McGraw-Hill Yearbook of Science & Technology

1984

D1116110

McGraw-Hill
Yearbook of
Science &
Technology

COMPREHENSIVE COVERAGE OF
RECENT EVENTS AND RESEARCH AS
COMPILED BY THE STAFF OF THE
McGRAW-HILL ENCYCLOPEDIA OF
SCIENCE AND TECHNOLOGY

1984

McGRAW-HILL BOOK COMPANY

New York   St. Louis   San Francisco
Auckland   Bogotá   Guatemala   Hamburg
Johannesburg   Lisbon   London   Madrid   Mexico
Montreal   New Delhi   Panama   Paris   San Juan
São Paulo   Singapore   Sydney   Tokyo   Toronto

1234567890   DODO   89876543

The Library of Congress has cataloged this serial
publication as follows:

McGraw-Hill yearbook of science and technology.
1962–   . New York, McGraw-Hill Book Co.

  v.   illus.   26 cm.
  Vols. for 1962–   compiled by the staff of the
McGraw-Hill encyclopedia of science and
technology.

  1. Science—Yearbooks.   2. Technology—
Yearbooks.   I. McGraw-Hill encyclopedia of
science and technology.
Q1.M13   505.8   62–12028
Library of Congress   (10)

ISBN 0-07-045492-2
ISSN 0076-2016

Table of Contents

Consulting Editors

Contributors

A list of contributors, their affiliations, and the articles they wrote will be found on pages 487–490.

Preface

The 1984 *McGraw-Hill Yearbook of Science and Technology*, continuing in the tradition of its 21 predecessors, presents the outstanding recent achievements in science and technology. Thus it serves as an annual review and also as a supplement to the *McGraw-Hill Encyclopedia of Science and Technology*, updating the basic information in the fifth edition (1982) of the Encyclopedia.

The Yearbook contains articles reporting on those topics that were judged by the consulting editors and the editorial staff as being among the most significant recent developments. Each article is written by one or more authorities who are actively pursuing research or are specialists on the subject being discussed.

The Yearbook is organized in two independent sections. The first section includes seven feature articles, providing comprehensive, expanded coverage of subjects that have broad current interest and possible future significance. The second section comprises 156 alphabetically arranged articles on such topics as superconducting computers, cellular receptors, herpes, diamond-turned optics, anomalons, musical instrument acoustics, and extensional tectonics.

The *McGraw-Hill Yearbook of Science and Technology* provides librarians, students, teachers, the scientific community, and the general public with information needed to keep pace with scientific and technological progress throughout the world. The Yearbook has successfully served this need for the past 22 years through the ideas and efforts of the consulting editors and the contributions of eminent international specialists.

SYBIL P. PARKER
EDITOR IN CHIEF

McGraw-Hill Yearbook of Science & Technology

1984

Origin of

H₂O

NH₃

CH₄

H₂O

H₂

NH₃

CH₄

H₂

E.T. Steadman

Planetary Atmospheres

Tobias C. Owen

Tobias C. Owen teaches astronomy at the State University of New York at Stony Brook. He is a coauthor (with Donald Goldsmith) of "The Search for Life in the Universe." He was a member of the molecular analysis team on the NASA Viking mission to Mars in 1976 and is on the imaging science team for the Voyager missions to the Outer Planets.

A survey of the planets reveals a remarkable array of different atmospheres. Venus and Mars, just inside and just outside Earth's orbit, respectively, have atmospheres that are more than 95% carbon dioxide, with surface pressures that differ by a factor of nearly 10,000. The Earth's atmosphere clearly shows the influence of the life that inhabits its surface, which produced the 21% oxygen content, maintains the 78% nitrogen, and is largely responsible for the absence of all but a trace of carbon dioxide (Table 1). On the outer planets, atmospheres are found to be dominated by hydrogen, with helium the next most abundant gas on Jupiter and Saturn, and possibly on Uranus and Neptune as well. Carbon is present as methane, CH_4, instead of CO_2—fully reduced instead of completely oxidized.

The small bodies that have retained atmospheres exhibit a comparable diversity. Jupiter's rocky satellite Io appears to have a local, transient atmosphere of SO_2 that is associated with volcanic eruptions driven by the dissipation of tidal energy in Io's interior. Titan, the largest satellite of Saturn, has an atmosphere of nitrogen that is denser than Earth's. Methane appears as a minor constituent in Titan's atmosphere and is probably going through the same changes of phase as Earth's water. Both Pluto and Triton (Neptune's large moon) appear to possess tenuous atmospheres containing some methane.

How has this diversity come to exist? Scientists are interested in the answer to this question not just because of a desire to explain planetary atmospheres as they are found today, but also because of the recognition that the origin of life on Earth is a direct consequence of atmospheric chemistry during the early stages of Earth's his-

Table 1. Composition of the Earth's atmosphere

Gas	Molecular weight	Fraction of dry air ($\times\ 10^{-6}$)		Amount, atm/cm
		By volume	By weight	
N_2	28.013	780840	755230	624000
O_2	31.999	209470	231420	167400
H_2O	18.015	1000–28000	600–17000	800–22000†,§
Ar	39.948	9340	12900	7450
CO_2	44.010	320	500	260*
Ne	20.179	18.2	12.7	14.6
He	4.003	5.24	0.72	4.2
CH_4	16.043	1.8	1.0	1.4
Kr	83.80	1.14	3.3	0.91
CO	28.010	0.06–1	0.06–1	0.05–0.8*
SO_2	64.06	1	2	1*
H_2	2.016	0.5	0.04	0.4
N_2O [3]	44.012	0.27	0.5	0.2
O_3	47.998	0.01–0.1	0.02–0.2	0.25†,‡
Xe	131.30	0.087	0.39	0.07
NO_2	46.006	0.0005–0.02	0.0008–0.03	0.0004–0.02*
Rn	222	$0.0^{13}6$	$0.0^{12}5$	5×10^{-14}
NO	30.006	Trace	Trace	Trace*

SOURCE: C. W. Allen, *Astrophysical Quantities*, 3d ed., Athlone Press, University of London, 1973.
*Greater in industrial areas.   †Meteorological or geographical variations.   ‡Increases in ozone layer.   §Decreases with height.

tory. Conditions on Earth have changed irreversibly since the chemical events that led to the origin of life took place, but in the outer solar system there are planets whose atmospheres resemble in some important respects the conditions postulated for the primitive Earth. In the inner solar system there are examples of planets like Earth on which life has evidently not developed; these planets constitute experimental controls. Hence by studying the atmospheres of the other planets, scientists hope to be able to address the fundamental and fascinating question of the origin of life on Earth.

Traditionally there have been two principal ideas for the origin of atmospheres: capture of gases from the primordial solar nebula (producing a primary, unfractionated atmosphere) and degassing from the rocky or icy material composing the planet itself (leading to a secondary atmosphere). The first of these

Table 2. Solar abundances of the elements*

Element	Value†
H	2.66×10^4
He	1.8×10^3
O	18
C	11
Ne	2.6
N	2.3
Mg	1.1
Si	1.0
Fe	0.9
S	0.5
Ar	0.11

*The abundance of silicon is made equal to 1.0, and the list is arbitrarily terminated at argon.

†The units indicate numbers of atoms on a scale normalized to silicon (that is, for each silicon atom, there are 18 oxygen atoms).

processes was thought to be responsible for the atmospheres of the giant outer planets, the second has been customarily associated with the development of atmospheres on the small, rocky inner planets. A third possibility is obtained by combining the first two: capture plus degassing.

Given that the composition of an atmosphere is highly variable with time, how can a planet's early history be reconstructed to see which of these hypotheses for the origin of atmospheres is correct? Composition changes as a result of escape from the exosphere, photochemistry in the upper atmosphere, and chemical reactions with the surface. To minimize the confusion produced by these processes, the ideas can be tested by examining the abundances of the noble gases and the ratios of various isotopes as they are found today. This procedure provides the best hope for distinguishing among the various possible chemical and mechanical processes that could change the atmospheric composition of a planet during its evolution.

CAPTURED ATMOSPHERES: THE GIANT PLANETS

An atmosphere that has simply been captured from the solar nebula should have a composition that reflects the elemental composition of the Sun itself (Table 2). This is therefore a very easy hypothesis to test. It is merely necessary to discover which molecules will form under local equilibrium conditions from this starting mixture and then examine the atmospheres of the planets to see if these molecules are there and whether the elemental abundances match those in the Sun.

Jupiter and Saturn. The two planets in the solar system that are generally considered to have atmospheres of this type are Jupiter and Saturn (Table

Table 3. Composition of the atmospheres of Jupiter and Saturn*

Gas	Jupiter	Saturn
H_2	1	1
CH_4	$1 - 2 \times 10^{-3}$	$\sim 2 \times 10^{-3}$
NH_3	2×10^{-4}	$\sim 2 \times 10^{-5}$
H_2O	10^{-6}	—
C_2H_2	8×10^{-7}	1×10^{-7}
C_2H_6	4×10^{-5}	7.5×10^{-6}
CO	3×10^{-9}	—
HCN	2×10^{-9}	$<7 \times 10^{-9}$
GeH_4	6×10^{-10}	—
PH_3	4×10^{-7}	3×10^{-6}
C_3H_4	Yes	Yes
C_3H_8		Yes

*Mixing ratios are given that provide the abundance of a given constituent relative to the abundance of molecular hydrogen.

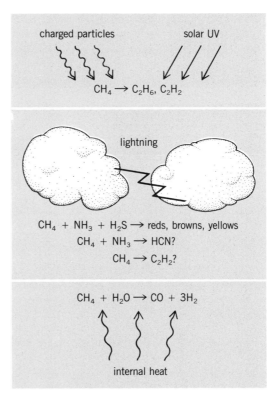

Fig. 1. Energy sources and molecules produced in Jupiter's atmosphere. (*After T. C. Owen, Planetary atmospheres and the search for life, Phys. Teacher, pp. 90–96, February 1982*)

3). In addition to the expected molecules—H_2, CH_4, H_2O, NH_3—a number of others—HCN, CO, C_2H_2, and so forth—are found that are not predicted by thermal equilibrium calculations. These molecules are either made at great depth where they are in equilibrium and then swept up to the region of the atmosphere where they can be detected (PH_3, GeH_4, CO), or they are produced high in the atmosphere by photochemical processes (C_2H_2, C_2H_6, C_3H_8) or by lightning discharges within the clouds of the troposphere (HCN, C_2H_2) [Fig. 1]. Thus chemistry is altering the composition of these atmospheres, even though they may have begun as a simple mixture of the same elements found in the Sun.

But did these atmospheres really begin that way? In Table 4 a comparison is made of the planetary ratios of helium and carbon to hydrogen with the solar values. Carbon is chosen for this analysis since it is present in the form of methane, which does not condense in the atmospheres of these planets. It is evident that there are real differences between the planets and the Sun. Although Jupiter appears to have a solar helium-hydrogen ratio, carbon is enriched by about a factor of 2, as it is on Saturn. This suggests a two-step process for the formation of these atmospheres: the accretion of a large, icy core followed by the capture of an envelope of gas from the nebula. This envelope contains the solar value of He/H, but methane outgassed from the core increases the observed abundance of methane on the planets.

In the case of Saturn, an additional process has acted to deplete the helium. Saturn, which is smaller than Jupiter, has cooled more rapidly during the 4.6 billion years since both planets formed, and has reached a stage at which helium is beginning to condense in the interior, thereby releasing gravitational potential energy and depleting the amount of helium found in the atmosphere. Observations of the degree of helium depletion and the amount of internal heat radiated to space are consistent with this theoretical model for Saturn's history.

The isotope ratios shown in Table 4 provide an-

other way of studying atmospheric histories. The presence of deuterium close to the interstellar abundance underlines the primitive nature of these atmospheres. This isotope has long been absent from the Sun since it undergoes nuclear reactions at relatively low temperatures. On the other hand, the high value of $^{12}C/^{13}C$ is difficult to understand, since interstellar values are two to four times lower. This new (1982) determination invites verification.

While all of these measurements represent the most recent work that has been done on these problems and include the excellent results obtained by the Voyager spacecraft, scientists are still dealing with

Table 4. Elemental abundances and isotope ratios in outer planet atmospheres*

	Elements	
	He/H	C/H
Jupiter	$\odot$	$2 \times \odot$
Saturn	$0.5 \times \odot$	$2 \times \odot$
Uranus	?	$\sim 20 \times \odot$
Neptune	?	$\sim 30 \times \odot$

	Isotopes
Jupiter	$^{12}C/^{13}C \sim 1.8 \times \odot$
	$D/H = 3.6 \times 10^{-5}$
Saturn	$D/H = 2.6 \times 10^{-5}$

*$\odot$ denotes the solar value.

interpretations of spectroscopic measurements made outside the planets' atmospheres. Such measurements are always subject to various errors and uncertainties, not all of which are known. It is therefore still within the realm of possibility that both Jupiter and Saturn do in fact contain solar proportions of the elements which have simply undergone the chemical and physical processing already described. In order to be completely certain about these alternatives, it would be useful to know the abundance ratios of many more elements. There will be an opportunity to make a much more definitive test in 1988 when the probe from the Galileo spacecraft enters the atmosphere of Jupiter; a mass spectrometer will be used to obtain direct measurements of molecular and elemental abundances.

Uranus and Neptune. Uranus and Neptune are much more difficult to observe from Earth than either Jupiter or Saturn, and they have not yet been visited by spacecraft. Only hydrogen and methane have been positively identified in their atmospheres, although acetylene and ethane are suspected on Neptune (see Table 4). Here the values of C/H are very definitely nonsolar, beyond the realm of possible errors in the data. The mean densities of these planets support this finding; both Uranus and Neptune are clearly deficient in hydrogen and helium compared to the solar mixture of these elements.

The principal question remaining for these two planets is whether or not their atmospheres contain a solar value of He/H. Helium can only be detected indirectly, by its effect on the hydrogen thermal emission spectrum. This measurement cannot be made from Earth, but there is hope that it can be accomplished from the Voyager spacecraft when they reach Uranus in 1986 and Neptune in 1989. At present the possibility must be considered that even the hydrogen observed in the atmospheres of these planets may be contributed in large part by degassing of their very large cores, thereby making the atmospheric values of He/H much smaller than the solar value. Such a possibility would be consistent with a theory for the origin of Uranus and Neptune which holds that they were built up from cometlike objects to form their massive cores, with subsequent capture of a thin envelope of gas from the solar nebula.

It is already obvious that Uranus and Neptune are not identical twins, but differ as much from each other as Jupiter from Saturn or Earth from Venus. For example, Neptune has both an internal source of energy and a thermal inversion in its stratosphere, while Uranus has neither. Hence, both the compositions and the histories of these two atmospheres may differ as well.

Summary. To summarize, it must be said that there is no clear evidence for a purely captured atmosphere. Jupiter and Saturn may possess such atmospheres, but the data available at present do not favor that hypothesis. Even in the realm of the giant planets, it seems that some degassing from a solid

core has played an important role in determining the compositions of the present atmospheres.

SECONDARY ATMOSPHERES

Traditional thinking about the formation of secondary atmospheres has concentrated on the Earth, Mars, and Venus. On these rocky bodies, one could see that impacts, erosion, and various forms of tectonic activity (volcanism, mountain building, crystallization of extruded magmas, and so on) would lead to the release of gases trapped or chemically bound in the materials making up the planet. The released gases formed the planet's atmosphere, whose composition could be expected to change with time as a result of escape, deposition, chemical reactions, and the changing composition of the released gases. This last-mentioned change occurs with the maturing of the planet as differentiation leads to a change in the chemical composition of the mantle, impacts become unimportant sources of volatiles, and extensive recycling of gases begins.

In just the last few years, it has become evident that one must consider analogs of these processes when discussing atmospheres on icy bodies in the outer solar system also. As mentioned above, Titan, Triton, and Pluto are all members of this group at the present time, but there is evidence for early, transient atmospheres on some of Saturn's smaller satellites as well. The present discussion of secondary atmospheres will begin with a consideration of the object in this new category that has received the most study.

Titan. One of the great triumphs of the Voyager missions was the new information they provided about the atmosphere of Titan, Saturn's largest satellite (Table 5). It contains more total nitrogen than the Earth's atmosphere, plus an intriguing variety of organic molecules, including an aerosol that is evidently produced by further reactions among these and other species. The aerosol absorbs strongly in ultraviolet light, exhibits a hemispheric change in

Table 5. Atmosphere of Titan

Constituents	Relative abundance‡
Major	
Nitrogen (N_2)	82–94%
Argon (Ar)	12%*
Methane (CH_4)	6%†
Trace	
Hydrogen (H_2)	2000 ppm
Acetylene (C_2H_2)	2 ppm
Ethylene (C_2H_4)	0.4 ppm
Ethane (C_2H_6)	20 ppm
Diacetylene (C_4H_2)	0.1–0.01 ppm
Methylacetylene (C_3H_4)	0.03 ppm
Propane (C_3H_8)	20 ppm
Cyanogen (C_2N_2)	0.1–0.01 ppm
Hydrogen cyanide (HCN)	0.2 ppm
Cyanoacetylene (HC_3N)	0.1–0.01 ppm
Carbon dioxide (CO_2)	0.007 ppm

*Deduced from $\bar{\mu} = 28.6$.
†Varies with altitude.
‡By number of molecules.

albedo (south is brightest as northern spring begins), and shows distinct layering plus a dark northern polar hood. Methane has been known to be present since its discovery in 1944, but all of the other constituents have been discovered within the last decade, with most being positively identified (including nitrogen and hydrogen) only by the Voyager spacecraft. The chemistry of Titan's atmosphere is being driven by solar ultraviolet light and precipitating electrons from Saturn's magnetosphere. It offers interesting parallels for some of the chemical reactions hypothesized for the primitive Earth, which were the first steps toward the origin of life. But how did the atmosphere itself originate?

It is immediately apparent that Titan's atmosphere is not a modified version of a captured atmosphere because it contains so little neon. In referring to Table 2, neon would be expected to have about the same abundance as nitrogen in the solar nebula. Thus the atmosphere of Titan should be about 66% Ne and 33% N_2. Instead less than 1% Ne is found, according to Voyager measurements. As a noble gas, neon will not form chemical compounds; it is too heavy to escape and it will not condense at Titan's temperatures. Hence the absence of neon in large quantities is a strong argument for a secondary atmosphere.

It is still not certain how the nitrogen in Earth's atmosphere originated, but for Titan's nitrogen the possibilities can be narrowed down to two: either it came to the satellite in the form of ammonia that was subsequently photodissociated with the escape of hydrogen and the formation of molecules of N_2, or it was trapped as N_2 in the ices that formed the satellite. In the first case, a surface temperature of 150 K would be required during the early phases of Titan's history to allow the ammonia to have a significant vapor pressure. Otherwise there would not be enough ammonia in the atmosphere to permit the photodissociation to occur. [Titan, with a surface temperature of 95 K and a surface pressure of 1.5 bars (1.5×10^5 pascals) has no detectable ammonia at present.] This is not a prohibitive requirement, in that one can readily assume that the release of gravitational energy during accretion would have warmed the surface layers of Titan, and an atmospheric greenhouse set up by the ammonia and the H_2 could have maintained sufficiently high temperatures to permit the postulated photochemistry to occur. There is no detailed model for this scenario, however, and in any case it does not account for a high abundance of argon in Titan's atmosphere, if indeed this gas is present.

The evidence for argon is indirect: When the pressure-temperature profiles from both the infrared spectrometer and the radio occultation experiments on the Voyager spacecraft are combined, the mean molecular weight ($\bar{\mu}$) that provides the best fit is slightly larger than 28. This requires the presence of a cosmically abundant gas with a higher molecular weight that will not condense at Titan's low temperatures. The gas must also be undetectable spec-

troscopically. Argon satisfies all of these criteria.

At the conditions postulated for the proto-Saturnian nebula [$P = 0.1$ bar (10^4 pascals), and $T = 60$ K], the saturation vapor pressure of argon is over two orders of magnitude above the pressure of this gas in the nebula. Hence it will not condense. How then did it appear in the atmosphere? The answer lies in the possible formation of clathrate hydrates as the ice in the satellite condenses out of the cloud. These hydrates consist of lattices of H_2O molecules in which spaces exist that can accept "guest" atoms or molecules. At the ambient pressure and temperature assumed for the forming Saturn system, a mixed clathrate hydrate of methane, nitrogen, and argon could form. But neon would not be trapped at these temperatures; hence its absence from the resulting atmosphere is consistent with this mode of origin.

Now there is a new way to provide the observed amount of atmospheric nitrogen: it is assumed that nitrogen was present in the proto-Saturnian nebula as N_2, and was trapped in the ices along with the argon and methane. During the accretion and differentiation of Titan, the trapped gases were released to form the atmosphere.

It is not yet possible to distinguish between these two alternatives. It can be said that the ratio of nitrogen to argon degassed by Titan is consistent with the solar ratio if 12% argon is indeed present. This calculation allows for nitrogen escape and deposition. The presence of 7 ppb of CO_2 in Titan's atmosphere implies the existence of 0.1–0.5% CO as well. The presence of N_2 in the proto-Saturnian nebula would imply the existence of some CO as well, reflecting a lower state of reduction of the nebula gases. In fact, CO has just been discovered on Titan by ground-based observations. Thus there is general consistency between the available data and this point of view, but the possibility that some of the nitrogen was in fact brought to Titan as ammonia cannot be ruled out.

To summarize, Titan offers the unique case of a highly evolved, reducing atmosphere. The origin of that atmosphere can be understood in terms of a model that includes the trapping of gases from the proto-Saturnian nebula in the form of clathrate hydrates. It seems that the gases in that nebula included both CO and N_2. The present surface temperature and pressure on this satellite are consistent with the existence of lakes or seas of liquid methane (Fig. 2). The presence of a variety of organic molecules, a thick aerosol layer, and a cold surface on which the chemicals formed in the atmosphere could collect and be preserved all make Titan an important natural laboratory for testing ideas about chemical evolution on the primitive Earth.

Inner planets. The study of Titan offers a new perspective from which to approach the problem of the origin of the inner planet atmospheres. First, it seems likely that the solar nebula was less reducing than is commonly thought—containing N_2 and CO instead of (or in addition to) NH_3 and CH_4. Second, Titan can be used as a giant analog of a comet in

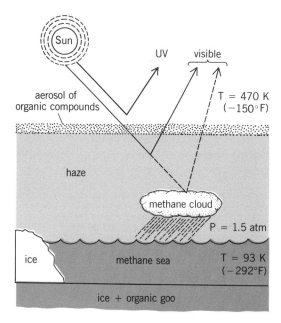

Fig. 2. Cross section of Titan's atmosphere showing the penetration of light of different wavelengths, with hypothetical configurations of haze and surface.

order to show what kind of volatiles these icy messengers from the outer solar system may have delivered to the inner planets. This source of atmospheric constituents must be considered, along with the meteorites and local condensates, since it is not yet clear which of these three provided most of the volatile elements in the present atmospheric gases.

It has long been recognized that the composition of the terrestrial atmosphere is dominated by the biological processes of the planet, and strongly influenced by nonbiological processes of erosion and chemical reactions with surface materials. If an attempt is made to reconstruct the Earth's composition in the absence of these processes, carbon dioxide would be the most abundant gas, with nitrogen only a minor constituent, and oxygen reduced to a few tenths of a percent. This reconstruction bears a

Table 6. Volatile inventories of main atmospheric constituents of Venus, Earth, and Mars

Gas	Venus (now)	Earth (total)*	Mars (total)*
CO_2	96.5%	98%	98%
N_2	3.5%	1.9%	1.7%
^{40}Ar	33 ppm	190 ppm	300 ppm
H_2O	Trace	3 km†	30 m†
Total atm = $10^6 \times \dfrac{M_{atm}}{M_{planet}}$‡	100	60	2

*No life, no weathering (total amount of volatiles degassed by the planet over geologic time).

†Thickness of a condensed layer of liquid water spread over the planet.

‡M represents mass.

striking resemblance to the present atmosphere of Venus (Table 6), except that a much greater amount of water is found on Earth.

This abundance of water is a direct consequence of Earth's position in the solar system. It is highly probable that Mars contains a large store of water beneath its surface in the form of permafrost, and if the planet were warmer, this subsurface water would become available. Erosional features on the Martian surface bear witness to a time when this condition was met, perhaps 3.5 billion years ago (Fig. 3). On the other hand, it is clear that the total amount of atmosphere (per gram of rock) degassed by Mars over geologic time is much smaller than that on either Earth or Venus (Table 6).

In dramatic contrast to Mars, Venus is too hot to maintain water on its surface or in its subsurface layers. How this happened can be demonstrated with a Gedanken experiment in which the Earth is moved to the orbit of Venus. The resulting increase in Earth's global temperature would lead to increased evaporation of water from the oceans, which in turn would increase the atmospheric greenhouse effect caused by water vapor, leading to a higher surface temperature, more evapoation, and so on. This positive feedback loop would lead to a "runaway greenhouse," in which the oceans would literally boil away. At that point the atmosphere would be so hot that there would be no cold trap to prevent water vapor from mixing high into the stratosphere. Dissociation would be rapid under the influence of solar ultraviolet light, with a massive escape of hydrogen. The residual oxygen would then be free to combine with the crust of the planet.

To apply this hypothetical scenario to Venus, it must be assumed that this planet started out with water, as did the Earth. Until recently, this assumption was in competition with the hypothesis that Venus has no water now since it began with no water, having formed too close to the Sun for this highly volatile substance to condense. Some of the scientists who favored the runaway greenhouse hypothesis argued that Venus, like Earth and Mars, received most of its volatiles from sources farther out in the solar system—the primordial equivalent of contemporary comets and meteorites. It now seems that this latter version is in fact correct, in that a huge enrichment in deuterium has been discovered in the residual water vapor in the atmosphere of Venus. The deuterium-hydrogen ratio on Venus is approximately 1×10^{-2}, compared with 1.6×10^{-4} on Earth and 2.5×10^{-5} in the interstellar medium. Such a large enrichment implies the escape of an enormous quantity of hydrogen, corresponding to an amount of water equivalent to at least 0.3% of the Earth's oceans. This is only a lower limit to the quantity of water originally on Venus, since the early stages of hydrogen escape would have been so rapid that no fractionation between hydrogen and deuterium would have taken place.

Thus, there appears to be support for a common source of volatiles for all three inner planets. But

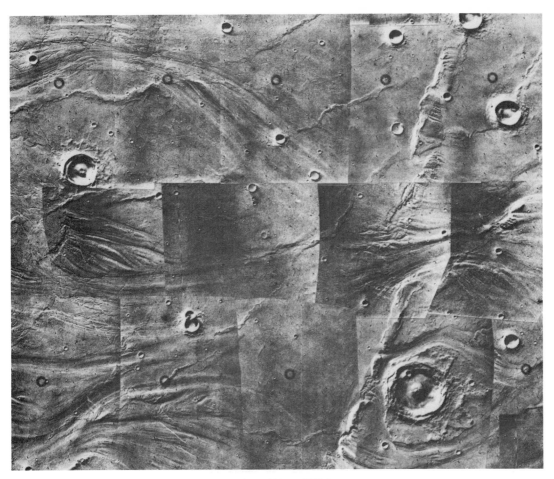

Fig. 3.   Evidence of ancient fluvial erosion on the surface of Mars. (*NASA*)

the picture is actually not quite so simple, as a detailed comparison of the noble gas abundances on these three planets indicates (Fig. 4).

Inspection of Fig. 4 reveals that Mars and Earth have noble gas abundance patterns that are similar to one another and that also match the pattern found in carbonaceous chondrite meteorites. This pattern is marked by a distinct depletion of neon and argon relative to krypton and xenon. Furthermore, neon is 1.8 times less abundant than argon, whereas on the Sun it is 24 times more abundant (Table 2). From this comparison, it appears that the fractionation of the noble gases from a solar mixture took place in the nebula before the planets formed. Both the meteorites and the material that accreted to form the planets accumulated this fractionated gas that was subsequently released into the planets' atmospheres.

In the case of Venus, however, the picture is not so clear. Here the argon-krypton ratio is closer to the solar value, while the neon-argon ratio again resembles the Earth, Mars, and the meteorites. Furthermore, the absolute amounts of neon and argon in the atmosphere of Venus are much greater (per gram of rock) than the abundances found in the atmospheres of the other two planets, or even in the carbonaceous chondrites. As indicated in Fig. 4, a class of meteorites called enstatite chondrites is

roughly similar to Venus, but the argon-neon ratio in these objects is not a good match. How did this enrichment and change of abundance pattern occur on Venus?

One suggestion is that the early solar wind imprinted a solar pattern of abundances on the material that accreted to form Venus, with much of the neon subsequently escaping by diffusion. In this model, the gas atoms are implanted in small dust particles. It is suggested that there was enough material in orbit around the Sun at the distance of Venus to shield the forming Earth from this same implantation process, so that only Venus shows this effect. This idea gains some support from the derived value for the ratio of the isotopes of neon on Venus, which was determined by the Soviet *Venera 13* and *14* spacecraft as $^{20}Ne/^{22}Ne = 11.6 \pm 0.6$. This is intermediate between the terrestrial value of 10.1 and the solar value of 13.6, suggesting the presence of a solar wind component.

In other words, there is evidence for at least two discrete sources of volatiles for Venus: one source to supply the early complement of water—the same combination of meteoritic and cometary bodies that brought volatiles to Mars and Earth—and a second source that selectively enriched the argon and neon, either through solar wind irradiation or by some other

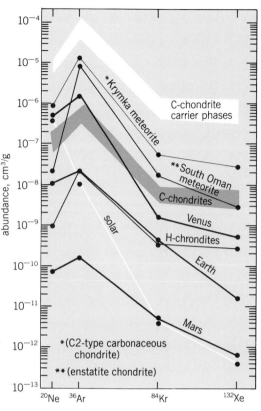

Fig. 4. Comparison of noble gas abundances in the atmospheres of Venus, Earth, and Mars with meteoritic and normalized solar abundances. (*Courtesy of E. Anders, Enrico Fermi Institute, University of Chicago*)

local process peculiar to Venus that must have been related to the way in which this planet formed.

Given that all three inner planets began with such a similar inventory of volatiles, why are their atmospheres so different today? The importance of being at the correct distance from the Sun in order to maintain a temperature that allows liquid water to exist on a planet's surface has already been explained. The presence of liquid water in turn has allowed life to originate on Earth, and life has totally altered the composition of its atmosphere. Moving the Earth closer to the Sun would preclude the existence of water. But what about the opposite maneuver? Is the dry, thin, cold atmosphere of Mars the only possible answer?

It seems that this case is not quite so clear. The tremendous CO_2 greenhouse found on Venus offers an alternative. Table 6 and Fig. 4 both show that Mars has outgassed much less than Earth and Venus. If Mars had been a larger planet, then its initial atmosphere might have been much denser and persisted longer, as it was fed by tectonic activity. If this atmosphere contained a high proportion of carbon dioxide, the resulting CO_2/H_2O greenhouse would have been capable of maintaining clement conditions over the globe. A CO_2 atmosphere with the same surface pressure as Earth's at the distance of Mars would keep the surface temperature above

the freezing point of water. However, could these conditions be maintained for 4.6 billion years? A planet with large polar caps of frozen water, with a temperate region only near the equator would be possible with this model, but it would still be a far cry from the Mars seen today.

CONCLUSIONS

At the beginning of this article, three possible models were offered for the formation of planetary atmospheres. It appears that one of these—simple condensation from the solar nebula—is excluded by the available evidence. Nevertheless, judgment must be withheld until in-place measurements from the atmosphere of Jupiter are taken to see whether or not this giant planet may in fact retain the elemental abundances exhibited by the Sun. If it does not, as current evidence suggests, then it must be concluded that the giant outer planets owe their atmospheres to a compound process of accretion, outgassing, and capture from the nebula.

The smaller bodies in the outer solar system may be rich in volatiles trapped directly from the nebula by the ice that constitutes about half of their individual masses. Subsequent degassing has allowed Titan to possess a substantial atmosphere. Pluto and Triton appear to have tenuous, methane-containing atmospheres. The reworking of the surfaces of Saturn's small, icy satellites testifies to the release of gas from the interiors of these moons, whose masses are too small to retain atmospheres.

In the inner solar system, there are secondary atmospheres that have been degassed by the planet-forming material. It appears that Venus, Earth, and Mars have all acquired a large share of their volatiles from a late bombardment by cometary and meteoritic material that condensed from the solar nebula at much greater distances from the Sun than the present orbits of these planets. But Venus exhibits an enrichment of argon and neon which indicates that this planet has had a history which sets it apart from the others. It is not yet clear what the essential features of this history have been.

The origin of life on Earth appears to be a unique phenomenon in the solar system. This can be understood in terms of Earth's distance from the Sun and its size, both of which favor the development of a global climate within which liquid water is stable. But how did life begin and in what kind of atmosphere? The erosion of the Earth's crust has removed most of the ancient rock record, making reconstruction of the early period of Earth's history extremely difficult. But ancient rocks should be much more common on Mars, judging from the preservation of impact craters over much Martian terrain. The possible similarity in early atmospheric history between Earth and Mars suggests that investigation of ancient Martian rocks—especially sedimentary rocks—might supply useful information about the early history of the Earth and its atmosphere. At the present stage of knowledge, the possibility that life actually began on Mars, too, and then died out as the global climate changed, cannot be ruled out. If so, some of those sedimentary rocks might even contain microfossils.

Long before the elaborate mission strategies required for the investigtion of such questions on Mars are developed, it is hoped that there will be new insights into atmospheric origins with simpler missions and new, Earth-based observations. Further investigations of Titan's atmosphere have high priority, owing to the interesting similarities between some of the chemistry currently taking place there and reactions hypothesized as initial steps toward the ori-

gin of life on Earth. The Galileo Project to explore the atmosphere of Jupiter and the Vega and Giotto missions to study Halley's Comet all promise a rich harvest of relevant new results before the end of the present decade.   [TOBIAS C. OWEN]

Bibliography: H. O. Holland, *The Chemistry of the Atmosphere and Oceans*, 1978; T. Owen, Titan, *Sci. Amer.*, 246(2):98–109, 1982; J. B. Pollack and Y. L. Yung, Origin and evolution of planetary atmospheres, *Annu. Rev. Earth Planet. Sci.*, 8:425–488, 1980; J. C. G. Walker, *Evolution of the Atmosphere*, 1977.

Intelligent

E.T. Steadman

Machines

Keith L. Doty

A professor in the Electrical Engineering Department at the University of Florida, Keith L. Doty is also associate director of the Center for Intelligent Machines and Robotics (CIMAR) there. He is investigating distributed microcomputer networks and exploring ways to incorporate them in mechanical robotic manipulators and data-base management systems. He is author of the textbook "Fundamental Principles of Microcomputer Architecture."

The idea that a particular machine exhibits intelligence may stimulate positive responses of excitement, interest, and hope or it may create negative responses of anxiety, hostility, and denial. This behavior is understandable when it is considered that, to most individuals, the most human characteristic is intelligence. Great philosophical debate revolves around the issue of whether the alleged intelligence exhibited by machines is of the same nature as human intelligence. The issues are quite involved and, since the implications strike at the core of the human self-image, arguments pro and con are clouded by emotional, rational, and religious biases.

INTELLIGENCE

The debate concerning intelligent machines centers on some basic philosophical questions, such as: Why are some machines referred to as intelligent? What are the characteristics of intelligence that leads to the designation of one machine as intelligent and another as not intelligent? Is it possible to build machines that are as intelligent as humans? Can machines really be intelligent at all, in any sense of the word?

Certainly, this brief survey of intelligent machines will not resolve the philosophical issues, but this article seeks to establish the viewpoint that machine intelligence is possible. However, this article does not attempt to equate machine intelligence with human intelligence, although the latter constantly serves as a model.

To get a clearer understanding of machine intelligence, it is helpful to consider the following dictionary definition of intelligence and then attempt to derive a plausible definition for machine intelligence: "The ability to learn or understand from experience; ability to acquire and retain knowledge from experience; mental ability; the ability to respond quickly and successfully to a new situation; use of the faculty of reason in solving problems, directing conduct, etc. effectively" [*Webster's New World Dictionary*].

It is notable that intelligence is an ability and not an agent. Implied, of course, is that some agent exhibits this ability. The presence of intelligence should not, however, immediately be taken as evidence that an agent with that ability is necessarily conscious or aware of itself.

The preceding definition presents a number of difficulties, however, if an attempt is made to apply it directly to machines. A number of terms associated with intelligence, such as learn, understand, mental ability, experience, and reason, have such broad meanings that semantic arguments could arise. To avoid this pitfall intelligence can be defined with the specific goal of being able to identify terms with known machine capabilities. With this approach machine intelligence can then be discussed on an operational and engineering level, and the relative degree of intelligence demonstrated by a particular machine can be assessed.

Machine intelligence is: the ability of a machine to modify its observable behavior based upon present and past responses to its environment as perceived by sensors that provide the machine with information; the ability to acquire sensory information, store that information, and compute results which can be retained to affect current and future behavior of the machine; and the ability to respond quickly and successfully to a new situation within the scope of its performance objectives. From this definition it may be observed that an intelligent machine is capable of acquiring information, storing information over time, computing and retaining results, and responding to new situations in a timely and observable manner based upon the first three capabilities.

All four of these activities can be observed in an electronic digital computer. For example, the computer might acquire information from a keyboard, store the resulting characters in its memory unit, interpret those characters as BASIC program commands, and perform the operations dictated by the commands, such as printing text on paper. It can be justifiably argued that this behavior does not even approach the complexity of human behavior. The primary point is that machine intelligence, like human intelligence, has degrees or levels of sophistication. Consequently, the phrase "within the scope of its performance objectives" in the above definition of machine intelligence is extremely important. For practical purposes the scope of machine behavior and its perceived universe is highly constrained to reduce complexity. It means that any intelligent machine may have the defined characteristics to a lesser or greater degree than some other machine. Examples of this will be discussed later in the article.

Artificial intelligence. At this point the concept of machine intelligence can be related to the earlier concept of artificial intelligence to try to see the differences. In P. H. Winston's definition, artificial intelligence is the study of ideas which enable computers to do the things that make people seem intelligent.

This definition is considered suspect by Winston himself, as well as by the present author, since it invokes, as an operational definition of intelligence, a comparison of machine behavior with human behavior and a judgment about whether any particular human behavior, and hence similar machine behavior, is intelligent. There are many difficulties in such an approach. People may not consider scratching their nose to be intelligent behavior, yet the amount of sensory and motor capability, backed by the elaborate control necessary to perform that task, is extraordinary. A machine with such capabilities would be considered intelligent.

Thus, a more direct characterization of intelligent machines is preferable, even if it does not present all the dimensions of human intelligence. As understanding of the principles that make intelligent behavior possible for machines increases, the position taken in this article will have to be modified.

Machine intelligence might be properly understood as applied artificial intelligence. Artificial intelligence is concerned primarily with the scientific aspects of intelligence, while applied artificial intelligence is more strictly engineering-oriented. The boundary between these two areas is indistinct, since the activity of one domain merges with the other.

Work in artificial intelligence lends direction to the development of practical engineering instances of intelligent machines. The enthusiasm of early researchers in artificial intelligence created what are now known to be unrealistic expectations in developing intelligent machines, and as a result the credibility of artificial intelligence has suffered. However, the potential economic value of even modest gains in artificial intelligence, appear to be on the verge of fruition. This is due to the significant development of low-cost, reliable microprocessor technology coupled with the desire to increase productivity through robots and flexible automation. The consumer marketplace, too, has changed radically because of the availability of low-cost computation and decision logic, as witnessed by the proliferation of intelligent machines such as microcomputer-controlled automobile engine combustion, microwave oven control, and microcomputer-driven video games. Most of these consumer applications have not come from developments in artificial intelligence, but future product enhancements will undoubtedly incorporate some artificial intelligence technology. The next major advance in machine intelligence will hinge on the development of small computers 1000 to 10,000 times faster than today's microcomputers. This computational capability will permit commercialization of some of the current esoteric work in knowledge based systems and stereoscopic vision.

Modeling machine intelligence. In order to distinguish an intelligent machine from other types of machines, a model must be formulated based upon the definition given previously. First, it is necessary that an intelligent machine exhibit one or more of three capabilities: (1) cognitive machine intelligence (refers to the ability to plan and establish goals, model the environment based upon sensory input, and change plans, goals, and environmental models based upon sensory input and the machine's current knowledge base); (2) sensory machine intelligence (the ability to perceive the environment and identify elements in that environment); (3) motor machine intelligence (the ability to adaptively move objects in the machine environment). The second and third areas require cognitive capabilities. Machine sensory subsystems must be able to sort and analyze sense data into patterns and identify them with respect to the knowledge base. Motor intelligent functions must manipulate objects in the environment without disrupting the machine's world in an undesirable manner.

The main emphasis here is that intelligent machines vary widely in their capabilities and do not necessarily exhibit these three aspects of intelligence in equal measure. For example, a chess-playing computer requires only a small amount of sensory input to monitor its world. Classifying the input character string "2–4 3–4," which designates a move of a piece from row 2, column 4, of a chess grid to row 3, column 4, as sensory input might appear a bit strange at first. Further consideration, however, reveals that inputs of this nature can be construed as highly structured sense data. To illustrate, suppose there is a vision system that monitors a real chessboard with pieces on it. When the machine's opponent makes a move, the vision system will analyze the new state of the chessboard and determine what piece was moved and where. Since this is the only information extracted from the large amounts of vision data produced by the camera, possibly in the order of a million bits, the whole vision system can be bypassed by having the opponent enter the move as direct sense data.

The output of a chess-playing computer can be on a video display unit, so the moving of chess pieces is not a motor requirement. This dimension of intelligence, therefore, can be bypassed altogether here. Many intelligent machines, in fact, do not require the ability to manipulate objects, as in this case.

Machines that exhibit intelligence. It is virtually impossible to find a machine that can perform the spectrum of sophisticated functions performed by humans. Today, no machine comes close to the general manipulative and problem-solving ability of even the most learning-disabled individual. This observation does not imply that machines are unintelligent, or that intelligent machines cannot do their assigned tasks as well as humans. From a pragmatic engineering viewpoint, intelligent machines that perform highly specialized tasks of economic or social value already exist and their further develop-

ment for economic and practical purposes is justified. Whether the performance level of human intelligence can ever by achieved by a machine will be clarified by continuing research and engineering developments. More important than the outcome, however, is the fact that the development of intelligent machines will continue as long as they provide support for economic progress or stimulate intellectual curiosity.

SPECTRUM OF MACHINES

With the above formulation and caveats of machine intelligence as a guide, the concept of intelligent machines can be looked at more closely, along with the meaning of machine intelligence, and a spectrum of machines can be examined to see why some are considered to be intelligent and others are not.

Video game machines. Most of the arcade-type video games lack any of the characteristics of intelligence. Their programs generate fixed patterns of actions determined solely by the player's success at playing the game. Sometimes these games generate psuedorandom changes of the machine's response, but these usually do not relate in any meaningful way to the player's actions. Although the human player may require great skill and dexterity to beat such machines, this does not mean that the machine is intelligent. The video game machines that display intelligence do not passively react to the player's responses, but actually generate counterstrategies. A computer chess game such as SARGON II illustrates this type of intelligence, which is referred to as cognitive machine intelligence.

Since a personal computer with sufficient memory can be programmed to play any existing intelligent game, it may be deduced that machine intelligence derives from the computer program and not the machine itself. Programmability, therefore, seems to be a requisite property of intelligent machines. In terms of current technology, this means that an intelligent machine usually contains a digital computer, which is often a microcomputer, as a subsystem. To increase a machine's intelligence requires, among other things, greater program size and complexity and the storage of massive amounts of information. This suggests that another aspect of intelligence is the ability to organize, reduce, and analyze great quantities of data.

Feedback control systems. Automobile cruise controls, central air conditioners, refrigerators, ice makers, autopilots, and many kinds of manufacturing equipment employ feedback control techniques to improve machine performance. A machine that incorporates feedback compares the actual machine response to a desired response and performs corrective action if the two do not correspond.

Figure 1 illustrates a simplified feedback arrangement for controlling a dc motor. In this system, if the output shaft position differs from the reference shaft position, the machine attempts to drive the motor in the direction that minimizes the error be-

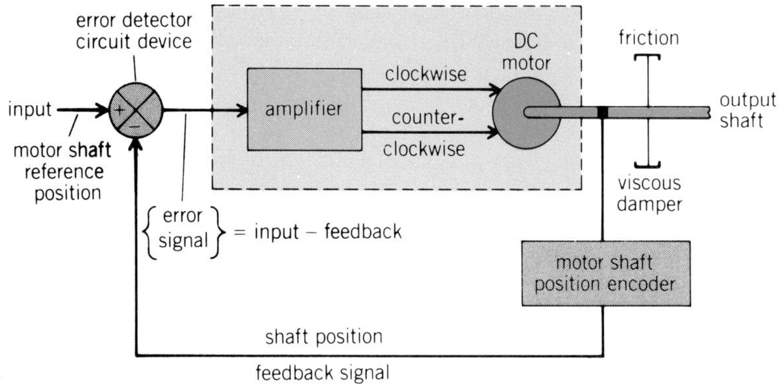

Fig. 1. Servomechanism illustrating the use of the feedback principle for positioning the output shaft of a dc motor.

tween the input shaft reference and the measured shaft position fed back to the input. There is no digital computer or program here, but the organization and arrangement of the system elements can be seen as a hard-wired analog computer program.

To consider the simple feedback system of Fig. 1 as intelligent would be inaccurate, but it does have some of the rudimentary features of intelligence. The machine responds to its environment (handles unpredictable loads); makes decisions (turns the output shaft toward a preferred position); and exhibits goal-directed behavior (drives the motor to correct for the load and move the output shaft back to the desired location).

Adaptive feedback control systems come one step closer to intelligent behavior than simple feedback systems. In adaptive control, the controller monitors the input-output behavior of the subsystem and periodically updates estimates of the parameters used to model the controlled subsystem. The controller then modifies the control algorithm to reflect its new knowledge about the system, hence the name adaptive control.

Machines with feedback, a large number of input-output parameters, and involved rules for ascertaining whether the outputs constitute a desirable configuration might easily suggest intelligence to a human observer. Somehow, then, complexity relates to the intuitive notion of intelligence. Unfortunately, any closed sequence of cause-and-effect relationships among physical phenomena may be described mathematically as a feedback system, even if the engineer would not identify it as such. This prevents an unequivocal statement that the presence of feedback implies machine intelligence, although an intelligent machine must necessarily employ feedback of some nature in order to respond to its environment.

Smart terminals and smart instruments. New product developments in computer terminals, point-of-sale terminals, laboratory instruments, and the like are often referred to by engineers as "smart." This designation usually means that the product incorporates a microcomputer subsystem which the user may program to some degree to increase the ma-

chine's flexibility and versatility, two characteristics associated with intelligence. The machines themselves, however, typically do not show any intelligence, since they cannot provide the flexibility and versatility automatically; in other words, human intervention is required to reconfigure them. While smart machines may be programmable, most are not intelligent because they lack programs that perceive and analyze environmental changes before taking action. Thus, programmability alone is not sufficient to make a machine intelligent.

Talking machines. Machine-generated speech applications are increasing. Early applications included telephone voice response systems, in which subscribers to a service, for example, stock quotations, would call a service number, enter appropriate codes by pushing tone-generating buttons, and then receive a voice reply. Machine speech technology also serves to aid the blind, enhance video games, and teach foreign languages. Some automobiles now provide voice announcements to remind occupants to fasten their seat belts, to alert the driver to potential danger, or to indicate when scheduled maintenance is due. In general, any situation where a machine communicates information to a person represents a potential application of a talking machine.

The ability of a machine to generate speech does not necessarily mean the machine has intelligence. The capability of stereo sound systems to play back audio recording illustrates this point. A spectrum of sophisticated techniques may be employed to generate speech, but no matter how complex the techniques, these techniques will not produce an intelligent machine unless the machine can be programmed to respond in a variety of ways to changing environmental conditions. Present developments in computer speech generation have dealt with the problem of efficiently representing and storing speech signals in electronic storage and regenerating them with high quality. The process of adding intelligence has been of secondary importance.

The simplest way to represent and store speech within a computer is to convert the analog acoustic signal to a sampled digital one and place the resulting digital form directly into computer memory for storage. Unfortunately, about 8000 samples per second need to be stored in order to produce high-quality speech. If each digital samples takes 8 bits, 64,000 bits would be required to store 1 second of speech. Since 1 second of speech translates to about three to six spoken words, it can easily be understood that this is far more storage than necessary for such a small amount of information. This scheme has other drawbacks, too: recording digital speech does not make the component parts amenable to computer algorithmic processing, an essential prerequisite for intelligence.

A second approach to machine-generated speech takes advantage of the known properties of speech and compresses the information found in 1 s of signal into about 1000 bits. Speech produced from compressed-data is of lower quality than that produced from direct recording. The resulting representation of speech data does not easily permit algorithmic manipulation, either. Texas Instruments' "Speak and Spell" educational toy, which helps teach the user to spell 200 English words, employs data-compression techniques. National Semiconductor has recently announced a speech processor which Ford Motor Company may use in some of its models. It controls both pitch and inflection of a word in order to combine isolated vocabulary words more naturally.

Speech synthesis directly from text appears to be the most suitable approach for efficient utilization of computer storage while representing data in a manner that makes algorithmic manipulation possible. Direct recording and data-compression techniques often have severely limited vocabularies, but speech synthesis from text essentially provides an unlimited vocabulary. Several firms hope to develop a system that integrates optical character recognition devices with a text speech synthesizer to produce a reading machine.

Synthesizing speech from pure English text, however, is quite difficult for general and language-specific reasons. Since English spelling is not phonetic, software must be written to convert English text to a phonetic representation. In addition, sound intonation, stress, elision, and other acoustic interactions resulting from connected speech cause wide variability in the way that speech is heard. Duplication of these sounds through software analysis of character text strings presents a formidable problem, and current intelligent machines are incapable of fully handling it. Nonetheless, the idea has merit, and with human preprocessing of the text data into a phonetic description, commercialization is possible.

Votrax Interface Division of the Federal Screw Works, for example, produces the ML-1 multilingual voice system (Fig. 2), which converts low-speed digital inputs into synthesized words and phrases by using phonemes as building blocks. The device performs this function by electronically simulating the vocal tract system to produce the sound associated with a particular phoneme. When the ML-1 is sent a 12-bit command word, it configures the vocal tract circuit to generate 1 of 128 different sounds with 1 of 8 pitch levels at 1 of 4 possible speaking rates. A string of these 12-bit commands can be programmed to generate any desired speech sequence with reasonable quality at the cost of about 120 bits per second of speech. Although the creation of these sequences burdens the programmer who must generate them, doing so greatly simplifies the process

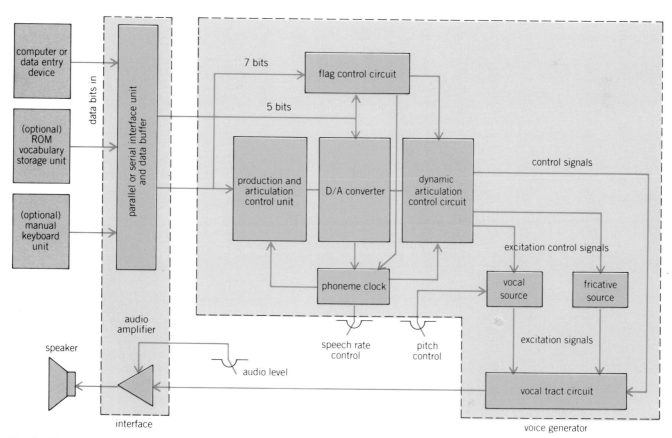

Fig. 2. Votrax ML-1 multilingual voice system, on computer command, generates speech phonemes and integrates them to produce synthesized speech. (*After B. A. Sherwood, The computer speaks, IEEE Spectrum, p. 23, August 1979*)

of speech synthesis. A truely intelligent machine would generate the sequences of 12-bit commands automatically and directly from the English text.

Machines that recognize and understand speech. While the synthesis of voiced phrases may be difficult, that task pales in comparison to the problem of constructing machines that allow natural language communication between humans and machines. Natural speech recognition by intelligent machines, let alone understanding, is extremely limited, and there is some doubt that a machine recognition system with human capabilities will ever be achievable. Others argue that limiting technology, cost constraints, and inadequate knowledge about the recognition process are hindrances to achieving this goal. Whatever the outcome in this debate, the availability of useful automatic speech recognition systems demonstrates the feasibility of this technology when the scope and intent of the application area are suitably constrained. In other words, it is not necessary to duplicate human functions to the last detail in order to have commercially useful intelligent machines.

Automatic speech recognition machines (Fig. 3) analyze incoming, microphone-transduced acoustic signal variations and energy content as a function of both frequency and time in order to recognize speech. Characteristic features extracted from this analysis are compared to the stored features of different vocabulary words. The vocabulary word that matches the most features found in the signal is presumed to be the word that was spoken. More advanced research machines employ other sources of information, such as language syntax and semantics, to help in the recognition process. For example, many people pronounce "pin" and "pen" the same way; even if they are enunciated correctly, an automatic speech recognition machine designed to distinguish the two words in isolation would have some difficulty. Context can help; for example, the use of the qualifier "ballpoint" leaves no doubt as to which of the two words was said.

In assessing the scope and performance of automatic speech recognition machines, a number of characteristics must be examined: (1) Does the machine recognize isolated or connected words? In the case of connected word recognition, does the speaker have to pause somewhat between words? (2) Is recognition speaker-dependent or -independent? Will the machine respond equally well to male, female, adult, or child? (3) Will the machine operate in a noisy environment or over a telephone? (Some recognizers require quiet environments and more expensive, wide-band channels for signal transmission.) (4) How much training is required for the system to recognize a particular individual's voice? Can the system automatically adapt to another speaker? (Sometimes the speaker has to adapt to the machine to ensure reliable recognition.) (5) What is the size of the system's vocabulary? (6) How accurate is the recognition? Two types of errors occur in recognition: rejection, where the machine indicates that the word cannot be identified; and substitution, where the machine misidentifies the word. Substitution errors cause the most trouble since further action might be initiated by a misunderstood verbal command. (7) How fast does the machine respond? The response time varies from 0.05 s for simple recognizers to minutes for sophisticated research machines that recognize complex vocabularies and sentence structures.

Nippon Electric's DP-100 is an automatic speech recognition machine that recognizes a large vocabulary of 1000 words with an accuracy of about 99%. The system was designed for routing and inventory control in a warehouse. It responds to a specific speaker in about 0.3 s and recognizes phrases of up to five connected words.

The Verbex Corporation 1800, designed to handle automatic telephone transactions, has a medium-sized vocabulary of 128 words, which it recognizes in real time with a reliability of over 99%. The system is self-training and speaker-independent (a mark of intelligence) and it recognizes connected speech.

Computer vision. With computer vision, meaningful descriptions of physical objects are constructed from images. This sharply contrasts with image processing, where different filtering techniques and signal conditioning produce a transformed image but no information about the meaning of the image. For example, if a color photograph of a red barn is examined, the viewer will identify the

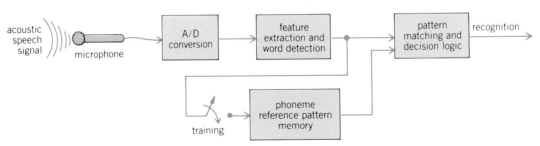

Fig. 3. Functional structure of a speech-recognition system. Incoming speech is transduced into analog form, sampled 8000 times per second, converted to a sequence of 12-bit digital values, and analyzed in both the time and frequency domains to determine characteristic features. These features are compared to prototype features in the phoneme pattern memory to see if a match can be made. If an appropriate number of features match with a prototype, the phoneme is recognized.

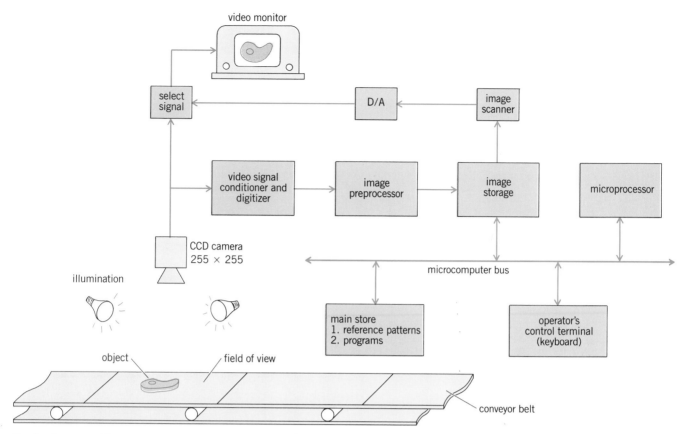

Fig. 4.   Computer vision system inspecting parts on a conveyor belt. Image signals sensed by the charge-coupled-device camera undergo preprocessing to make them amenable to analysis and recognition algorithms.

building, surrounding trees, and white puffs of clouds in a blue sky. These objects have meaning to the perceiver and result from a complex analysis of the image. In contrast to this, the color image may be transformed into a black-and-white image. Although this transformation changes the image, it does not produce a meaningful description of the objects.

Image meaning may be in the eye of the viewer. If the viewer is a machine dedicated to a specific set of visual tasks, it is expected, based upon the principle of economy of means, that the machine would derive only image meaning that is relevant to its job function. This is fortunate, since it allows engineers to constrain the analysis problem to manageable proportions for a number of applications, such as visual inspection, identification, and object location. An intelligent machine vision system, therefore, is one that extracts from images meaningful information useful to the performance of some task.

A typical vision machine (Fig. 4) consists of a camera coupled with some image-preprocessing hardware and a digital computer with image analysis programs. Cameras are either of the Vidicon raster scan variety or the newer charge-coupled device array imagers. The camera image must be converted into a two-dimensional grid of discrete points to make it compatible for computer storage and analysis.

The number of grid points or picture elements, called pixels, determine the amount of image data provided to the computer algorithms for analysis (Fig. 5). For some applications, 128×128 pixel arrays appear sufficient, while others require 512×512 (somewhat better than television). Typically, black-and-white pixel intensity is represented by an 8-bit number with 00000000 representing a black pixel and 11111111 a white pixel. In-between values yield 254 different gray levels.

Even with a modest 128×128 array of 8-bit pixels, the storage requirement for one image exceeds 128,000 bits. Processing such large quantities of information in a timely manner can tax even large mainframe computers. Considerable reduction of pixel information, the amount of processing per pixel, and faster technology are prerequisites for current applications. For example, in many instances recognition of an object on a conveyor belt or a platform requires only a two-dimensional outline of the object. If the object and the background have high contrast, a simple black-white binary code for each pixel reduces storage requirements by a factor of 8. Algorithms for processing two-dimensional binary images are also faster than gray-level image processing. Together these advantages translate into feasible real-time processing in the manufacturing environment.

Industrial applications of computer vision have

Fig. 5. Picture quality increases with the number of pixels: (a) 4 × 4 pixels, (b) 8 × 8, (c) 32 × 32, (d) 64 × 64, (e) 128 × 128, (f) 256 × 256. (*From D. H. Ballard and C. M. Brown, Computer Vision, Prentice Hall, 1982*)

Fig. 6.   Commercial computer vision system. (*Automatix*)

begun. Since the late 1950s considerable research effort has been expended on automating the sense of sight, resulting in an increase in knowledge about the physical, psychological, and cognitive dimensions of vision. Until the mid 1970s, applications of this knowledge to image analysis and pattern recognition were primarily restricted to military and space programs because of the cost and complexity of the equipment and the typically long computer-processing times. These cost factors have decreased rapidly in the past few years, stimulating new applications and more extensive research. The development of low-cost microcomputer and digital memory technologies are responsible for making cost- and time-sensitive industrial applications feasible.

General Motors Sight-I system determines the pose (position and orientation) of an integrated circuit chip assembly, inspects the chip for broken pieces, rejects defective assemblies, and aligns test probes with the electrical contact pads of a properly posed, nondefective chip. All this is accomplished in less than a second.

Automatix (Fig. 6) and General Electric make two-dimensional computer vision systems for parts identification, visual inspection, and pose determination. Designed to coordinate with robotic manipulation or to stand alone, these systems employ several microcomputer subsystems, along with special vision preprocessing units to speed up the analysis. Prototype objects to be recognized are presented to the system to train it. Features such as area, perimeter, centroid, moments, color, and number of holes are extracted for each prototype. Enough features are selected during training so that the vision system can reliably differentiate the prototype objects from each other and reject objects which do not have enough features of any one prototype.

When a vision machine is in operation (Fig. 6), a conveyor belt, robot, or other positioning device presents an object to the camera's field of view. A picture of the object is taken, digitized, and analyzed. The features computed from the image are compared to those of the prototypes in "feature space" and a "best" match determined.

Machines with tactile sense. Producing a machine with touch capabilities appears less difficult and exotic than producing machines with vision, but the range of applications seems to be of less economic value. However, industrial and manufacturing applications abound for machines with tactile capabilities. Touch allows both the detection of an object grasped in a robot end effector (or hand) and the control of the grasping force to prevent breakage and slippage. There is an adaptable touch sensor on the market that allows a robot to guide a deburring tool along the contours of a work piece whose surface is only partially known. Requirements for touch sensors to "feel" machined surfaces for cracks or mars exist, but such sensors have not yet been developed. Simple limit switches and whisker probes detect objects through direct contact, so they can be considered low-level touch sensors without intelligence. Such switches find wide application in all types of machines, from robots to lathes and milling machines.

If force and torque sensing are considered to be a form of touch, then that sense plays an essential role in automatic assembly. A robot wrist sensor (Fig. 7) has been developed that measures the forces experienced by the robot hand as it mates parts in the assembly process. The robot's knowledge of these forces indicates whether the parts have jammed or have mated smoothly. Computer algorithms can utilize this touch data to implement corrective procedures when required.

More sophisticated touch sensors, consisting of binary switch arrays, have been investigated in research laboratories. Each binary switch, or touch element (touchel), in the array indicates whether there is contact with the object at that point. From the spatial arrangement of on and off touchels, information about the object's shape can be determined. Intelligent touch machines of this nature are in research and in the early stages of commercial use. As industrial robots become more sophisticated, machines with touch capabilities will augment computer vision and provide the robot with a more complete view of its world.

Problems. Training, feature extraction, storage, and comparison characterize all forms of machine perception and pattern recognition, whether vision, speech recognition, or tactile sensing. While conceptually simple, machine perception presents difficult problems. What features are important? How can they be extracted from the raw sense data? Does the physical sense data have enough intrinsic information for recognition, or must other sources of knowledge be employed to provide a greater context? How should features be compared; for example, how red is red? What is an effective, reliable way of classifying an object? What is meant by "best match"? Once an image has been analyzed and objects recognized, how are their functional and spatial relationships represented on the computer so that

Fig. 7.   Robot wrist sensor for detecting forces on the hand.

Fig. 8.   Small industrial robot, the Unimation Puma 250. (*Unimation*)

these relationships can be manipulated to derive higher levels of information? The list goes on, but this should suffice to illustrate that while it is relatively easy to acquire a general grasp of the principles of machine perception, designing engineering machines which can perform the various perceptual tasks is by no means a simple process.

Intelligent robots. Industrial robots do not presently possess the overall total and coordinated mechanical dexterity of human articulation. The design and construction of a mechanical hand comparable to the human hand may serve as the ultimate goal in robot mechanism design. Mechanical dexterity alone, however, does not make a robot intelligent. To be intelligent a robot must sense environmental conditions and create motion to respond to those conditions through a decision-making process based upon some underlying task objectives. The broader the scope of such decision making, the more intelligence required.

Mobile robots. Intelligent mobile robots can be used in space as planetary exploratory vehicles. Since the round-trip time of flight for radio signals between

Earth and Mars is about 20 minutes, tight Earth-based control of a vehicle roving on the surface of Mars would be inefficient and would place the machine at great risk of falling into a crevice. Such vehicles must have their own intelligence in order to cope with environmental conditions that threaten their existence or mission. Earth uses of mobile robots include activities such as aiding the handicapped, harvesting, mining, nuclear power plant maintenance, and mobile military robots (which might serve as scouts to assess enemy strengths and weaknesses).

Industrial robots. Most industrial robots are stationary, and applications for them have advanced from the execution of simple, prestored, pick-and-place motion trajectories to complex assembly operations requiring real-time decision making, a feature of intelligent behavior. Early applications in materials handling and spot welding have led to more difficult tasks, such as paint spraying, machine tool loading and unloading, arc welding, and parts assembly. Implementation of these applications and an industrial robot's ability to produce any useful work

Fig. 9. Large industrial robot, the Cincinnati Milicron T³.

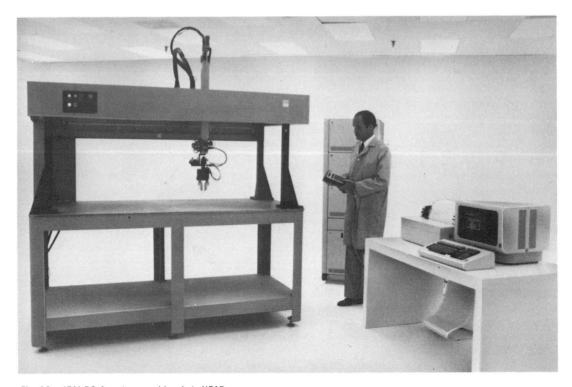

Fig. 10. IBM RS-1 parts assembly robot. (*IBM*)

still demand a highly structured environment. Adding intelligence to robots increases flexibility and reduces the need for special jigs and fixtures. The increased cost of intelligent robots and the associated potential loss in reliability due to their inherent complexity, on the other hand, tend to work against these advantages.

Industrial robots come in a variety of configurations, sizes, and load capacities. The anthropomorphic-style manipulators, such as the Unimation Puma 250 (Fig. 8) and the Cincinnati Milicron T^3 (Fig. 9), manipulate loads of 1 kilogram and 45 kilograms and have maximum reaches of 0.5 meter and 2.5 meters, respectively. Both of these machines have six revolute axes under computer control. Each axis can be set separately to position the robot end effector anywhere within its reach. Specially designed computer programming languages allow users to develop motion programs, either off-line or through a teach pendant, and to interface with the sensors. The Puma 250 is designed for light industrial applications, such as printed circuit board assembly, while the T^3 handles heavier-duty tasks.

The six-axis computer-controlled IBM RS-1 robot (Fig. 10) has three sliding joints that produce cartesian motion along the x, y, and z axes of a right-handed coordinate system, and three rotating joints (roll, pitch, and yaw) at the gripper's wrist. These six degrees of freedom allow the robot to position and orient the gripper in almost any configuration within its three-dimensional, rectangular work envelope. The RS-1 has been configured within IBM to assemble one of the company's tape drive transport mechanisms and to correct many error conditions detected during the assembly process by using a sensitive, tactile force sensor integrated into the gripper. One assembly error corrected by the robot system illustrates this capability. When the RS-1 attempts to insert a gear into a gear train on the tape transport and the two gears do not exactly mesh, the tactile sensor on the gripper detects a large force. By rotating the gear to be inserted slightly, the robot is able to mesh the gears and slide the added gear down its mounting pin. Only 10% of the programming for this particular assembly function actually involves direct control of the robot; the other 90% provides the robot with enough intelligence to interpret the gripper tactile sense data and invoke the appropriate manipulative responses.

The nascent field of intelligent machines spans a number of disciplines. Future advances will require engineering teams working cooperatively on mechanism and sensor design, machine intelligence programming techniques, and total systems integration.

[KEITH L. DOTY]

Bibliography: D. H. Ballard and C. M. Brown, *Computer Vision*, 1982; N. R. Dixon and T. B. Martin (eds.), *Automatic Speech and Speaker Recognition*, 1979; K. L. Doty, *Fundamental Principles of Microcomputer Architecture*, 1979; N. J. Nilsson, *Principles of Artificial Intelligence*, 1980; R. Reddy, Words into action, II: A task oriented system, *IEEE Spectrum*, pp. 26–28, June 1980; P. H. Winston, *Artificial Intelligence*, 1977.

Evolution of

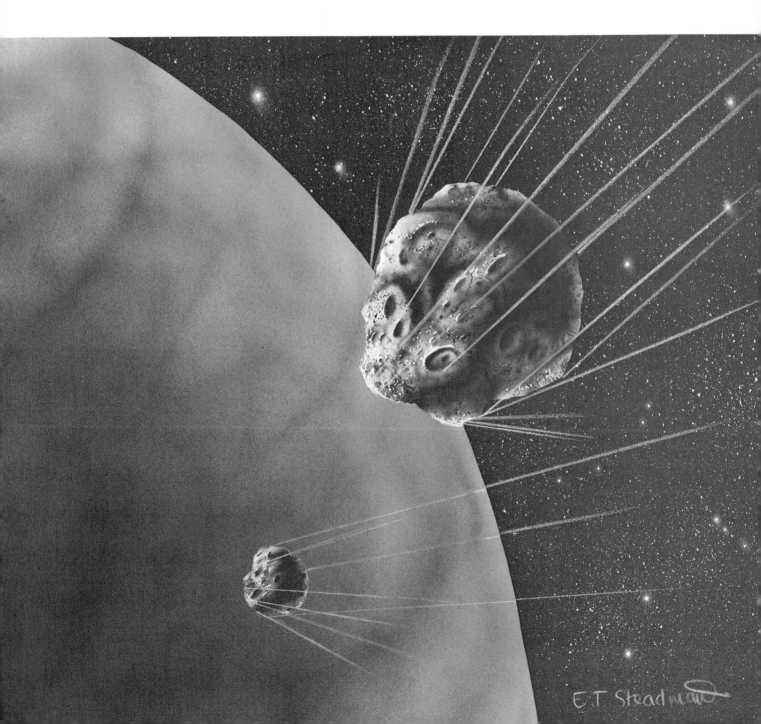

E.T. Steadman

the Earth's Crust

W. G. Ernst

After receiving a doctorate in geology from Johns Hopkins University, W. G. Ernst in 1960 joined the Earth and Space Sciences faculty at the University of California, Los Angeles, eventually becoming professor of geology and geophysics and a member of the Institute of Geophysics and Space Physics. Author of many journal articles as well as books, he concentrates his interests in rock-forming silicates, high-pressure metamorphic terranes, and the petrogenesis of oceanic crust and mantle.

The initial formation and chemical evolution of the Earth are incompletely understood. The interior of the planet—mantle plus core—has undergone profound thermal mineralogic and chemical changes over the course of geologic time, but direct evidence concerning these earliest stages is not obtainable because of the inability to sample and measure even the uppermost mantle. The atmosphere and hydrosphere also have evolved with time, but circulation and mixing, extremely rapid on a geologic time scale, have thoroughly destroyed evidence of precursor stages. Accordingly, scientists turn to the rocky outer shell, or sialic crust, of the Earth for information regarding the earliest phases of planetary development. Because the modern type of sea-floor spreading and subduction result in the recycling of the oceanic basement and its underpinnings (at current rates, every 50–200 million years), information regarding the nature of the early Earth resides chiefly in the continents.

The sialic crust, however, represents only a very small portion of this chemically differentiated planet. To understand the production of this surficial slag, the inferred history of the core and mantle must briefly be considered. On a grander scale, differentiation of the Earth as a whole can be placed in better perspective through consideration of the origin and general evolution of the terrestrial planets. Therefore, to begin with, a brief, necessarily speculative account of the formation of Earth, Moon, Mars, and Venus is needed before the terrestrial rock record is examined. The paleomagnetic evidence of ancient continental motion, and some constraints on past plate tectonic regimes will then be addressed.

A few definitions regarding geologic time are in

order. Three principal intervals are recognized: the Archean (from the formation of the planet to 2.5 Ga ago; Ga = billion years); the Proterozoic (from the close of the Archean to the Precambrian-Cambrian boundary); and the Phanerozoic (the time of megascopic life, commencing about 0.6 Ga ago). The time divisions employed in this article are as follows:

$$\left.\begin{array}{l} \text{Archean, 4.5–2.5 Ga} \\ \quad \text{Hadean, 4.5–3.9 Ga} \\ \quad \text{Early Archean, 3.9–3.3 Ga} \\ \quad \text{Middle Archean, 3.3–2.9 Ga} \\ \quad \text{Late Archean, 2.9–2.5 Ga} \\ \text{Proterozoic, 2.5–0.6 Ga} \\ \quad \text{Early Proterozoic, 2.5–1.6 Ga} \\ \quad \text{Middle Proterozoic, 1.6–0.9 Ga} \\ \quad \text{Late Proterozoic, 0.9–0.6 Ga} \\ \text{Phanerozoic, 0.6 Ga–present} \end{array}\right\} \text{Precambrian}$$

THE FIRST 600 MILLION YEARS

Isotopic evidence from meteorites and the Moon indicates that the age of the solar system is approximately 4.6 Ga. Gravitational collapse of a locally high density of interstellar gases in the solar nebula resulted in the formation of a rotating disk. The greatest mass of this proto–solar system became concentrated toward the center of the disk, resulting in ignition of thermonuclear reactions and consequent formation of the Sun. The accretion of planetesimals in more peripheral regions of the disk gave rise to discrete planetary bodies orbiting this newly formed central star. Although initially such condensates may have been rather cold, the burial of heat during successive impacts probably resulted in increasingly hot accretion. This primordial phase of formation of the solar system was probably completed by about 4.5 Ga ago, when the story of the Earth really began.

Planetary heating. Because of the relative abundance of the radioactive isotopes of such elements as uranium, potassium, and thorium in the newly formed Earth, a significant amount of planetary self-heating would have occurred subsequent to the accretionary stage. Moreover, most planetary materials, except for metals, are poor thermal conductors; radioactive and impact heating thus would have promoted incipient melting of the smaller terrestrial bodies and a substantial degree of fusion for the more massive ones. Because iron melts at a lower temperature than do silicates under high confining pressure (iron and silicate liquids being immiscible), a dense, iron-rich melt would have formed and migrated down the gravity gradient, displacing silicates upward and thus producing a molten metal core. The conversion of kinetic energy to heat during this process would have liberated additional thermal energy, contributing to the fusion and separation of metal core and silicate mantle. The slight cooling that has occurred since this earliest stage of planetary differentiation has resulted in formation of the Earth's solid inner core. Differential flow in the outer, liquid portion of the core is regarded as the source of the dynamo responsible for generation of the Earth's magnetic field.

Rock formation. No terrestrial rocks have survived to the present from this early, or Hadean, stage of planetary evolution; indeed, the most ancient samples now known, from the Isua metasedimentary belt of western Greenland, and perhaps the tonalitic gneisses of eastern peninsular India, are no older than about 3.8 Ga. However, evidence of this stage of development is preserved on the Moon: The anorthositic impact breccias of the lunar highlands contain rock fragments at least as old as 4.4–4.2 Ga, and the highlands are heavily cratered, attesting to continued, intense meteorite bombardment there—and probably also on Mercury, Venus, Earth, and Mars—to about 3.9 Ga. Such meteoritic accretion was undoubtedly most intense during the earliest stages of planetary formation, and tapered off toward the end of Hadean time. The formation of the highlands appears to have resulted from the gradual accumulation of fragments and blocks of plagioclase, minerals that floated on a hypothesized lunar magmatic ocean 4.4–4.2 Ga ago or even earlier. The lunar maria are floored by younger (Early Archean) basalts, indicating that the thermal budget of the Moon was such that at least local production of magma occurred as late as 3.8–3.3 Ga ago.

Hydrosphere and lithosphere. By 3.8 Ga ago, water-laid sediments containing silicic, fusible, continental crust–type debris were being deposited on the Earth. Clearly, the temperature of the surface of Earth was less than a few hundred degrees Celsius, as required for the existence of liquid water. Granitic crust was also present. Thus, no mat-

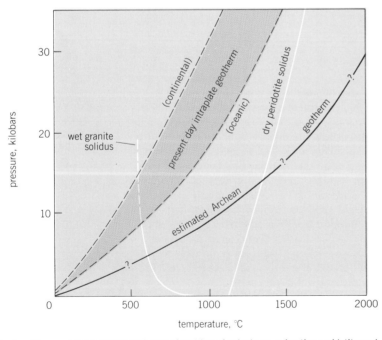

Fig. 1. Phase relations for the initiation of melting of anhydrous, primitive peridotite and of granite in the presence of excess H_2O. In the absence of H_2O, increased pressure elevates the (dry) melting temperatures of anhydrous phase assemblages, whereas aqueous fluids promote a lowering of (wet) fusion temperatures as pressure is raised. Also shown are modern intraplate geothermal gradients, and an estimated 3.6-Ga-old geotherm, assuming approximately three times the present-day heat generation and a stratified Earth. 1 kilobar = 100 megapascals.

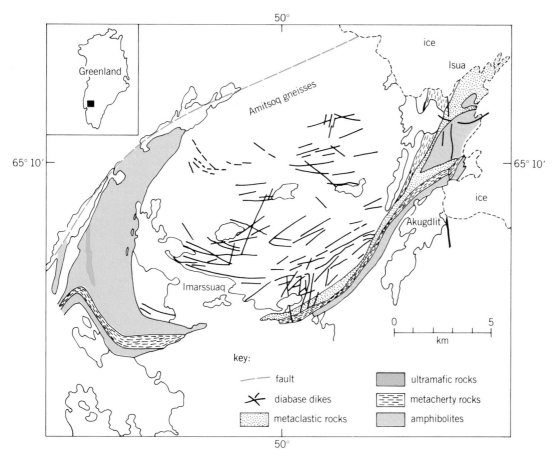

Fig. 2. Generalized geologic map of the Isua supracrustal belt of western Greenland. (*After J. H. Allaart, The pre-3760 Myr old supracrustal rocks of the Isua area, central West Greenland and the associated occurrence of quartz banded ironstone, in B. F. Windley, ed., The Early History of the Earth pp. 177–189, John Wiley & Sons, 1976*)

ter how the initial differentiation of the Earth was accomplished, a surficial layer of crust and overlying hydrosphere must have existed at least locally by latest Hadean time. And because the melting of mantle peridotite requires higher temperatures than those required for formation of oceanic and continental crust, at least a thin layer of solid mantle basement must also have been present by this time. Clearly, the chief components of the lithosphere—as well as both hydrosphere and atmosphere—were present on Earth by 3.8 Ga ago. This condition succeeded, and in part overlapped with, the earlier Hadean stage of Earth history, a primordial phase in the development of the planet. By analogy with the Moon, this precursor stage probably was characterized by: intense meteoritic infall during fractional crystallization of the hypothesized terrestrial magmatic ocean; widespread volcanism; abundant release of volatile constituents; sialic recycling (crustal foundering); and meteoritic reworking, or "gardening."

Geothermal gradients. Phase relations illustrating conditions of partial fusion, and inferred geothermal gradients for the present day and for the Early Archean Earth, are shown in Fig. 1. Although calculated temperature/depth relationships depend on the assumptions employed, the early Earth unquestion-

ably exhibited a higher average geothermal gradient than that which currently exists. The proportion of heat lost through a more efficient convective overturn of small, rapidly flowing cells probably would have been far greater during the Archean than at present. In any case, wet sialic crust—and even dry peridotite—would have begun to melt at shallower depths than at present (Fig. 1). For this reason, Archean continents and especially their solid mantle underpinnings, on the average, may have been relatively thin (that is, both the Mohorovičić discontinuity and the top of the asthenosphere probably were situated at shallower depths than they are today). As will be discussed later, mineralogic evidence suggests the local existence of Precambrian crustal thicknesses approaching those of present-day cratonal areas, but no evidence for ancient, comparably thick mantle lithosphere exists.

ARCHEAN ROCK RECORD

The oldest known preserved terrestrial rocks belong to the Isua supracrustal belt of western Greenland; they consist of amphibolites, ultramafic lenses, metacherts, carbonate-bearing mica schists, and metaconglomerates which contain volcanic clasts. The Isua supracrustals are engulfed by the 3.7-Ga-old Amîtsoq tonalitic and granodioritic gneisses (Fig. 2).

Granites and feldspathic gneisses of nearly comparable age crop out in the Minnesota River valley of the north-central United States and in eastern India.

Gray gneisses and greenstones. Archean complexes exposed in South Africa, Rhodesia, Western Australia, peninsular India, Canada, Brazil, and Fennoscandia bear testimony to the abundant occurrence of belts of greenstone and superjacent volcanogenic graywackes as old as 3.6–3.0 Ga. In general, these relatively small, linear complexes occur as feebly metamorphosed synclinoria surrounded by higher-grade tonalitic gneisses, pyroxene-bearing amphibolites, and granulites. The encompassing tonalitic terranes appear to truncate and deform the greenstone-graywacke belts, and are therefore probably of younger age (commonly 2.7–3.4 Ga old). However, the so-called gray gneisses of such complexes may represent rifted continental blocks between which mafic complexes were later extruded, or they may be remnants of the sialic substrate upon which plateau basalts and associated sediments were initially deposited (this appears to be the situation in Western Australia, Rhodesia, and parts of the Canadian Shield). For such gray gneisses, the geochronologic data would merely reflect a time of isotopic homogenization—and a resulting resetting of the isotopic clock—rather than their true geologic age. Such gneisses are antiformal, and their cross-cutting relationships to the feebly metamorphosed greenstone keels, as well as their much higher metamorphic grade, are probably indicative of buoyant, upward plastic flow from deeper crustal levels. These gray gneisses and associated granulites in turn are commonly intruded by migmatites and pink potassic postorogenic granites approximately 2.5–3.0 Ga old.

Archean assemblages of this sort suggest a model of downward movement of dense volcanogenic (mafic- and ultramafic-rich) supracrustal materials that displaced laterally, and in an upward direction, a complex of thermally softened, relatively buoyant, quartzofeldspathic rocks that were subsequently intruded by postorogenic granites. Figure 3 presents an example of this type from the South African Shield. Generally lacking from the geologic section in this area are mature platform-type strata such as orthoquartzites, thick layers of monomineralic limestones or dolomites, and widespread evaporitic sequences; peraluminous shales, redbeds, impure mixed carbonates, and banded iron formations of great lateral extent are also uncommon. Recumbent folds and low-angle thrust faults occur but are not common in Archean greenstone-granite terranes (Fig. 4).

Igneous rocks. The mafic igneous rocks so prominent in Archean greenstone belts consist chiefly of tholeiitic, commonly pillowed basalts, and are thought to represent submarine extrusions; these grade upward into andesitic lavas and pyroclastics. Such volcanic cycles are typically capped by immature clastic sediments, chert, or jaspilite. Komatiite and basaltic komatiite flows are the most distinctive rock types in the lower portions of many greenstone belts. These ultramafic lavas are characterized by MgO contents greater than 18% by weight, and CaO/Al_2O_3 weight ratios normally exceeding unity. That these rocks were completely molten is demonstrated by the occurrence of spinifex texture, that is, intermeshed bladelike crystals of olivine and orthopyroxene that in some instances exceed 1 meter in length. These crystals evidently grew perpendicular to the surface of nuclei-free, highly fluid, magnesium-rich volcanic flows. The occurrence of komatiite is significant, because anhydrous melts of such magnesian composition can be liquid only at temperatures approaching 1600°C, even at very low pressures. This rock type closely approaches the primordial mantle in composition, thus the occurrence of a komatiitic melt must reflect a high degree of mantle partial fusion—probably at considerable depth. The virtual restriction of such refractory lavas to Archean terranes seems to be a consequence of the high geothermal gradient of the early Earth and a declining heat flow over the course of geologic time. Extensive melting of the mantle therefore must have accompanied Hadean and Early Archean crustal differentiation, diminishing as the undecayed radioactive nuclides decreased in abundance and the Earth cooled.

Sedimentary units. Archean metasedimentary associates of the dominantly ultramafic, mafic, and more felsic igneous rocks of greenstone terranes are

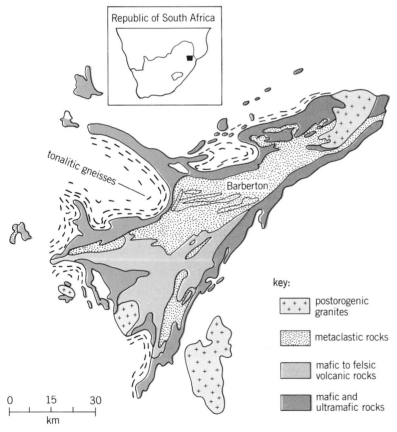

Republic of South Africa

tonalitic gneisses

Barberton

key:

+ + + postorogenic granites

metaclastic rocks

mafic to felsic volcanic rocks

mafic and ultramafic rocks

0   15   30
km

Fig. 3.   Map of the Barberton supracrustal belt, South Africa. (*After C. R. Anhaeusser, The Barberton Mountain Land, South Africa: A guide to the understanding of the Archean geology of western Australia; and also R. P. Viljoen and M. J. Viljoen, The geological and geochemical evolution of the Onverwacht volcanic group of the Barberton Mountain Land, South Africa, in J. E. Glover, ed., Symposium on Archean Rocks, Geol. Soc. Austral. Spec. Publ. no. 3, 1971*)

commonly interpreted as representing poorly sorted, first-cycle clastics; units rich in volcanogenic debris; and interlayered siliceous precipitates. Local derivation from an emergent volcanic belt seems required. The occurrence of cross-bedding, ripple marks, and other sedimentary features indicates that many such units were deposited in relatively shallow water.

Source of gneisses and granulites. The ultimate origin of the encompassing gray gneisses and granulites is obscure. Except for charnockites, these rocks are low in potash and have isotopic ratios indicating a mantle source. Some workers regard these largely tonalitic complexes as ancient, deeply eroded analogs of the batholithic belts that occur at modern convergent continental margins (for example, the Andes). If this interpretation is correct, the "postorogenic," Late Archean potassium-rich granites which intrude such high-grade metamorphic terranes would appear to represent mobilized sialic basement, with the interpreted upward buoyant movement presumably being a consequence of a higher geothermal gradient and of the partial fusion attending the local concentration of relatively fusible elements involved in mantle devolatilization. The reworked, secondary nature of such crosscutting plutons seems to be reflected by their enrichment in alkalis, high K_2O/Na_2O ratios, and disturbed isotopic natures.

Attendant physical conditions. The physical conditions suggested by the mineralogy and petrology of Archean crustal rocks must be considered. The presence in Archean sediments of water-worn clastic sulfide grains and, especially, of detrital uraninite—minerals which are resistant to long-term transport and attendant solution weathering only under essentially anoxic conditions—is consistent with the attendance of an early atmosphere/hydrosphere characterized by low oxygen fugacities. The occurrence of discontinuous, thin, but indisputably sedimentary, layers of Archean banded-iron formation is also consistent with a low oxygen fugacity, apparently required for transport of ferrous iron dissolved in aqueous solution across shallow basins for precipitation in regions typified by slightly more oxidizing environments. Low oxygen fugacity values apparently were maintained until about 2.0 Ga ago, permitting the formation of the great banded-iron formations of South Africa, Western Australia, Labrador, Minnesota-Ontario, and elsewhere during the Early Proterozoic.

Ferruginous cherts and related chemically precipitated sediments from the Archean of western Greenland and South Africa seem to have crystallized at temperatures approaching 100°C, whereas Proterozoic analogs from Canada, western Australia, and California yield apparent temperatures of 22–55°C, according to $^{18}O/^{16}O$ data. Exact values may be questioned, because oxygen-isotope paleotemperature determinations depend critically on the assumed isotopic composition of seawater and on the absence of postdepositional exchange, but the oceans evidently have been cooling over geologic time.

The virtual restriction of komatiitic lavas to the

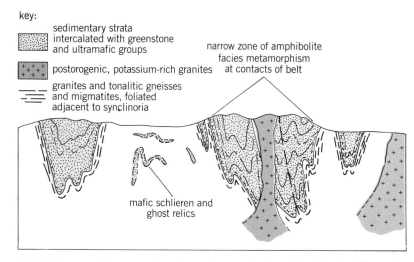

key:
- sedimentary strata intercalated with greenstone and ultramafic groups
- postorogenic, potassium-rich granites
- granites and tonalitic gneisses and migmatites, foliated adjacent to synclinoria

narrow zone of amphibolite facies metamorphism at contacts of belt

mafic schlieren and ghost relics

Fig. 4. Diagrammatic cross section of greenstone synclinoria enveloped in a tonalitic gneiss terrane. (*After C. R. Anhaeusser et al., A reappraisal of some aspects of Precambrian shield geology, Geol. Soc. Amer. Bull., 80:2175–2200, 1969*)

Archean suggests that, at least in the vicinity of the greenstone belts, geothermal gradients of the early Earth were higher than at present. The total absence of blueschists, eclogites, and other high-pressure, low-temperature metamorphic rocks from such terranes also attests to this ancient high heat-flow regime. This appears to be a consequence of the fact that heat production due to radioactive decay at and before 3.8 Ga was roughly three times greater than at present. Although the thickness of the early continental crust is a matter of some debate, disproportionately high Archean abundances of pyroxenebearing amphibolitic gneisses and sillimanitic granulites likewise suggest relatively elevated crustal temperatures.

Lithologic evidence thus indicates that the early Earth was characterized by: high geothermal gradients; a variable, but locally substantial degree of mantle partial melting; igneous rocks that were predominantly extrusive; an essentially anoxic atmosphere and ocean; and a compositionally immature veneer of locally derived volcanogenic sediments. The thermal structure for the Archean Earth necessitated both by calculated heat flow and by the occurrence of preserved granulitic and komatiitic terranes requires the presence of thin lithospheric plates and the generation of both oceanic- and continental-type protocrust as relatively near-surface phenomena. The maximum thickness of this early crust is uncertain, but was probably less than that of the present day.

The dimensions of Archean lithospheric plates are poorly constrained, but judging from the fact that preserved greenstone belts are relatively small linear features—typically on the order of 20×150 km in size—lithospheric plates may have been much smaller than modern ones. A large number of small, rapidly circulating upper-mantle convective cells would also be in accord with the high heat loss required by the inferred thermal structure of the early Earth.

CRATON FORMATION AND CONTINENTAL DRIFT

In general, available paleomagnetic data suggest that, with the notable exception of a proto–North Atlantic rifting/converging couple which produced the Caledonian orogenic belt, a Gondwana supercontinent existed from more than 800 million years ago up to the initiation during the Late Triassic of the current cycle of continental drift, at which time this megacontinent was disrupted by rifting. For rocks lithified during the Early Proterozoic, apparent polar wander curves for South America and most of southern and western Africa approximately coincide; at that time, then, at least these sialic masses appear to have been joined together as a supercontinent. A similar relation probably holds for the cratons of northern Africa and western Australia. Archean paleomagnetic data, however, are much less definitive. Nevertheless, it seems plausible that dispersal of the cratons may be a geologically recent development. During early portions of the Precambrian, accreting sialic segments may have been rapidly swept against one or more enlarging supercontinents, with more extensive areas of the Earth's surface being typified by open seas. In any case, accreting, presumably relatively rigid, continental crust-capped plates, as documented by polar wander curves, were apparently extant as early as about 2.6 and perhaps 2.8 Ga ago. During the Proterozoic, intracratonic orogenic belts formed across preexisting shield areas, as indicated by abundant radiometric data. Their large size must have helped in some manner to stabilize the supercratons.

Oceanic versus continental crust. Unfortunately, neither paleomagnetic nor geochronologic data are adequate to quantitatively estimate the area ratio of Archean ocean basin to continent. The proportion of the Earth floored by oceanic versus continental crust is a complex function of the amount and thickness of oceanic crust, with the lateral extent and thickness of the continents being determined by factors such as: (1) the geothermal gradient and the depth at which melting of fusible quartzofeldspathic material occurs; (2) the volume of the hydrosphere (hence the "freeboard" of the continents and their potential erosion to wave base); (3) the rate of sweeping together of continental debris by sea-floor spreading; and (4) the rate of production of sialic material through chemical differentiation of undepleted mantle. The Mohorovičić discontinuity may have been at relatively shallow depths during the early Precambrian, because of the relatively high geothermal gradient and the consequent fusion of deeply buried wet granitic crust (Fig. 1). The volumes of both hydrosphere and continental crust apparently have increased over the course of geologic time as a result of continued, but waning, mantle evolution. Convective overturn of the hot Archean asthenosphere undoubtedly was much more vigorous than at present. Accordingly, it can be concluded that the sialic crust was initially somewhat thinner, and that it was rather rapidly accreted into cratonal areas. Thus, it has been estimated that perhaps two-thirds, or as much as three-quarters, of the total volume of continental crust was produced during the Hadean and Archean, although prior to 3.8 Ga some of this material may have been returned to the mantle; isotopic data also support this concept of waning continental growth.

Continentality. Judging from the known rock record, the emergence of continental shelves and shallow epeiric seas of great areal extent may not have occurred until about 2.5 Ga. Archean sediments are typically first-cycle, chemically primitive, locally derived, and volcanogenic, whereas Proterozoic strata commonly include reworked, mechanically and chemically recycled units of much greater lateral continuity and chemical diversity. It is conceivable that the Archean microcontinental assemblies were largely submerged, as suggested by the apparent scarcity of platform sediments, with freeboard being widely attained only in the Early Proterozoic.

Continentality combined with transpolar drift allows for climatic extremes, and it is probably not a coincidence that the most ancient glacial deposits date from Early Proterozoic time.

Basaltic intrusives. Another piece of evidence supporting an increase in the thickness of sialic crustal material with the passage of time involves the nature of emplacement of basaltic magmas. Archean mafic igneous rocks are predominantly extrusive; huge floored intrusives of gabbroic magma appear in the geologic record only in units about 2.7 Ga old or younger. During the earlier Archean, sial may have been repeatedly breached by fluid ferromagnesian melts derived from depth, supporting the hypothesis that the continental crust was relatively thin. By the beginning of Proterozoic time, however, locally thick granitic crust apparently provided a gravitatively stable shield which could not be surmounted by dense mafic liquid (that is, the hydrostatic head on some such liquids was insufficient to permit their rise to the continental surface).

PLATE TECTONIC EVOLUTION OF THE EARTH

The configuration and present motions of lithospheric plates, a plate tectonic regime initiated during the Early Mesozoic opening of the Atlantic Ocean, and the fragmentation of Pangea, provide data regarding the current mechanisms by which oceanic crust is formed and destroyed. Effects of this modern cycle of plate tectonics can be reliably traced back into Paleozoic and latest Precambrian terranes, where structural evidence of convergent plate boundaries exists in the form of overthrust zones and allochthonous nappes. Sea-floor spreading as it occurs today produces a distinctive oceanic petrostratigraphic sequence that includes: tectonite peridotites; cumulate ultramafic-mafic complexes; sheeted diabase dikes and sills; and an overlying series of massive tholeiitic lavas, pillow basalts, breccias, and pelagic sediments. Such assemblages are found in Phanerozoic terranes, but similar suites are not recognized unambiguously from any but the youngest Precambrian sections. Paired metamorphic belts, also a typical product of Phanerozoic plate tectonics, are also rare or unknown in the Precambrian.

Preplate stage. The lithotectonic associations produced by plate motions in the Phanerozoic may thus be absent from all but the latest Precambrian. The possibility that plate tectonic processes did not operate on the early Earth must therefore be investigated. Certainly, this may have been the case on other terrestrial planets; the preservation of heavily cratered terranes on the Moon, Mercury, and Mars shows that these bodies were not subjected to major crustal reworking through the subduction of lithospheric plates, at least subsequent to the marked decline in intensity of meteoritic bombardment some 3.9 Ga ago. On the other hand, all of these planets have considerably smaller masses than the Earth and thus could not be expected to have sustained the impact-induced and radioactively induced high internal temperatures presumably required for the onset of plate tectonic processes for as extended a period. Thus, although primordial melting no doubt occurred on other terrestrial planets, the Moon, Mercury, and Mars evidently cooled to the stage at which the thermally driven gravitative instability required for mantle convection ceased prior to termination of the sweep-up of planetesimals. The larger mass of the Earth (and possibly Venus) allowed sustained retention of high temperatures and of the strong thermal gradients which undoubtedly served to fuel the heat engine that drives the lithospheric plates.

Platelet stage. Because this heat engine has existed since formation of the Earth's core in Hadean time, it seems evident that at least small, thin platelets must have bounded the Earth's dynamic surface beginning early in Earth history, a supposition that is consistent with the preserved Archean rock record and with the phase diagram in Fig. 1. Rapid convective overturn of the mantle probably would have driven such platelets against and beneath one another, but because of the elevated near-surface temperatures and the thinness of the plates, together with the small magnitude of the lithospheric/asthenospheric density inversion and the attendant minor gravitative instability, the lithosphere would not have been subducted to appreciable depths. This accounts for the lack of ancient high-pressure metamorphic rocks such as blueschists and eclogites, and for the absence of alkalic igneous suites (thought by most petrologists to be derived by extremely small degrees of partial fusion at substantial mantle depths), as well as for the local occurrence in Archean terranes of the highly refractory komatiites and related magnesian lavas. Because of the high rate of outgassing and high heat-flow regime of the primitive Earth, mafic volcanic activity would have been much more voluminous than at present, resulting in the concomitant rapid early development of quartzofeldspathic crust and the hydrosphere/atmosphere. Although Archean gray gneisses were no doubt remobilized during accretion and thickening of the cratons, their mantlelike geochemistries could well reflect their derivation from fertile peridotitic, basaltic, and amphibolitic (hydrated metabasaltic) precursors.

Supercratonal stage. Accordingly, the Archean–to–Early Proterozoic transition would have been characterized by a gradual change from small, thin, hot Archean platelets, driven about by numerous rapidly convecting asthenospheric cells, to more modern, relatively thick, cooler, laterally extensive, coherent plates, the motions of which may have been a function partly of negative buoyancy and partly of asthenospheric flow on a grander scale. By the Early Proterozoic, small sialic masses, produced chiefly during early mantle overturn and chemical segregation, had become assembled into major, thicker continental cratons. As freeboard thus became increasingly important, the processes of mechanical erosion and sedimentation began to play major roles in the production of chemically differentiated sedimentary facies.

Summary of stages. Thus, four transitional stages in the plate tectonic evolution of the Earth seem to be recognizable based on the work of P. E. Cloud, J. V. Smith, G. W. Wetherill, B. F. Windley, A. Kröner, and A. M. Goodwin: (1) a Hadean (preplate) stage, characterized by a partly molten planet that was intensely bombarded by meteorites and that underwent profound gravitative separation to form a metallic core, a ferromagnesian silicate mantle, and a continuously reworked crustal scum; (2) an Archean (platelet) stage, during which the Earth's surface was dominated by hot, soft, relatively thin, more or less unsubductable platelets that aggregated sialic material to form protocontinents; (3) a Proterozoic (supercratonal) stage, characterized by emergence of large continental shields, development of freeboard, intracratonal orogeny, and drift of supercontinents; and (4) a latest Precambrian-Phanerozoic (modern-type) cycle of plate tectonics, involving rifting, dispersal of continental fragments, subduction, and generation of long, linear, paired mobile belts at plate margins.   [W. G. ERNST]

Bibliography: J. H. Allaart, The pre-3760 Myr old supracrustal rocks of the Isua area, central West Greenland . . . , in B. F. Windley (ed.), *The Early History of the Earth*, pp. 177–189, 1976; C. R. Anhaeusser, The Barberton Mountain Land, South Africa . . . , in J. E. Glover (ed.), *Symposium on Archean Rocks*, Geol. Soc. Austral. Spec. Publ. no. 3, 1971; C. R. Anhaeusser et al., A reappraisal of some aspects of Precambrian shield geology, *Geol. Soc. Amer. Bull.*, 80:2175–2200, 1969; P. E. Cloud, Major features of crustal evolution, *Geol. Soc. S. Afr. Annexure*, 79:1–33, 1976; A. M. Goodwin, Precambrian perspectives, *Science*, 213(4503):55–61, 1981; A. Kröner (ed.), *Precambrian Plate Tectonics*, pp. 57–90, 1981; J. V. Smith, Mineralogy of the planets . . . , *Mineral. Mag.*, 43:1–89, 1979; R. P. and M. J. Viljoen, The geological and geochemical evolution of the Onverwacht volcanic group . . . , in J. E. Glover (ed.), *Symposium on Archean Rocks*, Geol. Soc. Austral. Spec. Publ. no. 3, 1971; G. W. Wetherill, The role of large bodies in the formation of the Earth and Moon, *Proc. Lunar Sci. Conf.*, vol. 7, pp. 3245–3257, 1976; B. F. Windley, *The Evolving Continents*, 1977.

Quality

Circles

Donald L. Dewar

Donald L. Dewar is president of the Quality Circle Institute in Red Bluff, CA. A pioneer in the quality circle concept, he has authored or collaborated on nearly 100 articles and papers, 19 books, and almost 2000 audiovisuals, and has implemented quality circles in hundreds of organizations throughout the world. He holds an M.B.A. degree from Santa Clara University.

Worker participation activities, under the stimulus of keen competition and a consumer demand for quality products and services, are assuming an increasingly important role in every kind of business and governmental organization. Quality circles, one of the important movements in the area of worker participation, is such an activity. Although relatively new in the Western world, it has grown to phenomenal proportions (with over 10 million participants) in Japan since its shaky beginning in 1962. It is credited with having contributed greatly to Japan's ascendency to economic and industrial strength. Quality circles, also known by a number of other names such as QC (quality control) circles and teams, is rapidly becoming a genuine movement throughout the industrialized world. This concept is concerned with an organization's most valuable (and most expensive) resource—one that is virtually untapped—its work force.

One definition of a quality circle is: a group of workers from the same area who usually meet for an hour each week to discuss their job problems, investigate causes, recommend solutions, and take corrective actions when authority is in their purview. The objectives are to effect improvements in quality, productivity, and motivation. Quality is a thought that resides in the mind of every executive. Participative management is the vehicle that puts it in the mind of every employee in an organization. Without attention to quality, sooner or later any organization will fail. As for productivity, clients and customers want quality services and products. Management alone cannot carry the burden of responding to these basic tenets of organizational survival. The intelligence and creative capacity of employees must be tapped to stem their complacency about poor quality and its inherent high cost. Every organization that uses quality circles reports improvement in employee morale (motivation). Commonly, an organization will recognize the indicators of improvement in employee morale: reduced absenteeism and lower employee turnover.

The key to achieving these objectives is participative management, and this approach should be encouraged from top to bottom of every organization. Responsibility for the quality of work should be returned to the worker; and the worker should be able to influence how the job can best be done. The worker's ideas and resourcefulness can often make the difference between success or failure of the best engineering concepts. The worker, being closest to the work itself, *can* make a good idea better. Success is assured when participation is encouraged, because it results in a strong commitment to attainment. Employees who are treated as important resources, and who are truly part of the team, will make "their" organizations formidable competitors in any field.

For decades the Japanese were viewed as producers of inferior merchandise. Today, Japanese exports are sought because of their reputation for high standards of design and quality. This transition was due in part to General Douglas MacArthur. In the postwar period, he knew that Japan's reputation for shoddy manufacturing would limit the acceptance of their merchandise, and so, with the help of Japanese leaders in government and industry, he obtained the services of various Americans to assist the Japanese in raising the quality levels of their products. Years later, in 1961, a series of exploratory meetings were sponsored by the Union of Japanese Scientists and Engineers (JUSE). When JUSE first suggested quality circles to its member companies, there was an enormous amount of skepticism. However, three circles were registered with JUSE during May of 1962, and by the end of the year there were twenty.

Even though quality circles were so successful in Japan, they did not reach the Western world until the 1970s. Quality circles were started at Lockheed in 1974, and their success drew the attention of many other companies in the United States and Europe. However, little occurred until 1977 when five other companies in the United States started circles; from there, the activity has grown steadily. In 1982 the number of companies in the United States using quality circles was conservatively estimated at 3000. The International Association of Quality Circles (IAQC), an organization established to provide guidance to those who wanted to initiate quality circle activities, was formed in late 1977.

OPERATION

Figure 1 illustrates the various processes that take place in the operation of quality circles. The first step is problem identification. The problems usually identified by circle members are uninteresting and mundane. These are problems that either no one else is aware of; or if others are aware of them they are too busy to give the problems attention or not interested enough to care. There is probably no one more qualified to identify, analyze, and solve these problems than the people who put up with them every day. Problems (often referred to as projects or themes) are suggested mostly by circle members. However, the circle should encourage ideas from as many other sources as possible, such as management personnel, staff members, and even nonmembers. There are various techniques used to identify problems, but the most effective one seems to be the brainstorming approach.

The list of problems developed in the previous step is carefully reviewed by the circle members and prioritized—the process of problem selection. The number one problem they wish to adopt as their circle project is selected by the leaders and the members.

Circle members have the responsibility of conducting the analysis of the problem selected. They may occasionally request temporary assistance from someone outside the circle, perhaps an industrial engineer, a technician, or a staff specialist. It is important that individuals brought in to give this momentary assistance do not try to take over the problems. It is preferable that only the assistance asked for be given; then the experts should step aside and let the circles continue as before.

The process of communicating the recommendations to the manager is one of the most exciting phases of quality circle operation. The management presentation is conducted as a stand-up function in which the members take part, using charts and graphs that they have prepared. It is common that preparations for the management presentation take place on the employees' own time at break, during lunch, and after work. Although nobody may ask them to use their own time, they do so because it is important to them to do the best possible job; it is a powerful and exciting way to communicate what they have done; and it is also an emphatic method of demonstrating the circle's value.

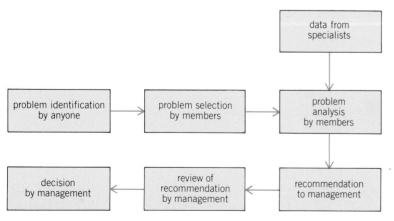

Fig. 1.   Flow process diagram for quality circle operations. (*After D. L. Dewar, The Quality Circle Handbook, Quality Circle Institute, p. F4–1, 1980*)

Quality circles operate through the normal management channels. The presentation is made to the individual to whom the supervisor (normally the circle leader) reports, rather than to the steering committee or to somebody on a high executive level. The normal channels are followed precisely, although higher-level management may be present at a management presentation. In fact, it is likely that higher-level management will often be present, but as observers, not as decision makers.

There is nothing automatic about implementing a quality circle solution. The decision of whether or not to do it has to come first, and the manager of the organization decides if it is the proper thing to do.

THE ORGANIZATION

There is no circle organization in the true sense of the word. There is only the normal management hierarchy and its relationship to the entire organization. The circle simply fits into it in a most complementary manner, as shown in Fig. 2. (The role of the facilitator, the coordinator of circle activities within an organization, is indicated in the block to the right.) The management personnel provide the support and guidance for circle activities through their participation on the steering committee. The supervisor is the circle leader, at least when the circle is new. The members are the employees in the area who desire to participate in the circle activities. Bringing it all together is the function of the facilitator, who coordinates all aspects of the operation.

Steering committee. The steering committee should be formed as soon as a decision has been made to implement quality circle activities. Fundamentally, the steering committee must set goals and objectives for quality circle activities. It should establish operational guidelines and control the rate of expansion. It should be presided over by a chairperson, and decisions should be reached by a democratic process.

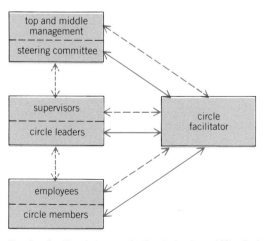

Fig. 2. Quality circle organizational structure. (*After D. L. Dewar, The Quality Circle Handbook, Quality Circle Institute, p. F10–1, 1980*)

Representatives from major departments within the company should be members of the steering committee. For example, in a factory it is common to have representatives from such diverse areas as manufacturing, quality control, personnel, education and training, engineering, finance, marketing, and the union. The facilitator should be a member also. Managers or top-level staff members are normally on the steering committee.

Establishing circle objectives. This is the process of identifying areas where improvement is possible. It could be as simple as one objective, or as complex as a dozen. The following is a list of objectives that might be considered for adoption by the steering committee:

Reduce errors and enhance quality
Inspire more effective teamwork
Increase employee motivation
Create a problem-solving capability
Build an attitude of problem prevention
Improve company communications
Develop harmonious manager/worker relationships
Promote personal and leadership development
Develop a greater safety awareness

Determining actions outside circle charter. Experience has shown that some items should be considered outside of the circle charter. These items, which generally include benefits and salaries, employment practices, policies on discharging employees, personalities, and grievances, should be identified and clearly spelled out so that the circles understand their boundaries.

Facilitator. One of several key ingredients for successful quality circle activity is the facilitator, sometimes called the quality circle coordinator, who coordinates the overall program. Although usually a full-time position, it may be part-time in smaller organizations. The facilitator should be selected by the steering committee as soon as a definite decision has been made to implement quality circles.

The facilitator trains the circle leaders; acts as backup to the leaders during member training; forms the link between the circles and the rest of the organization; works closely with the steering committee; attends most circle meetings, especially when a circle is new; and continually coaches the circle leader.

The facilitator is responsible for seeing that records are maintained on improvements in quality, reduction of costs, scheduled improvements, energy conservation, achievements, and improvements in safety as well as in attitudes. Most of these can be translated into dollar savings—a very understandable common denominator.

The leader. Experience demonstrates that circle activities will have a greater chance of success when the supervisor is the initial leader. The supervisor is already designated to perform a leadership role in that structure, and the quality circle concept gains quicker acceptance when it fits into the existing or-

Check Sheet

Error	June				total
	1	2	3	4	
addition	卌	卌 ⁄⁄⁄⁄	卌 ⁄	卌 ⁄⁄	27
multiplication	卌 卌 卌 ⁄⁄⁄⁄	卌 卌 卌 卌 ⁄⁄	卌 卌 ⁄⁄⁄⁄	卌 卌 卌	70
omission	⁄⁄⁄	卌 ⁄	⁄⁄	⁄	12
routing	⁄⁄		⁄	⁄⁄⁄	6
typing	卌 ⁄	卌	卌 ⁄⁄	卌 ⁄⁄	25
Total	35	42	30	33	140

Fig. 3. A check sheet designed by circle members has been used to collect data. (*After D. L. Dewar, The Quality Circle Handbook, Quality Circle Institute, p. 4–8, 1980*)

ganizational structure. If a quality circle did not operate within the existing organizational setup, it might be viewed by some as a competing organization.

If someone other than the supervisor is a leader, the quality circle would probably evolve in the following manner. The supervisor becomes the first leader. Later, the leader identifies another individual, usually a lead person, to act as an assistant leader. Eventually, the operation of the circle becomes smooth and efficient, and the supervisor feels no threat by not being personally in command. It is at this stage that the circle may consider electing its leader. The circle will likely be one or two years old when this occurs.

After the leader has received training, the circle is organized, and the leader provides training to the members. The facilitator is present, but only as a backup, during the first few training sessions.

Assuring successful activities. Successful quality circle activities are rarely due to chance. They are the result of the presence of certain essential elements, and the absence of one or more of these elements will impair the degree of success. The steering committee should be aware that, as a body, it can contribute immensely toward ensuring that these basic foundation blocks are in place:

Management is supportive.
Participation is voluntary.
There is a people-building attitude.
Training is provided.
Teamwork is encouraged.
Recognition is provided.
Problems are selected in members' areas of expertise.
Problems are solved, not just identified, by the circles.

TRAINING

Many problem-analysis programs fail because the participants are not adequately equipped to analyze, but there are tools to increase analytical effectiveness. Some worker-participation programs equip only supervisory personnel with problem-analysis skills since the supervisors are supposed to be the problem solvers. In quality circles, however, those most knowledgeable, the members, are very involved as problem solvers.

The basic analytical building blocks are the quality circle techniques. They are brainstorming, data gathering, Pareto analysis, cause-and-effect analysis, management presentation, histograms, control charts, stratification, and scatter diagrams.

Brainstorming. The technique used to bring everyone's ideas out into the open is brainstorming. Each member, in turn, voices one possible cause of the problem. These ideas spark enthusiasm and originality. Wild ideas are safe to offer because the rules of brainstorming do not permit criticism or ridicule. All ideas that are presented are recorded for later analysis.

Data gathering. A major function of the circles is to analyze problems. Usually, before analysis can begin, data must be accumulated. Since this is frequently done by the circle members, training in data-gathering and sampling techniques is necessary to ensure accuracy and save time. Check sheets are one of the convenient and economical ways to collect data. They are designed by the members and may take on the appearance illustrated in Fig. 3.

Pareto analysis. This is a technique that separates the few important problems from the many trivial problems. A completed analysis is depicted in graph form in Fig. 4. Each column represents a different problem, and the tallest is the problem that will be solved first because it is the most important.

Cause-and-effect analysis. This is a widely used quality circle technique. A diagram with the appearance of a fish bone is constructed by the members, as illustrated in Fig. 5. The problem is stated in the box to the right. The possible causes are

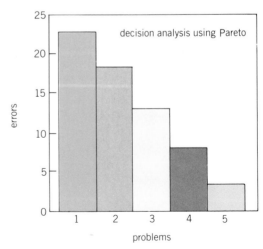

Fig. 4. A Pareto chart arranges the various problems in descending order of importance. (*After D. L. Dewar, The Quality Circle Handbook, Quality Circle Institute, p. 5–5, 1980*)

identified by the members, who circle the most likely causes and then prioritize them. The most likely cause is then verified in some manner.

Management presentation. The best plan will fail unless it is properly sold. Several times a year each circle has to do just that, by using a presentation setting to make recommendations or to indicate status to their manager. Their training in presentation techniques includes the basics of public speaking and the fundamentals of preparing and using graphs and charts such as histograms, scatter diagrams, and control charts. Each member usually takes part in this activity, despite the fact that many of them are extremely uncomfortable. It is viewed by members as a powerful form of recognition.

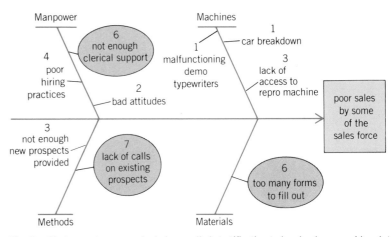

Fig. 6.   Circle members use a technique called stratification to break a larger problem into its component parts. (*After D. L. Dewar, Stratification: A Leader Manual, Quality Circle Institute, 1982*)

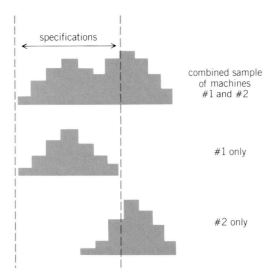

Fig. 5.   Completed cause-and-effect problem analysis. (*After D. L. Dewar, The Quality Circle Handbook, Quality Circle Institute, p. 6–10, 1980*)

Stratification. Sometimes a problem is best analyzed by taking it apart and examining each piece separately. For example, an excessive number of errors are occurring in one large department, as illustrated in the top portion of Fig. 6. It may be best to separately analyze what the error rate is within each group in that department. Perhaps the problem exists in only one small area; for example, machine number 1 in Fig. 6 is producing parts within the required specification, but number 2 is not.

Other training provided to members may include case studies of successful solutions worked out by other circles, communications, human relations, and group dynamics.

SUMMARY

Quality circles have demonstrated impressive gains in organizations where they have been used. It is common for organizations to realize from $3 to $6 in cost savings and cost avoidance for every dollar invested in quality circle solutions. Literally tens of thousands of organizations around the world are involved in quality circles; the greatest concentration of circles is in Japan, but their spread to the Western world involves thousands of companies and the list is rapidly growing. Organizations engaged in manufacturing, petroleum, chemicals, agriculture, medical care, banking, insurance, government, merchandising, the military, and so on—wherever people are involved in producing goods and services—are using quality circles. Also, the size of the organizations using quality circles varies tremendously, extending down to those that employ fewer than 100 people, or indeed sometimes as few as 10 people.

More than just a program, quality circles is a philosophy, a way of life which directly involves employees at the shop level in analysis and solution of product-quality problems. It does not just involve asking employees for suggestions—suggestions that someone else might analyze and implement. It is a bold step forward in reestablishing the dignity of the workers and the jobs they perform. It also provides workers with training in the basic tools of problem analysis so that they have the means to study and solve problems. In short, the people doing the work learn that they, not inspectors, actually control quality.   [DONALD L. DEWAR]

Cloning

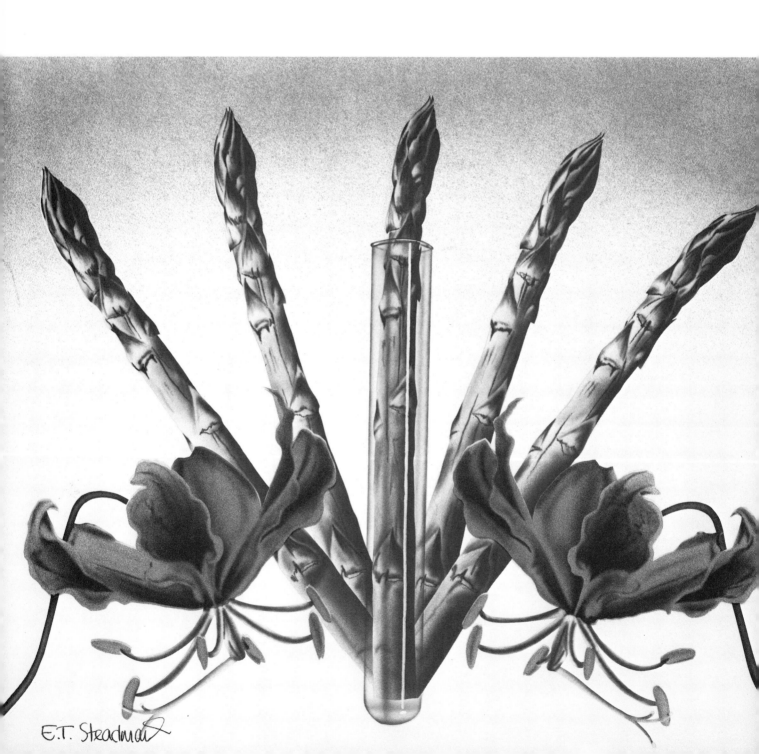

E.T. Steadman

Agricultural Plants

Bob V. Conger

Bob V. Conger is a professor of plant and soil science at the University of Tennessee. Recipient of a doctorate in genetics from Washington State University, he has published more than 60 scientific papers as well as a book on plant tissue culture, and serves as an editor of one journal and as an associate editor of another. His current research interests are in plant cell and tissue culture and the breeding of forage grasses.

Clone was defined by the Greeks as a twig or slip suitable for propagation. A current definition is the aggregate of individual organisms descended by asexual reproduction from a single individual, including all life forms from microorganisms to multicellular plants and animals. Indeed, now the term cloning is even used for the reproduction of deoxyribonucleic acid (DNA) from plants and animals in bacterial cells. In this article, the term cloning will be restricted to the application of cell and tissue culture technology and its potential for the improvement of agricultural plants. This is essentially correct usage since asexual reproduction is involved at some stage in the various manipulations of cultured cells to be discussed. A primary exception may be embryo culture, in which zygotic embryos (formed from sperm plus egg) are rescued from the mother plant and cultured aseptically on an artificial medium.

Manipulation of plant cells and tissues in culture constitutes a part of the new plant biotechnology, or genetic engineering, research that is expected to contribute to plant improvement efforts. The rate of genetic improvement of important crops has decreased in recent years, and yield plateaus, or ceilings, appear to have been reached for many species. The potential for study and manipulation of crop plants in test-tubes and petri (culture) dishes has created exciting possibilities in both fundamental and applied plant science for mass propagation, creation of new genetic combinations, development of disease-free plants, shortening the time period to produce new varieties, and so on. It has also attracted the interest and investment of several industrial firms, ranging from small, new enterprises to major oil, chemical, and seed companies.

Plant tissue culture is based on the concept of cellular totipotency; that is, individual cells possess the genetic information and physiological machinery to divide, differentiate, and develop into entire plants. In a broad sense, plant tissue culture may be defined as the growing of plant organs, tissues, cells, and protoplasts on a sterile synthetic medium containing hormones, mineral salts, vitamins, and an energy source (a form of sugar) under controlled conditions of light and temperature. Cells are produced from tissue or other cells which will later develop into plants. Tissue culture allows the techniques available for studying microbial organisms to be applied to higher plants. For example, the screening of 1 million cells for a particular trait, such as cold tolerance, requires a petri dish of only 2.5-in. (10-cm) diameter, whereas screening 1 million plants spaced on 1-ft (30.5-cm) centers requires almost 23 acres (9.3 hectares) in the field. Furthermore, the time required to grow and screen plants at the cellular, compared to whole plant, level may be reduced from months to a few weeks or even days.

Unfortunately, not all plant species respond similarly to tissue culture techniques. Much of the technology which was developed for so-called model plant systems, such as tobacco (*Nicotiana* species) and jimsonweed (*Datura* species) has been difficult to apply to important food and feed crops in the Gramineae (grasses and cereals) and Leguminosae (peas, beans, clovers, and so forth) families. Significant progress is now being made toward understanding basic cellular processes which will allow application of these techniques to the improvement of important crop species. The remainder of this article will discuss the progress and potential of cell and tissue culture technology in crop improvement. Examples with various plant species, both agricultural and nonagricultural, will be considered where appropriate.

HISTORICAL BACKGROUND

Interest and recognition for the technology's potential in crop improvement has flourished especially since the early 1970s. However, it is generally agreed that plant tissue culture began with the work of the German scientist G. Haberlandt. Working at the turn of the century, Haberlandt recognized the possibility of culturing isolated vegetative cells in simple nutrient solutions, demonstrating totipotency and studying the development from cell to multicellular whole organism. Although he was not successful in obtaining cell divisions, over the next several years others attempted similar experiments based on his work. Some of the significant events in plant cell and tissue culture since the beginning of the century are listed in the table.

The discovery of auxin (indoleacetic acid, IAA) by F. W. Went in 1928 was one of the most important finds in plant physiology and had a tremendous impact on plant tissue culture. Recognizing the importance of this hormone, the American scientist Philip R. White, and French scientists P. Nobécourt and R. J. Gautheret made significant discoveries in the 1930s. In 1934 White established an actively growing clone of tomato roots. During the same year Gautheret reported that pieces of tree cambium proliferated into an algalike growth when placed on a solidified (agar) medium containing a sugar (glucose) and an amino acid (cysteine hydrochloride). In 1937 White discovered the importance of B vitamins in the growth of cultured roots, and Nobécourt obtained cell proliferation from carrot roots. Also in the late 1930s Gautheret's work with carrot and White's work with a tobacco hybrid revealed that plant tissue would proliferate undifferentiated cells that could be repeatedly subcultured.

Kinetin, the first cytokinin, was discovered in 1956 by C. O. Miller and colleagues working in the laboratory of Folke Skoog at the University of Wisconsin. Skoog and Miller subsequently found that auxin-cytokinin interactions controlled differentiation in tobacco tissue cultures. A high auxin-to-cytokinin ratio caused root formation, whereas the reverse relationship caused shoot initiation.

One of the most significant developments was the growing of carrot cell cultures to entire plants by F. C. Steward at Cornell University in 1958. This event was one of the very first to affirm Haberlandt's totipotency concept. During the early to mid 1960s Skoog and his students, T. Murashige and E. Linsmaier, continued to make significant advances in the development of artificial growth media. The Murashige and Skoog formulation, published in 1962, and derivations and modifications thereof, is probably still the most widely used medium today.

The formation of haploid embryos from cultured anthers of jimsonweed was reported by Indian scientists S. Guha and S. C. Maheshwari in the mid

Some important discoveries in plant tissue culture

Event	Year	Scientists
First attempts to culture isolated cells	1902	G. Haberlandt
Discovery of auxin	1928	F. W. Went
Establishment of actively growing and dividing cultures	1930–1939	P. R. White, R. J. Gautheret, P. Nobécourt
Discovery of kinetin and the role of auxin-cytokinin interactions in control of differentiation	1955–1957	F. Skoog and C. O. Miller
Generation of entire plants from carrot cell cultures	1958	F. C. Steward
Enzymatic isolation of protoplasts	1960	E. C. Cocking
Mass production of haploid tobacco plants by pollen culture	1969	J. P. Nitsch and C. Nitsch
Regeneration of whole plants from tobacco protoplasts	1971	I. Takebe and coworkers
Creation of hybrid tobacco plant by protoplast fusion	1972	P. Carlson and coworkers

1960s. Later in the same decade J. P. Nitsch and Colette Nitsch described a method for producing hundreds of haploid plants from pollen grains of tobacco. Haploid plants in which the cells contain only the chromosome number of the gamete and one-half that of somatic cells are of great interest to plant breeders.

One of the most exciting areas of plant tissue culture, and one which has created the most interest, is protoplast culture. Protoplasts are plant cells in which the wall has been removed, usually by digestion with special enzymes. In 1960 E. C. Cocking reported the release of protoplasts from tomato root-tip cells by using a fungal cellulase. By using a sequential process of crude pectinase to release the cells and then cellulase to digest the cell wall, J. Nagata and I. Takebe obtained protoplasts from tobacco leaf cells. Takebe and coworkers then reported the regeneration of whole plants from tobacco leaf protoplasts in 1971. These events paved the way for the development of a hybrid between two tobacco species by fusing leaf protoplasts from the two species and then growing the hybrid (fused) cell to a plant. This was accomplished by Peter Carlson and colleagues in 1972.

Research during the past 10 years has led mainly to the improvement of culture and fusion techniques, the regeneration of plants from tissues and cells from a wider range of species, and the creation of hybrids by somatic cell fusion between more, including sexually incompatible, species. However, there is much progress yet to be made, especially with the culture of species in the Gramineae and Leguminosae families, before the technology can be routinely used in the improvement of important agricultural plants.

MASS PROPAGATION

The use of culture techniques for asexual (vegetative) propagation is probably the most advanced area of plant tissue culture, and is currently the one which has the most practical application. Commercial establishments engaged in propagating a variety of plants by this method are increasing in number each year. The primary advantages compared to conventional methods for clonal propagation are savings in space and time and the greater numbers which can be produced.

Cloning by either conventional or tissue culture techniques is especially valuable for heterozygous, sexually self-incompatible, and sterile plants. These characteristics occur in many asexually propagated ornamental, vegetable, and fruit crops.

Methods of tissue culture propagation. Multiplication in tissue culture can occur through: (1) enhanced formation of axillary shoots on shoot tips,

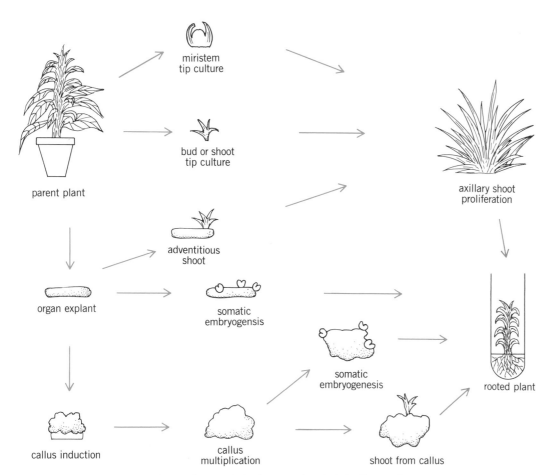

parent plant

miristem tip culture

bud or shoot tip culture

adventitious shoot

organ explant

somatic embryogensis

axillary shoot proliferation

somatic embryogenesis

callus induction

callus multiplication

shoot from callus

rooted plant

Fig. 1.   Schematic presentation of the various methods of propagating plants by tissue culture.

buds, or meristem tips followed by rooting of individual shoots; (2) production of adventitious shoots on plant parts which do not usually produce shoots, for example, leaves, also followed by rooting of individual shoots; and (3) somatic cell embryogenesis, that is, production of embryos from vegetative cells rather than from cells resulting directly from sexual fertilization (Fig. 1). These embryos will germinate and produce plants.

Production of plants by the axillary shoot method has an advantage, since aberrant plants occur at a frequency no higher than the normal rate. Production of abnormal (variant) plants in tissue cultures is undesirable and can be a major problem when the aim is to produce genetically and otherwise identical plants. Even though axillary shooting is slower than other tissue culture methods, it enables multiplication rates that are a million times faster than traditional methods. Adventitious shoots or asexual embryos can arise directly on the explant tissue (portion of plant used for culture) or can originate in the callus (undifferentiated cell mass from tissue culture). Both methods offer potentially much higher multiplication rates than axillary shooting. However, there are problems with maintaining chromosomal and genetic stability in callus and liquid cell suspension cultures started from callus. Somatic embryogenesis is potentially the most rapid method of cloning plants. However, much progress is needed to obtain synchronous development and devise protective coatings for prolonged storage, easy transport, and sowing of such embryos in soil. Because of the great interest in the technology, several commercial firms are investigating one or more of these problems, for example, the development of fluid drilling machines to sow somatic embryos.

Applications. Orchids were the first plants to be propagated by tissue culture. This research arose from Georges Morel's elimination of mosaic virus from meristem culture of orchid plants in 1960. He noted that the procedure produced a multitude of plants within a short time. It was estimated that as many as 4 million plants could be produced from a single explant within a single year. From a commercial standpoint, ornamentals are still the most commonly propagated plants by tissue culture procedures.

Asparagus is a primary vegetable crop which can be propagated through tissue culture. According to H.-J. Yang, asparagus is a cross-pollinated dioecious crop in which the staminate plants are more productive than the pistillate plants. Increased yields might result if a field of just staminate plants could be established. However, propagation by seed results in almost equal numbers of staminate and pistillate plants, and propagation by stem cuttings has not been possible. Division of individual crowns also has not been commercially feasible since only a few genetically identical plants can be developed at any one time. Tissue culture may be a useful method of propagating the high-yielding staminate plants. In theory, it has been estimated that 300,000 transplantable plants could be produced from a single shoot apex in 1 year. In actual practice, however,

70,000 plants are produced by one person working 200 days a year. This depends on an adequate supply of aseptic plants and the culture of 500 bud segments per day.

A large-scale application for mass propagation of the pyrethrum plant, which is a source of the broad-spectrum pyrethrum insecticide, was developed by Luis Levy (Fig. 2). The problem was the need to mass-propagate 57 superior clones which had been selected over a 10-year period, with less than 50 plants of each clone available. A multiplication rate of approximately 12 million plants per year was needed for 1000 hectares over 4 years. By starting with 100 explants and a combination of tissue culture with the last step of vegetative propagation, about 930,000 plants were produced each month, or about 11 million plants in 1 year. Thus, the target set was achieved. In addition to savings in time, labor, and money, the plants from tissue cultures were more vigorous and stress-resistant than plants originating from conventional means of propagation. Also, the tissue culture process could be designed to produce a predictable number of plants at a particular time, thereby allowing for greater ease of plantation planning.

Stages of tissue culture propagation. Tissue culture propagation has been outlined in three major stages (Fig. 3). Stage 1 is the establishment of an aseptic culture of the plant to be propagated (Fig. 3a and b). Various plant parts, such as leaves, root tips or shoot tips, inflorescence sections, and buds, may be explanted onto culture medium. It is absolutely necessary that the culture be free from infection and that explants be surface-sterilized to remove saprophytic bacterial and fungal contaminants. In some instances, cell suspensions initiated from callus may also be used to regenerate plants.

In stage 2 the objective is to obtain an increase in organs, especially shoots, which will give rise to plants. In many species the usual method of multiplication is by shoot formation from an intermediary callus (Fig. 3c). This method results in a rapid increase of propagable shoots, but has the serious disadvantage in some cases of producing genetically aberrant plants. In some species, axillary shoots may form directly on the explant. This is a slower method of shoot multiplication, but genetically deviant plants are virtually absent. Also, somatic embryos capable of giving rise to plants may form directly on the explant or from callus.

Stage 3 is the establishment of regenerated plants in soil. This involves rooting of shoots if roots are not present. For example, shoots may be placed in culture tubes (Fig. 3d) on a medium with no hormones and a reduced concentration of salts. A rooting hormone may or may not be needed to promote root growth. Plants may be established in small peat pots (Fig. 3e). A high-humidity environment is maintained during plant establishment by placing the peat pot with the seedling in a polyethylene bag. Plants may then be transferred to larger pots (Fig. 3f) or to the field. Plants may also be hardened to give some degree of resistance to stress from water

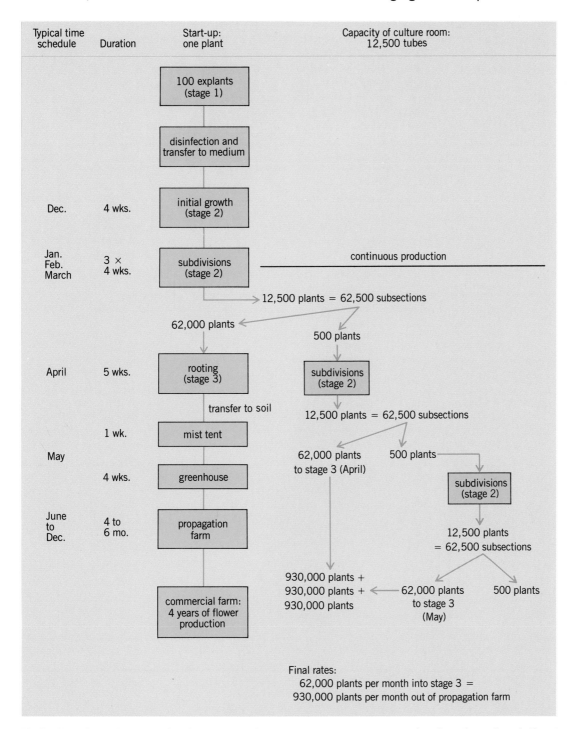

Typical time schedule	Duration	Start-up: one plant	Capacity of culture room: 12,500 tubes

Fig. 2. Flow diagram for propagation of pyrethrum by tissue culture. (*After L. W. Levy, A large-scale application of tissue culture: The mass propagation of pyrethrum clones in Ecuador, Environ. Exp. Bot., 21:389–395, 1981*)

and temperature before transfer to the field if desired.

Meristem tip culture. Meristem tip culture is a special method of tissue culture cloning whose primary purpose is usually the eradication of viruses from infected, vegetatively propagated plants. Viruses usually do not pass from plants to seeds and therefore are less of a problem in seed-propagated plants. However, in vegetatively propagated plants, viruses may be passed on continuously through the propagation of infected plant parts. It was discovered that cells in the most actively growing tissues, such as in the shoot apical meristem, are usually devoid of viruses.

The culture of meristem tips to produce virus-free plants was first used in orchids. This technique is, and will continue to be, a very important, commercially used tool to free vegetatively propagated species, especially ornamentals, from viruses. It has also been used to a limited extent in perennial, out-

Fig. 3. Stages of tissue culture in orchardgrass (*Dactylis glomerata*). (*a*) Leaf sections on medium in petri plate (stage 1). (*b*) Callus forming from leaf section (also stage 1). (*c*) Shoots differentiating from callus (stage 2). (*d*) Establishment and rooting of seedlings in culture tubes (stage 3). (*e*) Young plant after transfer to peat pot, allowing for further development before establishment in greenhouse pots or field. (*f*) Large plants after transfer to pots.

crossing agronomic crops such as red clover, white clover, and ryegrass, where clones must be maintained in a healthy and vigorous condition for the production of new varieties.

The technique involves explanting only a small portion of the meristem tip (Fig. 1). Smaller explants increase the probability of obtaining virus-free plants. However, shoot formation decreases with decreasing size of the explants. Larger explants are more likely to produce shoots, but the possibility of transferring the virus also increases. Therefore, the proper balance between plantlet formation and virus eradication from meristem tips must be obtained.

INDUCTION AND SELECTION OF VARIANTS

Genetic and cytogenetic (chromosomal) variability, which sometimes occurs in plants during the period of dedifferentiated cell proliferation (callus or liquid cell suspensions) between the explant and the next plant generation, was mentioned above. Regardless of whether plants are propagated by tissue culture or by conventional vegetative procedures, it is desired and expected that all resulting plants be genetically and visually identical to the parent plant. Phenotypic variants from tissue cultures were previously dismissed as artifacts, and they are obviously unwanted if the purpose is to produce exact copies of the parent plant. Currently, they are viewed as a rich and novel source of genetic variability to be exploited in plant improvement programs, and have been termed somaclonal variation by P. J. Larkin and W. R. Scowcroft. Somaclonal variation has been observed and sought in a wide range of agricultural crops. A few examples in specific crops and potential for use in other crops are presented below.

Sugarcane. The potential usefulness of variant plants generated in tissue cultures was first realized in sugarcane. Commercial sugarcane is heterozygous and contains germ plasm (genes) from three to five different sugarcane species. Conventional breeding and selection programs require 10–15 years to develop and release a new clone which, when released, can be vegetatively propagated as long as desired. Naturally occurring variability is low for some traits, for example, resistance to certain diseases. Plants from tissue culture experiments conducted by D. J. Heinz and independently by M. C. Liu showed variation in several morphological, cytogenetic, and biochemical traits. This variation was used to identify and isolate plants resistant to various important diseases. Resistance was obtained for Fiji disease (caused by a virus transmitted by leafhoppers), eyespot disease (caused by the fungus *Helminthosporium sacchari*, which produces a toxin causing lesions in leaf tissue), and downy mildew (a systemic disease caused by the fungus *Selerospora sacchari*). Resistant subclones were then utilized in the development of new varieties.

Significant variations in stalk length, number, diameter, volume, density, and weight; percent fiber; cane yield; and several other agronomic characters were also found. Some of the variants, when placed in field trials, showed significant improvements over the normally grown commercial varieties. Attempts to increase variation even further by applying radiation and chemical mutagenic treatments were not successful. Tissue culture methods by themselves seem adequate for inducing a high amount of variation in sugarcane.

Potato. One of the most interesting examples of somaclonal variation has resulted from James Shepard's work with potato. He has termed his technique protoplast cloning and the resulting plants which arise from individual protoplasts as protoclones. Potato is another crop which is asexually propagated for commercial production. The tuber is the organ most commonly used for propagation. Despite extensive breeding programs in North America and Europe, the most widely grown varieties are old varieties, many of them dating back more than 50 years. In the United States only four varieties constitute 72% of the total acreage; the most significant variety is Russet Burbank, selected in the early 1900s. This variety comprises 39% of Canadian and American potatoes. It arose as a sport (variant) in field plantings of the variety Burbank which date back to the original selection made by the renowned botanist Luther Burbank in 1871.

Conventional breeding of potatoes is complicated by low pollen fertility, tetraploidy, complex inheritance, and a fragmentary knowledge of potato genetics. Improvement in many traits is of interest, including resistance to various diseases. Each year 22% of the world potato crop is lost to disease. Spontaneous changes for disease resistance and other characters occur too infrequently to be of consistent value, and the use of induced mutations has resulted in a few improved varieties.

Shepard and colleagues reasoned that it might be easier to improve a current popular variety than to create a new one. They isolated protoplasts from leaf mesophyll of Russet Burbank and from these grew cell masses which regenerated into whole plants (Fig. 4). Unlike earlier experiments with tobacco, they found that most of the potato plants regenerated from protoplasts were not identical with the parent plant or with each other. A screening of more than 10,000 protoclones showed significant and stable variation for compactness of growth habit (some with a fuller and more efficient canopy), maturity date (some which set tubers earlier than the parent variety), tuber uniformity (some with much more uniformity in size and shape), tuber skin color (white as compared to brown in the parent), photoperiod requirements (13 hours required to set flowers as compared to 16 hours in Russet Burbank), and fruit production (some with 100 times more berries than the parent).

Of special significance is that disease-resistant protoclones were also recovered. The parent is susceptible to both early blight (*Alternaria solani*) and late blight (*Phytophthora infestans*) [late blight is the disease which caused the Irish potato famine in the 1840s]. Protoclones were isolated which were more resistant than the parent to early blight, and others were found that were resistant to multiple races of late blight. One protoclone outyielded the parent, but the difference was not consistent or statistically significant.

Corn. In 1970 the southern corn leaf blight epidemic, in which a specific race (race T) of *Helminthosporium maydis* attacked large areas of the United States crop, resulted in great losses. Susceptible corn types carried a factor in their cytoplasm associated with male sterility. Male sterile types are used in hybrid seed production to reduce labor and costs associated with hand detasseling of plants. Corn plants

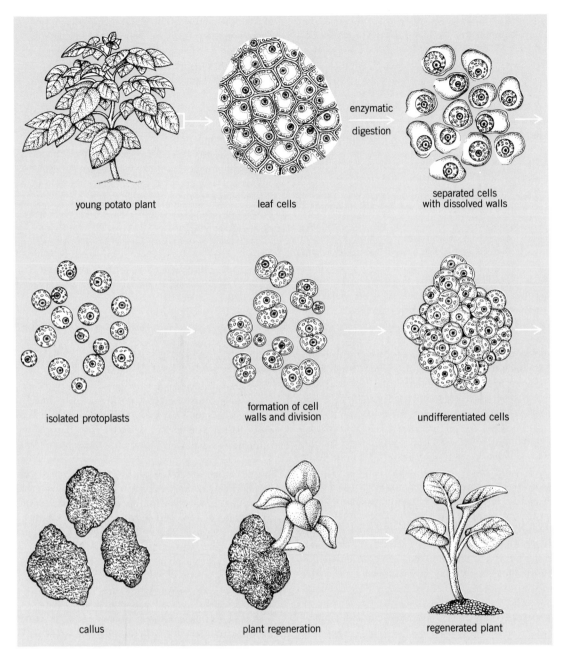

young potato plant   leaf cells

enzymatic
digestion

separated cells
with dissolved walls

isolated protoplasts

formation of cell
walls and division

undifferentiated cells

callus   plant regeneration   regenerated plant

Fig. 4.   Diagram of procedure to create protoclones in potato. (*After J. F. Shepard, The regeneration of potato plants from leaf-cell protoplasts, Sci. Amer., 246(5):154–166, May 1982*)

with normal (N) cytoplasm are resistant.

Somaclonal variation was found by Burle Gengenbach for resistance to the southern corn leaf blight. Tissue cultures initiated from plants with the T (Texas) male sterile cytoplasm are susceptible to the toxin produced by the fungus organism, while those from plants with N cytoplasm are resistant. The pathotoxin was used to select resistant cell lines from the susceptible T cytoplasm cultures. After five cycles of increasing the toxin concentration and selecting resistant cultures, cell lines were recovered which were resistant to lethal levels of the toxin. Plants regenerated from the cultures were also resistant.

Other examples. Currently, the most immediate practical application of variant selection in tissue

cultures would seem to be with vegetatively propagated crops in which variants, or sports, can be identified and then increased by direct vegetative propagation. Such examples were given above for sugarcane and potatoes. There is, however, much interest in mutant selection for all important crop species, for example, in the development of drought- and salt-tolerant plants for arid regions of the world; in herbicide resistance so that certain herbicides can be used to eliminate all but the desired crop species; and in cold tolerance. Biochemical mutants are of interest in cereal grains and seed legumes, especially those that will produce large amounts of certain essential amino acids which normally occur in very low quantities. This would improve the protein quality of these crops and provide more nutritionally balanced diets for that segment of the world population which depends so heavily on direct consumption of grain. Interesting somaclonal variation has been observed in several other economically important species, including rice, alfalfa, and ryegrass; many other examples also exist.

A major limiting factor in the use of mutant and variant selection technology in tissue and cell cultures is the inability to regenerate plants from single cells and protoplasts in many important crop species.

CREATION AND USE OF HAPLOIDS

Haploid cells possess the gametic or one-half the somatic (normal body cell) chromosome number. Such chromosome numbers in plants are found in egg and pollen cells. Plant breeders are interested in haploid plants which may be derived directly from these cells rather than from the fertilized egg. Recessive mutations are expressed immediately in haploids and are not masked by dominant genes from the other parent. For example, if gene A is mutated to gene a in a pollen cell, its expression would likely be masked by the dominant A gene from the egg cell, that is, Aa, in the resulting offspring. If the pollen grain could be induced to divide and differentiate shoots and roots and the chromosomes doubled, then the new plant would be homozygous for a and the gene would be expressed. The most common method of doubling chromosomes is by colchicine treatment.

Another important use for haploid cells is in the development of homozygous lines. After crossing two varieties of a self-fertilizing diploid species, three to five generations or more of selfing may be required to obtain the desired genes in a homozygous condition. For example, assume that the genetic makeup in an F_1 (first generation offspring from a cross between two parents) for one pair of chromosomes is Aa Bb cC Dd eE fF gG Hh Ii, and it is desired that genes a, b, d, h, and i be homozygous. Because of genetic recombination and segregation, several generations of selfing and evaluation may be required to obtain this genetic constitution. However, if the haploid plant, possessing the chromosome with makeup a b C d E F G h i could be isolated and doubled, homozygosity for the desired genes could be obtained in only one generation. The above example is for only one chromosome pair. Since plants possess several pairs of chromosomes, homozygosity may also be desired for different genes on other chromosomes.

Heterozygosity. In crop species that are naturally outcrossing (as opposed to self-fertilizing), there is also an interest in developing homozygous breeding lines so that controlled crosses might be made that would maximize heterozygosity (hybrid vigor), such as was accomplished many years ago with corn. Homozygous breeding lines are often difficult to obtain in these species because of self-incompatibility and inbreeding depression (opposite effect of hybrid vigor). E. T. Bingham has described a unique approach for maximizing heterozygosity in alfalfa. This involves:

scaling the natural tetraploids down to diploids (gametic chromosome number in this case); producing diploid hybrids by crossing F_1's; and doubling the diploid hybrids by colchicine treatment to obtain maximum heterozygosity in the autotetraploid (four identical sets of chromosomes). Theoretical models developed for alfalfa should also be applicable to other crop species with similar ploidy levels and type of reproductive behavior.

Production methods. Efficient methods for producing haploids do not exist for many crop species. In some species, they may be obtained by identification of those that occur naturally. In other species, such as barley, they may be obtained by very specific methods. The discovery that haploids could be obtained by anther (Fig. 5) and pollen culture has created much interest among plant breeders. Haploids may be produced from embryos arising directly from pollen (embryogenesis) or from callus (organogenesis). Embryogenesis is preferred because it usually produces a lower frequency of aberrant plants. Haploids can be obtained easily by anther culture of some species, such as tobacco and jimsonweed. In these species, haploids usually form from embryos. In important cereal and legume species, progress has been hampered by low frequencies of plantlet formation, production of chlorophyll-deficient plants, aneuploid and polyploid plants, and other aberrant plant forms. Usually haploids form by organogenesis from callus, which probably accounts for the higher frequency of abnormal plants. However, recent progress with rice, wheat, and barley has resulted in the development of haploid plants from anther cultures, and embryo formation has been reported in some cases. Some of these have been developed into homozygous lines which are now under evaluation in breeding programs in various parts of the world.

This is clearly one aspect of plant tissue culture in which the application is very real and near. It may have tremendous impact on plant breeding programs by greatly shortening the period required for variety development and obtaining a greater degree of hybrid vigor in a wider range of crop species.

PROTOPLAST TECHNOLOGY

Since the discovery that whole plants could be grown from isolated protoplasts and that hybrid plants could be produced by protoplast fusion, both scientists and nonscientists have speculated about various manipulations that might result in new plant types. Possible manipulations include the transfer of genes, for example, for nitrogen fixation or for changing protein composition, and creation of hybrids between widely unrelated species. However, this exciting area of plant cell and tissue culture may also be the farthest removed from practical application.

The main problems and limitations of this technology include: the failure to regenerate plants from protoplasts in a wide range of important crop species, and the lack of methods to identify and isolate desired altered cells. Within the two most important

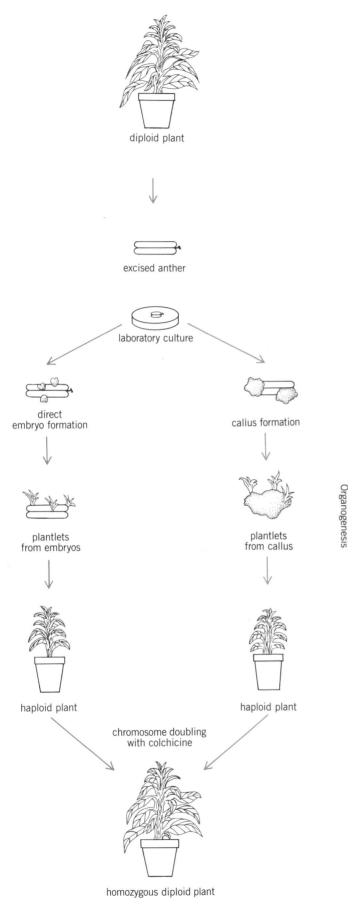

diploid plant

excised anther

laboratory culture

direct embryo formation

callus formation

plantlets from embryos

plantlets from callus

haploid plant

haploid plant

chromosome doubling with colchicine

homozygous diploid plant

Embryogenesis

Organogenesis

Fig. 5. Anther culture showing development of haploid plants either by embryogenesis directly from the anther or by organogenesis from callus.

crop families, the Gramineae (cereals and grasses) and Leguminosae (legumes), repeatable results for growing plants from protoplasts have occurred only with alfalfa, a forage legume. No plants have been regenerated from protoplasts of any major cereal (corn, wheat, rice) or grain legume (soybeans, peas, beans). Nevertheless, because of the far-reaching significance and interest of this technology, some of the results obtained to date with somatic hybridization and genetic transformation, and their future potential, deserve mention.

Somatic hybridization. The first scientists to create a hybrid by fusing protoplasts between two different tobacco species, Peter Carlson and coworkers, termed this process parasexual hybridization. Progress in this field was rather slow during the first few years following the initial success. Most examples included hybrids of plants within a species or between species which could also be hybridized sexually. Since 1977, however, numerous examples of parasexual hybrids have been published, including many between species which are sexually incompatible. Hybrids between wild and domesticated species of carrot, between woody and herbaceous species of jimsonweed, and, more interestingly, between tomato and potato have been reported. Yet, the number of plant species and genera in which parasexual hybrids have been created is still primarily limited to so-called model systems, that is, to genera within the Solonaceae family (tobacco, petunia, jimsonweed, and so on) and species of *Daucus* (carrot). Basically, the process includes protoplast isolation, fusion, identification, or selection of the hybrid fusion product and regeneration of a plant from the hybrid cell (Fig. 6). Also, it is desirable that the regenerated hybrid plant be genetically stable and fertile.

Protoplast isolation. Protoplast isolation represents no major problem. Various enzymes that separate cells from each other and dissolve cell walls are readily available, and protoplasts have been isolated from a wide range of tissues and species. Enzymes may be used singly or in combination, depending on the plant species and tissue. A major consideration is that the osmotic pressure (determined by solute, primarily salt and sugar, concentration) of the isolation and incubation solutions be similar to that of the plant cells to keep them from collapsing or bursting. Leaf mesophyll has been frequently used as a source of protoplasts in many species. The cells are relatively uniform and easy to isolate. Protoplasts have also been isolated from liquid suspension cultures initiated from callus. Walls of these cells are often more difficult to dissolve, and the possibility of genetic and chromosome abnormalities may be greater than in leaf mesophyll cells.

Protoplast fusion. Fusion also presents no major problem, especially since the discovery of polyethylene glycol as a fusing agent. Protoplasts have been fused between widely divergent plant species, such as barley and soybean, and even between plant and animal cells. A major problem after fusion is that of obtaining sustained cell divisions. If the species are too widely divergent, they are likely to differ in length of time required to complete a cell cycle; that is, chromosome divisions will not be synchronized in cells of the two species. This results in few or no divisions of the fusion product, or complete elimination of chromosomes from one of the species.

Identification of hybrid cell. If protoplasts from two different species are compatible and the hybrid fusion product can divide, form a cell mass, and eventually a plant, the problem then becomes one of identifying the hybrid cell. This is because protoplasts from a particular species will fuse not only with those from a different species but also with each other. In the initial hybrid between two tobacco species, preferential recovery of the fused hybrid cells was obtained with a selective growth medium. Protoplasts from each parental species were unable to regenerate into a plant on a medium possessing certain nutritional deficiencies which the hybrid cells were able to overcome. Callus masses were subjected to further selection pressure by placing them on a solid medium without adding hormones. Neither parental species was able to grow on this medium, while the hybrid grew vigorously. Shoots and leaves were obtained from cell masses. Since roots failed to form, the shoots were grafted onto a freshly cut stem surface of one of the parent plants. Grafted plants grew, flowered, and produced normal seeds.

Another method of selecting hybrid plants is to use parents that possess different genes for chlorophyll deficiency. When cells from different parents are fused the genes are complemented in the hybrid, resulting in normal green plants. This technique was used for producing a parasexual hybrid between two varieties of cultivated tobacco. Chlorophyll-deficient mutants are common or easily induced in a wide range of crop species; thus this method of hybrid identification is certainly one which is receiving consideration.

The use of mutants with different nutritional requirements for hybrid cell selection was mentioned above. This characteristic, in combination with drug resistance and susceptibility, was used to create somatic hybrids between two petunia species. Scientists in England produced a purple-flowered hybrid between red-flowered *Petunia hybrida* and white-flowered *P. parodii* by utilizing the actinomycin D susceptibility of *P. hybrida* and failure of *P. parodii* to grow on a special medium lacking a nutritional requirement. The fused hybrid cells were resistant to actinomycin D and also grew on the medium that would not support the growth of *P. parodii*.

Hybrids produced between two tobacco species and between two petunia species possessed the characteristics of those produced by sexual hybridization. For example, with petunia, the flower color of red × white was purple for both the parasexual and sexual hybrids. Also, seed planted from both kinds of hybrids produced plants which segregated for flower color (red, purple, and white) in a similar manner.

Various other selection procedures and modifications of those mentioned above have been used or

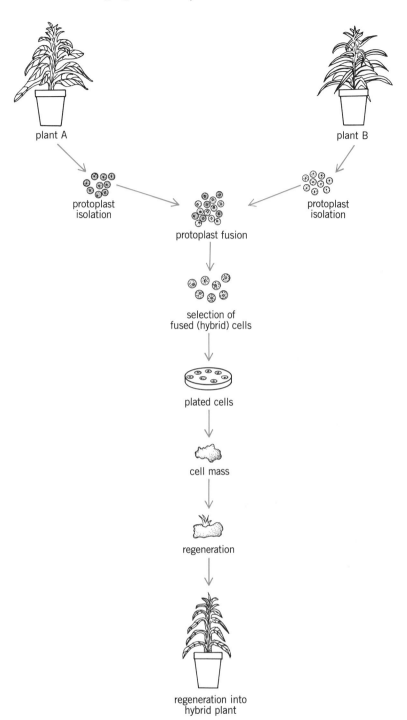

Fig. 6. Procedure for producing a new hybrid plant from two different plant species.

proposed. The scheme described for petunia represents an almost ideal situation in which large numbers of hybrid fusion products were easily isolated. However, nutrient-requiring or drug-sensitive mutants have been identified or developed in only a few species. Fused hybrid cells have also been identified microscopically and then physically removed and placed on a separate medium. More recently, identification has been made easier through the use of fluorescent microscopy, where the cells from the different parental types appear in different colors. The number of hybrid cells which can be recovered is definitely limited by this method. However, in a practical breeding program, the recovery of only one or a few hybrid cells may be adequate if the resulting plant can be easily cloned by vegetative propagation.

Potato-tomato hybrid. One of the more interesting parasexual hybrids is potato-tomato, developed by Georg Melchers. Hybrids containing the cytoplasm of potato were termed pomatoes and those containing the cytoplasm of tomato were termed topatoes. Leaf mesophyll protoplasts from a yellow-green tomato mutant were fused with those isolated from a liquid callus culture of potato. A total of nine hybrid plants were positively identified by analyzing a specific protein, ribulose bisphosphate carboxylase. The hybrids contained a peptide that was specific for both potato and tomato. Generally, the hybrid plants grew slowly and produced a limited number of leaves. The chromosome numbers varied from 48 (the expected number for the hybrid) to 74. Some plants produced flowers and small aborting fruits (Fig. 7a). Others produced thickened stolons resembling elongated, small tubers (Fig. 7b) which would sprout into new plants. No seed were obtained from any of the plants. While the interest in producing hybrids between widely divergent plant species is obviously great, it

Fig. 7. "Topato" plant. (a) Fruit, obtained after pollination with a wild potato species, *Solanum stenotomum*. (b) Tubers. (*Courtesy of Prof. G. Melchers, Max-Planck-Institut für Biologie, Tübingen, West Germany*)

may be too early to expect spectacular results, such as a plant that will produce abundant tomatoes above the gound and potatoes below the ground.

Potential. Problems with hybridization between widely unrelated species include failure of division synchrony, less vigorous and chlorophyll-deficient plants, sexual infertility, and chromosome instability. The first tobacco hybrids possessed the expected chromosome number of 42. However, later studies with hybrids from the same two species revealed wide variation in chromosome numbers. This suggests that chromosome numbers may not be stable even in somatic hybrids of plants which can be crossed sexually. Of course, the lack of sexual fertility would not be a problem in vegetatively propagated species. Thus, the potential for the technology may be realized first in plants such as ornamentals.

Protoplast fusion may also have great potential for transferring single chromosomes or parts of chromosomes and organelles, such as chloroplasts and mitochondira, from one plant to another. For example, a chromosome segment conferring disease resistance might be transferred from a wild relative to a domesticated species by irradiating cells to break chromosomes, fusing the protoplasts, and then selecting the desired hybrid cells with recombined chromosomes. Chloroplasts from species with high photosynthetic efficiency may be transferred to cells of species with lower efficiency. The same rationale would hold for transfer of mitochondria between plants. The transfer of functioning chloroplasts to cells from albino segments of leaves has already been accomplished in tobacco.

Gene transfer. Several reports appeared during the early 1970s describing the uptake and expression of foreign DNA in plant cells. For example, D. Hess reported using DNA from a red-flowered petunia to transform a white-flowered form into a red-flowered form. This "transformation" of plant cells, as had been demonstrated many years ago with microbial systems, created much interest among plant scientists concerned with genetic engineering of important crops. However, the failure of other scientists to repeat those experiments led to a certain amount of pessimism concerning the ultimate use of such technology. The recent and continuing discoveries on the nature and makeup of genetic material in higher plants and animals, the ability to clone (replicate) genes of higher plants in bacterial cells, the successful transformation of yeast and animal cells, and the discovery of new methods to introduce foreign DNA into plant chromosomes have renewed enthusiasm for this technology. Even so, plant genetic engineering using recombinant DNA technology is probably many years from practical application. However, some of the modifications proposed, methods of achieving them, and a few examples of the limited successes to date can be discussed.

Procedures. The concept is based on the insertion of DNA from a microorganism, or from another plant, into that of a recipient crop plant. Steps required in the process include: (1) uptake of the exogenous DNA by the recipient cell; (2) integration of the foreign

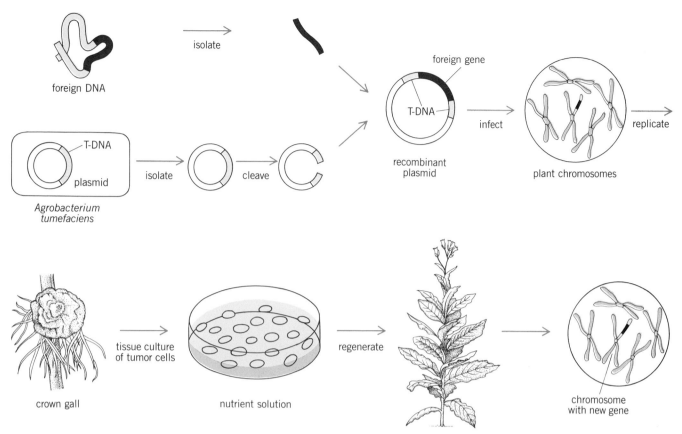

Fig. 8. Introduction of foreign DNA into plant cells via a Ti plasmid from *Agrobacterium tumefaciens*. The bacterium carries a plasmid which causes crown gall tumors in dicotyledonous plants. A section of the plasmid (T region) combines with the chromosomal DNA in the nucleus of plant cells. The T region is cut and a piece of foreign DNA (carrying desired genes) is inserted. The plant is infected with the bacterium and the T-DNA is replicated in the tumor. The tumor cells are isolated, grown in tissue culture, and regenerated into plants which might carry the new genes in their chromosomes. (*After W. J. Brill, Agricultural microbiology, Sci. Amer., 245(3):199–215, September 1981*)

DNA into either the chromosomes of the recipient plant or into the DNA of cytoplasmic organelles such as chloroplasts or mitochondria; expression of the new genes in the modified recipient cells and especially in the plant regenerated from that cell; replication of the extra DNA along with the recipient cell's own genetic material; and transmission of the new genes to daughter cells and offspring through the parent plant's pollen or egg cells.

Recently, the use of protoplasts for transformation experiments has received considerable attention. Absence of the wall makes the cell more akin to animal cells or yeast cells without walls in which successful transformations have been made. Apparently, the actual uptake of foreign DNA by plant cells is feasible; plant protoplasts seem to have an affinity for taking up particles, ranging from organelles and pieces of DNA to foreign objects such as small pieces of latex. The problems of inserting small pieces of DNA and having them expressed lie with the later steps listed above.

Genetic vehicle. A general strategy for plant cell transformation is the use of a genetic vehicle which is capable of either integrating itself into the plant genetic material (chromosomes or organelles) or independently replicating in the plant cell. Transmis-

sion to daughter cells and subsequent generations would also be required. A molecular vehicle is a segment of recombinant DNA (DNA in which a piece of plant DNA has been combined into a segment of bacterial or viral DNA) used to deliver a desired gene into a plant. A good example is virus particles, which invade the host cell and induce the cell to replicate the virus genes. If desired (useful) genes could be attached to a virus and then the plant cell infected with that virus, the desired genes could be replicated and expressed in the host cells. This approach is being explored with cauliflower mosaic virus. This virus is often cited as a potential vehicle because it is one of the few plant viruses known to contain double-stranded DNA (most plant viruses contain a different nucleic acid, RNA).

Several transforming vehicles have been proposed. The one most often suggested is a plasmid, that is, an extranuclear piece of DNA, residing in the bacterium *Agrobacterium tumefaciens*. This bacterium incites the formation of crown gall tumors in a wide range of dicotyledonous plants. The causal agent of the tumors is borne by a plasmid termed Ti, for tumor-inducing. Researchers postulated that if the bacterium could use the plasmid to genetically engineer tumors they might also be used to transfer useful genes. The segment of the plasmid responsible for infection is called the T region. It is the genes within this region that are integrated into the plant nuclear genetic material, replicated, and expressed. A scheme by which foreign genes may be introduced into plant cells with a Ti plasmid is shown in Fig. 8.

The theory for using Ti plasmids in plant transformation is to splice the T region of the plasmid, insert the desired genes, for example, for nitrogen fixation, and then introduce the modified plasmid into the host plant cell. Then the modified plasmid either would be incorporated into the plant genetic material and express and transmit its characteristics along with the DNA in the plant chromosomes, or would be carried as an extranuclear body and would perform these functions independently.

Concentrations of the desired genes to be transmitted may be increased by cloning. This involves isolating the DNA from the donor plant; splicing it into a plasmid from plant viral DNA; inserting the plasmid into a bacterium, for example, *Escherichia coli*; multiplication of the bacterium to produce multiple copies of the desired plant gene; and then infecting the recipient plant with the viral DNA.

Applications. Many experiments have been conducted to transfer genes from one plant species to another or from microorganisms to plant cells. One of the most interesting was conducted by John Kemp and Timothy Hall, who succeeded in transferring a gene which codes for the bean protein phaseolin into sunflower cells. The transferred gene directed the synthesis of messenger RNA in the resulting cell mass. However, the messenger RNA did not code for a protein, and the cells were not regenerated into a whole plant. Nevertheless, this was considered a major breakthrough since it involved the transfer of a gene from one plant to another and the coding from DNA into messenger RNA. This achievement has been popularly called the sunbean.

The transfer of genetic information or useful genes between sexually incompatible plant species has been of interest to plant breeders for many years. The possibility of such transfers seemed remote until a few years ago. The explosion of knowledge concerning the nature and makeup of genetic material in higher plants and animals has been phenomenal during recent years. Recombinant DNA and genetic engineering technology have allowed manipulations that were once thought to be nearly impossible. There is still a long way to go, however, and several major limitations still need to be overcome, including the ability to regenerate plants from cells in a wide range of important crop plants. Also, it must be remembered that even if it were possible to transform cells, it does not always mean that the character will be expressed in the whole plants regenerated from those cells or that the character will be transmitted through the seed to the next generation. If such seed transmission were not possible, the technology would be restricted to vegetatively propagated crops.

Nevertheless, there is cause for optimism. Even though practical applications of somatic cell fusion and DNA transfers are probably some years away, there is little doubt that continued progress in basic research in these areas will continue. It may eventually be possible to transfer genes controlling nitrogen fixation from bacteria to important cereal grains and thus enable them to grow independently of nitrogen fertilizers, or to make photosynthesis more efficient to greatly increase yields, or to modify the storage proteins of cereal grains to provide for more balanced human nutrition in developing countries. It must not be forgotten, however, that the development and application of any and all of this new technology will require the cooperation and collaboration of plant scientists in many disciplines, including the plant breeder, who will ultimately release the products for public consumption.

EMBRYO CULTURE

Although fertilization is normal in sexual crosses between many plants, seed is sometimes not obtained because the resulting embryos either abort or do not develop normally. This may occur when attempting hybridizations between both closely related and unrelated species. The possibility of rescuing embryos shortly after fertilization and culturing them under sterile conditions on artificial media was recognized many years ago, and many successful experiments were performed at that time. Recent advances in tissue culture techniques have increased the interest in embryo rescue and culture, producing hybrids that were once thought to be impossible.

Since embryos may abort at various times after fertilization, depending on the cross fertilization attempted, the need to rescue the embryo for tissue culture may exist very early. In some cases, fertilized ovules are rescued immediately after fertilization and cultured. In other cases, the ovules may be placed in a test-tube or petri dish and then fertilized. The latter method is termed *in vitro* fertilization and is somewhat analogous to the "test-tube baby" technique in humans and other mammals. The main difference is that the fertilized human or mammal egg is returned to the uterus for further development, whereas the fertilized plant ovule is not returned to the mother plant, and the embryo continues to develop and eventually germinate in the test-tube or petri dish.

Successful applications of this technique are too extensive to list here. The primary interest has been to make hybrids between different species. An example might be to transfer the perennial nature of a wild species to a cultivated species. Other examples include acceleration of seed development, overcoming self-incompatibility barriers, bypassing dormancy requirements, and transferring nuclear genes from one species into the cytoplasm (extranuclear portion of cell) of another. The procedure may involve culture of ovaries or egg cells as well as embryos of different stages.

This technique in cell and tissue culture has tremendous potential for practical application in plant breeding, and the technology is available now. The technique is likely to receive more and more attention as researchers try to combine desirable characteristics of plant species and introduce new ones from wild species.

GERM PLASM PRESERVATION

Germ plasm is the source of genetic diversity of living organisms, which allows them to adapt to changing environments and tolerate the onslaught of new disease and insect epidemics. Extinction of a species or group of organisms possessing a unique genetic constitution represents an irreversible loss of a valuable resource. Erosion of genetic resources is becoming an increasingly important concern. The southern corn leaf blight epidemic in 1970 is a recent vivid example of crop vulnerability to disease epidemics. Since the changing of wild species to cultivation has resulted in only about 150 plant species used to meet human food needs, many wild species which may have possessed disease resistance transferable to cultivated species have become extinct. Germ plasm collections of seed-propagated crops are maintained at the National Seed Storage Laboratory in Fort Collins, Colorado. However, even with these species many scientists claim that the collections are inadequate and that the stores of variability are being lost at an alarming rate.

The situation with vegetatively propagated crops is probably even more serious, since germ plasm cannot be easily stored and preserved. For these species, conventional methods of germ plasm preservation are very labor-intensive and expensive. Also, the material is subject to attacks by pathogens and pests. Tissue culture techniques are being explored to preserve genetic stocks of several crop species for future use. The technology is still in its infancy, but it basically involves freezing of cells, tissues, or plant organs at very low temperature and then thawing and regenerating whole plants from this material when needed. Problems and research areas for investigation include selection of the material for preservation; selection of a protective agent to treat the cells before freezing; and determination of rate of cooling, storage temperature, length of storage, and method and rate of thawing.

Liquid nitrogen ($-196°C$) or liquid nitrogen vapor ($-140°C$ or below) has been most commonly used for low-temperature preservation of plant tissue. Plant tissues are usually treated with a protective agent which protects against freeze-thaw damage by ice crystals forming within cells and must be removed after thawing. One of the most common and widely used agents is dimethyl sulfoxide (DMSO).

Although many plant systems may be used for preservation, for example, cells, somatic embryos, and pollen, meristems probably offer the best possibility for recovering virus-free plants that are genetically stable. There is still difficulty, however, in regenerating plants from cells in all but a few species, and the cell and tissue culture process tends to induce genetic aberrations. Additional research is needed on the feasibility of using this technique to preserve various plant species for future use. In species such as rare desert cacti, which are threatened by theft, this may be the most feasible or only way of ensuring their survival.

CONCLUSION

Cloning as discussed in this article is restricted to cell and tissue culture technology and its potential use in the improvement of agricultural crops. Certain manipulations, such as mass propagation, embryo culture, and selection of variants, are already in use and are contributing to plant-breeding efforts. Creation of haploids by anther and pollen culture may be just on the horizon for making significant contributions to crop improvement. The use of protoplast technology, both to create new hybrids by somatic cell fusion and to introduce foreign DNA into plant cells, seems to be further in the future in terms of actual application for important food and feed crops.

Limitations and problems include the inability to regenerate plants from cells and protoplasts of many important crop species, the presence of genetic and chromosomal abnormalities in callus and cell cultures, and the lack of good cell-selection schemes for recovering somatic-hybrid or genetically transformed cells. Scientists in many research laboratories throughout the world have made considerable progress in overcoming these problems and undoubtedly will continue to do so. Recent advances in the manipulation of plant cells and tissues inspire optimism for its role in supplementing and complementing conventional plant-breeding techniques to increase future food production.

[BOB V. CONGER]

Bibliography: E. C. Cocking et al., Aspects of plant genetic manipulation, *Nature*, 293:265–270, 1981; B. V. Conger (ed.), *Cloning Agricultural Plants via In Vitro Techniques*, 1981; S. H. Howell, Plant molecular vehicles: Potential vectors for introducing foreign DNA into plants, *Annu. Rev. Plant Physiol.*, 33:609–650, 1982; P. J. Larkin and W. R. Scowcroft, Somaclonal variation: A novel source of variability from cell cultures for plant improvement, *Theor. Appl. Genet.*, 60:197–214, 1982; T. Murashige, Plant propagation through tissue cultures, *Annu. Rev. Plant Physiol.*, 25:135–166, 1974; K. O. Rachie and J. M. Lyman (eds.), *Genetic Engineering for Crop Improvement*, Rockefeller Foundation Conference (May 12–15, 1980), 1981; J. Reinert and Y. P. S. Bajaj (eds.), *Applied and Fundamental Aspects of Plant Cell, Tissue, and Organ Culture*, 1977; J. F. Shepard, D. Bidney, and E. Shahin, Potato protoplasts in crop improvement, *Science*, 208:17–24, 1980; I. K. Vasil (ed.), *Perspectives in plant cell and tissue culture*, *Int. Rev. Cytol.*, suppls. 11A and 11B, 1980.

Electronic

E.T. Steadman

Publishing

David H. Goodstein

David H. Goodstein is director of Inter/Consult. He holds an appointment as Research Affiliate at the Visible Language Workshop of MIT. Contributing editor on new technology for "Graphic Arts Monthly" and the 11 member magazines of the Eurographic Press Association, he is also a popular lecturer on graphic arts and computer graphics.

At the intersection of microprocessors, graphic display technology, communications networks, laser nonimpact imaging, and the traditional print-oriented publishing industry, a new phenomenon known as electronic publishing is emerging. Over the last few years, significant strategic moves have been made by the top 100 media companies. Groups like Dow Jones, Grolier, Time Incorporated, CBS, and the New York Times have moved firmly into position to benefit from the distribution of their information in nonprint formats. At present there are few who claim profits from their activities in this area, but a great deal of work is under way in developing the technologies and market positions which will ultimately fuel the economics of this new industry segment.

Electronic publishing has become a reality over the last decade partly because of progress in digitization of both text and images and partly because of the availability of new media for information or communications. The availability of new media for delivery of soft copy depends on both technology and changes in the regulatory environment. Once both areas coincide, these new media will intensify the potential for nonprint publishing. Direct-home-delivery channels like cable TV or direct broadcast satellite (DBS), and retail distribution media like videodisks, are examples of information vehicles unavailable prior to 1978. The recent decision by the Federal Communications Commission to allow broadcasting of data on the 650 local frequencies previously reserved for voice telecommunications (for example, pocket message transmitters for doctors) is an example of a regulatory decision which opens new horizons for the use of portable computer terminals to both originate and access electronic information.

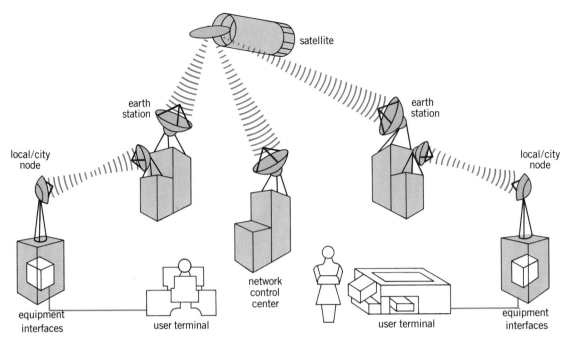

Fig. 1. Satellite-based network. (*Harris Corporation*)

Publishing environment. The immense growth of interest in electronic publishing has resulted primarily from technological advances affecting the areas of information distribution, display, and storage, combined with a dramatic decrease in the cost, and increase in the speed and power, of telecommunications.

In the area of communications, satellites have already begun to revolutionize commercial long-distance data transmissions. In the late 1970s and early 1980s the United States witnessed the rise of a new industry, dedicated to creation of nationwide data communication networks using packet switching technology, such as GTE's Telenet and Tymeshare's Tymenet networks. These networks (Fig. 1) are easy and inexpensive to use when compared to long-distance telephone lines. For most users, access to the network is through a local telephone call on a regular telephone. The networks are designed to allow data communications and computing devices manufactured by many different companies to connect with one another intelligently, that is, without the users having to concern themselves with the compatibility of individual devices. Due to their low cost, ease of use, and effect on the device compatibility issue, packet switching networks have created a strong and stable data telecommunications industry. This industry now seeks to encourage the growth of all kinds of electronic information services to increase its own revenues and profits.

The technical differences between packet switching networks and long-distance telephone lines show up in the rate structures. Long-distance telephone charges are a function of the distance between parties and the connect time, while satellite-linked, packet-switching network rates are a function of the volume of data transmitted and connect time. Since

this type of communication link depends for most of its operation on satellites which are in a geostationary orbit 22,500 mi (36,000 km) above the Earth, it is not difficult to see why surface distances are relatively immaterial to the cost of telecommunications using packet switching networks. Other resources used are equivalent to those used in a local telephone call.

In this environment, electronic publishers can reach a national market of potential subscribers without the costs and logistical problems associated with the nationwide distribution of printed material (Fig. 2). Since electronic publishing is based on a pay-per-use concept and the data base of text or images is continuously available, there are dramatically different economics from those associated with traditional publishing, where manufacturing and distribution of books, magazines, or newspapers represent by far the majority of costs. The effectiveness of the distribution network for the printed artifact was also the major determinant of economic success for publishing operations. Paper has long been the single highest cost in the entire complex set of operational steps which make up the publisher's business.

Some of the main issues unresolved in the electronic publishing industry relate to the cost and human factors engineering of the users' terminal equipment. In this regard, several significant advances were made in 1982. One such advance is the introduction of the pocket-sized IXO Telecomputing System. It has a built-in phone modem with autodialing and autologging so that users only need to enter a password to gain entry to an on-line data base. Its ease of use is indicated by the keyboard, which includes YES, NO, DON'T KNOW, and HELP keys. As the price of pocket-sized devices drops, they will be perceived as consumer, rather than business or luxury, items. This will have an enormous impact on the progress of market demand for electronic publishing products. Innovative information display devices, coupled with the continued proliferation of personal computers fostered by such giants as IBM and DEC, also will have a favorable impact on the future development of the electronic information industry.

Definitions of electronic publishing. Electronic publishing is actually a catchall phrase covering a wide variety of activities. The most important common feature of these activities is that they contain or convey information with a high editorial and value-added content in a form other than print. Included in the list of presently practiced electronic publishing activities are: on-line data bases, videotext, teletext, videotape cassettes, videodisks, cable TV programming (DBS delivery soon), and electronic mail and messaging.

For convenience and ease of understanding, these may be classified according to the amount of interactivity the user is allowed. One-way systems deliver information in a continuous manner and allow the user only passive viewing and crude control over sequence or position. All television and sequential-access image-archiving media, like videotape, are one-way. One-way-plus systems are continuous but offer some option for sequencing and selective access, with teletext being the best example. Two-way systems access information randomly under direct user control. Table 1 shows a matrix of the functional characteristics of various electronic publishing technologies.

Production of electronic information products. Crucial to the economical production of new information packages is the ability to manipulate images and type in totally digital formats. This was available only experimentally and at great expense until mid-1981. Moreover, software and engineering expertise for type- and image-processing systems, most commonly referred to as pagination or image-assembly devices, are maturing at a rapid rate (Figs. 3 and 4). This is occurring just as the new generation of high-powered microprocessors of both 16- and 32-bit varieties are becoming realistic components for lower-cost systems.

Publishing organizations who currently have a need and justification for completely integrated pagination to automate production of their print products may find themselves in the most favorable position to exploit new market opportunities. Once master images have been composed for the production of printing plates, it is (or will be) a simple matter to reprocess them into the formats used for the electronic publishing market. The standard home television set or

Table 1. Functional characteristics of the electronic publishing universe*

Mode	Activity	Suppliers	Products	Response	Advantages	Disadvantages
One-way	Videotape		Training Newsletters Magazines	Poor	Records broadcast TV	Editing poor; sequential access
	Cable TV	*American Baby*	Magazines News	Poor	Strong need; known ad format	Limited reuse value; no print recycling
One-and-a-half-way	Teletext	Telecable Teleprompter	News Classified ads Directories	Poor	Low-cost production and delivery; recycles data from text system; low capital cost	No graphics; poor ad medium; slow access; low data volume; boring
	On-line data base	CompuServe PIRA NYTIB	News Finance Technical abstract bibliography or full text	Good	Existing technology; recycles print; powerful tool	No graphic content; complex user protocol
Two-way	Videotext	Telidon Prestel	Information retrieval Transaction services	Excellent	Graphic content; recycles print; electronic mail and transactions	High capital cost; limited graphics; regulatory issues; user interface
	Videodisk	Pioneer RCA Philips	Book Catalog Training	Excellent	Interactivity; graphics; low cost per copy	Small on-line data base; high production cost; standardization

*© Inter/Consult, 1982.

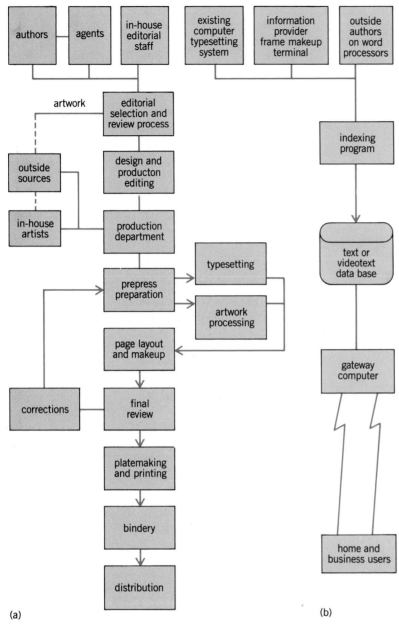

Fig. 2.   Comparison of work flows for (a) traditional and (b) electronic publishing for on-line data base or videotext.

ings at one end, and content-oriented offerings at the other. The other axis places products at one end and services at the other (Fig. 5).

Electronic mail characterizes the conduit end of the spectrum. It is a person-to-person communication system, without any information content inherent in the service. Each user is assigned an electronic mailbox, the equivalent of a computer directory or queue, to which other people can send messages, data, or documents instantaneously. The message remains in the recipient's mailbox until it is retrieved. A user can establish standing distribution or mailing lists and send the same message to everyone on the list instantaneously. In more sophisticated systems, a user's mailbox may also be intelligent, and know which subjects are most important, flagging certain items for priority attention or forwarding messages on certain subjects to the user at other locations, depending on the time of day or week.

The content end of the spectrum is well characterized by on-line data bases such as the New York Times Information System. A subscriber to this electronic publisher's offering can retrieve news stories based on key words and combinations contained either in a master index or in the story contents.

Until recently, content-rich media meant only collections of alphanumeric information. Now, however, the new technologies of video image digitization, storage, manipulation, and compression are bringing pictures into the domain of electronic publishing. Although the time delay inherent in creation and manufacturing of available formats of optical videodisks makes them less than a real-time medium, they are a very effective means by which interactive video publishing can be accomplished.

On-line data bases. Most mature of the electronic publishing activities is that of on-line data base publishing. On-line data bases utilize standard magnetic media and are composed of text characters stored in ASCII. They are accessed remotely by using standard data terminals via existing public telecommunications facilities at low to moderate speed, with most hard-copy output being done on line printers at central facilities. Direct output of full-text data from electronic text-processing systems to on-line data bases is already a reality in highly automated newspaper or magazine publishing environments.

This segment of the industry has been active since 1975. There are presently more than 500 producers offering more than 1400 individual products on a pay-per-use or subscription basis. Growth has held steady at 30% per year since 1975, and this growth trend is expected to continue at a 25–40% annual growth rate for the next few years.

Bibliographic data bases accounted for only 10% of revenues in 1980, a figure which is disproportionate to the percentage of products in this category. These data bases are published predominantly by government, quasigovernment, and not-for-profit suppliers. Source data bases (or factual packages) provide better revenue than bibliographic data bases.

its equivalent, incorporated into a computer terminal, will probably remain the standard for delivery of electronic information until at least 1988.

Electronic publishing may be characterized as being in a phase 2 state of evolution, and the supporting pagination technology as being in the early part of a phase 3 state. These stages are given in Table 2.

Conduit versus content. Several approaches to an analysis of electronic publishing have been made which are useful in providing different perspectives on this multifaceted subject. Tony Oettinger has developed an analysis of the information business which seems especially useful. This is a grid along which the various electronic publishing offerings can be plotted. It places conduit or media-oriented offer-

Table 2. Phases of electronic publishing

Phase number	Phase title	Implementation date*		Comments
		Videotext/ videodisk	Print-oriented pagination	
1	Conceptualization	1975–1976	1973	
2	Prototype construction	1978	1975	
3	Product availability	1983	1982	Requires at least two suppliers and the ability to deliver 5% of total market demand in acceptable production-ready condition.
4	Active market	1985	1983	Mature product; increasing number of products and suppliers developing segment orientation over time. Delivery of 65% of total market within 5 years from start.
5	Mature market	1992	1987	Decline in product development; concentration of ownership among suppliers; changing price and purchaser demographics; newspaper text processing is a good example.

*All dates are approximate.

However, they are more expensive to market than bibliographic data bases, which rely on well-trained librarians who understand and use them as professional tools. Source data bases, like Value Line or Standard and Poor's, with usage costs of over $100 per hour, require knowledgeable sales people and considerable customer training. They also demand some preplanning to obtain most effective results, eliminating casual usage. From a producer stand-point, factual packages also require a high degree of currency, accuracy, and powerful indexing. These are all expensive in terms of time and human judgment. However, factual data bases focus more on current data and can have less expensive on-line storage requirements than bibliographic varieties.

Inability to incorporate graphic material has been a major limitation and block to greater acceptance of on-line data bases. In 1982 the first attempt was

Fig. 3. Portion of master image of a fully composed page on Scitex's Vista system.

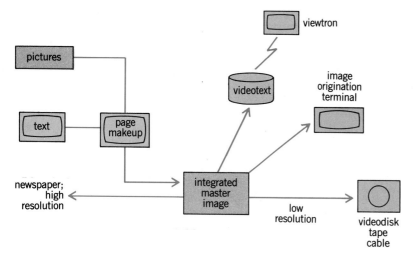

Fig. 4. Migration of text/image masters. Images and text are placed into machine-readable form and transmitted to various output devices. (*Inter/Consult*)

108,000 images, necessitating multiple disks for storage of a complete archive. This becomes especially inconvenient when disk volumes must be changed during a single search sequence.

A major impediment to industry growth and data base proliferation is lack of standardization in the user access protocol—the group of directives for specifying what operations are to be performed and which variable data are to be used. Nonstandard command codes to perform the same operations on different data bases create a Tower of Babel situation in which it can take as much as 6 months to train a fully qualified searcher, and each new data base has its own specialized commands which must be learned.

Availability of a new generation of self-contained data base processors will make entrance to this field easier for publishers. A multiprotocol processor, and batch search techniques which use intelligent microprocessor terminals and store-and-forward to reduce communications charges, will improve the user's situation and decrease costs.

Videotex. Videotex is the generic term for electronic home information delivery systems. Within this broad term there are two specific approaches, called teletext and viewdata, or videotext. Teletext is a one-

made to use videodisk, attached locally to the access terminal, as a way of integrating image material into the on-line data base presentation. This is exciting theoretically but may be difficult in practice, since each disk contains only 54,000 or

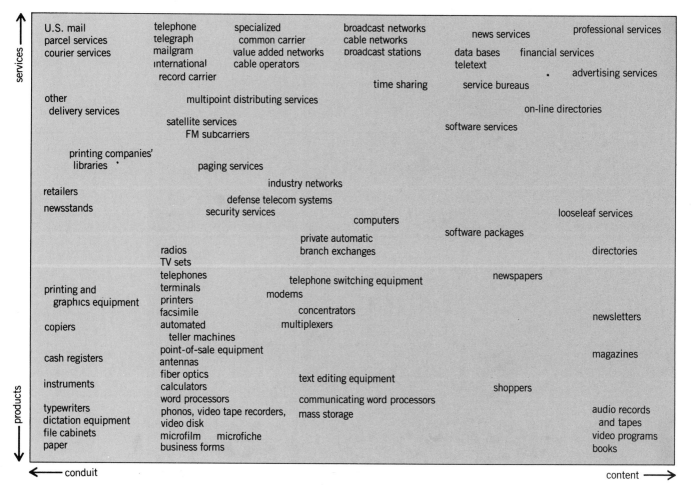

Fig. 5. The information business characterized in terms of conduit versus content as a function of available products and services.

way communications medium. Images, each constituting a single frame of TV data in a special, compressed format, are transmitted in a continuous sequence. Users indicate which frame they would like to see by interaction with the decoding unit in their local TV set. This "grabs" the desired frame or sequence of frames on its next cycle. Users have no way of transmitting information to the central broadcast facility. Individual teletext services have brand names like Ceefax, the BBC's pilot program, or Teletel, the French service.

Videotext, or viewdata, is a potentially more powerful and lucrative service since it involves users' ability to directly access a central data base interactively (Fig. 6) from their local TV. Users may request specific frames of information. More importantly, they can directly access what they want, meaning quicker response time and a more structured usage of the medium. Most exciting of all, users can communicate with other users via the system. They can also utilize transactional services, including banking or shopping based on information provided by the system.

The enormous potential of videotext services has not been realized as yet in any activity above the level of experiment, despite the avalanche of publicity which is heralding the arrival of a new era in home information delivery. Many fundamental problems exist in making the concept truly useful and commercially viable. Most of these relate to needed production capabilities, which requires competent digital type/image manipulation, cheap hard copy, and delivery systems which need adequate bandwidth to deal with pictures or productization of present information resources into videotext data bases. Recent estimates indicate that by 1990 a total of 10% of United States homes (and a much larger percentage of businesses) will be utilizing videotext in a significant way.

Unfortunately, the strategic and regulatory environment around videotext services has become the battleground for both national telecommunications policy and national technology prestige. This has overshadowed deficiencies in present product and production technology. Lack of acceptance by information providers and users has been unheard beneath the din of congressional arguments between the newspaper industry and AT&T over rights to revenues from home-accessed information bases. The economics of the new medium may prove to be wildly overestimated, with claims based on subsidized probes which rarely approach real-life market conditions.

Even for those systems which are actually operational, capital cost is high on both the publishers' and users' side. A new generation of data base and computing machinery will help the industry, as will new terminals designed for use by graphic designers and visual artists. In June 1982, AT&T introduced its new Frame Creation Terminal (FCT). This is a color-page or frame-makeup terminal which can operate either as part of an on-line system or as a stand-alone work station. It serves the needs of informa-

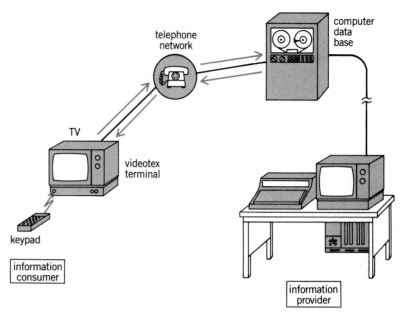

Fig. 6. Videotex information network.

tion providers for composing and formatting frames to be placed into the videotext data base. The FCT is equipped with a computer, color video display, keyboard, and graphics tablet. Storage of up to 600 frames on dual floppy drives is supported.

Videodisk. Laser videodisk is the most exciting of the new visual media. It has rich potential since it employs a random-access technique. Available videodisks store single-frame television images in the form of analog FM-encoded TV signals. These are decoded and played back by using a low-power laser. Future optical disks will be memory systems rather than frame storage devices. However, because of the enormous quantities of digital data they will be capable of storing in binary form, the creation of all-digital image archives will be their prime application.

These kinds of picture data bases in digital format are impractical with present technologies because of the enormous amount of data contained in images. As an example, the number of ASCII characters in a standard office document in 8½ × 11 in. (216 × 279 mm) format is about 3840 characters. This page could be transmitted over a 9600-baud phone line connection in 3.4 seconds. The same page in a digital scanned format would take 3,800,000 bits of storage, or 475,000 characters. This data would require 63 seconds of transmission time. A color picture would have four or five times the amount of information.

Laser-format videodisks contain 54,000 color video frames per side of the disk, and each frame is directly accessible by its frame number. Some player devices allow simultaneous access to both sides of the disk, making all 104,000 frames accessible simultaneously. When played at 30 frames per second, the laser disk functions much like a 30-minute video cassette, with improved video and two-channel stereo sound. However, the unique aspect of laser disks lies in their ability to display any individual frame, or sequence of frames, as a result of the interaction between a viewer and a computer linked to the laser disk player. Further sophistication may be added by use of a data-base management system in the control computer. This allows users to retrieve images based on their content, historical usage, source, or any other factor deemed to be important, and encoded during construction of the videodisk or its retrieval index base.

Videodisks are produced in large quantities by a stamping process, much like the manufacture of LP records. They are similar to books from a publishing perspective in that they nicely accommodate the situation where static information needs to be distributed to many different locations.

The laser consumer disk with its low cost and high on-line storage capability has been perceived as a suitable medium for several publishing products, including catalogs, training programs, and archive documentation, and as a source for locally stored graphics. However, all these applications accept the idea of a content-static package with a long production lead time being acceptable. An interesting prototype for industrial, research, or training applications of analog disks has been introduced by Vision Machine Research. This consists of two color-display screens controlled by a third data terminal accessing from four to ten laser disk players simultaneously. Multiple microprocessors offer data base management for search and retrieval of desired images. Full audio output for sound tracks and data communications makes this appear to be a powerful tool for a medium where access is the problem.

Most applications where image archiving is desired need a dynamically updatable and expandable storage capability. Picture morgues for newspapers or magazines, electronic document-on-demand systems, and electronic color systems all share this requirement. However, no systems for such applications are currently available, and even when the problems of recordability, erasability, data reliability, and archival durability are solved, an enormous amount of research and development on packaging of such large data bases will be required. However, the development of the videodisk as an alternative to print will blossom on the arrival (in 1983) of disks which store digital rather than analog information. This device will solve most of the problems of present analog-form disks.

Case studies. CompuServe is a model for the electronic information utility of the future. Access is made by a local telephone call in over 300 United States cities presently concentrated in the northeastern seaboard, Silicon Valley (the region in California where many electronics manufacturers are based), Los Angeles, and Chicago. CompuServe offers users 301 categories of information, from African weather to used-car purchase procedures (Fig. 7).

Telecable. Telecable is a teletext operation recycling the print operation's news over a channel leased from Wisconsin CATV. Brief news stories are displayed on the screen for 25 seconds with 25-second advertising slots between stories. On important stories, viewers are encouraged to become readers by referring them to the print product for more in-depth coverage. However, poor understanding of how to use the teletext medium for advertising seems to be a major problem. For this reason, Telecable recently introduced an hourly 5-minute live anchored

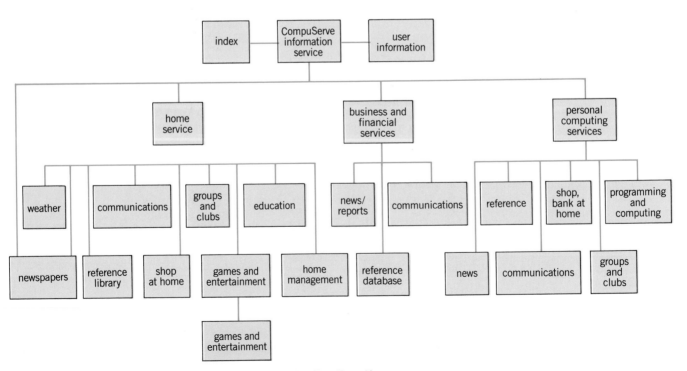

Fig. 7. CompuServe information utility. (*Inter/Consult*)

newscast for which it sells traditional 30-second advertising spots. Surveys now show frequency of use rising, with 73% of cable subscribers presently viewing Telecable 1.5 times daily, a 57% increase over the first year of operation.

Videodisk for how-to literature. One strong electronic publishing opportunity for laser disks is the traditional how-to book or magazine. Not only can the textual material and static pictures be presented, but animated sequences of a process can also be incorporated. The viewer has the option of viewing the film sequences at regular speeds or in slow motion, in both forward and reverse directions. The viewer also can stop the action at any point without additional wear on the disk. Users can follow different levels of instruction through the disk according to their level of skill or knowledge of the tools required for the task. The first commercially available interactive videodisk ever produced was an instructional package called "How To Watch Pro Football."

Applications in this mode of use range from the teaching of surgical procedures to physicians, to the teaching of origami (Japanese art of paper folding) to children. Tests by the U.S. Army have found that while experts can repair a piece of complex machinery as quickly using a printed manual, novices did far better using interactive videodisks in analyzing and fixing complex problems on unfamiliar machinery or systems.

Sears videodisk catalog. The first attempt to replace or augment full-color merchandise catalogs with videodisk was undertaken by Sears Roebuck in their Summer 1981 video edition. The catalog contained not only traditional text and graphic information frames, but TV-like action sequences and user-originated search/query capability. Action sequences included fashion modeling and operation of mechanical equipment like lawn mowers. Production values were extremely high, with TV advertising standards prevailing throughout. There was also a heavy mix of computer-generated graphic effects.

The catalog was made available to the public for home use or at public-access facilities in Sears retail locations. Reactions were extremely favorable, although specific sales data or comparisons with traditional catalog merchandising remained proprietary in nature. Unfortunately, slow acceptance by consumers of videodisk as a component in home entertainment systems will be a gating factor in use of the disk for catalog purposes.

Electronic magazines. Nonprint formats for magazines are an appealing and seemingly successful application of electronic publishing. Magazines are presently being published in videotape format, and expansion into videodisk is immanent. Several magazines are using their print product as the basis for a cable or broadcast television program. Videotex seems very suitable for certain kinds of magazines, especially those that are how-to—oriented.

American Baby magazine has now piloted a series of programs for cable distribution in major markets, making use of their name recognition and in some cases offering video versions of articles which originally appeared in print. Classic TV ad spots are used and heavy cross promotion of the print product is done as well.

Time Incorporated launched a series of videotext news and feature magazines in 1982. *Popular Science* has already started making content available through CompuServe. Some material is specially edited for the on-line format, including certain headline/capsule-oriented features. Automobile road tests and car repair articles are also offered. *Popular Science* encourages its readers to offer suggestions or pass on requests for more information via CompuServe's electronic messaging services. This is a forerunner of the kind of transactional services in responding to advertisers which will make video magazines an attractive and powerful medium.

As pagination of the traditional print product matures, electronic recycling of material will give publishers with such multifunctional equipment strong incentives for content recycling in other forms.

Opportunities and obstacles. It appears that a new electronic publishing industry is at the beginning stages of its evolution. Many of the larger participants in the traditional sectors have already made strategic moves to position themselves strongly in the new sector. Activity by traditional equipment suppliers has been less enthusiastic. Most capital-intensive product development for new equipment needed to supply this sector is being undertaken by data processing suppliers.

Is electronic publishing a threat to traditional printed information distribution? For the next 5 years, the answer is probably no. Production of printed matter will probably experience a short-term increase as a result of the packaging and collateral materials needed to deliver, sell, or use electronic information. In a 10-year time frame, however, color magazine and catalog printers may begin to experience a shift in their revenue base. Certainly vendors of graphic arts equipment and supplies, especially plates, films, and inks, are expected to be strongly impacted by 1990. This may seem remote until the payback periods on a film-coating alley or a large rotogravure printing facility are considered.

Electronic publishing is the most important sector of the emerging information industry. It is on the threshold of recognition as a major industry. When seen in combination with traditional publishing, it could, by the year 2000, represent the largest single sector of the United States industrial economy.

[DAVID H. GOODSTEIN]

Bibliography: T. S. Dunn (ed.), *Proceedings of the 1981 Lasers in Graphics, Electronic Publishing in the 80's Conference*, Vista, CA, vols. 1 and 2, 1981; D. Goodstein, Output alternatives, *Datamation*, 26(2):122–130, February 1980; D. H. Goodstein, The outlook for optical videodisk in graphic arts applications, *TAGA Proceedings 1981*, Rochester, NY, pp. 204–221, 1981; D. H. Goodstein, Videodisk, in *Auerbach Electronic Office: Management and Technology*, 1981; *The Seybold Report on Publishing Systems*, 1981.

Archeological

Cu Zn Ga Ge As

| 47 Ag | 48 | 49 In | 50 Sn | 51 Sb |

| 79 | 80 | 81 | 8 | 83 |

Metallurgy

Dennis Heskel

Recipient of a doctorate in anthropology from Harvard University, Dennis Heskel is an assistant professor in the Department of Anthropology at the University of Utah. He has worked on the development of metallurgy in the ancient Middle East and on the role of technology in the development of civilization.

The importance of metallurgy—the techniques of smelting, melting, and working of metals—in the description of ancient history is evidenced by the use of the terms Copper Age, Bronze Age, and Iron Age in the writings of ancient Chinese and Roman historians, and in the modern-day ordering of the development of human societies by archeologists. While these terms represent a number of changes and characteristics in the societies they describe, the metal they used is seen as an integral component of their level of organization.

Because of this perception, there has been interest since the nineteenth century in the use of a variety of scientific techniques to analyze metal objects from ancient societies. This article will review these major analytical techniques and provide examples of the different types of information produced by these studies.

ANALYTICAL TECHNIQUES

Since the nineteenth century, metal objects recovered from archeological sites, that is, the material remains of the behavior of past societies, have been examined by wet chemical analysis. This involves the use of a sequence of chemical reactions to identify the constituents of the metal used to produce the object and to determine the percentage of each element present. While this technique is very accurate, it is rarely used today because of two disadvantages: the series of chemical reactions require that a large sample be taken from the object, and wet chemical analyses are more time-consuming and expensive than other techniques now available.

The information obtained from wet chemical analysis has provided some useful information on the composition of ancient metal objects and on shifts in the types of metal used by past societies. One early and influential work demonstrated a shift from the use of copper-arsenic metal to the use of copper-tin bronze during the Bronze Age in central Asia. This shift in material has since been demonstrated to have occurred throughout the Middle East, Europe, and Peru, and is a focus of current research in archeological metallurgy.

Nondestructive methods. At present, less destructive physical methods of analysis are used to determine the chemical composition of ancient metals. These analyses determine the concentrations of select major elements (concentrations greater than 1%), minor elements (concentrations of 0.1–1%), and trace elements (concentrations less than 0.1%). The major elements define the character of the metal used in antiquity, and changes in concentrations of an element can reflect changes in technological developments, such as the use of copper-tin bronze. The minor and trace element concentrations provide information on the raw materials used to make the metal of the object, and thus can be used to identify the geographical source or mine of the raw materials, the type of ore exploited and the smelting techniques employed, and the competence of the metalworker.

The principal techniques used to obtain the composition of ancient metal objects are optical emission spectroscopy and atomic absorption spectroscopy. Both methods identify the distinct wavelength emitted by an element when its atoms are excited, and measure the concentration of these elements by the intensity of the wavelengths. To increase the accuracy of the quantitative analysis, comparative analysis with standards of known composition is required.

There are problems inherent in the technique in obtaining accurate composition data, particularly in the use of emission spectrographic analysis, primarily caused by the segregation of different elements during the cooling of the molten metal. In order to help ensure an accurate result it is necessary to obtain a sample large enough to minimize the effect of inhomogeneity of the materials but not so large as to damage the object. Taking multiple samples from different parts of the object is also a useful strategy to increase the accuracy of the results.

Information provided. The results of these compositional analyses have been used to provide two basic types of information: correlations of ancient mines and production centers with finished objects; and establishment of the development of metallurgy through time, based on technological changes in the metal produced, and of the distribution of objects produced by a specific group of smiths. Both of these sets of information provide valuable data on the interaction of different societies, or the importance of trade and contact, and on the spread and impact of knowledge in the ancient world.

Austrian Bronze Age. One of the most successful correlations of ancient mines, production centers, and finished metal objects was done by R. Pittroni and an Austrian team. This research documented the locale of production for copper objects in the Bronze Age of Austria by identifying as many minor and trace elements as possible in the artifacts of each stage of production. Since the small quantities present, less than 1%, would not have been intentional additions to the metal, the similarity of trace and minor elements present at each stage was used to correlate objects to production centers and mines. Since ores vary a great deal within a particular mine, only the presence or absence of an element was noted. This method has the advantage of providing reasonably accurate source and production locations for metal artifacts that cannot be determined by a distribution map of the objects themselves. This information provides an understanding of the complexity of resource procurement and scale of production by the ancient societies of Austria.

Andean metallurgy. H. Lechtman has established not only a description of the development of metallurgy in the Andes of South America, but also an understanding of how and why these developments occurred by using, in part, the results of spectrographic analysis.

The earliest metal used in Peru was worked gold, which from 1000 B.C. to 400 B.C. was probably associated with the presence of the Chavín religious cult. The use of copper and silver, however, also appears to have developed concurrently, but the presence of these metals represents a different line of technological development—one associated with trade in materials and ideas.

The skill and craft evident in the production of these gold and copper objects indicate that a sophisticated knowledge of a number of metalworking techniques was present in Peru by 400 B.C. Both of these metals remained the major sources for further developments of Andean metallurgy.

One of the most striking of the developments in Andean metallurgy is the production of tumbaga (gold-copper) alloys and gold-copper-silver alloys by the Mochica culture of the north coast of Peru, dating from 200 B.C. to A.D. 600. When cast, tumbaga was coppery in color and, as in the gold-copper-silver alloys, became golden when the metal was hammered into thin sheets, as copper is lost from the surface of the alloy through oxidation upon annealing. This concern for the golden color of metal led not only to the tumbaga alloys but also to the development of complex systems of gilding copper.

This knowledge of the sophistication of Mochica metallurgy provides significant insight into the technical skills of the early Peruvians; the importance of metallurgy to them; and their attitude toward materials—views and beliefs that were adopted and became important in the values of succeeding Andean cultures.

METALLOGRAPHIC ANALYSIS

Besides techniques that allow for the determination of the composition of ancient metal objects, metallographic examination of archeological materials is also important in studying the methods used to make metal artifacts.

Metallography. Metallographic examination consists of viewing a section of an artifact by using the reflected light of a metallurgical microscope (Fig. 1). The analysis is based on the identification of the mechanical and thermal history of the metal by the revealed internal structures. This information is useful for determining the technical skills of the metalworkers of a particular society at a specific time and also for establishing a sequence of changes and developments in the history of metallurgy.

A metallurgical microscope consists essentially of a compound microscope and an optical illuminating system with a strong light source, an objective lens system, and an eyepiece lens system. The total magnification is determined by multiplying the power of the objective by the power of the eyepiece. Typically, the magnification used to examine archeological samples is in the range $\times 50$ to $\times 500$.

The procedure of preparing an object for metallographic examination begins with the cutting of one or more sections from the artifact. The selection of the sample depends on the type of information sought. Since this is a destructive procedure, it is important that the section selected represent the object being investigated.

Polished samples. The section is then usually mounted in a hot or cold setting plastic, often Bakelite, to increase ease of further preparation. The upper surface is then polished by using a series of increasingly fine abrasive papers, diamond, and then finally alumina paste. The resulting polished surface is examined under a metallurgical or reflected light source microscope.

The microscope is used to distinguish the internal structures of the object. In the polished samples, primarily grains, grain boundaries, and the type and amount of corrosion can be identified. This information, particularly about grain size, is used to help determine the rate of cooling of the cast metal and thus something about the manufacture of the object. Inclusions in the metal, such as cuprous oxide in melted and cast copper and in copper alloy objects, also provide valuable information on the formation of the object. Research by C. S. Smith in 1965, and later by D. L. Heskel, used the absence and presence of these cuprous oxide inclusions to document the use of hammered native or naturally occurring copper and the shift to the use of melted and cast copper in Iran during the fourth millennium B.C.

Etched samples. Additional structures in the metal sample can be distinguished by etching the polished surface. Etching, or the submerging or swabbing of the surface with a chemical reagent, attacks the grain boundaries and secondary phases (preferentially). The etch dissolves some of the metal and thus reveals its internal structure. There are many etchants used for examing metal structures, each being most suitable for metal of a specific composition. Two of the most commonly used etchants are ferrous chloride with hydrochloric acid for nonferrous metals and alloys, and nitric acid in alcohol for iron and steel.

Once etched, the sample can be examined macroscopically at low magnification (typically below

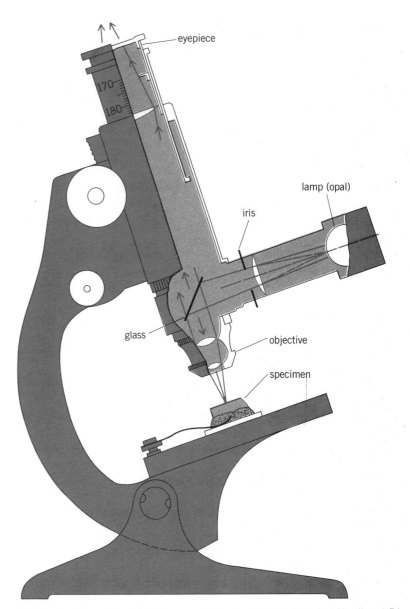

Fig. 1. Typical metallurgical microscope. (*After D. R. Cliffe, Technical Metallurgy, Edward Arnold Publishers, 1968*)

$\times 50$) and microscopically. Macroscopic examination can reveal flow lines, or the direction of the flow of metal when it is cold-worked, or hammered into shape (Fig. 2). Microscopic examination of the etched sample reveals additional internal structures that provide information on the mechanical and thermal history of the metal.

Nonferrous metals. Most of the ancient artifacts of nonferrous metals and alloys were cast or formed partly by pouring molten metal into a mold. When molten metal is poured, it absorbs oxygen and forms cuprous oxide. As the molten metal cools, it forms dendrites, or treelike growths. In a pure metal these dendrites meet and, with the spaces filled by the cooling liquid metal, form grains. With more than one metal, the process is the same, but the resultant structure is much different. In an alloy (a mixture of two or more metals which have different

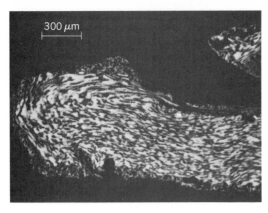

Fig. 2. Flow lines from a section of a copper spatula from Tepe Yahya, Iran, etched with K$_2$CrO$_7$.

characteristics than the metals themselves) which forms a solid solution, there will be a single phase of varying composition. These variations in microcomposition depend upon the melting point of the metals and which constituent solidifies first. In alloys which do not form solid solutions, such as copper-lead, the constituent metals will separate on solidification.

Some alloys form solid solutions within a limited range of compositions, such as copper-arsenic and copper-tin. These alloys, upon cooling from their molten state, will form a series of phases. The phases vary in composition and are also dependent on the melting temperatures of the various metals. This results in the presence of a cored structure or observable dendritic structure in the etched section of the as-cast metal (Fig. 3).

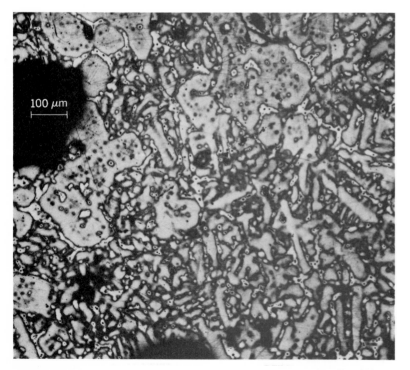

Fig. 3. Cored structure from a polished section of a cast copper piece from Tepe Yahya.

The size and shape of the grains or dendrites formed from pure and alloyed molten metal provide information on the rate of cooling of the metal and thus the method of casting. Rapid cooling produces small dendrites and grains, perhaps indicating the use of a metal mold. Slow cooling produces larger dendrites and grains, indicating the use of a clay mold or sand casting.

Often an object is worked from a cast shape into its final form. Cold working of the metal distorts and elongates the grains or dendrites. It shifts the crystal lattice of the metal, producing slip bands which appear as lines inside the grains, and also makes the metal brittle (Fig. 4). Annealing, or the heating of the metal at a temperature below the melting point, removes the internal stress caused by the distortion of the crystal lattice. It results in the formation of a different grain structure, characterized by the presence of equiaxial grains and annealing twins (Fig. 5). Several cycles of cold working and annealing are often used to produce the final shape of the object, but the resulting microstructures are similar. If the object is hammered after the final anneal, the twins become bent and irregular.

Cold working a metal makes it harder. This method of strengthening was recognized by ancient smiths. Edge hardening, or the hammering of edges (particularly blade edges) to make them harder, is a common technique to form metal objects in antiquity.

Iron and steel. The thermal and mechanical properties of iron and steel are very different from those of nonferrous metals and alloys. Thus, the macrostructures and microstructures of ancient iron and steel objects are quite different. Throughout most of antiquity, the smelting of iron oxide ore produced bloomery iron, a spongy mass of metallic iron mixed with slag and charcoal. This iron has a very high melting point; thus, unlike nonferrous metals, such iron was not used for casting objects. A notable exception is found in China, where as a result of knowledge from the production of pottery using very high temperatures, cast iron was produced by an early date—during the first millennium B.C.

In order to produce a usable object from an iron bloom, it is necessary to forge it, or heat and hammer the metal to expel the impurities. The resulting iron or wrought iron has a low carbon content (less than 0.5%) and is soft. The wrought iron is then forged into the desired shape. Wrought iron is softer than bronze, and thus the replacement of bronze by iron throughout the ancient western world must be viewed as a result of economic priorities—iron ore is more readily available than copper, and especially tin—and not the result of the material properties of the metal.

The addition of carbon to wrought iron in the range of 0.5–2% produces steel, a harder and thus more useful metal. The process of adding carbon to wrought iron is known as carburization and consists of heating the iron in charcoal at temperatures above 900°C. The microstructure of carbonized iron contains pearlite and cementite rather than the ferrite of wrought iron.

The steel produced by carburization can be hardened further by quenching, that is, heating the metal in a reducing atmosphere above 750°C and then plunging it into cold water. Quenching eliminates the pearlite phase and produces martensite, which is a carbon-rich solution. The quenched steel is very hard and brittle and must be tempered or reheated to remove some of the brittleness, although with a concomitant loss of hardness.

Evaluating ancient metals. Metallographic analysis of objects made of nonferrous metal and alloys, and of iron and steel, has provided valuable information on the development of metallurgy in antiquity.

China. A good example of the information obtained from macroscopic examination are the studies of the magnificent ceremonial bronze vessels produced during the Shang and Chou dynasties of China during the second and first millennia B.C. Recent research has shown that these vessels were not produced by the lost-wax casting technique—the molding of a wax model between clay cores and then melting away the wax, leaving the space to be filled with molten metal—but by an elaborate piece-molding technique (Fig. 6). These vessels were made by assembling a series of clay molds in which the decorative detail was first carved or incised around a clay core. The vessels are made of thin metal, 0.2–0.4 cm thick, and the positioning of the core and skill of the joints indicate the sophisticated techniques of the ancient Chinese smiths.

The reconstruction of the stages used to produce these vessels also required analytic techniques besides macroscopic examination, primarily the use of radiography. This is an important reminder that multiple analytical techniques must be used in the investigation of metal artifacts.

Iran. An interesting example of the type of information obtained from microscopic metallography is the recent research on the development of metallurgy in Iran during the fourth and third millennia B.C. Besides the basic elaboration of the technologies used to produce metal objects at the major fourth- and third-millennia sites in Iran, this study identified the importance of cultural factors in the use of these technologies.

Metallographic analysis of a number of objects from Tepe Hissar and Shahr-i-Sokhta, two Bronze Age Iranian sites, demonstrated that: the smiths of these sites shared a common, sophisticated set of metallurgical techniques, despite a distance of 500 mi (800 km) between the sites and the absence of any shared stylistic tradition; and the use of the metal technology differed considerably at each site, the result of cultural factors and priorities rather than of technical limitations. At Tepe Hissar emphasis was on the creation of beautiful finished objects, and not on the functional properties of the metal used. The emphasis of the smiths at Shahr-i-Sokhta was on advanced smelting techniques to produce quality metal combined with less advanced hammering and annealing techniques to produce simple forms.

Fig. 4. Slip bands from a copper pin from Tepe Yahya, etched with K_2CrO_7.

Role of culture. While it remains difficult to determine the specific cultural factors that affect the way a technology is used, it is important to understand that cultural choice and priority have a crucial role in the development of metallurgy, and that technological innovations do not just occur. This knowledge is important in recognizing that social, political, and economic conditions of a society determine the nature of technological developments, and that society is not totally dependent on technological innovations for change.

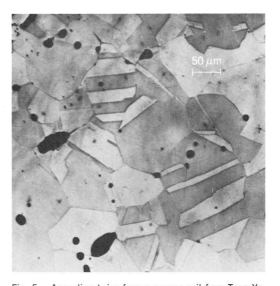

Fig. 5. Annealing twins from a copper nail from Tepe Yahya, etched with K_2CrO_7.

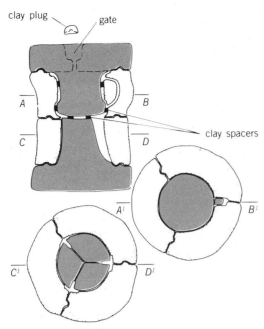

Fig. 6. Probable mold assembly (side view and cross sections) for casting a ceremonial bronze vessel. The gate, in which the melt is poured, is left as part of the finished casting with the placement of a clay plug. Clay spacers are used to separate the mold and core. (*After C. Stanley, A Search for Structure: Selected Essays on Science, Art, and History, MIT Press, 1981*)

CONCLUSIONS

As the investigation into the history of metallurgy and the development of human societies progresses, the role of archeological metallurgy, using the basic techniques mentioned, and several others, should increase in importance. The few examples presented of the type of information obtained and its significance to understanding the past should indicate the value of metallurgical examinations of archeological artifacts. In addition to the increased knowledge from, and inherent interest in, this research it is hoped that continued study will provide answers to how and why past societies produced and used the metal objects that formed such an important part of their material culture, and thus provide information on the behavior of these ancient societies.

[DENNIS HESKEL]

Bibliography: N. Barnard, *Bronze Casting and Bronze Alloys in Ancient China*, 1961; E. R. Caley, *The Analysis of Ancient Metals*, 1964; D. R. Cliffe, *Technical Metallurgy*, 1968; R. J. Gettens, *The Freer Chinese Bronzes*, vol. 11, 1969; D. L. Heskel, The development of pyrotechnology in Iran during the fourth and third millennia B.C., Ph.D. dissertation, Harvard University; G. L. Kehl, *The Principles of Metallographic Laboratory Practice*, 1943; H. Lechtman, The central Andes: Metallurgy without iron, in T. Wertime and J. Muhly (eds.), *Coming of the Age of Iron*, 1980; R. Pittroni, The applications of quantitative methods in archaeology, *Viking Fund Publications in Anthropology*, no. 28, pp. 21–27, 1960; R. Pumpelly, *Explorations in Turkistan*, vols. 1 and 2, 1908; C. S. Smith, Metallographic study of early artifacts made from native copper, *Actes du XIᵉ Congrès International d'Histoires des Sciences*, 6:237–243, 1965; C. S. Smith, *A Search for Structure: Selected Essays on Science, Art and History*, 1981; M. S. Tite, *Methods of Physical Examination in Archaeology*, 1972.

A-Z

Acoustic relaxation

Acoustic relaxation occurs in liquids or solutions when the quantity α/f^2 is found to decrease with increasing frequency according to Eq. (1). Here α is the

$$\frac{\alpha}{f^2} = \sum_i \frac{A_i}{1 + \left(\dfrac{f}{f_{c_i}}\right)^2} + B \qquad (1)$$

measured sound absorption coefficient at frequency f; A_i and f_{c_i} are the respective amplitude and relaxation frequency for the ith relaxation; and B represents all contributions to α/f^2 which are not frequency-dependent. In kinetic studies of fast reactions in solution, acoustic relaxation is one of a whole range of experimental methods, termed chemical relaxation techniques, which enable the experimenter to "tune into" the time scale of fast chemical equilibria which occur in the range $1-10^{-10}$ s. M. Eigen coined the term chemical relaxation spectrometry to describe these time resolution studies. Over the years many acoustic methods, based on a variety of different principles, have been developed and used to carry sound absorption and velocity measurements from a few kilohertz to gigahertz. The most versatile of these techniques operate in the ultrasonic frequency range 1–200 MHz, which enable the kinetics of fast reactions with half-lives of the order 100–1 nanoseconds to be studied.

In recent years acoustic techniques have been applied in many branches of the natural sciences, and notable contributions have been made in colloid and macromolecular chemistry, metal ligand interactions, and dynamic studies at phase boundaries in various systems, including liquid mixtures and liquid crystals. In addition, absorption and velocity measurements at different frequencies have been made on a range of mammalian tissues, mainly in connection with ultrasonic scanning and diagnosis in medicine.

Kinetic studies of micellar solutions. In the application of chemical relaxation methods to kinetic studies, the most significant advances of the 1970s were in connection with the kinetics of micellar solutions of surfactants. As a result of substantial experimental measurements involving many types of fast reaction techniques, complemented by a comprehensive theoretical treatment, the kinetics of micelle formation is now well understood. It is generally believed that solutions of surface-active agents above the critical micellar concentration contain monomers and also micellar aggregates which have a narrow distribution about a mean micellar number of the order 50–100. The surfactant aggregates, which are intermediate between monomers and micelles proper, are present in such small quantities that they cannot be detected by analytical techniques. Micelles are formed from monomers through a sequence of stepwise bimolecular reactions of the kind shown in reactions (2).

$$\begin{aligned}
A_1 + A_1 &\rightleftharpoons A_2 \\
A_1 + A_2 &\rightleftharpoons A_3 \\
A_1 + A_3 &\rightleftharpoons A_4 \qquad (2) \\
&\vdots \\
&\vdots \\
A_1 + A_{n-1} &\rightleftharpoons A_n
\end{aligned}$$

The chemical relaxation spectrum of a micellar solution is characterized by two relaxation times, a fast process of the order of nanoseconds, which is associated with the equilibrium involving the monomer exchanging with a micellar aggregate, and the slow relaxation time of the order of milliseconds, associated with the dissociation of a micelle to the monomer through all the elementary steps in reactions (2) [or the reverse process]. Ultrasonic relaxation has been used extensively to study the fast relaxation time in a number of surface-active agents and also in drugs which exhibit micellar properties. Recently these studies have been extended to investigate the kinetics of inverse micelles in hydrocarbon solvents. Micelles are known to solubilize material such as oils and alcohols. In certain cases, for example, in alcohols, the material is distributed between the bulk aqueous solution and the micellar phase. These solutions are important in connection with microemulsions, which are tertiary systems comprising water, oil, surfactant, and alcohol. Microemulsions have many potential industrial applications, especially in connection with tertiary oil recovery. Information concerning the dynamic exchange process of small molecules with micellar aggregates in these systems has been obtained from ultrasonic relaxation experiments. The method has also been used to investigate the kinetics associated with the exchange of water molecules having different environments in different lyotropic liquid crystals.

Phenomenological methods. In many examples of the application of ultrasonic relaxation in kinetic studies, a single relaxation has been observed experimentally, even though the process being investigated is known to be described by a series of stepwise elementary chemical reactions similar to the scheme of reactions (2). In these circumstances a conventional treatment of the data can be made only by trial and error, and becomes very difficult because of the many unknown parameters. An alternative approach to this problem, involving a phenomenological argument, has been proposed in an attempt to tackle this problem. From the amplitude and relaxation frequency a phenomenological coefficient, denoted l, can be evaluated and in favorable circumstances can be identified with the forward and backward rates of the process. This approach has been used successfully in several studies associated with the exchange process of small molecules with large aggregates.

Studies of conformational processes. In many cyclic, alicyclic, and acyclic molecules, internal rotation about single bonds, such as C—C, C—N, and C—O, can give rise to stable conformational isomers having different energies. When the potential barrier hindering the internal rotation is of the order of magnitude from a few kilojoules to 40 kilojoules, the rate at which one isomer is converted to another becomes extremely fast and in many cases can be studied only by ultrasonic relaxation. Many such studies have been carried out on a variety of simple molecules. Recently these studies have been extended to solutions of macromolecules, where the relaxation data are often described by a distribution of relaxation times. Despite this complexity, two specific conformational processes have been identified: the viscoelastic normal-mode motion of the whole polymer chain; and local internal rotation involving segments of the polymer chain. In biological macromolecules other specific conformational changes have been studied, including the kinetics of the helix-coil conformational transition in polypeptides.

Studies of metal ligand interactions. The ultrasonic method also is used to study the kinetics associated with metal ligand interactions and in particular the involvement of solvent. The method is ideal for these kinetic studies, since there is very little restriction on the range of solvent that can be used. Relaxation time measurements have also been carried out near the phase boundary in many systems, including the critical point in binary and tertiary liquid mixtures and various phase boundaries in thermotropic and lyotropic liquid crystals, as well as the thermotropic phase transition which occurs in phospholipid bilayer systems. In general, the relaxation is much broader than that described by a single relaxation time, and the amplitude of the process increases as the phase boundary is approached. The main reason for carrying out these measurements is to complement thermodynamic studies.

Although these studies have shown that processes such as concentration fluctuations and conformational changes are important at phase boundaries, at present there is no theoretical treatment describing these events in such a way that detailed information on the molecular level can be obtained from the ultrasonic relaxation experiments.

Studies of biological tissues. There has also been considerable progress in the use of ultrasonic scanning methods in medical diagnosis. The principle of ultrasonic scanning involves propagating ultrasonic pulses in the human body and detecting the echoes reflected from the internal organs. By means of suitable scanning, the detected echoes can be stored and displayed on a cathode-ray oscilloscope in such a way that they effectively form an acoustic map of the corresponding organs and tissue boundaries, which in effect is a black-and-white outline representing the gross anatomy. The ability of scanning devices to visualize internal organs directly by using the principle of ultrasonic echography is dependent upon the partial reflection of acoustic energy at tissue interfaces. The fact that partial reflection of this kind depends on the acoustic impedance of the various tissues has prompted many workers to carry out ultrasonic absorption and velocity measurements in different biological tissues over a range of frequencies. The motivation for a systematic study of different substances, such as biological tissue and solutions of globular proteins, involving ultrasonic measurements at different levels, is to attempt to understand the details of the processes responsible for the frequency dependence of sound absorption, including relaxation processes. These studies may lead to better methods of direct diagnosis, especially in the identification of different diseases within certain organs.

For background information *see* CHEMICAL DYNAMICS; CONFORMATIONAL ANALYSIS; SOUND ABSORPTION; ULTRASONICS in the McGraw-Hill Encyclopedia of Science and Technology.

[EVAN WYN-JONES]

Bibliography: D. Hall, J. Gormally, and E. Wyn-Jones, *J. Chem. Soc. Faraday II*, in press; E. Wyn-Jones and W. J. Gettins (eds.), *Techniques and Applications of Fast Reactions in Solution*, 1979.

Acoustics

The acoustic signals produced by charged particles traversing a medium offer a new technique for particle detection. Rapid thermal expansion of the medium produces an impulsive acoustic signal with a shape similar to one cycle of a sine wave. For charged particles this expansion results from thermalization of the energy lost by the particles ionizing a small fraction of the molecules as the particles traverse the medium.

Acoustic signals produced by electron beams traversing aluminum were observed in 1969, but there was little interest in these signals until 1976, when calculations presented by T. Bowen at the DUMAND (Deep Underwater Muon and Neutrino Detector) summer workshop predicted that charged particle beams or cascades induced by single particles with total energy deposited in a medium greater than about 10^{14} eV (about 160 ergs or 1.6×10^{-5} joules) should produce detectable acoustic signals. As an outcome of this workshop, three experiments were conducted in 1979 on the acoustic signals produced by proton beams traversing water. The results of these experiments, discussed below, agreed very well with Bowen's calculations, except that the acoustic signal did not vanish at 4°C, where the thermal expansion coefficient of water is zero, but at 6°C. This deviation from an expected pure thermal mechanism prompted another experiment in 1980 which showed that in addition to the thermal signal there is also a nonthermal signal produced by proton beams traversing water.

Proton-beam signals. The experiments conducted by the DUMAND group demonstrated that the acoustic signal period depended on the length of time the proton-beam pulse lasted, the diameter of the proton beam, the length of the proton beam in the water, and the location of the hydrophone with respect to the beam. The signal amplitude was observed to be linearly proportional to the total energy deposition over a range of 10^{15}–10^{21} eV. By using several different media, the signal amplitude was found to vary linearly with β/C_p (the thermal expansion coefficient divided by the specific heat at constant pressure for the medium) over an order of magnitude change in β/C_p. This result supports the assumption of a thermal mechanism, which is best demonstrated by the variation of the signal amplitude with temperature. At about 4°C water has its maximum density and $\beta = 0$. Thus at 4°C no thermal signal can be generated. Below 4°C, β is negative and the leading half cycle of the acoustic signal is rarefractive followed by a compressive trailing half cycle, and above 4°C the leading half cycle is compressive followed by a rarefractive trailing half cycle. For water, over the temperature range $T = $ 0–50°C, the variation in C_p is insignificant compared to the change in β. Thus the temperature dependence of the signal amplitude is described by $\beta(T)$. The results of the DUMAND experiments found that above about 10°C the signal amplitude was proportional to β and that below 4°C the leading half cycle amplitude was rarefractional. The zero signal amplitude which was expected at 4°C was observed at 6°C, in contradiction to a pure thermal mechanism for acoustic signal generation.

Nonthermal signals. This anomalous result prompted another experiment to study the acoustic signals produced by proton beams traversing water. Figure 1 shows the acoustic signals produced in water at several temperatures. At high temperatures (above 11°C) the bipolar acoustic signal is clearly evident after a delay of about 80 μs corresponding to the travel time of the acoustic signal from the beam path to the hydrophone. The small signal at 0 μs is due to electromagnetic interactions between the proton beam and the hydrophone. The factor given for each signal is the attenuation needed to keep the signal amplitude within the ±256-mV range of the digitizer used to record the signals. The squared-off portions of the signals, where the digitizer range is exceeded, do not affect the acoustic signal but only

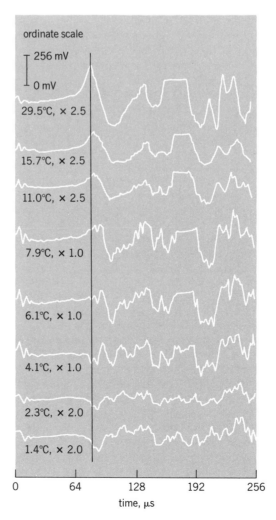

Fig. 1. Acoustic signals produced by proton beams traversing water at various temperatures.

that portion after about 128 μs which is a sum of several reflections of the acoustic signal. In Fig. 1, as the temperature decreases, the amplitude of the leading half cycle of the acoustic signal decreases and the amplitude maximum shifts in time to the right. Following down the vertical line, fixed in time coincident with the 29.5°C amplitude maximum, shows that the acoustic signal amplitude vanishes at about 6°C. Below 4°C, at this same time, the leading half-cycle amplitude is clearly negative. This constant-time amplitude analysis is similar to the method used to obtain the results discussed above, in which the acoustic signal amplitude was reported to vanish at 6°C. Near 4°C, where no thermal signal can exist, the data in Fig. 1 show a tripolar nonthermal signal with a rarefractive leading half cycle.

Figure 2 shows the effect of subtracting the 4°C nonthermal signal from the acoustic signal at various temperatures, and computer-filtering the result to remove high-frequency noise. At each temperature a bipolar signal can be seen with reflections overlapping the second half cycle. With the nonthermal signal removed, the maximum signal amplitude does not shift in time as the temperature varies. The amplitude of the leading half cycle as a function of the water temperature is shown in Fig. 3. The solid line is proportional to β for water. The amplitude of the signal remaining after subtracting the nonthermal signal is proportional to β and can be considered to have a pure thermal dependence. In addition, this analysis demonstrates that the acoustic signal generated by proton beams traversing water is a linear superposition of a thermal signal and a nonthermal signal, the amplitude of which is independent of temperature. It is this superposition of signals which causes the shift in time of leading half-cycle amplitude maximums seen in Fig. 1. The anomalous result discussed above occurs because at 6°C the ther-

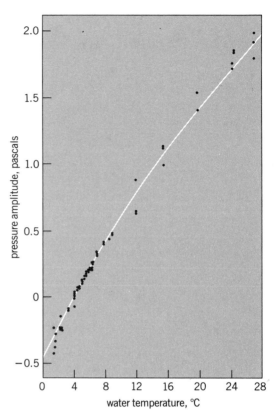

Fig. 3. Leading half-cycle amplitude of the thermal acoustic signal produced by proton beams traversing water as a function of the water temperature.

mal and nonthermal signal amplitudes are equal but opposite and the amplitude of the superposition is zero.

Tripolar acoustic signals would be produced by any mechanism which quickly expands or contracts a volume of a medium and then, in a time shorter than the sound transit time across the volume, allows the volume to relax. Three possible nonthermal mechanisms associated with charged particle beams have been suggested. These are microbubble production, molecular dissociation, and electrostriction. The first two of these mechanisms cause expansion of the affected volume and produce tripolar signals with compressive leading cycles, the opposite of what was observed. Electrostriction, produced by nonuniform electric fields embedded in a polarizable medium, causes contraction of the affected volume and would produce a tripolar signal with a rarefractive leading half cycle.

When a charged particle beam traverses a medium, there are two electric fields produced. One is from the charges in the beam and the other is from the charges freed by ionization of the medium. Calculations show that the electrostriction due to the freed charges is the dominant of the two sources and probably the source of the nonthermal signal produced by proton beams traversing water.

Applications. Some of the many possible applications of acoustic particle detection include particle-beam monitors, cosmic-ray detectors, massive-

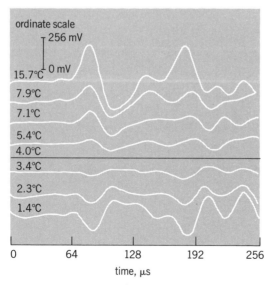

Fig. 2. Thermal acoustic signals produced by proton beams traversing water at various temperatures.

shower detectors, and cosmic neutrino and muon detectors.

Particle-beam monitors. At accelerators, the acoustic signal produced by charged particle beams would provide a means of measuring the diameter, position, and intensity of the beams. Beam intensities in the range of 10^8–10^{12} particles per beam pulse could be monitored acoustically.

Cosmic-ray detectors. Since acoustic signals propagate through most liquids and solids with little attenuation, large volumes could be monitored with a few hydrophones to detect the cascades produced by cosmic-ray nuclei. Analysis of the acoustic signals would give the charge and energy of the nuclei. The high acoustic threshold energy would automatically limit the acoustically detected cosmic rays to only those with very high energy (greater than 10^{14} eV).

Massive-shower detectors. The next generation of neutrino experiments at accelerators will require massive (on the order of 10^7 kg) detectors which must necessarily be made from inexpensive materials (for example, water). The acoustic signals produced by neutrino-induced cascades in the water could be analyzed to give a measurement of the energy deposited in the electromagnetic (γ, e^+, e^-) and the hadronic (p, $\pi^\pm$, . . .) portions of the cascade.

Cosmic neutrino and muon detectors. A similar application would be acoustic detection of cascades produced by high-energy cosmic-ray neutrinos and muons in supermassive (on the order of 10^{12} kg) detectors. A possible source of such a large volume (on the order of 1 km^3 of water) is deep in the ocean. This volume would be shielded from all cosmic-ray particles except neutrinos and muons by the overlying ocean. The existence of nonthermal signals is important to this application since the water temperature deep in the ocean is about 4°C.

Many more applications will be envisioned as the thermal and nonthermal acoustic signals produced by charged particles are studied.

For background information *see* Cosmic rays; Particle dectector; Sound in the McGraw-Hill Encyclopedia of Science and Technology.

[STANLEY D. HUNTER]

Bibliography: B. L. Beron et al., Mechanical oscillations induced by penetrating particles, *IEEE Trans. Nucl. Sci.*, 17(3):65–66, 1970; S. D. Hunter et al., Acoustic signals of non-thermal origin from high energy protons in water, *J. Acoust. Soc. Amer.* 69:1557–1562, 1981; A. Roberts (ed.), *Proceedings of the 1976 DUMAND Summer Workshop*; L. Sulak et al., Experimental studies of the acoustic signature of proton beams traversing fluid media, *Nucl. Instrum. Meth.* 161:203–217, 1979.

Aerosol

An aerosol is a two-phase system in which the particulate phase (either liquid or solid) is suspended in the gaseous phase. The beneficial uses of powders and particles have long been recognized, but when particles become suspended in air to form an aerosol, they are generally considered a nuisance. These nuisance aerosols cause problems such as air pollution, respiratory illnesses, and corrosive destruction of property. Recently, new beneficial uses for aerosols have been developed, and new techniques for their implementation and production are rapidly being discovered.

A variety of applications of aerosols have been developed, including agricultural, food production, medicine, military, household, and industrial. Indeed, many of the powdered materials in common use (such as powdered milk) are produced by an aerosol generation process.

Aerosols are produced either by mechanical dissemination of a bulk material into a gas or by physicochemical techniques in which a material is vaporized and condensed or reacted in the vapor phase to form a dispersed particulate. The most important characteristics of an aerosol that determine its properties (both for beneficial or detrimental reasons) are chemical species, mass or number concentration, particle size, and size distribution. These characteristics can be measured by a number of techniques that rely on some characteristic physical behavior that is dependent on the size or mass of the particulate.

Aerosol generation processes. Aerosols may be classified as either dispersion aerosols or condensation aerosols. These two types of aerosols are formed by mechanical or physicochemical means as outlined in Table 1. The table lists several typical generation methods and the particle diameter that is likely to be produced. Mechanical methods of generation generally yield particles on a mass basis that are larger than 1 μm, whereas physicochemical techniques produce particles that are predominantly smaller than 1 μm. One exception is the atomization of solutions and subsequent evaporation of the solvent. The final particle size is controlled not only by the initial drop size from the atomizer (mechanical generator) but also by the concentration of dissolved solute in the solution being sprayed. Practical considerations of solvent purity usually limit this technique to aerosols of particulate larger than 0.3 μm. The term nebulizer is often applied to atomization devices which deal with dilute solutions to produce very fine particles.

Many aerosol production techniques make use of combinations of the techniques listed in Table 1. For example, an electrically exploded wire can generate aerosols by mechanical separation of solid and liquid (molten) particles from the wire due to rapid heating, by vaporization and condensation of the metal, and by gas-phase reactions to form metal oxides or hydrides.

Gravimetric processes. Gravimetric aerosol generation in its most elementary form is illustrated by droplets of liquid issuing from an orifice such as an eye dropper, or dropping off a continuous or discontinuous surface. This technique yields low aerosol production and coarse particle size. The principle that governs the formation of droplets by gravimetric

Table 1. Aerosol generation processes

Aerosol type	Dissemination process	Typical technique	Aerosol material	Typical particle diameter, μm
Dispersion (mechanical)	Gravimetric	Eye dropper	Liquid	>500
	Rotary	Spinning disk	Liquid	>5
			Soluble solid	>0.3
	Hydraulic	Hydraulic nozzle	Liquid	>5
			Soluble solid	>0.3
	Pneumatic	Two-fluid nozzle	Liquid	>1
			Soluble solid	>0.3
		Film bursting	Liquid	>1
		Redispersive	Solid	>1
		Explosive	Liquid	>0.5
			Solid	>0.5
	Electrostatic	Charged needles	Liquid	>1
	Vibrational	Vibrating crystal	Liquid	>10
		Vibrating rod	Liquid	>5
Condensation (physiochemical)	Condensation	Vaporization/ condensation	Volatile	<2
		Pyrolysis	Nonvolatile	<0.5
	Gas-phase reaction	Chemical	Volatile	<0.5
		Photolytic	Volatile	<0.5

generation involves force balances due to gravity, liquid surface tension, viscosity, density, and feed rate. These parameters also govern the formation of aerosol from a rotary-type atomizer, which increases gravitational forces by introducing a centrifugal field. The spinning disk or spinning cup generator is used commonly in spray drying, fuel atomization, and agricultural applications. The disintegration of liquid takes place either by unstable breakup of jets or films, or by dropwise formation. Liquid films or ligaments are produced by introducing the liquid at the center of a rotating disk or cup. The disk is rotated at speeds of 10,000 to 100,000 revolutions per minute. The resulting velocity difference between the

Generation of a test aerosol by redispersion of a powdered silica. The powder is disseminated by compressed air through a pneumatic nozzle. Air and particles are mixed at sonic velocity.

liquid and the surrounding atmosphere produces an instability which shatters the liquid into fine droplets. The smallest drop size produced is limited by the structural strength of the rotating disk. Spinning disk atomizers are used in spray dryers with slurries (as in the production of powdered milk) because of the lack of plugging problems.

Hydraulic processes. Atomization by a hydraulic nozzle is the result of liquid jet instabilities due to a relative velocity difference between the liquid and the surrounding atmosphere. The fluid velocity is produced by forcing a liquid through an orifice or nozzle from a pressurized reservoir. Fine sprays can be produced by using very high pressure, although in general the particle size is rather coarse and polydisperse (with a wide range of sizes). There are several variations of hydraulic nozzles, such as swirl jets, impinging jets, and axial jets. Various spray patterns can be produced by shaping the nozzle outlet. The parameters affecting hydraulic atomization are the same as for gravimetric aerosol generation but include operating pressure. Hydraulic nozzles are used extensively for spraying agricultural pesticides, fuel atomization, spray drying, and watering lawns.

Pneumatic processes. The high relative velocity between gas and liquid is achieved with a two-fluid pneumatic nozzle by accelerating a gas to high velocity rather than the liquid. Pneumatic nozzles are sometimes called air-blast nozzles, and they are used extensively for fuel atomization and medical inhalation therapy. A familiar example of a pneumatic nozzle is the aerosol spray can, which produces an atomizing gas by chemical reaction or vaporization of a volatile liquid. Pneumatic nozzles will generally produce a finer spray than hydraulic or rotary atomizers.

Solid-particle aerosol generation by explosive techniques or by redispersion of a powder is also considered to be a pneumatic dissemination process because a high velocity of expanding gas is required. Film bursting generates an aerosol from the sudden failure of a stressed film of liquid. This is a low-capacity aerosol generation technique, but one which nature uses extensively. Resuspension of a dust from the ground by a strong wind is a good example of a powder redispersion aerosol. The van der Waals forces between small particles prevent redispersion of particles that are much finer than 1 μm diameter. Explosive dissemination can produce very fine particles, but coarse particles will also be present because of the difficult of uniformly distributing the explosive force and rapidly expanding gases. Explosive dissemination is used by the military to produce signal or screening smokes and for chemical weapons. The illustration shows powder redispersion to produce a smoke.

Electrostatic processes. Electrostatic dissemination produces a force on a liquid by exposing it to an electric field. Powders may be aerosolized, but this technique is much less developed. Forces on the aerosolized material are produced by free charges, or by polarizing the material and creating repelling forces. Electrostatics are sometimes used in conjunction with other aerosol generation techniques, such as in spray painting or fuel atomization.

Vibrational processes. By imposing mechanical vibrations on a liquid, instabilities are created which strip off ligaments of fluid and subsequently shatter the droplets produced. The major parameters for particle production are frequency of vibration, surface tension, and liquid density. Relatively uniform particle sizes can be produced by the technique, and it has been used to produce aerosols for medical therapy and for testing purposes.

Condensation and gas-phase reaction processes. The growth of particles under supersaturated conditions by vapor diffusion and condensation is a method of producing fairly uniform and usually submicrometer aerosols. Homogeneous nucleation occurs when a vapor material is sufficiently supersaturated such that the vapor self-nucleates and forms a molecular cluster on its own. Heterogeneous nucleation requires a very small particle (such as salt nuclei) upon which the vapor can condense. Vaporization-condensation and gas-phase reaction aerosol generation play a major role in the formation of atmospheric air pollutants. Gas-phase reaction can produce a liquid or solid product which then remains suspended in the atmosphere. The gas phase reaction of sulfur dioxide (SO_2) and water vapor contributes to the formation of sulfuric acid aerosol and acid rain in the Earth's atmosphere. In industry, nonvolatile materials can be pyrolyzed (reacted with heat in the absence of air) to produce liquid or solid aerosols. The formation of carbon black is a good example.

Aerosol characteristics. The most important physical property that gives an aerosol its unique characteristics is particle size. Other properties such as concentration, shape, density, and composition may also be important. The size of an aerosol particle can sometimes be ambiguous, and many times the size is actually defined by some characteristic physical consequence of the apparent size and reported as an effective size. The aerodynamic characteristics of the particulate generally determine whether an aerosol is stable or not. In the atmosphere, for example, the wind may suspend soils and pollens, but unless its diameter is below about 20 μm, the particulate will settle under the influence of gravity and not remain airborne. Particles smaller than 0.1 μm tend to flocculate or deposit on surfaces by diffusion. Stable aerosols are therefore generally in the size range of 0.1 to 20 μm in diameter. In industrial production or agricultural applications the particle sizes may be much larger in order to enhance collection or deposition on a surface. In these cases, stable aerosols are undesirable because of inefficient product utilization or subsequent air-pollution problems.

The concentration of particles in an aerosol can be expressed in many ways, although the two most common are by mass and by number. For a given mass of material suspended in the air, the mass concentration (for example, milligrams per cubic meter) remains constant as the average particle size decreases. However, the number concentration (for example, particles per cubic centimeter) and the total particulate surface area will increase dramatically for the same mass concentration as the average particle size decreases. For most aerosol applications the mass concentration is a more useful value, although in certain situations, such as health-related or optical properties, the number or surface concentration may be more suitable.

Typical aerosol concentrations of ambient particles are 10–250 $\mu g/m^3$, and the particle size is mostly below 1 μm. In an industrial situation, powders may be conveyed in air at concentrations up to 50 kg/m^3. A typical paint-spraying application from an aerosol can produces overspray breathing-zone concentrations of 50–100 mg/m^3 of particles less than 6 μm diameter.

Role of aerosols. Aerosols in the environment generally cause nuisance, health, and property destruction problems. Aerosols are produced naturally in the form of rain, snow, fog, pollens, volcanic ash, and many other materials. Air pollution caused by direct emission of particulate from power plants, automobiles, forest fires, and atmospheric reactions cause billions of dollars worth of damage annually. Health problems due to exposure to dusts and chemicals in aerosols in the work place have caused much misery and death. Long-term exposures of mine workers and dock workers to silica and asbestos, which caused respiratory diseases such as silicosis and asbestosis, are classic examples. Aerosols that contain particulate below approximately 5 μm diameter can penetrate the body's natural removal mechanisms in the upper respiratory tract and deposit in the lower respiratory tract. For this reason extensive studies of particulate air pollution, indoor air pollution, and inhalation toxicology are being conducted.

Aerosols can serve many useful purposes also, as outlined in Table 2. Some of the newer developments, such as ink jet printing, have greatly expanded capabilities within the field. In ink jet printing, droplets of ink are produced through a hydraulic nozzle, electrostatically charged, and then directed to the paper by means of an applied electric field. Extremely rapid and precise printing results.

The production of finer and finer metal powders for metallurgical sintering has been made possible by metal atomization. Metal powders produced by atomizing molten metal produces essentially spherical particles. After the aerosol is collected, the powders flow very freely, and sintered metal parts can be produced with little or no machining.

The food industry produces more and more convenience foods by the process of spray drying. By tailoring the particle size of the product in the atomization process, different properties of the food can be enhanced. Instant milk is the result of very fine particulate and loosely packed agglomerates. The fine sizes promote faster solubility. Flavor or sweetness of certain products can be enhanced by altering the particle size.

In the medical field, the same properties of aerosols that cause lung disease can be used to advantage in application of medications. Drugs can be dissolved in an aqueous solution and administered to the patient through an inhaler that atomizes the solution to fine enough particle sizes that there is efficient deposition on the lung surfaces. Direct application of drugs to the respiratory tract greatly increases effectiveness of the drugs and decreases the quantities required. New developments in the application of aerosols will greatly enhance the quality of life.

For background information *see* ATOMIZATION; PARTICLE PROPERTIES; PARTICLE SIZE ANALYSIS in the McGraw-Hill Encyclopedia of Science and Technology.   [CLYDE L. WITHAM]

Bibliography: R. Dennis (ed.), *Handbook on Aerosols*, Tech Information Center, ERDA, 1976; S. K. Friedlander, *Smoke, Dust and Haze*, 1977; N. A. Fuchs, *The Mechanics of Aerosols*, 1964; C. E. Lapple, J. P. Henry, and D. E. Blake, *Atomization: A Survey and Critique of the Literature*, SRI Int. Tech. Rep. no. 6, AD821–314, 1967.

Agricultural science (animal)

The rate and efficiency of animal production are determined by factors such as genetics, health, and feed resources. An important factor that determines the efficiency of animals in producing food and fiber for consumers is the climatic environment. Temperature is considered to be the major climatic factor affecting animal production. The thermal environ-

Table 2. Role of aerosols

Field	Typical applications	Type of aerosol generators used
Agriculture	Pesticide application, fertilizer application	Hydraulic sprayers, spinning disk, pneumatic
Environmental control	Particulate scrubbers	Pneumatic, hydraulic, condensation
	Cloud seeding	
Communications	Ink jet printing	Electrostatic and hydraulic
	Signal flares	Gas phase reaction
Laboratory or field testing	Particulate control technology	All generators
	Air-pollution sampling	
	Health studies	
Food production	Milk, coffee, convenience foods	Rotary
	Separations (for example, wheat chaff)	Pneumatic
Household	Cleaners, personal care	Pneumatic, hydraulic
Industrial	Coatings (paint, metal, polymer)	Hydraulic, pneumatic, electrostatic
	Fuel atomization	Hydraulic, pneumatic, electrostatic, vibratory
	Product formation (for example, fertilizer, carbon black)	Gas-phase reaction, hydraulic, rotary
	Metal atomization	Hydraulic, pneumatic, electrostatic
Medicine	Pharmaceuticals	Rotary, pneumatic
	Aerosol therapy	Pneumatic, vibratory (ultrasonic)
Military	Screening smoke, chemical weapons	Explosive, redispersive
	Flares	Gas-phase reaction

ment can be described in terms of effective ambient temperature, that is, the combination of air temperature, radiation, wind, precipitation, and humidity.

Livestock responds to wide variations in effective ambient temperature by altering feed intake, rate of energy loss, or rate of product output. This response lowers the rate of performance—the rate that animals grow, reproduce, or accomplish their desired function—and the efficiency of converting feed to animal product. A basic understanding of the relationship between animals and their thermal environment is necessary in order to assess the environment's impact on livestock performance.

Animal-environment relationships. The basic relationship between an animal and the thermal environment is shown in Fig. 1. The thermoneutral zone is defined as the range in effective ambient temperature in which rate and efficiency of performance are maximized. At temperatures below optimum, but still within the thermoneutral zone, a cool zone exists where animals use mechanisms to conserve body heat, such as postural adjustments, changes in hair or feathers, and vasoconstriction of peripheral blood vessels. The effectiveness of various insulative and behavioral responses to cold stress are maximal at the lower boundary of the thermoneutral zone, a point called the lower critical temperature. Below this point is the cold zone where an animal must increase its rate of metabolic heat production to maintain constant body temperature. In practice, lower critical temperatures may vary considerably depending upon specific housing and pen conditions, age, breed type, lactational state, nutrition, time after feeding, history of thermal acclimation, hair or wool coat, and behavior. For example, the predicted lower critical temperature for large ruminants on high feeding levels is considerably lower than for poultry, swine, or young ruminants (Fig. 2).

As temperature rises, the animal comes into the warm zone where thermoregulatory reactions are limited. Decreasing tissue insulation by vasodilation and increasing surface area by changing posture are

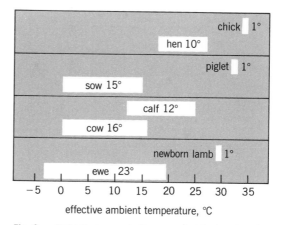

Fig. 2.   Estimated range in thermoneutral temperature for newborn and mature animals of different species. (*After National Research Council, Effect of environment on nutrient requirements of Domestic Animals, National Academy Press, 1981*)

mechanisms used to increase the rate of heat loss. In a hot environment, animals have to dissipate metabolic heat in a situation where there is a reduced thermal gradient between the core of the body and the environment. The higher-producing animals with greater metabolic heat (from product synthesis) tend to be the most susceptible to heat stress. This is different from cold conditions where high-producing animals with their higher metabolic heat production are in a more advantageous position than low- or nonproducing animals. Evaporation of moisture from the skin surface or respiratory tract is a mechanism used by animals to lose excess body heat in a hot environment.

The relationships between cold stress, critical temperatures, thermoneutral zone, and heat stress remain consistent; that is, cold temperatures always refer to those temperatures below the thermoneutral zone, even when this zone changes. The lower critical temperature may change from 0°C for a steer with a dry winter coat to 15°C if the coat is wet. Consequently, 10°C would not be cold stress for the dry steer but would be a cold environment when the same steer is wet. Thus, temperature alone is insufficient for assessing the potential impact of the total climatic environment.

Environment and performance. Both rate of animal performance, such as daily gain, or milk production, and efficiency of converting feed to animal product are functions of feed intake and feed energy required for maintenance. (Maintenance refers to energy expended to keep animals in energy equilibrium while maintaining normal body functions, with no growth or production of wool or milk.) The thermal environment affects both feed intake and maintenance requirement and in turn reduces both rate and efficiency of animal weight gain in the cold and heat (Fig. 3).

Voluntary intake is inversely related to ambient temperature, increasing moderately during cold but

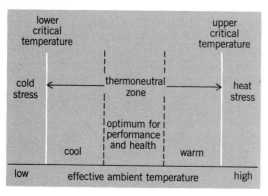

Fig. 1.   Schematic representation showing relationship of thermal zones and temperatures. (*After National Research Council, Effect of environment on nutrient requirements of Domestic Animals, National Academy Press, 1981*)

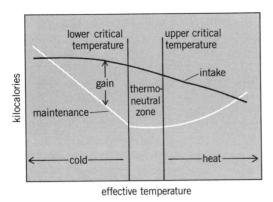

Fig. 3. Effect of temperature on rate of intake, maintenance energy requirement, and energy retained as product (gain). (*After D. R. Ames, Thermal environment affects production efficiency of livestock, Bioscience, 30:457, 1980*)

rapidly decreasing during heat stress. Metabolic heat production increases linearly during cold stress but nonlinearly during heat stress. The nonlinear increase is due to reduced efficiency of sweating and panting as heat stress progresses and to increased rate of heat production resulting from the Q_{10} (the approximate doubling of chemical reactions with each increase of 10°C in temperature) effect of elevated core temperature. Predicting the effects of heat stress on maintenance energy requirements, and consequently on the rate and efficiency of performance for livestock, is more difficult as compared with the effects of cold stress.

Numerous trials involving all species of livestock have shown that rate of performance by animals is reduced when exposed to adverse environments. In each case, maintenance requirement increases, with less energy available for production. In cold conditions, animals increase their intake, but not enough to keep pace with their rapidly increasing maintenance energy requirement. When exposed to heat stress, reduced intake combined with increased maintenance requirement results in lowered performance.

The relationship between performance (output) and intake (input) results in lower efficiency when livestock are exposed to either cold stress or heat stress (Fig. 3). This is most obvious during cold stress, when intake increases and performance (gain) de-

creases; during heat stress, the response is less obvious because both intake and performance are lower when compared with the thermoneutral zone. However, output (performance) is reduced because the maintenance requirement increases and intake decreases, resulting in a lower efficiency.

An example of temperature effect on rate of performance and efficiency of livestock is illustrated by data shown in the table, collected from swine grown in temperatures ranging from cold stress (0°C) to heat stress (35°C). Efficiency was highest in the thermoneutral zone and was reduced during both cold and heat stress. While the temperature and efficiency values may differ for animals with different insulation, diets, and so on, or for different species and products, the same general pattern of reduced efficiency is consistent among animals exposed to thermal stress.

Reproductive inefficiency is one of the most costly and production-limiting problems facing the livestock industry. Reproductive processes in both the male and female are sensitive to heat stress. As a general rule, increased temperature decreases ovulation rates, shortens intensity and duration of estrus, increases embryonic mortality, and decreases male fertility. Seasonal variations in reproductive activity are well documented but may be attributed to either temperature or photoperiodism. For example, seasonal activity in sheep is primarily controlled by the light-dark ratio, but heat (32°C) exposure of ewes during different phases of the reproductive cycle indicate that the embryo is most susceptible to heat stress shortly after conception, although significant losses may occur in later stages of pregnancy. The impact of thermal stress on the reproductive performance of sows has been identified as a major problem in the swine industry. Reports indicate increased embryonic mortality when gilts are subjected to heat stress immediately following conception, and substantial increases in stillborn pigs when gilts are exposed to heat in late gestation. Cattle are also susceptible to thermal stress. For example, reports indicate conception rates lowered by 48% in cows exposed to 32°C compared with 21°C. An example of response of the male to thermal stress shows that only 59% of gilts mated with heat-stressed boars were pregnant 30 days after breeding compared with 82% mated with control

Effect of temperature on intake, growth rate, and efficiency of energy conversion for swine (70–100 kg)*

Temperature, °C	Caloric intake (digestible energy), kcal/day	Growth rate kg/day	Product (gross energy), kcal/day	Caloric efficiency, %
0	15,377	0.54	2991	19.4
5	11,404	0.53	2936	25.7
10	10,616	0.80	4432	41.7
15	9,554	0.79	4376	45.8
20	9,766	0.85	4709	48.2
25	7,976	0.72	3988	50.1
30	6,703	0.45	2493	37.1
35	4,579	0.31	1717	37.4

*From D. R. Ames, Thermal environment affects production efficiency of livestock, *Bioscience*, 30:457, 1980.

boars. Specific physiological mechanisms which result in lowered reproductive efficiency are not totally understood, but may be due to either the direct effect on high temperature or the indirect effects mediated by physiological, nutritional, or endocrine responses to thermal stress.

In summary, the thermal environment can have a drastic effect on animal performance. Fortunately, the impact of both cold and heat can be tempered by the ability of animals to adjust physiologically and behaviorally to temperature extremes. Yet the cost of producing food and fiber of animal origin is increased significantly by exposure to the climatic environment.

Thermal environment management. Livestock producers are usually willing to incorporate management systems to improve energetic efficiency when it is economically advantageous, for example by modifying the existing environment to reduce the impact of thermal stress and improve energetic efficiency. Among the many possibilities for improving livestock environment to reduce cold stress are windbreaks (natural and constructed), sheds, confinement buildings without supplemental heat, and heated buildings. Modifications that provide the optimum environment from an efficiency viewpoint require the highest input (buildings with supplemental heat). Basically, the decision on the degree to which animal environments should be modified depends on the cost of improving the environment compared with the value of improved performance. In essence, the cost of energy in feed is weighed against energy costs of buildings and supplemental heat. Of course, such factors as cold tolerance of the animals, effect of diseases, and other determinants must be considered as well. Similar decisions regarding use of shades, sprinklers, and air-conditioned environments during heat stress are based on return on investment. Maximizing rate of performance of livestock is important, but not necessarily the major goal. Economic considerations largely determine the level of efficiency selected for livestock systems.

For background information *see* AGRICULTURAL SCIENCE (ANIMAL) in the McGraw-Hill Encyclopedia of Science and Technology.

[DAVID R. AMES]

Bibliography: D. R. Ames, Thermal environment affects production efficiency of livestock, *Bioscience*, 30:457, 1980; W. Bianca, Animal responses to meteorological stress as a function of age, *Biometeorology*, 4:119, 1970; National Research Council, *Effect of Environment on Nutrient Requirements of Domestic Animals*, National Academy Press, 1981.

Agricultural science (plant)

During recent years the productivity of crops has been increased by improved cultivars and by the extensive use of bees for crop pollination. This article discusses recent developments in seed production in new grass and legume cultivars, and current research on the value of different bee species as pollinators.

Seed production in grasses and legumes. Seed production is the process of increasing the small quantities of seed of improved cultivars developed by the plant breeder to supplies adequate for market demands. Plant breeding has been strengthened recently by improved methods of breeding and genetics. Commercial plant-breeding efforts have been encouraged by the passage of the Plant Variety Protection Act by Congress in 1970. This act provided plant breeders in the United States the same type of protection of rights to newly developed cultivars as was previously offered in other countries. These two factors have resulted in greater numbers of new cultivars entering seed production channels. *See* the feature article CLONING AGRICULTURAL PLANTS.

The process of developing a new cultivar is technically complicated and expensive. It requires a minimum of 5 to 10 years of breeding and testing, followed by a period of 2 to 4 years to increase seed quantities to an amount adequate for marketing. The American seed industry is organized to efficiently increase the few grams of stock seed of a new cultivar provided by the plant breeder to the millions of kilograms required by the forage and turf markets. The system includes the moving of stock seed to areas of specialized seed production and the use of modern techniques for maximum yields and quality. The marketing system makes the correct cultivar available to the consumer in adequate quantities.

Economic importance. Because of the economic benefit of modern grass and legume seed production, the results of plant genetic improvement are quickly made available to consumers. Improved cultivars provide greater forage yields for livestock; improved resistance against many diseases and insects; superior turf for home lawns, sports fields, and parks; and better soil stabilization, including improved revegetation of mine spoils. The resulting increases in meat and milk, and in environmental quality, can be measured in billions of dollars.

The United States is a major seed producer for export. In 1980 more than 40,000 metric tons, or 16%, of the seed crop was exported. Many foreign cultivars are introduced for seed production under the Organization for Economic Cooperation and Development (OECD) Certification Scheme, and are then returned to the country of origin for use. The United States production area is selected because of grower experience, reliability of performance, and seed quality. Major foreign consumers of these seeds include Europe, Japan, Canada, and Mexico.

Areas of production. The closing out of livestock from selected meadows and hay fields and the harvesting of the seed was once the major source of forage seed. The quality and yields from a meadow are extremely variable because of seasonal conditions and grazing intensity. This makes it difficult to balance seed supply with demand. The development of improved cultivars, requiring isolation from other cultivars to preserve genetic purity, also encouraged the establishment of fields specifically for seed production.

A major environmental requirement for high-quality seed production is a dry harvest season which favors natural field drying of seed. Artificial drying is costly and, if it is not carefully controlled, reduces germination and storage life. During the past 25 years there has been a major shift in seed production to the western states, since they offer a better environment for natural field drying of seed. California is the major production area of alfalfa seed, followed by Idaho and Nevada. Oregon is the leading producer of red and crimson clover seed, annual ryegrass, perennial ryegrass, red creeping and chewings fescue, orchardgrass, and bentgrass. Oregon, Washington, and Idaho are major bluegrass-producing states. Minnesota produces the largest quantity of timothy seed, and Missouri leads the nation in tall fescue seed.

Cultural practices and management. Specialist seed production, which consists of growing the crop on fields without grazing during the growing season, usually produces higher seed yields with better weed control. Catch-crop seed production, which consists of harvesting a pasture by removing the livestock long enough for the development of a seed crop, may be of lower purity and yield, but often produces a higher gross return because of the added income from livestock grazing. To be competitive, specialist seed producers must depend upon high average seed yields. Specialist seed fields are planted at a lower seeding rate, and thus greater care must be given to weed control than if the same cultivar is planted for pasture.

The seed certification label attached to a bag of seed assures the purchaser that the contents have been produced under specific requirements established by the Association of Official Seed Certifying Agencies and by the Seed Branch of the U.S. Department of Agriculture to protect the genetic purity of the cultivar. Seed fields must meet rigid standards to qualify for the certification program. These standards include field-use history, isolation from contaminants, and field inspections. After meeting the field requirements and postharvest seed tests, the seed is granted certification.

Control of weeds and other plants in seed fields is essential to meet contractual and market standards. The use of modern herbicides and application methods has made it possible to maintain seed quality. Selective herbicides will control certain types of weeds without crop injury. However, certain plants cannot be controlled selectively and special methods must be used. A technique widely used in the establishment of turf perennial ryegrass fields is the application of activated charcoal directly over the seeded row as a spray at planting time. Another unique method of weed control is based on differences in plant height. Weeds taller than the crop can be controlled by using a sponge-covered bar saturated with a systemic herbicide. The sponge is transported slightly above the crop canopy, wiping the chemical on the taller weed plants. The herbicide is translocated to the roots and affects only the plants contacted by the sponge. This method of weed control greatly reduces the amount of herbicide used and does not expose the crop plant to the chemical.

Certain legumes, such as alfalfa, require insect pollination for seed formation. The honeybee (*Apis mellifera*) is the major pollinating insect in many areas because it is easily managed. However, the honeybee prefers to visit blossoms other than alfalfa and will drift away from seed fields. Alfalfa producers in the Pacific Northwest states have made effective use of wild insects, such as the alkali bee (*Nomia melanderi*) or the alfalfa leafcutting bee (*Megachile rotundata*). Management methods have been developed to increase bee numbers and protect bees from predators. The increase in their numbers near seed fields has significantly augmented the yields of alfalfa seed.

Seed storage and marketing. In normal seed production and marketing, seed is moved rapidly from production to market and is planted 12–18 months after harvest. Under certain circumstances, it must be kept for a longer period, and special care may be needed to maintain an acceptable germination level.

Commercial storage life of seed depends on the species. Most grasses and legumes will retain viability for 3 to 10 years under good storage conditions. The most damaging combination of conditions affecting storage life is high temperature and high humidity. The condition of the seed at the beginning of the storage period is the most important factor affecting potential storage life. Seed that is free from damage, with a high viability and a moisture content below 12%, will have the maximum potential storage life, and the maintenance of a low temperature and relative humidity in the storage facility will prolong storage life. Packaging dry seed in moisture-proof bags will help maintain a low moisture level.

Artificially altering the temperature and humidity may not be economical for commercial seed stored for short periods, but is widely used for stock seed and germplasm preservation.

Forage and turf seed is marketed by firms or associations with production and consumption branches, or through brokers who deal with growers and wholesalers. The marketing network is international in scope. Some firms have seed production fields in several areas to protect themselves against crop failure due to weather conditions.

The trend toward private ownership of new cultivars makes possible better control of seed production and marketing. A stabilized production and seed supply will avoid the price variations that have caused marketing difficulties in the past.

[HAROLD YOUNGBERG]

Pollination. Insect-pollinated crops provide about one-third of the average American diet, either directly as fruits and vegetables or indirectly as oil seeds and livestock feed. In areas of the world where the staple food is rice, or another wind- or self-pollinated grain, possibly no more than 1% of the diet is from insect-pollinated crops. While home gardens may receive adequate pollination from existing populations of wild bees and honeybees, large acreages of insect-pollinated crops will produce low yields unless additional populations of honeybees or other pollinators are provided.

The value of honeybees as pollinators is over 20 times that of their value as producers of honey and beeswax. The honeybee (*Apis mellifera*) is the most important and manageable insect pollinator. However, it is threatened by the spread of the African bee, various parasites, insecticide poisoning, and lack of adequate summer forage for large numbers of colonies. A current trend is the utilization of other species of bees for pollination of specific crops. Notable successes have been achieved with the alkali bee (*Nomia melanderi*) and the alfalfa leafcutting bee (*Megachile rotundata*) for alfalfa pollination, and the orchard mason bees (*Osmia cornifrons* and *O. lignaria*) for fruit pollination.

Honeybees. The bee industry is threatened by the spread of the African bee, the so-called killer bee, from its established populations in Brazil. The African bee, a race of the honeybee, interbreeds with the Italian and other honeybee races, resulting in an aggressive hybrid that is prone to stinging and is difficult to manage, although generally the hybrid produces more honey. The African bee is now found in Panama, and at its current rate of spread, it could invade the United States by the next century. Importation of live adult honeybees into the United States is prohibited in order to prevent the spread of the African bee, the honeybee mite, and other parasites and diseases.

Beekeepers traditionally depended upon honey and wax, but rental of hives for pollination now forms a substantial part of their income. Because of the large number of bees required per acre, honeybees generally do not produce surplus honey when they are used for pollinating a crop. In commercial beekeeping, honeybees may be overwintered in southern states to be used in early crops such as almonds, and then trucked north as other insect-pollinated crops come into bloom. Bees are left in the crop only during bloom—usually less than 3 weeks. A major problem of large-scale beekeeping is locating summer forage for the hundreds of colonies of bees after crops requiring pollination have finished blooming. Large acreages of suitable honey plants, such as sweet clover, are needed if bees are to produce a honey crop and develop sufficient stores for winter.

The demand for honeybees for pollination is expected to increase. Additional apple trees and other fruits are being planted in the United States, and hybrid varieties of crops such as cotton, soybean, and sunflower are being developed by using insects to transfer pollen from male plants to female plants. The cotton currently grown in the United States does not require cross pollination, but hybrids produce 20% higher yields and have other advantages, such as pest resistance. If the whole cotton belt grew hybrid cotton, there would be a shortage of bees to pollinate the crop. With 500,000 colonies, California is the state with the largest number of honeybees, yet it has to import an additional 100,000 colonies a year to pollinate just the 300,000 acres (120,000 hectares) of almonds. However, the mass transport of bees provides a means of spreading bee diseases and parasites.

Improving pollination. One way to alleviate the shortage of honeybees is to improve the efficiency of the bee. Honeybee queens can be artificially inseminated, and progress has been made in developing strains of honeybees for special purposes. Through selection and artificial insemination, strains of honeybees were developed that collected 87% alfalfa pollen, as compared with only 6% collected by unselected strains. An alternative to improving the honeybee is to improve the plants. Many plants have intricate pollination mechanisms to ensure cross pollination. Some crop varieties, such as Red Delicious apples, are self-infertile and must receive pollen from a different variety in order to set fruit. Other mechanisms involve physical characteristics of the flower that affect their attractiveness to bees, such as color and odor. Self-infertility and other characteristics can be altered by selective plant breeding to produce varieties which are more readily pollinated.

Bee poisoning. Insecticides are needed to control crop pests, but they may adversely affect honeybees. Bees may be killed when sprayed in the field, when hives are accidentally sprayed, or when they forage on recently treated fields. Due to their smaller size, wild bees, such as the alfalfa leafcutting bee, are often more susceptible to insecticides than honeybees. Often the source of poisoning cannot be determined. Foraging bees may be affected and die before returning to the hive, but some slow-acting insecticides may be brought back to the hive with pollen. This problem was accentuated by the development of microencapsulated insecticides. These tiny plastic capsules containing insecticide adhere to the bee hairs much like pollen. They may not affect the foragers, but they can kill the brood when the contaminated pollen is fed to the young. Serious losses of bees due to several insecticides led to the development of the indemnity program by the U.S. Department of Agriculture, whereby beekeepers suffering loss of colonies due to insecticides could receive payment for killed hives. This program was dropped in 1981, and the Environmental Protection Agency, universities, and state apiculturists have tried to alleviate the bee-poisoning problem through regulatory and educational efforts.

Insecticide labels list potential hazards to bees, and highly toxic materials are not permitted on blooming crops. Insect pest control recommendations advocate the use of materials least toxic to bees at times and in a manner least likely to cause poisoning. For example, aerial application of insecticides to alfalfa seed fields in Washington state are commonly made after dark, when bees do not fly.

Such measures, and an increasing awareness of the need for pollinators, have reduced the bee-poisoning problem, but have not eliminated it.

Wild bees. An advantage of honeybees is their ability to forage on a large number of crops, but other species of bees can be used for pollinating specific crops. Except for bumblebees and a few others, such species are not social insects but are gregarious, with large populations of individual females, each constructing and provisioning her own nests. Unlike honeybees, they provide no honey, wax, or other products useful to humans, and they are useful only as pollinators. A good example is the alfalfa leafcutting bee, whose potential as an alfalfa pollinator was first noted in Utah in 1957. Following experimental work by several entomologists, seed growers provided nests for the bees, usually holes 5 mm (0.2 in.) in diameter and 10 cm (4 in.) deep drilled in wooden boards (see illustration). In such nests, a female constructs a cell from pieces she cuts from leaves, and provisions the cell with alfalfa nectar and pollen before laying an egg and sealing the cell with more leaf pieces. She makes additional cells until the hole is full. Each female can pollinate enough flowers in her lifetime of 1 month to produce nearly a pound of alfalfa seed. Seed growers use thousands of females per acre (1 acre = 0.4 hectare) and may obtain seed yields of over 1000 pounds per acre (1100 kilograms per hectare). Millions of these bees have been produced in the western United States and in other countries, resulting in a new industry which supplies bees, nesting materials, and shelters. Unfortunately, this bee has been subject to several predator, parasite, and disease problems that have been costly to control.

Orchard mason bee nests placed in an apple orchard.

The orchard mason bee is a distantly related species that plasters mud in nesting holes and pollinates apples and other fruit trees. Efforts are being made in the United States to develop populations for use as almond and apple pollinators, following successful use of a similar species in Japan. There are over 20,000 species of bees alone, as well as many other insects such as beetles, flies, and butterflies, that pollinate plant species. There is now a worldwide search being conducted for species that could be managed like the leafcutting bee for pollinating such crops as melons, hybrid sunflowers, onions, and cotton. The need for such insects will increase if honeybees cannot meet the rising demand for insect pollinators.

Pollination ecology. There is an increased interest in pollination ecology, and many scientists are engaged in studying pollination dynamics and bee behavior. Most plants attract pollinators by rewarding them with nectar. To promote maximum cross pollination, flowers produce nectar in proportion to the amount of energy expended by a specific pollinator, excluding other visitors by flower size, shape, or other characters. Recognition and exploitation of such relationships through plant breeding and other means may lead to improved use of honeybees and other insect pollinators.

For background information *see* AGRICULTURAL SCIENCE (PLANT); BEE; BEEKEEPING; BREEDING (PLANT); POLLINATION; SEED; SEED GERMINATION in the McGraw-Hill Encyclopedia of Science and Technology.

[E. C. KLOSTERMEYER]

Bibliography: R. Bekey and E. C. Klostermeyer, *Orchard Mason Bee,* Washington Ext. Bull. 0922, 1982; D. E. Caron (ed.), *Increasing Production of Agricultural Crops Through Increased Insect Pollination,* Proc. 4th Int. Symp. Pollination, Maryland Agr. Exp. Sta. SMP-1, 1979; L. O. Copeland, *Principles of Seed Science and Technology,* 1976; C. A. Johansen, *How to Reduce Bee Poisoning from Pesticides,* Western Regional Ext. Publ. 15, 1982; C. A. Johansen, D. F. Mayer, and J. D. Eves, Biology and management of the alkali bee, *Nomia melanderi* Cockerell (Hymenoptera: Halictidae), *Melanoleria,* 28:25–46, 1978; W. O. Lee, Clean grass seed crops established with activated carbon bands and herbicides, *Weed Sci.,* 21:537–541, 1973; S. E. McGregor, *Insect Pollination of Cultivated Crop Plants,* USDA Agr. Handb. 496, 1976.

Agricultural soil and crop practices

Tillage is an agricultural soil-management practice commonly employed by crop producers. In recent years, reduced tillage (including no tillage) has been advocated to reduce soil loss by erosion, water loss by evaporation and runoff, and cost of production.

Effects of tillage. Through the selection of a tillage method for a given soil in a given climate, the agricultural producer regulates the physical environment of the soil—primarily temperature, water content, oxygen diffusion rate, and the placement of

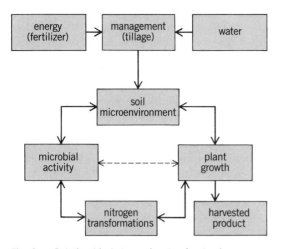

Fig. 1. Relationship between inputs of natural resources combined with management (including tillage) decisions upon biological activity in a soil. (*After J. F. Power and J. W. Doran, Nitrogen use in organic farming, in R. D. Hauck and D. A. Russell, eds., Nitrogen in Crop Production, American Society of Agronomy, 1983*)

oxidizable organic matter (chiefly crop residues). These factors in turn regulate the kind and rate of soil microorganism activity, as well as the environment for root growth and assimilation of nutrients and water (Fig. 1). The level of soil microorganism activity controls the decomposition rate of organic matter in the soil and, eventually, the availability of many chemical elements required for plant nutrition, that is, nitrogen and, to a lesser extent, phosphorus, sulfur, and other elements which are required in small quantities.

Reduced-tillage systems. Historically, the major reasons for tillage were to control the growth of weeds and to bury residues from the harvest of the previous crop. Soils tilled for these purposes were bare and subject to wind and water erosion, surface crusting, and evaporative loss of water. During the last several decades, herbicides have become available for weed control, and equipment has been improved to allow planting through crop residues that remain on the soil surface. Thus, technology now permits crop production with little or no tillage involved. Reduced-tillage systems have the potential advantage

of reducing soil erosion and production costs, conserving soil water, and increasing crop yields, as well as permitting cultivation of soils that were previously too erosive for crop production.

Intensive research has been conducted in recent years to achieve a better understanding of the effects of tillage methods upon changes in soil properties and subsequent plant growth. It was discovered that frequently no-tillage and reduced-tillage methods of cultivation increased soil water storage as much as 100 mm, primarily through reduced evaporation and runoff. In water-deficient climates, this extra soil water could potentially increase crop yields up to 1000 kg/ha annually. However, because of the increased possibility of below-optimum soil temperatures, greater disease or insect damage, restriction of soil nutrients, or other unknown reasons, the potential for increased yield through reduced tillage was not realized. Also, in poorly drained soils, crop yields produced by reduced- and no-tillage practices were often less than those produced by bare tillage.

Soil environment. More recently, research efforts have been intensified in order to gain a better understanding of the biological processes affected by tillage methods. Humans influence soil environment through the selection of soil- and crop-management practices, especially tillage practices. Tillage has two major effects upon the soil environment: a direct effect on the soil's physical environment caused by disturbance of the soil; and changes in the soil's chemical environment caused by the placement of crop residues. Soil disturbance generally increases soil porosity, which in turn affects content and movement of water and air (oxygen) in the soil and, subsequently, soil temperature. The soil organisms (including plant roots) respond to this soil environment in a predictable manner according to the known ecology of the organisms. If the environment is favorable to the characteristics of that organism, their activities are enhanced over those of organisms less adapted to that environment.

Numerous examples could be given of the differential responses of organisms to soil environments. For example, if approximately 60% of the soil pore space is filled with water, the activity of aerobic organisms will be near maximum. This condition favors the decomposition of crop residues and other

Ratios expressing no tillage compared to bare tillage on soil properties and microbial populations at seven United States locations*

Parameter	Soil depth, mm	
	0–75	75–150
Soil organic carbon	1.25	0.96
Total soil nitrogen	1.20	1.01
Soil water content	1.47	0.98
Aerobic microorganisms	1.35	0.71
Facultative anaerobes	1.57	1.23
Denitrifiers	7.31	1.77
Mineralizable soil nitrogen	1.35	0.96

*After J. W. Doran, Soil microbial and biochemical changes associated with reduced tillage, *Soil Sci. Soc. Amer. J.*, 44:765–771, 1980.

organic materials, resulting in the formation of maximum quantities of carbon dioxide and nitrate-nitrogen. However, as the soil dries and the average water-filled pore space drops below 50%, much of the soil water will become a thin film around the circumference of the soil pore. Such an environment is better suited for the larger cells of fungi than for bacteria. Consequently, exchangeable ammonium produced by fungal activity may accumulate in the soil because the activity of nitrifying bacteria is restricted. On the other hand, if water-filled pore space exceeds 70%, oxygen diffusion will be retarded and aerobic activity will be suppressed, while anaerobic processes (such as denitrification) will be enhanced.

Comparison of bare and no tillage. The effects of tillage practices upon the soil environment and upon subsequent biological activity have recently been documented. J. W. Doran sampled seven tillage experiments from West Virginia to Oregon to compare the effects of no tillage with bare tillage on several soil properties, water content, and microbial populations and their activity. Both maize (*Zea mays*) and wheat (*Triticum aestivum*) crops were included. Average values for the comparison of some parameters are presented in the table. Results from all locations showed the same trends as these average values. Compared with bare tillage, no tillage, with its absence of soil disturbance and placement of crop res-

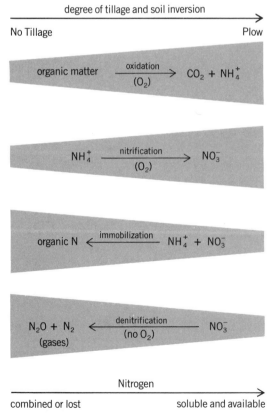

Fig. 2.   Effects of tillage intensity upon magnitude of soil carbon and nitrogen transformations. (*After J. W. Doran, Tilling changes Soil, Crops Soils Mag., 34(9):10–12, 1982*)

idues on the surface, resulted in greater soil surface organic-matter levels and greater soil water contents, a soil environment more conducive for microbial growth. Below the 75-mm depth, bare tillage generally increased aeration within the soil and oxidation of the organic matter buried by tillage. Thus, greater aerobic microbial activity extended to deeper depths on tilled as compared to nontilled soils.

Intensity of tillage. The effects of reducing the intensity of tillage are summarized in Fig. 2. As intensity of tillage increased from no tillage at one extreme to plowing at the other extreme, the rate of oxidation of organic matter to carbon dioxide and production of ammonium and the rate of nitrification of ammonium to nitrate-nitrogen were increased. However, immobilization of inorganic nitrogen into organic forms and denitrification both decreased. Thus, reduced-tillage practices tend to reduce rates of oxidative processes, resulting in an accumulation of organic matter and mineralizable nitrogen, and also reduce loss of topsoil by erosion.

For background information *see* AGRICULTURAL SOIL AND CROP PRACTICES; NITROGEN CYCLE; SOIL MICROBIOLOGY in the McGraw-Hill Encyclopedia of Science and Technology.

[J. F. POWER]

Bibliography: J. W. Doran, Soil microbial and biochemical changes associated with reduced tillage, *Soil Sci. Soc. Amer. J.*, 44:765–771, 1980; J. W. Doran, Tilling changes soil, *Crops Soils Mag.* 34(9):10–12, 1982; W. R. Oschwald (ed.), *Crop Residue Management Systems*, Amer. Soc. Agron. Spec. Publ. 31, 1978; J. F. Power and J. W. Doran, Nitrogen use in organic farming, in R. D. Hauck and D. A. Russell (eds.), *Nitrogen in Crop Production*, American Society of Agronomy, 1983.

Air navigation

Generally speaking, air navigation devices may be considered to be of two types: those devices which use measurements made on electromagnetic paths are called the classical type, and those which obtain their information by doubly integrating acceleration are called the self-contained type. This article describes the problem involved in attempting to integrate the communication, navigation, and identification equipments—a system of the classical type; and how microprocessors enhance the operation of integrated inertial systems—systems of the self-contained type.

INTEGRATED CNI SYSTEMS

The integration of airborne communication, navigation, and identification systems has been perceived as advantageous for several decades. Properly implemented, it would save spectrum space and airborne size, weight, and cost, while improving performance. However, a number of factors, mainly political rather than technical, have prevented most proposed systems from becoming operational. There are signs that this situation may be about to change. Two systems are receiving relatively large funding from military sources, and there are some signs of support from the civil sector.

Present nonintegrated systems. Since the end of World War II, most civil and military aircraft have carried three basic sets of airborne equipment: two-way communication, primarily by voice; radio navigation, for en route flying and for landing; and identity replies, by transporders, to interrogations from secondary radars on the ground. These pieces of equipment are commonly referred to as CNI (communication, navigation, and identification) equipment. The nonintegrated systems in most common use or in development are listed in Table 1. All these systems make use of a radio transmitter at one end of a pair and a radio receiver at the opposite end. Obviously, not many aircraft carry all of these systems. However, a large body of civil aircraft carry ILS, VOR, VHF communication, DME, and ATCRBS, and many military aircraft carry UHF communication, Tacan, and IFF. It is in these areas that most efforts at integration have been concentrated.

Philosophies of integration. There are at least three schools of thought on the problem of integration.

The first school advocates leaving the present system alone. This school claims that separate systems, independent of each other, each with its own antenna, electronics, and display, offer through their redundancy the greatest safety and reliability. Furthermore, it is possible to upgrade one system at a time, as state-of-the-art technology advances. This philosophy is the same as that which has led to multiengine transports, each engine with its own fuel supply and controls. Admittedly this type of system costs more, but the advantages are claimed to justify the increased expense.

A second alternative is to invent new signal formats which combine communication, navigation, and identification. This concept was the popular approach immediately after World War II; for example, in the years 1946 to 1949 the U.S. Air Force sponsored a half dozen contractual efforts in which this philosophy was pursued. In almost all cases, a single frequency band was chosen to allow a single antenna system. In some cases voice communication was replaced with digital communication by keyboard. All of these worked, but none has survived. However, the concept is not dead; it is in evidence in the military's Joint Tactical Information Distribution System (JTIDS) program.

A third school advocates keeping present signal formats but integrating the hardware. This school, recognizing the political tasks involved in changing frequency bands, aims at doing at least the whole receiving job after the signal has arrived inside the airplane (using existing signal formats). The U.S. Air Force's Integrated CNI Avionics (ICNIA) program is probably the best example of this approach.

Status of integrated systems. While not commonly referred to as a CNI system, the civil very-high-frequency (VHF) system comes as close to being operational as anything that is actually flying. Conceived in the 1940s, it placed the instrument landing system (ILS) localizer, VHF omnidirectional radio range (VOR), and VHF communication in adjacent frequency bands such that a single 108–136-megahertz receiver, like the thousands of general-

Table 1. Nonintegrated CNI systems in use and under development

Type*	System	Frequency	Function	Number of aircraft
		Systems in use		
N	Omega	10–13 kHz	Oceanic navigation	10,000
N	Loran C	90–110 kHz	Medium-range navigation	1,000
N	ADF	200–1600 kHz	Medium-range navigation	120,000
C	HF communication	3–30 MHz	Oceanic communication	3,000
N	Markers	75 MHz	ILS approach	90,000
C	VHF FM	30–88 MHz	Army communication	10,000
N	ILS localizer	108–112 MHz	Left-right guidance	100,000
N	VOR	108–118 MHz	Short-range navigation	179,000
C	VHF communication	118–136 MHz	Civil communication	250,000
C	VHF communication	225–400 MHz	Air Force/Navy communication	20,000
N	ILS glide-slope	329–335 MHz	Up-down guidance	90,000
N	Tacan	960–1215 MHz	Military navigation	25,000
N	DME	960–1215 MHz	Civil navigation	60,000
I	ATCRBS/IFF	1030, 1090 MHz	Identification, military and civil	100,000
		Systems under development		
	Sincgars	30–88 MHz	Antijam Army communication	
C	Seek Talk	225–400 MHz	Antijam Air Force	
C			communication	
N	Navstar/GPS	1227, 1575 MHz	Navigation by satellites	
N	MLS	5.0–5.25 GHz	Precision landing	

*C = communication; N = navigation; I = identification.

aviation "com.-nav." receivers, could be tuned to any one of the three facilities. Furthermore, by employing two com.-nav. receivers, communication can be carried on by one while the other furnishes navigation. Thus each function is provided with a backup. Unfortunately, vertical polarization is used for communication and horizontal polarization is used for navigation. The antenna system is consequently a hybrid.

Joint Tactical Information Distribution System. This is a United States military system under development which operates in the 960–1215-MHz band. This band is already occupied by Tacan, DME, and ATCRBS/IFF. JTIDS adds communication capabilities (both voice and data). Communications is effected by the use of pulses which are simultaneously hopped in frequency and biphase-modulated so that they not only are highly resistant to interference, but also produce negligible interference for the existing occupants of the band. (Such spectrum-spreading techniques are gaining favor in many other applications where the addition of new services to an existing frequency band is desirable.) Thus, in one system, JTIDS offers all three CNI functions through the use of a new signal format for communication and existing formats for navigation and identification. The system has completed the advanced development phase and is currently in engineering development; it is, however, still a line-of-sight system, and thus is limited to the capabilities of the 960–1215-MHz band. This system makes no pretense of covering all of the functions listed in Table 1.

Integrated CNI Avionics (ICNIA). This is a U.S. Air Force system still in the early stage of development as compared with JTIDS. It takes most of the existing CNI equipments, with their signal formats, and compresses them into approximately half the volume of their present format.

Two parallel approaches are currently being funded. In one of them, the radio-frequency–intermediate-frequency (rf-i.f.) sections are miniaturized, while certain commonalities are applied to the signal-processing sections. In the other one, an "all-band" receiver is constructed in which, by using a sampling rate of several hundred megahertz per second, any frequency can be selected, and several frequencies can even be used simultaneously.

Whichever system wins out, ICNIA aims at being a CNI system which covers most current functions plus some new ones, such as the satellite-based Global Positioning System. Since ICNIA uses existing signal formats, it may encounter less political opposition. Also it may be less vulnerable to the "all-eggs-in-one-basket" criticism, which may have prevented the implementation of many previous integrated systems, since it preserves the individual characteristics of the existing systems.

Discrete Address Beacon System (DABS). This is a civil system in which the Air Traffic Control Radar Beacon System (ATCRBS) is provided with a discrete address for each aircraft. This facility is pro-

vided by the use of interrogation codes for each aircraft. Since a unique data link is then available between individual aircraft and the ground, it becomes possible to add a two-way data link and, in so doing, a two-way communications capability to an identification system. The resulting system does not quite constitute a complete CNI system (the system lacks a navigation function), yet it integrates two of the important functions. In the United Kingdom the system is termed ADSEL (for Selective Address), and it is called Mode-S by the International Civil Aviation Organization. The Federal Aviation Administration plans the widespread implementation of DABS after 1990. *See* AIR-TRAFFIC CONTROL.

[SVEN H. DODINGTON]

INTEGRATED INERTIAL SYSTEMS

Technological advances in microprocessors, inertial sensors, and digital communications have ushered in a new era in air navigation, integrated inertial navigation systems. In an integrated navigation system all the avionics subsystems associated with navigation are interconnected so that each subsystem receives what it requires from the others on a timely basis. Their collective outputs are automatically blended and interpreted, resulting in the control data needed to operate aircraft efficiently and safely. Because so much data must be absorbed rapidly in a highly dynamic environment, it would be virtually impossible for the flight crew to perform all the necessary tasks without the integrated navigation system. At the heart of such a system is the inertial reference subsystem (IRS), which senses aircraft attitude and motion. Other sensor, display, and control subsystems are integrated with the IRS to constitute a modern integrated navigation system for high-performance aircraft.

System requirements. Integrated navigation systems must provide more than simple point A to point B guidance; they must do it with efficiency and high precision in all phases of operation from takeoff through climb, cruise, descent, approach, touchdown, rollout, and taxi. Although the IRS plays the key role in all of these functions, it is augmented by other sensors to provide optimum accuracy. The data required for aircraft operation and their sources are shown in Table 2.

Also required are the means for processing these data to derive the guidance steering commands and status information for the automatic flight control system and flight instruments. In the newest commercial jets, such as the Boeing 757/767 series and the Airbus Industrie 310, this processing is done by the flight management computer (FMC), which stores flight plans, airways and approach chart data, and airplane and engine data. The FMC also mixes position data from two or more sensors to assure the highest accuracy, and processes fuel flow and thrust data to compute minimum time/fuel consumption profiles. Not all integrated navigation systems have an FMC, however. In a system without an FMC the navigational computations are done by the inertial subsystem; hence, it is called an inertial navigation system (INS) to distinguish it from the IRS, which does not perform all the navigational computations. The trend in commercial aircraft is toward the IRS with an FMC to reduce crew work loads, save fuel, and improve schedule adherence. The INS approach is used in commercial aircraft placed in service before fuel costs soared. In the military the emphasis is on standardization to achieve economies of scale and to minimize training and logistics costs. The United States Air Force standard INS is defined in terms of form, fit, and function (F^3) requirements to assure INS compatibility and interchangeability with different aircraft types. Similarly, the U.S. Navy standardized on the carrier aircraft inertial navigation system (CAINS), which has special features to permit alignment on a moving carrier deck.

System description. Prior to the advent of inertial systems, aircraft attitude, heading, and motion were sensed by vertical and directional gyros, airspeed sensors, and barometric altimeters. These sensors are still installed on aircraft either as primary instruments or as backup for integrated navigation systems. Although these sensors enable the pilot to fly the aircraft and perform dead-reckoning navigation, they have serious limitations because the gyros are gravity-erected and thus cannot maintain level in the presence of accelerations, such as those caused by turns. Also, the air data sensors do not provide motion sensing with respect to the earth. An inertial system solves both these problems: it does not suffer from accelerations and furnishes velocity with respect to the earth. (Fundamentally, an inertial system senses motion with respect to space; but, when Schuler-tuned and torqued at the earth's rate, it will maintain level and sense motion with respect to the earth.) Therefore, at the heart of an integrated navigation system is an inertial system.

Inertial systems can be mechanized either as gimbaled or strapdown. In a gimbaled system the inertial sensing components (gyros and accelerometers) are mounted in a four-gimbal structure to isolate the sensors from attitude and heading changes. As the technology of gyros and computers improved, it became practical to build gyros with very precisely controlled torquing rates and computers fast enough to follow attitude and heading changes. This practice led to the strapdown inertial system which has no need for gimbals. In a strapdown system the gyros and accelerometers are bolted directly to the airframe, and they therefore sense directly airframe body rates and accelerations. This eliminates the need for the separate body-mounted rate gyros and accelerometers required when a gimbaled inertial system is used. Although this is an important advantage, large inventories of gimbaled systems will ensure their continued use for years to come. Eventually they will likely be superceded by strapdown systems.

There are two types of strapdown gyros in use: the conventional two degree of freedom spinning-wheel gyro and the newly developed single degree of freedom ring-laser gyro (RLG). Spinning-wheel gyros,

Table 2. Data required for aircraft operation

Data	Source
Position in three dimensions (3D)	IRS (inertial reference system)
Aircraft body rates (3D)	IRS
Aircraft accelerations (3D)	IRS
Pitch and roll attitude	IRS
Magnetic heading	IRS
Ground speed and ground track	IRS
Flight path with respect to the earth	IRS
Vertical velocity	IRS
Airspeed and barometric altitude	DADC (digital air data computer)
En route position updates (overland)	VOR/DME (VHF omnidirectional range/distance measuring equipment) or for military use Tacan (tactical air navigation)
En route position updates (global)	Omega
Approach guidance	ILS (instrument landing system) or its successor MLS (microwave landing system)
Weather avoidance	Weather radar

by virtue of the angular momentum of the spinning mass, tend to resist changes in direction caused by aircraft attitude changes. The sensitive axes of these gyros must be electronically torqued to follow pitch, roll, and heading changes. The torquing current is a direct measure of the aircraft body rates. The RLG, on the other hand, is a solid-state sensor and not actually a gyro. It is an angular rate sensor which uses two beams of laser light counterrotating in a closed path to measure angular rotation. When the RLG is rotated, the frequency of the laser beam traveling in one direction is either increased or decreased, depending on the direction of rotation, and the opposite change in frequency occurs to the laser beam traveling in the opposite direction. The difference in frequencies is directly proportional to the angular (aircraft body) rate. The RLG received an important vote of confidence when Boeing selected a strapdown ring-laser gyro IRS for the 757/767 series. This choice was significant because it marked the first major application of RLGs and sparked more intense competition which will lead to broader use of RLGs in aircraft of all types, especially if experience bears out their promise of higher reliability. The military has supported extensive RLG development and plans their use if current tests succeed.

With either the conventional spinning-wheel or laser gyro the process of obtaining attitude and position data is the same. By integrating the aircraft body rates, the attitude is measured. Three orthogonally disposed accelerometers, also firmly bolted to the airframe, measure acceleration in pitch, roll, and yaw. The attitude data supplied by the gyros are used to transform these acceleration measurements into north, east, and vertical components which are

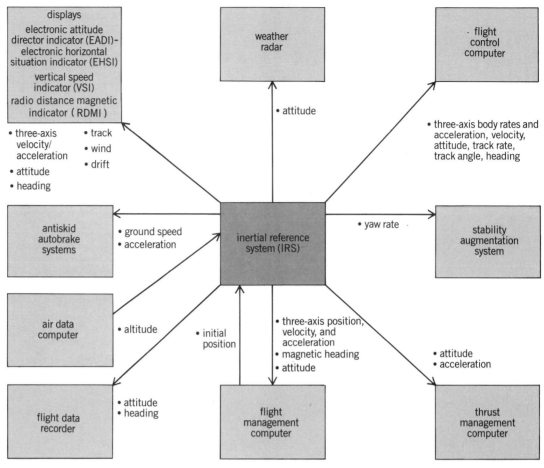

Relation of the inertial reference system (IRS) in the Boeing 757/767 series to other systems. (*After P. J. Fenner and C. R. McClary, The 757/767 inertial reference system (IRS), 14th Joint Service Data Exchange for Inertial Systems, November 18–20, 1980*)

then double-integrated to yield a three-dimensional change in position. From these basic measurements the IRS generates the data shown in Table 2.

Because these IRS outputs are comprehensive and superior in quality to vertical and directional gyros, overall performance is improved. For example, the automatic flight control system provides better response and more precise aircraft control because the IRS provides ground-referenced velocity in addition to attitude data that are not degraded by acceleration. These dynamically exact attitude data are also used to stabilize the weather-ground mapping radar image (and the "heads up" display in military aircraft) so that it exactly matches the real world as viewed from the cockpit. This is an important confidence-builder for the pilot. The IRS-furnished ground velocity and acceleration enable several avionics subsystems to function with greater precision than air data would permit. For example, the FMC can make better estimates of time of arrival and improve operation in areas with time-controlled flow. It also computes wind continuously. This is useful in landing, especially when wind shear conditions exist, and in energy management to conserve fuel. In addition, ground speed and accelera-

tion data aid in controlling the antiskid/automatic braking system, automatic thrust reversal, nose-wheel steering gain, and taxi speed. The illustration shows the central role of the IRS in the Boeing 757/767.

Supporting sensors. The stand-alone capability of most civil inertial systems is about 2 nautical miles per hour (with 95% probability of achieving this accuracy; 1 nmi = 1852 m). Federal Aviation Administration accuracy requirements for area navigation in the United States' national airspace are: 1.5 nmi en route, 1.1 nmi in the terminal area, and 0.3 nmi in nonprecision approach. Therefore, the inertial reference needs a source of position updates from other navigational aids.

For en route operations the VOR/DME or Tacan ground-based radio transmitters provide range and bearing position fixes. The integrated navigation system mixes this radio data with inertial data to provide improved accuracy.

For transocean operations aircraft are spaced further apart. The Federal Radionavigation Plan (July 1980) specifies that position accuracy must be maintained to within 12.6 nmi. The inertial system normally meets this requirement, but redundancy is required to protect against inertial system failure. Frequently, this redundancy is satisfied by the addition of an Omega navigation system which provides continuous worldwide position accurate to about 2 nmi (daytime) and 4 nmi (nighttime) from eight ground-based transmitters located in various places around the world. In an integrated navigation system the Omega data are used to limit the inertial error to the nominal Omega accuracy.

For en route vertical positioning, barometric altitude is adequate to meet Federal Aviation Regulations for vertical separation which is 1000 ft (305 m) below 29,000 ft (8839 m), and 2000 ft (610 m) at and above 29,000 ft. The output of the vertical axis of the inertial reference in an integrated navigation system is mixed with barometric altitude and rate of altitude change. This allows tighter vertical control of the aircraft and detection of vertical windshear. The barometric altimeter also helps the inertial sensor by providing a long-term reference to mix with short-term (fast response) vertical acceleration measurements of the inertial sensor, which would be unstable without the damping provided by the altimeter.

For approach operations precise guidance is provided by the ground-based localizer (lateral) and glide slope (vertical) transmitters, which constitute the instrument landing system (ILS). The ILS provides a single azimuth path along the runway centerline and a fixed angle of descent (typically 3° above horizontal). According to the Federal Radionavigation Plan the newer microwave landing system (MLS) will be phased in beginning in 1985. The MLS will offer multiple azimuth paths and descent angles, providing precision guidance over a much larger volume of airspace and allowing aircraft of different types to use optimum glide slope angles. International agreements have been made to establish an all-weather landing capability. These agreements will be implemented in a series of step reductions in allowable weather minimums for landing, consistent with demonstrable advances in equipment capability. The MLS will support category IIIC landings which represent zero–zero weather conditions (zero-feet decision height and zero-feet runway visual range). Integrated navigation systems with inertial reference sensors will provide the necessary attitude precision together with ground-referenced velocity and aircraft body rates to allow category IIIC landing when augmented with guidance signals from an MLS receiver.

Systems integration. Depending on the particular aircraft and mission, different degrees of systems integration may be mechanized. In most aircraft the integration is done on a federated basis; that is, the individual subsystems are autonomous and their computational needs are furnished by their own self-contained digital processors. The subsystems are then interlinked via a digital bus network that provides the input/output communications between the inertial reference and the other subsystems. In military aircraft higher degrees of systems integration may be mechanized to improve the probability of mission success. Weapons delivery and aerial cargo drop aircraft require dynamically exact and very precise attitude and velocity data to accurately determine the release point. Reconnaissance aircraft need precise and lag-free position and attitude data for radar and camera records. High-G acceleration maneuvers demand that the attitude sensors be decoupled from acceleration-induced errors. All these requirements place stringent performance demands on the inertial and radio-positioning subsystems.

To meet these requirements, systems integration designers frequently take advantage of the complementary nature of the various sensors to achieve synergistic benefits. For example, inertial systems have virtually instantaneous response, but radio-positioning systems, such as Omega and loran C, suffer acceleration-induced measurement (position) lags. This is because the receiver which tracks loop bandwidth must be narrowed (inducing a long time constant) to accommodate low signal-to-noise conditions which can occur in weak signal areas, or during maneuvers because of antenna shielding, or as a result of interference jamming. By proper mixing using real-time adaptive filter algorithms, the radio-positioning data can be used to bound the inertial position errors and calibrate the biases in the inertial sensors to upgrade their performance. This provides a more accurate (calibrated) inertial navigation capability to carry out a mission in case the radio signals become unusable (out of range or jammed by the enemy). At the same time, the inertial reference can provide dynamic aid to the radio receiver, thereby eliminating the lag in position measurements and allowing receiver operation with extraordinarily narrow bandwidths to improve its resistance to jamming. Hence, by blending the complementary characteristics of the inertial and radio-positioning subsystems, both elements exhibit performance which is superior to that which either could achieve operating alone. The combined result is a hybrid system with synergistic properties.

Future systems. Integrated navigation systems will be improved in the very near future by the addition of a collision avoidance system and the new satellite-based radio navigation system. In 1981 the FAA defined a specific technical approach to collision avoidance and established the guidelines for implementing the system before the end of 1984. The approach that was selected is the threat collision avoidance system (TCAS), which issues an evasive maneuver command when it senses a collision threat. It relies on existing air-traffic control transponders that are installed on all commercial and most other aircraft. It is entirely an airborne system—it does not depend on ground-based equipment. It is also a cooperative system analogous to the use of headlights on automobiles, and so all aircraft must be equipped for the system to work. To accommodate the owners of small private aircraft, for example, a bare-bones version, TCAS I, that provides only proximity warning, will be authorized. Airliners will carry the complete version, TCAS II, that outputs a directional interrogation signal and receives range, bearing, altitude, and angle-of-arrival data from which coordinated evasive maneuvers will be computed. TCAS II has a scanning antenna which uses the heading information furnished by the integrated navigation system.

The NAVSTAR Global Positioning System (GPS) is being implemented by the Defense Department to provide very precise three-dimensional position and velocity information for land, sea, and air use. Eighteen satellites will provide continuous worldwide signal coverage. Although some limited coverage is available today to support various test programs, fully operational coverage is not expected until 1987. Government policy regarding civil usage (whether it will be with or without cost to the users, the accuracy available to nongovernment users, and so forth) has not been established, but the Federal Radionavigation Plan highlights the prospect of civil use of GPS and has set 1986 as a target date for reaching a decision on the selection of navigation systems for the future. If GPS becomes a national standard, it will probably replace Omega and loran C, after a period of overlap service, and it may also replace VOR/DME and Tacan. Furthermore, GPS offers the designers of integrated navigation systems new opportunities to exploit the very precise position and velocity data it will provide. Simplified (lower cost) strapdown inertial systems that provide aircraft attitude, heading, rates, and accelerations, but not position, could be integrated with GPS to form a new hybrid system that would become the basis for the next generation of integrated navigation systems.

For background information *see* AIR-TRAFFIC CONTROL; ELECTRONIC NAVIGATION SYSTEMS; GYROSCOPE; INERTIAL GUIDANCE SYSTEM; SATELLITE NAVIGATION SYSTEMS in the McGraw-Hill Encyclopedia of Science and Technology.

[JOHN J. HOPKINS]

Bibliography: P. J. Fenner and C. R. McClary, The 757/767 inertial reference system (IRS), *14th Joint Service Data Exchange for Inertial Systems*, November 1980; J. J. Hopkins, Integrated satellite navigation and strapdown attitude and heading reference system, *Inst. Navig. J.*, 28:189–198, fall 1981; M. Kayton and W. Fried, *Avionics Navigation Systems*, 1969; P. J. Klass, Litton tests laser-gyro inertial system, *Aviat. Week Space Technol.*, 113(22):144–145, December 1, 1980; U.S. Departments of Defense and of Transportation, *Federal Radionavigation Plan*, March 1982.

Air-traffic control

Air-traffic control procedures make considerable use of secondary surveillance radar, known internationally as SSR but commonly termed the ATC radar beacon system in the United States. Secondary surveillance radar, which had its origins in World War II, requires each cooperating aircraft to carry an electronic device, the transponder. Although SSR has brought many benefits, the growth of air traffic has revealed significant problems with SSR in regions of high traffic density. To avoid these difficulties while ensuring worldwide compatibility with existing systems, an extended version of SSR, Mode S, has been developed for use internationally.

The SSR airborne transponder listens on a frequency of 1030 MHz for a suitably coded interrogation and replies on a frequency of 1090 MHz. The interrogator antenna is often comounted with the primary radar antenna of a mechanically scanning surveillance radar, and primary and secondary radars use similar methods to determine range and bearing of a target.

The "clutter" problems (due to returns from terrain or precipitation) that trouble primary radar are avoided by SSR, making it much easier to transmit radar data to a remote site. It can also detect targets at long range without expensive, high-power transmitters. The aircraft reply can be coded to give additional information. For example, to a Mode A interrogation the aircraft replies with one of 4096 identity codes set up on switches in the cockpit. A Mode C interrogation will receive a reply containing a coded version of the altimeter reading. It is therefore possible, if Mode A and Mode C interrogations are interlaced, to put height and identity data on the radar display.

One common frequency is used by SSR for all interrogations, and another is used for all replies. As air-traffic control sensors and equipped aircraft have proliferated, problems have arisen because of the reception of replies intended for other interrogators ("fruit") and because of replies to the sensor's own interrogations which overlap in time ("garble"). In particular, two aircraft on a given route at different heights may give replies which remain garbled, and therefore unintelligible to the sensor, for a period of several minutes.

Development of Mode S. In 1968 the U.S. Department of Transportation set up an Air-Traffic Control Advisory Committee, which concluded that current SSR was reaching the end of its useful life and that there was a need for a successor system which would use different frequencies. Parallel work in Britain resulted in a suggestion that seems likely to extend the life of SSR by many years. The proposed solution, the subject of cooperative effort in the United States and Britain, was known originally as DABS in the United States and ADSEL in Britain. It has now been agreed internationally that the system will be known as SSR Mode S.

An SSR ground station with Mode S facilities can transmit interrogations to draw Mode A or Mode C replies from regular transponders. An aircraft with a Mode S transponder will reply to Mode A or C messages from a standard air-traffic control sensor. The new facility comes into play only when both the ground station and the aircraft have Mode S capability. It is therefore possible to have a smooth transition from the existing to the future system. It has been demonstrated, both in the United Kingdom and in the United States, that airlines can replace existing transponders with a plug-in Mode S replacement without any need for other modifications; and that the Mode S equipment operates both with existing surveillance systems and with prototype Mode S interrogators.

Mode S surveillance. Mode S provides a new ability to transmit an interrogation message addressed specifically to a chosen aircraft. There are 16 million possible addresses, more than the total number of aircraft in the world. Each Mode S interrogator will normally maintain a file with the identity and approximate position of all Mode S–equipped aircraft within its coverage. It can then carry out a roll call of covered Mode S aircraft, each being interrogated in turn as the scanning antenna points in an appropriate direction. Mode S interrogators also have an all-call mechanism to enable surveillance of aircraft having only Mode A and C transponders, or of Mode S aircraft not yet on file. A Mode S transponder responds to an all-call interrogation with a reply which includes its own address. The sensor can then add this address to the roll call file, and by means of a discrete address message can temporarily lock out the specified transponder against later all-call interrogations.

If a Mode S station has been out of service for some reason, it may be necessary to use all-call first to rebuild the file of aircraft addresses. Until this is done, the sensor may face the garbling and other problems that Mode S is designed to avoid. Once the sensor has read the address of one of the aircraft within its coverage, it can be locked out from responding to subsequent all-calls, and the situation can be brought under control step by step. To aid this process, there is provision for a stochastic all-call message, to which the reply is given only at irregular intervals, the reply probability being specified by the ground sensor.

Compatability between Mode S and existing SSR is obtained by exploiting a feature of standard transponders which is intended to inhibit responses to the antenna side lobes of the interrogator. Conventional SSR interrogators radiate, in addition to signals in the main beam of the directional antenna, a comparison pulse from a more or less omnidirectional antenna. The side-lobe suppression pulse inhibits any reply when the main interrogation is less powerful than the reference signal. This facility has made it possible, without requiring any modifications to existing transponders, to devise a Mode S interrogation message which inhibits any reply by old-style transponders. A Mode S aircraft will disregard this suppression and react in a manner determined by the remainder of the interrogation message.

Since Mode S has thus given a new meaning to the old-style side-lobe suppression pulse, yet another side-lobe suppression mechanism is included in the Mode S interrogation message to handle side-lobe problems during Mode S target acquisition.

Multisite operation. To provide adequate low-level SSR cover, there must be considerable overlap between the cover at higher level provided by adjacent sensors. Given a suitable network management system, this overlap can provide valuable redundancy if a ground sensor breaks down or if cover is lost due to a hole in the aircraft antenna coverage. For sensors connected to a common management network, this overlap makes possible easy transfer of a target moving out of the coverage of one sensor and into another.

In areas like western Europe, where there are numerous autonomous air-traffic control organizations, the system must ensure that any all-call lockout is unlocked often enough to enable sensors which are not part of the same management network to obtain data on incoming traffic. This problem needs further study.

Mode S data link. Mode S was conceived as a solution to the surveillance problem, but the opportunity also arose to expand the original facilities used by SSR to telemeter identity and height. Thus, Mode S has a provision for more general messages of both the up-link and down-link type. A down-link message may be a response to instructions on the up-link from the ground or it may originate in the aircraft. Standard messages of up to 224 bits can be transmitted in either direction within 10–15 s without disrupting surveillance. Extended-length messages of up to 1280 bits require a more elaborate protocol. Unlike Mode A and Mode C, Mode S message formats use error-correcting codes, and there is a provision for repetition of any message which is too corrupted for automatic correction at the receiver.

The Mode S data link forms the basis of an airborne collision-avoidance system, TCAS (traffic alert and collision avoidance system), which is presently the subject of intensive research and development in the United States. TCAS uses Mode S in an air-to-air role. Threat detection relies mainly on range (measured directly) and relative height (derived from telemetered height data), and on the rate at which these two quantities change with time. The alarm criteria are such that no warning should be issued if the aircraft are separated in accordance with air-traffic control rules. If necessary, TCAS uses the data link to ensure that two or more TCAS aircraft adopt complementary escape maneuvers. Because of compatibility between Mode S and SSR, TCAS can protect its aircraft against a threat with old-style SSR, provided that the threat is being interrogated by a ground-based sensor. Because TCAS and air-traffic control are connected to the Mode S data link, there is an obvious potential for sophisticated future systems which combine ground- and air-derived data.

For background information *see* AIR-TRAFFIC CONTROL; ELECTRICAL COMMUNICATIONS; RADAR in the McGraw-Hill Encyclopedia of Science and Technology.

[STANLEY RATCLIFFE]

Bibliography: B. Alexander, *IEEE Trans. Aerosp. Electr. Sys.*, Department of Transportation Air Traffic Control Advisory Committee, AES-6(2):106–111, 1970; R. C. Bowes et al., ADSEL/DABS: A selective address secondary surveillance radar, *Agard Conf. Proc.*, vol. 188, pap. no. 9, 1975; U.S. National Standard for the Discrete Address Beacon System, *U.S. Federal Register*, 1981.

Animal virus

While serological methods permit the classification of virus isolates based upon amino acid sequence differences in virus-coded proteins, these methods are limited to detection of differences which occur in exposed antigenic determinants of virus proteins. A more complete comparison of virus genomes for purposes of identification and classification is based on nucleotide sequence analysis.

Nucleotide-sequence-specific cleavage. Until recently, only indirect methods for comparison of nucleotide sequence relationships were available. Immunological cross-reactivity, heat stability, and electrophoretic mobility are all properties which depend on the amino acid composition and arrangement in virus proteins and therefore reflect nucleotide sequences. Now, however, relatively simple methods are available which allow direct analysis and comparison of the genomes of viruses. The total nucleotide sequence of the entire genome is known for a very few small bacterial viruses and for some members of the papovavirus group. Methods for total nucleotide sequence determination are being developed rapidly, but complete sequence analysis is still a major undertaking. More limited but direct information about nucleotide sequences can be obtained by analysis of the size distributions and compositions of polynucleotide fragments produced by nucleotide-sequence-specific cleavage of virus genomes.

Two classes of nucleases have been employed as sequence-specific cleavage reagents. Ribonuclease, which cleaves RNA chains after a specific nucleotide [for example, ribonuclease T1 cleaves after every guanylate (G) residue], produces a set of fragments, the sizes of which are determined by the position of the specific residue in the viral genome. Such sets of fragments can be resolved by two-dimensional chromatographic separations to yield a "fingerprint." Comparison of these fingerprints for two RNA genome viruses can show if the two viruses have the same or different distribution of G residues in their genomes. This method has been invaluable in the study of oncornaviruses and the relationships between multiple endogenous retroviruses from the same animal.

For analysis of DNA-containing viruses, bacterial restriction endonucleases have been employed. This terminology reflects the known biological function of these nucleases; that is, they recognize and cleave at specific nucleotide sequences which are present in foreign DNA but absent in the DNA of the cell which produces the enzyme, so that entry of foreign DNA is restricted. These enzymes come from a variety of organisms and exhibit a number of DNA-sequence specificities, allowing the identification of many different DNA sequences.

The general method followed in the use of restriction endonuclease for virus identification can be described as follows. Isolated viral DNA is incubated with a specific endonuclease until all DNA sequences which are susceptible to the nuclease have been cleaved. The set of DNA fragments is then resolved on the basis of size by gel electrophoresis. The large fragments become the slowest because of the sieving effect of the gel, so an inverse relationship between size and migration is observed. The position of the DNA fragments can be determined by radioautography on x-ray film if the viral DNA is labeled; or the DNA can be visualized and photographed directly by the fluorescence of a dye, such as ethidium bromide, which may be included in the agarose gel and binds to the DNA. Less than 1 μg of DNA per sample is required for fluorescent detection, and 5000-counts-per-minute radioactive label is required for autoradiographic detection. One microgram of DNA is obtained from about 5×10^9 particles of herpesvirus or 1.5×10^{11} particles of papovavirus.

Classification of virus isolates. The first application of restriction endonuclease cleavage site analysis to comparative virology was the study of EcoR1 cleavage patterns of several adenoviruses (serotypes 2, 3, 5, 7, 12) and an adeno-SV40 hybrid virus (Ad2 + ND1). Each serotype was distinct, and it was found that two independent members of the same adeno-7 serotype gave related but distinguishable patterns of DNA fragments.

The same approach was applied to herpesvirus DNA analysis, and it was possible to classify herpes simplex virus isolates into two groups on the basis of endonuclease cleavage patterns. It was found that these groups were the same as those obtained by biological characterization (HSV-1 and HSV-2). Furthermore, intratypic variation was recognized; that is, various isolates of HSV-1 showed strain-specific differences which could be used to further subclassify these HSV-1 isolates. This intratypic variation has been used to follow the course of a hospital-based outbreak of HSV-1. The power and clarity of this approach was immediately apparent, and the use of restriction endonucleases for studies in molecular epidemiology has progressed rapidly in the past 5 years.

The relationship of varicella virus (chicken pox) to herpes zoster virus (shingles) has also been studied by restriction endonuclease cleavage patterns. Immunological studies suggested that the same virus, called varicella-zoster virus (VZV), caused both diseases. This conclusion was more firmly substantiated by direct analysis of DNA from varicella virus and zoster virus; both proved to be the same virus.

In the papovavirus group of DNA viruses, SV40 and polyoma virus have been extensively characterized by cleavage site analysis, and more recently the human papovavirus JCV and BKV have come under extensive study. This analysis provided rapid and definitive proof that SV40, JCV, and BKV are distinct viruses.

Restriction endonuclease cleavage site analysis of DNA virus genomes is a product of very recent progress in molecular biology, and is of considerable biological and clinical interest. The methods are sufficiently simple and reproducible to be of use in a variety of areas: virus identification and classification, epidemiological studies, and virus-host interactions.

For background information *see* ANIMAL VIRUS; GENETIC MAPPING; NUCLEIC ACID; VIRUS in the McGraw-Hill Encyclopedia of Science and Technology.

[WILLIAM C. SUMMERS]

Bibliography: T. G. Buchman et al., Restriction endonuclease fingerprinting of herpes simplex virus DNA: A novel epidemiological tool applied to a nosocomial outbreak, *J. Infect. Dis.*, 138:488, 1978; C. Mulder et al., Specific fragmentation of DNA of adenovirus serotypes 3, 5, 7, and 12, and adeno-simian virus 40 hybrid virus Ad2 + ND1 by restriction endonuclease R + EcoR1, *J. Virol.*, 14:68, 1974; J. E. Oakes et al., Analysis by restriction enzyme cleavage of human varicella-zoster virus DNAs, *Virology*, 82:353, 1977; J. E. Osborn et al., Comparison of JC and BK human papovaviruses with simian virus 40: Restriction endonuclease digestion and gel electrophoresis of resultant fragments, *J. Virol.*, 13:614, 1974; L. Rymo, T. Lindahl, and A. Adams, Sites of sequence variability in Epstein-Barr virus DNA from different sources, *Proc. Nat. Acad. Sci. USA*, 76:2794, 1979; J. Skare, W. P. Summers, and W. C. Summers, Structure and function of herpesvirus genomes, 1. Comparison of five HSV-1 and two HSV-2 strains by cleavage of their DNA with EcoR1 restriction endonuclease, *J. Virol.*, 15:726, 1975.

Anomalons

Startling evidence that nuclear matter may exist in a hitherto unexpected form has come from recent experiments carried out at the Lawrence Berkeley Laboratory's heavy-ion accelerator, the bevalac, and from cosmic-ray experiments. The experiments indicate that high-energy nuclear fragments associated with the fragmentation of nuclei at relativistic energies possess anomalously short interaction mean free paths, that is, large reaction cross sections. A surprising discovery from the latest experiments is that the shortening of the mean free paths of projectile fragments is compatible with the existence of nuclear entities—anomalons—that appear to exhibit dimensions comparable to those of uranium nuclei.

To date, no known particle or explanation within the framework of conventional nuclear physics has accounted for the anomalons. Theoretical speculations on the anomalon effect are focusing most intensely on two concepts: one is that quark bundles in nuclei might account for the observations; the other is that the anomalon effect may be due to extraordinary nuclear configurations and states of excitation.

Projectile fragments. Whenever a projectile nucleus at relativistic energy collides with a nucleus at rest, that is, the target nucleus, a large variety of nuclear fragments are produced. The phenomenon of projectile fragmentation, where nuclear fragments are emitted from the vertex of a nuclear collision essentially in the direction and velocity of the incident projectile nucleus, is relevant to anomalon experiments. An example of a chain of projectile fragmentation reactions observed in a nuclear emulsion is shown in Fig. 1. This figure is a microprojection drawing of an event produced by an ^{56}Fe (atomic number $Z = 26$) nucleus at an energy of 1.9 GeV/nucleon, where the initial collision generates a chain of projectile fragmentation reactions that emit, sequentially, projectile fragments of charges $Z = 24$, 20, and 11. All the interactions in Fig. 1 are characteristic of high-energy nuclear interactions, with significant particle production and target excitation indicated by the large number of target prongs emitted from the vertices of the nuclear stars.

Mean free paths. The collisions of beam nuclei as illustrated in Fig. 1 occur randomly along their paths in matter. However, the average distance between collisions, or the mean free path, for each nuclear species is a well-defined quantity and can be measured accurately. An important property of the mean free path of a relativistic nuclide in matter is that it is an invariant quantity and, in the present context, depends only on the nature, that is, the physical sizes, of the projectile and target nuclei.

Experimental evidence for anomalons. Since the discovery of heavy nuclei in primary cosmic rays in 1948, and particularly since the development of the bevalac in the 1970s, comprehensive studies on the mean free paths of relativistic nuclei have been carried out in nuclear emulsion detectors. Experiments designed to measure the mean free paths of nuclei and their fragmentation products by use of these

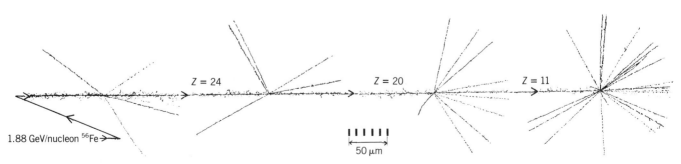

Fig. 1. Succession of four projectile fragmentation reactions initiated by a 1.88 GeV/nucleon ^{56}Fe nucleus (enters from left) as observed in nuclear emulsion. Actual distance between first and fourth interactions is 5.7 cm (2.2 in.).

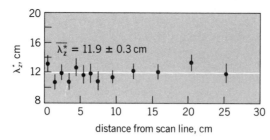

Fig. 2. Measured values of the mean free path λ_Z^* of 2.1 GeV/nucleon ^{16}O beam nuclei versus the distance from the scan line, or entry point, in an emulsion detector. 1 cm = 0.4 in.

sensitive photoemulsions take advantage of the fact that emulsions serve as both the target material and particle (track) detector.

Microscopic examination of the emulsions is performed by following each beam track upon its entry into the emulsion stack until it either interacts or leaves the detector. Whenever a fragmentation reaction is detected, the projectile fragments are likewise followed, and the charges and path lengths of interacting and noninteracting projectile fragments are measured and recorded. From such data, the observed mean free path of projectile fragments (as well as beam nuclei) of charge Z can be given by Eq. (1), where $\Sigma \ell_i$ is the total path length followed for both interacting and noninteracting tracks that leads to N_Z interactions.

$$\lambda_Z^* = \Sigma \ell_i / N_Z \qquad (1)$$

The mean free paths λ_Z^* of beam nuclei are exemplified by those shown in Fig. 2, where the value of λ_Z^* for ^{16}O nuclei at 2.1 GeV/nucleon is plotted as a function of the distance from their entry into the emulsion stack, that is, the scan line. These data are well accounted for by a constant value of $\overline{\lambda_Z^*} = 11.9$ cm (4.7 in), a quantity independent of the distance from the scan line. This behavior is expected from conventional nuclear theory and is characteristic of the mean free paths of all beam nuclei measured to date, ranging from ^{4}He to ^{56}Fe, which by definition are taken to be the mean free paths of normal nuclei.

In contrast to the ^{16}O beam data, the mean free paths of projectile fragments show quite dramatically that they are not constant with the distance D after their point of emission. This is shown in Fig. 3, which presents the results from three independent experiments on the mean free paths of projectile fragments. Two of these experiments used bevalac beams of ^{56}Fe, ^{16}O, and ^{40}Ar at approximately 2 GeV/nucleon, whereas the third experiment involved the reanalysis of data from an extensive series of cosmic-ray balloon flights. Plotted are the charge-independent mean free path parameters Λ^* for the projectile fragments versus the distance D, where Λ^* is defined in Eq. (2). This equation accounts well for the λ_Z^* versus Z dependence observed for the mean free paths of beam nuclei, where $\Lambda_{\text{beam}} \approx 30$ cm (12 in.) and $b \approx 0.4$. The data in Fig. 3 show that the mean free paths of projectile fragments are consistently lower than Λ_{beam} for the first few centimeters from their origins, becoming compatible with Λ_{beam} ($\Lambda^*/\Lambda_{\text{beam}} = 1$) for distances D greater than about 5 cm (2 in.).

$$\lambda_Z^* = \Lambda^* Z^{-b} \qquad (2)$$

The short mean free paths of projectile fragments at small distances D mean that there is an excess in the number of interactions at these distances, confirming the long-suspected anomalous behavior of projectile fragments observed in cosmic-ray experiments. The latest results not only indicate that the mean free paths of projectile fragments are short at distances D less than about 2–3 cm (1 in.), but that at larger distances they revert to the mean free paths of normal beam nuclei.

Interpretive models. To gain some insight into the nature of the excess of interactions of projectile fragments at short distances, one possible assumption is the following: In addition to normal nuclei, there is a component of anomalous projectile fragments, that is, anomalons, which are produced with probability a, having a constant, anomalously short mean free path λ_a. Estimates of a and λ_a from data in Fig. 3 give $a \approx 0.06$ and $\lambda_a \approx 2.5$ cm (1.0 in.). The astonishing conclusion is that the 2.5-cm mean free path of anomalons required by this model to account for the suppression of Λ^* at short distances is comparable to the mean free path expected for uranium nuclei. The curve in Fig. 3 is the computed Λ versus D relation based on the assumption that 6% of the projectile fragments are anomalons, the remaining 94% of the projectile fragments being normal nuclei. Although this primitive model reproduces the trend of the experimental data, it is by no means unique. For example, a much larger population of anomalons that attenuate by decay, rather than by nuclear interaction, with a lifetime that corresponds to a decay distance of a few centimeters could also

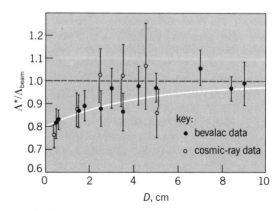

Fig. 3. The mean free path parameter Λ^* versus distance D from the origins of projectile fragments. The values of Λ^* are normalized by the values of Λ_{beam} observed for the same nuclei. The horizontal line at $\Lambda^*/\Lambda_{\text{beam}} = 1$ is the prediction for normal nuclei; the curve is the prediction assuming a 6% admixture of projectile fragments with $\lambda_a = 2.5$ cm (1.0 in.).

account for the data, provided 100% of the projectile fragments are anomalons and have cross sections about twice those of normal nuclei of the same charge. Irrespective of these interpretive models, the fundamental observations indicating the existence of nuclei 2 to 10 times larger than normal appears to be unexplainable by conventional nuclear theory.

The observations that point to the existence of nuclei with anomalously large interaction cross sections among the projectile fragments of relativistic nuclei must now be taken seriously. More experimental details on anomalons are vitally needed, particularly on their production and interaction mechanisms, their lifetimes, decay modes, and masses.

Theoretical speculations. The experimental guidelines needed for realistic theoretical interpretations of the anomalon effect are tenuous. Still, theoretical speculations abound. A theory that nuclear collisions can alter the quark structure in nuclei to produce color polarization inside the nucleus, thereby giving rise to larger nuclear collision cross sections, has been put forward. Alternatively, a lagrangian field theoretic approach leads to hadroid solutions that exhibit the appropriately long-range forces required to explain the large interaction cross sections of anomalons. So far there is no theoretical concept to explain both the enhancement of the cross section and the remarkably long lifetime of anomalons, estimated to be in excess of 10^{-10} s.

A true nuclear puzzle has thus emerged from relativistic heavy-ion experiments. Its solution presents an exciting challenge to both experiment and theory, from which a more fundamental description of nuclear matter may evolve.

For background information *see* NUCLEAR REACTION; NUCLEAR STRUCTURE; QUARKS in the McGraw-Hill Encyclopedia of Science and Technology.

[HARRY H. HECKMAN]

Bibliography: H. B. Barber, P. S. Freier, and C. J. Waddington, Confirmation of the anomalous behavior of energetic nuclear fragments, *Phys. Rev. Lett.*, 48:856–859, 1982; E. M. Friedlander et al., Evidence for anomalous nuclei among relativistic projectile fragments from heavy-ion collision at 2GeV/nucleon, *Phys. Rev. Lett.*, 45:1084–1087, 1980; A. L. Robinson, A nuclear puzzle emerges at Berkeley, *Science*, 210:174–175, 1980; B. M. Schwarzchild, New evidence for anomalously large nuclear fragments, *Phys. Today*, 35(4):17–19, 1982.

Antimicrobial resistance

Resistance of microorganisms to chemotherapeutic agents has been known since the first use of antimicrobial drugs in the specific therapy of infectious diseases. It was not long after the introduction of sulfonamides and antibiotics into medicine that resistant strains of bacteria began to appear. Despite warnings of serious complications in the future, the use of antimicrobials in medicine has increased continuously. Furthermore, antibiotics not only have been used for the treatment of human and animal diseases, but have also been applied in agriculture and given to livestock as feed additives. The uncontrolled use of antibiotics has resulted in the development and spread of bacteria resistant to these useful agents. Hospitals are plagued by pathogenic bacteria resistant to many antimicrobials. Multiresistant bacterial strains, causing epidemics of typhoid and dysentery, have also appeared outside hospitals. Several pathogenic species that had been uniformly sensitive to antibiotics have recently acquired resistance to certain drugs; this is the case for the organisms responsible for gonorrhea (*Neisseria gonorrhoeae*), pneumonia (*Streptococcus pneumoniae*), and infantile meningitis (*Haemophilus influenzae*). This article deals with the genetic mechanisms of development of resistant bacteria and describes the biochemistry of resistance to antimicrobial agents in these microorganisms.

Chromosome-mediated resistance. Drug resistance in bacteria refers not to the natural resistance of a species, but to acquired genotypic changes that persist during cultivation in the absence of the drug. These changes may be brought about by spontaneous mutation, followed by selection of the mutant.

Although bacteria are the smallest living cells, they contain all the machinery required for growth and self-replication. Bacteria possess a nuclear body (nucleoid), often referred to as the bacterial chromosome. This structure is a single, circular, tightly coiled DNA double helix. The *Escherichia coli* chromosome is composed of about 5000 different genes. During replication of the chromosomal DNA, spontaneous mutations in genes can occur. The frequency of such an event per gene—the mutation rate—is very low. Usually, only one antibiotic-resistant mutant cell is found among 1 million to 1 billion cells. An antibiotic-resistant population, therefore, arises only by the selection of the mutant cell. This selection is exerted by the respective drug. Development of resistance in bacteria by this classical darwinian evolution process (mutation followed by selection), however, seems to be a rather rare event in nature. This mechanism is of importance mainly in tubercle bacilli.

Plasmid-mediated resistance. Besides the chromosomal DNA, many bacteria contain other chromosomelike, circular DNA molecules known as plasmids. These autonomously replicating structures are 100 to 1000 times smaller than the chromosome. Most plasmids are not necessary for the survival of the bacterial cell; however, some can contain genetic information which alters the phenotype of the host cell significantly. Plasmids which carry antimicrobial-resistance genes are called R plasmids. Such plasmids can contain from one to many resistance markers; today it is not unusual to find single plasmids carrying genes for resistance to ten or more antibiotics.

Transmissible plasmids enable bacteria to mate with other bacteria and exchange genetic material, especially that of the plasmids themselves. Trans-

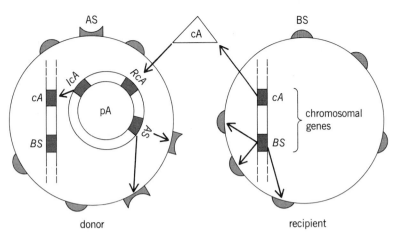

Fig. 1. A model showing specific pair formation in *Streptococcus faecalis*. Here cA and BS are the genetic determinants for the clumping-inducing agent cA for plasmid A (pA) and the binding substance BS, respectively. AS is the genetic determinant for the aggregation substance AS. RcA is the genetic determinant for a regulatory protein which responds to cA, resulting in a turning-on of the determinant AS. IcA is the genetic determinant for a substance which represses (or inactivates) the determinant for endogenous cA. (*After G. M. Dunny et al., Plasmic transfer in Streptococcus faecalis: Production of multiple sex pheromones by recipients, Plasmid, 2:454–465, 1979*)

missible plasmids are often observed in the gram-negative enterobacteria, in which the mating process may involve bacteria of unrelated species, and even of different genera and families. Plasmid exchange by mating consists of several conjugal steps: For pair formation, specific proteinaceous sex pili of plasmid-positive donor cells form conjugation bridges between the donors and recipients. Mobilization and transfer begins with an enzymatic "nick" in a single DNA strand of the plasmid which then enters the receptor via a cytoplasmic bridge. Establishment involves the synthesis of the complementary DNA strand in the recipient. Simultaneously, a strand complementary to the one left over in the donor is synthesized. After completion of the process, the donor cell, as well as the receptor cell, contains one copy of the plasmid, and both can become genetic vehicles for plasmid transmission again.

Not much is known about the mechanism of a conjugationlike mating process in the gram-positive *Streptococcus faecalis*. Pair formation in this organism begins with the excretion by receptor cells of a peptidelike substance of low molecular weight. This clumping-inducing agent, or sex pheromone, causes the synthesis of an aggregation substance in plasmid-containing donor cells. The aggregation substance locates itself on the surface of the donor cell, where it can now recognize a binding substance on the surface of receptor cells. Thus, the formation of mating aggregates resulting from random collision of nonmotile enterococci is facilitated, and plasmid exchange can begin (Fig. 1).

Nontransmissible plasmids are mainly observed in the gram-positive *Staphylococcus*. Staphylococcal plasmids are transferred from donor to receptor with the help of bacteriophages. During maturation of these viruses in plasmid-containing cells, a plasmid, instead of the phage genome or part of it, can be packaged in the phage head. This defective, nonlethal phage particle transfers the plasmid to an appropriate receptor cell. The process is called transduction (Fig. 2).

Another transfer mechanism of both transmissible and nontransmissible plasmids is transformation. Lysis of plasmid-containing cells results in the liberation of plasmids, which then are transferred to appropriate receptors as naked DNA molecules. Under physiological conditions, the transformation process for plasmid exchange plays a rather minor role.

Physiologically, exchange of plasmids has been observed in both basic groupings of bacteria, gram-positive and gram-negative. It is not known whether plasmids may be exchanged across the line between these groups. Nevertheless, it is obvious that the broad host range for R plasmids of gram-negative and of gram-positive bacteria favor the spread of antibiotic resistance in these organisms.

Transposition. One of the main questions regarding infectious multiple-drug resistance concerns the evolution of single plasmids that carry different genes for resistance to many antibiotics. An answer to this problem was recently provided by the discovery of an important new phenomenon in bacterial genet-

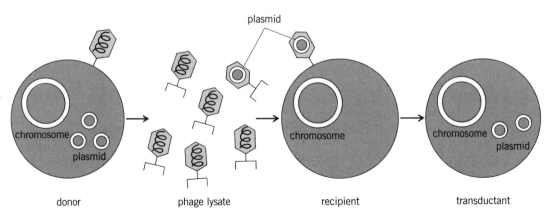

Fig. 2. Schematic illustration of the transfer of a plasmid by transduction.

ics, that of translocation (transposition). Resistance genes of plasmids are often attached to special genetic structures known as transposable DNA segments (transposons). These DNA pieces have the ability to jump from one genetic site to another within a cell. Transposition can occur between a plasmid and the chromosome, or vice versa. Evidence suggests, however, that transposition between plasmids is much more frequent (Fig. 3). All transposable DNA pieces have terminal repeat sequences, mostly in inverted orientation. These inverted repeats seem to be essential for excision and insertion of the sequence at specific sites on DNA molecules. Accumulation of resistance-determining transposons on plasmids has led to the development of plasmids with many resistance genes. Given these facts, the dramatic amplification and evolution of bacterial resistance in the past decades is no surprise.

Biochemistry of antimicrobial resistance. Several phenotypic mechanisms of antimicrobial resistance have been observed. The enzymatic detoxification of antibiotics is a mechanism often found in plasmid-borne resistance. Resistance to penicillins and cephalosporins is mainly due to the production of penicillinases and cephalosporinases, respectively, which hydrolyze the β-lactam ring of these antibiotics. Chloramphenicol resistance is predominantly the result of acetylation of this drug by chloramphenicol acetyltransferase.

Interference with drug transport is the mechanism of resistance to tetracyclines. Plasmid-coded resistance to this drug is correlated with the appearance of a tet (tetracycline-resistance) protein in the membrane of cells which prevents the transport of tetracycline through the membrane layer. Although resistance to aminoglycoside antibiotics had originally implicated detoxification mechanisms, it has become apparent that this is not the case. It can be assumed today that modification of aminoglycosides by *O*-phosphorylation, *O*-nucleotidylation, or *N*-acetylation either interferes with transport or blocks it directly by interacting with the transport mechanism or carrier.

Alteration in target sites involves the development of resistance of the target molecules to the respective drug. For example, resistance to erythromycin is due to dimethylated ribosomal RNA, and resistance to streptomycin depends on altered ribosomal proteins.

Metabolic bypass mechanisms account for resistance to sulfonamides and trimethoprim. In both cases, a plasmid provides the cell with resistant metabolic enzymes, dihydropteroate synthetase and dihydrofolate reductase, respectively, that substitute for the inhibited normal enzymes and allow continued functioning of the blocked pathway in the presence of these drugs.

Controlling resistance. It is obvious that bacteria possess an impressive genetic versatility enabling them to adapt continuously to a changing environment. The problem of resistance, therefore, will not be solved by developing or inventing new antibiot-

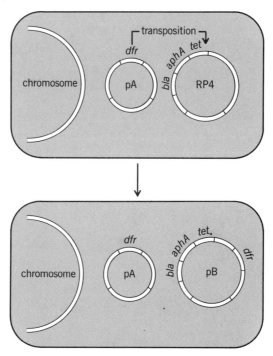

Fig. 3. Schematic illustration of the transposition of a transposon which contains the determinant for resistance to trimethoprim (dfr) from plasmid A (pA) to the R plasmid RP4. The plasmid B (pB) arose by addition of transposon (dfr) from pA to RP4. The resistance determinants and products involved are: dfr for dihydrofolate reductase, bla for β-lactamase, aphA for aminoglycoside phosphotransferase, and tet for tetracycline resistance protein.

ics. Nor will it be solved by developing chemicals that will interfere with the action or transfer of either plasmids or transposable genetic elements. The concept must be that bacteria can become resistant to every possible drug. Control of antibiotic resistance will be possible only by restricting the use of antibiotics, both in medicine and in agriculture.

For background information *see* BACTERIAL GENETICS; BACTERIOPHAGE; DRUG RESISTANCE; TRANSDUCTION (BACTERIA); TRANSFORMATION (BACTERIA) in the McGraw-Hill Encyclopedia of Science and Technology.   [FRITZ KAYSER]

Bibliography: G. M. Dunny et al., Plasmid transfer in *Streptococcus faecalis*: Production of multiple sex pheromones by recipients, *Plasmid*, 2:454–465, 1979.

Archaebacteria

Archaebacteria are a recently recognized group of organisms that look like typical bacteria under the microscope. Yet, they not related to the other bacteria any more than they are to animals or plants. The discovery of archaebacteria has caused biologists to rethink the basic categories of living organisms, and will give new insights into the nature of the universal ancestor of all existing life.

Categories of living organisms. A century ago, biologists believed that life on Earth was only of two kinds, plant and animal. Today, however, taxonomists recognize there are five (or, in some systems, four) kingdoms of living organisms: plants, animals, fungi, protists (for example, paramecium and ameba), and monera (bacteria). Viruses are usually discounted because they are not self-replicating. This five-kingdom classification is misleading, however, for on a more fundamental level there are only two groupings, the so-called eukaryotes and prokaryotes. The basis for the latter distinction is cell type. Eukaryotic cells tend to be relatively large (roughly 10 μm in linear dimension); have a nucleus defined by a nuclear membrane; and contain organelles, such as mitochondria and (in the case of plants) chloroplasts, as well as certain other structures visible under the light microscope. Prokaryotic cells, on the other hand, are smaller and contain no internal structure visible under the light microscope. All organisms that can be seen with the naked eye (animals, plants, and some fungi), as well as some microscopic forms (the remaining fungi and protists), are composed of eukaryotic cells. Bacteria comprise the prokaryotes.

Bacterial evolution. The Earth is 4.5 billion years old. Fossil eukaryotes older than 1 billion years have not been found. However, microscopic bacterial fossils are common in sedimentary rocks as old as 3.5 billion years. Thus it seems that an almost 2-billion-year period in Earth history existed in which the major, if not the only, living forms were bacteria. Until recently little was known about bacterial evolution or about the bacterial phylogenetic tree. Thus, the greatest extent of the evolutionary history of this planet was not understood.

Based on the fact that all life shares a common genetic code and certain other complex features, it is inferred that all life, eukaryotic and prokaryotic, has arisen from a common ancestor. This universal ancestor supposedly gave rise to a line of eukaryotes on the one hand and a line of prokaryotes on the other. While this is true for eukaryotes, it turns out not to be true for prokaryotes. Actually, there are two unrelated prokaryotic lines of descent, but this could not have been known before biologists were able to measure genealogical relationships among bacteria.

The reason so little was known about bacterial evolutionary relationships was that bacterial shapes (rods, spheres, spirals, and so on) are too simple to be used for the determination of relationships. A solution to this problem became possible through the development of molecular phylogenetic techniques which permit the measurement of evolutionary relationships between organisms on the basis of genetic similarity; that is, the extent of similarity in the sequences of a given protein (or gene) isolated from two different organisms is a measure of how recently the two shared a common ancestor. At the molecular level prokaryotes and eukaryotes are similar in general but different in detail. Most, if not all, molecular functions in cells are distinct for eukaryotes and prokaryotes. Unfortunately, generalizations about prokaryotes were premature because only a few bacteria were chosen for study, for it was believed that

all prokaryotes would be very similar at the molecular level. Were it not for this prejudice, the archaebacteria would have been discovered at least a decade sooner.

Molecular phylogenetic methods. Molecular phylogenetic approaches were seriously applied to bacteria for the first time during the 1970s. The most useful of these involved a ribosomal RNA. By comparing defined portions of the sequence of this molecule from one prokaryote with another, it was possible to measure evolutionary distances and genealogical relationships among the bacteria. To date, over 250 species of bacteria have been characterized in this way.

The most surprising finding of these studies was that the bacteria comprise two distinct major groups, one that is now called the eubacteria (or true bacteria), and the other the archaebacteria. The first of these groups contains the typical and well-known bacteria, for example, the actinomycetes, streptomycetes, cyanobacteria, other anaerobic photosynthetic bacteria (the so-called purple and green bacteria), pathogenetic bacteria, and so on.

Phenotypes of archaebacteria. The archaebacteria come from unusual, if not extreme, habitats. Four general phenotypes of archaebacteria are now known: methanogens, extreme halophiles, thermoacidophiles, and anaerobic thermoacidophiles. The methanogens live under highly reducing conditions (complete absence of oxygen), and all produce methane gas from carbon dioxide and hydrogen. They are almost the sole source of biologically produced methane on this planet. Methanogens occur in many places—the stomachs of cattle, sewage fermenters, hot springs, bogs (methane is called marsh gas), and the ocean bottom. Many methanogens can live on extremely simple media, for example, inorganic compounds, such as ammonium salts, phosphates, and sulfides; and carbon dioxide and hydrogen gases. Many also grow at very high temperatures, and some are not destroyed even by boiling.

The extreme halophiles are organisms that must live in very salty environments (for example, the Great Salt Lake and the Dead Sea), and they grow even in saturated brines. Salt evaporation ponds sometimes turn red because of the growth of halophiles in large numbers. They also have been known to discolor salted fish. The red pigment they possess, bacterial rhodopsin, allows them to derive energy from sunlight. It is very much like the visual pigment rhodopsin in the human eye.

The so-called thermoacidophiles include two recognized genera, *Sulfolobus* and *Thermoplasma*. These organisms grow under extremely acidic conditions (some at pH values near 1) and at very high temperatures (some at temperatures within 10° of boiling). Thermoacidophiles are found in such places as acidic hot springs (the odoriferous sulfur caldron type) and smoldering coal refuse piles. They contribute to the acidity of their environment by oxidizing sulfur compounds to sulfates (for example, sulfuric acid). The *Sulfolobus* species are distributed worldwide, including sites such as Yellowstone Park and the hot springs of Iceland and New Zealand.

The fourth type of archaebacteria are also thermoacidophiles, but grow under strictly anaerobic conditions. Since they have only recently been discovered, relatively little is known about these bacteria. Unlike *Sulfolobus* species, they do not oxidize sulfur compounds, but reduce them to hydrogen sulfide. Like some methanogens, some species will grow under extremely simple conditions, such as in solutions of inorganic salts.

Characteristics of archaebacteria. Archaebacteria differ from eubacteria (and from eukaryotes) in the details of their molecular organization. They have unique cell walls (most often constructed of protein) and lipids. (Normal lipids are composed of straight-chain fatty acids bonded chemically to glycerin through ester linkages. Archaebacterial lipids consist of branched hydrocarbons bonded to glycerin through ether linkages. The former are saponifiable lipids; the latter are not.) Archaebacteria contain certain unusual biochemicals; their enzymes tend to be slightly different from eubacterial enzymes, and a number of their genes seem to jump from one position to another in the bacterial chromosome (which is a relatively rare process in the eubacteria).

A pressing question, then, is whether the archaebacteria and the eubacteria are more closely related to each other (both are prokaryotes when defined by cell structure) than either are to the eukaryotes. Molecular phylogenetic measurement indicates that they are not. Thus, the category prokaryotes actually comprises two evolutionarily separate groups of bacteria, not one as had so long been believed.

The picture of evolution that emerges from these findings, the universal phylogenetic tree, is thus one in which the universal ancestor gave rise to three primary lines of descent, the archaebacteria, the eubacteria, and the eukaryotes. These are the phylogenetic groupings at the highest level and should probably be called the primary kingdoms (or urkingdoms)—to distinguish them from the eukaryotic kingdoms (animals, plants, and so on).

This discovery greatly increases the perspective on a number of major evolutionary problems, such as the nature of the universal ancestor, the evolution of biochemical pathways, and the origin of the eukaryotic cell. Some biologists believe that some characteristics of the eukaryotic cell are derived from archaebacteria. As more is learned about archaebacteria in the 1980s, biologist may acquire a new perspective in the study of evolution.

For background information *see* BACTERIA; BACTERIAL TAXONOMY; in the McGraw-Hill Encyclopedia of Science and Technology.   [CARL R. WOESE]

Bibliography: C. R. Woese, Archaebacteria, *Sci. Amer.* 244:98–122, June 1981.

Avalanche

Snow avalanches are known to result from certain combinations of snow, weather, and terrain conditions. A few experienced observers are able to predict avalanche occurrence with fair success by sub-jectively integrating these factors, but such skills are difficult to acquire and even more difficult to teach. More objective approaches, using statistical or deterministic models, appear promising and will be useful for forecasters concerned with providing real-time avalanche warnings for large mountain areas. In steep, snow-covered terrain where there are numerous year-round residences and extensive development, the emphasis is on the delineation of areas threatened by destructive avalanches. Of primary concern in such places is the relatively infrequent, randomly occurring avalanche that runs far beyond what is generally considered to be its normal boundary. Avalanche dynamics models that predict the runout distance, velocity, and impact force of such avalanches are simplistic approximations of actual avalanche motion. All such models require empirical coefficients to achieve reasonable results. More realistic models and better information on the coefficients would improve the accuracy and usefulness of the predictions.

Forecasting occurrence. A number of attempts have been made in recent years to use statistical methods to correlate weather and snow conditions with avalanche occurrence. Univariant and multiple linear regression, discriminate analysis, and time-series methods have all been tried with varying degrees of success. The most frequently used weather variables are precipitation, precipitation intensity, air temperature, wind speed and direction, snow-drift amount and intensity, global radiation, and hours of sunshine. Snow-cover variables which are commonly used either singly or in combination are: depth of snow on the ground, amounts of new snow, snow temperature, settlement of the snow cover, and hardness of surface layers. These variables may be entered as current data or as the sums or averages for past time intervals of from 6 to 96 hours. Lagged data, that is, data for previous periods, may be discounted or reduced in value since their contributions to instability decrease with time.

The use of weather and snow conditions to predict avalanches presents serious problems to any statistical approach that requires normally distributed independent variables. Weather and snow-cover factors seldom have normal distributions and are often strongly correlated. For example, high-density snow, heavy precipitation, relatively warm temperatures, and cloudy weather usually correlate.

Other problems stem from the relatively few avalanche days per year. Any forecast scheme that predicts persistence will achieve unrealistically high scores. One 12-year record used for modeling showed only 128 avalanche days—an average of only 10–11 days per year. Because all data sets show a preponderance of nonavalanche days, if disciminant analysis is being considered, some technique, such as random selection of nonavalanche days, must be used to provide approximately the same number of avalanche and nonavalanche days. Such balanced samples may then be too small to permit partitioning to get more homogeneous data. For example, conditions leading to dry avalanches are different from those leading to wet avalanches. Hence, most researchers divide their data into at least these two categories. It may also be useful to develop predictions for different degrees of avalanche intensity. To do so requires still further partitioning of the data, which further reduces sample size.

Another problem with developing a data set for statistical analysis is the difficulty of collecting a complete tally of all avalanches and of knowing when the events occurred. Because no reliable instrument exists for recording avalanche occurrence over large areas, visual observations must be employed. Because most avalanches occur during stormy weather, when visibility is very poor, a number of them are undetected. The time of occurrence of natural avalanches in many areas can be approximated only to the nearest 12 hours, because constant surveillance is impractical and detection during the night is impossible.

The biggest obstacle to avalanche forecasting, however, is determination of the entity to be forecast. This dependent variable should be unambiguous, easy to monitor, and easy to quantify. The most helpful information would be the size of the additional load needed to release an avalanche at the weakest place in the snowpack. So far, researchers have been unsuccessful in developing nondestructive techniques for measuring snowpack stability for single avalanche paths, much less for entire mountain areas. About the only unambiguous indication of snowpack stability is the release of an avalanche. Although it is not the preferred dependent variable, most models use an avalanche count or some modification of it.

Some of the current forecast models are based on the concept of an avalanche day, which is merely a day when at least one avalanche is observed in the forecast area. Some of these models predict whether or not a given day will yield an avalanche. Others predict the probability of a given day being an avalanche day; predict the total number of avalanches for a given sample of paths; or use an avalanche activity index that is usually based on some combination of number and size of avalanches.

Maximum accuracy of the statistical models seems to range from 60 to 80%. They do better predicting days with numerous avalanches than those with only a few.

Current research on this topic is directed toward replacing or supplementing statistical models with physical-process or deterministic models. The deterministic models attempt to simulate the physical processes that are thought to cause avalanches. Emphasis is given to precipitation and wind drifting of snow as major loading mechanisms, and to changes within the snowpack that determine its strength. Attempts are also being made to use so-called pattern recognition techniques to relate present weather and snow cover factors to past conditions that produced avalanches.

Efforts to enlarge existing data bases must continue, and new data bases incorporating a variety of climatic and terrain conditions must be developed. All types of models are dependent on such data for verification and calibration. An accurate estimate of avalanche activity for large mountainous areas is the most difficult portion of these data bases to acquire. Records of the number, size, type, location, and time of occurrence of avalanches in complex and often remote terrain, collected over long periods of time, are essential for continued research.

Forecasting velocity, runout distance, and impact force. The increase in year-round residences and developments in mountainous areas has emphasized the need to delineate areas threatened by destructive avalanches. A primary concern is the extent and intensity of the occasional avalanche that runs far beyond its normal boundaries. Because such events are relatively rare and appear to show a random time frame, the development of models for predicting and computing their runout distance, impact force, and velocity are highly desirable for zoning and planning purposes. All of the models proposed in the past decade seek a computational procedure that is simple enough to solve on existing computers, while still giving a reasonable approximation of the complex and highly variable motion exhibited by avalanches.

It is reasonably certain that avalanche motion varies from simple block sliding to highly turbulent flow, and that frontal velocity is highly variable, depending on snow and terrain conditions. The moving snow mass is made up of particles that range from individual ice grains to massive blocks of tightly bonded snow. In most cases, bulk density of the moving snow ranges from a dense layer near the sliding surface, to less-dense layers above, to an all encompassing snow-dust cloud that shows an airlike density. The moving mass can entrain additional snow at its leading edge and deposit snow at its trailing edge. It usually moves as an elongated mass with indistinct boundaries. At any given time the moving mass occupies only part of the avalanche path. The moving snow compresses with pressure, temperature, and especially upon impact.

None of the existing models attempts to cover all the conditions mentioned above. Instead they make a series of simplifying assumptions. Some models assume motion to be steady and uniform down a longitudinal profile of the path. None allows for horizontal spreading. Most assume the snow mass moves as a continuous, homogeneous, noncompressible, isotropic fluid. All depend on one or more friction or resistive terms to keep velocities and runout distances within observed limits. Many yield velocities only for segments of the profile where slope and other conditions can be considered uniform.

Major research efforts are directed toward modification of the simplifying assumptions to attain more realism and better evaluation of the friction and resistive parameters based on observed data, especially for unusually long-running avalanches. The earliest practical model, developed in the 1950s, was based on open channel flow of fluids. It assumed the moving snow was continuous, and required an arbitrary selection of the place on the longitudinal profile where deceleration set in. An alternative model that considers the avalanche to be a center of mass permits all calculations to be referenced to an easily identified starting point rather than to an arbitrarily selected point on the profile. Unfortunately, neither model gives an estimate of the depth or distribution of the debris in the runout zone other than the final location of its leading edge.

Another recent model uses a finite difference solution of the Navier-Stokes fluid equations to model avalanche motion. This model modifies the Navier-Stokes assumptions and uses three parameters (one viscosity and two friction) to control output. It provides a good estimate of the distribution of the moving snow and of the debris in the runout zone. A further modification permits the moving granular material to lock up and move as a block when velocity drops below a critical value. This lockup phenomenon is often observed in natural avalanches.

Field data from a variety of snow and terrain conditions are extremely important for verifying and improving these models. Data that include avalanche velocities, snow conditions, debris distribution, and runout distance are scarce—especially for the relatively infrequent long-running avalanches.

Future avalanche research should concentrate on measuring the velocity and bulk density of moving avalanches at several places along the path. One study planned for the near future will use radar to track avalanches in motion to learn more about avalanche velocity and its variations along the path.

Additional research that has a bearing on avalanche problems include studies of wind transport and deposition of snow and of the structure and strength of deposited snow.

For background information *see* AVALANCHE in the McGraw-Hill Encyclopedia of Science and Technology.

[MARIO MARTINELLI, JR.]

Bibliography: R. Perla and M. Martinelli, Jr., *Avalanche Handbook*, no. 489, USDA, 1975; Symposium on Snow in Motion, Fort Collins, CO, August 12–17, 1979, *J. Glaciol.*, vol. 26, no. 94, 1980.

Bacterial endotoxin

Few, if any, bacterial products cause such an extraordinary variety of biological effects in susceptible hosts as bacterial endotoxins. Chemically they are lipopolysaccharide macromolecules that are produced by numerous species of gram-negative bacteria, including *Neisseria*, *Haemophilus*, *Pseudomonas*, *Pasteurella*, and others belonging to the large family of Enterobacteriaceae, as well as by anaerobes. The endotoxins are macromolecules consisting of lipid A and a core polysaccharide bridged by KDO (2-keto-3-deoxy-D-mannooctulosonic acid). In microorganisms, protein associated with lipopolysaccharide is referred to as lipid A–associated or lipopolysaccharide-associated protein. The core poly-saccharide may be enlarged by the presence of O-specific side chains, characterizing serogroups and serotypes (O antigens) of many pathogens, such as shigellae and salmonellae.

The biologic effects of these endotoxins in susceptible hosts include fever, toxic shock, alteration of metabolism, changes in the hematopoietic system, the Shwartzman and Sanarelli phenomena (hemorrhagic lesions of skin and internal organs, respectively), tumor necrosis, alterations of the coagulation system, changes in nonspecific resistance, activation of complement, and both enhancement and suppression of immune responses. Many of these effects are due to the lipid A component. Effects on the immune system elicited by lipopolysaccharide as antigen include formation of antibodies and multiplication of immune lymphocytes with specificities directed against the components of lipopolysaccharide. Studies of the mode of action have been greatly facilitated by the recognition that certain strains of inbred mice are resistant to endotoxin. Endogenous mediators, produced in response to lipopolysaccharide by certain host cells, play a major role in its biological effects.

Sources of lipopolysaccharide. Numerous genera of bacteria produce lipopolysaccharide, which accounts for their endotoxicity, as part of the cell wall. They are gram-negative and include both cocci and bacilli, aerobes and anaerobes, and pathogens and nonpathogenic microorganisms. Rough (R) organisms derived from smooth (S) forms contain lipopolysaccharide, even when the change from S to R results in loss of virulence. Lipopolysaccharide-producing microorganisms include, among others, the following genera: *Neisseria*, *Branhamella*, *Acinetobacter*, *Escherichia*, *Salmonella*, *Shigella*, *Serratia*, *Proteus*, *Yersinia*, *Pseudomonas*, *Fusobacterium*, *Bacteroides*, *Veillonella*, and even species belonging to the photosynthetic prokaryotes.

Isolation of lipopolysaccharide. A variety of methods have been used to isolate lipopolysaccharide from bacteria, such as the procedures involving trichloracetic acid, phenol water, aqueous ether, aqueous *n*-butyl alcohol, and phenol–chloroform–petroleum ether. Significant differences exist among various lipopolysaccharide preparations: some still contain the lipopolysaccharide-associated protein and others do not, and some are highly immunogenic and others far less so. In addition, the biologic properties of the isolated products may be altered by heating and by other procedures. Depending on the strain of microorganisms used, different lipopolysaccharide preparations may be obtained. Some of these preparations, such as certain products obtained from photosynthetic prokaryotes, are nontoxic.

Chemical characteristics of lipopolysaccharide. The chemical structure of the endotoxic lipopolysaccharide molecule has been largely clarified. Essentially, it consists of lipid A linked to the core, which consists of sugars, by means of KDO. Rougher strains have less sugars, and mutants with increasing roughness serve as sources of lipopolysaccharide preparations. Extremely rough strains produce lipopolysaccharide consisting of only lipid A and KDO. In the lipopolysaccharides obtained from S forms, repeating polysaccharide units are attached to the core and give the entire molecule its O (group or type) specificity. Because these repeating units differ from group to group or from type to type, the numerous lipopolysaccharide preparations differ in antigenicity. In addition, protein is present in the cell walls of these microorganisms.

Lipopolysaccharide–host cell interactions. Lipopolysaccharide interacts with host cells both in vivo and in vitro. This latter interaction can be clearly demonstrated with lipopolysaccharide and erythrocytes. Lipopolysaccharide binds to red blood cells, and as a result these cells acquire a new antigenic reactivity, dependent upon the particular haptenic specificity of the lipopolysaccharide molecule. A cell membrane component has been isolated and identified which serves as receptor for lipopolysaccharide and related molecules. Current studies suggest that lipid A–specific receptors may be present on some lymphoreticular cells and thus could possibly contribute to immune regulation. A variety of substances, including certain proteins and antibiotics, notably polymyxin B, inhibit cell attachment of lipopolysaccharide. In fact, this antibiotic even prevents lipopolysaccharide toxicity in animals.

Host mediators. Recent studies on the mode of action of lipopolysaccharide have revealed that many of the overt changes produced in susceptible hosts are due to mediators produced by the host. For example, the characteristic pyrogenic response of humans, rabbits, and other animal species is due to the production or release of a protein, referred to as endogenous pyrogen, which is produced by macrophages. Probably two different kinds of protein molecules act in this capacity, and their relationship to other mediators is being investigated at the present time. Two particularly important mediators have been identified and are referred to as interleukin 1 and 2. Interleukin 1 is also known as LAF because it acts as a lymphocyte-activating factor, probably by means of helper cells. Interleukin 2 is known as TGF, or T-cell-derived growth factor. These two mediators affect the immune responses of certain hosts. Other important mediators are interferon, the colony-stimulating factor which modulates the granulopoietic system, and the tumor-necrotizing factor.

Lipopolysaccharide susceptibility of animal species. Different species of animals vary significantly in their susceptibility to lipopolysaccharide. This susceptibility pattern is not related to evolution. In fact, even crabs (for example, *Limulus polyphemus*) are highly susceptible, and amebocyte lysate obtained from these animals is used as an extraordinarily sensitive indicator for lipopolysaccharide. At the other end of the scale, humans are highly responsive and, upon intravenous injection of as little as 0.001 μg/kg, develop fever. In contrast, some monkeys are very resistant, tolerating even large

amounts of lipopolysaccharide without developing fever or other manifestations. Differences exist even within a species; a few mouse strains may be resistant while most others are susceptible. In view of the current interest in the staphylococcal toxic shock syndrome, it is noteworthy that staphylococcal enterotoxin and exotoxin C markedly increase susceptibility to the toxic effect of lipopolysaccharide. Repeated daily injections of endotoxin results in decreasing fever responses and eventually in tolerance to the toxic effects of lipopolysaccharide.

Biological effects of lipopolysaccharide. In susceptible hosts, including rabbits, dogs, and humans, lipopolysaccharide causes a large variety of reactions. Upon intravenous injection of lipopolysaccharide in minute quantities, rabbits develop a characteristic biphasic fever response; the second phase is due to the release of endogenous pyrogen.

Hemorrhagic necrosis develops in rabbits upon two properly spaced injections of lipopolysaccharide. In the Shwartzman reaction the first dose is given intracutaneously and the other intravenously. The resulting local reaction is referred to as the local Shwartzman phenomenon. Hemorrhages develop in many sites of the body when two doses of lipopolysaccharide are injected intravenously, a phenomenon referred to as the Sanarelli reaction. Related to these phenomena is the hemorrhagic reaction of the placenta in pregnant animals and of tumors, such as sarcoma, upon a single injection of lipopolysaccharide.

Lipopolysaccharide alters many metabolic activities; for example, it causes changes in blood glucose levels and enzymatic activities of the liver. It also has striking effects on the hematopoietic system, causing an initial drop in the number of white blood cells followed by a marked rise, and it increases susceptibility to epinephrine. Conversely, BCG immunization, adrenalectomy, and staphylococcal enterotoxin enhance susceptibility to lipopolysaccharide.

Lipopolysaccharide exerts a marked influence on the immune system. Depending upon the particular conditions, it acts as an adjuvant or as an immunosuppressive agent. It is also a mitogen and a polyclonal activator of lymphocytes. In addition, lipopolysaccharide activates the complement system by the alternative (sometimes referred to, erroneously, as alternate) pathway, as well as by the classical pathway, by means to two different mechanisms.

Applications. Hopes for the use of this material in humans arose with the discovery of the beneficial effects of lipopolysaccharide in experimental animals, notably the increased nonspecific resistance to infection. Few applications have emerged for the use of lipopolysaccharides and its components in medicine. At the present time research is concentrated on the immunogenic properties of core lipopolysaccharide for active immunization against infection caused by microorganisms producing an identical or similar core macromolecule, and the use of the corresponding antibodies. Lipopolysaccharide vaccine has been used for specific immunizations against *Pseudomonas aeruginosa* infections. Immunologic injury may follow when lipopolysaccharide is present in tissues and when the corresponding antibodies, such as those directed against lipid A, are produced and react with this antigen.

Numerous studies of the hemagglutination test for the identification and quantitation of lipopolysaccharide antibodies engendered by various pathogens in both experimental and natural infections have proved its usefulness. The hemagglutination test is based on the fact that lipopolysaccharide of different specificities readily becomes attached to indicator red blood cells; in turn, these modified erythrocytes become suitable reagents for the demonstration of the corresponding antibodies. Each erythrocyte can be antigenically modified by some 10 different antigens, thus providing a multivalent reagent.

Since lipopolysaccharide, even in minute quantities (> 0.001 µg/kg), causes febrile reactions in humans, and since this endotoxin is very widely distributed in nature and resists heating to a significant degree, it is important that lipopolysaccharide contamination of materials to be injected into patients be avoided. Its presence can be detected by two tests. One test measures the temperature response of rabbits and the other the clotting of amebocyte lysate of crabs (limulus test).

Immunization with R mutants or core lipopolysaccharide, as well as with the antiserum to the core lipopolysaccharide, protects experimental animals against infection and endotoxicity. R mutants from different microorganisms vary in their protective capacity against various pathogens. Even protection against *Haemophilus influenzae* has been demonstrated.

Conclusions. Much has been learned about the chemical composition and biologic properties of lipopolysaccharide. A single molecular mechanism responsible for the extraordinarily numerous biological reactions in susceptible hosts has not been identified. Unexplained, too, are the nature of the significant differences in susceptibility of various animal species and the sequential susceptibility and refractoriness of chick embryos. Finally, the role of lipopolysaccharide produced by gram-negative bacteria present in the host, notably on the development of the immune system and other defense mechanisms against infection in humans and animals from birth to old age, remains to be elucidated.

For background information *see* ANTIGEN; IMMUNITY; VIRULENCE in the McGraw-Hill Encyclopedia of Science and Technology.

[ERWIN NETER]

Bibliography: E. H. Kass and S. M. Wolff, Bacterial lipopolysaccharides: Chemistry, biology, and clinical significance of endotoxins, *J. Infect. Dis.*, 128(*supp.*):9–306, 1973; O. Lüderitz et al., Chemical structure and biological activities of lipid A's from various bacterial families, *Naturwissenschaften*, 65:578–585, 1978; D. C. Morrison and J. A. Rudbach, Endotoxin-cell-membrane interactions leading to transmembrane signaling, *Contemp. Topics Mol. Immunol.*, 8:187–218, 1981; D. C. Morrison and J. L. Ryan, Bacterial endotoxins and host immune responses, *Adv. Immunol.*, 28:293–450, 1979; A. Nowotny (ed.), *Beneficial Effects of Endotoxins*, (in press); D. Schlessinger, (ed.), Endogenous mediators in host responses to bacterial endotoxin, *Microbiology-1980*, pp. 3–167, American Society for Microbiology, 1980.

Bacterial taxonomy

As in the classification of other kingdoms, the hierarchically higher ranks of bacteria are divided into two or more lower ranks which are arranged in classes, orders, families, genera, and species. In bacterial taxonomy, the basal unit is the species, which is built up from individual strains that share a number of common characters. Related species unite as genera, similar genera unite as families, and so on. Although these groupings and arrangements are principally based on phylogenetic considerations, larger ranks in the bacterial classification can be considered to be artificial and speculative.

The classification of bacteria is based on morphological, physiological, and biochemical characters. Somatic components and metabolites are analyzed by using gas chromatography and infrared spectra, and zymograms (enzyme profiles generated by zone electrophoresis) have also been introduced into the classification of bacteria in recent years. More recently, the base composition of the deoxyribonucleic acid (DNA) and the homology between DNAs of bacteria have become fundamentally important in bacterial taxonomy.

Pathogenicity is a character of limited value in bacterial classification, because it is a variable and nonunitary character which is difficult to determine in cultures as well as in living organisms.

Conventional and numerical classification. Bacterial classification depends on the characterization of microorganisms. In former systematics, now termed conventional or traditional taxonomy, the detail derived from characterization varies with different organisms, and different taxonomists use different criteria for characterization. For example, a cytologist may attach great importance to morphologic and cytologic characters rather than physiological and biochemical characters; on the other hand, some taxonomists have emphasized antigenic structure in the classification of bacteria. In addition, there has been disagreement among taxonomists about the value of different characters. Conventional taxonomists search for characters that are regarded as important because of certain overall features and try to group organisms that share common important characters. The opinion of each taxonomist differs on which organisms are similar and which are different. Thus the classification is subjective, and the results may vary for different taxonomists. In general, however, conventional classifications have yielded relatively satisfactory results when they have been confirmed by numerical methods.

A recent development in bacterial classification has been a return to the theories and ideas of the plant taxonomist Michel Adanson. One of the essential elements of Adansonian principles is that each character is considered to be of equal value. P. H. A. Sneath first applied Adansonian principles to bacterial taxonomy, and further studies on a wide range of bacteria by many investigators showed that his principles were applicable to the classification of bacteria. The computer analysis of characterization is generally referred to as numerical classification because the correlation of characters is appraised statistically. While organisms which share common important characters are grouped together in the conventional classification, there is no need in the numerical classification for any test result to be uniformly positive on all organisms within a given group. The grouping is based on a measure of overall similarity between pairs of organisms. Numerical classification is simply concerned with grouping organisms on the basis of similarity of phenotypic characters. The methodology is objective, since the grouping of organisms is based on a statistical analysis of similarity without any knowledge of the nature of the characters or any working hypothesis· of the evolution of taxonomic groups.

In numerical classification studies, the number of unit characters to be analyzed should not be less than 60 in order to ensure a satisfactory minimization of sampling error. Also, the unit characters should be independent of one another. Taxonomic categories based on morphology, physiology, and biochemistry should be evenly distributed among a series of unit characters. Each unit character chosen is studied for every strain in the collection. To provide input to computers, the data represent qualities that are recordable as positive or negative or as traits which can be scored by unit designation. Although many different coefficients for the estimation of resemblance between pairs of organisms have been devised, two coefficients are commonly used: the simple matching coefficient (S_{SM}) and Jaccard's coefficient (S_J). The S_{SM} is used to express two-state data (positive or negative) only, and any character for which either or both pairs have a missing entry is ignored. S_J is identical to S_{SM}, but matching negative results are ignored.

The number of groups in a collection of organisms can be determined from tabulation of S_{SM} or S_J pairs recorded in a triangle matrix (see illustration). A hierarchical arrangement of groups of strains can be obtained by plotting the degrees of similarity in a dendrogram.

Each individual group is referred to as a phenon, and measurable degrees of difference between phenons are called taxonomic distances. It is difficult to define a level of similarity for taxonomic ranks such as species and genus. In general, a 70% single-linkage cluster in either S_{SM} or S_J gives an approximation to the genus and 80% to a species, but many exceptions exist.

Genetic approach to classification. Recently, taxonomists have oriented themselves toward the characterization of the genetic composition of microorganisms. Deoxyribonucleic acid (DNA) contains four bases, guanine (G), cytosine (C), thymine (T), and adenine (A). Base composition can usually be expressed as a percentage of G + C to total base (%GC, or mole % GC). It has been suggested that the G + C ratio of a single genus should be homogeneous, and different genera often have different G + C ratios. The G + C ration can be determined by several different techniques, but thermal melting (Tm) and buoyant-density techniques are most commonly used. In the thermal melting method, heating of DNA in solution leads to the denaturation of the double-stranded helix to single-stranded DNA. The denaturation temperature depends on the base composition of DNA; the greater the G + C content, the higher the denaturation temperature. The buoyant density is estimated by CsCl density gradient ultracentrifugation. DNA molecules band at an equilibium in the linear CsCl density gradient when they are ultracentrifuged in a CsCl solution. The density of the DNA at the midpoint of the band is called the buoyant density of DNA, and is related to the base composition of DNA.

The determination of the extent to which single strands of DNA from two different strains reassociate to form double strands has developed as a taxonomic tool. A given strand of DNA can reassociate with another strand from the same species or from a different species, if the two strands contain comple-

A	100								
B	56	100							
C	64	88	100						
D	52	96	96	100					
E	56	92	92	88	100				
F	100	56	64	52	56	100			
G	96	52	60	64	64	96	100		
H	48	88	92	96	84	48	60	100	
I	88	60	56	60	72	88	92	72	100
	A	B	C	D	E	F	G	H	I

(a)

A	100								
F	100	100							
G	96	96	100						
I	88	88	92	100					
E	56	56	64	72	100				
B	56	56	52	60	92	100			
C	64	64	60	56	92	88	100		
D	52	52	64	60	88	96	96	100	
H	48	48	60	72	84	88	92	96	100
	A	F	G	I	E	B	C	D	H

(b)

Determination of the number of groups in a collection of organisms. (a) Tabulation of similar unit characters between each strain and every other strain (A to I) in a triangle matrix. (b) Each strain is rearranged in order of higher level of similarity. Two clusters are formed at the 88% level of similarity.

mentary base sequences. The homology of DNA can be estimated by measuring the extent and stability of pairing between single-stranded nucleotides. Deoxyribonucleic acid reassociation is assumed to be a definitive method for establishing biological relatedness at the species level.

Although there is a variety of methods for monitoring DNA reassociation, two basic methods are usually applied in bacterial classification: binding of radiolabeled, denatured DNA in solution to an unlabeled single strand of DNA immobilized on membrane filters or in agar gels; and the free-solution method, including the hydroxyapatite method, the S1 nuclease method, and the renaturation-rate method. In the hydroxyapatite method, the double-stranded DNA is separated from the single strands by hydroxyapatite chromatography, and the radioactivity of liquid eluted from hydroxyapatite columns is measured. The observed homologous binding for each labeled DNA is designated 100% and the reassociation values of the heterologous DNA are determined relative to the binding ratio of homologous DNA and expressed in a percentage. The S1-nuclease method is carried out by precipitation of S1-resistant DNA by trichloroacetic acid and filtration on cellulose nitrate filters or by adsorption of the DNA on DE81 filters after treatment of the hybridization mixture by S1-nuclease. The radioactivity is measured in a vial containing a scintillator. The DNA reassociation is calculated by determining the ratio between the average counts in the nuclease-treated and the nontreated samples. Formed DNA-DNA duplexes can also be analyzed by thermal analysis and estimates of base mismatching can be made. In the renaturation rate method, the rate of reassociation of heterologous DNA is slower than that of homologous DNA. The reassociation of DNA can be determined by the spectrophotometer, for which radioactive labeled DNA is not needed.

Conclusions. Modern taxonomic methods form a great contribution to bacterial classification. However, each method has its utility and its limitations, and each method should be combined with other approaches in microbial classification.

According to the International Code of Nomenclature of Bacteria, certain principles have to be followed for the valid publication of new names or new combinations, including the designation of the type of a named taxon, and a detailed description of the taxon. The new name or new combination must be published in the *International Journal of Systematic Bacteriology*.

For background information *see* BACTERIAL TAXONOMY; DEOXYRIBONUCLEIC ACID (DNA); NUCLEIC ACID; NUMERICAL TAXONOMY; TAXONOMY in the McGraw-Hill Encyclopedia of Science and Technology.   [RIICHI SAKAZAKI]

Bibliography: J. De Ley, Hybridization of DNA, in *Methods in Microbiology*, 1971; J. Marmur and P. Doty, Determination of the base composition of deoxyribonucleic acid from its thermal denaturation temperature, *J. Mol. Biol.*, 5:109–118, 1962; C. L. Schildkraut et al., Determination of the base

composition of deoxyribonucleic acid from its buoyant density in CsCl, *J. Mol. Biol.*, 4:430–443, 1962; P. H. A. Sneath and R. R. Sokal, *Numerical Taxonomy: The Principles and Practice of Numerical Classification*, 1973.

Biliprotein

Phytochrome, the light-sensitive pigment responsible for many regulatory responses in green plants, exists in two forms, Pr [red absorbing (structure I*a*)] and Pfr [far-red absorbing (II)], which are intercon-

(I*a*) R = C_2H_3
(I*b*) R = C_2H_5
R′ = H or protein

R′ = H or protein

(II)

vertible by light. Recent research on the chemistry of phytochrome has resulted in the identification of the structure of the Pfr chromophore. This research relied heavily on chemical and theoretical models for two reasons. First, phytochrome is present in plants only in low concentrations and is still difficult and tedious to isolate in larger amounts; and second, analysis of phytochrome is impeded by the large apoprotein that is covalently bound to the chromophore.

Chromophore structure. Studies of chromophore structure involve the conjugation system, the substitution pattern, and covalent bonds to the protein.

Conjugation system. Even before Pr was isolated, the similarities of the absorption spectra of Pr to those of the photosynthetic antenna pigment phycocyanin (I*b*) led researchers to suggest that the chromophore structure for these two pigments was similar. The spectra of both native Pr, for which the suggested conformation is shown in structure (III), and phycocyanin are so strongly influenced by noncovalent interactions with the apoprotein that they cannot readily be related to free bile pigments like phycocyanobilin (IVa). The interactions can be uncoupled by denaturation—for example, treatment with urea

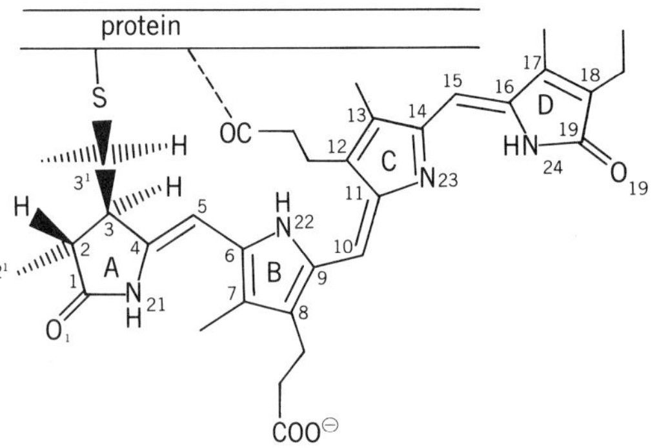

(III)

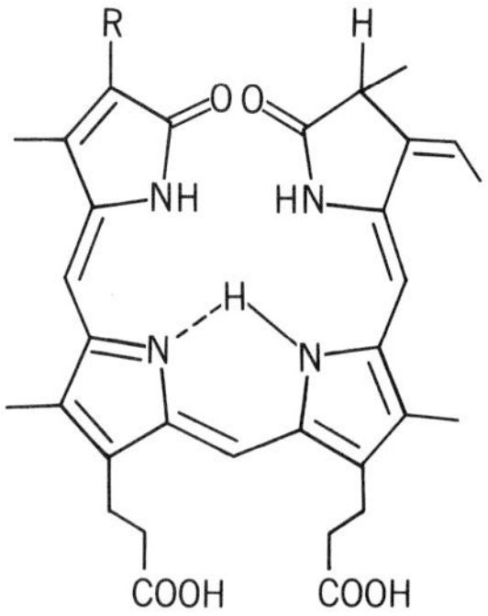

(IVa) R = C$_2$H$_5$
(IVb) R = C$_2$H$_3$

or heat—a process which in the case of phycocyanin is reversible. The spectra of the denatured biliproteins can then be compared to those of free bile pigments. In particular, denatured phycocyanin is very similar to the synthetic A-dihydrobilindione (V) not

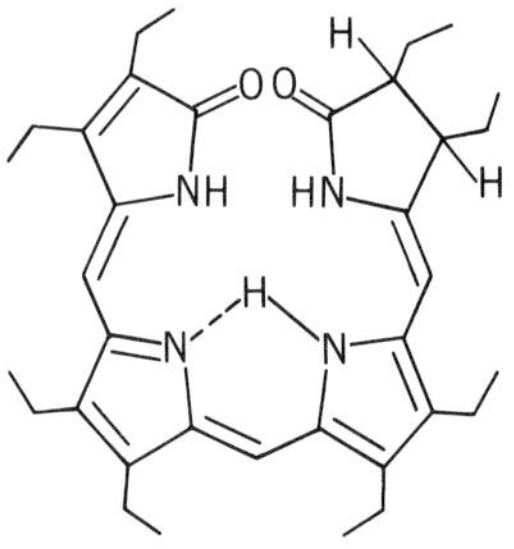

(V)

only as free base but also after protonation, deprotonation, and complexation with zinc, thus proving that the same conjugation system is present in both. The spectrum of denatured Pr is red-shifted by about 10 nanometers as compared to native phycocyanin, which can be taken to indicate the presence of an additional conjugated double bond in a β-pyrrolic position. Since the extinction coefficients of (V) are

well known, uncoupling from the protein also allowed the determination of the chromophore content to be one chromophore per peptide chain in both large and small phytochromes.

Denaturation of Pfr gave a surprising result: the conjugation system was apparently one double bond shorter than that of Pr, in contrast to the long-wavelength shift of native Pfr. Reactivity studies with a variety of bile pigments, for example, A-dihydrobilindione (V) and phytochromobilin (IVb), and bilipeptides of phycocyanin and Pr suggested that either an oxidation reaction at the methine bridge next to the reduced ring A or the geometric (Z, E) isomerization at the C-5 or C-15 methine bridge was responsible. In both instances the "missing" double bond is still present, but is no longer fully conjugated to the main π-system. This is due to steric hindrance, which twists the respective end ring out of the plane of the remaining three rings, as demonstrated by x-ray crystal analysis of related pigments and by molecular orbital calculations.

Substitution pattern and covalent bonds to the protein. The IX α-substitution pattern of Pr, Pfr, and phycocyanin has been derived, first, from chromic acid degradation, which cleaves the tetrapyrrole into a mixture of cyclic imides. These principally bear the original β-pyrrolic substituents, but side reactions are encountered at ring A and, in the case of phytochrome, at the vinyl groups of ring D. Detailed analysis of the reactions at ring A make it possible to determine not only the site of linkage to the protein, for example C-3, but also the stereochemistry at all three of the asymmetric centers at C-2, C-3, and C-3^1. These reactions have also been applied to the 3-ethylidenebilindiones (IVa) and (IVb) and to the protein-bound chromophores. The degradation of both Pr and Pfr lead to identical mixtures of imides. Less extensive and milder degradation methods have recently been developed that allow a selective cleavage of the tetrapyrrole at either the C-5 or the C-10 methine bridge. These reactions have been used to study a potential second and more labile bond to the protein via one of the propionic ester groups. Preliminary results have shown that at least some of the chromophores in phycocyanin do not contain such a bond.

The proposed structures have been confirmed by two other methods. One is the cleavage of the phycocyanin chromophore with a mixture of hydrobromic acid and trifluoroacetic acid to yield (IVa). Although this treatment did not yield (IVb) from Pr because of side reactions of the C-18 vinyl group, correlation was made by the observation of identical products from both Pr and (IVb) under the same reaction conditions. The structures of (IVa) and (IVb) have been confirmed by total synthesis.

The other method used to establish the molecular structures is nuclear magnetic resonance (NMR) spectroscopy of the bilipeptides. The assignment of all chromophore signals in the phycocyanin and Pr peptides was done by comparison with the respective chromophore-free synthetic peptides. The nuclear magnetic resonance spectroscopic analysis of Pfr posed a special problem. The Pfr peptides are at least moderately stable only under acidic conditions, and these facilitate exchange of protons essential to the analysis, for example, at position C-5. This problem was ameliorated by carefully balancing the $^1H/^2H$ ratio and the acidity of the solvent and by the use of the more readily accessible (E, Z, Z)-(V) and phycocyanin peptides as models. These isomers can be prepared from the respective (Z, Z, Z) isomers by a general method involving the photoisomerization of reversibly formed addition products at C-10.

Noncovalent interactions of chromophore and protein. These consist of static interactions and dynamic interactions.

Static interactions. The properties of the phytochrome chromophores are profoundly altered by noncovalent interactions with the native protein, which are only partly understood. In the case of Pr, these interactions are rather similar to those seen in the photosynthetic antenna pigments, phycocyanin, and allophycocyanin, which are therefore good models for Pr. They show the same increased chemical stability as free bile pigments such as A-dihydrobilindione or the denatured or proteolytically degraded biliproteins, and they show the same pronounced increase of the long-wavelength absorption and decrease in the near-ultraviolet absorption. These changes are good examples of molecular ecology, that is, the adaptation to environmental conditions on the molecular level.

The absorption changes are most likely due to a conformational change of the chromophore. Free bile pigments of the biliverdin type are present in a predominantly cyclic-helical, porphyrin-type conformation, and the similar absorption spectra of the biliprotein chromophores uncoupled from the protein suggest a similar conformation. Molecular orbital calculations predict just this type of spectrum for a cyclic conformation and suggest that the spectra of native biliproteins reflect an extended conformation as shown in structure (III). This view is supported by conformationally restricted bile pigments, for example, isophorcabilin (VI), which show the

(VI)

spectral features expected for an extended conformation. Extended conformations are energetically less favorable than cyclic ones, and this energy has to be derived from the protein. It is interesting to note

that phycocyanin has indeed a rather small stabilization energy as compared to other proteins of similar size.

The decreased reactivity of native biliproteins has been studied by isotope-exchange experiments and by chromophore modifications, again with A-dihydrobilindione as a model. It has been shown that the chromophore is accessible to reagents in the aqueous phase and is therefore thermodynamically rather than kinetically stabilized by the protein. It is as yet unclear whether this stabilization is due to conformational changes.

The chromophore-protein interactions are less well understood in Pfr. Its chromophore conformation is probably similar to that of Pr, but the exceptionally large red shift of the native as compared to the denatured pigment cannot be rationalized this way. The only factor known to produce similarly large optical shifts in free bile pigments is deprotonation. Since the pK of the chromophores is within the physiological range, it has been suggested that the native Pfr chromophore is present as an anion.

Dynamic interactions. An important property of linear tetrapyrrole bile pigments is their flexibility. It has been shown that the two forms of cyclic-helical (Z,Z,Z)-bilindiones with opposite chirality interconvert rapidly at ambient temperatures. Similarly, the broad, unstructured absorption bands of free bile pigments are thought to arise from a superposition of several rapidly interconverting conformers with similar energies but different electronic transition energies. The absorption maxima remain broad even at 77 K. The chromophore absorptions of phytochromes are, by contrast, much narrower. This is likewise true for phycobiliproteins such as phycocyanin. The optical spectra of these compounds are complicated because of the presence of more than one chromophore in the monomer which absorb light at distinctly different wavelengths. Nevertheless, the low-temperature spectra can be resolved into a set of narrow lines. It is therefore likely that the chromophores of biliproteins have lost their flexibility and are fixed by the proteins in defined conformations, which is an important factor in establishing their photochemical and physiologically crucial properties.

As known from other systems, one important consequence of an immobilization of structure is a general decrease in radiationless deactivation of excited states. The latter process is dominant ($> 99\%$) in free bile pigments like (V), but is largely reduced in biliproteins, thus resulting in either high fluorescence (like phycocyanin) or high quantum yields of photochemical events (like phytochrome) to an extent that depends on the specific environment of the chromophores. It is noteworthy in this connection that partially denatured biliproteins acquire photochromic properties similar to partially denatured phytochrome.

The profound similarities of phycocyanin and Pr with regard to chromophore structure and protein interactions make phycocyanin the best available model for phytochrome and raise the possibility of a common evolutionary origin of these functionally quite different biliproteins.

For background information *see* PHYCOBILIN; PHYTOCHROME in the McGraw-Hill Encyclopedia of Science and Technology.

[HUGO SCHEER]

Bibliography: H. Falk, N. Müller, and T. Schlederer, *Monatsh. Chem.*, 111:159, 1980; A. Gossauer, R.-D. Hinze, and R. Kutschan, *Chem. Ber.*, 114:132, 1981; W. Rüdiger, *Struct. Bond. (Berlin)*, 40:101, 1980; H. Scheer, *Angew. Chem.*, 93:230, 1981; H. Scheer, *Angew. Chem. Int. Ed. Engl.*, 20:241, 1981; J. P. Weller and A. Gossauer, *Chem. Ber.*, 113:1603, 1980.

Biophysics

Use of ultrasound in both diagnostic and therapeutic medical applications continues to increase in scope. In diagnostic uses, accurate and complete information is desired about the structures through which the sound is propagating and about how these structures are changed (for example, in pathological cases) without producing any alterations of the structures. In therapeutic applications, it is important to be able to predict the response of the medium to ultrasonic exposure, and to determine accurately the deposition of energy in the medium needed to produce the desired effects. The safe and effective use of ultrasound for both purposes depends on a detailed knowledge of the manner in which ultrasonic energy propagates through various tissues and organs.

Sound propagation in nonlinear media. In the derivation of the quantitative relations which describe acoustic wave propagation in material media, it is necessary to invoke three constitutive relationships. First, an equation of continuity expresses the conservation of momentum. Second, a dynamic equation relates the motion of the particles of the medium to the forces influencing them. Third, an equation of state involves variables descriptive only of the medium itself, such as pressure, density, and temperature. The wave equation results from these three relations with the elimination of all but one of the dependent variables.

If the equation of state is a more general relation, allowing for complete description of the medium, the resulting wave equation becomes intractable. The simplest of all equations of state comprises a linear relationship between changes in pressure P in the medium and changes in density ρ of the medium, Eq. (1), where the subscripted quantities represent

$$p - p_0 = A \left(\frac{\rho - \rho_0}{\rho_0} \right) \qquad (1)$$

ambient values and the unsubscripted ones are instantaneous quantities, and A is a constant. Using this relation, along with the other constitutive equations, leads to the wave equation which, though simple and not descriptive of more involved phenomena, does describe much about acoustic wave propagation in such simple media.

An approximation to the general equation of state which provides for description of more details of acoustic wave propagation is accomplished by the addition of the quadratic term, an in Eq. (2). This

$$p - p_0 = A \left(\frac{\rho - \rho_0}{\rho_0} \right) + B \left(\frac{\rho - \rho_0}{\rho_0} \right)^2 \qquad (2)$$

equation of state expresses the deviation of the medium from linear pressure-density behavior and leads to nonlinear phenomena (which also result, in part, from increase in the amplitude of the wave). As a result of this nonlinearity, an originally sinusoidal wave becomes distorted as it propagates away from the source because regions of higher pressure travel at greater speeds than do regions of lower pressure. The amount of distortion increases with the intensity of the wave, the distance traveled, and the frequency of the fundamental. In the frequency domain, the distortion process implies the generation of harmonics at the expense of energy extracted from the fundamental. Thus, as an originally monochromatic wave travels away from the source, energy is transferred from the fundamental to the harmonics. The harmonics have zero amplitude at the source, but increase in amplitude with propagation distance until the rate of energy input becomes less than the energy output because of absorption and other loss phenomena. Since high-frequency components of an acoustic wave are absorbed more readily than are the lower-frequency components, the effective absorption of the distorted wave is greater than that of the monochromatic wave of the fundamental frequency. Because of the greater absorption, the rate of energy deposition within the propagating medium and the consequent heat development from a distorted wave is greater than that from an undistorted wave. For very intense waves the increased absorption of the wave leads to acoustic saturation, limiting the amplitude of the propagating acoustic radiation.

The relative magnitudes of the constants A and B in the equation of state provide a measure of the degree of nonlinearity of the medium. Measurement of the fundamental of the acoustic wave process and of the harmonics as a function of distance from the source provides a methodology for determining the ratio B/A.

Nonlinearity of biological media. It has been determined that the rate of absorption of ultrasound by biological media is substantial and that it is the macromolecular content, largely the protein content, that is mostly responsible for this high absorption. Thus, a convenient tissue model is an aqueous solution of easily available proteins such as hemoglobin and serum albumin. Determinations of the nonlinearity parameter B/A for such solutions show that it increases nearly linearly from a value of 5.2 for water (at 30°C) to 7.3 for a 40% solution (40 g of solute per 100 cm³ of solution) of bovine serum albumin and 7.6 for a 50% solution of hemoglobin. Whole blood contains about 12% hemoglobin and about 7% plasma proteins, together wth a small quantity of other solutes. It is more structured than aqueous protein solutions because of the cellular structures which enclose the hemoglobin, but may be considered homogeneous and isotropic on a macroscopic scale. The B/A value of whole blood is found to be very similar to that of a hemoglobin solution of the same dry-weight content, suggesting that the cellular structures do not contribute significantly to the nonlinearity of the suspension.

An excised tissue and its homogenate provide further opportunity to assess the influence of tissue architecture upon B/A. Determinations of B/A for porcine liver have yielded a value approximately 15% greater than the homogenized preparations of this same tissue. Homogenized liver has a B/A value which may be slightly higher than a protein solution of the same dry-weight content.

Finally, the excess B/A, that is, the contribution to the value of B/A due to the presence of the solute, per unit concentration, has been examined as a function of the molecular weight of the solute. The B/A value for solutions of dextrose, and also its polymer dextran, in the molecular weight range of about 10^2 to 10^6 daltons shows the B/A value to exhibit very little dependence on molecular weight.

The B/A values of the various biological media investigated thus far seem to suggest that the nonlinearity depends upon the amount of solute in a solution, and upon some architectural features of the structure, but not upon moleclar component size. Whether this comprises additional information that will alter with various pathological states to provide for more sophisticated diagnoses remains a question for further study and development.

For background information *see* SOUND; THERMOTHERAPY; ULTRASONICS in the McGraw-Hill Encyclopedia of Science and Technology.

[FLOYD DUNN]

Bibliography: R. T. Beyer and S. V. Letcher, *Physical Ultrasonics*, 1969; F. Dunn, W. K. Law, and L. A. Frizzell, Nonlinear ultrasonic wave propagation in biological materials, *Proceedings of the IEEE Ultrasonics Symposium*, pp. 527–532, 1981; L. E. Kinsler and A. R. Frey, *Fundamentals of Acoustics*, 2d ed., 1962; W. K. Law, L. A. Frizzell, and F. Dunn, Ultrasonic determination of the nonlinearity parameter B/A for biological media, *J. Acoust. Soc. Amer.*, 69:1210–1212, 1981.

Buildings

The major consideration in the structural design of high-rise buildings is the resistance to lateral forces (wind or earthquake) and the control of drift (lateral building movement). In recent years the use of a tube structural system has become popular for accomplishing these goals. The tube system consists of horizontal beams (spandrels) rigidly connected to closely spaced columns (10–12 ft or 3.0–3.7 m apart) around the perimeter of the building. The columns cause a frame action, with minimum floor-to-floor horizontal drift, simulating a shell type of structure which resists lateral forces.

The skin-tube-framed building behaves like a cantilevered hollow tube, with all sides of the building resisting forces regardless of the direction of lateral force, as opposed to a more flexible framework where only the frames in the direction of the applied force are activated. This is possible because the more rigid tube permits sufficient distribution of vertical shear forces to activate all the columns, to some degree, to resist loads in any direction.

Stressed-skin tube. A new concept has been developed which takes the tube system a step further. The development of the stressed-skin tube utilizes the facade of the building as a structural element which acts with the framed tube to make the building structure behave like a cantilevered hollow tube.

Because of the contribution of the stressed-skin facade, the framed members of the tube require less mass, and are thus lighter and less expensive. All the typical columns and spandrel beams are standard rolled shapes, minimizing the use and cost of special builtup members. The depth requirement for the perimeter spandrel beams is also reduced, and the need for upset beams above floors, which would

Fig. 1.   Stressed-skin panels being connected to the tube framing of the Dravo Tower, Pittsburgh, designed by Welton Becket Ass. (*Turner Construction Co.*)

encroach on valuable space, is minimized.

This structural system has been used on the 54-story Dravo Tower in Pittsburgh (Fig. 1). The exterior framed-steel tube works together with a unique exposed-steel stressed skin, eliminating the need for a separate, nonstructural exterior wall. In addition to reducing the sizes of primary-frame members, this system requires no additional lateral resistance from the core structure in the center of the building. The simplified core structure and the elimination of upset beams add more than 4% interior space to the building. The exterior framed steel tube, with columns spaced 10 ft (3.0 m) on center, provides all the necessary lateral resistance to maintain stresses within code allowables. However, a high-rise building with members which are sized to resist lateral loads only within allowable stress tolerances will have an unacceptable amount of drift; in this case, a drift equal to the building height divided by 275.

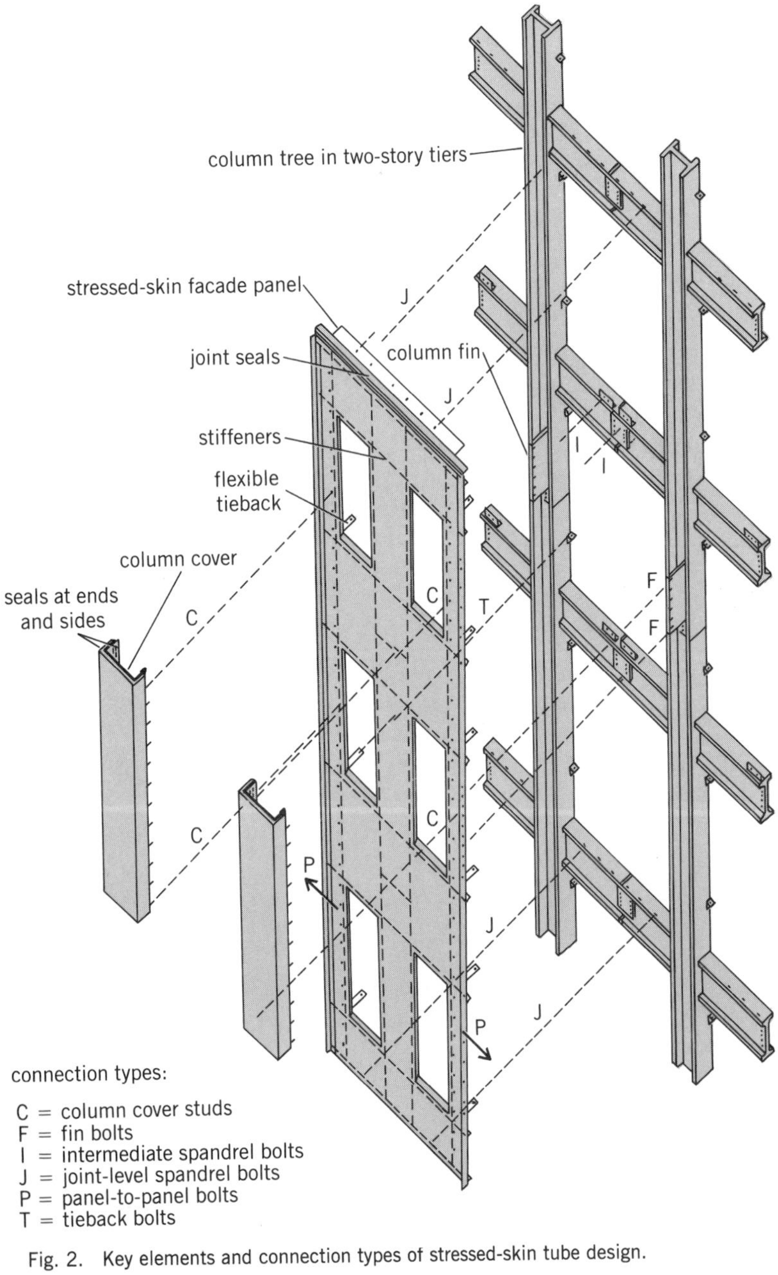

In a conventional framed steel tube, steel is added to the spandrels and columns, and often to the core, to provide the additional lateral stiffness required for acceptable drift, on the order of the height divided by 500. This lateral stiffness is accomplished in the Dravo Tower with the stressed-skin facade. Since the degree of drift is not related to structural safety, there are no code requirements dictating the amount of building drift. The Dravo Tower's core resists some of the gravity loads and the primary tube structure resists the rest of them, plus all lateral loads within code allowables, so the building is structurally sound even without the stressed skin. The steel skin does, however, add the additional structural stiffness necessary to eliminate an unacceptable amount of drift. Therefore, by maintaining this functional distinction between the steel skin and the rest of the structural framing in both design and analysis, the steel skin can be exposed, without fire protection.

Each stressed-skin panel is ¼–⁵⁄₁₆ in. (6–8 mm) thick, 36 ft (11.0 m) high, and 10 ft (3.0 m) wide, and typically contains six window openings that are 6 ft (1.8 m) high and nearly 2 ft (0.6 m) wide. The window detailing is similar to that used on airplanes: the glass is retained by H-shaped neoprene structural gaskets pressed into the edges of the window openings, and the corners of the window openings are curved to reduce local stresses. This is accomplished automatically in the factory by a computerized steel-cutting process, avoiding extra costs. Attention to detailing permits rectangular sheets of glass to be utilized even though there is a slight curvature in the corners of the window openings.

The structural system did not inhibit the architectural design of the Dravo Tower. Continuous rows of windows were not a design option because the skin serves a structural function. This was not a serious constraint, however, as glass areas are now typically held to about 30% of a building's facade to maximize energy conservation. The location of the window openings in the Dravo Tower is compatible with the architectural layouts for office space throughout the building.

Development of the stressed-skin tube concept required a considerable amount of research. Numerous mathematical models were necessary to determine the exact behavior of the stressed-skin panel and its interaction with the tubular frame. A full-scale test panel confirmed the complicated overall and local buckling analysis which was conducted on the stressed-skin panels. This concept will provide architects and engineers with another choice for an efficient way of resisting lateral loads in high-rise buildings.

Between the core and the exterior tube, 47-ft (14.3-m) clear spans are achieved with 24-in.-wide (61-cm) flange beams which contain mechanical distribution systems to minimize floor-to-floor heights. The

connection types:

C = column cover studs
F = fin bolts
I = intermediate spandrel bolts
J = joint-level spandrel bolts
P = panel-to-panel bolts
T = tieback bolts

Fig. 2.   Key elements and connection types of stressed-skin tube design.

floor typically consists of 2-in. (5-cm) metal decking, spanning 10 ft (3.0 m) between the beams, with 2½-in. (6-cm) stone concrete on top.

The net result of this integrated architectural and structural design is to provide cost-effective column-free interior space with a high ratio of net to gross floor area. The heart of the cost effectiveness lies in the dual function of the stressed-skin exterior wall, which acts as a facade and provides the stiffness required to minimize drift.

Design and construction. This approach requires an in-depth analysis. The entire structure must first be analyzed as a tube system without the skin to be sure that all stress levels are within code allowables. The tube structure must be analyzed again with the addition of the skin to be sure that an acceptable level of drift is obtained. In addition, the skin must be designed for a combination of in-plane forces, wind forces perpendicular to the skin, some compressive forces transferred from the columns due to column shortening, residual fabrication stresses due to cutting the window openings, thermal forces, and overall and local buckling. A unique connection system was developed to minimize thermal and compressive stresses in the skin, thereby optimizing its use as an efficient stressed skin.

Figure 2 shows an exposed view of the key components of the stressed-skin tube design. The framed steel tube is fabricated from two-story-high tree sections which are then connected on the building site. These consist of 24-ft-high (7.3-m), 14-in.-wide (36-cm) flange columns with 5-ft-long (1.5-m) spandrel beams, which are 30-in.-deep (76-cm), rolled, wide-flange shapes welded to each side of the columns.

Connection types are shown in Fig. 2. The stressed skin is connected by fin bolts only at its center. At the top and bottom of each stressed-skin panel, connections are made to the spandrels through a bent plate. This connection method minimizes the transmission of gravity forces from the columns into the panel as the columns shorten under load. It also permits the panels to expand and contract with temperature changes without developing any significant temperature stresses.

Because the panels do not accept any significant compressive forces due to column shortening, they can be erected at any time during building construction. If stresses could be transferred due to column shortening, the panels could not be erected until a significant amount of the frame was constructed. The design provides for efficient construction, with the erection of each panel directly following the erection of the frame section. In addition, the type of connections and their locations are designed so that they can all be made from the inside of the building without any scaffolding.

As shown in Fig. 2, there are additional connections: P, which transfer shear forces between panels, and T, which are flexible tiebacks that transfer horizontal loads between the panel and the frame without permitting the transfer of vertical loads. Stif-

feners, similar to those utilized on light-aerospace-type structures, are used to reinforce openings and prevent panel buckling.

For background information *see* Buildings; Structural analysis; Wall construction in the McGraw-Hill Encyclopedia of Science and Technology.

[RICHARD L. TOMASETTI]

Cellular receptors

Cellular receptors play an important role in the interaction of nerve cells with certain chemical compounds. This article describes the role of receptors in the brain in mediating the activity of opioids and in controlling appetite. The article also discusses the effects of alcohol on the functioning of brain receptors.

Opioid receptors. Opium has been used for medical and other purposes for thousands of years. Morphine, the active ingredient of opium, is an opiate alkaloid which, together with many synthetic derivatives, has been studied for almost 200 years with the goal of distinguishing the analgesic properties from the addictive properties. The marked stereospecificity of the action of these drugs, and their extremely high potency, suggested the existence of specific opiate receptors in brain tissue. The presence of such receptors was demonstrated in 1973 almost simultaneously in several independent laboratories.

The question of why there should be specific opiate receptor sites in brain tissue prompted the discovery less than 2 years later, of endogenous peptide ligands that bind to these receptors in the brain. It should be noted that since the term opiate refers only to alkaloids derived from opium, the receptors under discussion will be designated here as opioid, a term which encompasses all compounds with properties like the opiate morphine and therefore includes the endogenous peptide ligands.

The first endogenous opioids to be identified were methionine- and leucine-enkephalin, both of which are pentapeptides; this was followed by the discovery of other larger opioid peptides, such as β-endorphin and dynorphin. However, even before the demonstration of stereospecific receptors and endogenous ligands in the brain, the complex pharmacology of the opiate and opiatelike synthetic analgesics suggested the presence of more than one type of opiate receptor. The existence of mu (μ), kappa (K), and sigma (σ) opioid receptors was proposed, based on the pharmacological properties of different types of analgesic drugs. Subsequent to the discovery of endogenous ligands, another receptor type, the delta (δ) receptor, was proposed. Strong evidence now exists for the presence of distinct μ- and δ-receptors in the brain; comparable evidence for distinct K- and σ-receptors is not yet available.

Properties. Opioid receptors, like those for other neurotransmitters or hormones, are generally considered to be high-molecular-weight proteinaceous components of the plasma membrane of cells. These receptors are associated with membrane fractions of nerve tissue; their concentrations in synaptosomal cell fractions implies a localization in synapses.

The first phase of receptor activation is the bind-

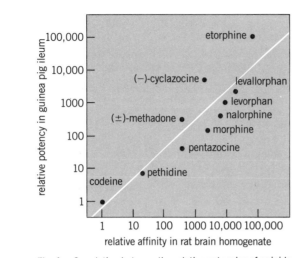

Fig. 1. Correlation between the relative potencies of opioids to reduce stereospecific binding of ³H-naloxone in rat brain homogenates (relative affinities for rat brain receptors) and the relative agonist potencies in the guinea pig ileum (relative ability to inhibit the electrically induced contractions of this tissue). Values are plotted on a logarithmic scale; correlation coefficient is $r = 0.974$ ($n = 11$). (After H. W. Kosterlitz and A. A. Waterfield, In vitro models in the study of structure-activity relationships of narcotic analgesics, Annu. Rev. Pharmacol., 15:29–47, 1975)

ing of an opioid to the receptors. This phase can be studied by binding radiolabeled opioid ligands (such as ³H-dihydromorphine or ³H-naloxone) to tissue preparations and measuring the results quantitatively in terms of the affinity of the ligand for the binding site. This process is saturable and also stereospecific in that the (−)isomers are generally many times more potent than the (+)isomers. Unlabeled compounds can be tested for receptor affinity by their ability to competitively inhibit the binding of labeled ligands to the receptors. Evidence for the participation of protein in the binding site is provided by the fact that proteolytic enzymes and sulfhydryl reagents disrupt the binding.

For functional receptor activation to occur, binding of the ligand to the receptor must be followed by an effector event which alters the state of the target cell. This event is quantified in terms of the efficacy of the ligand, and isolated strips of guinea pig ileum or mouse vas deferens provide simple systems for these measurements. Opioids are found to inhibit electrically induced contractions of these tissues, which contain both μ- and δ-receptors; the guinea pig ileum has a predominance of μ-receptors while the mouse vas deferens has a predominance of δ-receptors. Morphine has greater affinity for μ-receptors and is therefore more effective at inhibiting contractions of the guinea pig ileum; the enkephalins have greater affinity for δ-receptors and are more effective at inhibiting contractions of the mouse vas deferens.

Both morphine and the enkephalins are full agonists at these receptors; that is, they are able to produce 100% inhibition of the contractions. Partial agonists such as pentazocine are unable to produce a full effect regardless of dose. Antagonists are agents that have an affinity for the receptor (that is, bind to the receptor) but do not activate the effector event. By binding to the receptor they can prevent an agonist from activating the effector event. Naloxone is such an antagonist of opioid receptors. It will block the effect of the opioids on the vas deferens and ileum tissue preparations and will also block opioid-induced analgesia in the whole animal. Sodium ions inhibit the binding of most opioid agonists to receptors but increase the binding of antagonists such as naloxone. The effects of receptor activation may be mediated either by inhibition of adenylate cyclase or by inhibition of the membrane permeability to sodium.

One way to assess whether the binding represents physiologically functional receptors of interest is to determine whether the addition of a series of unlabeled ligands to the binding assay displaces the labeled ligand with relative potencies that are proportional to their relative biological potencies in the living organism. In many cases such a correlation has been observed for the opioid receptors (Figs. 1 and 2), but there are exceptions. Such exceptions can be explained in terms of differing efficacies or differing bioavailabilities which are not measured in binding assays.

Distribution and function. Both direct receptor

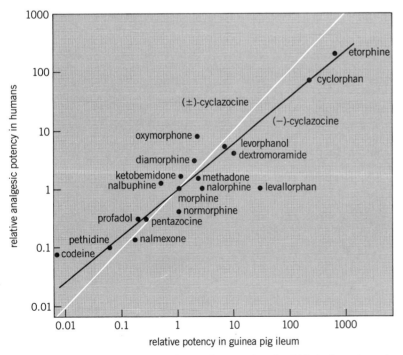

Fig. 2. Correlation between the relative agonist potencies of opioids in the guinea pig ileum and the relative potencies to produce analgesia in humans. The values are plotted on a logarithmic scale. The correlation coefficient is $r = 0.926$ ($n = 19$) minus codeine. (After H. W. Kosterlitz and A. A. Waterfield, In vitro models in the study of structure-activity relationships of narcotic analgesics, Annu. Rev. Pharmacol., 15:29–47, 1975)

binding and autoradiographic procedures have been employed to study the distribution of opioid receptors in the brains of several species, including human, monkey, and rat. The receptors are widely distributed throughout the brain but are not uniform in number or type among different areas, indicating multiple hormonal or transmitter roles subserving different functions. This is demonstrated by a few examples. The μ- and δ-receptors are for the most part localized in different areas throughout the brain, although several regions are enriched in both types. While there is evidence that both μ- and δ-receptors may mediate analgesic effects, the differential distribution suggests various other functions. A high level of binding is associated with the limbic system, in regions such as amygdala (mainly δ-receptors) and nucleus accumbens (mainly δ-receptors), which probably play a role in regulating emotional phenomena; opioids are known to influence the emotional response to pain. However, receptors in these regions may also be involved in euphoria and reward behavior. Relatively high levels of opioid receptors (mainly μ-receptors) and opioid peptides are also found in the hypothalamus. Many of these receptors probably reflect the role of opioid peptides in regulating neuroendocrine mechanisms. The high density of receptors found in thalamic regions (mainly μ-receptors), in periaqueductal gray matter (mainly μ-receptors), and in the substantia gelatinosa of the spinal cord (both μ- and δ-receptors) most likely reflect functions of the opioid peptides in the modulation of pain. The localization of opioid receptors in the nucleus tractus solitarius (both μ- and δ-receptors), vagal fibers (both μ- and δ-receptors), and area postrema probably reflect the role of opioids in the regulation of cardiovascular, visceral, and respiratory functions and in the induction of nausea and vomiting. Although the basal ganglia are generally associated with motor control, functions of high levels of opioid peptides and receptors in the corpus striatum (both μ- and δ-receptors), especially the globus pallidus, are not presently known.

The mixed agonist-antagonist benzomorphan compounds (for example, pentazocine) which appear to have low addiction liability have been thought to utilize K-receptors to produce analgesia. Correlation with K-activation was made, however, before recognition of δ-receptors, and these supposed K-selective agents are now also known to have significant δ-activity. Indeed, there is growing evidence that spinal analgesia not mediated by μ-receptors, previously associated with K-receptors, is more likely mediated by δ-receptors.

Opioid receptors are most likely intended to receive the various endogenous opioid ligands as neurotransmitters or neurohormones. Ligand-receptor specificity is not yet established, but it has been suggested that the enkephalins and a fragment of dynorphin may be natural ligands for δ-receptors and K-receptors, respectively.

The most obvious useful activity of opioid receptor activation is analgesia, and a corollary to this has been tolerance and physical dependence. Since the discovery of opioid receptors and their endogenous ligands, however, the possible role of these systems in neuroendocrine regulation, cardiovascular regulation, and the etiology of various psychiatric diseases has received greater attention. There is also a possibility that one or more of the endorphins play a certain role in regulating the immune system.

The existence of several distinct receptors provides promise for the development of new analgesics with lesser addiction liability and also for the development of selective new treatments for various diseases. Attempts are being made to develop chemical structures specific for each receptor type. For example the met-enkephalin analog, metkephamid, now being developed may be the first useful peptide-based drug to emerge from this work, although it is not an entirely specific ligand. Metkephamid has been shown to produce at least part of its analgesic effect by action on the δ-receptor; this has not been demonstrated for any other useful or potentially useful analgesic. It may also have lesser addiction liability than drugs of equivalent analgesic potency. The most unique aspect of the drug would be that it may not cross the placental barrier as readily as conventional analgesics, a fact which could give it an important place in obstetrics.

[ROBERT C. A. FREDERICKSON]

Brain glucoreceptors and appetite control. Food intake in mammals appears to be partly controlled for the maintenance of energy balance; that is, there appear to be mechanisms which increase eating in response to signals of metabolic need. Consequently, physiologists have suggested that the body must monitor the metabolism of a certain nutrient and that variation in the metabolism of this nutrient generates signals which alter appetite and food intake. Glucose (blood sugar) is the only nutrient for which unequivocal evidence of such a signal role exists. Drug- and hormone-induced reductions in the availability or the cellular metabolism of glucose (glucoprivation) cause increased appetite and eating in most mammals, including humans. There is evidence that the brain contains special nerve cells (glucoreceptors) which monitor the availability or utilization of glucose. Furthermore, recent experiments have shown that the glucoreceptors that mediate feeding and the release of epinephrine are located in an area of the brain far from where they were thought to be. Uncovering the location of these glucoreceptors should make it possible to determine how they participate in the control of normal eating.

Evidence for glucoreceptors. Electrical recordings have shown that there are many nerve cells in the brain which change their activity in response to increases or decreases in glucose availability. Such glucose-sensitive cells are located throughout the brain, but have been studied mostly in the hypothalamus. Unfortunately, simply finding cells whose electrical activity changes in response to altered glucose availability does not prove that these glucose-sensitive cells are the glucoreceptors which bring about increased eating and epinephrine secretion in response to glucoprivation.

Experiments performed during the late 1960s and 1970s showed that certain toxic analogs of glucose, such as gold thioglucose, damaged specific regions of the hypothalamus and that this damage was accompanied by overeating and obesity. Consequently, it was proposed that the obesity syndrome produced by gold thioglucose was the result of damage to glucoreceptors which control feeding. However, it has been subsequently found that the same kinds of lesions were also produced by chemicals unrelated to glucose. Furthermore, animals with lesions produced by gold thioglucose still increase their eating in response to reduced glucose availability and are able to secrete epinephrine in response to this stimulus. Therefore, if gold thioglucose does in fact destroy glucoreceptors, they are not the receptors which mediate increased eating or epinephrine secretion in response to glucoprivation.

The best evidence for the existance of brain glucoreceptors for control of eating and epinephrine secretion comes from behavioral and physiological studies utilizing glucose analogs which inhibit glucose metabolism. The most powerful of these analogs are 2-deoxyglucose (2DG) and 5-thioglucose (5TG). When either of these substances is injected into the cerebroventricular system, animals immediately increase their food intake and epinephrine secretion. When minute doses of 2DG or 5TG are injected into the brain they elicit the same physiological responses as larger doses that are injected intravenously, and it is logical to conclude that the glucoreceptors which mediate eating and epinephrine secretion are located in the brain.

Location of glucoreceptors. The hypothalamus is clearly an important participant in the control of eating and energy balance and, as previously mentioned, it does contain glucose-sensitive cells. However, direct injection of 2DG into various hypothalamic regions does not elicit eating. Therefore, there is no direct evidence to support the existence of hypothalamic glucoreceptors for feeding. In fact, the most recent evidence demonstrates that the glucoreceptors which mediate both increased feeding and epinephrine secretion are located in the hindmost portions of the brain (pons-medulla) and not in the hypothalamus or other forebrain structures. For example, in one experiment silicone plugs were placed in the cerebral aqueduct of rats, functionally isolating the forebrain ventricles (lateral and third ventricle) from the fourth ventricle of the hindbrain. Although 5TG-elicited eating and epinephrine secretion when infused into any brain ventricle prior to placement of the plug, only infusion into the fourth ventricle caused eating after plugging the cerebral aqueduct. Therefore, the receptors which mediate these responses must be located near the fourth ventricle in the pons-medulla. Additional data which support a hindbrain site for glucoreceptors mediating feeding and epinephrine secretion come from work with rats in which the forebrain is separated from the hindbrain by a cut passing through the brain, just caudal to the hypothalamus. In this preparation, called the decerebrate preparation, any neurally mediated response to any stimulus must be mediated by nerve cells caudal to the cut, that is, caudal to the hypothalamus. It was found that decerebrate rats secrete epinephrine in response to systemic 2DG administration. Furthermore, a recent report suggests that decerebrate rats increase the amount of a sugar solution swallowed if they have been made hypoglycemic with insulin. These findings, taken together with the results of the aqueduct obstruction and ventricular infusion experiments, clearly indicate that there are glucoreceptors in the hindbrain which mediate eating and epinephrine secretion. The existence of hypothalamic glucoreceptors for these two responses is not supported, and the precise location of the hindbrain glucoreceptors has not yet been determined.

Mechanisms. There are two different hypotheses regarding the events which serve as stimuli for the brain glucoreceptors. The first, a metabolic hypothesis, predicts that the glucoreceptors are sensitive to their own energy metabolism. According to this hypothesis, agents such as 2DG and 5TG cause feeding because they inhibit glycolysis within the glucoreceptor cells. The second and more recent hypothesis, the membrane receptor hypothesis, suggests that the glucoreceptor membrane interacts with glucose in the extracellular fluid. Reduced glucose interaction with the membrane of the glucoreceptor cells causes activation of these nerve cells and results in feeding.

Evidence for the metabolic hypothesis is compelling if not fully convincing. Firstly, all of the substances which have been shown to activate the glucoreceptors for feeding are antiglycolytic agents at their effective concentrations. Secondly, glucose analogs which are transported by the nerve cell membrane but which do not inhibit glycolysis are not effective for eliciting feeding and epinephrine secretion. Finally, ketone bodies, such as β-hydroxybutyrate, which are metabolized by enzymes independent of those responsible for glucose breakdown can, under certain circumstances, prevent feeding elicited by 2DG. Since the structures of glucose and β-hydroxybutyrate are very different, it seems unlikely that β-hydroxybutyrate would interact with a membrane glucoreceptor. It is more likely that β-hydroxybutyrate prevents 2DG-induced feeding by replacing calories lost due to inhibition of glucose utilization.

Support for the membrane glucoreceptor hypothesis rests mainly upon the finding that injection of alloxan into the brain ventricles permanently attenuates or abolishes feeding in response to 2DG. Alloxan temporarily antagonizes activation of taste-bud sugar receptors by glucose. The activation of these taste receptors is thought to occur through a membrane-receptor interaction. Furthermore, alloxan destroys the insulin-secreting B cells in the pancreas and this action is antagonized by glucose. One mechanism by which alloxan may destroy B cells is through toxic interaction with proposed membrane glucoreceptors in B-cells. By analogy, it has been proposed that brain glucoreceptor cells are destroyed by interaction of alloxan with their membrane receptors. In support of this proposal, it has been shown that infusion of glucose prevents alloxan from abolishing eating in response to 2DG. Although evidence supporting the metabolic hypothesis is currently stronger, neither hypothesis can be ruled out until the brain glucoreceptors are precisely located and their cellular characteristics studied. Demonstration that the glucoreceptors which mediate eating are in the hindbrain dramatically reduces the area of search for these receptors and brings closer the goal of understanding how brain glucoreceptors participate in normal and abnormal eating behavior.

[ROBERT C. RITTER]

Alcohol effects on brain receptors. Acute ethanol intoxication leads to inebriation, caused by the impairment of central nervous system activities. On the other hand, chronic alcoholism, in addition to inducing certain behavioral changes, is characterized by tolerance and physical dependence on alcohol. Moreover, alcohol exhibits a potent interaction with many psychoactive drugs. Acute ethanol intoxication often potentiates the effects of agents which depress the activity of the central nervous system, whereas chronic alcoholism generally induces resistance to these effects. In view of the fact that, to a large extent, brain functions seem to be modulated by neurotransmitters, the effect of acute and chronic ethanol intoxication on receptors for endogenous and exogenous ligands has been the subject of numerous studies. However, at the present time much of the data is contradictory or incomplete, and a clear understanding of the functional and anatomic relationships of these receptors has not yet emerged. It is therefore appropriate to view studies of the effects of ethanol on membrane receptors of the central nervous system with caution.

Because the function of many membrane-associated proteins is affected by the physical state of membrane lipids, receptor function in the brain might similarly be affected, since receptors reside in the plasma membrane. Acute exposure to alcohol perturbs all biological membranes, including those of the brain; specifically, the presence of ethanol increases membrane fluidity. By contrast, chronic ethanol intoxication leads to adaptive changes in membrane lipids, such that these membranes become more rigid, that is, less fluid. Since the functions of membrane proteins, such as enzyme activity, ion transport, and receptor binding, are thought to be optimal at the normal membrane fluidity, any change in fluidity might interfere with such functions. Moreover, ethanol intoxication has been shown to decrease protein synthesis in a number of systems and thus may affect the synthesis of membrane receptors. The sensitivity of membrane receptors may also be influenced either directly by the presence of ethanol or indirectly by changes in the concentration of neurotransmitters. This sensitivity may be modulated by the density or number of receptors, their binding affinity, or the coupling of receptor binding to intracellular functional activity.

Adrenergic receptors. The sensitivity of norepinephrine receptors has been measured as the norepinephrine-sensitive adenylate cyclase activity. Chronic ethanol treatment of rats causes an increase in the basal activity of adenylate cyclase, and a decrease in its sensitivity to norepinephrine. After withdrawal of ethanol, the sensitivity of the β-adrenergic receptors was reported to be enhanced. The binding of the β-adrenergic receptor ligand, dihydroalprenolol, was decreased in brains of ethanol-treated rats immediately after withdrawal, but was increased 2–3 days later, the change being attributed to an increase in the number of receptors. It was suggested that the increase in β-adrenergic receptor binding contributes to the hyperactive state observed in later stages of alcohol withdrawal in rats. The affinity and number of α-adrenergic receptors in the brain were unchanged in rat cortex during withdrawal from chronic alcohol treatment, although minor changes were found in the midbrain and brainstem.

Cholinergic receptors. Acute ethanol intoxication has been reported to inhibit the release of acetylcholine in the brain. Changes in the availability of this neurotransmitter within the synapse can lead to changes in the receptor density or the kinetic characteristics of the receptor. During ethanol withdrawal, the specific binding of the acetylcholine analog quinuclidinyl benzylate (QNB) was increased, apparently owing to an increase in QNB binding sites rather than to a change in binding affinity. However, this increase in the number of muscarinic cholinergic receptors was very short-lived. Chronic ethanol treatment did not lead to any change in muscarinic cholinergic receptors in mouse striatum, but dic lead to an increased number of such receptors in other areas of the brain. Functional correlations in this area are not well understood, and further work is clearly required.

Dopamine receptors. Ethanol, in concentrations up to 500 mM, did not change the density or binding affinity of the dopamine receptor in mouse striatum in tissue culture. GTP increased the affinity of the receptor for dopamine fivefold in the presence and absence of ethanol. The effect of chronic ethanol intoxication on dopamine receptors in the brain is not clear; subsensitivity, no change, and supersensitivity have all been reported. On the basis of the response of ethanol-withdrawn animals to apomorphine, it was suggested that the decreased effects of neuroleptics and apomorphine on dopamine synthesis after ethanol administration result from a change in postsynaptic receptors, which reside on neurons of the feedback loop controlling dopamine synthesis. On the other hand, ethanol treatment may alter the dopaminergic neurons themselves, so that they no longer respond in a normal fashion to regulatory stimuli.

Opiate receptors. The opiate antagonist, naloxone, has been reported to block certain ethanol-induced behavioral changes and the effects of ethanol on calcium ion metabolism. Therefore, changes in opiate receptor function might be related to the development of tolerance to certain effects of ethanol. High (nonphysiological) concentrations of ethanol inhibited high-affinity binding of dihydromorphine (DHM), whereas concentrations of ethanol which can be achieved in the blood of humans or experimental animals were reported to have a slight stimulatory effect on such binding. Thus, the concentration of the ligand used to assess the effects of ethanol is very important. Because other alcohols inhibit DHM binding in a way which is not linearly related to their membrane-water partition coefficient, the ability of ethanol to alter opiate binding may not be exclusively dependent on its membrane-perturbing activity. During withdrawal from chronic ethanol intoxication, the dissociation constant (K_d) of DHM was increased for the high-affinity morphine-binding site. The effects of chronic ethanol treatment on the high-affinity morphine-binding site may involve an effect caused by sodium ions. In the brains of control animals, sodium ions tended to increase the K_d value of the high-affinity morphine receptor, whereas in comparable tissue from ethanol-treated animals there was no change in K_d. Neither the affinity nor the number of low-affinity morphine-binding sites was changed. The low-affinity morphine-binding site has been suggested to be identical to the high-affinity binding site for enkephalin. The fact that chronic ethanol treatment was reported not to have any effect on high-affinity binding or the number of receptors for enkephalin in the caudate nucleus of the mouse supports this concept. Naloxone was also reported to antagonize the analgesia produced by salsolinol, the condensation product of dopamine and acetaldehyde, the latter being the first metabolic product of ethanol oxidation. Both salsolinol and tetrahydropapaveroline, another condensation product, inhibit naloxone binding in brain homogenates. These data suggests that a mechanism involving opiate receptors may be involved in the possible effects of these condensation products.

Benzodiazapine–γ-aminobutyric acid receptor complex. Whereas the interaction of ethanol and benzodiazapines is well known, the effects of ethanol on the benzodiazapine receptor are less well understood. The complex can be visualized as a γ-aminobutyric acid (GABA) receptor, a benzodiazapine receptor, and a picrotoxinin-barbiturate receptor. The chloride ion channel, whose opening is regulated by the binding of GABA agonists to the GABA receptor, is associated with these receptors. Depressant drugs potentiate chloride ion permeability, while excitants block this channel. One group has reported that acute ethanol treatment increases the density of GABA-receptor sites and that after chronic ethanol treatment a tolerance to this effect occurs. It was speculated that the effect of ethanol on the complex might be mediated indirectly through the picrotoxinin site. Acute treatment of mice with ethanol

resulted in an increased density of low-affinity GABA-binding sites in whole brain, whereas chronic ethanol treatment was said to decrease the number of low-affinity sites. Ethanol had no effect on benzodiazapine receptors in an isolated membrane fraction, but increased the binding of benzodiazapine to the detergent-solubilized receptor. However, these studies were done at temperatures of 0–4°C. Other studies indicate that at physiological temperatures the presence of ethanol increases the binding of benzodiazapine to the membrane-bound receptor in tissue culture. Some investigators have reported that chronic ethanol administration does not lead to any changes in the affinity or density of benzodiazapine-binding sites in any area of the brain, whereas others have found decreased benzodiazapine binding after chronic ethanol administration to experimental animals. Again, these studies were conducted at low temperature. Studies at physiological temperatures indicate that chronic ethanol treatment leads to decreased benzodiazapine binding in the membrane-bound receptor. Enhancement of the binding of benzodiazapine might lead to facilitation of GABA-mediated inhibitory transmission and might be responsible for some of the central results of alcohol ingestion, including sedative, anxiolytic, and muscle-relaxing effects.

Glutamate binding. Glutamate has been suggested as an excitatory neurotransmitter in the brain. Chronic ethanol administration to rats was reported to increase glutamate binding in a brain synaptosomal fraction because of an increase in maximal binding capacity, not because of a change in binding affinity. This increase was said to represent an adaptive response of the brain to the depressant effects of ethanol.

It is likely that both acute and chronic alcohol intoxication exert significant effects on neurotransmitter receptors in the brain. Data are still fragmentary and controversial, and much additional work remains to be done before a coherent picture emerges.

For background information *see* ALCOHOLISM; ANALGESIC; ENDORPHINS; HUNGER; MORPHINE; NEUROBIOLOGY; OPIATES; SYNAPTIC TRANSMISSION in the McGraw-Hill Encyclopedia of Science and Technology.   [EMANUEL RUBIN]

Bibliography: R. C. A. Frederickson, Endogenous opioids and related derivatives, in M. Kuhar and G. W. Pasternak (eds.), *Analgesics: Pharmacological and Clinical Perspectives*, in press; R. C. Ritter, P. G. Slusser, and S. Stone, Glucoreceptors controlling feeding and blood glucose: Location in the hindbrain, *Science*, 213:451–453, 1981; S. Ritter, J. M. Murnane, and B. L. Ladenheim, Glucoprivic feeding is impaired by lateral, or fourth ventricular alloxan injection, *Amer. J. Physiol.*, 243:R312–317, 1982; E. Rubin and H. Rottenberg, Ethanol-induced injury and adaptation in biological membranes, *Fed. Proc.*, 41:2465–2471, 1982; E. J. Simon and J. M. Hiller, Opioid peptides and opiate receptors, in G. J. Siegel et al. (eds.), *Basic Neurochemistry*, 3d ed., pp. 255–268,

1981; P. G. Slusser and R. C. Ritter, Increased feeding and hyperglycemia elicited by intracerebroventricular 5-thioglucose, *Brain Res.*, 202:474–478, 1980; R. Tabakoff and P. L. Hoffman, Alcohol and neurotransmitters, *Alcohol Tolerance and Dependence*, pp. 201–226, 1980.

Chlorophyll

Although the vast majority of research on the chemical and photochemical properties of chlorophyll has been undertaken in organic solvents, this is somewhat inappropriate since essentially all chlorophyll in nature is associated with proteins and is not free in solution. Until recently this approach was due to the lack of well-characterized chlorophyll-protein complexes isolated from plants or photosynthetic bacteria. Recently, many natural chlorophyll-protein complexes have been isolated and characterized, and a number of synthetic chlorophyll-protein complexes have been prepared as models for the natural system.

The chemical structures of chlorophyll and bacteriochlorophyll are shown in Fig. 1. These intensely colored macrocyclic compounds have characteristic electronic absorption and circular dichroism spectra, and redox properties in organic solvents. These properties are usually very different for the identical chromophores in natural chlorophyll-protein complexes. In this context, chlorophyll should be viewed as a prosthetic group in a protein, much like heme, retinal, iron-sulfur clusters, and so on. In each of these, the protein environment modulates the properties of the prosthetic group, without altering it chemically. At the present time a number of possible molecular mechanisms can be identified which can account for some or all of the characteristic properties of chlorophyll or bacteriochlorophyll when it is associated with proteins; these will be discussed in this article.

Natural chlorophyll-protein complexes. The photosynthetic apparatus and its associated chlorophyll are functionally divided into antenna chlorophyll and reaction-center chlorophyll. In both systems, and in all organisms studied to date, most if not all of the chlorophyll is intimately associated with proteins. Most chlorophyll-protein complexes are integral membrane proteins; that is, they tend to be highly hydrophobic. The greatest portion of the chlorophyll (95% or more) in any organism is associated with the antenna complex, and serves to absorb and transport excitation energy passively to the reaction center, where the actual photochemistry of photosynthesis takes place. In this way, the organism maximizes the absorption of incident light while minimizing the number of reaction-center complexes. The latter involve many components which participate in electron transport, and it would be inefficient to duplicate this complicated system any more than necessary. The functional distinction between antenna and reaction-center chlorophyll is not to be confused with a chemical distinction: the chlorophyll or bacteriochlorophyll chromophores are chemically identical in antenna and reaction-center complexes, but their properties are altered by the differences in their environment.

Antenna chlorophyll. There do not appear to be universal chlorophyll-containing proteins, conserved from species to species. At the present time,

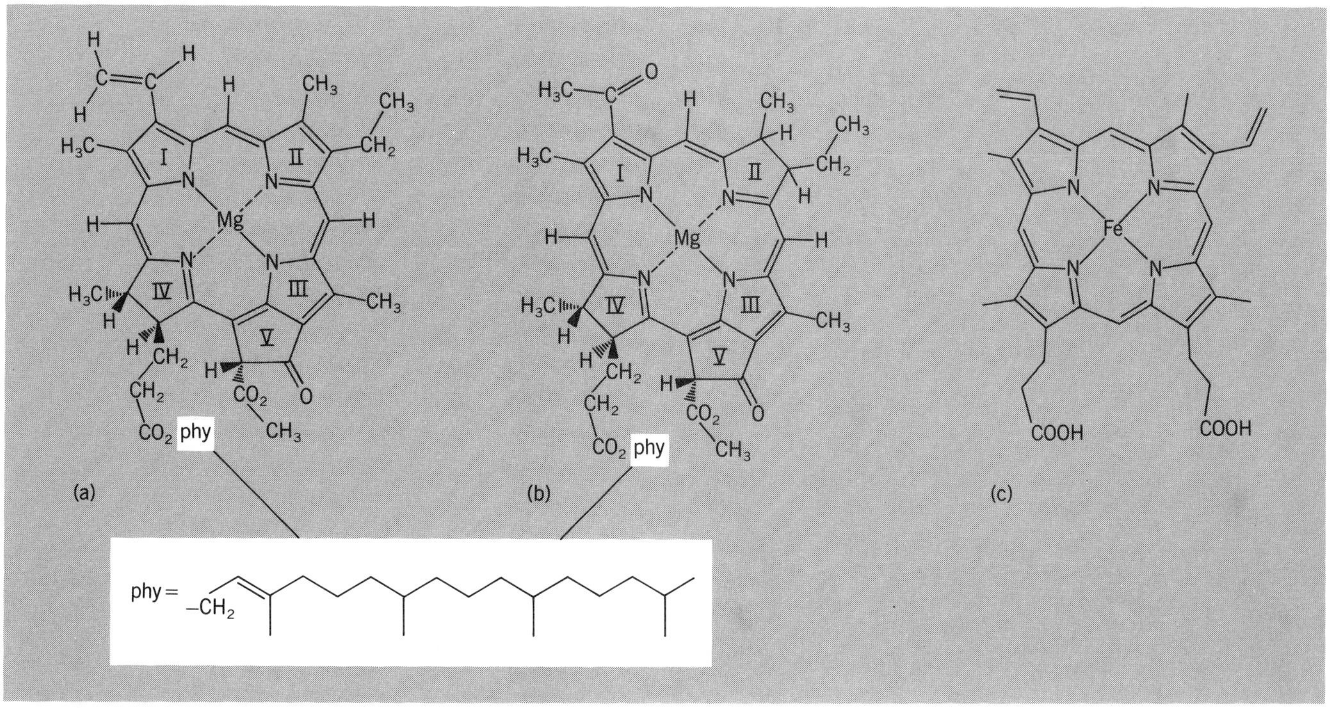

Fig. 1. Chemical structures of (a) chlorophyll *a*; (b) bacteriochlorophyll *a*; (c) heme.

several different complexes have been isolated from a variety of species at various levels of purity. The best-characterized complex is the unusual water-soluble protein from *Prosthecochloris aestuarii*, which is believed to be an antenna complex in this green bacterium. This protein has been crystallized and studied in detail by x-ray diffraction. The protein is a trimer; each monomer unit contains seven bacteriochlorophylls. These bacteriochlorophylls have no obvious order in the protein; that is, there is no evidence for translational or rotational symmetry among the bacteriochlorophylls. This is interesting because it might have been expected that an array of bacteriochlorophylls with some obvious structure would lead to particularly efficient transport of excitation energy. The bacteriochlorophylls in this complex all have nearest-neighbor bacteriochlorophylls within 1.2 nanometers. The electronic absorption maxima for the characteristic lowest energy bands are considerably red-shifted relative to those of the chromophore itself in organic solvents. Unfor-

tunately, it is very difficult to understand this spectrum because all seven bacteriochlorophylls are observed simultaneously, and there is no reason to believe they are spectroscopically identical.

Reaction-center chlorophyll. Reaction-center protein complexes have been studied in greatest detail, as these complexes are the site of the primary energy-transducing steps of photosynthesis. Reaction-center complexes have been isolated from many species of bacteria, as well as from green plants. These complexes from the purple photosynthetic bacteria are by far the best characterized, and will be used to illustrate features of bacteriochlorophyll-protein complexes. The absorption spectrum of the best-characterized reaction center from *Rhodopseudomonas spheroides*, R-26 mutant, is shown in Fig. 2a. This reaction center contains four bacteriochlorophylls and two bacteriopheophytins (bacteriochlorophyll in which the central magnesium atom has been replaced by two hydrogens) associated with two proteins. This spectrum is compared with that of an equimolar mixture of these two chromophores in an organic solvent (Fig. 2b).

It is immediately apparent that the reaction-center proteins substantially modify the absorption characteristics of both chromophores. The origins of these shifts have been the subject of great interest. At the simplest level, the absorption feature at 870 nm (Fig. 2a) has approximately the intensity associated with two bacteriochlorophylls and is often described as a dimer or a special pair. This feature has been shown to be associated with the primary electron donor in these reaction centers. In Fig. 2a, the feature at 800 nm is also associated with two bacteriochlorophylls, and the feature at 760 nm is associated with the two bacteriopheophytins. Although this assignment is widely agreed upon, there is relatively little direct evidence to support it. This is especially true because none of the absorption features corresponds closely to features of known chemical species in organic solvents. At the present time there is no structural information about chlorophyll- or bacteriochlorophyll-protein interactions in reaction centers from any organism.

Synthetic chlorophyll-protein complexes. One way to assess experimentally the role that proteins may play is to build model systems which duplicate some of the properties of the natural system. Unfortunately, it has not proved to be possible at this time to extract chlorophyll or bacteriochlorophyll from a native complex and reconstitute it, but it has been possible to prepare synthetic chlorophyll-protein complexes. Inspection of the chemical structures of the chlorophylls in Fig. 1 shows that they are not dissimilar to porphyrins (for example, heme), especially if the long hydrocarbon tail (phy) is removed. Heme is an intensely red chromophore which is responsible for the red color of blood, in which it is complexed with the protein apohemoglobin, and for the color of muscle, in which it is complexed with the protein apomyoglobin. These oxygen-carrying proteins are perhaps the most studied proteins,

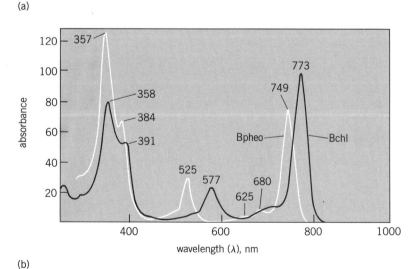

(a)

(b)

Fig. 2. Comparison of the electronic absorption spectrum of photosynthetic reaction centers with the absorption spectra in organic solvents of the bacteriochlorophyll components. (*a*) Spectrum of reaction centers from *Rhodopseudomonas spheroides*, R-26 mutant. (*b*) Spectra in diethyl ether of an equimolar mixture of bacteriochlorophyll *a* (Bchl) and bacteriopheophytin *a* (Bpheo).

and have been characterized by every conceivable method, including x-ray crystal structures. Noting the similarity between the chlorophylls and heme, S. Boxer and K. Wright have substituted many chlorophylls (with the long hydrocarbon tail removed) for heme in both apomyoglobin and apohemoglobin. These intensely green, semisynthetic proteins, called chloroglobins, provide the first structurally defined 1:1 complex between chlorophyll and a protein.

The electronic absorption spectra of chlorophylls and bacteriochlorophylls in the chloroglobin complexes are nearly indistinguishable from those in organic solvents. Thus, in this particular protein environment, there is very little perturbation by the protein of the electronic structure of the chromophore. Surprisingly, the circular dichroism in the chloroglobin complexes is very different from the intrinsic chlorophyll circular dichroism. Circular dichroism is a measure of the difference of absorption of right and left circularly polarized light by a sample. In order for a molecule to exhibit circular dichroism activity it must be asymmetric. This can arise by the presence of one or more asymmetric centers in the molecule (as is the case for chlorophyll) or by asymmetry in the environment around the chromophore (as may arise in a protein). These effects are essentially additive. Apparently the asymmetry imposed by the protein environment can overwhelm the intrinsic circular dichroism. This result shows that the circular dichroism of natural chlorophyll-protein complexes depends largely on the particular protein, and it is consequently very difficult to interpret. Single crystals of the chloroglobins have also been prepared and studied spectroscopically in detail.

Mechanisms of protein-induced effects. Several possible mechanisms can be suggested to explain the spectral shifts of chlorophyll in proteins.

Specific ligation. The central magnesium atom in the chlorophylls is always five-coordinate, and can be six-coordinate in some cases. The fifth or sixth ligand, or both, may be provided by the protein, and this ligation can alter the position of the central magnesium atom, perturbing the electronic structure.

Exciton interactions. When two or more chromophores are in proximity, there can be a variety of interactions between molecules. One of the simplest is the resonance or exciton interaction in a molecular dimer, leading to a splitting of each monomer energy level into two dimer levels. This gives rise to two absorption bands in the dimer in place of single bands in the monomer, though the dimer bands may not be well resolved. Furthermore, depending on the specific geometry of the dimer, either the higher or lower energy band may be very weak, giving rise to an apparent shift in the absorption. Because all natural chlorophyll-protein complexes studied to date have more than one chlorophyll and bands are always shifted, exciton effects may be very important in determining the absorption properties. This effect is often invoked to explain the extremely red-shifted absorption of the special-pair dimeric electron donor at 870 nm in reaction centers, as shown in Fig. 2a, though this is not the only explanation for the band shift. Although this is not really a direct effect of the protein, the exciton interaction requires specific positioning of the chromophores by the protein.

Electrostatic interactions. Proteins contain many ionizable amino acid side chains, whose charges depend on pH, ionic strength, and local interactions. When a molecule like chlorophyll is excited, there is a change in the charge-density distribution within the chromophore. This change may be either stabilized or destabilized by the presence of a local charge in the protein, depending on the sign of the change in charge density, the sign of the local charge, and their relative locations. Electrostatic effects of this sort have been invoked to explain band shifts in the visual pigments by some workers, and similar effects may prove to be very important in chlorophyll-protein complexes as well. This is an especially attractive mechanism because the wide range of spectral shifts can be easily accommodated by positioning charged groups at different distances and positions around the chromophore. Furthermore, there may be important chemical consequences, as the primary photochemistry of photosynthesis generates charged intermediates which may be stabilized by nearby charged groups.

Any one, or all, of these mechanisms may be operative in any particular chlorophyll-protein complex. The precise mechanism is not yet understood in any complex which has been isolated to date; however, the role of each mechanism will become clear as detailed structures become available for these complexes.

For background information *see* CHLOROPHYLL; OPTICAL ACTIVITY; PHOTOSYNTHESIS in the McGraw-Hill Encyclopedia of Science and Technology.

[STEVEN G. BOXER]

Bibliography: R. K. Clayton and W. R. Sistrom (eds.), *The Photosynthetic Bacteria*, 1978; B. W. Mathews and R. E. Fenna, Structure of a green bacteriochlorophyll protein, *Acct. Chem. Res.*, 13:309–317, 1980; K. Nakanishi et al., An external point-charge model for bacteria rhodopsin to account for its purple color, *J. Amer. Chem. Soc.*, 102:7945–7946, 1980; K. A. Wright and S. G. Boxer, Solution properties of synthetic chlorophyllide- and bacteriochlorophyllide-apomyoglobin complexes, *Biochemistry*, 20:7546–7556, 1981.

Clinical pathology

Fine-needle aspiration biopsy, a diagnostic method in which cells and small tissue fragments are obtained for microscopic examination by means of a small-bore needle, has gained wide acceptance as an accurate, reliable, and safe technique. It is utilized principally in the evaluation of neoplastic disease. Initially considered a method of limited application and often used only in evaluating lesions that could be directly visualized and palpated, it is now employed at virtually any body site and has found particular application in the diagnosis of intrathoracic and intraabdominal tumors. The expanded application of this simple technique has been made possible by the development of advanced methods of localizing lesions, including biplanar fluorography, ultrasound, and computer-assisted tomography. In addition, improved biopsy instruments have been developed which provide better samples. The expertise necessary for interpretation of these samples has been made possible through the recently expanded knowledge of the cytopathologic manifestations of disease.

Historical perspective. The concept of obtaining material for diagnosis through needle aspiration is not new. Isolated reports of successful needle biopsy date back to the nineteenth century, and the first reports of patients in whom a diagnosis was made by this means were published in 1930. For many years the method made little progress toward becoming part of standard diagnostic methods. Pathologists were reluctant to accept the idea that accurate diagnoses could be rendered on such small samples, and clinicians were fearful that the procedure would promote the spread of tumor or result in tumor seeding of the needle tract. The technique was more widely accepted in Europe than in North America, and it was there, particularly in Scandinavia, that the method utilizing small-bore needles was developed.

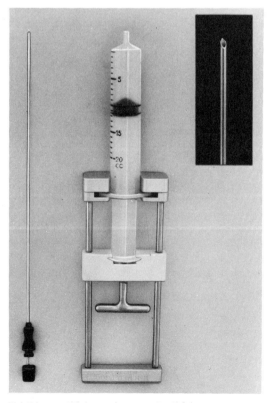

Fig. 1. Equipment utilized for fine-needle aspiration biopsy, including syringe, syringe holder, and "chiba" needle. Inset shows magnified view of needle tip.

Numerous cases were reported in which diagnostic needle biopsy was performed without serious complications, and concerns regarding accuracy and safety gradually dissipated.

Biopsy methods. The biopsy needle that is commonly utilized is thin-walled and flexible. It has an inner stylet and an outside diameter of 0.06–0.07 mm (0.024–0.028 in.) (Fig. 1). The length of the needle varies with the site to be sampled: short needles are utilized for superficial lesions and long needles for deep-seated ones. The needle, with stylet, is placed into the lesion to be sampled, and is guided by palpation or radiologic means, depending on the tissue to be biopsied. After placement, the stylet is removed, a syringe is attached to the hub, and negative pressure is applied to aspirate cellular material into the needle barrel. The needle and the attached syringe are removed, and the aspirated material, which consists of usually one or two drops, is expelled onto microscopic slides. It is immediately fixed and stained for microscopic examination.

Indications and advantages. The most frequent application of needle aspiration biopsy is in the evaluation of tumors for the diagnosis or exclusion of cancer. It also is applied in the evaluation of inflammatory disease to establish a diagnosis and to obtain material for identification of infectious organisms. Several advantages are offered by this diagnostic method. The technique provides a highly accurate diagnosis which can be performed rapidly, often on an outpatient basis, thereby saving unnecessary hospitalizations. There is minimal risk and insignificant trauma to the patient. The procedure is low in cost, particularly when compared to alternative methods for biopsy of intrathoracic or abdominal lesions, which often require thoracotomy or laparotomy and a prolonged hospital stay. The technique requires no anesthesia, and patients who are poor surgical risks can be safely biopsied. Also, patients who are clinically inoperable can receive appropriate chemotherapy or radiotherapy without needing major surgery performed solely for diagnostic purposes.

Complications. The complication rate for needle aspiration biopsy with current techniques and instruments is quite low. One study, for example, reports no serious complication in over 8000 aspirations. Significant hemorrhage is seldom encountered, even if major blood vessels are punctured, because of the small diameter of the biopsy needle. Pneumothorax, the accumulation of air in the pleural space, is a relatively common complication of lung biopsy and occurs in about 25% of patients. In needle biopsy, however, only about 5% of cases will exhibit symptomatic pneumothorax and require treatment. Biopsy-induced infection is a theoretical complication; it is avoided by utilizing sterile technique. Implantation of tumor cells along the needle tract after biopsy is exceedingly rare and long-term studies indicate minimal risk. Morbidity following needle biopsy is also rare.

Diagnostic interpretation. The accurate diagnostic interpretation of needle aspiration biopsies requires close cooperation between the radiologist, clinician, and pathologist. A representative sample must be obtained from the lesion in question, and the cellular material adequately preserved and stained for microscopic examination. The interpreting pathologist must be knowledgeable in both the cellular and tissue manifestations of neoplastic and non-neoplastic disease, and evaluate the specimen in relationship to the patient's entire medical situation.

The aspirated cellular material is prepared for examination in a variety of ways. The most frequently utilized technique involves the spreading of a drop of aspirated material directly onto glass microscope slides. These are then immediately placed in 95% ethanol for fixation, or are air-dried and stained, either with Papanicolaou stain or Giemsa stain. Many laboratories utilize both stains on separate slides. If the particular case warrants it, the wide variety of histochemical techniques applicable to tissue analysis can be applied. In addition, aspirated material can be prepared for electron microscopic or immunohistochemical evaluation.

The diagnosis is based on well-defined cytomorphologic characteristics. In the presence of malignancy the biopsy sample is generally highly cellular. Malignant cells, being less cohesive than normal cells, are present as numerous single cells and as aggregates. The frequent presence of necrosis within malignant tumors is often reflected by the occurrence of cellular debris. Other criteria of malignancy include disorganization of cell arrangement; marked variation in cell and nuclear size and shape; increased nuclear-cytoplasmic ratio; abnormal, irregular, or thickened nuclear envelopes; irregularities of nuclear chromatin distribution; and enlarged, often irregular nucleoli. Increased numbers of mitoses or abnormal mitoses may also be evident. The actual cytologic features vary with the organ and type of tumor biopsied (Fig. 2).

Accuracy. With currently applied methods, needle aspiration biopsies are highly accurate in the diagnosis of malignant tumors. In most reported series of lung biopsies, for example, 85–90% of patients with cancer were accurately diagnosed. No cytologic diagnosis is obtained in 10–15% of patients with malignancy. These false negative results are almost always on the basis of technical sampling problems and not due to a failure to recognize abnormal cells which are present. It is, therefore, important to emphasize that in the absence of a definitive diagnosis of a benign lesion by needle aspiration, the lack of malignant cells does not exclude the possibility of cancer. The patient with a nondiagnostic fine-needle aspiration biopsy usually requires further diagnostic evaluation.

Summary. Fine-needle aspiration biopsy is a highly accurate, rapid, safe, relatively inexpensive, and expeditious diagnostic technique. It has become widely accepted as a valuable addition to the meth-

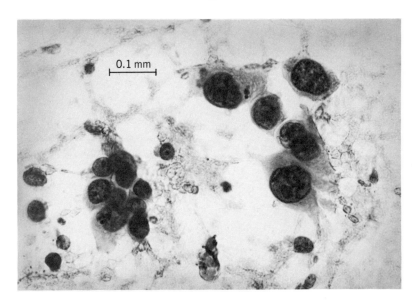

Fig. 2. Malignant tumor cells (stained with Papanicolaou stain) in aspiration biopsy of squamous cell carcinoma of lung.

ods available for the evaluation and management of both neoplastic and non-neoplastic disease.

For background information *see* CLINICAL PATHOLOGY in the McGraw-Hill Encyclopedia of Science and Technology.

[THOMAS A. BONFIGLIO]

Bibliography: T. A. Bonfiglio, Fine needle aspiration biopsy of the lung, *Pathol. Annu.*, 16:159–180, 1981; W. J. Frable, *Thin-Needle Aspiration Biopsy*, 1982; T. S. Kline, *Handbook of Fine Needle Aspiration Biopsy Cytology*, 1981.

Coal

Over 800 million tons (725 million metric tons) of coal is mined annually in the United States, of which approximately 85% is used to supply one-fourth of the nation's energy requirement. Most of this coal is consumed in electric power plants and industrial furnaces, where the coal is combined with air and burned, liberating energy but also generating by-products which can pollute the air.

Coal combustion. This involves processes that span a wide range of designs from small steam boilers with combustion chamber dimensions of several meters to large electric power plant furnaces 10 to 15 stories high. Methods of preparing the coal for fuel and the conditions under which it burns also vary widely. To illustrate the basic features of one important type of coal combustion system, consider the combustion process of a large, modern electric power plant. First, the coal is pulverized to a consistency resembling talcum powder. The coal particles are then blown into the furnace, where they are mixed with air and burned in suspension, generating peak flame temperatures near 1650°C (3000°F). During the combustion process the organic coal constituents burn, and the noncombustible coal mineral matter is released as small fly ash particles which remain mostly in suspension and are carried from the furnace by the gaseous exhaust products. During the residence in the furnace and subsequent passage through various heat-extraction stages, the gases are cooled to about 150°C (300°F). Afterward they may pass through one or more exhaust-gas cleanup devices before they exit through the smokestack.

The complex physical and chemical phenomena that make up the coal combustion process, and the various solid and gaseous by-products, have been the subject of extensive research in recent years. This research has been motivated by a variety of factors related to improving the design and operating characteristics of furnaces and associated combustion equipment. Another major incentive has been the national commitment to minimize the potential adverse environmental impact of increased coal utilization. The major combustion-generated air pollutants, whose emission levels are now regulated in many industries, are intimately related to the thermal and chemical conditions inherent to the combustion process and the by-products which these conditions produce. The most notable of these by-products are sulfur oxides, nitrogen oxides, and fly ash particulate matter.

Sulfur oxides (SO_x). These are gases that result from the oxidation of sulfur compounds contained in the coal. This conversion of sulfur appears to be an unavoidable feature of most commercial coal combustion processes. (An exception now rarely in use is the fluidized-bed boiler, where the sulfur oxides can be retained in the solid residues of the combustion process by reacting them with calcium oxide or carbonate to form sulfates.) Consequently, to minimize the emission of SO_x to the atmosphere, many combustion processes are equipped with scrubbers to remove most of the SO_x from the exhaust gases. Also, burning coals with lower sulfur content results in less SO_x. In another approach, coal cleaning, some of the sulfur is removed from the coal prior to combustion. Recent years have seen a growth in the development of coal-cleaning technology and a resurgence of interest in fluidized-bed boilers as a means of reducing the overall cost of controlling SO_x emissions.

Nitrogen oxides (NO_x). These are gases that originate from two sources of nitrogen during combustion of coal. The first source is the nitrogen contained in the air suppled to the combustion process. The high temperatures and oxidizing conditions encountered during combustion readily convert a small portion of this nitrogen to NO_x. Reducing peak flame temperatures and minimizing the amount of air in contact with the fuel when these highest temperatures occur form the basis for many NO_x emission abatement techniques. The second and most significant source of NO_x during coal combustion is the nitrogen contained in the coal itself. This source of NO_x has been the subject of most NO_x-control research in recent years. Laboratory tests have shown that this fuel-bound nitrogen can account for up to

75% or more of the NO_x generated during coal combustion. The actual chemical mechanism by which the fuel nitrogen is converted to NO_x is not fully understood. However, a highly complex picture is emerging in which the actual combustion temperatures and fuel-air mixing conditions are important, as well as the way in which the nitrogen is bound in the coal.

The fuel nitrogen can be divided generally into two parts: volatile nitrogen, associated with the more volatile coal fractions that are released from the coal and burned during the early stages of combustion; and char nitrogen, which is contained in the slower-burning fractions of the coal. In laboratory experiments the conversion of volatile nitrogen to NO_x has been shown to be larger than that of char nitrogen, the former accounting for more than half of the total NO_x from fuel nitrogen. Additionally, the volatile NO_x formation can be significantly affected by changes in the mixing between coal and air during early combustion stages. Thus, design of improved coal burners to produce favorable mixing patterns appears to offer a promising avenue to control this source of NO_x. The char NO_x appears relatively insensitive to early combustion conditions and may prove more difficult to control. Minimizing NO_x simultaneously from both fuel nitrogen sources and nitrogen contained in the air, while still meeting the various practical design and operating constraints of the combustion system, is the major thrust of current efforts to develop improved low-NO_x combustion systems.

Fly ash. This is the residue remaining after the combustible organic materials in the coal are burned. It consists of a vast number of small particles that are formed in suspension within the furnace. In pulverized coal combustion as much as 70–90% of this material remains entrained in the exhaust gases at the end of the combustion process. The remainder is retained in the furnace as slag or ash deposits which are removed during the course of normal furnace operation.

There has been wide interest in the origin and evolution of fly ash during coal combustion and in the physical and chemical characteristics of fly ash particles. Such information is crucial in a variety of areas, which includes: the development and design of particulate emission controls for combustion systems; the understanding of the health effects of particulate matter; and the design and operation of combustion and heat-transfer equipment. To understand the characteristics of fly ash, the nature of the mineral composition of the coal must first be considered.

The mineral constituents of coal consist of a majority of the naturally occurring elements found in the periodic table. These elements originated from the plant material that composed the coal seams and from exposure to a variety of other factors during the long process of coal formation. These other factors include waters and soils that contained the seams, groundwater that flowed through the seams, and the influence of adjacent mineral bodies. Distinct mineral types, such as aluminosilicate clays, sulfides and sulfates, carbonates, and silica, have been identified in coal. Specific elements have also been found to be associated with the organic matter in the coal, that is, chemically bonded with the organic coal matrix rather than occurring in the identifiable mineral structures. These coal elemental and mineral constituents in various chemical forms compose the fly ash emitted from the combustion process.

Fly ash particles vary dramatically in size, morphology, and composition within any given combustion exhaust stream. This variation is influenced by both the nonhomogeneous nature of coal composition and the characteristics of the coal combustion environment. Only within the last decade have the first detailed size distributions and chemical compositions of fly ash been reported. Most of this information is associated with the combustion of pulverized coal.

Particle sizes and shapes. The sizes of fly ash particulate matter have been found to span a wide range from about 0.01 μm in diameter to several hundred micrometers. Fine particles with diameters less than about 0.5 μm are now believed to have formed in a fashion distinctly different from that of the larger particles. Although the exact formation mechanisms are debated, it is generally accepted that most fine particles result from a process involving the high-temperature volatilization of ash components during combustion, followed by condensation and subsequent coagulation of the volatile material as the combustion products cool. Thus, the temperature history of the combustion process seems to play a key role in the formation of these small particles.

The formation of the larger particles is thought to result from the melting of the mineral inclusions in the coal as the organic coal matrix burns. These molten inclusions may coalesce and grow in size to form large droplets. The final result is predominantly fused glassy spheres (see illustration). Broken, crystalline, or irregularly shaped particles, hollow spheres, and even spherical shells encapsulating smaller spheres have been observed, emphasizing the complexity of large-particle evolution.

Chemical composition. A number of studies have determined the chemical composition of fly ash particles and the distribution of specific chemical elements as a function of particle size. Overall, the particulate chemical composition reflects the mineral content of the coal, but individual particles may contain significantly different composition and structure. While numerous mineral forms have been found, one recent investigation identified three major mineral matrices: glass, mullite-quartz, and magnetic spinel.

The distribution of specific mineral elements over the fly-ash-particle size range has been measured by a few investigators. It is difficult to generalize these findings since they involve a variety of particle-sampling techniques, analytical procedures, coal com-

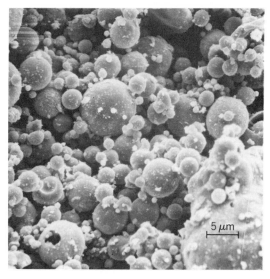

Scanning electron photomicrograph of fly ash particles from pulverized coal combustion.

positions, and coal combustion system designs. However, it does appear that the size distribution of specific elements can be explained at least in part by their high-temperature chemical behavior during the combustion process. For instance, many of the more volatile elements, such as arsenic and selenium, have been found to be enriched on the smaller fly ash particles. This appears to be consistent with the volatilization-condensation model of small-particle formation. In contrast, other more refractory elements, such as iron and aluminum, exhibit a more or less uniform distribution across the particle size range.

For background information *see* AIR-POLLUTION CONTROL; COAL; COMBUSTION in the McGraw-Hill Encyclopedia of Science and Technology.

[MICHAEL W. McELROY]

Bibliography: M. W. McElroy et al., Size distribution of fine particles from coal combustion, *Science*, 215:13–19, January 1, 1982; E. J. Mezey, S. Singh, and D. W. Hissong, *Fuel Contaminants*, vol. 1, *Chemistry*, EPA 600/2–76–177a, July 1976; D. W. Pershing and J. O. L. Wendt, Relative contributions of volatile and char nitrogen to NO_x emissions from pulverized coal flames, *Ind. Eng. Chem. Proc. Des. Dev.*, 18(1):60–67, 1979; R. D. Smith, The trace element chemistry of coal during combustion and the emissions from coal-fired plants, *Prog. Energy Combust. Sci.*, 6:53–119, 1980.

Cogeneration

Cogeneration refers to the on-site, concurrent generation of electricity with process steam or heat as a more efficient use of fuel to meet both the electrical and thermal energy requirements of an industrial process plant. Historically, the use of cogeneration was a common practice in industrial plants before low-cost electricity was readily available from the vast network of public utility grids that exist today.

Recent changes in fuel costs, electricity prices, and regulations governing the utility companies, however, have resulted in a renewed interest in cogeneration. It is becoming an increasingly attractive investment for industries requiring process heat or steam.

Energy-saving potential. The potential for saving energy stems from the basic inefficiency that is inherent in the generation of electricity alone. In a typical power plant, electric generators are driven by gas or steam turbines with such pressure requirements that the exhaust heat or steam from these turbines contains approximately two-thirds of the energy originally delivered to the turbine. Since fuel is burned to provide the energy needed to drive these turbogenerators, roughly one-third of fuel energy can be converted to electrical energy. Unless the utility can sell some of its exhaust steam or heat to a process plant located close by, this inefficiency must be reflected in the price of the electric power sold to its customers.

Frequently, this inefficiently produced electricity is purchased by industrial process plants to meet their electric power demand, while they are burning fuel themselves just to generate steam for their process heat requirements. In many industrial processing plants these requirements can be more than adequately met by the exhaust steam from a turbogen-

erator, and since little additional fuel is required to produce the necessary power steam at higher pressure, industrial processing plants are being encouraged to examine on-site cogeneration of their own electric power. In this way, fuel already being consumed for industrial process heating can do double duty by producing electricity as well, while reducing total industrial demand for purchased electricity.

Figure 1 illustrates the economic rationale for the use of industrial cogeneration. Figure 1a suggests a hypothetical processing plant which purchases electricity to meet its electric power demand and purchases fuel to generate its process steam requirement. The costs per hour for these purchases are shown as $35.00 and $100.00 for electricity and fuel, respectively, for a total energy cost of $135.00 per hour. If the same processing plant could meet both its electrical and process steam demand from an on-site cogeneration system, as shown in Fig. 1b, then the plant would purchase no electricity, and total energy cost would be approximately $105.00 per hour for purchased fuel only, suggesting that more than 20% in total energy costs could be saved.

Available systems. Various schemes have been devised for cogeneration systems in order to provide flexibility in accommodating differnt processing plant situations. The three most commonly used systems include the gas turbine, steam turbine, and reciprocating internal combustion engine (most often a diesel engine). Of these, the gas turbine (Brayton engine) and the steam turbine (Rankine engine) are the two systems most widely used. In the Brayton engine system (Fig. 2), fuel is used by the gas turbine to generate electricity while steam is raised in a waste heat boiler by using the exhaust heat from the turbine. In the Rankine engine system (Fig. 3), fuel is used to raise steam in a high-pressure boiler to drive the steam turbine for generating electricity. The exhaust steam from the turbine is then delivered throughout the plant to meet process steam demand. In the case of the reciprocating internal combustion engine, the engine is used to drive the electric generator while the engine exhaust, usually supercharged with compressed air, is used as the heat source in a low-pressure waste heat boiler to provide process steam for the plant.

Food industry applications. Until now, the use of cogeneration in the food industry has been generally confined to the processing of bulk commodities where highly energy-intensive operations are required, such as in beet sugar processing, corn wet milling, and cane sugar processing and refining. Many of the processing plants in these industry subsectors are quite large and require huge amounts of steam to evaporate the millions of pounds of water from dilute sugar and starch solutions that are processed daily, as well as electric power to drive heavy machinery.

In the rest of the food industry, where most processing operations are considerably less energy-intensive, the historic availability of low-cost fuels and electricity has created a large dependence on the

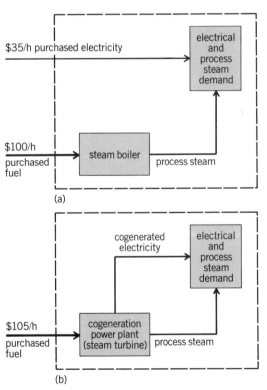

Fig. 1. Economic rationale for the use of industrial cogeneration. (a) Separate purchases of electric power and fuel for process steam. (b) Purchase of fuel for cogeneration of electrical and process steam. (*After A. A. Teixeira, Energy cogeneration: Something for almost nothing, Food Eng., 51(12):72–74, 1979*)

use of purchased electricity and small package boilers to provide process steam. Therefore, with the exception of a few scattered processing plants involving breakfast cereal, gelatin, and canning operations, there has been little use of cogeneration in the rest of the food industry. In almost all cases where cogeneration is being used, the systems are based on steam-driven turbines with high-pressure steam boilers.

Because of the dramatic changes occurring in the cost and availability of energy, the future outlook for cogeneration in the food industry may be quite different from the present situation. The energy and cost savings potential of cogeneration systems can be expected to take on new economic significance in the choice of industrial utility systems for replacement or plant-expansion programs in the United States food industry. Recent studies indicate that cogeneration is already being closely examined in the soybean oil, malt beverage, and citrus industries. The fairly steady demand profile (hourly, daily, and monthly variations in steam and power demand), and appropriate balance between demand for steam and electricity that is found in existing process plants within these subsectors makes them likely candidates for the future use of cogeneration. In some cases, key decision makers have already anticipated their future use of cogeneration by specifying high-pressure steam boilers with a power steam option in their new plant-expansion programs. With these boilers already in place, the turbogenerators can be installed at a later date when the decision to cogenerate is made.

Current research. The opportunity for using cogeneration in any food processing plant should be evaluated by examining three basic criteria: appropriate balance between demand profiles for steam and electricity; economic justification based on price for purchased electricity and capital investment required; and procurement of a satisfactory standby agreement with the local utility.

Much of the recent research on opportunities for cogeneration in the food industry has been supported by the U.S. Department of Energy (DOE). As a result of evaluating the third criterion, the DOE and the Energy Regulatory Commission have taken steps toward establishing new rules and guidelines regarding rates and exemptions for qualifying cogeneration facilities.

Other studies sponsored by the DOE have focused on the technical and economic feasibility of site-specific opportunities for industrial cogeneration. A study made by the DOE in 1980 includes an economic analysis of optimum cogeneration systems designed for a soybean-crushing mill and a malt beverage brewery as examples of food industry applications for cogeneration. A 1982 study compared the economic feasibility of three different cogeneration systems for potential application to the citrus industry. These studies have shown that there can be wide differences in economic feasibility for cogeneration depending on the many factors that surround an in-

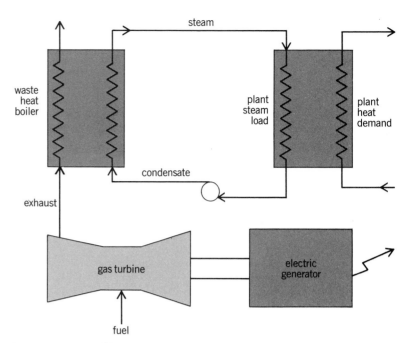

Fig. 2.   Schematic of a gas turbine (Brayton cycle) cogeneration system. (*After M. A. Leo, Energy conservaton in citrus processing, Food Technol., 36(5):231–244, 1982*)

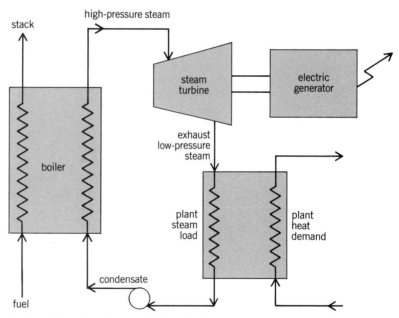

Fig. 3.   Schematic of a steam turbine (Rankine cycle) cogeneration system. (*After M. A. Leo, Energy conservation in citrus processing, Food Technol., 36(5):231–244, 1982*)

dividual plant situation. Economic feasibility is highly site-specific and should be evaluated on a case-by-case basis.

For background information *see* ELECTRIC POWER GENERATION; FOOD ENGINEERING; STEAM–GENERATING UNIT in the McGraw-Hill Encyclopedia of Science and Technology.

[ARTHUR A. TEIXEIRA]

Bibliography: M. A. Leo, Energy conservation in citrus processing, *Food Technol.*, 36(5):231–244, 1982; A. A. Teixeira, Energy cogeneration: Something for almost nothing, *Food Eng.*, 51(12):72–74, 1979; A. A. Teixeira, Congeneration of electricity in food processing plants, *Agr. Eng.*, 61(1):26–29, 1980; U.S. Department of Energy, *Industrial Cogeneration Optimization Program*, Rep. no. DOE/CS/05310-1, 1980.

Concrete

Today, as increased emphasis is being placed upon the conservation of energy and materials, greater attention is being given to improving the properties of conventional building materials or to finding replacements that exhibit a superior cost-property balance. As a result, the technology of composite materials is being applied to concrete, the most widely used construction material in the world.

A concrete-polymer composite is a combination of portland cement concrete and polymers which has properties superior to either material. Portland cement concrete suffers from serious drawbacks, such as little or no resistance to chemical attack; freeze-thaw deterioration; low tensile, shear, and bond strengths; and inherent microstructural problems such as airvoids and shrinkage cracks. Structural applications of polymers are limited as a result of their time- and temperature-dependent properties.

International interest in the use of polymers in concrete dates back to the early 1950s. This work expanded into large-scale research in the 1960s and early 1970s. Since then, many attempts have been made to incorporate polymers into concrete. These polymers have been used to modify concrete in three different ways: adjustment of the rheology of the uncured concrete mix, acceleration of strength development during the setting process, and improvement of the ultimate properties of hardened concrete. Due to the availability of various kinds of monomers and polymerization techniques, the technology of concrete-polymer composites has now matured, and even with high material expenses, the cost/benefit ratio is becoming comparable to conventional materials. Such concrete composite materials

are being tailor-made to obtain improved mechanical, electrical, and thermal-insulating properties.

In an attempt to find a suitable substitute for cement, progress has been made toward developing new, lightweight polymer concretes. Such concretes, while costing more per unit volume than normal-weight concrete, result in reduced costs. Waste products from industry, agriculture, and also naturally occurring materials have been successfully recycled to make such lightweight polymer concretes. Additionally, the increased use of such wastes might prove to be ecologically beneficial

Three prinicipal classes of concrete-polymer composites exist: polymer-impregnated concrete, polymer concrete, and polymer–portland cement concrete. The distinctions among these classes are very important in the selection of materials and in design.

Polymer-impregnated concretes are formed by allowing organic monomers to permeate hardened, precast portland cement concrete. The monomers are subsequently polymerized within the mineral structure, and the polymer adds certain properties to the original concrete.

Polymer concrete is prepared by the addition of a monomer system to particular materials. Polymerization is initiated after efficient mixing of the monomer with the inorganic solids and occurs within a few minutes after the mixing operation. Thus, polymer concrete consists of an aggregate and a polymeric binder.

Polymer–portland cement concretes are formed by the addition of a monomer, prepolymer, or dispersed polymer into a portland cement mix. A polymer network is formed in place during curing of the concrete. The resultant properties of the composites depend on the type of monomer and the selected method of polymerization.

Early developments. Although polymer concretes and polymer–portland cement concretes were first used during the 1950s, major interest in these and in polymer-impregnated concrete began to develop in the mid-1960s. Interest on a worldwide scale is accelerating rapidly. Some concrete polymer materials have been in use for many years, others are undergoing their first applications, and some are awaiting acceptance by a justifiably conservative technological world. After about 25 years of research and development, enthusiasm for the fascinating new combinations of properties that the materials possess has been tempered with a healthy skepticism about whether or not these materials have a place in engineering applications. Fortunately, even under the cold and impartial eye of cost effectiveness, there are applications which are suitable for the unique properties of such materials.

Polymer-impregnated concrete. The basic method for producing polymer-impregnated concrete, shown schematically in the illustration, consists of the fabrication of precast concrete, oven drying, saturation with a monomer, and in-place polymerization. A low-viscosity monomer such as methylmethacylate (com-

Basic method of producing polymer-impregnated concrete.

mercially used to produce Plexiglas or Lucite), is commonly used.

The principal result of impregnation is the sealing of the continuous capillary pore system in concrete, which results in large decreases in permeability. Significant improvements in structural properties also occur; for example, the impregnation of a standard-weight construction-grade concrete will increase the compressive strength from approximately 4000 psi (27.6 megapascals) to greater than 20,000 psi (138 MPa). Similar improvements in tensile and flexural strengths are obtained, and resistance to chemicals and weathering is also markedly improved. The degree of improvement is dependent upon the fraction of the voids in the concrete that are filled with polymer. Other beneficial effects may be due to the ability of the polymer to: act as a continuous, randomly oriented reinforcing network; increase the cement paste-aggregate bond; absorb energy during deformation; and increase the strength of the aggregate.

As a result of its superior properties, polymer-impregnated concrete has been considered for applications requiring high strength and corrosion resistance. End uses in bridge decks, structural elements, and pipes have been examined. Unfortunately, the impregnation process is complex, and monomer costs are high. Despite a great deal of optimism, relatively few applications appear to have achieved commercial acceptance.

Polymer concrete. In polymer concrete, the polymer is the only binder of the aggregate. Therefore, the properties of the product are more dependent upon the characteristics of the polymer than are those of polymer-impregnated concrete.

A wide variety of monomers and aggregates have been used to produce polymer concrete. Examples of the former are epoxy, polyester, methymethacrylate, and furane derivatives. Silica and limestone aggregates are commonly used, but lightweight materials such as perlite and foamed glass can be used when low weight and thermal-insulating properties are needed.

The time for curing and for development of a high proportion of maximum strength can be readily varied from a few minutes to hours. The bonding strength of polymer concrete to substrates (metals, portland cement concrete, and so on) is also usually high. In spite of the cost, polymer concrete is particularly useful for maintenance and repairs, especially when delay and inconvenience must be avoided. Thus the cost/benefit ratio is favorable.

By carefully selecting the particle size of the aggregate, as little as 7–8 wt % polymer is needed. Compressive strength as high as 30,000 psi (206 MPa) can be obtained. Typical flexure strength is 3000 psi (20.6 MPa).

The versatility of polymer concrete has led to many applications throughout the world, including electrical insulators, columns for mine supports, sewer pipes, flooring, concrete patching, and overlays for highway bridge decks.

The production of polymer concrete normally requires the use of dried aggregate since the presence of water significantly reduces the strength of the product. This disadvantage has recently been overcome by the use of furfuryl alcohol monomer. Mixtures of this material with water-saturated aggregate, cured at $-4°F$ ($-20°C$), have yielded compressive strengths of 2500 psi (17 MPa) 1 hour after mixing. The material appears to be suitable for the rapid repair of water-saturated concrete and asphalt pavements under all weather conditions.

Polymer–portland cement concrete. From the standpoint of process technology, the addition of a monomer or dispersed polymer to a portland cement concrete mix is an attractive idea. Unfortunately, most common monomers either interfere with the cement-curing reactions or are chemically degraded by the cement. The inclusion of epoxies or polyesters can be effective, although usually rather high concentrations are required to improve the mechanical properties.

The incorporation of polymeric latexes has received the most attention. These latexes consist of very small spherical particles of high-molecular-weight polymers held in suspension in water by the use of surface-active agents, and typically contain about 50% solids by weight. Acrylics, styrene-butadiene copolymers, polyvinyl acetates, or natural rubbers are commonly used.

The mixing and handling of polymer–portland cement concrete is very similar to that of conventional concrete. Concrete mixers are used, with cement, aggregate, and water; conventional trowels and screeds are employed for finishing. Curing, however, is different.

Whereas conventional concrete requires extended periods of 100% moist conditions for ideal cure, the film-forming feature of the composite is such that after 1 day of moisture cure, the surface can be uncovered. By then a film has formed on the surface, retaining internal moisture for continued cement hydration, while the exposed surface air dries. Generally, after several days of air cure at 59 to 80°F (15 to 27°C), the concrete can be put into service.

In general, latex-type polymer–portland cement concrete exhibits excellent bonding to steel reinforcement and to old concrete; good ductility; resistance to penetration by water and salt; and excellent durability under freezing and thawing conditions. Not surprisingly, the precise balance of properties depends on the nature of the polymer and its concentration. While flexural strength and toughness are usually increased, the modulus of elasticity may or may not be increased; the more rubbery the polymer, the lower the modulus.

The first latex-modified overlay on a bridge in the United States was installed in 1960. The material is now accepted as a standard construction material, covering 9×10^7 ft^2 (8.4×10^6 m^2) of bridge decks, and is being installed at an estimated 2.2×10^6 ft^3/year (6.2×10^4 m^3/year). It has proved itself not only in the laboratory but also in the field.

Summary. In scientific research and development and in the introduction of concrete-polymer materials to commercial applications, it is apparent that significant progress has been made and that worldwide interest in the materials exists. Commercial applications for polymer concrete and polymer–portland cement concrete are in effect on a relatively large scale and markets are developing rapidly. Few commercial applications for polymer-impregnated concrete exist, primarily because of the complex and costly impregnation process, high monomer costs, and the rapid development of polymer concrete technology, which has produced lower-cost composites with similar properties.

For background information *see* COMPOSITE MATERIAL; CONCRETE; POLYMER; PORTLAND CEMENT in the McGraw-Hill Encyclopedia of Science and Technology.

[L. E. KUKACKA]

Cosmology

The past decade has seen tremendous progress in understanding the distribution of galaxies in space and the origin and evolution of large agglomerations of galaxies. Cosmography, the mapping of the distribution of galaxies, makes it possible to confront cosmological theories with observations. Large-scale agglomerations of galaxies are usually referred to, perhaps improperly, as superclusters. These are groups of galaxies, whose longer dimensions extend over 100 to 300 million light-years (1 ly = 9.46 $\times$ 10^{12} km). The structures are rather irregular, generally with no central condensation and with patchy variations of density. They are probably the fundamental unit cell of the universe. Clusters and groups of galaxies are embedded in them. These large structures are essentially regions which define positive density fluctuations; that is, the number of galaxies per unit volume exceeds the mean number density of galaxies in the universe.

A remarkable and unexpected discovery of the last decade is that there exist not only regions in which galaxies are clustered on a very large scale, but also space which is void of galaxies. Such voids appear to be an important characteristic of the universe and put further constraints on the theoretical models. They also influence the dynamics of galaxies over large scales.

Observation of voids. The peculiar motion of a distant galaxy, outside the great clusters of galaxies, is small in respect to the motion due to the expansion of the universe. It is therefore possible, by using the relation between distance and red shift discovered by E. P. Hubble, to determine the distance from red-shift measurements. The distance, together with the positional coordinates on the celestial sphere, give the position in space of an extragalactic object, and thus mapping in three dimensions is made possible.

To date, various regions of the sky have been surveyed relatively completely. Among these are a region in Coma, the Coma-A1367 supercluster (Fig. 1); a region in the constellation of Hercules, the Hercules supercluster (Fig. 2); and the Perseus-Pisces supercluster (Fig. 3). [In Figs. 1, 2, and 3 a red shift of 10^3 km/s corresponds to a distance of approximately 20×10^6 parsecs (1 pc = 3.1 $\times$ 10^{13} km), or approximately 60×10^6 ly.] It is clear from the figures that galaxies tend to be found in certain regions of space rather than in others and that there are vast regions which are devoid of objects. In Hercules, for instance, the void extends over a depth of about 300 million ly and covers an area of at least 8 $\times$ 5 degrees.

At present the statistics of voids—their number and size distribution—is unknown. It is even uncertain what the largest size is. Detection of a vast void in the direction of the constellation of Boötes has been contradicted by the detection of a sizable number of galaxies in that region. Nevertheless, the fundamental facts of the existence of such voids and the need to better understand their cosmological significance remain.

Irregularities in the distribution of matter perturb locally the cosmic velocity field which is regulated by the mean mass of the universe. The density excess of a supercluster exerts a braking action on the surrounding matter so that it locally decelerates the expansion of the universe. Such an effect has been measured in the local supercluster. Dynamically the void acts as a negative density fluctuation so that it should have the equivalent effect of further pushing matter away and locally accelerating the expansion. Such local properties must be avoided when the properties of the universe as a whole are sampled.

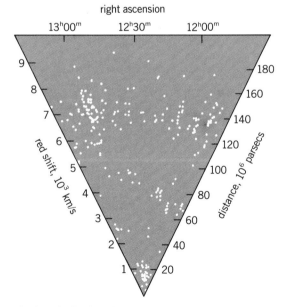

Fig. 1. Distribution of galaxies in the Coma-A1367 region. Coma cluster is at approximately right ascension = 13^h, and A1367 cluster at approximately right ascension = 11^{h}30^m. Region void of galaxies at red shifts between 5000 and 7000 km/s is clearly visible. (*After S. A. Gregory and L. A. Thompson, The Coma/A1367 supercluster and its environs, Astrophys. J., 222:784–797, 1978*)

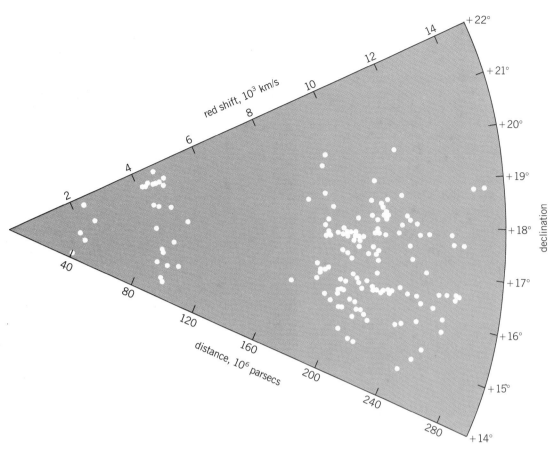

Fig. 2. Distribution of galaxies in the observed field of the Hercules supercluster. (*After M. Tarenghi et al., The Hercules supercluster: I. Basic data, Astrophys. J., 243:793–801, 1979*)

It becomes important, therefore, to determine the scale length of the fluctuations and study some properties of the universe, such as expansion and density, over larger scales. In a conservative view such a characteristic size is larger than 60 million ly and probably smaller than about 300 million ly.

Theory of cluster and void formation. Many-body simulations show that the gravitational interaction among galaxies can create clumpiness on all scales up to about 30 million ly. Such a process of gravitational instability certainly plays an important role on a relatively small scale. It has, however, serious difficulties in reproducing the elongated structures seen in Figs. 1, 2, and 3 and detected also in other surveys.

Some astronomers therefore favor the hypothesis that the present irregularities in the distribution of matter originated from large-scale density and velocity fluctuations in the early universe. In this theory only large-scale (adiabatic) fluctuations survived the recombination era, the time when decreasing temperature ($T \sim 4000$ K) allowed the recombination of protons and electrons to form hydrogen, and galaxies resulted from the fragmentation of these large-scale density enhancements. The theory is capable of explaining the formation of structures of various shapes and can account for the presence of spaces which are practically empty. In an alternative idea

for the formation of superclusters, and related voids, the effect of the birth of a massive galaxy on the surrounding intergalatic medium is considered. It has been shown that a slowly expanding shell is formed by the energy released by the massive stars of the galaxy. The cooling shell is unstable, and after some time it fragments and forms new bound objects. The mass of the shell and the size of the void are in agreement with observational values. It has been pointed out, however, that such a process requires "seed" galaxies, and these may be formed, for instance, by the mechanism of fluctuations in the early universe mentioned above. The alternative theory therefore postulates the same origin and initial evolution for the large-scale fluctuations, but a somewhat different final product.

Galaxies and matter. The question of whether a void of galaxies is also a void of matter has not yet been answered observationally. In fact the possibility that intrinsically faint galaxies could exist in the voids cannot be excluded. However, the existence of such galaxies would imply that the luminosity function of galaxies, that is, the number of galaxies per unit volume and per interval of brightness, changes over scales of the order of 30 million ly. There is no evidence of such variation; on the contrary, the part of the luminosity function which can be observed is the same in different regions of space.

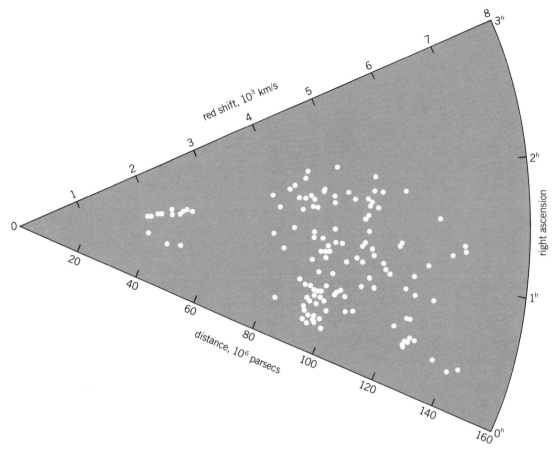

Fig. 3. Distribution of galaxies in the Perseus-Pisces region.

Unfortunately, it is impossible to observe faint galaxies over large distances because of their low brightnesses and so a direct observational determination is not feasible. The voids may also be regions where conditions for galaxy formation were unfavorable, in which case they could be filled with invisible matter.

On the other hand, it is also possible to assume the reasonable view that galaxies, bright and faint, are good tracers of the distribution of matter, so that lack of fairly bright galaxies means also lack of faint galaxies and of matter in general.

A problem and another possibility exist, however. One of the main goals of observational cosmology is the determination of the mean mass of the universe, since such measurements determine the cosmological model. In a big bang cosmology, the theory which is most widely accepted, helium is formed soon after the beginning of the universe, and its abundance is very closely related to the present mean mass density. According to modern observations of the helium abundance in young galaxies, that is, in galaxies where there has been little processing of material through stellar nuclear reactions, the mean mass density of the universe must be very low. This is in contradiction to the larger mean mass density observed for galaxies and clusters of galaxies. The helium abundance, however, sets constraints on the amount of matter that exists in the form of neutrons and protons (baryons); it does not exclude the possibility of large amounts of nonbaryonic matter which, naturally, can be detected by its gravitational effects. Perhaps the latter contributes to the masses of galaxies and clusters.

The least exotic and most discussed possibility is the existence of matter in the form of massive neutrinos. In fact, it is known that neutrinos are very abundant in the universe, and there are theoretical and some experimental reasons to believe they are particles of nonzero rest mass. The definite answer to the existence of massive neutrinos will come from experiments which, at the present time, are either in progress or being planned by various physicists.

Since the confinement of neutrinos by a gravitational field is a function of their mass, knowledge of their mass (assuming they have one) will determine whether they can solve only local astronomical problems, such as the mass discrepancy observed in clusters of galaxies, or whether they are spread throughout the universe. In this case, they would also fill the voids detected in the distribution of visible galaxies. In that case, most of the matter in the universe would be in a form which easily escapes detection.

For background information *see* COSMOLOGY; NEUTRINO; SUPERCLUSTERS in the McGraw-Hill Encyclopedia of Science and Technology.

[GUIDO CHINCARINI]

Bibliography: G. Chincarini, Clumpy structure of the universe and general field, *Nature*, 272:515–516, 1978; G. Chincarini and H. J. Rood, The cosmic tapestry, *Sky Telesc.*, 59:364–371, 1980; S. A. Gregory and L. A. Thompson, Superclusters and voids in the distribution of galaxies, *Sci. Amer.*, 246(3):106–114, March 1982.

Data flow systems

Data flow systems are an important alternative to conventional programming languages and architectures and offer key advantages over them. They have the potential to realize large amounts of parallelism (present in many applications) and effectively utilize very-large-scale integration (VLSI) technology. Recent years have witnessed an increasing interest in data flow systems. Today there are at least a dozen major projects in the United States, Europe, and Japan where research is being actively pursued on languages, compilers, architectures, and performance evaluation. Several prototype systems are now operational. Much progress has been made, but several important questions remain unanswered. Active pursuit of research and development in this area will probably continue for several years.

Basic concepts. Data flow systems use an underlying execution model which differs substantially from the conventional one. The model deals with values, not names of value-containers. There is no notion of assigning different values to an object which is held in a global updatable memory location. A statement such as $X: = B + C$ in a data flow language is only syntactically similar to an assignment statement. The meaning of $X: = B + C$ in data flow is "compute the value $B + C$ and bind this value to the name X." Other operators can use this value by referring to the name X. The statement has a precise mathematical meaning and defines X. This definition remains constant within the scope in which the statement occurs. Languages with this property are sometimes referred to as single assignment languages. The second property of the model is that all processing is achieved by applying functions to values to produce new values. The inputs and results are clearly defined, and there are no side effects. Languages with this property are called applicative. Value-oriented, applicative languages do not impose any sequencing constraints in addition to the basic data dependencies present in the algorithm. Functions must wait for all input values to be computed, but the order in which the functions are evaluated does not affect the final results. There is no notion of a central controller which initiates one statement at a time sequentially. The model described above can be applied to languages and architectures.

Data dependence graphs. The computation specified by a program in a data flow language can be represented as a data dependence graph, where each node is a function and each arc carries a value. Very efficient execution of a data flow program can be achieved on a stored program computer which has the properties of the data flow model. The machine language for such a computer is a dependence graph rather than the conventional sequence of instructions. There is no program counter in a data flow computer. Instead, a mechanism is provided to detect when an instruction is enabled (all required input values are present). Enabled instructions together with input values and destination addresses (for the result) are sent to processing elements. Results are routed to destinations, and this may enable other instructions. This mode of execution is called data-driven.

The illustration is an example of a machine language program for a data flow computer. Values are carried on tokens which flow on arcs of the graph. The graph in the illustration has four tokens initially. Tokens x and y carry input values; two control tokens have the value F. Iteration and conditional execution are achieved by using the SELECT and DISTRIBUTE operators. SELECT routes a token to its output arc either from arc T or from arc F depending on the value of the control token on the horizontal input arc. DISTRIBUTE routes a token on its input arc either to the T or F output depending on the value of the control token. All operators remove the input tokens used and produce a number of identical result tokens, one for each destination. In the illustration, the initial output of the upper SELECT is a token with value x since the control to-

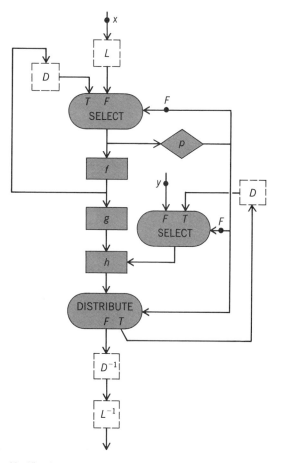

Machine language program for a data flow computer, represented as a data dependence graph.

ken has a value F. The input tokens x and F are removed and two output tokens are produced, one for the function f and the other for the predicate p. It is a useful exercise to follow through the execution of the graph assuming that each arc is a first-in-first-out queue and can hold an unbounded number of tokens, and that L, L^{-1}, D, and D^{-1} are the identity operations.

Static and dynamic architectures. The execution model of a data flow computer as described above, though radically different from conventional processing, is the basis for most of the data flow machines currently being investigated. Individual differences arise due to the amount of parallelism that can be realized, the mechanisms for detecting and scheduling enabled instructions, and the handling of data structures. Two different architectures, static and dynamic, are explained below.

In static architecutres, an instruction (like h in the illustration) is represented in memory by a packet which has an operation code and space to hold two input values and one or more destination addresses. Hardware is provided with the memory to detect the arrival of both input operands. One restriction in static data flow computers is that no arc in the graph can carry more than one token during execution. Control signals are sent from destination nodes to source nodes to indicate the consumption of previous values. Nevertheless, static machines can realize several different forms of parallelism. For example, f and p can be evaluated simultaneously. Also, g and h can be executed in parallel since they form two stages in a pipeline. However, one form of parallelism cannot be realized. Assume f and h are simple functions and compute very fast and g is relatively slow. If the single-token-per-arc limitation is removed, then several tokens would accumulate at the input of g allowing the possibility of invoking multiple, simultaneous instantiations of g. Dynamic data flow architectures can realize this form of parallelism.

Dynamic data flow architectures allow multiple tokens per arc. A token carries a value and a label. The label specifies a context, an iteration number, and a destination address. Each instruction knows its successors and sets the destination field of the result token appropriately. In addition, the L, D, D^{-1}, and L^{-1} operators (ignored so far in the illustration) modify the context and iteration number. The L operator creates a new context by stacking the previous context and iteration number and also sets the new iteration number to 1. The D operator increments the iteration number. The D^{-1} operator resets the iteration number to 1, and the L^{-1} operator restores the old context and iteration number (stacked by L). Dynamic data flow computers use an associative memory to hold tokens which are waiting for their partners to be produced. This mechanism is used to bring together tokens with identical labels. When this event occurs, the destination instruction is fetched and, together with the input values, is sent to a processing element. On completion, result tokens are produced with appropriate values. With

this mechanism, simultaneous evaluations of g are possible. Since h may not be associative, successive evaluations of h must proceed in the specified order. The token-labeling mechanism guarantees this, irrespective of the order in which the simultaneous evaluations of g complete. The token-labeling concept can be extended to handle recursion and generalized procedure calls.

Data-driven and demand-driven execution. In data-driven execution, the sources of a node N produce the input values and execution of N is triggered when all input values are produced. Another execution rule, called demand-driven, is sometimes used. With this rule, nothing happens until a result is demanded at a primary output of a graph. The corresponding node then demands its inputs. These demands flow opposite to the arcs in the graph until the primary inputs are reached. A node executes only if its result has been demanded and its own demands satisfied. An advantage of this approach is that the computations which occur are exactly those that are required. This rule is also called lazy evaluation.

Comparison with conventional systems. Conventional languages and architectures are characterized by the existence of a sequential controller and a global addressable memory which holds objects. Languages such as FORTRAN and PL/1 allow aliasing and side effects and impose sequencing constraints not present in the original algorithm. The compile step attempts to recover the parallelism obscured by the language and generate a data dependence graph. Depending on the parallelism to be exposed, this can be a complex step. Code is then generated for a scalar processor, a vector processor, or a multiprocessor consisting of several uniprocessors sharing storage. Each alternative is examined briefly below.

In uniprocessors, sequential decoding of instructions is necessary to place appropriate interlocks on storage and thereby guarantee the logical correctness of results. Techniques such as overlapping, pipelining, and out-of-sequence execution are used to design high-performance processors. However, because of the sequential decode, concurrency can be obtained only in a small window around the program counter. Furthermore, high-performance uniprocessors cannot effectively utilize VLSI technology.

The decoder limitation can be circumvented if a single instruction can initiate multiple operations on a data structure. This leads quite naturally to vector architectures. Portions of a program coded in instructions which have vector operands can be executed at very high speeds limited only by hardware and memory bandwidth. However, not all applications with parallelism can be vectorized, and very sophisticated compiler analysis is needed to generate vector code automatically from sequential programs.

Conventional multiprocessing can utilize VLSI technology well. The key problem is the execution model with its global updatable memory. Since the processors execute asynchronously, race conditions can arise and are prevented by embedding synchronization primitives in the code. Usually, the overhead for synchronization is large, and low-level parallelism cannot be realized. The code must be partitioned into relatively large blocks of computation with few synchronizing primitives. Moreover, if large amounts of parallelism are not obtained, the performance of the entire system can be critically dependent on processor-processor and processor-memory communications latency. The current state of development does not support compilation of sequential programs on a multiprocessor system. Also, multiprocessor code with embedded synchronization is extremely difficult to verify.

The complexity of the compile step for both vector and multiprocessor architectures can be reduced by extensions to sequential languages. However, this forces the programmer to consider parallelism explicitly, an added complexity.

Advantages and limitations. Data flow systems can overcome many of the disadvantages of conventional approaches. In principle, all the parallelism in the algorithm is exposed in the program without the programmer having to deal with parallelism explicitly. Since programs have mathematical properties, verification is simpler. Generating the dependence graph from the program is a simple step. Systems of large numbers of slow-speed processors are possible, and the approach therefore exploits VLSI technology. If large amounts of parallelism are realized, then processor-memory and processor-processor communications latency is not as critical. Since there are constraints on the production and use of information, protection and security can be more naturally enforced.

Several important problems remain to be solved in data flow systems. Handling complex data-structures as values is inefficient. There is no complete solution to this problem yet. Data flow computers tend to have long pipelines, and this causes degraded performance if the application does not have sufficient parallelism. Since the programmer does not have explicit control over memory, separate "garbage collection" mechanisms must be implemented. The space-time overheads of managing low levels of parallelism have not been quantified. Thus, though the parallelism is exposed to the hardware, it has not been demonstrated that it can be effectively realized. The machine state is large, and without the notion of a program counter, hardware debugging and maintenance can be complex. Data flow also shares problems with conventional multiprocessor approaches: program decomposition, scheduling of parallel activities, establishing the potential of utilizing large numbers of slow processors over a variety of applications, and system issues such as storage hierarchy management and disk seek-time limitations.

For background information *see* DIGITAL COMPUTER; DIGITAL COMPUTER PROGRAMMING in the McGraw-Hill Encyclopedia of Science and Technology.   [TILAK AGERWALA]

Bibliography: Data Flow Systems, *Computer*, special issue ed. by T. Agerwala and Arvind, 15(2):10–69, February 1982; P. C. Treleaven et al., Data-driven and demand-driven computer architecture, *Comput. Surv.*, 14(1):93–143, March 1982.

Desertification

Large stretches of land in various parts of the world, which until recently supported crops and livestock, are now barren desert, and desertlike conditions are increasing at an alarming rate. The formation of these conditions, a process known as desertification, is in most places the result of human activity. Evidence for recent climatic changes as a cause of desertification is scanty or nonexistent, although there is a strong suspicion that at least in some areas human activity can induce climatic change. The most striking examples of desertification occur during and just after a prolonged drought, when the impact of people and livestock on the land is most severe. However, drought is not the cause of desertification; it simply accelerates the process.

Causes of desertification. Desertification sets in when the cover of vegetation is reduced and the biological effectiveness of scarce or irregular rainfall is decreased because of high evaporation rates. Loss of vegetation may occur for a variety of reasons, including overgrazing by livestock, clearing and attempts to replace the original vegetation with crops, felling of trees for firewood, and burning of grassland in the hope of obtaining better pasture. Removal of vegetation can result in soil erosion by wind and water. Each year, millions of tons of topsoil from Africa are desposited by wind in the Atlantic Ocean.

Desertification has been going on for a long time, possibly since the beginnings of agriculture 10,000 years ago, but only in the last 20 years has it been recognized as one of the most serious threats to humans. Civilizations from the past declined and disappeared because the land supporting them was rendered useless, for example, the concentration of salts in land irrigated by the Babylonians and Sumerians eventually ruined their agriculture. North Africa was much greener and more productive in Roman times than it is now, and parts of India that were once forested are now barren desert because the trees were cut down for firewood. There are examples of humanly induced desertification in virtually every part of the world, including the southwestern United States. However, nowhere is the problem more acute than in the arid savanna of Africa south of the Sahara, an area which first directed the world's attention on the seriousness of desertification.

Food shortages. The Sahel, the region where the southern edge of the Sahara gives way to arid, grassy savanna, has been inhabited by nomadic pastoralists for thousands of years. The pastoralists and their animals had come to terms with a risky environment where rainfall was scarce and unpredictable and where periodic food shortages were accepted as part of the normal way of life. In 1911 and 1940 there were severe droughts which had disastrous effects, although what happened to the people and their animals is not known with any degree of certainty.

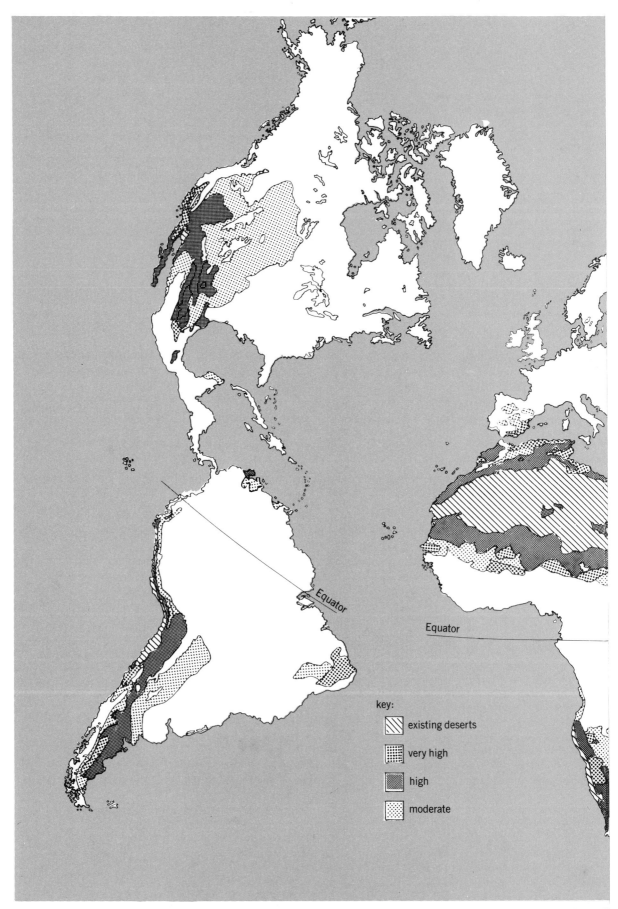

Desertification map of the world, showing the risk of desertification as assessed by climate, changes in vegetation, soil erosion, and human population trends. (*After UNCOD, 1977*)

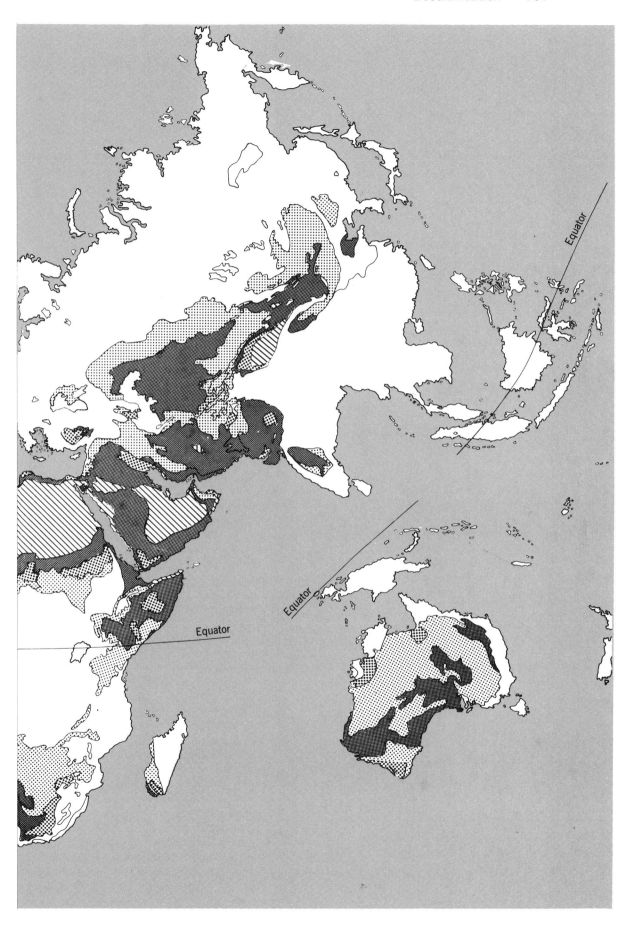

At Rosso, Mauritania, the average annual rainfall is 284 mm. In 1968 it was only 122 mm, which is within the normal range of variability, and hence there was no cause for alarm, especially since in 1969 there was 295 mm of rain, just above the average. In 1970 the rainfall was 149 mm, followed by 126 mm in 1971 and a mere 54 mm in 1972. By 1973 it became apparent that the rainfall throughout the Sahel was exceptionally low and the people destitute and starving. The six countries most affected—Mauritania, Upper Volta, Mali, Niger, Chad, and Senegal—were in a desperate state, and parts of some other countries, including much of Gambia, northern Ghana, and northern Nigeria, also suffered. It is believed that 100,000 people died in 1973 alone, and about 7 million became dependent on food handouts from richer countries and from international agencies which, by 1973, had at last discovered the extent of the disaster. About $200 million were donated to help the people of the Sahel, but it was too late to save pasture and crops. The 1968–1973 drought was certainly severe, but it must be assumed that comparable droughts had occurred before. However, this time the pressure of too many people and animals on the land caused desertification on a massive scale.

The pastoralists suffered most, but in the south of the area, where there is usually just enough rainfall to cultivate some crops, millions of people went hungry and many starved. By 1973, Lake Chad was one-third its normal size and the Niger and Senegal rivers had failed to flood, leaving vast plains arid with no prospect of crop production. Two million nomadic pastoralists lost half their livestock and 15 million cultivators lost half their crops. Many moved south, seeking food and shelter in towns. Many never returned because the drought left the land damaged and useless or because they had lost the will to try again in such a harsh environment.

UNCOD. The great drought was, however, a stimulus for action. In 1977 the United Nations Conference on Desertification (UNCOD) was assembled in Nairobi to examine how recurrent drought and human impact on the land can result in long-term or permanent land degradation. Thousands of pages of documentation were produced by UNCOD, and estimates showed that about 50,000 km^2 of the world's land surface degrade by desertification each year. Many schemes to combat desertification were put forward. Some were impractical; for example, the proposition to create a vast green belt of trees and shrubs to hem in the Sahara. Perhaps the most valuable contribution from UNCOD is the map they produced showing the risk of desertification throughout the world (see illustration). The degree of risk was assessed by climate (especially the amount and seasonal distribution of rainfall and the likelihood of drought), vegetational changes brought about by human activities, soil erosion, population density, and the rate of increase in numbers of both people and livestock. The risk is greatest in and around arid and semiarid areas within the tropics, and is non-existent in extremely arid areas, which are already natural deserts. The most vulnerable areas appear to be in Africa, north of the Equator and south of the Sahara. Here live many of the world's poorest people, who have few or no resources to combat desertification and whose numbers are rapidly increasing.

There has never been a proper ecological study of desertification.

Such a study would have to continue over decades, during which time there would have to be at least one drought, before any significant results would emerge. As a consequence, desertification cannot be defined with precision; rather it is perceived. Pastoralists and cultivators in dry lands have a remarkable intuitive ability to assess the carrying capacity of the land, and so an ecological research project would have to include an assessment of the people's knowledge and behavior as well as an analysis of the interactions between climate, soil, vegetation and animals.

Overgrazing. Some of the most striking examples of desertification occur where livestock are moved from place to place and allowed to graze on wild grasses. Provided there is sufficient rainfall, no great damage is done, because grasses, especially the perennial species, have the capacity to withstand and even to prosper as a result of grazing. But once movement is restricted and numbers exceed the carrying capacity, grass is destroyed—especially during periods of scarce rainfall.

Grasses first appeared in the Miocene, about 25 million years ago, and at the same time wild grazing mammals, with teeth adapted for cropping grass, made their appearance. It is almost certain that grazing pressure from wild mammals did not cause desertification. Nowadays, however, nonnative cattle, sheep, and other livestock are found in those areas where the grasses are not specifically adapted to the effects of their grazing. They are also maintained at densities far in excess of the wild grazers, and the result has been a breakdown in the coevolutionary relationship between eater and eaten. Much the same has happened in the national parks of East Africa and elsewhere, where wild grazers are confined at high density to relatively small areas and are thus no longer able to undertake long-distance migrations in response to the annual cycles of rain and drought.

Overgrazing is not the only cause of desertification in pastoral lands. Animals need water and concentrate at water holes where the trampling of hooves can cause far more damage to vegetation than cropping. Some species of plants can withstand trampling better than others. These tougher plants are often unpalatable, even poisonous, and because of a lack of competition they tend to take over the ever-increasing patches of bare soil. When it is dry, trampling breaks up the soil making it subject to wind erosion. In wetter conditions the soil becomes a sea of mud and eventually dries out to form a hard crust, leaving little prospect for rapid colonization by grasses. The provision of water boreholes in the

Sahel by outside agencies resulted in considerable local damage by trampling, and many cattle died of starvation; water alone cannot keep a cow alive.

Old, dying grass is not good food for livestock, and so pastoralists set fire to it during the dry season. The effect is generally beneficial as the land is cleared of unwanted vegetation, and nutrients are rapidly recycled. Burning also stimulates the growth of new grass, even before the arrival of the first rain, and this provides nutritious food for animals. However, severe, extremely hot fire can destroy the humus and the nitrogen-fixing bacteria in the soil. Fine particles deposited after a fire may reduce the ability of the soil to absorb water and hence retard the regrowth of grass. On the whole, regular burning increases the amount of grass at the expense of other vegetation, although the role of fire in areas undergoing desertification is not fully understood.

Cultivation on arid land. Rain-fed cultivation is possible in places with very little rain. It becomes increasingly possible, on a seasonal basis, where the annual rainfall is not less than 250 mm and where there is a single wet season. If there are two wet seasons in the year, the minimum rainfall requirement is about 500 mm. Cultivation in arid places is by trial and error, with successes more obvious but not necessarily more frequent than failures. The failures are not always caused by insufficient effective rainfall; depletion of soil nutrients and soil erosion are often responsible for forcing people to abandon an area they are trying to cultivate. Shifting cultivation, in which vegetation is cut and burned, providing a rich but temporary source of nutrients, and light tilling of the land are commonly practiced throughout the dry tropics. After two, three, sometimes up to ten growing seasons, yields fall abruptly and the land is abandoned and allowed to revert to its original vegetation.

In dry regions, shifting cultivation can support about 25 people per square kilometer, but the system comes under stress when rotations are shortened and soil fertility is depleted. The carrying capacity is easily exceeded as soon as numbers of people are too high for the land to support. With continued use there is little chance for the natural vegetation to reestablish itself, and once again, erosion by wind and rain become the main hazard. Large areas, especially in the southern Sudan and in the Sahel, that supported shifting cultivation a few years ago are now barren and totally unproductive.

Irrigated agriculture is a more permanent proposition for arid-land agriculture. Once the decision to irrigate has been made, the people must decide to settle in the area. Irrigation is labor-intensive and also requires capital investment which is not always readily available. The main problem is maintaining flow of water which goes directly to crops and is not lost by seepage and evaporation. Another problem is the buildup of high salinities which can completely ruin an area for cultivation; then the land is usually abandoned and becomes a desert.

Virtually all the areas of the world at risk from desertification are, by Western standards, highly marginal. In Europe and North America such land would not be used for crops or for livestock other than sheep; rather, it would probably be set aside as wilderness where people could go for certain recreations. But in poor tropical countries people have to do the best they can in these highly marginal and risky places; there is simply nowhere else to go.

For bakground information *see* DESERT ECOSYSTEM in the McGraw-Hill Encyclopedia of Science and Technology.   [DENIS OWEN]

Diamond-turned optics

Diamond-turned optics are optical elements that have been machined to a specular finish on precise metalworking lathes. The precision of these lathes is so great that the optics produced on them can be used in the near infrared without further optical working, and many smaller diamond-turned optics (less than 10 cm in diameter) with simple shapes can be used even in the visible region.

While diamond turning is a useful and cost-effective technology wherever there is a need to make nonspherical or discontinuous surfaces, such as multifaceted scanning mirrors, the advantages of diamond turning are best used to solve the unique problems of high-energy laser optics. Diamond turning has permitted optical systems to be designed with totally unconventional surfaces. At the same time, these systems appear to work so effectively that they are influencing diamond-turning technology to produce machines of larger capacity, higher accuracy, and better surface finish. The availability of unconventional surfaces will encourage their further use.

Machining process. Although the diamond-turning lathe is responsible for the accurate geometric shape of the optics, the diamond-tipped cutting tool creates the specular or mirror finish on the parts. The unique properties of diamond—its hardness, high thermal conductivity, and low coefficient of friction against many materials—produce the extremely low surface roughness that accounts for the mirror finish on the optics.

The lathes on which the optics are turned have several basic components. There is an air-bearing spindle and faceplate on which the part is mounted. Two linear slides run perpendicular and parallel to the spindle axis. A lead screw–driven carriage rides on these slides and carries the diamond-tipped cutting tool. A computer-driven numerical control unit drives servos on the lead screws to continuously position the carriage to produce virtually any desired figure of revolution. This leads to the principal advantage of diamond turning—any mirror shape that is a figure of revolution can be created without resorting to traditional optical-polishing methods.

In addition to the components of the machine that are responsible for creating the geometry of the optics, there is an equally important group of items that ensures the accuracy of the components. First, the spindle and slides are mounted on a massive base with good damping properties. This in turn is supported on an active vibration isolation system to mechanically decouple the machine from its surroundings. A thermal shroud and coolant temperature control unit isolate the machine thermally. Most machines also have a tool position feedback system that includes a distance-measuring interferometer sensitive to less than 25 nanometers. A coolant is generally used to help the diamond-cutting tool produce the best possible surface finish. Lastly, a vacuum chip extractor is used to ensure that the fine finish is not damaged if a chip becomes lodged between the tool and the work. This could easily ruin an otherwise satisfactory optic.

Application to high-energy lasers. In a high-energy chemical laser, it is necessary to have optics that are highly reflecting, have a high laser damage threshold, and are resistant to corrosion. The surfaces produced by diamond turning have these characteristics, whereas those made by conventional polishing methods do not. Diamond-turned surfaces have reflectivities matched only by similar metals evaporated on polished glass substrates. Because the diamond-turned surface is free of embedded polishing compounds (dielectric materials) and seemingly free of a work-hardened layer due to polishing, they exhibit a high threshold to damage by high-power laser emission. They are also less subject to corrosion than conventionally polished mirrors.

In addition to these positive attributes relating to surface finish, diamond turning permits a degree of dimensional control that is impossible to maintain by conventional optical polishing methods. First, there is the overall shape of the optics required for high-energy lasers. To extract energy efficiently from an annular gain region laser and yet provide for feedback, surfaces of revolution with a high degree of cone (axicon surfaces) are required. These surfaces depart drastically from spherical shapes and have slope discontinuities where they cross the axis of symmetry. They are virtually impossible to produce by traditional methods, which can make only spherical or near-spherical surfaces.

High-energy lasers also need scraper mirrors to direct part of the energy out of the laser cavity, and they need aperture stops, or stray light–dumping optics. Both these elements require optically good figures out to the edge of the part, and the edge must be sharp and of the correct geometrical shape. Here again, conventional polishing leaves a rounded surface or rolled edge. However, because of the low cutting forces and exceedingly clean cut of the diamond, the diamond-turned surfaces are flat to within micrometers of the physical edge of the part.

To ensure a high-quality output beam from the laser, the optics must be well aligned to within a few seconds of arc or better. Diamond turning permits the addition of reference flat surfaces on the parts. These are turned as a part of the same cut that produces the conical surface. In this way the plane of the reference surface is perpendicular to the optical axis of the conical surface to the accuracy of the spindle, which is far less than a second of arc.

Reference surfaces must be maintained accurately in tilt relative to other surfaces, and longitudinal spacing and centering are required as well. True bores are as easy to make as the plane reference surfaces, and the diameters of bores may be held to less than a micrometer. Overall dimensions are a critical factor in high-energy laser optics as well, because many optics have cooling passages just a few tenths of a millimeter below the optical surface. It is necessary to hold the separation between the cooling passages and the optical surface to a few tens of micrometers, and to hold parallelism better than this, so that the cooling is uniform over the active surface. Conventional optical polishing techniques are simply not adequate to maintain these types of angles and dimensional tolerances. On the other hand, they are relatively easy to maintain with diamond turning.

Problems. In spite of the advantages that diamond turning has for high-energy laser optics, there are three major problems involving materials, mechanical design, and surface finish. First, not all materials can be diamond-turned. From a strength and thermal standpoint, some of the most desirable materials, such as molybdenum, titanium, and silicon carbide, are impossible to diamond-turn. The softer, nonferrous metals, like aluminum, copper, and silver, turn beautifully but are structurally unsatisfactory. A compromise is usually made by using a high-strength substrate material that is plated with electroless nickel. It turns nicely and is hard enough that it can also be polished well if the diamond-turned surface finish is not good enough.

Second, the mechanical problem is related to how parts are supported on the diamond-turning machine and on the optical laser mount. Real physical materials are not stiff enough to hold their shape to optical tolerances under gravity and centrifugal loadings without extreme precautions and excellent mechanical design. This is a nontrivial problem and will affect high-energy laser optics independently of their method of manufacture.

Third, current diamond-turning machines are not capable of producing a sufficiently good surface finish that the optics can be adequately tested with visible light. There is a combination of high and moderately high spatial frequency roughness on the order of 30 nm root-mean-square in height that causes a substantial amount of light to be diffracted out of the desired optical path. Because these errors are high in spatial frequency, they are easily removed by conventional polishing perpendicular to the diamond-turning marks. Improvements in surface roughness by a factor of 10 are not hard to achieve by this method, and they then make the optics useful for their intended purposes. It is expected that as the art of diamond turning matures, solutions will be found to the surface finish problem, and the post-diamond-turning polishing step will be eliminated.

For background information *see* LASER; LATHE; OPTICAL SURFACES in the McGraw-Hill Encyclopedia of Science and Technology.

[ROBERT E. PARKS]

Bibliography: W. P. Barnes, Jr., Basic properties of metal optics, *Opt. Eng.*, 16:320–23, 1977; S. R. Lange and R. E. Parks, Characterization of scattering from diamond-turned surfaces, *Proc. SPIE*, 257:169, 1980; T. T. Saito (ed.), Precision machining of optics, *Opt. Eng.*, 17:569–626, 1978.

Double beta decay

The existence of a new subatomic particle, the electron neutrino, was postulated 50 years ago in order to conserve energy in nuclear beta decay, the weak process by which a nucleus of mass number A and charge Z decays into a daughter nucleus with $Z + 1$ or $Z - 1$. Later experiments in beta decay helped demonstrate unexpected properties of the weak interaction, including the maximal violation of parity and the interrelation of the charge-changing weak and electromagnetic interactions. These studies were the basis for a major theoretical advance in modern physics, the unified description of the weak and electromagnetic interactions in the model of S. L. Glashow, S. Weinberg, and A. Salam. Attempts are being made to generalize this model to encompass the strong interaction responsible for the binding of nuclei. Such grand unified theories suggest that some of the exact symmetries of the Glashow-Weinberg-Salam model, such as those responsible for massless neutrinos, maximal parity violation, and the existence of a conserved quantity called lepton number, may in fact be violated to some small degree. An important test of these symmetries may be provided by a second nuclear process, double beta decay.

Double beta decay is the radioactive decay of a parent nucleus of mass number A and charge Z to a daughter nucleus with $Z + 2$ or $Z - 2$ by emission of two electrons or two positrons. This decay will occur if the energy of the parent nucleus exceeds that of the daughter by the equivalent of at least two electron masses, $2m_ec^2$ (where c is the speed of light). As double beta decay is a second-order process of the weak interaction, it proceeds much more slowly

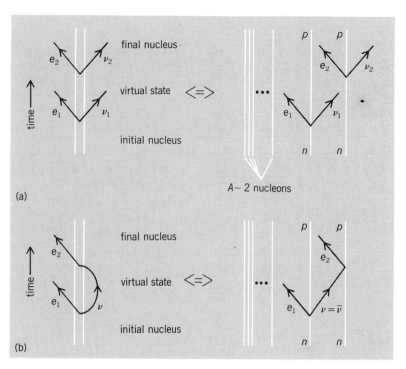

Fig. 2.   Mechanisms for (a) 2ν and (b) 0ν double beta decay. In each part of the figure, the two vertical lines in the diagram on the left represent the whole nucleus, while the diagram on the right is an expanded version in which each vertical line represents a single nucleon.

than ordinary first-order beta decay. Consequently, the study of double beta decay is feasible only in cases where the beta-decay transition to the daughter nucleus $(A, Z \pm 1)$ is either strongly hindered or energetically forbidden. This occurs frequently for parent nuclei with even Z and A, where the nuclear pairing force may render the (A, Z) and $(A, Z \pm 2)$ nuclei more stable than the odd-odd nucleus $(A, Z \pm 1)$ [Fig. 1].

Lepton number conservation. The underlying motivation for the study of double beta decay is the connection with lepton number conservation and the behavior of the electron neutrino under charge conjugation. In ordinary beta decay the neutrino and its charge conjugate, the antineutrino, are produced, together with their respective charged partners, the positron and its antiparticle, the electron, as in reactions (1). There also exist corresponding neutrino- and antineutrino-induced reactions (2). Double beta decay can then occur as a result of correlated beta decays of a pair of protons, for example, within a nucleus. This reaction (3) proceeds through a short-

$$p \rightarrow n + \beta^+ + \nu$$
$$n \rightarrow p + \beta^- + \bar{\nu} \qquad (1)$$

$$\nu + n \rightarrow p + \beta^-$$
$$\bar{\nu} + p \rightarrow n + \beta^+ \qquad (2)$$

$$2p \rightarrow p + n + \beta^+ + \nu \rightarrow 2n + 2\beta^+ + 2\nu \qquad (3)$$

lived, virtual excitation of the intermediate nucleus $(A, Z - 1)$, whose decay produces a final state with two neutrinos and two positrons (Fig. 2a). The decay rate can be measured by detecting the posi-

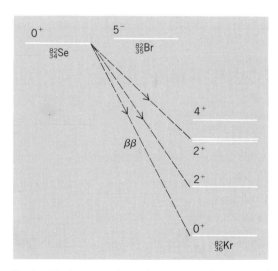

Fig. 1.   Nuclear energy levels for the double beta decay of ^{82}Se.

trons, whose summed energy has the distribution shown in Fig. 3.

It is not clear whether the neutrino and antineutrino "defined" by the beta decay reactions in reactions (1) and (2) are actually distinct. As the neutrino has no charge or measurable magnetic moment, it differs from other fermions in lacking an obvious behavior under the operation of charge conjugation. If $\nu \equiv \bar{\nu}$, that is, if the neutrino is a Majorana particle, double beta decay may also occur by exchange of a virtual neutrino between two nucleons with no neutrino emission (Fig. 2b), as in reaction (4).

$$2p \rightarrow p + n + \beta^+ + \nu$$
$$\equiv p + n + \beta^+ + \bar{\nu} \rightarrow 2n + 2\beta^+ \quad (4)$$

The lifetime for this decay is approximately $10^{13 \pm 2}$ years for Majorana neutrinos and infinite in the Dirac case, where the neutrino and antineutrino are completely distinct. In contrast, typical two-neutrino decay lifetimes are $10^{21 \pm 2}$ years, regardless of whether the neutrino is a Dirac or Majorana particle.

Early geochemical measurements showed that the double beta decay lifetimes of several nuclei exceeded 10^{17} years. Prior to 1957 it was believed that this established the Dirac character of the neutrino, and lepton number was introduced as the quantum number distinguishing ν and $\bar{\nu}$. The neutrino and electron are assigned $\ell = +1$, the antineutrino and positron $\ell = -1$. The assumption that additive lepton number is conserved then permits two-neutrino decay, but forbids neutrinoless double beta decay, for which the change in $\Sigma \ell$ is ± 2.

Following a suggestion by T. D. Lee and C. N. Yang, it was demonstrated in 1957 that the neutrino produced in beta decay is almost entirely left-handed, so that its intrinsic spin is aligned antiparallel to its direction of motion, while the antineutrino is right-handed (parallel alignment). This unexpected discovery of nearly maximal parity violation in the weak interaction exposed a flaw in the double beta decay argument for Dirac neutrinos. Even if $\nu \equiv \bar{\nu}$, double beta decay would be prohibited, as in reaction (5),

$$2p \rightarrow p + n + \beta^+ + \nu^{\mathrm{LH}}$$
$$\equiv p + n + \beta^+ + \bar{\nu}^{\mathrm{LH}} \nrightarrow 2n + 2\beta^+ \quad (5)$$

because it is a right-handed, not left-handed, antineutrino that participates in the familiar weak interactions. Therefore, the absence of neutrinoless double beta decay can be attributed to helicity suppression, irrespective of the Majorana/Dirac character of the neutrino, so that the introduction of lepton number to distinguish ν from $\bar{\nu}$ is not required by experiment.

The current interest in double beta decay stems from the predictions of modern gauge theories that the helicity of the weak leptonic current is not exact, being broken either by a small right-handed current admixture of strength η or by a nonzero neutrino mass m_ν. Experiment demands only that η be less than a few percent, while a very recent measurement of the beta decay spectrum of the triton suggests a nonzero neutrino mass in the range of 14 to 46 eV. Thus, careful searches for neutrinoless beta decay may yet provide important tests of lepton number conservation.

Experimental results. The existing experiments on double beta decay are of two types. Half-lives for three reactions, $^{130}\mathrm{Te} \rightarrow {}^{130}\mathrm{Xe}$, $^{128}\mathrm{Te} \rightarrow {}^{128}\mathrm{Xe}$, and $^{82}\mathrm{Se} \rightarrow {}^{82}\mathrm{Kr}$, have been inferred from geochemical measurements of the concentrations of noble gas daughter isotopes in old ore samples. These measurements are made possible by the high sensitivity of noble gas mass spectrometry, by the ease with which the product isotopes may be outgassed from ore samples, and by the low abundance of ambient noble gases in the Earth's crust. Unfortunately such measurements provide no direct information on the mechanism (2ν or 0ν) of double beta decay and suffer from uncertainties in ore age determination.

Laboratory experiments are extremely difficult because of the long half-lives for double beta decay and the presence of background radiation due to cosmic rays and trace quantities of radioactive isotopes in the experimental apparatus. A number of stringent limits on half-lives for neutrinoless double beta decay have been obtained by searching for two beta particles in coincidence carrying off a total energy equal to that released in the nuclear decay (Fig. 3). Because of the continuous distribution of the energy of the two beta particles, the lepton-number-conserving two-neutrino mode is even more subject to background uncertainties. Recently, the first observation of two-neutrino double beta decay was made for the transition $^{82}\mathrm{Se} \rightarrow {}^{82}\mathrm{Kr}$, yielding a half-life of $(1.0 \pm 0.4) \times 10^{19}$ years. A partial summary of experimental results is given in the table.

The experimental status of double beta decay is nebulous because of the conflict between the laboratory measurement of the two-neutrino half-life of $^{82}\mathrm{Se}$ and the earlier total half-life determined geo-

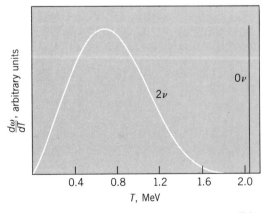

Fig. 3. Comparison of the differential decay rate $d\omega/dT$ for 0ν and 2ν double beta decay in the transition $^{76}\mathrm{Ge} \rightarrow {}^{76}\mathrm{Se}$, where T is the sum of the kinetic energies carried off by the electrons, and $d\omega/dT$ is the number of decays per unit energy interval. The 0ν spectrum is a line at $T = T_0$, the total kinetic energy released in the decay.

Partial summary of beta decay results

Reaction	T_0, $m_e c^2$*	Half-life $\tau_{1/2}$, years	Type of experiment
$^{130}\text{Te} \rightarrow {}^{130}\text{Xe}$	5.0	$10^{21.4}$, total	Geochemical
$^{128}\text{Te} \rightarrow {}^{128}\text{Xe}$	1.7	$10^{24.6}$, total	Geochemical
$^{82}\text{Se} \rightarrow {}^{82}\text{Kr}$	5.9	$10^{20.4}$, total	Geochemical
		$10^{19.0 \pm 0.2}$, 2ν	Laboratory
		$\geqq 10^{21.5}$, 0ν	Laboratory
$^{876}\text{Ge} \rightarrow {}^{76}\text{Se}$	4.0	$\geqq 10^{21.7}$, 0ν	Laboratory
$^{48}\text{Ca} \rightarrow {}^{48}\text{Ti}$	8.4	$\geqq 10^{19.5}$, 2ν	Laboratory
		$\geqq 10^{21.3}$, 0ν	Laboratory

*Total kinetic energy carried off by electrons or neutrinos, or both, in units of the electron mass.

chemically, which is longer by a factor of about 25. Theoretical estimates of the two-neutrino half-life are roughly in accord with the laboratory measurement, but are highly uncertain because of the difficult nuclear structure calculations required. New laboratory experiments are planned which should resolve this conflict.

Limits on lepton number violation. Implicit in these experimental results are stringent limits on the mass and right-handed coupling of Majorana electron neutrinos. Perhaps the most interesting case is the decay of the two tellurium isotopes, ^{130}Te and ^{128}Te. Because of the different energies released in these decays, equivalent to 7.0 and 3.7 electron masses, respectively, it can be shown that the ratio of the total geochemical half-lives is quite sensitive to the mode of double beta decay. If it is assumed that the nuclear matrix elements determining the decay amplitudes of these similar nuclei are equal, an assumption that nuclear structure calculations support, this ratio (6) can be calculated easily for two-

$$\frac{\tau_{1/2}(^{128}\text{Te})}{\tau_{1/2}(^{130}\text{Te})} = \begin{cases} 25, & 0\nu \\ 5100, & 2\nu \end{cases} \qquad (6)$$

neutrino and no-neutrino decay. Although the experimental situation is unsettled, the most recently published value for this ratio, 1590, is bracketed by these limits and suggests that both 2ν and 0ν double beta decay is occurring. Detailed calculations show that a Majorana neutrino with a mass of 10 eV or a right-handed coupling of strength $\eta = 5 \times 10^{-5}$ is consistent with this measurement. In the absence of direct detection of neutrinoless double beta decay, however, this result should be interpreted cautiously as establishing rough upper limits on the strength of lepton number nonconservation. Slightly less stringent limits on η and the Majorana neutrino mass can be derived from laboratory limits on the neutrinoless decay of ^{76}Ge and ^{82}Se, the derivations of which require calculation of nuclear matrix elements and thus have unknown uncertainties.

Recently, a measurement of the end-point spectrum in the ordinary beta decay of the triton provided evidence for a nonzero neutrino mass between 14 and 46 eV. If this result is confirmed, double beta decay measurements have just reached the sensitivity where the behavior of Dirac and Majorana neutrinos can be distinguished. This exciting result

explains the present enthusiasm for extending the sensitivity of neutrinoless double beta decay measurements by one or two orders of magnitude, as under the most cautious interpretation such results would test the charge-conjugation properties of neutrinos in the relevant mass range.

For background information *see* FUNDAMENTAL INTERACTIONS; NEUTRINO; PARITY (QUANTUM MECHANICS); RADIOACTIVITY; SYMMETRY LAWS (PHYSICS); WEAK NUCLEAR INTERACTIONS in the McGraw-Hill Encyclopedia of Science and Technology.

[W. C. HAXTON]

Bibliography: D. Bryman and C. Picciotto, Double beta decay, *Rev. Mod. Phys.*, 50:11–21, 1978; H. Primakoff and S. P. Rosen, Double beta decay, *Rep. Prog. Phys.*, 22:121–166, 1950; S. P. Rosen, Lepton nonconservation and double beta decay, *Proceedings of the 1981 International Conference on Neutrino Physics and Astrophysics*, pp. 76–92.

Earth, age of

The age of the Earth is presently measured by the same radiometric methods used to date rocks and minerals, but the application of these methods is necessarily more complex. In the simplest case, the age of a mineral is found by determining the quantity of radioactive parent isotope present in a sample and the amount of radiogenic daughter isotope that has been produced by radioactive decay of the parent. Since the daughter isotope is commonly not entirely of radiogenic origin, a correction must be made for the amount of nonradiogenic daughter isotope present in the mineral when it first crystallized. The age, calculated from the known rate of decay of the parent, will be correct if the mineral has behaved as a closed chemical system (no gain or loss of parent or daughter by processes other than radioactive decay) since crystallization.

Determining the age of the Earth also involves the same problems encountered in measuring the age of a rock or mineral. Basically these are to establish the amount of radiogenic daughter produced by the radioactive parent, which requires a proper correction for initial nonradiogenic daughter, and evidence for closed-system behavior. The recent developments in this field have led mainly to increased confidence in the validity of the corrections used for nonradiogenic initial daughter isotopes, with no significant change in the formerly accepted value for the age of the Earth.

Terrestrial-meteorite lead age. The following discussions consider the method for determining the Earth's age based on comparison of terrestrial with meteorite data. The most reliable measure of the age of the Earth is obtained by comparing terrestrial and meteorite lead-isotope data.

This approach is necessary because the Earth has undergone such extensive geochemical changes since formation that no record exists for its initial state, in particular for the isotopic composition of its initial lead. On the other hand, the parent bodies of meteorites were small in comparison with the Earth, so that geochemical differentiation terminated shortly after their formation, preserving their initial geochemical record.

The uranium-lead system is used to date meteorites, as shown in the table. Then, for meteorites

Uranium-lead system for dating meteorites

Parent	Daughter	Half-life, 10^9 years
^{238}U	^{206}Pb	4.47
^{235}U	^{207}Pb	0.704
—	^{204}Pb	No long-lived parent

that have existed as closed chemical systems since the time of crystallization, Eq.(1) can be applied, where prim denotes primordial, or initial, ratios and rad denotes the radiogenic components. Since radiogenic ^{207}Pb and ^{206}Pb are produced at different rates, the ratio of the two isotopes depends on the age of the samples, as shown in Eq.(2), where T is the age, ^{235}U and ^{238}U are the present-day abundances of the two isotopes, whose decay constants are λ^5 and λ^8, respectively.

$$\frac{\dfrac{^{207}Pb_{rad} + {}^{207}Pb_{prim}}{^{204}Pb_{prim}} - \left(\dfrac{^{207}Pb}{^{204}Pb}\right)_{prim}}{\dfrac{^{206}Pb_{rad} + {}^{206}Pb_{prim}}{^{204}Pb_{prim}} - \left(\dfrac{^{206}Pb}{^{204}Pb}\right)_{prim}} = \frac{^{207}Pb_{rad}}{^{206}Pb_{rad}} \quad (1)$$

$$\frac{^{207}Pb_{rad}}{^{206}Pb_{rad}} = \frac{^{235}U\,(e^{\lambda^5 T} - 1)}{^{238}U\,(e^{\lambda^8 T} - 1)} \quad (2)$$

Since investigations show that the isotopic abundances to the two uranium isotopes are constant in nature, with $^{235}U/^{238}U = 1/137.9$, the ratio $^{207}Pb_{rad}/^{206}Pb_{rad}$ determines an age T. The primordial lead ratios are usually taken as those measured directly from the troilite (iron sulfide) phase of the iron meteorite, Canyon Diablo. The troilite has such high lead and low uranium concentrations that the lead ratios have remained unchanged since the time of crystallization. With the primordial ratios established, it is possible to show that most (though not

all) stone meteorites crystallized 4.56 billion years ago.

If it is assumed that the primordial lead ratios from the troilite also apply to the Earth, an age of the Earth can be calculated from the radiogenic $^{207}Pb/^{206}Pb$ ratio of terrestrial lead. Clair Patterson first used this method, selecting a basalt and ocean sediment as sources for modern terrestrial lead. Those data yielded an age of 4.43 billion years for the Earth (adjusted to present values for uranium decay constants and primordial lead ratios). In this case, the geochemical evolution of the Earth complicates the choice of the terrestrial lead because the closed chemical system requirement is almost certainly violated. This complication led to the suggestion that the age calculated from young rocks was strongly influenced by geochemical differentiation in the source. An older age of 4.56 billion years (adjusted for present-day values of decay constants and newer measurements of the primordial lead ratios) was calculated when a 2.7-billion-year-old lead was substituted for the young terrestrial leads. This is logical since the aberration due to geochemical differentiation should be less for the older rocks. The age of the Earth calculated this way also agrees very closely with the stone meteorite ages.

A recent calculation of the age of the Earth was done by using a modified approach that does not require exact knowledge of the Earth's initial lead, but only that the isotopic ratios were initially uniform, and that U/Pb ratios were initially uniform in the Earth. Five leads ranging in age from 3.4 to 2.65 billion years old from three continents yielded a value of 4.54 billion years by this method. Four younger leads, 0.4 to 2.60 billion years old, yielded a younger apparent age of the Earth of 4.46 billion years, presumably due to the effects of geochemical differentiation. These findings agree well with those using a 2.7-billion-year-old lead in place of young terrestrial leads in the $^{207}Pb/^{206}Pb$ ratio.

The use of troilite from iron meteorites to determine primordial lead ratios has sometimes been criticized on the basis that the parent material of meteorites had to experience considerable differentiation to produce troilite. If this process took place over a long period of time, the ratios might conceivably have increased due to added radiogenic lead. A recent detailed study of lead isotope ratios in three stone meteorites for which the corrections for radiogenic lead are small—a few percent, owing to the lead-rich character of the stones—indicates that none of these meteorites, whose differentiation history is much simpler than that of the troilite, gives any evidence for $^{206}Pb/^{204}Pb$ or $^{207}Pb/^{204}Pb$ ratios that are lower than those in the troilite. It appears that the ratios had not evolved in the source materials before the troilite separated, and that the troilite ratios are suitable values for primordial lead of meteorites.

Chondrite-neodymium age. Another uncertainty in the age calculation is the extent to which data from meteorites can be applied to the Earth. The agreement in the lead-lead ages of stone meteorites and the Earth suggests, but does not prove, that the primordial ratios assumed for the earth are correct. New supporting evidence for the general applicability of the meteorite model to the Earth comes from another decay system, that based on the decay of samarium-147 to neodymium-143. It is difficult to determine an age of the Earth directly from a single decay system such as this in as much as the Sm/Nd ratio in the sampled rock is different from that in the source because of the differentiation accompanying the production of the magma. However, it is possible to determine the initial $^{143}Nd/^{144}Nd$ ratios of igneous rocks of various ages and to test whether the ratios appear to fit evolution in a source having a Sm/Nd ratio similar to that in chondritic meteorites. The Sm/Nd ratio is very uniform in chondrites, much more uniform than the U/Pb ratio. From this observation, plus measurement of the $^{143}Nd/^{144}Nd$ ratio can be calculated as a function of time. Such an evolution curve is shown in the illustration, which also gives a comparison of the initial ratios in terrestrial igneous rocks from around the world with the chondrite evolution lines. For the most part the agreement is very good for basaltic rocks over a time span reaching back 3.6 billion years. Basaltic rocks generally originate from the Earth's outer mantle and contain little or no crustal material, so that they are likely to be the least differentiated material. Granitic rocks, which often contain a component of recycled older crustal materials, fit the chondrite evolution line quite well for rocks older than 2 billion years, but show much scatter in younger rocks. This is expected in areas where the Earth's crustal rocks grow increasingly heterogeneous over geological time because of geochemical differentiation processes. The net result of the recent Sm-Nd studies is to underscore the general similarities between isotope ratios in the primordial Earth and meteorite parent bodies.

Future estimates. An interesting question concerns how estimates of the age of the Earth are likely to change in the future. The age derived from the

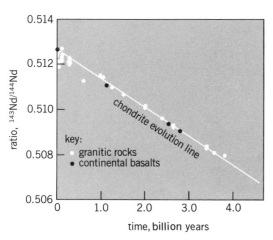

Isotopic composition of neodymium in igneous rocks as a function of age. The age of the Earth is calculated as 4.56 billion years.

meteorite dating method, pioneered by Patterson, has remained essentially unchanged over the past 30 years. Improved precisions in measurements and new geochemical methodologies (for example, the Sm-Nd studies cited above) have all served only to further establish the validity of the 4.55-billion-year age. It seems safe to assume that, barring radical changes in the understanding of fundamental concepts such as the laws of radioactivity, the present value will not change significantly in the future.

For background information *see* EARTH, AGE OF; LEAD ISOTOPES (GEOCHEMISTRY); METEORITE; ROCK AGE DETERMINATION in the McGraw-Hill Encyclopedia of Science and Technology.

[GEORGE R. TILTON]

Bibliography: C. J. Allegre and D. B. Ben Othman, Nd/Sr isotope relationship in granitoid rocks and continental crust development: A chemical approach to orogenesis, *Nature*, 286:335–342, 1980; B. B. Hanan and G. R. Tilton, Pb-Sr isotope study on unequilibrated chondrites, *EOS*, 63:363 (abstr.), 1982; F. Tera, Aspects of isochronism in Pb isotope systematics: Applications to planetary evolution, *Geochim. Cosmochim. Acta*, 45:1439–1448, 1981.

Echinodermata

Echinoderms are conspicuous and convenient animals for study in the deep sea, where knowledge of the nature and rates of life processes of bottom-dwelling animals is particularly sparse. There are, however, technological difficulties in collecting live deep-sea animals for study in the laboratory and also difficulties in making long-term observations from deep-diving submersibles. Sampling populations at a fixed station every few months therefore remains a good way to understand both population turnover and the processes involved in breeding, growth, and survivorship of animals in this habitat.

Two assumptions are generally made concerning deep-sea fauna. First, because of the supposed stability of physical conditions, particularly temperature, which serves as a cue for seasonal breeding in coastal populations, it is assumed that biological rhythms are absent. Second, because of the low fallout of edible material, indications of low rates of degradation of organic material by bacteria, and low density of animal life on the deep ocean floor, biologists have assumed that there may be a general slowing in the tempo of life at great depths. Studies of brittle stars and starfish from sample time series from the Rockall Trough, located west of the British Isles, have cast doubt on the validity of both of these generalizations.

Reproductive patterns (brittle stars). Two species of brittle star show annual periodicity in breeding; *Ophiura ljungmani*, and *Ophiocten gracilis*. *Ophiura*, which is found throughout the deep Atlantic Ocean at depths of up to 4 km (2.5 mi), is a small species whose body form appears very similar to closely related species from shallow water. Slices of gonadal tissue taken from individuals at different times of the year show that both male and female gonads undergo a rapid, synchronized development in autumn and become ripe in winter, with spawn-out probably occurring in late January or early February. The large number (up to 5700 per individual) and small size (maximum diameter 90 μm) of the ripe eggs suggest a prolonged development of the free-swimming larvae.

Ophiocten is found on the upper continental slope around the northern perimeter of the North Atlantic. Reproductive development is similarly synchronous, with spawning probably occurring in early spring. Both egg size and fecundity of *Ophiocten* resemble that of *Ophiura*. However, in the case of *Ophiocten* a long-armed larval form called *Ophiopluteus ramosus*, which was previously described from the plankton of the surface, has been almost certainly linked to *Ophiocten gracilis*. This larva, collected from mid-water plankton hauls from the Rockall Trough, would appear to change to a miniature adult form in late spring and then fall to the bottom. That this occurs over a wide area and at depths adverse to their subsequent survival is evident from fine-meshed bottom trawlings taken from deep water. In terms of numbers of smaller animals, such samples may be dominated by these postlarvae in summer-time. Yet many may be dead on reaching the bottom, and the sample time series makes it clear that none survives into the following winter. Such wasteful overdispersal of larvae is now well known among coastal species. However, it is somewhat unexpected in the deep sea, where the low fallout rate of detrital material to the bottom favors a high degree of adaptive refinement in deep-sea animals in order to make the best use of the sparse resources. Another consequence of long planktonic development in coastal waters is a sporadic or variable level of recruitment from year to year. Such variability is also evident from the time series of samples of *Ophiura ljungmani*.

However, two other species studies, the spiny-armed *Ophiacantha bidentata* and the larger *Ophiomusium lymani*, show reproductive strategies closer to those expected in the deep sea. *Ophiacantha* has large egg size (about 650 μm), lower fecundity (around 200 eggs per individual), and absence of any seasonal periodicity in gametogenic development, features that were previously considered characteristic of deep-sea animals. Populations of *Ophiomusium* are found on the sediment surface of the lower continental slope at around 2 km (1.2 mi) depth. Analysis of egg-size frequencies in the gonads of females has also failed to detect any kind of synchrony in reproductive development. The egg size of *Ophiomusium* (maximum 460 μm), fecundity (1200 eggs per individual), and lack of seasonal breeding indicate a short external development with the larva subsisting on yolk reserves before changing to the adult form.

Size-frequency distributions (brittle stars). As expected, *Ophiacantha* shows no sign of seasonal variation in recruitment, and shows a bimodal dis-

tribution of size frequencies. A left-hand peak represents fast-growing juveniles that probably experience high mortality before they reach adult size, at which time growth slows, forming a right-hand peak of breeding adults. A similar size structure is shown by *Ophiomusium*, but in this case low peaks are discernible in the frequencies between the two peaks that must represent annual pulses or recruitment. It seems possible that this pulsed recruitment results from a higher survival of baby brittle stars after their settlement in summer, a time when rapidly sinking organic aggregates, such as fecal pellets from the springtime burst in planktonic production at the surface, become available as food. With a slowing of growth on reaching breeding size, these year-class frequencies stack up on one another to give rise to the apparently unimodal peak of adults. The size of this mode relative to juvenile frequencies clearly will depend on how fast adults die off, while the degree to which growth is slowed (as a result of the burden of reproduction) will determine the mean body size of adults making up this peak.

By using a computer it is possible to simulate size-frequency distributions for *Ophiomusium* conforming

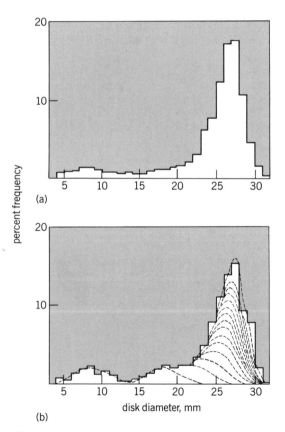

(a)

(b)

Fig. 1. Frequency distributions of the size of the central disk of the body of the brittle star *Ophiomusium lymani*. (a) Measurements of a population sample from the Rockall Trough. (b) Computer simulation of age structure (dotted lines) to show that as growth slows, year classes form a stack of frequencies. (*After J. D. Gage and P. A. Tyler, Growth strategies in deep-sea ophiuroids, Proceedings of the International echinoderm Conference, Tampa, 1981, Balkema, in press*)

to different rates of mortality and growth in order to compare samples obtained from different parts of its worldwide range. These results suggest that *Ophiomusium* reaches breeding size by its second or third year after settlement, but that the oldest adults in the stock of adult frequencies in Rockall may be up to 45 years old (Fig. 1). Elsewhere a large adult stack is absent; the lower survival rate of adults inferred is perhaps the result of greater competition for resources among very dense populations.

Growth and survivorship (brittle stars). It is possible to follow the growth in body size of modes in size frequencies that correspond to successive animal recruitments of *Ophiura ljungmani* only for juveniles. The rapidly reducing, and hence more randomly scattered size frequencies of adults probably are the result of a high mortality throughout the lifetime of the brittle star. Growth, although similar to that for *Ophiomusium* in that there is slowing toward an asymptotic size, is slower than that of *Ophiomusium*.

When growth and survivorship curves for the two species are compared (Fig. 2), it is possible to interpret them in terms of a dichotomous model. The model is driven by mortality pressure on a theoretical population *a*; the particular type of growth curve evolved, *ac* or *ab*, depends on the predictability of recruitment success. Species with a long, free-living larval life, such as *Ophiura ljungmani* and *Ophiocten gracilis*, probably are at the greatest survival risk, leading to variable recruitment success as a result of the vicissitudes of life in the plankton. Features of particular value to such type *ab* species, in addition to high fecundity, are longevity and early maturity in order to spread the risk of consecutive bad years in recruitment (that might result in extinction of the population) over as long a period as possible. It is possible that such populations never reach a full equilibrium with the carrying capacity of their environment, so that competition for resources may not be so keenly felt as for two *ac* species, such as *Ophiomusium* and *Ophiacantha*. Type *ac* species suffer less variability in survival of their yolk-fed larvae, although the size frequencies show that high mortality nevertheless still occurs. The population is able to respond to this by rapid growth to a large size, or a spiny body form. Breeding adults only grow slowly since greater diversion of resources is required to produce even a modest number of such large, yolky eggs. Life may be short in dense populations, where any impairment in performance—resulting, for example, from age—cannot be tolerated in a habitat where resources are exploited to the full and where rapidly growing young require every scrap of available food.

Starfish. Such variety in life histories is found also among deep-sea starfish. Species from Rockall show a wide variation in fecundity, accessory cell development, and resorptive processes in the gonad. Most species have the large eggs suggestive of nonfeeding larval development while being carried by currents for a short time near the ocean floor. However, two

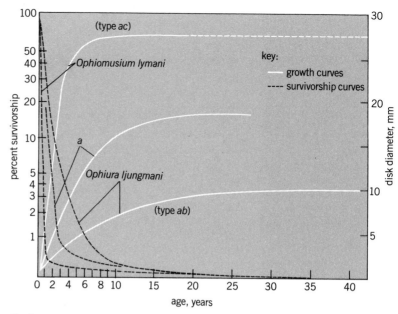

Fig. 2.   A growth and survivorship model to explain the evolution of contrasting life histories exemplified by *Ophiomusium lymani* and *Ophiura ljungmani*. There are two possible responses of a theoretical population *a* to mortality pressure: toward a prolonged breeding life and slowing of growth if mortality is heavy at all stages but variable only among young (type *ab*), or toward growing and breeding as quickly as possible to minimize risk of extinction where there is constant mortality loss of young (type *ac*). (*After J. D. Gage and P. A. Tyler, Growth strategies in deep-sea ophiuroids, Proceedings of the International Echionderm Conference, Tampa, 1981, Balkema, in press*)

species have the small eggs and high fecundity characteristic of prolonged and independent larval life. It is particularly significant that these two species show evidence of seasonal breeding timed to coincide with the annual cycle in planktonic production at the surface, although their larval forms await identification.

Conclusion. Because there seem to be clear advantages for some deep-sea animals to synchronize their life histories to the annual productivity cycles at the surface, it may be surprising that more species do not do so. In fact, deep-sea animals appear as variable in these respects as those in shallow water. However, it is possible that many species respond in less obvious, and so far undetected, ways to periodicities in their environment, perhaps by subtle shifts in biochemical mechanisms involving storage of food reserves.

For background information *see* DEEP-SEA FAUNA; ECHINODERMATA; PERIODICITY IN ORGANISMS in the McGraw-Hill Encyclopedia of Science and Technology.

[J. D. GAGE]

Bibliography: J. D. Gage and P. A. Tyler, Growth strategies in deep-sea ophiuroids, *Proceedings of the International Echinoderm Conference*, Tampa, 1981, in press; F. J. Rokop, Reproductive patterns in the deep-sea benthos, *Science*, 186:743–745, 1974; P. A. Tyler and J. D. Gage, Gametogenic cycles in deep-sea phanerozoic asteroids from the N.E. Atlantic, *Proceedings of the International Echinoderm Conference*, Tampa, 1981, in press.

Ecological interactions

Many marine invertebrates, especially in tropical waters, harbor endosymbiotic zooxanthellae, unicellular dinoflagellates which are able to forgo a free-living existence for symbiotic life inside the tissues of an animal. Reef-forming scleractinian corals and the giant clam *Tridacna* are well-known examples of hosts in a symbiosis which results from free-living dinoflagellates directly infecting or entering the animal host, most probably through the gut, either in the larval stage of the animal or soon after settlement. Recently a number of nudibranchs have been discovered that harbor symbiotic zooxanthellae obtained from eating zooxanthellae-containing coelenterates. The coelenterate tissue is digested in the gut, and the intact, functional zooxanthellae are removed and relocated in specialized nudibranch cells where they begin a new life in a new host. The nudibranch-zooxanthellae symbiosis is quite different in that the zooxanthellae are already endosymbiotic when ingested by the nudibranch.

Morphological adaptations. This secondary symbiosis has been found to have evolved independently at least six times within the nudibranch suborders Aeolidacea and Arminacea, and it has led to major morphological changes in the animals concerned. Studies of genera with both symbiotic and nonsymbiotic members show that the changes involve the development of mechanisms to place maximum numbers of captive zooxanthellae in the best positions for efficient photosynthesis. Most aeolids are hydroid feeders and are usually found living on their food. Their basic anatomy is fairly uniform, and the most characteristic feature is the presence of cerata—tubular outgrowths of the body wall which occur in clusters or rows on the back or the sides of the elongate body. Each ceras is a blood-filled papilla containing an unbranched tubule of the digestive gland. In the alcyonarian-feeding genus *Phyllodesmium*, one species, *P. poindimiei* (Fig. 1), feeds on a telestacean alcyonarian with few zooxanthellae; it is always found with its food, and has no symbiosis and no major morphological adaptations. However, in *P. cryptica* and two other species which feed on *Xenia*, an alcyonarian with many symbiotic zooxanthellae packed in its tentacles, the cerata have become flattened, and the single digestive gland tubule has been replaced by a ramifying network of fine tubules throughout the cerata. The *Xenia* feeders are normally found on or near *Xenia*, and histological evidence suggests a high turnover and low retention rate of zooxanthellae. Two other species, *P. macphersonae* and *P. longicirra*, have never been found feeding which indicates that they are getting sufficient nutrients from the zooxanthellae. In particular, *P. longicirra* (Fig. 2) has spectacular morphological adaptations for zooxanthellae symbiosis. The cerata have become large, flattened, paddle-shaped structures, and the digestive gland forms a ramifying network of fine tubules not only in the cerata but in all parts of the dorsal and lateral body wall. Healthy zooxanthellae are found in cells of the digestive gland, especially where the gland is closest to the outer body wall and thus nearest to the light. *Phyllodesmium longicirra* is truly a solar-pow-

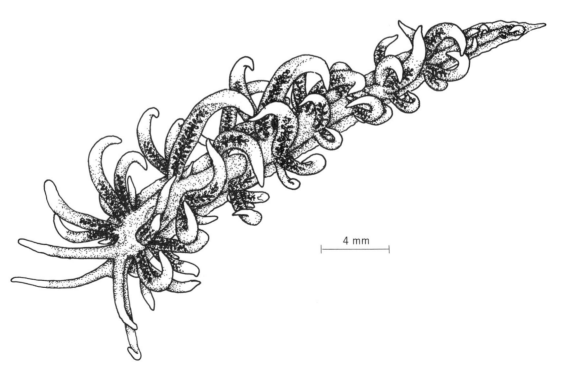

4 mm

Fig. 1. *Phyllodesmium poindimiei*, a nonsymbiotic alcyonarian-feeding aeolid. (*From W. B. Rudman, The anatomy and biology of alcyonarian-feeding aeolid opisthobranch mol-luscs and their development of symbiosis with zooxanthellae, Zool. J. Linn. Soc., 72(3):219–262, 1981*)

Fig. 2. *Phyllodesmium longicirra*; dark irregular rings are gardens of zooxanthelae.

ered nudibranch, with large light-collecting paddles containing gardens of zooxanthellae which appear as irregular brown rings in the live animal.

The pattern of flattening of cerata and breaking up of the digestive gland into a network of fine tubules throughout the body wall, as well as the cerata, is also found in the anthozoan-feeding family Aeolidiidae, with *Aeolidiopsis* feeding on the colonial anthozoan *Palythoa*, and *Spurilla* feeding on solitary sea anemones. In another family, the Glaucidae, *Pteraeolidia ianthina* has increased its light-collecting surfaces not by flattening the cerata but by arranging them in fan-shaped clusters along the sides of a greatly elongated body. Zooxanthellae are found both in the cerata and in ramifications of the digestive gland throughout the body wall, and like *Phyllodesmium longicirra*, *Pteraeolidia ianthina* (Fig. 3) has never been found feeding.

In the suborder Arminacea, two apparently unrelated nudibranchs have evolved zooxanthellae symbioses, and in each case they have developed morphological changes to place zooxanthellae in optimal positions for photosynthesis. In most arminaceans the digestive gland is a large, discrete mass within the visceral cavity, but in *Pinufius rebus*, which lives and feeds on the scleractinian coral *Porites*, and in *Doridomorpha gardineri*, which feeds on the alcyonacean blue coral *Heliopora*, the digestive gland is broken up into a network of tubules close to the dorsal body wall. In *Pinufius*, ceratalike papillae have evolved, and in *Doridomorpha*, a wide mantle skirt has developed; in both cases the dorsal body surface has been thus increased.

Evidence. Most evidence at present comes from comparative anatomy, histology, and behavioral ob-

servations, and initial physiological studies are confirming the existence of a functional zooxanthellae-nudibranch symbiosis. The morphological changes are all involved in placing zooxanthellae in the largest possible area of lighted surfaces. In those aeolids with the most complex morphological adaptations the animal no longer needs to be constantly near its coelenterate food. A parallel can be seen among the herbivorous sacoglossan opisthobranchs, some species of which remove intact algal plastids from the algae they feed on and retain them as functional symbionts in the cells of their digestive glands. The major morphological adaptations involve mechanisms which place the maximum number of chloroplasts in the best positions for efficient photosynthesis. The digestive gland has broken up into a network of fine tubules and the body wall has developed either large flaplike parapodia or dorsal papillae. Also, those species with the greatest ability to retain plastids are seldom found near their algal food.

Physiological studies on the zooxanthellae-nudibranch symbiosis have begun in Australia and Hawaii using *Pteraeolidia* and *Spurilla*. These preliminary studies show that a substantial proportion of carbon fixed by the zooxanthellae in photosynthesis goes to the animal, and that homogenates of animal tissue cause leakage of photosynthetic products from

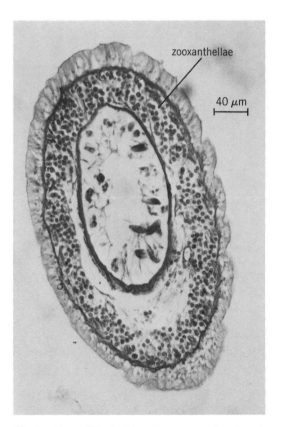

Fig. 3. *Pteraeolidia ianthina*; transverse section through upper part of ceras showing band of zooxanthellae in subepithelial layer. (*From W. B. Rudman, The taxonomy and biology of further aeolidacean and arminacean nudibranch molluscs with symbiotic zooxanthellae, Zool. J. Linn. Soc., 74(2):147–196, 1982*)

the zooxanthellae. *Pteraeolidia* with few zooxanthellae, starved in the light, can greatly increase its zooxanthellae population, and a comparison between specimens of *Pteraeolidia* starved in light and dark conditions show that those in the dark have a significantly greater loss of weight than those in the light.

Significance. All the available evidence suggests that the nudibranchs have evolved mechanisms to maintain healthy populations of zooxanthellae in their tissues and are able to extract photosynthetic products from the dinoflagellate partner. In scleractinian corals the zooxanthellae symbiosis is known to enhance calcium carbonate deposition, but this is of no significance to a shell-less animal. The seemingly most advanced zooxanthellae symbionts have apparently lost the dependence most aeolids have on their coelenterate food. This at first may appear advantageous, but as dependence on a specialized food source also greatly enhances the chances of meeting and mating, the advantage of the symbiosis remains obscure. In terms of general studies on zooxanthellae symbiosis, the secondary symbiosis described here provides an interesting insight. Recent studies have suggested that because zooxanthellae in different hosts have different reactions, the zooxanthellae species *Gymnodinium microadriaticum* may in fact be a complex of species. The discovery of zooxanthellae that can be removed from a symbiotic relationship with corals, anthozoans, and alcyonarians, and relocated in nudibranchs in another phylum, suggests that symbiotic zooxanthellae are capable of great physiological plasticity. The reported physiological differences may reflect a temporary physiological adaptation to an immediate host rather than a deeper genetic difference.

For background information *see* ALGAE; ECOLOGICAL INTERACTIONS; NUDIBRANCHIA in the McGraw-Hill Encyclopedia of Science and Technology.

[W. B. RUDMAN]

Bibliography: W. B. Rudman, The anatomy and biology of alcyonarian-feeding aeolid opisthobranch molluscs and their development of symbiosis with zooxanthellae, *Zool. J. Linn. Soc.*, 72(3):219–262, 1981; W. B. Rudman, Further studies on the anatomy and ecology of opisthobranch molluscs feeding on the scleractinian coral *Porites*, *Zool. J. Linn. Soc.*, 71:373–412, 1981; W. B. Rudman, The taxonomy and biology of further aeolidacean and arminacean nudibranch molluscs with symbiotic zooxanthellae, *Zool. J. Linn. Soc.*, 74(2):147–196, 1982.

Ecosystem

The answer to the question of why there are so many kinds of plants and animals depends upon the scale being considered. On a large scale of time and space, the number of all species on Earth at any time is a function of the rates of extinction and of evolution of new species. The history of this process is contained in the fossil record, which reveals a pattern of changes in evolutionary time: sudden mass extinctions followed by bouts of rapid evolution of new species are interspersed with periods of little or no change.

A similar pattern of events is seen on a smaller scale of time (ecological time) and space. Rapid changes in species diversity can be observed in a natural community over a few years. A disturbance (for example, a storm, fire, or flood) may cause the reduction or extinction of local populations in parts of the area. As in the fossil record, this is followed by a rise in diversity, not through evolution, but by an invasion of species from the surrounding undisturbed community. With time, the increase in diversity may stop and either level off or even decrease. Examples of such changes are shown in Figs. 1 and 2. The initial increases in diversity are due to gradual invasion of more and more species. The latter reductions are due to chance local extinctions or to competitive elimination of some species by others that are more efficient in exploiting the resources or more effective in interfering with their neighbors. If this process continues long enough, the diversity may stabilize because the resident species, being the superior competitors, cannot be replaced by others and are capable of self-replacement.

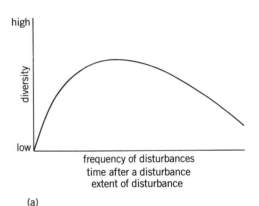

(a)

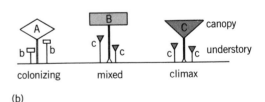

(b)

Fig. 1. Graphic representations of the intermediate disturbance hypothesis for the maintenance of species diversity. (a) Species diversity reaches a maximum for a certain frequency of disturbances. After a disturbance the diversity increases, reaches a maximum, and then decreases to a stable value. Diversity also depends on the extent of the disturbance and is usually low for small or very large disturbances. (b) Pattern of changes in species composition of adult and juvenile trees at three different successional stages of the Budongo forest in Uganda. First stage is dominated by a few species (class A), and juveniles are from a different species (class B). In the second stage, adults of class B dominate, and juveniles are again from a different species (class C). In the third stage, species of class C occur as juveniles and adults. (*After J. H. Connell. Diversity in tropical rain forests and coral reefs, Science, 199:1302–1310, 1978*)

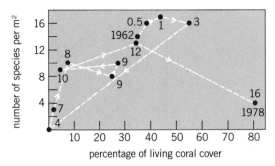

Fig. 2. Changes in species diversity and percent cover of living corals on a permanently marked site on the reef at Heron Island, Queensland, Australia (1962–1978). Sequence of observations is shown by the arrows; number of years elapsed since the original observation in 1962 is indicated at each point.

Maintenance of high diversity. Several possible mechanisms can operate to maintain high diversity under equilibrium conditions. In the first, called niche differentiation, each species is the most efficient competitor on a particular range of resource types. The number of species in the local community is then determined by the heterogeneity of resources and by the degree to which they can be divided between the different specialist species.

In the second mechanism, called compensation, the process of competitive elimination is not allowed to proceed to completion. For example, if the mortality rate is highest among the most common species, then, as another species begins to win in the competition for resources (and therefore becomes more common) its advantage will be compensated by an increase in its mortality. The same would apply if the reproductive or growth rates were reduced more by crowding from its own species than from other species. If these compensatory processes operate strongly, no species will become so common as to eliminate its competitors.

A third possible mechanism is called circular networks. Here, instead of the existence of a single competitor that can eliminate all others, the competitive network is circular: A can eliminate B, B can eliminate C, and C can eliminate A. If competition occurs between pairs of species, they will coexist without elimination.

These three mechanisms can operate while the community is in a stable state of equilibrium. If it is not at equilibrium, at least two other mechanisms can operate. During the process of initial invasion and competitive elimination the species composition is changing toward an equilibrium. If another disturbance occurs, this process is interrupted and set back. Since the diversity starts low, increases, and then drops toward the final equilibrium state, interruptions tend to keep the community in the intermediate stages, in which diversity is higher than before or later. This has been called the intermediate disturbance mechanism of species maintenance (Fig. 1). Lastly, under certain circumstances (perennial species, changing environments that favor the recruitment of offspring of different species at different times) diversity can be maintained by chance; hence this mechanism has been termed a lottery.

Species fluctuations in nature. For the first set of mechanisms to apply, the populations are assumed to have reached a stable equilibrium state. To find out how often this happens in nature the population have to be observed over at least one complete turnover (replacement of one generation by a new generation) to see whether they are maintaining the same population size. A recent survey of studies in which this had been done showed that a few populations changed very little over one or more turnovers, whereas most changed considerably. Within the few communities with sufficient data to judge, some species changed little while others changed a great deal, so that the overall species composition fluctuated. In general, few populations or communities appeared to have reached a stable equilibrium state.

Several examples of the nonequilibrium maintenance of species diversity will show how this happens in nature. One of the first observations of this pattern was made in the Budongo rain forest in Uganda: when the forest was spreading into grassland, the diversity was low; only a few species colonized the grassland. However, they did not replace themselves; in the colonizing forest the young trees were of different species and of a high diversity. When these grew up they formed the highly diverse mixed rain forest. However, the young trees in this forest were mainly a single species, ironwood; most of the mixed forest species did not replace themselves. The final stage was a self-replacing stand of low diversity, mainly ironwoods (Fig. 1). Thus, diversity was highest at an intermediate stage of succession after invasion of large openings.

A second example comes from a wave-beaten seashore in California, where boulders are often overturned, burying the plants and animals that live on top of the boulders and exposing those that ordinarily are sheltered by living on the bottom. This disturbance kills both types and thus provides empty space which new species can invade. Since small boulders overturn more often than larger ones, observations can be made over a continuum of disturbance frequency. This study showed that on the tops of small boulders the community consists of only the few weedy species that are adapted to quick invasion and rapid growth to maturity. On middle-sized boulders this happens at first, but then other species that grow more slowly begin to invade. They gradually fill in the spaces vacated when the earlier ones die. Thus, the species diversity rises gradually. On very large boulders, which are overturned only in very severe storms, this process of succession may continue for several years. After 1 or 2 years one particular seaweed species begins to predominate, spreading into vacated spaces, persisting, and excluding any other invaders. By simply holding the space and gradually spreading, it finally covers almost all the space. Thus, high diversity occurs only on middle-sized boulders, which are disturbed at intermediate frequencies; smaller and larger boulders have lower diversity because they are disturbed more or less frequently, respectively.

Another study concerns the diversity of corals on tropical reefs. By following the events on permanently marked quadrats for many years, two episodes of destruction of corals by hurricanes were witnessed. The first killed all the corals (Fig. 2, fourth year). New species of corals invaded and the diversity again increased. A second hurricane in the tenth year pruned off the tops but did not kill many of the colonies, which quickly regrew. During a 4-year interval of no disturbance, a few colonies overgrew all the others, which then died. As in the study of boulders in temperate latitudes, the diversity was highest at the intermediate stages of recovery from disturbance.

In instances such as these, the mechanisms of niche differentiation, compensation, or circular networks probably contribute little toward the maintenance of high diversity. Instead, periodic disturbances keep them in the intermediate stages of high diversity, continually interrupting the process of competitive elimination. This situation also favors the operation of the lottery mechanism, although there are few instances with the requisite information to decide whether it was operating or not.

Periodic disturbances by storms, floods, fires, landslides, and so on, are ubiquitous; given enough time, all points on the Earth will experience a disturbance strong enough to affect all the resident organisms. Although nature is seldom, if ever, in balance, organisms have evolved to cope with this natural situation of being perpetually in the process of recovering their ecological balance.

For background information *see* CLIMAX COMMUNITY; ECOLOGICAL INTERACTIONS; ECOLOGICAL SUCCESSION; ECOSYSTEM in the McGraw-Hill Encyclopedia of Science and Technology.

[JOSEPH H. CONNELL]

Bibliography: J. H. Connell, Diversity in tropical rain forests and coral reefs, *Science*, 199:1302–1310, 1978; W. J. Eggeling, Observations on the ecology of the Budongo Rain Forest, Uganda, *J. Ecol.*, 34:20–87, 1947; S. A. Levin, Population dynamic models in heterogeneous environments, *Annu. Rev. Ecol. Systemat.*, 7:287–310, 1976; W. P. Sousa, Disturbance in marine intertidal boulder fields: The nonequilibrium maintenance of species diversity, *Ecology*, 60:1225–1239, 1979.

Electrical utility industry

This article reviews the major developments in the United States' electrical utility industry in 1982, and then discusses the impact of computers on the industry.

1982 DEVELOPMENTS

Two developments set 1982 apart from all other years in the history of the electrical utility industry.

For the first time, peak summer megawatt demand was actually lower than that of the previous year, and not a single new steam-generating unit was ordered. The lack of steam generator orders and the drop in peak demand are not unconnected conditions. Though the drop in peak summer demand resulted entirely from the severe decline in industrial demand caused by the economic recession, it delays the date by which utilities must construct new facilities to meet future demands with adequate reliability. The actual construction time for a modern unit is about 5–6 years, and utilities currently find themselves with a large surplus of capacity either already in service or under construction. Even if future growth averages the same 3.2% rate that it has over the last 10 years, utilities will not need new capacity until 8–10 years from now.

This past year was also marked by the newly demonstrated willingness of utilities to cancel construction of units which have already cost extremely large amounts of capital. Two nuclear plants which already cost $200 million and $400 million, respectively, were canceled because of a predicted lack of need in the future and because of severe financial difficulties that continue to plague the utilities.

Ownership. Ownership of electric utilities in the United States is pluralistic; that is, it is shared by investor-owned corporations; customer-owned cooperatives; and public groups on the municipal, district, state, and federal levels of government. The industry, however, is essentially dominated by the investor-owned sector. Investor-owned utilities serve 77.4% of the 95.1 million electricity customers in the United States. Municipal-, district-, and state-owned utilities serve only 12.4% of the total customers, slightly more than the cooperatives, which serve 10.2%. Although federal utilities serve a few customers on the retail level, they are primarily wholesalers who sell to other utilities.

Investor-owned utilities also own and operate 77.8% of the nation's installed generating capacity. Publicly owned utilities own 10.1% of all installed capacity, and cooperatives own only 2.5%. Federal entities own and operate 9.6% of the installed capacity.

The large discrepancy between the percentage of customers served by cooperatives and the much smaller percentage of these cooperatives' capacity occurs because such organizations are mostly distribution companies which buy their power from others at wholesale rates and distribute it to their member customers. Some cooperatives do generate; and some, for example, the Basin Electric Cooperative, do not serve distribution customers at all, but act as generating and transmission companies to produce and transmit wholesale power to other distribution cooperatives.

The extraordinary financial pressures (due to a combination of escalation in fuel costs, high inflation, and the reluctance of regulators to permit cost-recovering rate increases) on investor-owned companies in recent years have led them to seek partnerships with others. This has produced a clear trend of selling shares of large new generating units to cooperatives and publicly owned utilities. These two types of utilities are eager to buy into these large, efficient units to gain access to the lower-cost energy produced by the economies of scale of these units. By themselves, their financial resources generally could not permit the building of such units, especially nuclear ones, but their access to lower-cost funding makes them an attractive partner for the investor-owned utilities.

Duke Power Company, one of the largest investor-owned utilities, recently sold a 75% share of a 1250-MW nuclear unit to a consortium of ten cooperatives from North Carolina and five from South Carolina. Duke will construct the plant and then operate it under contract from the other owners.

Some cooperatives and municipals now see the financial plight of the privately owned companies as an opportunity to build large units themselves and sell the power to the investor-owned companies. Seminole Electric Cooperative in Florida, for example, with no retail customers, is currently building two 600-MW coal-fired units, the first of which is due to be in service in 1983.

Capacity additions. Utilities had a total generating capacity at the end of 1982 of 652, 175 MW, having added 17, 716 MW during the year (see table).

The composition of the capacity additions made during 1982 was 613 MW of conventional hydroelectric, 8606 MW of fossil-fired steam, 1067 MW

of pumped-storage hydroelectric, 6961 MW of nuclear power, and 469 MW of diesel and combustion turbines. Of this additional capacity, 12,383 MW was installed by investor-owned utilities, 2265 MW by publicly owned utilities, 1193 MW by cooperatives, and 1875 MW by federal agencies.

The composition of total plant by type of generation at the end of 1982 was 448, 185 MW (68.7%) fossil-fired steam; 65,584 MW (10.1%) conventional hydroelectric; 13,416 MW (2%) pumped-storage hydroelectric; 67,697 MW (10.3%) nuclear; 51,730 MW (8%) combustion turbines; and 5563 MW (0.9%) internal combustion engines, essentially all diesels (Fig. 1).

Fossil-fueled capacity. Only one oil-fired unit entered service in 1982, and there are no others planned for the future. There are a considerable number of older oil-fired units in operation that could be converted to coal firing. These are heavily concentrated (62%) in the northeast, with some in the east-central (22%) and southeast (13%). Utilities plan to convert 9976 MW from oil to coal in the 1982–1990 period.

Because of the environmental problems associated with coal burning, utilities have launched a number of developmental projects to perfect fluidized-bed combustion. This is a technique in which the coal is burned with limestone in a bed that is kept turbulent, or fluidized, by passing air through it. This provides efficient combustion and pollution control simultaneously. A number of developmental pro-

United States electric power industry statistics for 1982*

Parameter	Amount	Increase or decrease compared with 1981, %
Generating capability, $\times 10^3$ kW		
Conventional hydro	65,584	1.0
Pumped-storage hydro	13,416	1.1
Fossil-fueled steam	448,185	1.0
Nuclear steam	67,697	1.1
Combustion turbine and internal combustion	57,293	1.0
TOTAL	652,175	1.0
Energy production, $\times 10^9$ kWh	2,317.6	−1.3
Energy sales, $\times 10^6$ kWh		
Residential	747,700	1.6
Commercial	554,900	2.5
Industrial	743,400	−7.1
Miscellaneous	76,300	−0.6
TOTAL	2,122,300	−1.5
Revenues, total, $\times 10^6$ dollars	125,400	12.4
Capital expenditures, total; $\times 10^6$ dollars	45,260	8.0
Customers, $\times 10^3$		
Residential	84,300	1.0
TOTAL	95,100	1.0
Residential usage, kWh per customer	8,911	1.0
Residential bill, cents/kWh (average)	6.6	1.1

*From 33d annual electric utility forecast, *Elec. World*, 196(9)75–86, September 1982; and extrapolations from monthly data of the Edison Electric Institute.

cesses to gasify coal and use the clean gas as a boiler fuel are also under way. A 40-MW prototype gasifier was expected to be in service at the Wood River plant of the Illinois Power Company before the end of 1982.

Nuclear power. The year 1982 marked the fourth straight year that no new orders for nuclear units have been placed. The cancellation of units already announced or under construction continued, with nine units totaling 11,200 megavolt-amperes (MVA) joining the list of those canceled. Because the average completion time from date of Nuclear Regulatory Commission approval to entrance into commercial service has now stretched out to more than

10 years, units ordered and begun in previous years continue to come on line. During 1982, two units with a combined capacity of 1948 MVA began commercial service. The total capacity of all nuclear units now in service in the United States is 67,697 MW, provided by 77 units in 55 individual generating plants. One of the units which entered service in 1982 was a boiling water reactor (BWR), and one was a pressurized water reactor (PWR). This brings the totals in the United States to 27 BWRs and 50 PWRs.

Utilities currently have 18 units under construction but less than 50% complete, with total power of 20,439 MW. There are an additional 42 units under construction but more than 50% complete, totaling 44,618 MW. Altogether, plants totaling 65,057 MW are in construction and scheduled for completion by 1991.

The consortium of 750 utilities which are sponsoring the construction of the sodium-cooled 375-MW Clinch River breeder reactor were granted permission in 1982 by the Nuclear Regulatory Commission to begin site clearing for the project. The breeder reactor, which is designed as a test bed for breeder technology, has been threatened by withdrawal of federal financial support, though components costing close to $100 million have already been fabricated for the project.

Combustion turbines. Combustion turbines have historically been installed by utilities for use at times of peak demand, supplying 10–15% of that peak. The low cost of such machines, currently about $290/kW compared to $660/kW for coal-fired units, more than offsets their high fuel consumption when they are used during peak demand for approximately 200 h/yr. Combustion turbines, which use either natural gas or oil as fuel, have heat rates of about 13,000 Btu/kWh (3.8 joules of heat per joule of electric energy) compared to 9000 Btu/kWh (2.6 joules of heat per joule of electric energy) for a modern fossil-fuel unit. Further, they have quick-start capability, which permits them to be brought up to full load in 2–3 min and provides flexible capacity for emergency conditions. They are also used to provide start-up power for generating stations that have experienced a complete shutdown, as during a blackout of an entire area or system. Because of government restrictions on fuel and because there are already excess reserves at peak periods, the percentage of peak represented by gas turbines is decreasing, and currently stands at 8%.

Although individual units as large as 125 MW have been installed, most units are in the 25–50-MW range. In 1982 utilities added 467 MW of combustion turbine capacity, raising the total industry capacity to 51,730 MW, which is 8% of total installed capacity.

Combustion turbines are also used in combination with steam turbines in a highly efficient combined-cycle operation. In this mode, the 900–1000°F (482–538°C) gas-turbine exhaust produces steam in a heat-recovery boiler which supplies a steam turbine. This cycle may have efficiencies as high as 60%, com-

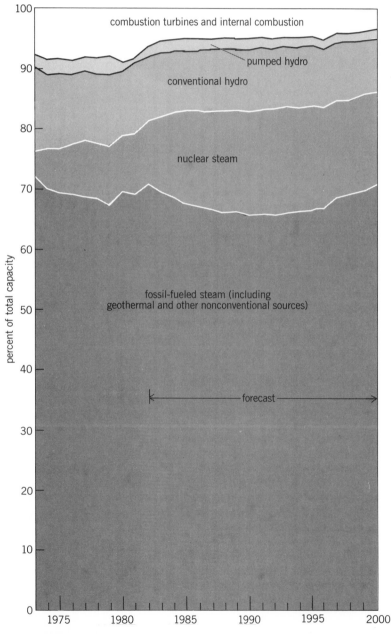

Fig. 1.   Probable mix of net generating capacity. (*After 33d annual electrical industry forecast, Elec. World, 196(9):75–86, September 1982*)

pared to 34–35% for the most efficient steam cycles. In this mode the combustion turbine will contribute about 70% of the combined output.

Utilities spent $243 million on combustion turbines in 1982.

Hydroelectric installations. Utilities brought 613 MW of conventional hydroelectric capacity into service in 1982, raising the total now installed to 65,584 MW. Hydro units of this type, that is, hydroelectric turbines driven by water impounded behind a dam or by the natural flow of a river, constitute 10.2% of total installed capacity of all types of utility systems in the United States. Future plans call for an additional 5400 MW of capacity to be built over the next 10 years, though locations for the required dams are becoming increasingly difficult to find and license. The few major sites suitable for high dams are, in general, in areas where the environmental effects of the resulting lake are unacceptable. Utilities spent $520 million on this type of installation during 1982.

The future of hydroelectricity may rest with small-scale projects that can either be incorporated in existing dams or be powered by the normal flow of the river. These units, carrying ratings of 1.5–50 MW, have become especially popular in the west, where the extensive irrigation canal systems often have locks that create heads of 15–25 ft (4.5–7.5 m), adequate to power small tube- or bulb-type hydroelectric units.

Pumped storage. Pumped storage represents one of the few possible methods for storing large amounts of energy from electric generators. Water is pumped from a body of water on which the generating unit is located into a reservoir elevated some distance above. This is normally done during off-peak periods when large, efficient base-load units are available. During the subsequent peak demand period, this water is released through the pumps, which can be reversed to act as turbines, recovering only about 65% of the energy originally expended but reducing the need for the equivalent capacity to be provided at peak.

Utilities installed an additional 1067 MW of pumped-storage capacity during 1982, raising total installed capacity in the United States to 13,416 MW, or about 2% of the nation's total capacity.

Utilities spent $224 million for pumped-storage facilities in 1982. The investor-owned companies spent $195 million, federal agencies spent $6.2 million, cooperatives $11 million, and publicly owned utilities $11.9 million.

Renewable energy sources. Utilities have actively sought alternatives to the combustion of coal, oil, and gas, and the use of uranium. At the end of 1982, utilities had a total of 1293 MW of capacity other than the conventional types listed above. These included geothermal, wind, and solar sources, plus waste-, and refuse-fired units. (Because of the generally small size of these units, the total is included in the fossil-fuel category for convenience.) Utilities have planned units which will provide an additional 4055 MW for the 1982–1990 period. These consist of 2389 MW of geothermal, 385 MW of solar, 359

MW of wind, and the rest waste- or refuse-fired capacity. More than 80% of this new capacity will be located in California.

Rate of growth. The year 1982 brought into sharp focus the growing polarity of those forecasting the rate of growth of electrical energy in the United States. Because of the intermittent years of low growth, such as 1982, some credible forecasters have begun to project extremely low or even negative growth prospects over the next 10–20 years. These projections are predicated on the rise in electricity prices as rates adjust to absorb the large additions to rate base that the new, expensive units represent as they come into service, coupled with stable prices for competitive energy sources such as coal and oil. Others, however, point to the fact that the normalized rate of growth, even throughout the past 10 years has tracked the growth in the gross national product. These forecasters are predicting growth of peak demand in the range of 2.5–3%.

A long-term decrease naturally arises from the mix of demographic factors that characterize a maturing society such as the United States. Population is growing at a decreasing rate and is thus passing beyond the years of peak consumption. This effect will be compounded by price-induced conservation. Because of the high technological content of the utility industry's plant, the preponderance of skilled labor employed, the high cost of capital engendered by unresponsive regulation, and escalating fuel costs, electricity prices should rise slightly more than inflation over the coming decade—somewhat moderating the utility industry's growth.

Industry and commerce have also embraced the concept of demand control. In this technique, electrical equipment is computer-controlled to minimize peak demand on the entire manufacturing plant or building within the constraints of required production. This reduces the demand charge from the utility, which is a substantial part of energy cost to industry. Load-shedding agreements have become popular between utilities and groups of industrial customers acting as a load block who, in return for a reduced rate for electricity, agree to reduce demand at peak upon request.

The overall declining pattern of growth has a major effect on reserve margins—that is, the excess of installed capacity over demand—on a national basis. A rule of thumb is that the average reserve margin should be about 25%, and it is now about 40%. Though utilities are delaying the construction of, or canceling, many major generating units, some that were started years ago will continue to come into service, supporting an elevated level of reserve which will not decline to the 25% level until 1990.

Usage. Sales of electricity also declined in 1982, dropping 1.5% from 1981. Total national usage was 2122×10^9 kWh. The drop was due almost entirely to sagging industrial sales, but a record low number of housing starts and a cool, wet summer also contributed.

Commercial sales held up best again in 1982,

rising 2.5% to a total of 554.9×10^9 kWh. This rise, however, was moderating at year's end.

Industrial sales dropped 7.1% from a poor 1981, finishing the year at about 743.4×10^9 kWh. Residential sales rose slightly from 1981, recording a rise of 1.6%, with 747.7 kWh consumed. Poor housing figures and a cool summer held sales in this category down when a substantial rebound had been expected after a very poor showing in 1981.

Despite the continued increases in the cost of electricity, electric heating for residences continues to make gains. Slightly over one-half of all new homes constructed in the United States in each of the last 10 years have been electrically heated, and in 1982 heating energy sales topped 156.4×10^9 kWh.

Residential use per customer rose slightly to 8911 kWh from 8863 kWh in 1981. Continued rate increases boosted the average residential rate to 6.6 cents/kWh and resulted in an annual average bill per residential customer of $588. Total revenue for the utilities industry was $125.4 billion.

Fuels. The effect of escalating prices for gas and oil and the long-range threat of the Fuels Use Act of 1974 have moved utilities strongly toward the use of coal as a fuel. Consumption in 1981, the last year for which good figures are available, rose 4.7%, to 596.4×10^6 tons (540.9×10^6 metric tons). Oil dropped 16.5% to 350.9×10^6 bbl (55.7×10^3 m³). Essentially all the shift from oil use was absorbed by coal, since consumption of natural gas also dropped 1.1% to 3642.7×10^{12} ft³ (103.1×10^{12} m³).

The same type of shift was seen in actual energy generated. Coal generated 1202.8×10^9 kWh, or 59% of the total; oil accounted for 205.8×10^9 kWh, or 10.1%; gas was used to generate 345.2×10^9 kWh, or 17%, and nuclear power put out 272.3×10^9 kWh, representing 13.4% of all generation. The rest was primarily hydroelectric, but with a very small contribution from other sources such as geothermal.

Distribution. Distribution capital expenditures for 1982 amounted to $6.5 billion, and an additional $235 billion was spent to maintain existing plant. During the year, 18,500 mi (29,800 km) of three-phase equivalent overhead lines and 10,500 three-phase equivalent mi (16,914 km) of underground lines came into service, at voltages ranging from 4.16 to 35 kV. The majority of this mileage was at 15 kV, which accounted for 12,358 three-phase equivalent mi (19,906 km) or 66.8% of overhead and 7350 three-phase equivalent mi (11,800 km) or 69.5% of underground circuitry. The percentages for overhead construction held by other voltage classes were 13.5, 15.2, and 4.5% for 35, 25, and 4 kV, respectively. For underground construction, the equivalent percentages were 9.3, 17.1, and 4.2. During 1982, utilities energized 18,830 MVA of distribution substation capacity and expended $1.01 billion in capital for substation construction.

Transmission. Utilities spent $3.5 billion in capital accounts for transmission lines in 1982. During the year, they spent $1.26 billion for overhead lines at 345 kV and above, and $963 million for overhead circuits of 220 kV and below. For underground transmission construction, which can cost an average of eight times more than equivalent overhead construction, capital expenditures amounted to $127.6 million to voltages of 220 kV and higher, and $23 million for circuits at 161 kV and below. Utilities installed 2680 mi (4317 km) of overhead lines at 345 kV and above, but 5545 mi (8932 km) at 220 kV and below. Looking at the capacity of those lines rather than the mileage gives a different picture of the place of the different voltage classes. Of the overhead lines installed in 1982, lines at 345 kV and higher totaled 2130 gigawatt-miles (3439 GW-km), while lower voltage lines contributed only 570 GW-mi (917 GW-km). The picture is completely different in the underground sector, because present cable technology costs favor the lower voltages. In 1982, only 28 mi (45 km) of cables operating at or above 230 kV and 73 mi (117 km) at or below 161 kV came into commercial service.

Utilities brought 44.2 GVA of transmission substation capacity into service in 1982 and spent $998.6 million for substation construction. The maintenance of existing transmission plants cost $682.5 million.

Capital expenditures. Total capital expenditures in 1982 rose to $45.3 billion, despite utility efforts to curtail construction programs. Of this total, $33.5 billion went for generating facilities, $3.5 billion for transmission, $6.5 billion for distribution, and $1.7 billion for miscellaneous facilities, such as headquarters buildings and vehicles, which cannot be directly associated with other categories. Total assets held by the investor-owned segment of the industry were $289.5 billion at the end of 1981. Municipals held about $30 billion in assets, and cooperatives about $33 billion.

[WILLIAM C. HAYES]

IMPACT OF COMPUTERS

Computer hardware and software applications for the power industry have been evolving into complex computer and communication systems encompassing all the major functions of the utility business (Fig. 2). These systems are being implemented to model more closely the utility operating procedures and policies in four integrated and interfaced categories: corporate information systems, energy management systems, power plant systems, and facility information systems. These systems are essential for the effective management and control of utility operations and will become even more important for the survival of the electrical utility industry in the future. This section discusses the key planning and development considerations associated with the implementation of each of the above functional systems, and explores their future effectiveness and productivity in electrical utility operations.

Corporate information systems. Corporate information systems encompass the traditional utility

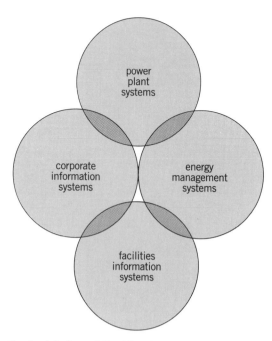

Fig. 2.   Interface relationships of electrical utility industry systems.

computer applications in the areas of accounting, financial, engineering, customer service, and administrative systems, which until recently represented the bulk of utility data processing and involved the mechanization of existing manual systems. With the advancement of data base and data communication technology, these systems, which generally operated in batch mode, have been replaced with completely new on-line systems. The planning and development of these systems has relied heavily on the application of structured design methodologies, project management disciplines, and user involvement. In the future, in addition to these development approaches, more advanced programming languages will be applied with more end-user participation and extensive networking. The overall structure of corporate information systems will continue to evolve from computerization of manual functions to modeling of the procedures and policies of all utility functions. In the utility customer-service area, by 1985 integrated configurations of computer and communication networks, in various stages of development, will be implemented into system architectures comparable to those of large-scale airline reservation systems. In the engineering and construction areas, the computer-aided design technology now maturing will have significant productivity impact on the design and construction of utility generation, transmission, and substation facilities. The application of data base and data communication technology will facilitate the development of the engineering information system concept, which will integrate engineering design with the construction and materials functions of the utility. In the accounting area, by 1984 the shift from financial accounting to

operational accounting will have been completed and will close the loop between construction management and materials management through the work order control system interface. In the area of power system planning and financial models, the impact of macro and micro modeling of the utility corporate planning functions will be felt in terms of high utilization of computer resources. This will advance the implementation of parallel computing, microprocessor technology, and data highway computer architectures.

Energy management systems. Decisions and commitments are required many years in advance for the implementation of large-scale energy management systems needed for the control and efficient operation of electrical energy systems (Fig. 3). Power system operations have a direct impact on generation and transmission expansion plans, as power systems are operated at increasingly closer limits. This mode of operation makes energy management systems a major component of utility management strategic planning and demands management's direct involvement in every step of the decision process. Computers are essential to energy management systems for control of the power system, optimization of fuel, and increased productivity. However, computers also have a high potential for unproductive utilization if system design does not faithfully model the procedures and policies of the function it represents and if operating management does not understand the basic functions and associated relationships modeled by the computer.

The basic functional objectives of energy management systems are to aid operating personnel in the operation of the power system and to maintain acceptable levels of power system security. These objectives must be conceptualized into operating philosophies by management and translated into precise function, clearly understood by operating personnel, to be performed in the areas of: automatic generation control, unit commitment, economic dispatch, system security, supervisory control, system monitoring and alarming, spinning reserve pricing, interchange transactions, generation status reporting, generation reserve calculations, system load forecasting, automatic time correction, incremental pricing, and construction and maintenance scheduling.

In order to follow these operating philosophies, energy management systems must perform the following supporting functions, which need to be identified and explained to operating personnel: data acquisition, human-machine interface, energy accounting, system test and failover, external data exchange, and regional coordination.

The human-machine interface function has an enormous impact on the successful implementation and operation of energy management systems and must be defined in terms which are familiar to operating personnel and utility management. This function comprises the hardware and software devices which enable operating personnel to obtain in-

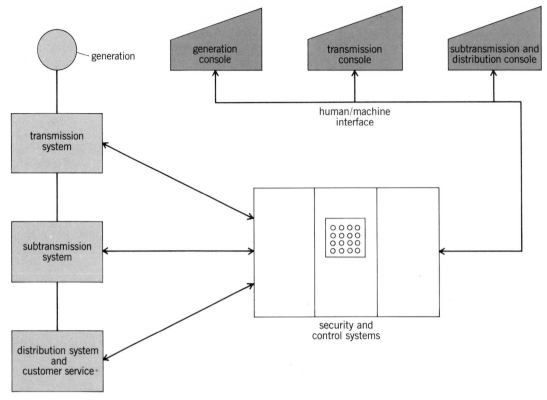

Fig. 3.   Centralized energy management system design.

formation and issue commands for the efficient control of power systems. Particular attention should be given to implementing the human-machine interface through staged console configurations in order to minimize human error and organizational problems. Also, a comprehensive cost/benefit study should be made of the functions that are to be modeled by the computer system and explained to operating personnel so that benefits and cost justification can be realized.

The relationship of the bulk power system (generation and transmission) to the distribution system, including the service connection to the customer, should be taken into account. The distribution system should not be regarded as separate from the all-important bulk power system, but should be seen as an integral part of the energy management system. From the point of view of the energy management system, transmission, subtransmission, and distribution should be centralized.

Power plant systems. The fossil and nuclear power plant computer requirements impact all power plant functions and encompass application areas in administration, operation, and maintenance; power system control (as interfaced with the energy management system); and production monitoring and control (fossil boiler/turbine control, nuclear reactor/turbine control). The power plant administration, operation, and maintenance applications are directly related to the corporate information systems. The power system control is related to the energy management system, and the production mon-

itoring and control is related to the process control standalone systems not connected to corporate information or energy management systems. These similarities provide a clear opportunity for interfaces between corporate information, energy management, and administration, operation, and maintenance systems for efficient utilization of computer resources in electrical utility operations.

The power plant administration, operation, and maintenance functions can be implemented by a plant information system with a single data base. This concept can be expanded for fossil and nuclear plants in a station information management system (Fig. 4) which has the following objectives: maximization of plant availability; utilization of work force at the highest level possible; streamlined information flow and auditable record-keeping; and provision of history for better resource planning. The major functions of a station information management system would be: management action control; station work management; and analysis and support. The station information management system interfaces directly with accounting, payroll time activity, and material management information systems.

Facilities information systems. A major development effort in the electrical and gas utility industry, which is still in the initial stages of feasibility and cost/benefit studies, is the implementation of a utility geographic data base concept which would support mapping, networking, and tax accounting applications. The concept would be to store information on utility facilities and land as a model of

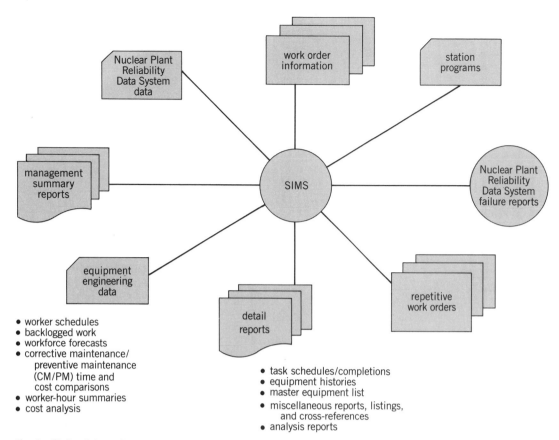

- worker schedules
- backlogged work
- workforce forecasts
- corrective maintenance/
  preventive maintenance
  (CM/PM) time and
  cost comparisons
- worker-hour summaries
- cost analysis

- task schedules/completions
- equipment histories
- master equipment list
- miscellaneous reports, listings,
  and cross-references
- analysis reports

Fig. 4. Station information management system (SIMS) for a nuclear power plant.

the existing procedures and policies of utility assets. The development effort is comparable in magnitude of investment and implementation time to that of large-scale customer information and energy management systems. The facilities information system would have direct interfaces with both customer and energy management systems and should be planned and designed with the appropriate strategic impact in perspective. The data base concept allows the modeling of the facilities as they exist in the real utility world, that is, in terms of their location and connectivity. In this fashion, facilities and land data can be retrieved by geographic areas and by network. This represents a radical departure from traditional manual drafting techniques and the many automated mapping systems. The impact of this approach will revolutionize the use of computers in the electrical and gas utility industry in the future. When fully interconnected with third-generation customer information and energy management systems, it will represent a new dimension in the productive utilization of computers throughout the power industry.

For background information *see* ELECTRIC POWER GENERATION; ELECTRIC POWER SYSTEMS; ELECTRIC POWER SYSTEMS ENGINEERING; ENERGY SOURCES; FLUIDIZED-BED COMBUSTION; TRANSMISSION LINES in the McGraw-Hill Encyclopedia of Science and Technology.

[V. CONVERTI]

Bibliography: Edison Electric Institute, *Statisti-*

cal Yearbook of the Electric Utility Industry, 1981; 1982 annual statistical report, *Elec. World*, 196(3):65–96, March 1982; 33d annual electrical industry forecast, *Elec. World*, 196(9):75–86, September 1982; 21st annual steam station cost survey, *Elec. World*, 195(11):69–84, November 1981.

Electronics

New applications of electronics technology have produced a growing need for electronic components capable of operating at high temperatures (above 300°C). The need for electronic devices for geothermal well probes, planetary space probes, jet-engine controls, and nuclear power plant instruments is supplying the impetus for advancements in this area, since conventional silicon diodes, transistors, and integrated circuits will not function in this temperature range. Recent research activities have explored several new technologies to fill this need. The research has been carried out in three primary areas: new silicon devices, compound semiconductor devices, and integrated thermionic circuits. Some of these devices have the potential of extending the operational range of electronic circuits up to 800°C.

Silicon devices. Semiconductor devices can be divided into two categories: minority carrier devices, such as diodes and bipolar transistors, and majority carrier devices, such as the field-effect transistor (FET). Minority carrier devices made from silicon do not work well above 250°C. This is because the energy bandgap in silicon is only 1.1 eV, and at temperatures above 250°C a large number of minority carriers are generated by thermal excitation.

The density of these thermally generated carriers in silicon is approximately 10^{10}/cm^3 at room temperature and increases by about 6 orders of magnitude at 400°C (to approximately 10^{16}/cm^3). The large density increase in thermally generated carriers causes a large increase in reverse leakage current of *pn* junctions and makes the performance of minority carrier silicon devices, such as bipolar transistors and diodes, marginal above 250°C.

Majority carrier devices in silicon (such as FETs) do not depend on minority carriers for their operation and therefore can be made to operate at higher temperatures. Enhancement-mode metal-oxide semiconductor FET (MOSFET) devices have been made to operate at 350°C. To achieve this operating temperature it is necessary to remove the input-gate-protection bypass diodes and use dielectrically isolated (rather than junction-isolated) devices because of diode leakage. The upper limit of operating temperature for the silicon MOSFET is not known definitely. Problems that occur at high temperatures are oxide degradation in the gate and leakage at the source-drain *pn* junctions. It is generally felt that small-area devices may be eventually made to operate at 400°C.

Compound semiconductor devices. The basic need for a simple high-temperature rectifier diode and moderate power devices with gain have led to high-temperature research on compound semiconductors. Many of the group III–V compound semiconductors have energy bandgaps larger than silicon. For example, the semiconductor bandgap in gallium phos-

phide (GaP) is 2.2 eV. The thermally generated minority carrier density in GaP at 400°C is approximately 10^{12}/cm^3 (4 orders of magnitude less than silicon). This level is small enough that it is possible to build minority carrier devices at high temperatures by using this and other wide-bandgap semiconductor materials. Diodes and transistors have been demonstrated at 450°C in GaP material. Figure 1 shows a scanning electron microscope photograph of a GaP bipolar transistor. The device in the photograph is a grown-junction *pnp* transistor made by liquid-phase epitaxy. The center contact is the emitter, the two outside pads are base contacts, and the collector contact is the bottom of the chip. Compound semiconductor devices of this type are estimated to be operational up to approximately 525°C, with moderate power-handling ability. Allowable current densities in these devices can be up to 100 A/cm^2, even at 400°C. While this current-carrying capacity is about 1 order of magnitude lower than that used in room-temperature silicon devices, it is still adequate for most high-temperature applications.

Research is also being carried out on compound semiconductor heterojunction devices. These are *pn* junctions made from dissimilar semiconductor materials such as gallium arsenide and gallium aluminum arsenide (GaAs and GaAlAs). The physics of junctions of dissimilar semiconductors can be exploited to reduce the leakage currents which plague high-temperature devices, making successful high-temperature minority carrier devices possible.

Integrated thermionic circuits. A dramatic departure from semiconductor technology is the integrated thermionic circuit (ITC). This technology combines photolithographically defined subminiature thin-film metal patterns with planar vacuum-tube technology. The result is a technology that allows fabrication of active circuits with a density approaching that of present room-temperature silicon integrated circuits. Figure 2 shows a cut away view of a single ITC active device. In this structure the grids (*G*) are coplanar with the cathode (*K*) on the lower plane; the anode (*A*) is on the upper plane, and the space between the cathode and anode is evacuated. The grid metal and oxide-coated cathode are photolithographically defined on a sapphire substrate. During normal operation a resistance heater is used to heat this substrate to approximately 800°C to produce electron emission from the cathode. A positive voltage is applied to the anode to collect these electrons, and a voltage is applied to the grid lines to set up an electric field which controls this current flow in a manner similar to that of a conventional triode vacuum tube. The resulting devices produce current densities of approximately 0.1 A/cm^2. Because of this limitation these devices are low-power devices, operating at power densities about 3 orders of magnitude less than room-temperature silicon devices. Thus, they are not suitable for use in power-conditioning or line-driving applications. The small size of these small-signal devices does allow the integration of analog and digital circuits with de-

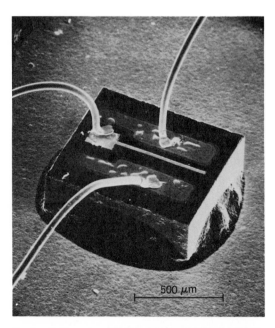

Fig. 1. Gallium phosphide bipolar transistor for use at temperatures up to 450°C. Center contact is the emitter, base contacts are two outside pads, and collector contact is the bottom of the chip. Chip size is 500 × 750 μm.

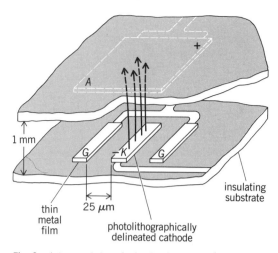

1 mm

thin
metal
film

25 μm

photolithographically
delineated cathode

insulating
substrate

Fig. 2. Integrated thermionic circuit (ITC). (*After Proceedings of the High Temperature Electronics and Instrumentation Conference, December 1981, Sandia National Laboratories Rep. SAND82-0425*)

vice densities of several thousand per square centimeter. ITCs have been life-tested at 500°C for several thousand hours with no detectable degradation. An upper temperature limit is not known, but it is estimated that, with suitable packaging techniques to reduce outgassing problems, device temperatures should be able to approach the cathode temperature of 800°C. These devices are also capable of surviving severe radiation environments and may be applicable to nuclear power reactor instrumentation.

Use of amorphous metal films. One of the weak links in semiconductor technology at high temperatures is the metallization used. All semiconductor metallizations are presently made from polycrystalline metals. Failures that occur with room-temperature semiconductor devices are usually attributed to metal failures due to diffusion, corrosion, or electromigration of the thin metal films used to contact and interconnect the semiconductor devices. Since these failures are due to thermally activated processes, they occur even more rapidly at elevated temperatures. In order to make reliable high-temperature semiconductor devices with reasonable lives this problem must be solved. It is known that most of these metal failures occur along grain boundaries in the polycrystalline metals. A potential solution to this problem is the use of amorphous metal films (which have no grain boundaries) in place of the polycrystalline metals. Recent experimental results have demonstrated that amorphous metal films on semiconductors are several orders of magnitude less susceptible to these failures at high temperatures, and these films can be produced by processes compatible with the semiconductor industry.

For background information *see* INTEGRATED CIRCUITS; SEMICONDUCTOR HETEROSTRUCTURES; TRANSISTOR; VACUUM TUBE in the McGraw-Hill Encyclopedia of Science and Technology.

[ROGER J. CHAFFIN]

Bibliography: Proceedings of the High Temperature Electronics and Instrumentation Conference, Houston, December 1981, Sandia National Laboratories Rep. SAND82-0425; *Proceedings of the Conference on High Temperature Electronics,* Tucson, March 1981, IEEE Rep. 81CH1658-4.

Elementary particle

The search for the most basic constituents of matter has progressed through the sequence atom, nucleus, and nucleon, to the interior of the nucleon. Quarks and leptons are now considered to be the elementary particles, along with the gauge particles that mediate their interactions. Quarks (q) are the constituents of the proton, neutron, mesons, and other particles which interact strongly—such particles are called hadrons. The quark model, first proposed by M. Gell-Mann and G. Zweig in 1964, explains the numerous hadronic particles as composite objects. According to current, but still unproved, theories, quarks are permanently confined inside hadrons and can never be isolated as free particles. Nonetheless, compelling evidence for the existence of quarks is derived from experiments that probe hadrons at very short distances. Leptons are particles which are not subject to the strong force; the electron and its neutrino are examples. Both quarks and leptons have spin angular momentum of ½ in the quantum unit of $h/2\pi$, where h is Planck's constant.

Major progress has been made in understanding the four basic forces of nature—strong, electromagnetic, weak, and gravitational—in the framework of gauge theories, of which quantum electrodynamics (QED) was the forerunner. The color force of the strong interactions is due to the exchange of a massless spin-1 gluon (g) between quarks. The electromagnetic force is transmitted by the massless photon (γ), which couples to electric charge. The carriers of the weak force, charged ($W^{\pm}$) and neutral (Z^0) spin-1 bosons, are now being actively sought at particle colliders, in the mass range 80–95 GeV predicted by the unified gauge theory of electromagnetism and weak interactions. A spin-2 massless graviton is the carrier of the gravitational force, though a renormalizable gravitational theory remains to be constructed. Grand unified theories, which unite the strong and electroweak interactions, intimately relate quarks and leptons and make the dramatic prediction that the proton will decay with a lifetime just beyond present experimental limits. Left unexplained is why there are three, or perhaps more, families of quarks and leptons, distinguished only by mass. This repetition has been interpreted by some physicists as an indication that quarks and leptons may themselves be composed of still more elementary particles.

Color. The everday world is composed of just four particles—the up quark (u), the down quark (d), the electron (e), and its neutrino (ν_e). These form the lightest of the three known quark-lepton families. Their electric charges, given in terms of magnitude of the electron charge, are $Q_u = +\frac{2}{3}$, $Q_d = -\frac{1}{3}$, $Q_e = -1$, $Q_{\nu_e} = 0$. The quark content of the proton, neutron, and charged pi mesons are $p = uud$,

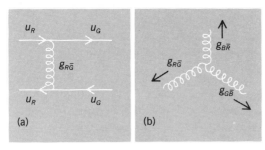

Fig. 1. Mediation of strong force by colored gluons. (*a*) Red and green *u* quarks exchanging colors via a red-antigreen (or green-antired) gluon. (*b*) Coupling of three gluons at a vertex.

$n = ddu$, $\pi^+ = u\bar{d}$, and $\pi^- = \bar{u}d$, where $\bar{u}$ and $\bar{d}$ denote antiquarks. The lowest nucleon resonance Δ^{++} with charge $+2$ and spin $\frac{3}{2}$ then has a wave function $\Delta^{++} = uuu$, which is symmetric in the three quarks. This is forbidden by the Pauli exclusion principle unless quarks possess a hidden quantum number, which is conventionally called color. Quarks come in the three primary colors—red, green, and blue—while antiquarks have comple-

mentary anticolors. Only colorless (that is, white) combinations are allowed to materialize as hadrons. A proton contains a quark of each color, while mesons contain quarks and antiquarks of complementary colors. Quarks interact through the exchange of double-colored gluons (Fig. 1). There are eight nonwhite color combinations of these massless gluons. The gluons can also interact with each other through exchange of gluons. The gauge theory of colored quarks interacting through colored gluons has the symmetry of the group SU(3) [associated with the

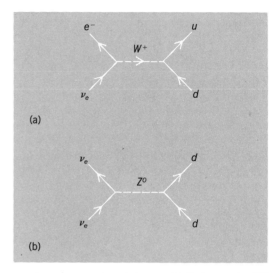

Fig. 3. Exchange of weak bosons. (*a*) Exchange of a $W^{\pm}$ boson, changing the flavors of quarks and leptons. (*b*) Flavor-conserving neutral current process resulting from Z^0 exchange.

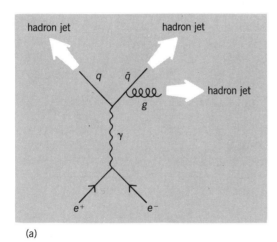

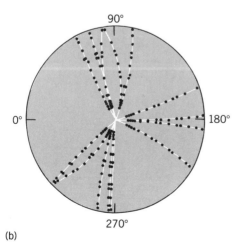

Fig. 2. Three-hadron jet arising in an e^+e^- collision from the production of a quark, an antiquark, and a gluon. (*a*) Mediation of the electromagnetic force between the e^+e^- and $q\bar{q}$ pairs by proton (γ) exchange. (*b*) Typical observed three-jet event.

three colors] and is known as quantum chromodynamics (QCD). It has the property of being asymptotically free, which means that very small probes inside the hadron see essentially free quarks because the gluon interactions become smaller at progressively shorter distances. Quarks and gluons are seen in high-energy collisions as jets (narrow sprays) of hadrons. In electron-positron collisions a two-jet structure occurs from the emission of a quark and an antiquark in opposite directions, which then fragment into back-to-back jets. Three-jet events arise when one of the emitted quarks radiates a gluon which also fragments into hadrons. Figure 2 shows an example of a three-jet event observed by the TASSO collaboration at the PETRA collider in Hamburg, Germany. *See* GLUONS.

Flavor. Different kinds of quarks, such as *u* and *d*, are called quark flavors. Each quark flavor comes in all three colors; the strong interaction is independent of flavor. Similarly, different types of leptons, such as ν_e and *e*, are referred to as lepton flavors. Weak interactions mediated by the $W^{\pm}$ boson can change the flavor from *d* to *u*, and ν_e to *e* (Fig. 3*a*). The weak force thus provides a link between quarks and leptons. The exchange of the neutral weak boson Z^0 does not change flavors (Fig. 3*b*).

Generations. For still unknown reasons, nature has provided two further quark-lepton families beyond (u, d, e, ν_e), with corresponding electric charges. The second family consists of the charm quark (c), the strange quark (s), the muon (μ), and its neutrino (ν_μ). The known members of the third generation are the charge $-\frac{1}{3}$ bottom quark (b) and the tau lepton (τ). The decays of the τ give indirect evidence for its neutrino (ν_τ). The missing top quark (t), with charge $+\frac{2}{3}$, is predicted to have a mass of 25 GeV, which is just beyond accessible energies of experimental searches.

The weak decays of hadrons containing a quark constituent of heavy flavor proceed through the flavor-changing $W^\pm$ couplings. Figure 4 illustrates basic quark mechanisms for charm decay.

Generation mixing. The charged weak bosons also cause transitions between quarks belonging to different generations. The $W^\pm$ couples equally to each of the quark and lepton doublets shown below. These

$$\begin{pmatrix} u \\ d' \end{pmatrix} \quad \begin{pmatrix} c \\ s' \end{pmatrix} \quad \begin{pmatrix} t \\ b' \end{pmatrix} \quad \begin{pmatrix} \nu_e \\ e \end{pmatrix} \quad \begin{pmatrix} \nu_\mu \\ \mu \end{pmatrix} \quad \begin{pmatrix} \nu_\tau \\ \tau \end{pmatrix}$$

are doublets of the weak-isospin SU(2) symmetry. The d', s', and b' quark states are not identical to the flavor states d, s, and b of the strong interac-

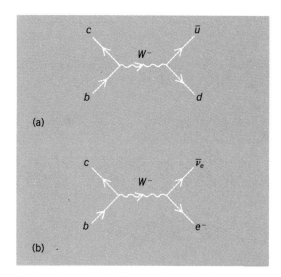

Fig. 5. Decay of the bottom quark b to the charm quark c and (a) $\bar{u}$ and d or (b) e^- and $\bar{\nu}_e$, via generation mixing.

tions. The mixing among different generations is small, with preferential mixing among nearest neighbors in mass. This mixing is responsible for the weak decays of strange particles and of particles with a bottom quark constituent. Figure 5 illustrates bottom quark decay through the $b \rightarrow c + W^-$ transition that arises from the b admixture in the s' state. The $W^\pm$ couples only to left-handed quark or lepton doublets (that is, the particle spin is correlated with the direction of motion, like a left-handed screw), which gives the observed parity-violating effects.

Quark and lepton masses. Since quarks cannot be isolated, their masses can only be found indirectly. Their effective masses depend on the means by which they are determined, especially for the light quarks. Constituent quark masses are deduced from the observed hadronic masses. The constituent masses of the u and d quarks are about 0.3 GeV (that is, about one-third of the nucleon mass). Lepton masses and constituent quark masses of the three generations are listed in the table. Neutrinos are expected to have masses which are considerably smaller than the masses of the charged leptons. One experiment has reported evidence for a ν_e mass in the range of 14 to 46 eV.

$W^\pm$, Z^0, and Higgs particles. The unified SU(2) $\times$ U(1) theory of electromagnetic and weak interactions by S. Weinberg, A. Salam, and S. Glashow gives a highly successful account of electroweak ex-

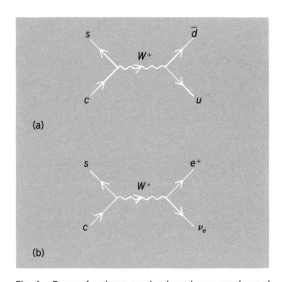

Fig. 4. Decay of a charm quark c to a strange quark s and first-generation members (a) u and $\bar{d}$, or (b) e^+ and ν_e, proceeding via the $W^\pm$ carrier of the weak force.

Charges and masses of leptons and quarks

Generation	Lepton flavor	Charge	Mass, GeV	Quark flavor	Charge	Mass, GeV
First generation	ν_e	0	$< 0.05 \times 10^{-6}$	u	$+\frac{2}{3}$	0.3
	e	-1	0.511×10^{-3}	d	$-\frac{1}{3}$	0.3
Second generation	ν_μ	0	$< 0.57 \times 10^{-3}$	c	$+\frac{2}{3}$	1.5
	μ	-1	0.106	s	$-\frac{1}{3}$	0.5
Third generation	ν_τ	0	< 0.250	t	$+\frac{2}{3}$	>19
	τ	-1	1.784	b	$-\frac{1}{3}$	4.7

periments at low energies in terms of a single parameter, $\sin^2 \Theta_W$ (related to the ratio of the SU(2) and U(1) couplings of gauge bosons). The theory predicted the existence of the weak neutral currents mediated by the Z^0, which were subsequently observed. The theory also predicts the masses of the weak bosons to be $M_{W^\pm} \simeq 83$ GeV and $M_Z \simeq 94$ GeV. These large masses account for the weakness of the weak force at low energy compared to the electromagnetic force. The discovery of weak bosons at these predicted masses in proton-antiproton experiments now under way at the CERN collider in Geneva, Switzerland, will be a crucial test of the theory.

The $W^\pm$ and Z bosons acquire their masses through spontaneous symmetry breakdown caused by a scalar field. The associated spin-0 physical particle, known as the Higgs particle, has very small couplings to quarks and leptons that are proportional to their masses. An argument has been given for a Higgs particle mass of 10 GeV, but it could be much larger, in the TeV (10^{12}-eV) range.

Leptoquark bosons. The gauge symmetries $SU(3)_c \times SU(2) \times U(1)$ of the color and electroweak forces were further "grand"-unified into a single group, SU(5), by H. Georgi and Glashow. The couplings of the gauge symmetries depend on the energy scale of the interaction in which they are measured. At low energies the $SU(3)_c$, SU(2), and U(1) couplings are quite different, but they depend on the mass scale of the interaction, and all converge to a common value around 10^{15} GeV, as needed for grand unification. The value of the parameter $\sin^2 \theta_W$ of the electroweak unification is predicted because all couplings are now related, and is in agreement with the experimental value. The ratio of the b-quark mass to the τ-lepton mass is also correctly predicted. The leptons and quarks of a given generation become companions in the multiplets of the SU(5) group, in a way analogous to that in which u and d form an SU(2) weak-isospin doublet. The additional charged vector gauge bosons (X) of the SU(5) group of mass 10^{15} GeV can change quarks to antileptons, or quarks to antiquarks (Fig. 6). These leptoquark transitions cause the proton to decay (for example, adding a spectator u quark in Fig. 6 represents the decay mode $p \rightarrow e^+ \pi^0$). The expected proton lifetime is on the order of 10^{31} years, just beyond present experimental lower limits. Deep mine experiments of greater sensitivity are under way to detect proton decay.

Composites. Unified theories leave the family pattern and the masses of quarks and leptons unexplained. Complicated spectroscopies in the past were eventually understood as the result of a binding force acting on a few constituents; efforts are being made to construct quarks and leptons as composites of still more elementary particles. To date there is no experimental evidence of composite structure, with a lower limit on the composite mass scale of leptons of 150 GeV from $e^+e^- \rightarrow \mu^+\mu^-$ experiments. Only shorter distance probes can decide the matter.

For background information *see* ELEMENTARY PARTICLE; FUNDAMENTAL INTERACTIONS; GLUONS; LEP-

TON; QUANTUM CHROMODYNAMICS; QUARKS in the McGraw-Hill Encyclopedia of Science and Technology.　　　　　　　　　　[VERNON BARGER]

Bibliography: H. Fritzsch and P. Minkowski, Flavor dynamics of quarks and leptons, *Phys. Rep.*, 73:67–173, 1981; P. Langacker, Grand unified theories and proton decay, *Phys. Rep.*, 72:185–385, 1981; A. Salam, Gauge unification of fundamental forces, *Rev. Mod. Phys.*, 52(3):525–538, 1980; G. 't Hooft, Gauge theories of the forces between elementary particles, *Sci. Amer.*, 242(6):104–138, June 1980.

Evolution

The last decade has witnessed a vigorous, and often heated, reassessment of the processes which govern organic evolution. The debate was opened by a paper, published in 1972 by N. Eldredge and S. J. Gould, advocating a controversial explanation for evolutionary change. These authors argued that evolution proceeds intermittently: species arise very quickly through rapid evolutionary change and then remain essentially unchanged for long periods. They coined the term punctuated equilibrium for this idea because rapid evolutionary advance punctuates long intervals of evolutionary stasis (equilibrium). Punctuated equilibrium stands in sharp contrast to the older, dominant graduational view of evolution, in which evolution is seen to occur slowly through the gradual accumulation of small changes within a species.

Graduational model. The graduational model of evolution, based largely on the views of Darwin but also incorporating a wealth of new information, especially in the fields of genetics, population biology, and biogeography, can be summarized as follows. Members of a population vary genetically (that is, genotypically) and morphologically (that is, phenotypically). Natural selection acts on the morphologic variation. Those individuals whose characters are best adapted to function effectively in the population's environment are more likely to survive to reproduce and to pass their genes to succeeding generations. Poorly adapted individuals will be culled from the population before reproducing and their genes will be lost. As a result, the gene pool and morphology of the population move toward that of the successful individuals. This process, continued generation by generation, will eventually yield a new species when the genetic composition of the evolving population becomes sufficiently different from that of the original population. New genes and phenotypes can be introduced into this mix spontaneously by mutation so that there is always genetic fodder for initiating evolutionary change. Since genetic modifications must sweep through the entire population to produce evolutionary movement of the population, speciation is viewed as a protracted, continuous process, composed of the sequentially accumulated results of numerous small evolutionary advancements.

Increase in the number of species through time

ELEMENTARY PARTICLE

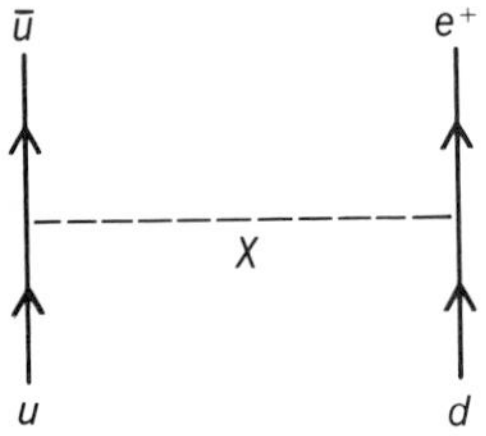

Fig. 6. Exchange of leptoquark vector boson X, causing transition of the d quark to the positron and the u quark to the ū antiquark.

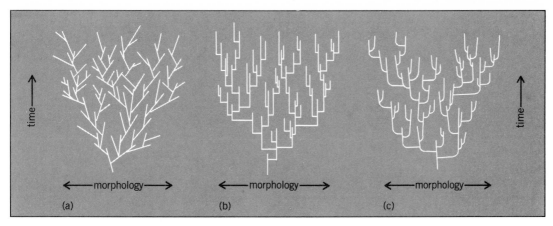

Fig. 1. Hypothetical phylogenies for alternative evolutionary models. Each nonhorizontal line represents a species. Multiplicative speciations are indicated by bifurcations. (*a*) Extreme graduational model. (*b*) Extreme punctuational model. (*c*) Punctuational composite view. (*After S. M. Stanley, Macroevolution: Pattern and Process, W. H. Freeman, 1979*)

can be accounted for in the graduational model by splitting a population into two or more subpopulations, each reproductively isolated from the others. This can occur by the development of some barrier, such as a mountain chain or sea, which comes to divide the originally continuous range of a species. The isolates then proceed independently along their own adaptive lines, eventually producing new species. The graduational model thus produces an evolutionary picture like that in Fig. 1*a*. Morphology evolves slowly with time, and the rate of morphologic change associated with multiplicative speciation events (bifurcations in Fig. 1*a*) is no greater than that occurring within a species, or a nonbranching lineage.

Punctuated equilibrium. Punctuated equilibrium devalues the evolutionary importance of long-term, gradual genetic and morphologic change. Although most advocates of the punctuational view do not deny that gradual change, as described above, can take place within a lineage, such change is not seen as a major factor in speciation or evolution. Instead, speciation is thought to occur as a rapid genetic and morphologic shift within very small, reproductively isolated groups of individuals. Such shifts are thought to occur in response to a variety of causes, including chromosomal alterations (major restructuring of genetic material); change in regulatory genes (genes controlling the growth, operation, and development of an organism); biogeographic phenomena such as

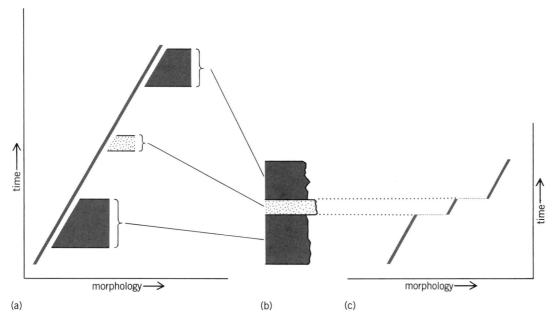

(a) (b) (c)

Fig. 2. Effect of gaps in rock record on gradual evolving lineage. (*a*) Change in morphology with time for hypothetical lineage. Dark bars show time intervals during which sediment is deposited, and segments of the phylogeny can be preserved as fossils. (*b*) Stratigraphic sequence produced by deposition noted in *a*. (*c*) Apparent punctuated evolutionary pattern of fossils preserved in rocks.

the founder effect; and the fixation in a small population of one or more macromutations (mutations having profound structural effects). The founder effect is encountered in situations where a small number of individuals colonize a new area. These founder individuals carry only a small fraction of the original population's gene pool, and give rise to a population with a gene pool that is very different from the original population.

Once formed, these small genetic isolates—incipient species, actually—become subject to natural selection. If successful, they expand, often replacing the parent species, and become full-fledged species themselves. The actual process of speciation occurs so quickly, and with such small numbers of individuals, that natural selection cannot act until the new species begins to expand. Thus, natural selection, in the punctuational view, acts on fully formed species, determining which will survive and which will not; but it plays no role in creating species.

On this point gradualism and punctuated equilibrium differ most profoundly. The graduational model invokes natural selection as the prime molding force in speciation, while the punctuational model assigns natural selection the essentially negative role of weeding out species once they have formed.

Punctuated equilibrium produces a phylogenetic tree like that in Fig. 1b. All morphologic change, that is, all evolution, is concentrated in the act of speciation. Once a species is formed, it does not vary; it enters a period of evolutionary stasis which is only broken by new episodes of speciation. A more moderate view, allowing some gradual change during a species' existence, is shown in Fig. 1c. However, such gradual change is minor compared to that occurring in punctuational speciation events, and has no obligatory relation to the morphological change associated with speciation (sometimes in opposite direction to speciational change). To a punctuationalist, speciation and evolution are decoupled from traditional darwinian gradualism. This more moderate approach is favored by most proponents of the punctuationalist view.

Much of the debate about these alternative views of evolution has focused on a search for hard evidence. This search has taken two forms: paleontologic and neontologic.

Paleontologic evidence. The fossil record has always presented something of an enigma to paleontologists because impeccable examples of gradual evolution are virtually nonexistent. Instead, the evolutionary record, as preserved in rocks, is one of apparent stasis punctuated by episodes of rapid evolutionary change. Since Darwin's time, gradualists have reconciled this disconcerting fact by arguing that it results from the incompleteness of the fossil record, and that the discontinuous nature of stratigraphic sequences superimposes a punctuational veneer on a graduational framework. The essence of this argument is illustrated in Fig. 2.

Punctuationalists counter this argument by asserting that discontinuities in the fossil record are precisely what one would expect from an episodic evolutionary process. Since, in this view, speciation occurs with such swiftness, in groups so small numerically and so localized geographically, the chance of seeing a speciation event in the rocks is practically nil. Only when a new species expands and establishes itself is it likely to be preserved in the fossil record. The fossil record is by its nature a history of established species, not a speciation. It is too coarse an instrument to reveal fine evolutionary detail.

Most paleontologists recognize the inadequacies of established phylogenies for resolving the current graduational-punctuational debate. Much effort has therefore gone into researching stratigraphic sequences that are uncommonly complete (although there is some evidence that even these may not be complete enough). The results have not proved conclusive. Examples of both types of evolutionary processes seem to occur. A study of the Permian fusulinid *Lepidolina multiseptata* (Fig. 3) appears to of-

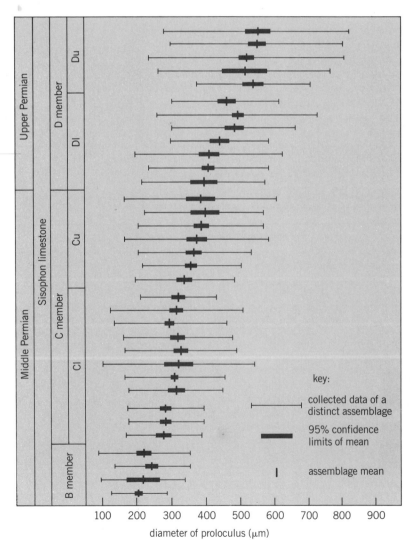

Fig. 3. Paleontologic evidence for graduational model. The graph shows the plot of increase in diameter of the initial chamber (proloculus) of the fusulinid *Lepidolina multiseptata* through time. (*After T. Ozawa, Evolution in Lepidolina multiseptata (Permian foraminifer) in East Asia, Mem. Faculty Sci., Kyushu Univ., 23:117–164, 1975*)

fer a fairly strong case of gradual evolution, while mollusks from the Plio-Pleistocene deposits of the Turkana Basin in the East African rift appear to be a good example of punctuated equilibrium (Fig. 4). Some lineages, like the Eocene condylarths, have been alternately cited as providing evidence for both views.

Punctuationalists have applied their theory of evolution with some success to paleontologic problems that have long evaded satisfactory explanation in a gradualist light. One of these problems is adaptive radiation—the rapid diversification of organisms from a single ancestral group. As an example, consider the appearance of the orders of placental mammals in the early Cenozoic. The gradualist view explains this diversification as the result of unusually rapid, but nevertheless gradual, transformation of established species under greatly increased selection pressures stemming from the availability of ecologic niches left vacant by the demise of the dinosaurs. Yet the average duration of mammalian species is about 1–2 million years, even in those lineages undergoing rapid change. Thus, the early Cenozoic mammalian diversification, which required about 10 million years, can be considered as the consequence of gradual evolution in about 10 species laid end to end. To go from mouselike, insectivorous ancestors to huge cetaceans (whales and porpoises) in 10 linear, gradually evolving species seems an impossible task. Punctuated equilibrium, in which speciation can occur in very short intervals (a few generations perhaps), and which can incorporate extensive change, accounts for this and other examples of adaptive radiation more economically and straightforwardly.

Yet many objections to punctuated equilibrium must be overcome before it can be generally accepted as an adequate explanation for evolutionary change. Chief among these concerns is the status of evolutionary stasis—is it real, or is it an artifact of the way researchers study fossils and of deficiencies in the rock record? The paleontologic concept of species usually involves a high degree of morphologic variety. In assigning individuals to a paleontologic species, this fact covertly forces the appearance that a given species is longer lasting than may actually be the case because individuals differing only slightly from the species norm are often included in the old species, when in fact they may represent the products of gradual evolution toward a new one. Morphologic stasis, and hence punctuational evolution, is artificially promoted as the process producing new species in such a framework.

In addition, the fact that only the most resistant parts of an organism's anatomy are generally preserved as fossils means that paleontologic species designations are dependent on only a small fraction of the total potential complement of evolutionary change. This can be a significant limitation when speciational change occurs in a part of the organism's morphology which is not fossilized. The effect is to augment the facade of evolutionary stasis, and to support the punctuational model.

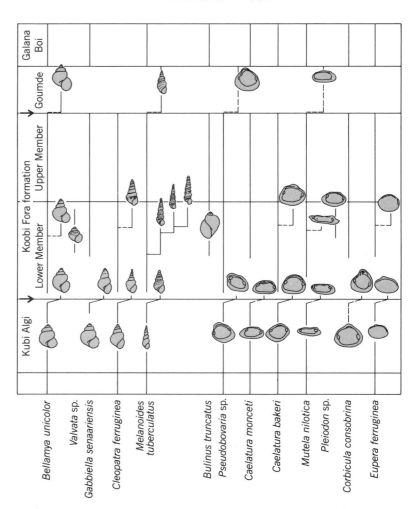

Fig. 4.   Paleontologic evidence for punctuational model. Plot of change in shell morphology with time for mollusks from the Turkana Basin deposits. The sketches show the shell form in each lineage. The principal evolutionary events illustrated are two episodes of simultaneous speciation (heavy arrows), and the adaptive radiation of several stocks in the middle of the sequence. (*After P. G. Williamson, Paleontological documentation of speciation in Cenozoic molluscs from Turkana Basin, Nature, 293:437–443, 1981*)

Neontologic evidence. As yet, no genetic mechanism for producing evolutionary stasis has been unequivocally identified. Punctuationalists have advanced the idea that stasis is the result of the genetic inertia characteristic of large populations. Stasis is thought to stem from the suppression of new mutations in the genome due to the difficulty of spreading such mutations through a large population. This would supposedly have the effect of maintaining a constant genome (stasis) until some unlocking mechanism, such as a chromosomal aberration or regulatory gene alteration, permitted a massive morphologic shift (speciation). But in many species, individuals of large populations are often scattered to such a degree that genetic interchange is not very effective in suppressing all new mutations. Moreover, populations often vary numerically with time, so that many species must endure periods when mutational genomic alterations cannot be suppressed. Most importantly, some recent work indicates that the genetic complement of some species is in constant turmoil, and that mutation is much more common than previously thought. Under these conditions it is difficult to find a basis for the long constancy in genetic composition of a species required for stasis and punctuational evolution.

Conclusion. The debate on punctuated equilibrium versus graduational evolution is far from over. Many more years of careful research in a wide array of subjects will probably be required to adequately explore the ramifications of these ideas, and to devise quantitative tests that can help distinguish the true nature of the evolutionary process. The key evidence in this case, comparable in its way to the discovery of magnetic anomalies which provided convincing documentation of the existence of sea-floor spreading, is not yet in hand. Pronouncements on the outcome of this controversy are still premature.

For background information *see* ANIMAL EVOLUTION; FOSSIL; PALEOBOTANY; PALEONTOLOGY; PLANT EVOLUTION in the McGraw-Hill Encyclopedia of Science and Technology.

[JOHN A. CHAMBERLAIN, JR.]

Bibliography: N. Eldredge, and S. J. Gould, Punctuated equilibria: An alternative to phyletic gradualism, in T. J. M. Schopf (ed.), *Models in Paleobiology*, pp. 82–115, 1972; P. D. Gingerich, Stratigraphic record of early Eocene *Hypsodus* and the geometry of mammalian phylogeny, *Nature*, 246:107–109, 1974; T. Ozawa, Evolution in *Lepidolina multiseptata* (Permian foraminifer) in East Asia, *Mem. Faculty Sci.*, *Kyushu Univ.*, 23:117–164, 1975; T. J. M. Schopf, Punctuated equilibrium and evolutionary stasis, *Paleobiology*, 7:156–166, 1981; S. M. Stanley, *Macroevolution: Pattern and Process*, 1979; P. G. Williamson, Paleontological documentation of speciation in Cenozoic molluscs from Turkana Basin, *Nature*, 293:437–443, 1981.

Extensional tectonics

The extension of the continental lithosphere without (or prior to) the formation of oceanic lithosphere is of central importance to tectonics, yet accurate determinations of large-scale intracontinental extension tested by several independent methods are sparse. For example, in the Basin and Range Province of the western United States, which has been cited as a region of Cenozoic extensional tectonics for over a century, estimates of provincewide extension have ranged from 10 to 100% increase over its original width. Only recently have workers there begun to establish a meaningful understanding of its magnitude and structural expression.

Magnitude. Large-magnitude (hundreds of kilometers) extension is supported by several independent arguments. Strike-slip faults in the southern part of the Great Basin region separate areas of Cenozoic upper-crustal extension from relatively stable tectonic blocks (Fig. 1). Linear geologic features, offset along the Garlock Fault, Las Vegas Valley shear zone, and Lake Mead fault system, allow reconstruction of this narrowest part of the province to a preextension configuration. In this reconstruction,

the Sierra Nevada, Mojave Desert, Spring Mountains, and Colorado Plateau are treated as stable upper-crustal blocks which have moved relative to each other in response to crustal extension, assuming the Spring Mountains are fixed relative to the Mojave block. The reconstruction indicates a minimum of 65% increase in the original width of the province, or about 140 km of total divergence between the Sierra Nevada and the Colorado Plateau. Other lines of evidence support large-magnitude extension: (1) the thickness of the Basin and Range crust is nearly half that of the Colorado Plateau and Sierra Nevada; (2) paleomagnetic evidence suggests large, clockwise rotation of the Oregon Coast ranges away from North America in late Cenozoic time, implying about 340 km (210 mi) of extension in the northern, wide portion of the Basin and Range; and (3) the apparent mismatch of Mesozoic tectonic elements between western Idaho–eastern Oregon and northern California suggests several hundred kilometers of extension in the northern Basin and Range.

Structural expression. The only reason for skepticism over such a large amount of extension of the continental lithosphere in the western United States is the apparent lack of evidence, until very recently, for structures in the geologic record by which it could be accommodated. The Basin and Range traditionally has been viewed as a system of horsts and grabens bounded by high-angle faults, in which the extension is accommodated by igneous intrusion and ductile stretching in the lower crust. This structural geometry is capable of producing only about 50–70 km of extension across the entire province. However, in the last 15 years, a number of workers have identified low-angle normal fault terranes in the province, and it is now clear that their presence on a provincewide scale is a rule rather than an exception.

More recently, a number of large, near-horizontal dislocation surfaces have been identified as normal faults, usually serving as lower boundaries to complex mosaics of blocks bounded by both high- and low-angle normal faults. These normal faults have many tens of kilometers of displacement on them, and because they are shallowly inclined, nearly all of their displacement accommodates crustal extension. Collectively, these faults could easily accommodate the large-magnitude extension of the Basin and Range Province deduced by other methods. Beneath these large, low-angle normal faults in some parts of the Basin and Range, the rocks have apparently been stretched and intruded in a direction parallel to that of the mosaics of normal fault blocks. Thus the low-angle normal fault terranes were believed by some workers to substantiate what is termed the classic-rift model, shown in Fig. 2, whereby the lithosphere extends by large-scale pure shear, similar to the stretching of a rubber band. In this model, extension associated with shallow, brittle normal faulting is accommodated directly below by intrusion and ductile stretching in the middle and lower crust. The pure-shear model implies that the crust

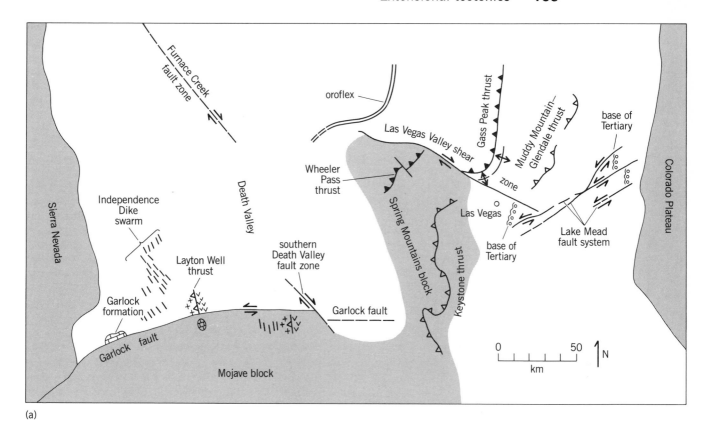

(a)

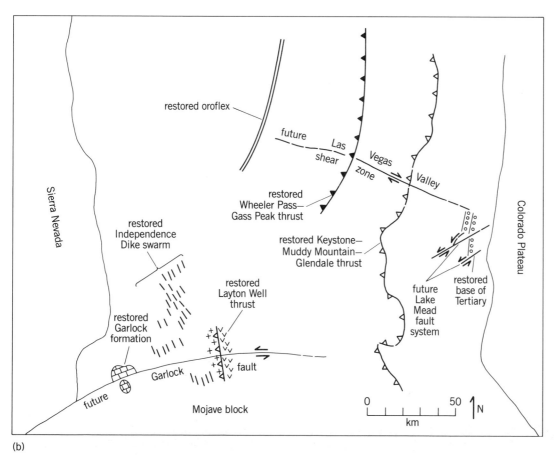

(b)

Fig. 1. Strike-slip faults and offset geologic features in a transect of the Basin and Range Province at the latitude of Las Vegas, Nevada: (a) present configuration and (b) preextension configuration (20 million years ago).

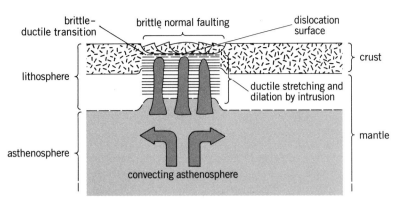

Fig. 2. Pure-shear, classic-rift model of extensional tectonics. (*After B. Wernicke, Insights from Basin and Range surface geology for the process of large-scale divergence of the continental lithosphere (abstract), in Papers Presented to the Conference on Processes of Planetary Rifting, Lunar and Planetary Institute, Houston, pp. 90–92, 1981*)

and mantle lithosphere thin by proportionate amounts during rifting.

As the results of detailed studies accumulated, however, it was found that most of the ductile deformation in these terranes largely predated extension in the normal fault mosaics and movement on the dislocation surfaces, and that a large percentage of these surfaces had no evidence of ductile deformation beneath them at all. The current data base indicates that the dislocation surfaces and overlying fault mosaics are the expression of large-scale simple shear of the upper crust, analogous to the well-

known fault geometries of compressional orogenic terranes.

In compressional mountain belts the amount of horizontal translation of thin sheets of rock at upper- and middle-crustal levels is typically several hundred kilometers or more, all of which must accommodate convergence of the lithosphere. Thus, the observation that extension and compression have highly analogous upper crustal geometries prompts the hypothesis that extension on a lithospheric scale is accommodated by the divergence of two rigid plates separated by a shallowly inclined fault, or, at depth, ductile shear zone. With this model, surficial extension in normal fault mosaics at the featheredge of the upper plate is transformed directly to convecting asthenosphere by the shear zone, as is indicated in Fig. 3.

The large magnitude of extension observed in the Basin and Range lends considerable credibility to a recently developed concept that the subsidence history of sedimentary basins on rifted continental margins reflects thermal reequilibration of mechanically extended continental lithosphere. However, such a model is not considered entirely satisfactory because it assumes uniform extension of crust and mantle lithosphere, whereas subsidence profiles of some basins that have been studied could be explained only by nonuniform extension. The markedly heterogeneous nature of lithospheric extension indicated by such studies is most easily explained by a simple-shear model of rifting, as shown in Fig. 3, as opposed to the pure-shear model of Fig. 2, in which crust and mantle lithosphere should be proportionately thinned.

For background information *see* FAULT AND FAULT STRUCTURES; PLATE TECTONICS; STRUCTURAL GEOLOGY in the McGraw-Hill Encyclopedia of Science and Technology.

[BRIAN WERNICKE]

Bibliography: W. Hamilton and W. B. Myers, Cenozoic tectonics of the western United States, *Rev. Geophys.*, 4:509–549, 1966; D. McKenzie, Some remarks on the development of sedimentary basins, *Earth Planet. Sci. Lett.*, 40:25–32, 1978; L. R. Royden and C. E. Keen, Rifting process and thermal evolution of the continental margin of eastern Canada determined from subsidence curves, *Earth Planet. Sci. Lett.*, 51:343–361, 1980; B. Wernicke, Low-angle normal faults in the Basin and Range province: Nappe tectonics in an extending orogen, *Nature*, 291(5817):645–648, 1981.

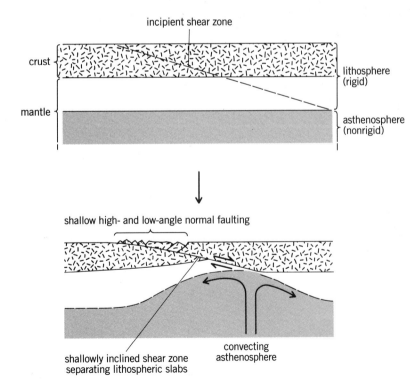

Fig. 3. Simple-shear model of extensional tectonics. (*After B. Wernicke, Insights from Basin and Range surface geology for the process of large-scale divergence of the continental lithosphere (abstract), in Papers Presented to the Conference on Processes of Planetary Rifting, Lunar and Planetary Institute, Houston, pp. 90–92, 1981*)

Extinction (biology)

The sudden extinction of formerly diverse, abundant, and widely distributed animals or plants is one of the unsolved problems of paleontology. Recent international interdisciplinary conferences have documented the nature, extent, timing and prior and subsequent global environmental conditions of these major extinction events. Attempts have been made to evaluate possible physical causes, ranging from climatic change to meteoritic impact.

Mass extinctions. Fossils record a continual diversification over 3500 million years (m.y.) of earth history, from the development of simple bacteria and blue-green algae to an impressive variety of marine plants by 800 m.y. ago. During the last 700 m.y., both animals and plants have left a relatively continuous record of increasing complexity and diversity. As the total diversity increased, some earlier forms evolved into slightly different ones, but others disappeared without descendants. These gradual changes through time resulted in recognizably distinct assemblages which can be used by the biostratigrapher to determine the geologic age of the enclosing rocks. However, at a few times in the geologic past, fossils recorded more sudden extinctions that affected a relatively large number of species and temporarily resulted in a lessened global diversity.

Significantly wide-ranging extinctions appear to have occurred during the Cambrian Period (500–570 m.y. ago), Late Ordovician (about 430 m.y.), Late Devonian (about 350 m.y.), late Permian (225 m.y.), Late Triassic (195 m.y.), and Late Cretaceous (64 m.y. ago). The first two of these extinctions reflected rapid evolutionary diversification and replacement of preexisting groups by new ones, but they did not decrease overall diversity. Like earlier extinctions, the Devonian and Permian extinctions affected the marine organisms, but their effect was more widespread. Major groups of phytoplankton, foraminifera (shell-bearing protozoans), passively suspension-feeding invertebrates (corals, bryozoans, brachiopods, crinoids), and primitive fish disappeared, and total global diversity was much reduced. Land animals and plants changed more slowly, following the pattern shown by the early Paleozoic invertebrates of continual diversification and replacement of primitive types. Unlike earlier events, the Late Cretaceous extinction affected a greater diversity of organisms and habitats, ranging from certain marine phytoplankton and zooplankton to the giant marine reptiles, terrestrial dinosaurs, and flying reptiles.

Recent discovery of an elevated level of iridium and other rare siderophile elements in rocks at the Cretaceous-Tertiary boundary has renewed support for the theory that collision of a large meteor or comet with the Earth caused these extinctions, either as a direct effect of the impact by shock wave or heat blast; by dust thrown into the upper atmosphere, blocking out sunlight for a period sufficient to kill off short-lived photosynthetic organisms; by cyanide poison released into the atmosphere or oceans by the comet; or by destruction of the ozone layer and production of massive amounts of nitric oxide (NO) that later washed out in rainfall as nitric acid, selectively causing extinction of calcareous-shelled marine plankton in the near-surface waters.

Cretaceous plankton. Two planktonic groups were most severely affected by the extinction event at the end of the Cretaceous: the coccolithophores and planktonic foraminifera. Coccolithophores are planktonic unicellular algae whose tiny flagellated cell is covered by calcareous platelets (coccoliths)

1–10 μm in diameter. They may be numerous, up to a few hundred thousand individuals occurring in a liter of surface water of the present North Atlantic. Their life span is short (a few hours to about 5 days), and the rapid turnover results in the accumulation of coccolith oozes on the sea floor. From their first appearance their diversity and numbers increased rapidly to a maximum number of species in the very latest stage of the Cretaceous. Their abundance in the warm, shallow Cretaceous seas resulted in the worldwide accumulation of coccolith limestones and chalks; and as from a half million to 69 million coccoliths are present in a single cubic millimeter of chalk, these abundant fossils gave the name to this geologic peiod (*creta* is Latin for chalk). Nevertheless, at the end of the Cretaceous, all but one or two species disappeared suddenly, about 90% becoming extinct. The extinction of these important phytoplanktonic primary producers had a domino effect on the food chain, affecting the largest marine animals and those at the end of the more complex food chains most severely. As important contributors to the calcareous sediments, the extinction of most coccolithophores also changed the character of the sediments, sharply decreased the rate of deposition, and left a gap in the rock record.

The other major group of calcite-depositing plankton of the Cretaceous was the planktonic foraminifera, marine protozoans whose tiny shells also contributed to the pelagic limestones of the Cretaceous, although in lesser amount. They were also diverse, geographically widespread, and abundant until the latest Cretaceous and similarly disappeared very suddenly. Only one or two species persisted and later gave rise to the planktonic assemblages of the Cenozoic.

The noncalcareous marine microplankton did not undergo as complete or as sudden a change. Thus, about 60% of the dinoflagellates disappeared (only the organic or calcareous resting cysts produced seasonally by a small fraction of the standing crop are fossilized), but the change was not as complete as for the two calcareous plankton groups.

The record is less complete for siliceous plankton. Diatoms, the dominant marine phytoplankton today, and radiolarians, the major silica-depositing protozoans, were present during the Cretaceous, and were locally abundant enough to produce diatomites or radiolarites. Other species occur in Cenozoic rocks. Because the opaline silica of the radiolarian shells and diatom frustules commonly dissolves in the water column or on the sea floor, a record of abrupt extinction would be obscured by the rarity of rock sequences that contain well-preserved siliceous fossils and that span this time interval.

Estimates of the actual time involved in the Cretaceous calcareous plankton extinctions range from a few thousand years to as little as 50 years. Regardless of its total duration, the time interval during which deposition of coccoliths and foraminiferal tests almost ceased has been exhaustively studied. Very detailed studies have been made at Zumaya

(Spain), Gubbio (Italy), Stevns Klint (Denmark), and in the Lattengebirge (Bavaria), as well as in the deep sea cored by the oceanographic Deep Sea Drilling Program. Additional early Paleocene faunal zones have been found that record the millions of years of slow and gradual evolutionary replacement of the once-abundant calcareous microplankton that had been so suddenly eliminated. The abruptness of the microplanktonic change has become increasingly apparent, whereas the extinction of the Cretaceous ammonites (distantly related to the living *Nautilus* and octopus) has been shown to be much more gradual. Thus, the marine calcareous microplankton suffered sudden and nearly complete extinction while other coexisting marine organisms showed much slower evolutionary change. The shallow-water benthos also was not as influenced by the extinction event as was the marine plankton.

Land extinctions. On land, dinosaurs also disappeared at the end of the Cretaceous. It is not certain whether or not the marine and terrestrial extinctions were contemporaneous, and hence possibly the result of a single catastrophic event, or if one or the other extinction was somewhat earlier, and their causes were unrelated. The dinosaur fossil record is much less complete than that of marine organisms, and in some localities plant fossil evidence suggests that the dinosaurs may have disappeared entirely somewhat before the close of the Cretaceous Period. If so, marine extinctions could not have caused the terrestrial extinctions, for example, by changes in atmospheric composition resulting from the extinction of the dominant photosynthetic phytoplankton. Nor could both extinctions have had a common cause, whether it was related to temperature, climate, or an impact event. The catastrophic events supposedly resulting from an asteroid or comet collision would have affected land animals and plants sooner and more drastically than marine ones, yet no sudden or major change is indicated by the land plant fossil record. Mammals and birds, as well as crocodiles, snakes, turtles, and other reptiles, also persisted while the dinosaurs disappeared.

The geochemical evidence of an asteroid or comet impact about 64 m.y. ago appears quite convincing. However, no biologic change of global extent, instantaneous occurrence, and universal effect has yet been demonstrated to have resulted from this impact. Instead, the Cretaceous extinction was, as the earlier ones, very selective. Rapid and all-inclusive extinction affected only the calcareous microplankton, while some organisms gradually decreased in importance over millions of years, and others showed relatively little, if any, change. Over the past few years, knowledge of the physical and biologic nature of the Cretaceous Earth has greatly increased, but some of the pieces to the puzzle of the Cretaceous extinctions may still be missing.

For background information *see* EXTINCTION (BIOLOGY) in the McGraw-Hill Encyclopedia of Science and Technology.

[HELEN TAPPAN]

Bibliography: W. K. Christensen and T. Birkelund (eds.), *Proceedings of the Cretaceous Tertiary Boundary Events Symposium*, II, University of Copenhagen, pp. 1–250, 1979; Conference on large body impacts and terrestrial evolution: Geological, climatological and biological implications, Snowbird, Utah, *Lunar Planet. Inst. Contrib.*, 449:1–69, 1981; H. Tappan, Extinction or survival: Selectivity and causes of Phanerozoic crisis, *Geol. Soc. Amer. Spec. Pap.*, no. 190, 1982; H. Tappan and A. R. Loeblich, Jr., Evolution of the oceanic plankton, *Earth Sci. Rev.*, 9:207–240, 1973.

Extrusion

Continuous extrusion is a metal-forming process whereby feedstock is pushed continuously through a die to form an indefinitely long product. In addition, the product cross section may vary considerably—being either smaller or larger—from that of the feedstock, necessitating the generation of high pressures in the region of the die. A continuous extrusion machine, therefore, has to be capable of taking feedstock at atmospheric pressure and passing it continuously into a highly pressurized environment and out again through a die which may considerably change its dimensions or shape.

Conform process. The Conform process achieves this objective by making effective use of the considerable frictional force which is generated between a billet and its container in a conventional extrusion press—a shortcoming in conventional extrusion. This frictional force is used to convey the feedstock—in either solid or particle form—into the extrusion chamber and to generate the pressures necessary for extrusion. The process is achieved by the use of a moving, continuous extrusion chamber operating in

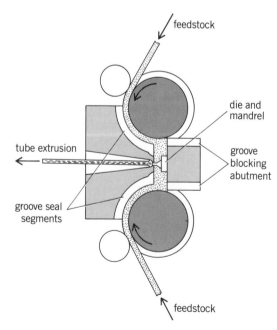

Fig. 1. Arrangement of wheels in the two-wheel conform machine. (*After B. Maddock, Company builds around new extrusion technology, Metallurgia, July 1981*)

combination with a stationary die. In considering the possible variants of such a machine, the simple concept of using a rotating, grooved wheel acting in conjunction with a stationary shoe was adopted. The shoe is arranged to overlap a portion of the wheel circumference and is designed to carry the die and tooling (Fig. 1). The groove may have a variety of shapes and may be in either the periphery or the face of the wheel, although all production machines to date use peripheral grooves. If a square groove is used, it will form three sides of a square-section extrusion container; the fourth, closing side is formed by tooling supported in the shoe and projecting a short way into the wheel groove. The extrusion container so formed is closed at the die zone by an abutment which is usually mounted in the shoe, completely filling the cross section of the groove. An extrusion die is normally located in either the abutment or the tooling adjacent to it. As the wheel is rotated, the feedstock is pulled forward by friction up to the abutment. Sufficient frictional force may be generated between wheel groove and feed to cause it to yield and deform plastically over a short length in front of the abutment, called the extrusion grip length. It should be noted that with three moving sides driving the feedstock, and one stationary side (the shoe tooling) resisting motion, the net effect is of two sides driving. Within the upset plastic zone the compressive stress builds up rapidly until it reaches a maximum in front of the abutment. Each increment of grip length adds to the stress in the material until the necessary flow stress is attained at the die face, causing extrusion to occur. The length of this upset zone is self-adjusting, depending on the extrusion pressure required, and it is only nec-essary to provide sufficient overall grip length in the shoe tooling for the process to operate successfully.

In the classical case, the feed is an interference fit in the groove. In practice this does not have to be the case: a wheel groove slightly oversized to feedstock diameter operates very successfully with many nonferrous metals such as aluminum. Indeed Conform machines are capable of accepting partic-ulate feeds, such as powders or granules, compact-ing them into a plastic mass in the zone in front of the die, and extruding a homogeneous product. In practice, wheel grooves are often fully radiused to match the feedstock diameter. Although the match-ing of groove geometry to feedstock shape is advan-tageous, an exact match is not essential since the Conform principle ensures a final precise fit over the extrusion grip length immediately before the die.

Basic design. A Conform machine consists of two basic elements, a rotating wheel and a stationary shoe. The shoe is held near the wheel, but can readily be removed for access to tooling.

The rotating components constitute a heavy-duty steel shaft on which is mounted the wheel assem-bly, the whole unit being carried on substantial journal bearings carrying the high radial loads de-veloped in the process. Rotational speeds tend to be low, typically below 60 rpm (revolutions per min-ute) on machines currently in operation, but the torque requirement may be 40,000 N-m (newton-meters) for a 300-mm-diameter (12-in.) wheel, so that high-torque, low-speed drives are necessary. This can be achieved either hydraulically or electrically, and machines with both types of drive are in opera-tion. The rotational axis may be vertical, horizon-tal, or at any angle of inclination.

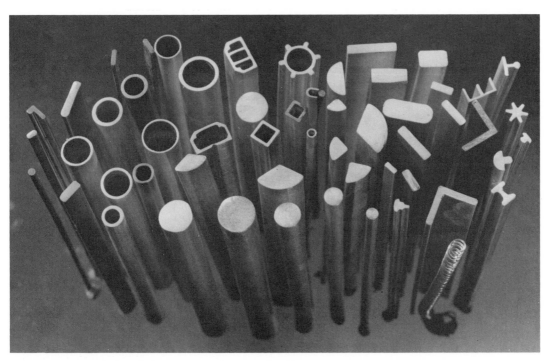

Fig. 2. Typical range of Conform aluminum products from 9.5-mm-diameter (3.74-in.) feedstock.

Wheels are usually made of high-strength, hot-working steels, often built up from two or three disks, and are axially preloaded by a hydraulic nut to simplify repair or replacement. The shoe, normally a massive, rigid steel section, is sized to provide a working zone of at least 90° for the inserted tooling segments. The frictional driving force and necessary tooling length depend on the coefficient of friction and yield strength of the feedstock. With most metals—particularly galling metals—grip lengths are remarkably short, even with particulate feed. For example, aluminum rod of 9.5-mm (3.74-in.) diameter extruding either to wire of 1-mm (0.04-in.) diameter, or tubing with 9.5-mm outside diameter and a 1-mm wall, requires an overall grip length of 100 mm (3.9 in.).

The die is the area of greatest possible variations in design, profile, and material selection. Conventional extrusion-die practice can be adapted to Conform operations, but due regard must be paid to the steadily increasing pressure profile along the working zone. Die technology has been developed which enables a wide selection of products to be made from one size of feedstock and machine by simply changing the die assembly (Fig. 2).

A typical large production machine is illustrated in Fig. 3. Available production machines range from those with wheels of 225-mm (9-in.) diameter, 9.5-mm (0.375-in.) groove width, and 70-kW power, up to machines with wheels of 500-mm (20-in.) diameter, 40-mm (1.5-in.) groove width, and 350-kW power. Design studies have been undertaken for machines with wheels of 1000-mm (39.4-in.) diameter, 50-mm (2-in.) groove width, and 500-kW power, and 2.2×10^6 N-m torque, which will be capable of extruding up to 2 metric tons (2.2 short tons) per hour of aluminum sections or 6 metric tons (6.6 short tons) per hour of copper sections.

A two-wheel Conform machine has been developed that has certain operational and product quality advantages (Fig. 1). Two-wheel Conform machines are used for the extrusion of aluminum cladding on steel wire. A linear version of the process, called Linex, has also been developed.

Production machines are currently in operation extruding typically 1 metric ton (1.1 short ton) per hour of solid aluminum conductor sections from 19-mm-diameter (0.75-in.) EC grade feedstock. To date, Conform technology has found its principal application in the extrusion of aluminum and its alloys into a wide range of sizes and sections (Fig. 2); however, machines are also being used on numerous other nonferrous metals and alloys, including copper, zinc, and precious metals, and the potential for ferrous metals and plastics is being explored.

Particulate feed and metal recycling. The potential for particulate metal feed was recognized in the original patent for Conform machines, and consid-

Fig. 3. Conform production machine. The wheel diameter is 400 mm (15.75 in.) and the applied power is 250 kW. (*Courtesy of Babcock Wire Equipment Co. Ltd.*)

erable attention has been given to this aspect of their application. Powders—of aluminum, copper, and their alloys—can be successfully extruded into homogeneous products. It is noticeable that the mechanical properties of the aluminum products are considerably greater than similar products extruded from rod feed and this can be attributed to the presence of finely divided metal oxide in the product (dispersion strengthening). The manufacture of electrical conductors from granulated copper cathode (copper obtained by electrolytic refining), extruded by the Conform process, has been developed. This process offers considerable energy and plant-investment savings.

Secondary metal products can be made from granulated scrap and reject materials suitably cleaned and segregated. While for some applications these products may not meet very tight technical specifications, there are many lower-grade products for which this process is acceptable, providing considerable potential for energy savings.

Conclusion. The particulate feed aspect of the Conform process perhaps offers the greatest potential for future developments, both in energy and materials conservation and in the prospects for the development of new metals and alloys, including fiber-reinforced products. The development of large machines capable of accepting continuously cast, nonferrous sections, such as those sections produced by modern casting wheel techniques, should be followed by the development of large friction-actuated, continuous extrusion machines. These machines could compete successfully with large conventional presses for the extrusion of a wide range of nonferrous products, offering considerable savings in space, capital investment, and overall operating costs.

For background information *see* EXTRUSION; METAL FORMING; POWDER METALLURGY; TOOLING in the McGraw-Hill Encyclopedia of Science and Technology.   [CLIFF ETHERINGTON]

Bibliography: D. Green, Continuous extrusion forming of wire sections, *J. Inst. Met.*, 100:295–300, 1972; J. A. Pardoe, The technology of reclaiming non-ferrous metal from waste by the continuous extrusion process, *Fourth World Recycling Congress*, April 1982; J. A. Pardoe, Wires and sections from non-ferrous metal powders, *Met. Power Rep.*, vol. 36, no. 8, 1981.

Fern

Recent studies in fern morphology have expanded the understanding of the mechanisms controlling the orientation of cell divisions in fern gametophytes and of the differentiation of fern shoot systems.

Gametophyte morphogenesis. An objective of the study of morphogenesis is the description of the origin of biological form in terms of such cellular phenomena as polarity, expansion, division, and differentiation. Of particular interest to botanists are the processes by which a plant cell selects the plane of its forthcoming division, that is, the orientation

in which the new cell wall is inserted between the future daughter cells. A change in the predominant division plane is often observed as the first step in the initiation of new structures or the transformation of existing structures in plants. The gametophytes of certain ferns have served as convenient experimental objects for studying how the plant cell controls its plane of division, because the two major growth stages of these organisms exhibit different patterns of cell wall insertion. Until recently, the morphogenetic research on fern gametophytes had assumed that specific gene products direct the dividing cell to select a particular plane of division. However, new experimental evidence shows that the incipient cell wall assumes the plane of minimal surface area within the dividing cell. This dependence on surface area seems to indicate that cells in the fern gametophyte may have the ability to translate the information inherent in some biophysical process into specific division planes.

Experimental organisms. The life cycle of ferns, like those of all plants, consists of alternating sporophyte and gametophyte generations (Fig. 1). Fern sporophytes are large, often perennial plants which are prominent members of many forest and field communities, whereas the gametophytes are minute, ephemeral structures. Specialized cells within the diploid sporophyte undergo meiosis to produce haploid spores. Each spore will then develop into a haploid gametophyte upon which the sexual organs are borne. A sperm from an antheridium fertilizes the egg within the archegonium; the resultant diploid zygote grows into a sporophyte, and thus the cycle has repeated itself.

The gametophyte of such ferns as *Onoclea sensibilis* (sensitive fern) and *Adiantum capillus-veneris* (maidenhair fern) have become favored experimental organisms for morphogenetic investigations. The ga-

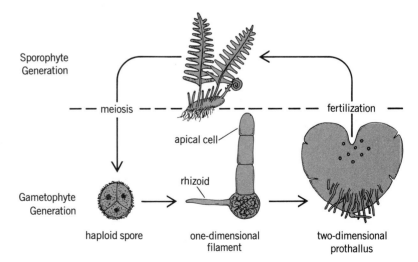

Fig. 1. Abbreviated diagram of the generalized fern life cycle which is composed of alternating gametophyte and sporophyte generations. Approximate sizes: spore 50 μm in diameter, filament 300 μm in length; prothallus 1–3 mm in width; sporophyte leaf 0.2–1.0 m in length. The reproductive organs where gamete formation, fertilization, and meiosis occur are not depicted.

metophytes of these ferns have two distinct developmental stages, each one with its own characteristic plane or planes of cell division. In the initial filamentous growth form, the apical cell undergoes repeated divisions in which the new cell walls are all made to adopt an orientation transverse (perpendicular) to the long axis of the filament. Eventually, the apical cell will, however, divide in the longitudinal (parallel) plane. This first longitudinal division marks the transition to the two-dimensional growth form, wherein the cells at the apex divide in various planes to generate the cordate shape of the mature prothallus. Since this transition in growth form involves the relaxation of prior restraints on cell division, these particular fern gametophytes have been intensively studied in the hope of elucidating the cellular mechanisms that control the plane of cell division.

Genetic hypothesis. It has long been appreciated that fern filaments exposed to blue wavelengths of light show a rapid transition from filamentous to two-dimensional growth. In contrast, red light tends to perpetuate indefinitely the filamentous growth stage. In the 1960s, these observations were interpreted in terms of the operon theory, which was first formulated to characterize the process of gene expression in bacteria. It was proposed that the morphogenetic effect of light is the regulation of the differential transcription of specific messenger RNAs responsible for each developmental stage. Blue light was thought to induce or depress a specific constellation of genes which codes for the proteins (enzymes, pigments, structural components) necessary to initiate and sustain two-dimensional growth, whereas red light was thought to restrict transcription to those genes that mediate filamentous growth (Fig. 2a). By inference, this hypothesis assumed that certain gene products could be found which have the ability to manipulate the orientation of the new cell wall.

Despite some early results that seemed to support this genetic hypothesis concerning the control of the division plane in fern gametophytes, careful studies have ultimately failed to relate the transition in growth form to the differential synthesis of macromolecules. For instance, the bulk RNAs of two-dimensional gametophytes contained higher percentages of the nucleic acid bases guanine and cytosine than did the RNAs of filamentous gametophytes. But an analysis of different RNA fractions showed that this shift in RNA base composition was totally attributable to an increased level of chloroplastic ribosomal RNA, which presumably does not participate in gametophyte morphogenesis. Although two-dimensional gametophytes grown under blue light exhibited higher rates of protein synthesis than filaments exposed to red light, this difference was similarly referable to an enhanced level of protein synthesis in the chloroplasts. Finally, it was alleged that various inhibitors of RNA and protein synthesis could selectively block the synthesis of the hypothetical gene products involved in the transition in growth form.

Further research has demonstrated that these synthesis inhibitors acted as general inhibitors of gametophyte growth rather than specific inhibitors of two-dimensional growth. Even though treated gametophytes remained filamentous for a longer period of time than the control plants, the inhibitors did not cause the young gametophytes to continue filamentous growth beyond the cell number at which the controls initiate two-dimensional growth. Given the equivocal nature of the evidence detailed above, botanists have been forced to reconsider whether specific genetic directives can determine the characteristic plane of cell division for each stage of gametophyte development.

Physical hypotheses. In the nineteenth century,

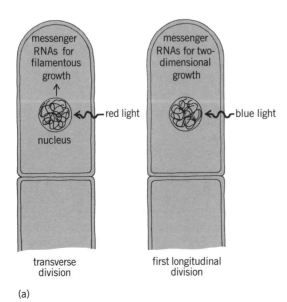

(a)

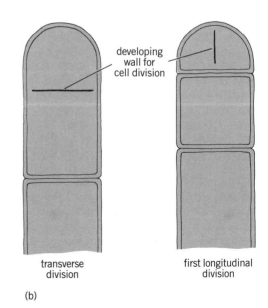

(b)

Fig. 2. Schematic representations of the (a) genetic and (b) minimal surface area hypothesis to explain the change in the plane of apical cell division that accompanies the transition from filamentous to two-dimensional growth.

Relationship between cell geometry and orientation of cell divisions

Apical cell dimensions (length/width ratio)	Calculated division place of minimal surface area	Observed number of cell divisions	
		Transverse divisions of filamentous growth	First longitudinal division of two-dimensional growth
≥ 2.00	Transverse	174	—
1.51–2.00	Transverse	139	—
1.01–1.50	Transverse	397	—
0.86–1.00	Transverse or longitudinal	170	21
0.71–0.85	Transverse or longitudinal	54	77
0.51–0.70	Longitudinal	3	151
≤ 0.50	Longitudinal	—	72
		937	321

Adapted from T. J. Cooke and D. J. Paolillo, Jr., The control of the orientation of cell divisions in fern gametophytes, *Amer. J. Bot.*, 67:1320–1333, 1980.

plant morphologists devoted considerable study to the geometric preconditions for cell division. This work culminated in the so-called principle of minimal surface area which states that a new cell wall will, in the absence of anisotropic external forces, occupy the plane of minimal surface area within the dividing cell. Several laboratories interested in fern gametophyte morphogenesis have recently applied this long-neglected principle to the problem of two-dimensional growth. This principle implies that the young gametophyte should continue filamentous growth as long as the transverse division plane represents the surface of minimal area in the apical cell, and two-dimensional growth should start once the longitudinal plane becomes the orientation of minimal surface area (Fig. 2b). To test this prediction, one group made time-lapse measurements of fern filaments exposed to various treatments, including different wavelengths of light, growth regulators, and RNA synthesis inhibitors, as the apical cells underwent the last series of transverse divisions in filamentous growth as well as the first longitudinal division in two-dimensional growth. Using various geometric models for the dividing apical cells, those workers converted their measurements of length/width ratios into relative internal surface areas and then noted whether the observed plane of cell division coincided with the calculated orientation of minimal surface area. Because the results from all the experimental treatments showed no essential difference, all the data can be pooled in the table. Depending on the geometric model used for the apical cell, the length/width ratio that makes the surface areas of the transverse and longitudinal planes equal falls somewhere between 0.7 and 1.0. A ratio higher than that value signifies that the transverse plane represents the orientation of minimal surface area within the dividing cell; a lower ratio makes the longitudinal plane the position of minimal area. This experiment has firmly established that the orientation of all divisions, and hence the growth form of the gametophyte, shows a striking dependence on the surface-area relations of the dividing apical cell.

The next step in this research is to find what this surface-area relationship could disclose about cellular mechanisms that might dictate the orientation of division. One can almost rule out the possibility of direct genetic control, because the activity of gene products, that is, proteins, depends on their concentration as well as the concentrations of substrates, products, and cofactors. Since concentration is a volume-dependent function, a cell could, in theory, use the activity of a particular gene product to determine its volume. For instance, the need to dilute an endogenous inhibitor below a threshold concentration has been proposed as a mechanism to explain why yeast cells do not divide until they have attained a certain minimal volume. But such volume information cannot reveal the internal plane of minimal surface area. Moreover, the fact that various growth regulators and RNA synthesis inhibitors do not affect the relationship between surface area and division plane argues against genetic control. On the other hand, many physical processes (surface free energy, stress-strain mechanics, light fluence, electric current density, and so on) are functions of the surfaces upon which they act, and thus a major emphasis of current research is to identify a biophysical process in the cell which might serve to direct the alignment of the new cell wall.

Two possible mechanisms for the physical control of the division plane have been suggested. One hypothesis views the developing cell wall as a weightless surface, much like the thin films that compose a mass of soap bubbles, where the thermodynamic constraint of minimal surface free energy forces the partitions within the bubbles to position themselves in the orientation of minimal surface area. The other hypothesis makes the assumption that a cell wall will be inserted in a plane free of shear stress; this shear-free plane shifts from transverse to longitudinal orientation as the length/width ratio declines below unity. Both these mechanisms seem unsatisfactory, at least in their present form. Observations of developing cell walls indicate that the incomplete wall adopts the plane of minimal surface area before it makes contact with existing walls, a situation incompatible with the thermodynamic mechanism which depends on mutual contact of all surfaces. The prediction of shear-free planes within the dividing apical cell is

Fig. 3. Habit of *Lygodium japonicum*. (a) Juvenile plant with two fully expanded leaves. (b) Rhizome segment of adult plant showing stem bifurcations and dorsal position of the leaves. (c) Twining rachis and expanding leaflets of an adult leaf climbing on a string.

based on deplasmolysis studies of how the shrunken protoplast restores its original volume upon the removal of the osmoticum. Since the stresses inside growing apical cells bear no relation to those observed in expanding protoplasts, it is inappropriate to use deplasmolysis experiments to determine shear-free planes. Clearly, further research will be nec-

essary to identify the physical mechanism whereby plant cells choose their plane of division.

Conclusion. The study of fern gametophyte morphogenesis has led to several insights about the control of cell division in plants, the most significant of which is the tendency for the new cell wall to occupy the orientation of minimal surface area. This dependence on surface area makes physical processes the most reasonable candidates for possible control mechanisms, even though the nature of the actual mechanism remains a question for additional research.

[TODD J. COOKE; RICHARD H. RACUSEN]

Shoot morphology. Ferns exist in nearly the same variety of habitats as the highly diversified flowering plants. They have adapted to life in mesic to arid terrestrial, epiphytic, and even aquatic environments. The vegetative body of the fern, like that of the flowering plant, consists of the following organs: leaf, stem, and root. However, in contrast to the flowering plants, which frequently possess complex, highly branched stems, fern shoots usually consist of an infrequently branched or totally unbranched stem which bears large structurally and functionally complex leaves. In addition to performing the light-gathering and gas-exchange functions of photosynthesis, the leaves of ferns also bear the spore-producing structures (sporangia). In many fern species the leaves are adapted for specific growth habits or functions which, when they occur in flowering plants, involve the modification of entire shoots. Examples of such fern leaves that are flowering plant shoot analogs include the stolonlike leaves of the walking fern *Camptosorus rhizophyllus*, the scrambling, thicket-forming leaves of *Gleichenia*, and the twining leaves of the climbing fern *Lygodium*.

The large final size of many fern leaves is partially due to a prolonged period of apical growth, during which time the growing leaf tip is protected within the crozier, or fiddlehead. In seed plants, although the shoot as a whole may undergo essentially indeterminate (unlimited) growth, the leaves typically have very limited apical growth, producing determinate (limited to a definite size and form) organs. For this reason the prolonged apical growth, large size, and complexity of some fern leaves suggest at least a superficial resemblance to a shoot.

It is a widely held tenet of comparative morphology that the leaves of ferns and seed plants, the so-called megaphylls, were similarly evolved from modified branch systems. Therefore the shootlike characteristics of some fern leaves have been taken as evidence that the leaf and stem are not as clearly differentiated as organ types in the former group; that is, ferns have a more primitive level of organography. This hypothesis was recently examined in a morphological investigation of the development of the shoot system of *Lygodium* (Schizaeaceae). *Lygodium* was chosen for this study since its indeterminate, climbing leaves apparently represent an extreme example of leaf elaboration resulting in a convergence to a shootlike form and function. In addi-

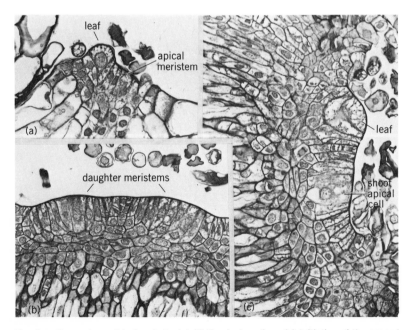

Fig. 4. Comparison of leaf and shoot initiation in *Lygodium*. (a) Initiation of the second leaf of the juvenile plant. (b) Terminal dichotomy of an adult shoot apical meristem. (c) Initiation of climbing leaf on dorsal (upper) flank of adult shoot apical meristem.

tion the juvenile plant of *Lygodium* produces a series of small, determinate leaves (Fig. 3*a*). The initiation and early development of these more typical leaves was studied and compared with that of the problematical climbing leaves of the adult plant.

Organ differentiation. The adult plant of *Lygodium* consists of a bifurcating, creeping rhizome which bears leaves on its upper surface and roots on its lower surface (Fig. 3*b*). The adult leaves circumnutate, climbing to a final length of from 2 to over 20 m, depending upon the species. In addition to their twining, indeterminate growth, these leaves resemble shoots in their formation of budlike, resting leaflet apices (leafbuds) which can grow out to ramify the leaf or replace the original leaf apex if it is damaged (Fig. 3*c*).

Information from early stages of development is often valuable in the determination of structural relationships (homologies) of plant organs. In spite of its superficial resemblance to a shoot, a critical analysis of the development of the climbing leaf revealed its actual foliar or leaflike nature. The adult shoot apical meristem of *Lygodium* resembles that of other ferns. A tetrahedral apical cell is surrounded by a group of elongate prismatic cells and an underlying zone of isodiametric subsurface cells (Fig. 4*c*). The first indication of leaf initiation is the enlargement of a single prismatic cell on the dorsal (upper) flank of the meristem (Fig. 4*c*). This leaf apical cell undergoes successive divisions, becoming lens-shaped and forming a protuberant, dorsiventral leaf primordium.

Although the shoot apical meristem of the juvenile plant is smaller and vertically oriented, its structure is similar to that of the adult. As in the initiation of the climbing leaves, the juvenile leaf is first evident by the enlargement of a single prismatic cell on the flank of the apical meristem. Subsequent divisions of this cell and its recent derivatives form an emergent primordium very similar to that of the climbing leaf (Fig. 4*a*). In their manner of initiation, the juvenile and adult leaves are very similar to each other as well as to the leaves of other ferns. This supports the argument that, in spite of their shootlike features, the climbing leaves are in fact foliar organs.

In the ferns branching may occur by outgrowth of lateral buds located on the stems and leaves, or by a terminal bifurcation of the shoot apical meristem. In order to examine the extent of differentiation between the shoot organs (leaf and stem), branch initiation was studied in *Lygodium* and compared with leaf initiation. In contrast to the lateral, appendage-like initiation of new leaves, new shoots arise only by a terminal bifurcation or dichotomy of the shoot apical meristem. No lateral buds are produced. Branching begins by a broadening of the apical meristem in the horizontal plane. The original apical cell and the prismatic cells adjacent to it cease producing tissue for the elongation of the stem and undergo cell divisions to form a packet of small cells at the center of the meristem (Fig. 4*b*). This isolates

two groups of still meristematic cells within the broadened apical meristem. These cells continue to divide and form two identical daughter meristems, each of which initiates a new apical cell by oblique divisions of one of the elongate prismatic cells.

The distinctiveness of the leaf and stem components of the *Lygodium* shoot is suggested by their differences in gross morphology. The foregoing descriptions of their development make it clear that new stem branches and new leaves are fundamentally different types of structures from the time of their initiation.

Leaf development. In spite of their initial similarity, the primordia of the juvenile and adult leaves of *Lygodium* produce a series of leaves of very different size and form. Early differences in amount of apical growth and manner of leaflet initiation account for these differences. The primordia of the first

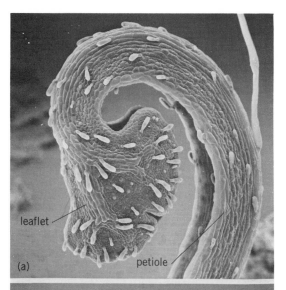

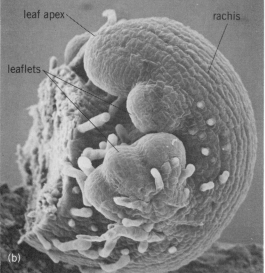

Fig. 5.   Scanning electron micrographs of leaf development. (*a*) Early stage of lamina development in the second juvenile leaf. (*b*) Apex of the climbing leaf dissected from the crozier.

juvenile leaves cease apical growth while very small, and undergo successive bifurcations of the primordial lamina to produce their determinate final form (Figs. 3a and 5a). Later juvenile leaf primordia have more prolonged apical growth and show a transition to lateral leaflet initiation producing a pinnate final form. The climbing leaves grow continuously at the apex and, like the later juvenile leaves, initiate leaflets laterally (Fig. 5b). In spite of their size and complexity, the climbing leaves are formed by the familiar basic processes of fern leaf morphogenesis and do not show any peculiar shootlike characteristics in their later development.

Conclusions. Analysis of the developmental morphology of the shoot of *Lygodium* showed that in spite of certain functional and morphological similarities to twining shoots, the climbing leaves are in fact fundamentally foliar organs. They are complex and highly modified structures, but these modifications represent specialization of the leaf organ and not a retention of shoot characteristics in a primitive leaf-shoot transitional organ. These conclusions are based purely on structural observations and comparisons. The actual evolutionary origin or strict homology of the leaves of higher vascular plants is unknown; however, the differentiation between leaf and stem is marked in the ferns and comparable to that seen in the shoots of seed plants.

The simple form and branching pattern of the stem and the complexity of the leaf of *Lygodium* demonstrates the integration of growth usually seen between these two organs in vascular plants. *Lygodium* is a striking example of the phenomenon of leaf elaboration and stem simplicity which demonstrates the integration of growth as it commonly occurs in the ferns.

For background information *see* APICAL MERISTEM; PLANT GROWTH; PLANT MORPHOGENESIS; POLYPODIALES in the McGraw-Hill Encyclopedia of Science and Technology.

[R. J. MUELLER]

Bibliography: D. W. Biehorst, On the stem apex, leaf initiation, and early leaf ontogeny in Filicalean ferns, *Amer. J. Bot.*, 64:125–152, 1977, T. J. Cooke and D. J. Paolillo, Jr., The control of the orientation of cell divisions in fern gametophytes, *Amer. J. Bot.*, 67:1320–1333, 1980; A. E. De-Maggio and V. Rahgavan, Photomorphogenesis and nucleic acid metabolism in fern gametophytes, *Adv. Morphogen.*, 10:227–263, 1973; G. P. Howland and M. E. Edwards, Photomorphogenesis of fern gametophytes, in A. F. Dyer, (ed.), *The Experimental Biology of Ferns*, pp. 393–434, 1979; D. R. Kaplan, Morphological status of the shoot systems of Psilotaceae, *Brittonia*, 29:30–53, 1977; J. H. Miller, Orientation of the plane of cell division in fern gametophytes: The role of cell shape and stress. *Amer. J. Bot.*, 67:534–542, 1980. R. J. Mueller, Shoot morphology of the climbing fern *Lygodium* (Schizaeaceae): General organography, leaf initiation, and branching, *Bot. Gaz.*, 143(3), 1982; R. J. Mueller, Shoot ontogeny and the comparative development of the heteroblastic leaf series in *Lygodium japonicum* (Thunb.) Sw., *Bot. Gaz.*, 143(4), 1982.

Fertilizer

Phosphorus occurs widely in the lithosphere and constitutes nearly 0.12% of the Earth's crust. Mineral deposits bearing minable proportions of phosphorus belong to one of two types of mineral groups: by far the most abundant, the apatite group; and the aluminum phosphate group derived from apatite by weathering. The term phosphate rock is often used to designate a rock that is naturally sufficiently high in phosphorus to be used directly in fertilizer manufacturing. The beneficiated concentrate of a phosphate deposit is also called phosphate rock.

Phosphate rock for direct application. In the phosphate industry, phosphate use is often expressed in terms of phosphorus pentoxide (P_2O_5) equivalent. In 1980, nearly 31 million metric tons of P_2O_5 (13.5 million metric tons of phosphorus) were consumed as fertilizers worldwide; almost 95% of that was derived from phosphate rock by chemical processing. The balance (1.65 million metric tons of P_2O_5) was in the form of directly applied phosphate rock, of which 51% was consumed in the Soviet Union, 39% in the People's Republic of China, and the rest primarily in Brazil, Malaysia, Sri Lanka, Indonesia, India, and the United Kingdom, in descending order. In the United States, 4.89 million metric tons of P_2O_5 was used in 1980, of which only 624 tons was in the form of directly applied phosphate rock. Although the amount of phosphate rock used for direct application has not changed appreciably over the last 10 years or so, world consumption of phosphate fertilizers has increased dramatically during the same period.

Mineralogy of phosphate rock. Apatitic phosphate rock occurs in sedimentary, metamorphic, and igneous deposits. However, sedimentary deposits account for nearly 85% of the phosphate rock mined worldwide. The physical and chemical characteristics of metamorphic and igneous rocks render them unreactive and unsuitable for direct application. The igneous materials are coarsely crystalline and have no internal surfaces. The metamorphic deposits are likewise highly consolidated. By contrast, sedimentary phosphate rocks consist of porous, loosely consolidated aggregates of microcrystals and possess a relatively large internal specific surface area.

The apatitic group of minerals is complex because of varying degrees and kinds of isomorphic substitution in the basic crystal lattice. Fluorapatite, $Ca_{10}(PO_4)_6F_2$, rarely occurs as such in nature, except in certain igneous deposits. Substitution of CO_3^{2-} for PO_4^{3-} and of Na^+ and Mg^{2+} for Ca^{2+} is very common in sedimentary rocks. Such deposits are referred to as francolites or carbonate apatites. Usually there is an increase in the number of fluorine atoms in the crystal lattice to balance the charge deficit incurred by CO_3^{2-} substitution. Carbonate apatites in nature have the following approximate formula:

$$Ca_{10-a-b}Na_aMg_b(PO_4)_{6-x}(CO_3)_xF_{2+y}$$

$$a = 1.327 \times \frac{x}{6-x} \qquad y \cong 0.4x$$

$$b = 0.515 \times \frac{x}{6-x} \qquad x < 1.38$$

There is also substitution at the fluorine position, where OH^- and Cl^- may substitute for F^- leading to the end members hydroxyapatite and chlorapatite, respectively. Mixed substitutions also occur.

The aluminum phosphate group of minerals also exhibits a complex pattern of isomorphic substitution. The end members of these series are wavellite [$Al_3(PO_4)_2(OH)_3 \cdot 5H_2O$], crandallite [$CaAl_3(PO_4)_2(OH)_5 \cdot H_2O$], and millisite [$CaAl_6(Na,K)(PO_4)_4(OH)_9 \cdot 3H_2O$]. Although most of the directly applied phosphate rock belongs to the apatitic group of minerals, a significant proportion is derived from the aluminum-phosphate deposits.

Agronomic and economic considerations. The profitability of direct application of phosphate rock is determined largely by agronomic and economic factors. However, agricultural production strategies and other socioeconomic factors may outweigh agronomic factors, especially in developing countries. The agronomic effectiveness of directly applied phosphate rock is determined by three interrelated factors: properties of the rock itself, soil properties, and crop characteristics.

Phosphate rock properties. Sedimentary apatitic phosphate rocks differ significantly in their reactivity, or their ability to provide enough phosphorus to ensure that plant growth is not limited by phosphorus supply. An index of reactivity, measured by solubility in citric acid, formic acid, or neutral ammonium citrate, is useful in ranking phosphate rocks, but is not linearly related to the available proportion of phosphorus in phosphate rocks. The index of reactivity is usually expressed as the percentage of phosphorus in phosphate rock which is soluble in the extracting agent. The index is not additive. For example, if two phosphate rock materials, A and B, both contain 30% P_2O_5 with 26% of the P_2O_5 in phosphate rock A and 13% of that in the phosphate rock B being citrate-soluble, then application of phosphate rock B at twice the amount of P_2O_5 as from phosphate rock A does not provide the same crop response. In fact, response to phosphate rock A will be still higher than to phosphate rock B. Reactivity of apatitic phosphate rocks is related to the degree of isomorphic substitution in the apatite mineral. Generally, the more the substitution, the higher the reactivity. Reactivity is also measured by citrate solubility for the aluminum phosphate minerals, but the indexing capability of such a mea-surement is on a different scale than that of apatitic rocks.

Availability of phosphorus in phosphate rock increases with the decreasing particle size of individual aggregates, and reaches a maximum at 150 μm (100% passing through U.S. standard 100-mesh screen). For particles finer than 150 μm, availability does not increase with decreasing particle size.

Sedimentary apatitic phosphate rocks are aggregates of microcrystallites loosely bound together. Extraneous components are often present as discrete mineral grains in the aggregate and sometimes act as cementing agents to hold the microcrystallites together. Accessory calcite and dolomite decrease the short-term availability of phosphorus in phosphate rocks whereas quartz is inconsequential because of its near inertness in soil. However, silica quartz and gypsum, which sometimes occur as cementing agents, may occlude the apatite surfaces or block the internal pores of the aggregates. In either case, there is a negative effect on reactivity. Silicate clays and halides are largely removed by beneficiation procedures. Oxides and hydroxides of iron and aluminum are not readily beneficiated out of a rock concentrate, but have little or no influence on phosphate rock reactivity in soil.

Soil properties. Of the many soil factors which affect availability of phosphorus in phosphate rocks, soil pH is probably the most widely used. The process of phosphate rock dissolution in soil consists of H^+ ions moving toward, and Ca^{2+}, $H_2PO_4^-/HPO_4^{2-}$, and HCO_3^- ions moving away from, the apatite surface. Therefore, soil properties which affect these processes will also affect the plant availability of phosphorus in phosphate rock. Thus, phosphorus availability in phosphate rock increases with lower soil pH, with higher affinity for calcium in the soil (for example, low base saturation, high cation-exchange capacity, heavy texture, and high soil organic-matter content), and with lower levels of soluble phosphorus in the soil solution. Since ionic movement to and from the apatite surfaces is diffusion-mediated, other soil factors that affect diffusion will also affect phosphate rock dissolution. Factors such as soil moisture, soil temperature, and distribution of phosphate rock particles in soil are therefore relevant factors, even though they are not intrinsic soil properties. As a general rule of thumb, reactive apatitic phosphate rock materials will give best results if soil pH is less than 6.5, will give mixed results if soil pH is 6.5–7.5, and will be nearly useless when soil pH is greater than 7.5. However, the effects of soil pH in some cases can be augmented or negated by the soil's affinity for calcium.

As for the aluminum phosphate type of phosphate rock materials, phosphorus availability is enhanced by higher soil pH because the activity of aluminum in the soil solution decreases with increasing soil pH. The optimum range of soil pH for this type of phosphate rock is 6.5–8.0.

Crop characteristics. Plants vary in their ability to

effectively use phosphorus derived from phosphate rock. This is the result of variation in growth rate of plants; in their demand patterns for phosphorus and calcium, and how these patterns alter the composition of the soil solution at the root-soil interface; in root morphology and extensity of root elongation and root hair formation; in the presence or absence of symbiotic mycorrhizal associations; in the functional relationship between rate of phosphorus uptake by plant roots and its concentration in the soil solution; and in the soil solution concentration of phosphorus that is required to maintain an optimum growth rate. Generally phosphate rocks are best utilized by crops which are slow-growing, have well-developed fibrous root systmes, and have high calcium requirements. Crop listings vary from one researcher to another, probably because of varietal differences within species. In general, legumes utilize phosphate rock better than cereals, dicots better than monocots, warm-season crops better than cool-season crops, and perennial crops better than annual crops.

Role in crop production. From a mechanistic point of view, utilization of phosphorus from directly applied phosphate rock depends on the concentration of phosphorus that can be maintained in the soil solution–phosphate rock particle interface; on root length and the probability of a rootlet coming near a phosphate rock particle; and on the root capacity to absorb and utilize phosphorus from that vicinity. Quite often, the concentration of phosphorus that can be maintained at the soil solution–phosphate rock particle interface is suboptimal, especially for annual, fast-growing crops. Yields will consequently be lower than the potential maximum for such crops. And the yield loss cannot be alleviated by increasing the rate of phosphate rock addition, as was shown in the example of phosphate rocks A and B. The practical implication is that maximum crop yields which can be attained with water-soluble phosphorus fertilizers (for example, superphosphate and ammonium phosphate) are not attainable with phosphate rock.

Economic analyses of phosphate rock use in the United States indicate that low phosphorus content; high cost of shipping, storage, and handling per unit of phosphorus; and low relative effectiveness argue against its use in most situations, especially when production strategies are designed to maximize returns on inputs and land resources. This explains the miniscule tonnage of phosphate rock used for direct application in the United States. However, when levels of production are constrained by factors other than phosphorus fertility, such as in many developing countries, crop production maxima may be within the limit attainable with phosphate rock. In such situations, economic analysis of phosphate rock use may be more favorable, especially during initial phases of development, or when use of domestic phosphate rock sources obviates the need to expend hard currency to import porcessed, water-soluble phosphate fertilizers. Under such conditions, direct application of phosphate rock is not only possible, but often the most feasible fertilization method both agronomically and economically.

For background information *see* FERTILIZER; PHOSPHATE MINERALS in the McGraw-Hill Encyclopedia of Science and Technology.

[F. E. KHASAWNEH]

Bibliography: G. W. Cooke, Experimental work in the United Kingdom on the agricultural value of rock phosphates, *Seminar on Phosphate Rock for Direct Application*, International Fertilizer Development Center, Muscle Shoals, AL, pp. 304–324, 1978; L. Gachon, Role of phosphate rocks in the phosphatic fertilization of French soils, *Seminar on Phosphate Rock for Direct Application*, International Fertilizer Development Center, Muscle Shoals, AL, pp. 343–348, 1978; F. E. Khasawneh and E. C. Doll, The use of phosphate rocks for direct application to soils, *Adv. Agron.*, 30:159–206, 1978; B. C. Marwaha and B. S. Kanwar, Utilization of ground rock phosphate as a direct phosphatic fertilizer: Review, *Fert. News*, 26(2):10–20, 1981.

Food engineering

Recent advances in food engineering have included development of food extrusion processes, instrumental food texture measurements, and optimization of food package design.

Extrusion processing. Food extrusion has become a widely applied technique for producing many convenience foods for human consumption, such as macaroni, snack foods, ready-to-eat cereals, pregelatinized starches, soup bases, full-fat soy flour, and textured plant protein. In addition, extruders are widely used to manufacture both dry chunk and semimoist pet foods.

Extruder operation. The food extruder performs several processing functions. It mixes granular feed components with water and other minor ingredients and then forces them through a die under pressure to yield specifically shaped pieces. The material is then cut into the desired lengths with a rotating knife. The extruder also functions in heating or cooking the ingredients to gelatinize (solubilize) the starchy components and to denature (insolubilize and texturize) the protein constituents. The final product is a convenience, or so-called instant, food.

The principal part of an extruder is a flighted Archimedes' screw which rotates within a cylindrical barrel (Fig. 1). The action of the flights on the tightly fitting screw conveys the food ingredients forward, compressing them within the flights with an action that results in thorough mixing. During this process, heating of the food mass occurs by converting the mechanical energy required to turn the screw into heat by friction (viscous dissipation), by heat transfer from the jackets surrounding the barrel, and, in some instances, by direct steam injection into the food.

The specific action of the food extruder is a function of the design of the screw, its speed of rotation, and the moisture content of the ingredients.

High-shear cooking extruders have shallow-flighted screws which revolve rapidly to dissipate more mechanical energy, improve mixing, and increase the temperature of the discharge product to 150–200°C. These units are used extensively in the manufacture of ready-to-eat cereals, snacks, textured plant protein, and dry pet foods. Low-shear cold-forming extruders have deep flights and relatively low turning speeds, resulting in a much smaller energy addition from mechanical dissipation, with product temperatures remaining under 100°C. This equipment is employed for manufacturing macaroni, chewing gum, snacks requiring precise formation, and ready-to-eat cereals which are subsequently puffed.

Extruder processes. Transformation of the free-flowing granular feed materials into a homogeneous plasticized mass occurs rapidly within the extruder.

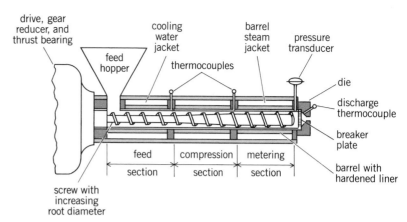

Fig. 1. Cross-sectional drawing of a food extruder showing important components. (*After J. M. Harper, Extrusion processing of foods, Food Technol., 32(7):67, 1978*)

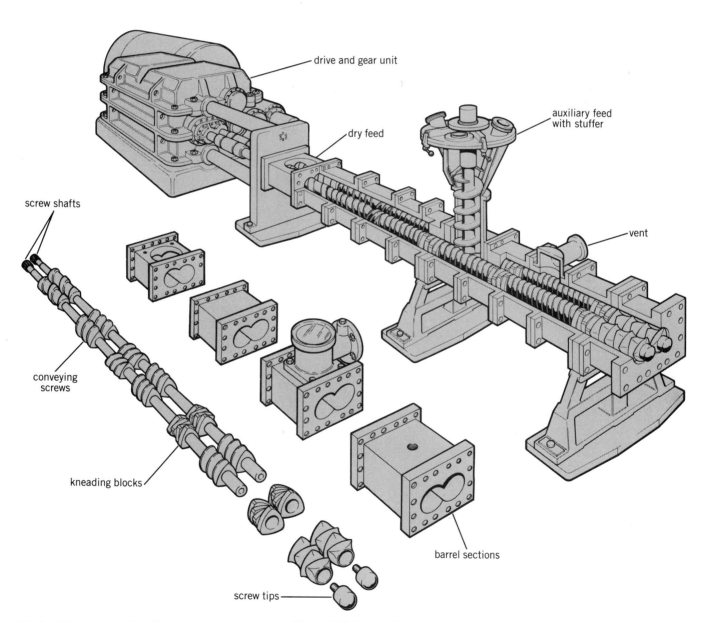

Fig. 2. Twin-screw extruder with cutaway exposing screw tips. (*Werner & Pfleiderer Co.*)

This is most pronounced in the cooking extruder, where a highly viscous dough is formed. While the temperature of the dough exceeds 100°C, the high pressure within the barrel prevents flashing or boiling of the available moisture content (18–35%). Once a cereal-based dough passes through the die, the pressure is released, and the moisture within the dough flashes, causing the product to puff. An expanded low-density material suitable for snacks and ready-to-eat cereals is the result. When high-protein doughs are extruded, cross-linking occurs between the long protein molecules that have been aligned by the shear within the screw, resulting in a meat-like, fibrous product called textured plant protein.

The average residence time of the food ingredients as they pass through the extruder varies between 60 and 240 s. A substantial portion of this time is spent in the feed section of the screw because of its deeper flights. Since high temperatures exist only near the discharge of the screw, residence time at these temperatures is normally less than 30 s. Consequently, extrusion cooking is referred to as a high-temperature short-time (HTST) process. Such processes have the advantage of beneficially cooking the feed ingredients, denaturing enzymes which exist in raw products, and pasteurizing the product, leaving it nearly free from microorganisms while retaining most of the vitamins.

Recent applications. Research and development in the food extrusion industry centers on elaborating processes which utilize a greater array of ingredients; for example, the density and texture of cereal-based foods are controlled by using purified starches and emulsifiers. High-density textured products, used as meat analogs, are produced from high-protein soy protein isolates with special extrusion dies. Additionally, mathematical analysis has been applied to improve control, scale up experimental results, and achieve maximum production capacity of extrusion lines.

Another recent development has been the use of twin-screw extruders (Fig. 2), in which two corotating screws with intermeshing flights are used. The basic advantage is improved mixing and residence-time control achieved with screws having different flight configurations along their length which promote conveying, mixing, or pressure increase. Such machines manufacture a variety of products, but one of the most interesting is Scandinavian crisp bread which is made by extruding rye and wheat at low moistures (less than 6%). In this instance, the one-step extrusion process replaces the conventional baking line, which is much larger, has higher capital costs, and consumes more energy.

[JUDSON M. HARPER]

Instrumental texture measurements. Since texture is a sensory attribute, strictly speaking it cannot be measured with an instrument. "Texture-measuring instruments" quantify mechanical properties of foods which must be related to the sensory perception of texture. The tests are performed by deforming the material under an applied force and measuring its response. The probes used and the rates of force (or deformation) application can be varied to suit specific purposes.

The most widely used instruments are universal testing units which can be fitted with different probes. The recorded force-deformation, force-time, or deformation-time curves can be quantified in terms of specific parameters, for example, firmness, force required to break, or energy required to break. This texture profiling technique can be used with a single step or a sequence of force-application steps on the same specimen.

Texture. Texture is defined broadly as the sensory manifestation of structural and mechanical (rheological) properties of food and involves the visual, tactile, kinesthetic, and aural senses. Vision detects structural or geometric elements which previous sensory experience indicates will affect the tactile or kinesthetic perception, for example, coarse versus fine breadcrumb or lumps in a smooth sauce. Hearing detects characteristic sounds produced on crushing of crisp or crunchy foods. Tactile and kinesthetic senses (in the mouth or in the hand) are the most important in perceiving texture. As a sensory attribute, texture can be perceived, analyzed, and interpreted only by a sensory/physiological apparatus. Because of problems in sensory measurement, much effort has been exerted in designing texture-testing instruments and correlating them with sensory perception.

General principles. Texture-testing devices consist of four basic elements: a probe for contacting the sample such as a flat plunger, a penetrating rod, or shearing jaws; a driving mechanism for imparting motion to the probe, either at a constant or at a variable rate; a sensing element for detecting the resistance of the foodstuffs to the applied forces (cutting, piercing, compressing, shearing, and so forth, depending on the probe and the type of motion); and a readout system, which can be a maximum force dial or a recorder tracing the force-distance or force-time curve.

Most texture tests are destructive but none duplicate the situation in the mouth. The main difference is the absence of the tongue which turns food in the mouth and permits the new surfaces to be exposed to the applied forces and facilitates mixing with saliva. The role of saliva as a solvent and lubricant can be approximated by adding water (or synthetic saliva) to the specimen, although this is not done as a rule. Because of the absence of the tongue, and of the confining environment provided by the cheeks, the action of texture-testing devices is most similar to the first chew or the first few chews; the best correlations with sensory assessment have therefore been obtained for the first bite. Nondestructive texture tests (for example, sonic resonance) have been proposed but in general do not correlate as well with sensory measurements.

Early work. Probably the earliest texture-testing

device was a puncture tester constructed in 1861 by A. Lipowitz in Germany. In 1905 J. von Hankoczy in Hungary designed an apparatus for measuring the strength of gluten, and in 1907 K. B. Lehmann in Germany described two instruments for testing the tenderness of meat. Since that time, over 100 instruments have been constructed and described, and over 60 have become commercially available. Early instruments were commodity-specific, for example, the Pea Tenderometer, the Tarr-Baker Delaware Jelly Tester, the Ball Compressor for cheese, and the Bloom Gelometer for gelatin. They can be classified into groups according to the test principle: penetrometers, compressometers, shearing devices, cutting devices, masticometers, consistometers, viscometers (for fluid and semifluid materials), extru-

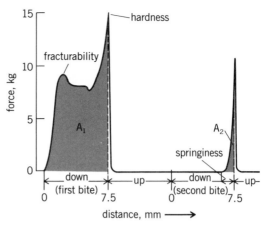

Fig. 4. Typical two-bite (two complete compression-decompression cycles) texture profile from which several textural characteristics can be quantified. The area ratio A2/A1 is a measure of cohesiveness. (*After M. C. Bourne, Texture profile of ripening pears, J. Food Sci., 33:223–226, 1968*)

Fig. 3. Example of a multipurpose texture-testing instrument shown operating in tensile mode. On the left is the crosshead to which different probes can be attached and the load cell; on the right, the recorder and control modules. (*Instron Corporation*)

sion devices, tensilometers, and miscellaneous devices which usually approximate the conditions to which the test material is subjected during preparation or handling.

Universal testing units. Since all texture-testing instruments exhibit common basic elements and since the probe contacting the food is the most crucial feature, multiple-purpose instruments equipped with a number of different probes have been constructed and commercialized (Fig. 3). They offer versatility, flexibility, accuracy, and other design advantages. They are generally limited to research because of their high cost. Simpler, less expensive models are available for quality control applications.

Texture profiling. Early instruments were limited to one-point measurements. The realization that tex-

ture is a multidimensional attribute led to the development of texture profiling, that is, quantifying several parameters from recorder tracings. Originally developed for tests involving two or more bites, texture profiling is now being applied to one-bite tests (Figs. 4 and 5).

Multiple-bite tests can be performed by keeping the height of the sample constant after deformation, or by increasing the extent of deformation, for example, from 10 to 50 to 90%. Texture-profiling tests are commonly performed by compressing the speci-

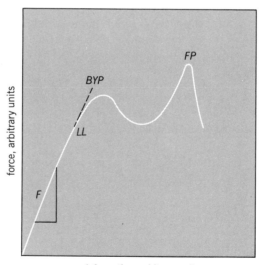

Fig. 5. Typical one-bite texture profile. *F* = firmness or Young's modulus; *LL* = limit of linear response; *BYP* = bioyield point or initial yield; *FP* = fracture point or massive failure. Both force and deformation parameters can be quantified. (*After J. G. Brennan, Food texture measurement, in R. D. King, ed., Development in Food Analysis Techniques, vol. 2, Applied Science Publishers, 1980*)

men between two flat plates. However, tests are also used in which heterogeneous material is pierced and curve deflections are related to structural elements, or the same material is forced several times through an extrusion plate and the drop in resistance is related to tenderness.

Although considerable literature has been accumulated on principles and applications of instrumental texture testing, it is still a developing field and much is left to the ingenuity of individual researchers. Key problems are correlations with sensory assessment and lack of standardization methods.

[ALINA S. SZCZESNIAK]

Food packaging. The food package serves multiple purposes. It functions as a materials-handling tool, an aid in promotion and sales, and a facilitator of food processing. However, its primary function is to protect food from contamination and from undesirable environmental factors. The package reduces the food's deterioration rate by means of various biological, chemical, and physical mechanisms, which are affected by environmental factors such as temperature, oxygen pressure, water-vapor pressure, and light intensity.

The food package must provide an adequate shelf life. This is defined as the period of time in which the desirable food qualities are maintained without the development of undesirable components. Collectively known as organoleptic factors, desirable qualities include nutrient content, taste, flavor, appearance, and texture. The level is specified by appropriate agents such as the manufacturers, buyers in both the wholesale and retail markets, or various governmental authorities.

Barrier properties. The key properties of packages that are effective in controlling the shelf life of foods are their barrier properties. These are permeability to gases, vapors, and liquids; transmission of light; resistance to penetration by insects; and thermal properties, which control the heating and cooling of packages. The other extremely important set of package properties is the mechanical strength and integrity of the package. Optimization of food-package design from the point of view of food protection means selection of materials and construction methods that will produce packages with these properties.

Modern packaging materials supply a large range of properties which can be utilized in package-design studies. The table shows examples of properties of selected packaging materials which are approved for use in food packaging.

Optimizing package design. Optimization procedures have not been standardized, but can be classified into several steps which form a logical framework. A determination of the limiting deteriorative mechanisms is made for the food to be protected, and a prediction of shelf life is completed by preparing a computer-assisted simulation. Then a choice of the least expensive packages can be made.

After these initial studies, optimization proceeds via the following steps: (1) Properties of food that determine quality depend on the initial condition of the food and on reactions that change these properties with time. These reactions, in turn, depend on the internal environment of the package. It is assumed that the deteriorative mechanisms limiting shelf life, and their dependence on environmental parameters such as oxygen pressure, water activity, or temperature, can be described by a mathematical (though not necessarily analytical) function. (2) Maximum acceptable deterioration level can be determined by correlating objective tests of deterioration with organoleptic or toxicological parameters. (3) The internal environment depends on the condition of the food, on package properties, and on the external environment. An assumption in optimization is that changes in environmental parameters can be related to food and package properties. (4) Barrier properties of the package can, in turn, be related to internal and external environments. (5) The various equations thus developed can be combined and solved with or without the aid of a digital computer. Solutions are generated which predict storage life or required package properties for a given form of storage. When an exact mathematical formulation of the course of deterioration is required in addition to the optimization objectives, it is possible to use a mathematical procedure known as dynamic optimization.

Typical barrier properties of selected food-packaging materials

Material	Oxygen permeability, cm^3/m^2 · day (1 mil film)	Water permeability, g/day · m^2 at 90% relative humidity (1 mil film)	Tensile strength, lb/in.^{2}	Heat sealability
Coated cellophane				
(at low humidities)	1–100	3–20	10,000	Depends on coating
(at high humidities)	Greater than 1000	High	10,000	Depends on coating
Polyethylene	3000–15,000	5–15	5000	Excellent
Polyester	100	15	20,000	Requires special equipment
Nylon				
(at low humidities)	100	5–10	10,000	Sealable
Saran	10	1	8000	Sealable
Laminations containing metal layers	Less than 1	Less than 0.1	Greater than 10,000	Excellent

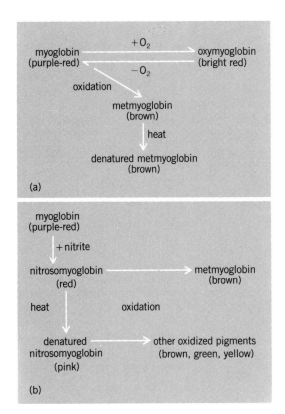

myoglobin
(purple-red) $\xrightarrow{+O_2}$ oxymyoglobin
(bright red)

$\xleftarrow{-O_2}$

oxidation

metmyoglobin
(brown)

heat

denatured metmyoglobin
(brown)

(a)

myoglobin
(purple-red)

+ nitrite

nitrosomyoglobin $\longrightarrow$ metmyoglobin
(red)   (brown)

heat   oxidation

denatured $\longrightarrow$ other oxidized pigments
nitrosomyoglobin   (brown, green, yellow)
(pink)

(b)

Fig. 6. Schematic of the denaturation mechanisms causing color changes in meats. (a) Fresh meats. (b) Cured meats.

Examples of optimization problems. A common problem which yields to a solution by optimization is the packaging of moisture-sensitive foods. Many of the reactions that cause quality deterioration depend on moisture content. There are two types of such dependence: the first involves reactions for which a definite, critical moisture content may be postulated below which the reaction rate is considered negligibly small; and the second is affected by those reactions that proceed over the entire range of water contents at significant rates. The analysis of the problem involving the first type of dependence is relatively simple. If a linear isotherm relating moisture content to partial pressure of water is assumed, and the storage conditions are considered to be constant with respect to temperature and relative humidity, then the safe storage time has a simple mathematical relationship to food and package properties.

Another packaging problem that can be handled by optimization involves color changes in prepackaged meats. Consumers are greatly influenced by the color of meat at the time of purchase. Before the advent of refrigeration and of protective meat packaging, the loss of red color was often associated with meat spoilage. Now, however, perfectly safe, nutritious, and palatable meat may lose its red color during refrigerated storage. But the fact remains that the consumer is attracted to meats which have a bright-red color. This consideration complicates the optimization of meat quality through packaging. The

color of meats is due to the protein pigment myoglobin. In the presence of oxygen this pigment forms the oxygenated oxymyoglobin. Both of these pigments contain heme iron in the reduced (ferrous) state. Upon oxidation, the iron pigment becomes the brown metmyoglobin, which becomes denatured when heated. In meats cured with nitrite, myoglobin is converted to nitrosomyoglobin, which can also be oxidized and denatured by heat (Fig. 6).

Oxidation to the undesirable brown pigments varies with oxygen concentration, and the mechanisms are different for fresh meats and cured meats. In cured meats oxygen always accelerates the color change, but in uncured meats the oxidation is more rapid at 2–5% oxygen. Oxidation is also greatly accelerated by light, especially ultraviolet light. Optimal packaging of meats is responsive to these influences.

Cured meats are best packed under vacuum or in inert gas, and in packages which have a low permeability to oxygen (around 10 $cm^3/m^2 \cdot day$) and also exclude the ultraviolet portion of the light spectrum. Fresh meats are packed optimally under conditions in which oxygen transmission is kept at levels up to 100 times higher than those suitable for cured meats.

For background information *see* FOOD ENGINEERING in the McGraw-Hill Encyclopedia of Science and Technology.

[MARCUS KAREL]

Bibliography: W. M. Breene, Aplication of texture profile analysis to instrumental food texture evaluation, *J. Texture Stud.*, 6:53–82, 1975; J. G. Brennan, Food texture measurement, chap. 1 in R. D. King (ed.), *Development in Food Analysis Techniques*, vol. 2, 1980; J. M. Harper, *Extrusion of Foods*, vols. 1 and 2, 1981; M. Karel, Packaging protection for oxygen-sensitive products, *Food Technol.*, 28(8):50, 52–53, 56–58, 1974; M. Karel, Protective packaging of foods, in M. Karel, O. Fennema, and D. B. Lung (eds.), *Physical Principles of Food Preservation*, 1975; A. Kramer and A. S. Szczesniak (eds.), *Texture Measurements of Foods*, 1973; J. de Man et al. (eds.), *Rheology and Texture in Food Quality*, 1976; I. Saguy and M. Karel, Modeling of quality deterioration during food processing and storage, *Food Technol.*, 34(2):81, 1980.

Food manufacturing

Food manufacturing operations can be grouped into three classes: separation processes, assembly processes, and preservation processes. The processes involved in separating edible food material from traditionally inedible parts are responsible for the greatest quantities of waste. Other sources of waste are wash waters, and water used to heat, transport, and cool edible and inedible materials. This article deals with research on upgrading inedible food process waste streams to useful sources of nutrients for human consumption.

Separation processes can be as simple as size-grading fresh-market fruit to remove undersized material which is used in juice, sauce, or dried pieces.

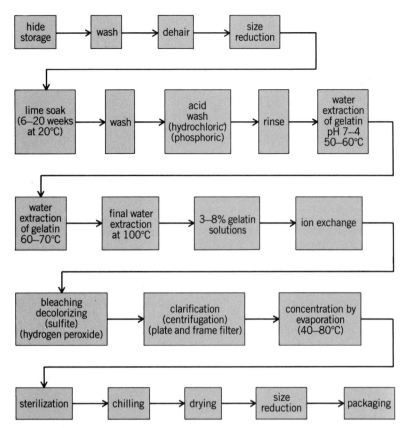

Flow chart for recovery of hide gelatin (alkaline process, type B).

By contrast, gelatin prepared from waste bones and hide is the product of a complex chemical extraction process in which inedible protein, through a number of carefully controlled extraction treatments, is made into a high-purity, highly functional protein isolate (see illustration). Most processes for recovering useful materials from liquid or solid streams generated during the separation of edible food products from skin, shell, bone, pits, dirt, and other contaminants fall between these two extremes. Recently, fermentation processes have been increasingly used in the upgrading of processing wastes that would normally go to the sewer or at best be used as animal feed or plant fertilizer.

Waste streams. Waste streams having useful quantities of protein, lipid, carbohydrate, fiber, pigment, or other desirable chemical contents originate from dry or mechanical milling of food or from chemical treatment or wet extraction of foods. Each commodity class yields waste streams characteristic of processing methods used in their preparation. There are five commodity classes which produce waste streams potentially useful for food material: (1) Dairy wastes, such as whey and acid whey, can yield protein, salts, lactose, hydrolyzed lactose (glucose and galactose), and lactose fermentation products. (2) Meat and poultry process wastes can yield biologically active hormones, isolates, and concentrates; enzymes; functional proteins; and proteins to enrich plant protein sources. (3) Cereal- and grain-milling wastes can yield functional protein isolates; refined fiber; fermentable carbohydrates; vitamins, oils, and waxes; and carbon for combustion to process energy. (4) Fruit and vegetable process waste streams can yield fermentable carbohydrates, functional plant proteins, starches, fiber, pigments, flavoring materials, complex polymers for structure modification, and carbon for combustion to process energy. (5) Marine products wastes such as chitin and fish protein isolates and concentrates can yield functional proteins, and binding and structuring agents.

It is important to note that waste streams are closely related to the process used to separate the edible food from inedible portions. As process technology develops to increase the direct yield of food from the raw material, the amount of by-product or waste decreases. Genetic engineering of raw materials can be used to eliminate wastes entirely by breeding out the undesirable part of the plant or animal separated by processing. For example, the connecting tissue area of processing tomatoes has been eliminated.

Tomato processing. Processing of tomatoes for canning as whole tomatoes is a good illustration of an industry's attempt to recover as much product as possible. Caustic has been used to peel tomatoes to reduce labor and unit process costs. However, the caustic peel waste represented a true waste stream in that it was highly alkaline and dilute since large quantities of water were used to wash the alkaline-softened skins from the tomato. An advanced chemical peeling system was developed, named dry caustic peeling, which yielded a more highly concentrated peel waste stream. This material was shown to be useful as a source of tomato solids which, when neutralized with hydrochloric acid, can be concentrated from 3% solids to a useful tomato concentrate product.

More recently the tomato industry has developed pressure-steam-peeling systems which eliminate the use of caustic and will allow the return of the steam-peeled skins to the pulping process to make concentrated tomato products. Thus the primary waste stream from the tomato industry, aside from soil and broken and decayed fruit washed away prior to entering the process, is tomato seeds. Tomato seeds, estimated to be about 4×10^5 metric tons from United States production, contain over 30% of their dry weight as protein. When ground tomato seed was added to bread, replacing 10% of the flour, it was found to improve protein quality and loaf volume without major changes in appearance.

Meat processing. The meat industry has developed many uses for by-product and waste streams rich in protein, fat, and minerals (as bone). Even with the efficient recovery of hides, trimmings, hair, and other wastes, water used in processing beef and in cleaning equipment has been shown to contain 2 to 5% of the protein of the carcass weight entering the process. An ion-exchange system has been designed for recovering these proteins from dilute streams in a form that is suitable for food use. Air flotation at pH 3 to 5 is recommended to remove fats and suspended solids prior to protein recovery by ion exchange.

Fish processing. There are methods for upgrading a wide variety of fishery wastes to useful food products. As with meat processing, mechanical deboning and forming is a key step in recovering protein normally lost by conventional separation methods. Chitin represents a rich source of protein and can be modified chemically to form a thickening or binding agent for foods. Chitin chemistry is being actively researched with the objective of producing safe food additives for emulsification, water binding, and structuring.

Fermentation. The need for pure L-form amino acids and the opportunity to produce "natural" vitamins, sweeteners, fatty acids, pigments, flavorings, and stabilizers has created renewed interest worldwide in fermentations and biochemical engineering to produce such food ingredients from food processing wastes. Microbial production of ethanol, vinegar, monosodium glutamate, and citric acid exemplifies useful food ingredients which can be manufactured from waste streams.

Single cell protein has been developed as a source of animal feed and as a potential human food. The production of single-cell protein from food processing and other waste streams has been studied. One problem is the seasonal nature of many food processing operations, such as fruit, vegetable, and fish processing. Dairy and meat plants are less subject to seasonal fluctuations but may operate only 5 days per week. Since continuous year-round fermentation operations would not be possible under these conditions, more expensive batch systems would be needed. Other requirements would be large, inexpensive holding tanks and microbes which will not be affected by contaminating organisms. Successful single-cell protein production would depend on finding microbes which: are able to grow rapidly on a variety of waste carbon sources, including cellulose; can provide a good quality protein; are free of toxic substances; are easy to harvest, refine, and package; have acceptable organoleptic quality and digestibility; and have the ability to overgrow or provide a stable ferment against chance contaminants (for example, by growing at low pH values).

Fungi may represent a useful class of microbes which meet these requirements. A protein yield of up to 25 g is possible for 100 g of carbohydrate in the substrate. Potentially useful fungi include *Aspergillus oryzae*, *Fusarium semitectum*, *Gliocladium deliquescens*, and *Trichoderma viride*.

The economics of protein production from waste streams must be evaluated on a source-by-source basis. However, considerable experience has been gained with yeast production from whey. Carbohydrate-rich wastes from potato processing operations could yield a year-round source of substrate for single cell protein production in the form of yeast or fungi.

Parameters. The use of food processing wastes to produce useful nutrients and functional additives for food processing is an emerging technology which is dependent on the processing methods generating a waste stream. Since wastes can be treated for disposal by established sewage systems at a known cost, upgrading of these wastes for food must be done in competition with these costs and with similar foods from conventional sources. The decision to develop technology to upgrade wastes is dependent on these parameters and also on the costs of the technology to change processing methods to eliminate the waste stream under consideration. Seasonal food processing operations would favor process modification to increase yields and reduce waste streams. Well-established year-round processes as are found in the meat, dairy, and cereal industries could favor investment in technology to upgrade dry and wet waste streams to valuable food ingredients.

For background information *see* FERMENTATION; FOOD MANUFACTURING; GELATIN in the McGraw-Hill Encyclopedia of Science and Technology.

[DANIEL F. FARKAS]

Bibliography: J. H. Green and A. Kramer (eds.), *Food Processing Waste Treatment*, 1979; G. M. Pigot, New approaches to marketing fish, in A. M. Altschul (ed.), *New Protein Foods*, vol. 2, *Technology*, part B, chap. 1, 1976; D. W. Stanley, E. D. Murray, and D. H. Lees (eds.), *Utilization of Protein Resources*, 1981; S. R. Tannenbaum and G. W. Pace, Food from waste: An overview, in G. G. Birch, K. J. Parker, and J. T. Worgan (eds.), *Food From Waste*, chap. 2, 1976.

Fourier series and integrals

Several decades ago J. E. Littlewood conjectured that there is a constant C such that whenever n_1, n_2, . . . , n_N are N distinct integers, inequality (1) is

$$\frac{1}{2\pi}\int_{-\pi}^{\pi}\left|\sum_{k=1}^{N}e^{in_k x}\right|dx > C\log N \qquad (1)$$

valid. Over the years, the esthetic appeal of this conjecture has led a number of mathematicians to try to prove it. Their efforts have generated useful insights and techniques in harmonic analysis, and in analytic number theory as well. Finally, in 1981, two proofs of the conjecture appeared, one by S. V. Konjagin in the Soviet Union and one by three United States authors. The latter proof is short and satisfying; it dispenses with the earlier combinatorial/counting approaches. The primary contribution to the work was that of B. P. Smith.

There is still no soft proof, from general structural considerations, that inequality (1) holds with some function on the right-hand side that tends to infinity with N. Apparently that was not known at all until the 1960 work of Paul Cohen, who obtained inequality (1) with $C(\log N/\log\log N)^{1/8}$ on the right-hand side. In the late 1970s, S. K. Pichorides and J. J. F. Fournier, with different methods, obtained $C(\log N)^{1/2}$. Pichorides's ideas underlie Konjagin's proof of the original conjucture.

Plausibility of Littlewood's conjecture. Littlewood's reasons for making the conjecture are easy to understand. For each integer n, the function $f(x) = e^{inx} = \cos nx + i\sin nx$ is defined on the circle group T (the real numbers modulo 2π) and is called a character of T. Inequality (1) means that the average absolute value on T of the sum of N distinct characters must be at least that big; in other words, there is a limitation on the extent to which the characters can cancel each other out. Some simple examples will serve to show why inequality (1) is so plausible. It seems reasonable to believe that it is when the set $E = \{n_1, n_2, \ldots, n_N\}$ consists of N consecutive integers that the most cancellation occurs. In that case, the left-hand side of inequality (1) is easily estimated to be approximately $(4/\pi^2) \log N$. It also seems reasonable that the least cancellation should occur when n_k is several times as large as n_{k-1} for each k; and in that case, the left-hand side of inequality (1) is easily shown to behave like $N^{1/2}$. If those two cases represent the extremes, it follows that inequality (1) should hold in all cases, for some C.

Relation to Fourier transforms. The Littlewood conjecture is an interesting part of the theory of Fourier transform behavior. To discuss its meaning in that context, it will be useful to introduce some further terminology. Let M denote the convolution algebra of all bounded complex-valued Borel measures on T. The norm of a measure μ in M is given by Eq. (2) as the total variation of μ. The Fourier (or Fourier-Stieltjes) transform of μ is a function defined on the group Z of integers by Eq. (3). The algebra M includes the subalgebra of Lebesgue-integrable functions f; thus if $d\mu(x) = (1/2\pi)f(x)dx$, then Eqs. (2) and (3) become Eqs. (4) and (5). Let B denote the algebra of all Fourier transforms of elements of M, and define the norm of a transform by Eq. (6).

$$\|\mu\| = \int_{-\pi}^{\pi} d|\mu|(x) \qquad (2)$$

$$\hat{\mu}(n) = \int_{-\pi}^{\pi} e^{-inx} d\mu(x) \qquad \text{for } n \text{ in } Z \qquad (3)$$

$$\|f\| = \frac{1}{2\pi} \int_{-\pi}^{\pi} |f(x)|dx \qquad (4)$$

$$\hat{f}(n) = \frac{1}{2\pi} \int_{-\pi}^{\pi} e^{-inx} f(x)dx \qquad (5)$$

$$\|\hat{\mu}\|_B = \|\mu\| \qquad (6)$$

Thus if g is a function defined on Z, and $g = \hat{\mu}$, then $\|g\|_B$ means $\|\mu\|$. If

$$f(x) = \sum_{n \in E} c_n e^{inx}$$

then $\hat{f}(n)$ equals c_n if n is in E, and 0 otherwise. In particular, if each $c_n = 1$, then f is the indicator function χ_E, which equals 1 on E, and 0 elsewhere. If $\#E$ denotes the number of elements of E, then inequality (1) can be rewritten as inequality (7).

$$\|\chi_E\|_B > C \log(\#E) \qquad (7)$$

The device of the Fourier transform provides a way to represent an object on T—like a sum of characters, or any measure—as a function on Z. Harmonic analysis is concerned with relations between properties of the objects and properties of their transforms; the transform is a tool for studying the object, and vice versa.

An indicator function is equal to its own square, which is to say that it is an idempotent. In the study of B, as with many Banach algebras, the identification of all its idempotent elements was an essential step toward understanding its structure. If E is a set of integers, then its indicator function χ_E is in B (that is, χ_E is a transform of some element of M) if and only if E is in the coset ring of Z, which means that E can be obtained by taking finite unions, intersections, and complements of cosets of subgroups of Z. That is the idempotent theorem, due to Henry Helson in the case of Z and to Cohen in more general settings. It has been shown that the theorem is stable, in that it still applies to functions that are close to indicator functions. In fact, if $f \epsilon B$ and $|f(n) - \chi_E(n)| < \varepsilon$ for all n, and if $\|f\|_B < (\log(1 + \sqrt{2}))^{-1} |\log \varepsilon| - 2$, then E is in the coset ring.

One objective in the study of B is to be able to tell which of the bounded functions on Z belong to B and which do not—what a transform can or cannot do. To that end, it is helpful to know that certain behavior implies a large norm. While $\|\chi_E\|_B = 1$ if E is a coset in Z—for example, the set of all odd integers, or a set with only one element—all results obtained to date suggest that the more complicated a combination of cosets E is, the larger $\|\chi_E\|_B$ must be. Inequality (7) is a manifestation of that principle, since an N-element set is the union of N cosets and has no simpler description as a member of the coset ring of Z.

Generalizations. The foregoing remarks are limited to objects on T and their transforms on Z, but the results extend to other dual pairs of locally compact abelian groups. The principle of Fourier transform behavior that has been discussed finds a most delicate and remarkable manifestation in the following problem: given a function on (for example) a finite subset E of Z, when and how can its definition be extended to all of Z so that the extension belongs to B, vanishes on most of the complement of E, and has a tolerably small norm in B? The answers suggest a deep relationship between the properties of a transform and the arithmetic of Z.

For background information *see* FOURIER SERIES AND INTEGRALS; GROUP THEORY; MEASURE THEORY in the McGraw-Hill Encyclopedia of Science and Technology.   [O. CARRUTH MCGEHEE]

Bibliography: C. C. Graham and O. C. McGehee, *Essays in Commutative Harmonic Analysis*, 1979; S. V. Konjagin, On the Littlewood problem (in Russian), *Isv. Akad. Nauk. SSSR*, 45:243–265, 1981; O. C. McGehee, L. Pigno, and B. Smith, Hardy's inequality and the L^1 norm of exponential sums, *Ann. Math.*, 113:613–618, 1981; B. P. Smith, Helson sets containing the identity are uniform Fatou-Zygmund sets, *Indiana Univ. Math. J.*, 27:331–347, 1978.

Fractals

Many systems in nature are not uniform in space. Instead, they exhibit geometrical features which appear self-similar under different magnifications, for example, the shapes of coastlines, the distribution of galaxies in the universe, the shapes of polymers, and the mass distributions in random alloys. Fractals are geometrical shapes which imitate these self-similar objects. Each such shape is characterized by a (usually noninteger) fractal dimensionality, which identifies quantitatively its self-similar properties. These notions have been increasingly employed in statistical physics, where the physical properties of systems are strongly dependent on their spatial dimensionalities.

Fractal dimensionality. Consider first the example of the coastline. It is very difficult to distinguish between two pictures of a portion of a coastline taken at different magnifications. In 1961 L. F. Richardson first estimated how the measured length of a coastline varies with the "yardstick" used. Later, in the 1970s, B. B. Mandelbrot interpreted these results through the introduction of fractal dimensionalities. In order to explain this interpretation, it is useful to construct a geometrical model of a coastline. Such a model is represented by the Koch curve (Fig. 1). The curve is constructed in a recursive way: begin with a straight segment, divide its length by 3, and then replace the (single) central piece by two similar pieces, as in Fig. 1a. The same procedure is now applied to each of the four new segments (Fig. 1b), and the procedure is repeated an infinite number of times. Clearly, the curve is self-similar; a magnification by a factor of 3 of any portion of it will look the same as the original curve. In the construction, the length-scale is changed by a factor of $b = 3$, and the number of basic units by a factor of $N = 4$. The fractal dimensionality D is now de-

Fig. 2. Three stages of the Cantor set.

fined by the equation $N = b^D$, that is, by

$$D = \frac{\ln N}{\ln b}$$

In Fig. 1, $D = \ln 4/\ln 3 \simeq 1.26$. A similar value applies to the west coast of Britain; that is, except for its regularity, the Koch curve imitates the self-similar features of this coast (for an appropriate range of length scales). Note that the equation reduces to that used for (integer dimensionality) hypercubic systems; for a regular cubic lattice, $N = b^3$.

Each geometrical shape is characterized by several additional dimensionalities: it is embedded in a euclidean space with dimensionality d_E, and it has a topological dimensionality d_T ($d_T - 1$ equals the dimensionality of the set needed in order to disconnect the shape into two pieces). In Fig. 1, $d_E = 2$ and $d_T = 1$. In general, $d_T \leq D \leq d_E$.

Another example in which fractals play an important role arises in the field of data transmission and noise. Consider a time interval of undisturbed transmission. An accurate examination of this interval reveals a short error burst, which divides the original interval into two unperturbed ones. These new intervals also contain error bursts, though much shorter. Eventually an infinite sequence of unperturbed (short) intervals is obtained, divided by a hierarchy of error bursts. A fractal model which describes this phenomenon is represented by the Cantor set (Fig. 2). The segments, separated by gaps, represent the unperturbed intervals. Only three steps of the construction are shown. According to the model, each time a segment is examined with a finer resolution (that is, the resolution is improved by a factor $b = 3$), a gap, that is, an error burst, is found in the middle. At each iteration two small pieces are obtained out of a large black segment. Hence $N = 2$, and, from the equation above, $D = \ln 2/\ln 3 \simeq 0.63$. The Cantor set should be embedded in a space with $d_E \geq 1$, and its topological dimension is $d_T = 0$ (it consists of a "dust" of isolated points).

A generalization of the Cantor set to a three-dimensional euclidean space was the basis of the fractal model of the universe, introduced by E. E. Fournier d'Albe in 1907. According to that model, the mass of the universe is concentrated in aggregates located in a certain hierarchical order. Present estimates of the fractal dimensionality of the mass distribution in the universe yield $D \simeq 1.3$. Apart

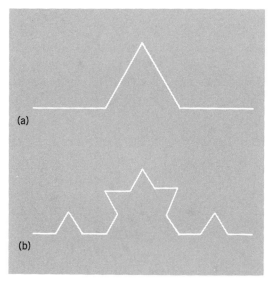

Fig. 1. Two stages of the Koch curve. (a) First stage. (b) Second stage.

from being a good descriptive model, the Fournier picture consisted of an explanation to the Olbers paradox. According to H. Olbers, the sky should always be bright, provided that the universe is static and infinite, with a homogeneous mass distribution. Since, according to Fournier, the mass scales with some power of the distance which is smaller than 2, the paradox is resolved. (However, both general relativity and the theory of the expanding universe are also capable of explaining this paradox.)

In the above examples nonrandom fractals are chosen to describe random phenomena. The fractal nature of the random systems is satisfied only in the statistical sense.

The rigorous mathematical generalization of dimensionality to noninteger values was originally developed by F. Hausdorff in 1919 and later by A. S. Besicovitch. The details of the Hausdorff-Besicovitch dimension will not be discussed, but the simple definition of D in the equation is consistent with it.

In addition to the fractal dimensionality D, fractals may be characterized by many other geometrical parameters. For example, one can construct fractal models of the universe which have the same value of D, but differ in the distribution of sizes of the empty volumes. These differences are measured by the lacunarity. Another important characteristic, the order of ramification, has to do with the number of links which should be cut in order to isolate an arbitrarily sized region around a given point.

Statistical physics. Probably the most significant progress in the application of fractal dimensionalities to science has been achieved in statistical physics. Anomalous dimensions of operators near critical points, as well as the analytic continuations of dimensions in the context of critical phenomena, emphasize the important role of noninteger dimensionalities. However, within the theory of critical phenomena, a lattice with noninteger dimensionality is considered to be an abstract generalization of a hypercubic lattice. By contrast, fractals are realizable, specific geometrical shapes.

Fractals are used frequently in statistical physics. A simple example is that of random walks. Since in this case the number of steps (the mass) scales as the square of the distance, it follows that random walks have $D = 2$, for all d_E. In contrast, the fractal dimensionality of self-avoiding walks (which do not intersect themselves) depends on the euclidean dimensionality in which they are embedded. They have the same D as nonbranching polymers (in dilute solution). The approximate formula of P. J. Flory, $D = (d_E + 2)/3$, gives excellent estimates for $d_E \leq 4$. At $d_E \geq 4$ self-avoiding walks are similar to common random walks. Self-avoiding walks may be modeled by certain Koch curves.

Another problem in which the concept of fractals enters naturally is that of percolation: a lattice is constructed of a random mixture of conducting and nonconducting links. At low concentrations of conducting links, these form only finite, isolated clusters, and the lattice is an insulator. Above a critical

threshold concentration p_c, an infinite cluster is formed, and the system becomes a conductor. Computer simulations by S. Kirkpatrick and others have shown that near p_c the infinite cluster is self-similar and can be characterized by $D \simeq 1.9$ at $d_E = 2$ and by $D \simeq 2.5$ at $d_E = 3$. The description of the infinite cluster as a fractal was earlier suggested by Mandelbrot and by H. E. Stanley and collaborators.

The conductivity of the infinite cluster is due only to its "backbone," without all the dangling links. Near p_c the backbone also turns out to be self-similar (with $D \simeq 1.6$ at $d_E = 2$ and $D \simeq 2.0$ at $d_E = 3$). Recently, Y. Gefen and collaborators proposed a simple geometrical model for the backbone. At $d_E = 2$, this model reduces to the Sierpinski gasket, a few iterations of which are shown in Fig. 3: at each iteration, the basic triangle is divided into four new triangles, and only three of these are kept for further iterations. Clearly, $N = 3$, $b = 2$, and $D = \ln 3/\ln 2 \simeq 1.58$. The model imitates many of the geometrical features of the backbone (for example, its order of ramification).

The advantage of regular models, like the one shown in Fig. 3, is that exact calculations of their

Fig. 3.   Four stages of the two-dimensional Sierpinski gasket.

physical properties can be performed. For example, Gefen and collaborators used their model to calculate the conductivity of the random resistor network by calculating that of the model fractal. The results give very good estimates of the measurable quantities.

Much of the progress in understanding phase transitions and critical phenomena comes from studies of the dependence of these phenomena on dimensionality. This dependence can be explicitly studied on fractals, which offer well-defined geometrical systems, with noninteger dimensionalities. Recently, Gefen and coworkers solved various phase transition problems on a variety of fractal structures. It was discovered that critical phenomena are not uniquely determined by D but also depend on other geometrical characteristics (lacunarity, order of ramification, and so forth). Many other phenomena in statistical physics involve self-similar features, including chaotic motions, diffusion, and roughening.

To summarize, fractals form a natural descriptive approach to many irregular, self-similar phenomena. In addition, by using simple fractal models, calculations which are extremely difficult to do on real systems can be performed, and the effect of various geometrical parameters can be controlled.

For background information *see* CRITICAL PHENOM-

ENA; PHASE TRANSITIONS in the McGraw-Hill Encyclopedia of Science and Technology.

[AMNON AHARONY; YUVAL GEFEN]

Bibliography: B. B. Mandelbrot, *The Fractal Geometry of Nature*, 1982; B. B. Mandelbrot, *Fractals: Form, Chance and Dimension*, 1977; Y. Gefen et al., Solvable fractal family and its possible relation to the backbone at percolation, *Phys. Rev. Lett.*, 47:1771–1774, 1981; Y. Gefen, B. B. Mandelbrot, and A. Aharony, Critical phenomena on fractal lattices, *Phys. Rev. Lett.*, 45:855–858, 1980.

Frontier molecular orbital theory

Frontier molecular orbital theory is a qualitative method for the understanding and prediction of structures and reactions of molecules. It has been applied primarily in the area of organic chemistry. The central assumption upon which frontier molecular orbital theory is based is that the highest occupied and lowest unoccupied molecular orbitals (HOMO and LUMO, respectively) of molecules have the greatest influence on the shapes and reactivities of molecules. Molecules of moderate complexity always have a large number of doubly occupied molecular orbitals. The highest-energy occupied orbital is the HOMO. Similarly, the lowest-energy vacant orbital from the large set of unoccupied orbitals is the LUMO. The HOMO and LUMO are the frontier molecular orbitals of a molecule. This assumption is related to the common assumption that the valence orbitals of atoms are the most influential in bonding.

The identification of the special significance of frontier molecular orbitals was made by Kenichi Fukui and coworkers in Japan in the late 1950s. Frontier molecular orbital theory is supported by perturbation theory arguments. When two molecules interact, their molecular orbitals overlap, leading to repulsion or stabilization. The interaction of filled orbitals is repulsive (closed-shell repulsion), while the interactions of doubly occupied orbitals of one molecule with vacant orbitals of the other molecule lead to stabilization. The greater these stabilization energies, the more facile the reaction.

According to second-order perturbation theory, with overlap neglected, the stabilization, ΔE, arising from mixing of a filled orbital, ϕ_i, on one molecule with a vacant orbital, ϕ_j, on the second molecule is given by Eq. (1)

$$\Delta E = -2 \frac{H_{ij}^2}{\epsilon_j - \epsilon_i} \qquad (1)$$

H_{ij} is the interaction matrix element for orbitals ϕ_i and ϕ_j, and ϵ_j and ϵ_i are the corresponding energies. Since H_{ij} is approximately proportional to the overlap in space of the two orbitals, those orbitals that overlap most lead to the largest stabilization. ΔE also increases as the energetic separation of the two orbitals decreases.

Among the stabilizing interactions, those involving the HOMO of one molecule and the LUMO of the other, and vice versa, are greatest, because these orbitals are separated by the smallest energy gap.

Frontier molecular orbital theory is successful not only because the frontier orbitals are closest in energy, but also because the interactions of remaining orbitals discriminate less between different reaction paths.

Although there are exceptions, a general principle enunciated by Fukui is remarkably successful: reactions occur so as to maximize the interaction between the frontier molecular orbitals. Although Fukui and coworkers applied this idea to many organic reactions, in 1965 Roald Hoffmann and R. B. Woodward brought Fukui's theory to the general attention of organic chemists through the discovery of the principle of conservation of orbital symmetry, which showed the predictive power of qualitative molecular orbital techniques, including frontier molecular orbital methods. For their advancements of organic theory, Fukui and Hoffmann were awarded the Nobel Prize in chemistry in 1981.

Cycloadditions. An example of the application of frontier molecular orbital theory to the determination of the allowed or forbidden nature of cycloadditions is demonstrated by the comparison of two reactions

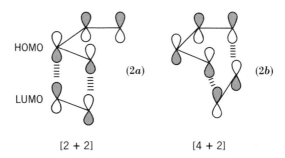

(2). Here the numbers in brackets refer to the number of π electrons which are involved in bonding changes from each molecule. One pair of frontier molecular orbitals, the butadiene HOMO and the ethylene LUMO, are represented. The filled and open *p*-orbital lobes represent the plus and minus signs of the wave functions in different regions in space. When ethylene approaches butadiene in a [2+2] fashion [reaction (2a)], which would give vinylcyclobutane, there is essentially zero overlap of the frontier molecular orbitals and no stabilization. However, approach in the [4+2] fashion [reaction (2b)] results in larger overlap and large stabilization yielding cyclohexene. Consideration of the diene LUMO and ethylene HOMO leads to the same conclusion. The [4+2] reaction occurs readily in a concerted fashion (no intermediates), while the [2+2] occurs with difficulty, and only via a stepwise mechanism involving intermediates. Similar ideas have been applied to understand the difference between allowed (relatively facile concerted) reactions and forbidden (disfavored by concerted mechanisms) reactions.

Reactivity differences. Reactivity differences for allowed reactions are also treated successfully by frontier molecular orbital theory. Since the stabilization arising from the HOMO-LUMO interaction is inversely proportional to the HOMO-LUMO gap, an increase in the energy of the HOMO or decrease in the energy of the LUMO lowers the activation energy of a reaction. For example, the activation energies of [4 + 2] reactions like that shown in reaction (2b) are decreased approximately linearly as the HOMO energy of the diene increases or as the LUMO energy of the alkene decreases. The energy of the HOMO of a molecule is thus a measure of the nucleophilicity of a molecule, while the LUMO energy is a measure of electrophilicity.

Regioselectivity. The orientational preference (regioselectivity) of organic reactions can also be predicted. Thus, as shown in reaction (3), substi-

tuents such as electron donors (D) on the diene and electron acceptors (A) on the alkene make the coefficient magnitudes (which determine the size of the p orbitals) unequal on the termini, so that a specific orientation is favored in order to maximize HOMO-LUMO overlap. These types of arguments have been used to rationalize the regioselectivity observed in a variety of thermal and photochemical reactions.

Variations in regioselectivity with reagents can be rationalized by consideration of both frontier molecular orbital coefficients and charges. Thus, enolates tend to protonate on oxygen, the site of highest negative charge as shown in reaction (4a) with a Brönsted acid (HA^+); but alkylate on carbon, the site of highest HOMO coefficient as shown in reaction (4b) with an alkyl halide (RX). Such reactions are called charge-controlled and frontier orbital–controlled reactions, respectively.

Molecular conformations. Stabilities and conformations of molecules can also be predicted by frontier molecular orbital theory. Thus, the aromaticity of benzene and the antiaromaticity of cyclobutadiene may be derived by considerations of the stabilizing interactions which occur upon union of π frontier molecular orbitals of fragments. Thus a 4π unit and a 2π unit can be united to form benzene with stabilizing HOMO-LUMO interactions, while the union of two 2π units to form cyclobutadiene occurs with no HOMO-LUMO stabilization. Similarly, the preferred staggered conformations of ethane and *gauche* conformation of 1,2-dihaloethane result from the

maximization of frontier orbital interactions between the two fragments formed by scission of the C-C bond.

For background information *see* MOLECULAR ORBITAL THEORY; QUANTUM CHEMISTRY; WOODWARD-HOFFMANN RULE in the McGraw-Hill Encyclopedia of Science and Technology.   [KENDALL N. HOUK]

Bibliography: I. Fleming, *Frontier Orbitals and Organic Chemical Reactions*, 1976; K. Fukui, *Theory of Orientation and Stereoselection*, 1975; K. N. Houk, Frontier molecular orbital theory of cycloaddition reactions, *Acc. Chem. Res.*, 8:361, 1975; K. N. Houk, Applications of frontier molecular orbital theory to pericyclic reactions, in A. P. Marchand and R. E. Lehr (eds.), *Pericyclic Reactions*, vol. 2, pp. 182–272, 1977; R. B. Woodward and R. Hoffmann, The conservation of orbital symmetry, *Angew. Chem. Int. Ed. Engl.*, 8:781, 1969.

Fungi

The filamentous fungus *Neurospora crassa* has been used for many years to study the basic relationships between genes and proteins. Recently, with the introduction of recombinant DNA techniques, new information has accumulated on the nature of the components of the ribosome, the site of protein synthesis in the cell; on the genes which code for these components; and on the mechanisms of synthesis, processing, and assembly of these components into a functional ribosome.

As with other eukaryotic organisms, *Neurospora* has two types of ribosomes. One type, located in the cytosol, is used to synthesize proteins coded by genes in the nucleus, while the other type, located in the mitochondria, synthesizes proteins coded by mitochondrial DNA.

Cytosolic ribosomes. Cytosolic ribosomes in *Neurospora* are the typical 80S ribosomes found in the cytosol of all other eukaryotic cells. Each 80S ribosome consists of a large (60S) and a small (37S) subunit containing about 70–75 proteins (the ribosomal proteins) and 4 RNA molecules (ribosomal RNA). Protein synthesis on cytosolic ribosomes can be inhibited by cycloheximide and other inhibitors of 80S protein synthesis, but not by chloramphenicol, tetracycline, or other substances which are known to selectively inhibit bacterial or organelle protein synthesis. Several mutants resistant to cycloheximide have been isolated and shown to have cycloheximide-resistant ribosomes in cell cultures.

The ribosomal RNA (rRNA) molecules are also typical 80S ribosomes and are identified by their size as 26S, 17S, 5.8S, and 5S. The 26S, 5.8S, and 5S rRNA molecules are components of the large subunit of the ribosome, while the 17S rRNA is a component of the small ribosomal subunit. These RNA molecules are encoded in nuclear DNA by genes (rDNA genes) which have been cloned and analyzed by recombinant DNA techniques. The 5.8S, 17S, and 26S rDNA molecules are found as part of a major rDNA locus of 6 million daltons, which contains 190 copies of a unit repeated in tandem and consisting of an untranscribed spacer, the 17S rRNA sequence, a transcribed spacer containing the 5.8S rRNA sequence, and the 26S rRNA sequence. The arrangement of these genes is typical of that found in a wide variety of eukaryotes. The multiple copies of the rDNA appear to be identical with respect to both length and restriction endonuclease cleavage sites. Such homogeneity is similar to what has been found for rDNA in other lower eukaryotes, such as the yeast *Saccharomyces cerevisiae* and the slime mold *Dictyostelium discoideum*, but not in higher eukaryotes.

The sequences which code for 5S rRNA are not located within the major rDNA repeat unit. Instead, the 100 copies of the DNA coding for this RNA are scattered throughout more than 30 different sites in the genome. This pattern differs from most other organisms, in which the 5S rRNA sequences are located within the large rDNA unit (although they are transcribed separately) or are clustered at a small number of sites in the genome. In addition to the 100 genes which are transcribed into 5S RNA, at least one pseudogene, whose sequence is related to the 5S genes but is not transcribed into RNA, has been identified.

The major rDNA unit is initially transcribed into a single 2.4 million–dalton ribosomal RNA precursor, which is then processed in several steps to produce each of the three RNA molecules encoded within the transcribed unit. A cold-sensitive mutant, which grows normally at 25°C (77°F) but not at 10°C (50°F), has been shown to have a defect in the processing steps which produce the functional 17S rRNA and, as a result, shows a significant underproduction of the small ribosomal subunit.

The mechanism of replication of rDNA that maintains sequence homogeneity is not understood. However, recent experiments have demonstrated that *Neurospora* also has a mechanism to regulate the number of copies of the major rDNA unit in the genome. A mutant has been used that contains a translocation of the nucleolus organizer region (the site of the major rDNA genes) to construct a strain carrying two nucleolus organizers. This strain, which initially has twice the normal number of copies of rDNA, gradually loses copies during subsequent vegetative transfers so that by the eleventh transfer the normal value is restored. This process acts as a regulatory mechanism to maintain the correct number of copies of rDNA in the genome, and is reminiscent of the process of demagnification found in *Drosophila* strains which contain extra copies of rDNA genes as a result of unequal crossing over. Magnification of rDNA genes in *Drosophila* mutants with a deficiency in the number of rDNA copies indicates that this regulatory process can operate in either direction. However, the specific mechanism by which either magnification or demagnification occurs has not been determined.

Mitochondrial ribosomes. Mitochondrial ribosomes in *Neurospora* are typical 70S ribosomes found in bacterial cells and in the mitochondria and chloroplasts of eukaryotic cells. They are smaller than the 80S ribosomes of the cytosol, contain fewer ribosomal proteins (about 53), and have only two ribosomal RNA molecules—25S and 19S. The RNA molecules are encoded in mitochondrial DNA, and the organization and transcription of these genes are known in considerable detail. They are located in a 20–kilobase pair (kb) region of mitochondrial DNA which also contains most of the transfer RNA (tRNA) genes used in mitochondrial protein synthesis. The two rRNA genes are separated by about 5 kb, and the 25S RNA contains an intervening sequence (intron) of approximately 2.3 kb. The tRNA genes are located primarily in two clusters, one between the two rRNA genes and the other adjacent to the large rRNA gene. Analysis of temperature-sensitive mutants defective in a factor required for processing the 25S RNA has shown that it is synthesized as a 35S precursor, with removal of the intron apparently occurring in a single cleavage reaction.

Although the mitochondrial rRNA molecules are encoded in mitochondrial DNA, nearly all of the genes encoding the mitochondrial ribosomal proteins are located in nuclear DNA. This means that the mitochondrial ribosomal proteins must be synthesized on cytosolic ribosomes and transported into the mitochondria, where they are assembled with the mitochondrial rRNA to form the mitochondrial ribosome. One exception is the mitochondrial ribosomal protein S-5, which is encoded in mitochondrial DNA and synthesized on mitochondrial ribosomes. This mechanism is similar to the ones found in yeast and protozoa, and it has been suggested that S-5 may be closely related to the yeast protein encoded by the yeast mitochondrial gene *var-1*.

The interaction of mitochondrial protein synthesis with other biochemical functions of the mitochondria has recently been examined by using a novel, extranuclear mutant of *Neurospora* known as [C93]. This is a temperature-sensitive mutant which has a normal phenotype at 25°C, but when it is grown at 37°C the rate of mitochondrial protein synthesis is reduced to about 25% of that of the wild type and the mitochondria are deficient in cytochromes *b* and *aa*$_3$. There is also a deficiency in the production of the small subunit of the mitochondrial ribosome at the higher temperature. Although these results may initially suggest that the defect is in the synthesis or processing of the mitochondrial rRNA, it has been found that the mitochondria of [C93] grown at 37°C (98.6°F) are deficient in one of the mitochondrially coded subunits of mitochondrial adenosine triphosphatase (ATPase), the enzyme responsible for generating ATP in the mitochondria. It has been suggested that the primary lesion in this mutant is in the assembly of the ATPase. The alterations in the ribosomes and in the rate of mitochondrial protein synthesis are thought to be secondary consequences of this mutation, either because the ATPase plays a direct role in the regulation of mitochondrial protein synthesis or because the lack of production of ATP indirectly results in a decreased rate of protein synthesis. These hypotheses emphasize the complexity of interrelationships that exist in the mitochondria which must be taken into account in the analysis of mitochondrial biochemistry.

An interesting feature of the protein-synthesizing systems of *Neurospora*, referred to above for the case of the mitochondrial ribosomal proteins, is the fact that many of the proteins found in the mitochondria are synthesized on cytosolic ribosomes and transported into the mitochondria. These proteins are apparently synthesized as precursor proteins containing leader sequences which may be used as recognition signals for receptors in the mitochondrial membrane. The proteins are transported into the mitochondria after completion of their synthesis, at which time the leader sequences are cleaved from the protein. Such proteins include apocytochrome *c*, an ADP/ATP carrier, several subunits of cytochrome oxidase, and subunit 9 of the oligomycin-sensitive ATPase. The ATPase subunit is an interesting protein for several reasons. First, it is a very hydrophobic protein, but the leader sequence apparently makes the protein sufficiently hydrophilic so that it can be transported through the aqueous cytosol to the mitochondrial membrane. Second, although in *Neurospora* the gene coding for this protein is located in nuclear DNA, in yeast the corresponding gene is found in mitochondrial DNA. Obviously, the mechanisms regulating the synthesis and assembly of this protein differ significantly between these two fungi.

For background information *see* CYTOPLASMIC INHERITANCE; DEOXYRIBONUCLEIC ACID (DNA); GENE; PROTEIN; RIBONUCLEIC ACID (RNA); RIBOSOMES in the McGraw-Hill Encyclopedia of Science and Technology.   [JERRY FELDMAN]

Bibliography: R. A. Collins et al., A novel extranuclear mutant of *Neurospora* with a temperature-sensitive defect in mitochondrial protein synthesis and mitochondrial ATPase, *Mol. Gen. Genet.*, 181:13–19, 1981; S. J. Free, P. W. Rice, and R. L. Metzenberg, Arrangement of the genes coding for ribosomal ribonucleic acids in *Neurospora crassa*, *J. Bacteriol.*, 137:1219–1226, 1979; K. D. Rodland and P. F. Russell, Regulation of ribosomal RNA cistron number in a strain of *Neurospora crassa* with a duplication of the nucleolus organizer region, *Biochim. Biophys. Acta*, 697:162–169, 1982; P. J. Russell, E. U. Selker, and J. A. Jackson, Cold-sensitive mutation in *Neurospora crassa* affecting the production of 17S ribosomal RNA from precursor RNA, *Curr. Genet.*, 4:1–5, 1981; E. U. Selker et al., B. Alzner-DeWeerd, and U. L. RajBhandary, Dispersed 5S RNA genes in *N. crassa*: Structure, expression and evolution, *Cell*, 24:819–828, 1981; R. Zimmermann, B. Hennig, and W. Neupert, Different transport pathways of individual precursor proteins in mitochondria, *Eur. J. Biochem.*, 116:455–460, 1981.

Geochemical prospecting

A new geochemical method which can aid in locating economically significant concentrations of molybdenite (MoS_2) ore is based on the fluorine content of micas associated with mineralization. So-called porphyry molybdenum deposits are formed when granitic melts of appropriate composition intrude other rocks near the surface. Under favorable conditions, a supercritical aqueous fluid containing small amounts of molybdenum in solution may separate from the silicate melt and penetrate the surrounding rocks, precipitating molybdenite in tiny cracks in the host rock in response to falling temperature and changing chemical parameters of the fluid. The hydrothermal fluid travels well beyond the zone of mineralization, reacting with, and changing the mineralogy of, most of the rocks it contacts. Because this alteration halo commonly extends far beyond any recognizable concentration of ore minerals (and is therefore more likely to be recognized either in drill core or in outcrop), the minerals formed as a product of alteration play an important role in prospecting. Fluorine-bearing minerals, such as fluorite, topaz, and micas, are associated with alteration in porphyry molybdenum deposits, but it is important to know how much fluorine enrichment is necessary for a given rock to be considered anomalously high and thus potentially in proximity to an ore deposit.

Distribution of fluorine in rocks. Fluorine is a minor element in most crustal rocks, ranging from about 400 to 1200 ppm. However, fluorine is strongly concentrated into a few important minerals: fluorite, CaF_2 (about 50% F); topaz, $Al_2SiO_4(F,OH)_2$ (17–20% F); apatite, $Ca_5(PO_4)_3(OH,F,Cl)$ (0 to about 4% F); and micas and amphiboles of widely varying composition (0 to about 10% F). In fluorite and other less common halides, the fluorine content is fixed by stoichiometry. In contrast, the fluoride ion in topaz, apatite, micas, and amphiboles substitutes for hydroxyl (OH) in the crystal structures of these minerals, and thus the amount of ion varies in response to the activity of fluoride present during crystallization. This variation allows these minerals to be used as so-called barometers of relative fluoride activity.

Fluoride-hydroxyl exchange. The thermodynamics of fluoride-hydroxyl ($F \leftrightarrows OH$) exchange in minerals such as micas can be described by the simple exchange reaction (1), where OH-mica and F-mica

$$OH\text{-mica} + HF \rightarrow F\text{-mica} + H_2O \qquad (1)$$

represent pure hydroxyl-mica and pure fluor-mica respectively (the cationic composition of the mica must also be specified), and HF and H_2O are gaseous components present in an aqueous supercritical fluid in contact with the mica. The fluid phase serves as an exchange reservoir for F and OH anions. The equilibrium constant for the reaction is expressed in Eq. (2), where a_F and a_{OH} are the ac-

$$K = \left(\frac{a_F}{a_{OH}}\right)_{mica} \left(\frac{a_{H_2O}}{a_{HF}}\right)_{fluid} \qquad (2)$$

tivities of fluor and hydroxyl components in the mica, and a_{H_2O} and a_{HF} are the activities of H_2O and HF in the fluid phase.

The chemical formula for mica may be simplified to $KM_{2-3}Si_3AlO_{10}(OH,F)_2$. The cations represented by M are most commonly Al^{3+} (muscovite), Mg^{2+} (phlogopite), and Fe^{2+} (annite), with lesser amounts of Fe^{3+}, Li^+, and Ti^{4+}. The constituent ions in all micas are arranged in distinct layers, which accounts for many of the most obvious physical properties of the micas, including their ability to be split into very thin sheets. Within these layers the M cations are located in the centers of octahedra formed by O, OH, and F anions; thus F anions are bonded most strongly to the M cations, and the equilibrium constant for the exchange reaction depends very strongly on the cation population in this octahedral sheet. Experimental calibration of the equilibrium constant involves equilibration of synthetic micas of known compositions with aqueous fluids in which a_{H_2O}/a_{HF} was controlled. The experimental conditions were total pressures of 1–2 kilobars (1–2 × 10^5 kilopascals) and temperatures of 400 to 700°C, which were chosen to be similar to physical conditions existing during natural hydrothermal processes. At the close of the experiment, the F/OH ratio of the mica was measured, and the equilibrium constant was calculated. Results showed that for the same fluid composition, Mg was much more effective than Fe in concentrating F into the mica structure, while Al was less effective than Fe. The difference could amount to several weight percent F; as a result of this, the F content of a mica cannot be used as a sole measure of the relative HF activity present during crystallization, but must be considered in light of the overall composition of the mica.

Application to rocks. In order to use these data as a relative HF barometer, the equation for the equilibrium constant can be rearranged in logarithmic form as Eq. (3). The right-hand side of Eq. (3)

$$\log \left(\frac{a_{H_2O}}{a_{HF}} \right)_{fluid} = \log K - \log \left(\frac{a_F}{a_{OH}} \right)_{mica} \qquad (3)$$

can be calculated for a specific mica both from the experimental calibrations and from a complete chemical analysis of the mica. The complete analysis is required to correct for the specific M cation population on F/OH partitioning. The left-hand side of Eq. (3) is an activity ratio. However, for hydrothermal fluids which are predominantly aqueous, any variation in that ratio reflects changes in HF activity because the activity of H_2O remains essentially constant, with decreasing activity ratios corresponding to higher HF activities. One complication is that the equilibrium constant depends on temperature, and the temperature of F ⇆ OH exchange can never be known precisely; nonetheless, if two micas in the same rock (presumably exposed to the same aqueous fluid and temperature) show different F/OH ratios after adjusting for possible cationic differences, then those ratios must reflect different HF activities which existed in the aqueous fluid. Another important factor is that the kinetics of F ⇆ OH exchange in micas is relatively rapid, so the fluorine imprint that can be read from micas is related only to the last fluid which was in contact with them. Micas which

were crystallized originally from a silicate melt and subsequently exposed to a circulating hydrothermal fluid as temperatures declined well below magmatic values will record only the fluorine signature of the last hydrothermal fluid present before the rocks solidified.

For example, micas from the Henderson molybdenite deposit from Clear Creek County, Colorado, were studied in terms of fluoride-hydroxyl exchange. More than 150 micas obtained from drill cores taken through the ore deposit were analyzed. Figure 1 is a cross section through part of the ore body. The white line encircles rock containing more than 0.1% MoS_2 and represents the lower limit of economically recoverable ore. The activity gradients in the hydrothermal fluids were calculated from micas in one of the major zones of alteration. Bearing in mind that lower ratios mean higher fluorine activities, it appears that HF activities as calculated from mica compositions increase consistently toward the center of most intense mineralization. These gradients were calculated for a constant temperature of 350°C. Assuming any other constant temperature would change the absolute numbers but not the relative spacing between the lines, the activity gradient is identical for any constant temperature. On the other hand, if a temperature gradient existed instead of an isothermal profile (a likely possibility), the fluo-

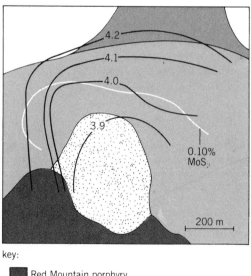

key:

■ Red Mountain porphyry

□ Urad porphyry

▣ Primos porphyry

▦ Henderson granite

□ Silver Plume granite

Fig. 1. Activity gradients (log a_{H_2O}/a_{HF}) for a vertical section through the Henderson molybdenite deposit. Black lines are activity ratio contours; white line marks the economic limit of molybdenite mineralization. The age of Silver Plume granite is 1.4 billion years; Red Mountain porphyry, Urad porphyry, Primos porphyry, and Henderson granite are different granitic phases of the igenous complex responsible for the ore deposit (28–23 million years). (*After A. J. Gunow, S. Ludington, and J. Munoz, Fluorine in micas from the Henderson molybdenite deposit, Colorado, Econ. Geol., 75:1127–1137, 1980*)

rine activity gradient will be increased, for example, from 0.3 log unit for the isothermal case to 1.3 log units for a 150° thermal gradient. In both cases, increasing HF activities point toward most intense concentrations of molybdenite.

Fluorine index. Thus, micas are effective indicators of changes in HF activity in geochemical environments. The above method is difficult to apply to exploration, however, because detailed geologic information must be available in order to decide how to treat the temperature dependence of the equilibrium constant. A simple adaptation involves a graphical technique in which micas are characterized according to a fluorine index (FI). In this technique, only the values for the weight percent MgO and weight percent F are needed, based on the assumption that Mg-F bonds are the most important contributors to fluorine fractionation in mica. Based on calculations from more rigorous thermodynamic expressions, lines of equal FI were drawn such that one FI unit equals 0.25 unit of log (a_{H_2O}/a_{HF}). Biotites from the Henderson deposit (Fig. 2) have FI's ranging from 9 to 11, with the strongest concentration around 10. In contrast, biotites from another important porphyry ore deposit at Santa Rita, New Mexico, in which the principal metal is copper rather than molybdenum, cluster around an FI of 6, forming a very distinct population. These micas do overlap partially with a group of biotites from typical (unmineralized) granites, however.

A more intriguing scientific question relates to the significance of the high fluorine activities associated with precipitation of large amounts of molybdenite, whether the fluorine is an essential ingredient related to complexing and transport of molybdenum in the hydrothermal fluid, or merely a fellow traveler which reflects more of the petrologic and geochemical evolution of the granite bodies which generated the ore deposit. The question is still open. Either way, it is clear that fluorine is an essential indicator, which opens the possibility that micas may be successfully used, in conjunction with other geologic and geochemical tools, as a guide to exploration for these deposits.

For background information *see* GEOCHEMICAL PROSPECTING; MICA; MOLYBDENITE in the McGraw-Hill Encyclopedia of Science and Technology.

[J. L. MUNOZ]

Bibliography: A. J. Gunow et al., Fluorine in micas from the Henderson molybdenite deposit, Colorado, *Econ. Geol.,* 75:1127–1137, 1980; J. Munoz and A. Swenson, Chloride-hydroxyl exchange in biotite and estimation of relative HCl/HF activities in hydrothermal fluids, *Econ. Geol.,* 76:2212–2221, 1981; W. H. White et al., Character and origin of climax-type molybdenum deposits, *Econ. Geol.,* 75th anniv. vol., pp. 270–316, 1981.

Glaciology

The motion of glaciers and ice sheets has been the object of scientific investigation for more than a century; to some extent it still remains a mystery because the varied velocity scales are not yet fully understood. It appears that several mechanisms may be responsible for very similar flow conditions.

Cold and temperate ice zones. In general, large ice masses are polythermal, that is, they consist of two zones: a cold zone, in which the temperature of the ice is below the melting point, and a temperate zone, in which the ice is at the melting point. In the cold zone, heat generated by internal friction affects the temperature distribution in the ice mass, and this in turn affects its motion. In the temperate zone, on the other hand, heat generated by friction melts some of the ice. Cold zones can be described in terms of a heat-conducting viscous model, but temperate zones can be described only in terms of a mixture of interacting ice and water.

The flow and temperature distribution in glaciers depends on what happens within the ice mass and its bounding surfaces—the free surface, the ice-rock interface, and the ice-water interface of possible floating portions. Cold and temperate ice are separated by an interface at which the temperature gradient is generally discontinuous. The body flow is governed by differential shearing; its value depends on the creep law of ice under slow plastic deformation. Older models relate deformation rate to stress with a stress-dependent coefficient so that isotropy of the ice is obtained in three dimensions, in partial disagreement with field observation. More realistic models of ice creep under plastic deformation therefore take into account the anisotropic properties of the ice.

Glacier velocities and surges. The overall velocities in a glacier are strongly dependent on what happens at the ice-rock interface. The boundary

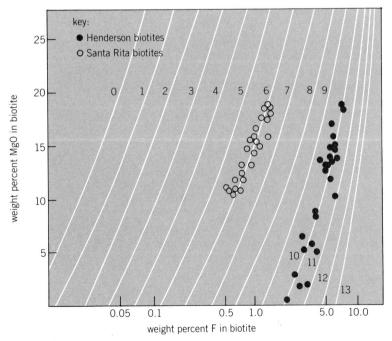

Fig. 2. Fluorine indexes for biotites from the Henderson molybdenum deposit and the Santa Rita porphyry copper deposit.

conditions depend on whether the ice is cold or temperate. At the cold portions of the base the no-slip condition applies; adhesive forces prevent the ice from sliding over the bedrock. When the ice at the base is temperate, however, the heat generated by deformation causes melting; hence a layer of water separates the bottom from the rock and acts as a lubricant, thereby allowing the ice to slide. The thickness of the lubricating film may be negligibly small, but it is sufficient to change the boundary condition from no-slip to perfect sliding. Bed resistance is then due only to pressure variations of the undulating bed.

In alpine regions when summer meltwater becomes abundant, interstitial water pressure increases and may partly lift the ice off its bedrock. Cavities are formed, the effective bed is smoothed, and the effective sliding resistivity is therefore reduced. The functional relationship connecting the basal sliding velocity u_b and the basal shear traction τ_b is qualitatively known for sliding without cavity formation, as shown in Fig. 1a. Here, u_b and τ_b are uniquely related. When cavities are formed, the behaviors shown in Fig. 1b or c have been hypothesized. Here, for a given basal traction, two or more values of the sliding velocity exist, giving rise to a bifurcation mechanism which may explain the surging motion of glaciers. In the lower part of the curve in Fig. 1c, u_b increases slowly with τ_b until an upper limit τ_u is reached, whereupon u_b jumps suddenly to a much higher value, corresponding to a glacier surge. Conversely, when starting at high values of τ_b, u_b decreases slowly with τ_b until a lower limit τ_l is reached, at which point u_b suddenly drops to a much lower value, corresponding to a sudden slowing of the ice.

The transition from the situation of Fig. 1a to that of c is described by a variable connected to the water content in the basal cavities. It is believed that this variable is the interstitial pressure, and field evidence has been collected in support of this hypothesis. A more complete picture of how basal sliding probably occurs is displayed in Fig. 2, which is a three-dimensional graph. The dependence of the basal sliding velocity on the two independent variables, basal shear and a variable describing the thickness of the lubricating film, is represented by

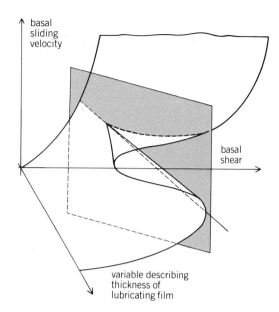

Fig. 2.   Three-dimensional graph showing dependence of basal sliding velocity on basal shear and a variable describing thickness of lubricating film. (*After K. Hutter, Glacier flow, Amer. Sci., 70(1):26–34, 1982*)

a curtain-folded surface. The curves cut out of this surface by planes parallel to the basal sliding velocity–basal shear stress plane are of the forms shown in Fig. 1a and c.

The multiplicity of flow conditions suggested by Fig. 2 is physically impossible. Under such circumstances, nature favors the stable configuration. Hence, if a glacier is in an unstable configuration, any perturbation will cause it to bifurcate rapidly into the stable configuration. Such a bifurcation may successfully explain surges. Surging glaciers move over a relatively short period of time—1 to 2 years— at velocities several orders of magnitude greater than those of normal glaciers. One cause of the surges is the peculiar sliding law of Fig. 2. However, there must also be other causes of surging motion, since some cold glaciers are known to surge occasionally.

In cold ice the velocity and the temperature distribution interact in a nonlinear fashion. By using a simplified mathematical model, it has been shown that under certain conditions of geothermal heat flux and atmospheric temperature the longitudinal surface velocity is nonuniquely related to these thermal boundary conditions. In the multivalued range, for a given geothermal heat flow there are small and large steamwise surface velocities, providing another possible bifurcating mechanism, but it is not firmly known yet whether the bifurcation would correspond to a transition from an unstable to a stable flow configuration.

Prediction of ice sheet flow. Prediction of ice sheet profile geometry is one of the most intriguing problems of glacioclimatology, because it will ultimately lead to understanding of the disintegration and formation of large ice sheets and will allow prediction of glacier advance and retreat on a more local scale.

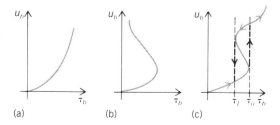

Fig. 1.   Three possible relationships between the component of glacier velocity parallel to the smoothed-out base u_b and the basal shear traction τ_b. (a) Sliding without cavity formation. (b, c) Sliding with cavity formation. (*After K. Hutter, Glacier flow, Amer. Sci., 70(1):26–34, 1982*)

Simple models frequently fail to match observations and bear the disadvantage of not being set on the rigorous footing of basic physical laws. When attempting to predict mathematically the surface profile of an ice mass under steady or transient conditions, the profile geometry must be determined, together with the associated flow and temperature fields. By incorporating into the governing equations the fact that ice sheets are long and wide in comparison to their maximum thickness, a workable scheme was developed, resulting in a nonlinear set of equations for the surface profile that incorporates differential creep and basal sliding, yet allows the accumulation- versus ablation-rate function and the geothermal heat to be freely and realistically assigned.

Localized flow. From a practical point of view these are often most significant. Glaciers are extensively observed, with a view to preventing catastrophes involving human beings. Ice avalanches need not be catastrophes, because they are often preceded by unusually large surface velocities, which increase in time and are confined to limited regions. Thus, the date of the avalanche can be forecast by fitting a hyperbola through the velocity-versus-time curve of a typical point and identifying the avalanche with the asymptote of the hyperbola.

For background information *see* GLACIOLOGY in the McGraw-Hill Encyclopedia of Science and Technology.

[KOLUMBAN HUTTER]

Bibliography: G. C. K. Clarke, U. Nitson, and W. S. B. Paterson, Strain heating and creep instability in glaciers and ice sheets, *Rep. Geophys. Space Phys.*, 15:235–247, 1977; K. Hutter, *Theoretical Glaciology*, 1983; A. Iken, The effect of the subglacial water pressure on the sliding velocity of a glacier in an idealized numerical model, *J. Glaciol.*, 27:407–421, 1981; L. A. Lliboutry, Local friction laws for glaciers: A critical review and new openings, *J. Glaciol.*, 23:67–96, 1979.

Gluons

In the past two decades physicists have come to understand that most subatomic particles, such as the neutron, the proton, and related particles, are not fundamental in themselves, but are composed of more elementary constituents called quarks. A new theory of the interaction of quarks, called quantum chromodynamics (QCD), has been developed in analogy with the well-known quantum theory of eletromagnetic forces—quantum electrodynamics (QED). Quantum electrodynamics describes the manner in which the attraction mediated by electric fields holds a proton and electron together to form a hydrogen atom. Similarly in quantum chromodynamics there are strong fields which hold the quarks together to form a proton or neutron. Moreover, in any quantum theory, these fields can sometimes be manifest as particles. In the case of QED the particle associated with the electromagnetic field is the photon. In quantum chromodynamics the particles are called gluons. In the past few years there has been evidence that gluons do in fact exist and that bound states which contain only gluons—called glueballs—may also exist.

Elementary particles. In the 1930s the basic particles which make up matter were known to be the electron, the proton, and the neutron, and the quantum of light, the photon. However, soon after this, other unexpected elementary particles were discovered. Aside from the neutrinos, the new particles were unstable in that they decayed into lighter particles within a very short lifetime (usually 10^{-23} to 10^{-8} s). In particular, when high-energy acclerators were developed in the 1960s the numbers of the known elementary particles grew tremendously. Present lists of these particles contain several hundred entries.

The particles in such a list fall into two categories. One class, called hadrons, interact with each other by the same strong forces that hold a nucleus together. The other class, called leptons, are not affected by these strong forces. The lepton family contains the electron, the electron's neutrino, and four other particles. The photon is generally not classified as either a hadron or a lepton; all other known particles are hadrons.

Quarks. In order to explain the multitude of hadrons, in the early 1960s M. Gell-Mann and G. Zweig independently proposed that all hadrons then known were made up of three types of quarks. This proved to be a great simplification. Just as the periodic table can be understood as the result of building atoms out of protons, neutrons, and electrons, so the list of strongly interacting particles can be understood as the result of building particles from quarks. During the 1970s new particles were discovered which increased the number of fundamental quarks from three to five. A sixth quark is predicted theoretically but has not been discovered.

However, in its early stages there were several major puzzles with the quark model. The most general was the question of how the quarks interact with each other. Among the more specific problems was the fact that several quark model calculations were wrong by a factor of 3. In addition, it appeared that particles existed in which two or more quarks shared a state with the same quantum numbers, a situation forbidden by the Pauli exclusion principle. It was gradually accepted that the solution to the last two of these puzzles was an idea suggested by O. W. Greenberg in 1965. He had proposed that each of the species of quarks comes in three types. These types are distinguished by an extra quantum number which has come to be called color, although it has nothing to do with the colors that are seen optically. This idea remedied the factor of 3 in the calculations and allowed all states to obey the Pauli exclusion principle.

At first the color quantum number was thought to be passive, in that it did not affect the way that quarks behaved. Quarks of different colors would be distinct but would act in the same manner, much as is the case with ordinary visual color (that is, the color of an automobile does not affects its driving characteristics). However, the idea of color turned out to be the clue for the general question of how quarks interact.

Quantum chromodynamics. Perhaps the best-understood interaction is electromagnetism. Particles which carry one or more units of charge will generate electric and magnetic fields. Another charged particle nearby will be influenced by these fields. In particular, it is the Coulomb force, due to the electric field, which is responsible for binding together electrons and nuclei to form atoms. This theory culminates in quantum electrodynamics, which accurately describes how electromagnetic forces behave in the subatomic world.

The theory which attempts to explain the strong interactions of quarks, quantum chromodynamics, uses color in much the same way that quantum electrodynamics uses charge. The important new ingredient of this theory is the existence of color fields. Particles which carry a color generate color fields around them and interact with each other through these fields. Quarks with opposite color attract each other. (Opposite has a more technical definition in the case of three colors than it does with the usual electric charge, but the resulting physics is similar.)

In quantum electrodynamics the quantum of the electromagnetic field is the photon. The photon behaves as a particle in the photoelectric effect and in gamma rays, among other situations. In quantum chromodynamics there are eight quanta corresponding to the color fields. They are called gluons because of their role in "gluing" the quarks together to form protons, neutrons, and so forth. In many ways gluons are similar to photons, but there is one very important distinction. Electromagnetic fields and photons are neutral, but the color fields and gluons carry the color quantum number. This means that there are strong forces between quarks and gluons, between gluons, and between quarks. Quantum chromodynamics is therefore very difficult to solve and indeed is still not completely understood, although many predictions of specific processes have been extracted.

Evidence for gluons, like evidence for quarks, has been obtained only indirectly, not by direct observation of the particles. One class of experiments involves the annihilation of very-high-energy electrons and positrons. Sometimes when this annihilation occurs, the energy reappears again in the form of an electron and a positron, and some of the time it appears as an electron and a positron plus an extra photon. The rates and distributions for these events are correctly predicted by quantum electrodynamics. Likewise, quantum chromodynamics predicts that, in the same reaction, a given fraction of the time the energy will reappear as a quark and an antiquark, and sometimes as a quark and an antiquark plus a gluon. Due to the strong forces between the quarks and gluons, these particles will not emerge to be detected as isolated particles, but jets of many hadrons will emerge that carry the original directions of the quark, antiquark, and gluon. Early experiments of this sort, performed at the Stanford Linear Accelerator Center (SLAC) in California and at the Deutsches Elektronen Synchrotron (DESY) in Hamburg, Germany, provided some of the most important evidence for quarks and for the existence of three colors. As the experiments have been refined and pushed to higher energies, they have uncovered the quark, antiquark, plus gluon reaction, providing further evidence for the existence of gluons.

Glueballs. One of the novel predictions of quantum chromodynamics is that bound states consisting only of gluons should exist. This can occur because the gluons carry the color quantum numbers, and in fact the forces between two gluons are stronger than those between quarks. These bound states are generally called glueballs, although sometimes they are referred to as gluonium. It is impossible at present to determine the properties of glueballs directly from quantum chromodynamics. However, several phenomenological models, based on the expected behavior of quantum chromodynamics, do exist and make predictions concerning glueball structure. They are expected to have masses and properties similar to the hadronic states already discovered, which are composed of a quark plus an antiquark (the mesons).

Very recent experiments using the electron-positron storage ring at the Stanford Linear Accelerator Center may have uncovered one such glueball. The experiments study the decay modes of a state of a charmed quark bound to its antiquark. The quark and antiquark will annihilate, and some of the time the energy appears in the form of two gluons plus a photon. Most frequently the gluons will each form a jet of hadrons, but it is possible for the two gluons to bind together into a glueball. This leads to a final state consisting of a glueball plus a photon. Experimenters have looked for such a configuration and have discovered a new hadron which they call $\iota(1440)$. (The symbol ι is the Greek iota; 1440 is a label which describes the mass of the particle in MeV/c^2.) It has the behavior expected of a glueball and does not appear to be interpretable as a state containing quarks. It is widely believed that this state is in fact the first glueball. Future experiments will be able to confirm this conclusion. The observations of gluons and glueballs support quantum chromodynamics as the correct theory of strong interactions and suggest that a major advance has been made in understanding the subatomic world.

For background information *see* COLOR (QUANTUM MECHANICS); ELEMENTARY PARTICLE; FUNDAMENTAL INTERACTIONS; PHOTON; QUANTUM ELECTRODYNAMICS; QUANTUM FIELD THEORY; QUARKS in the McGraw-Hill Encyclopedia of Science and Technology.

[JOHN F. DONOGHUE]

Bibliography: F. E. Close, *An Introduction to Quarks and Partons*, 1979; J. F. Donoghue, Glueballs, *Comments Nucl. Part. Phys.*, 10:277–285, 1982; S. L. Glashow, Quarks with color and flavor, *Sci. Amer.*, 233(4):38–50, 1975; G. 't Hooft, Gauge theories of the forces between elementary particles, *Sci. Amer.*, 224(6):104–138, 1980.

Grafting in plants

Plant grafting is an ancient practice. It was known to the Chinese as early as 1560 B.C., and Aristotle, Theophrastus, and the Apostle Paul each mentioned grafting in their writings. Today plant grafting is a widely used means of plant propagation and growth control that is of considerable economic importance. Grafting is used to: perpetuate clones of plants that cannot be readily reproduced by other propagative methods; obtain benefits of desirable rootstocks; change cultivars of established plants; and obtain special forms of plant growth. A recent renewal of interest in vegetative compatibility-incompatibility has led to a better appreciation for the complex events underlying graft formation. Research is now concentrated on the basis for the cohesion of the stock and scion, how the vascular system (that is, xylem and phloem) is reestablished between the tissues, and the causes of graft incompatibility.

Development of a compatible graft. The development of tensile strength in the compatible autograft in eggplant (*Solanum pennellii*) is presented in Fig. 1. The developmental pattern, which is typical for compatible grafts, consists of three phases. Phase I cohesion lasts 2–4 days and is characterized by the development of a weak cohesion between the stock and scion. Phase II cohesion lasts from days 4 to 16 after grafting and is characterized by a rapid increase in the tensile strength of the graft union. During phase III cohesion, the tensile strength of the graft union equals that of an ungrafted stem.

Initial cohesion. The initial cohesion of the stock and scion (phase I) is the result of a cellular wound response induced in the grafting cells by the graft incision. These cells secrete cell wall precursors into the graft interface. The cellular organelles responsible for this secretion are dictyosomes (Fig. 2). The subsequent polymerization of the cell wall precur-

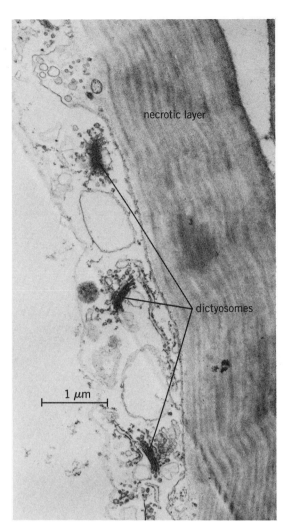

Fig. 2. Accumulation of dictyosomes along the cell wall adjacent to the graft interface at 6 hours after grafting in the compatible *Sedum* autograft. (*From R. Moore and D. B. Walker, Studies of vegetative compatibility-incompatibility in higher plants, I: A structural study of a compatible autograft in Sedum telephoides (Crassulaceae), Amer. J. Bot., 68:820–830; 1981*)

sors is responsible for the initial cohesion of the graft partners. The initial cohesion does not involve cellular recognition and is independent of other events in graft development. Furthermore, since the stock and scion of incompatible grafts also adhere during the early stages of graft development, the initial cohesion of graft partners is not directly related to graft compatibility-incompatibility. The continued increase in tensile strength of the graft union over time (phases II and III) is due to the interdigitation of callus cells at the graft interface, and the redifferentiation of a lignified strand of xylem linking the stock and scion.

Callus cells at graft interface. Cellular division results in the production of callus cells at the graft interface soon after grafting. Callus proliferation has the following features in common with the deposition of wall materials responsible for the initial cohesion of the graft partners: (1) callus proliferation is a

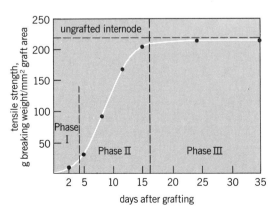

Fig. 1. Development of tensile strength in the compatible autograft in eggplant (*Solanum pennellii*).

characteristic wound response in plants; (2) callus proliferation is not directly related to graft compatibility-incompatibility; and (3) callus formation is independent of other events in graft formation.

The role of callus cells was shown in the graft development of monocots. Monocots are notoriously difficult to graft, and their tissues do not readily dedifferentiate to form callus. Also, unsuccessful plant grafts are often characterized by decreased callus formation. Therefore, adequate callus formation appears to be necessary for the subsequent expression of graft compatibility. Callus cells also (1) interdigitate at the graft interface, thereby increasing the tensile strength of the union; (2) rupture the necrotic layer (that is, the layer of cells killed in making the graft incision), thus establishing direct cellular contact between the stock and scion; and (3) are the cells through which vascular tissue subsequently redifferentiates.

Redifferentiation of vascular tissue. Redifferentiation of xylem and phloem across the graft interface is one of the last major events to occur during the development of a successful graft. The redifferentiating strands of vascular tissue originate from cells near the severed vascular bundles. The amount of vascular tissue joining the two graft partners is largely influenced by leaves on the scion. Auxins (a type of plant hormone) have been implicated in the redifferentiation of vascular tissue between the stock and scion. The formation of a lignified strand of xylem contributes significantly to the tensile strength of the graft union (phase II cohesion).

Graft incompatibility. While it appears that the general sequence of events outlined above (that is, cohesion of the stock and scion, production of callus cells, and redifferentiation of vascular tissue) occurs in response to grafting of compatible tissues, there are several different responses that result in incompatibility between two grafted tissues. One type of incompatibility response is characterized by the lack or incomplete redifferentiation of vascular tissue across a graft interface. In this type of response, both cut surfaces produce callus tissue that remains healthy, but subsequent redifferentiation of vascular tissue is abnormal or nonexistent. Most research reports on graft incompatibility fall into this category because they have been performed on fruit-tree grafts that are only partially compatible, or become incompatible at a later date.

These cases of delayed incompatibility may result from compounds associated with one partner having an adverse effect on the other partner. There are reports where a 20-year-old graft which appeared compatible became incompatible. Accumulation of toxic by-products or the production of a new secondary metabolite in response to aging are possible explanations for these cases of delayed incompatibility.

The movement of toxic substances between stock and scion can elicit graft incompatibility, as has been shown in reports of the translocation of metabolites and viruses from one graft partner to another, as well as in several studies involving reciprocal grafts. For example, although a particular graft combination of A/B (that is, stock over scion) may be compatible, the B/A orientation may result in an incompatible graft. This suggests that certain factors produced in one partner (in this case B) are translocated downward from the B scion and bring about incompatibility with A. If an interstock (C) is placed between B and A to produce B/C/A, the union may become compatible, suggesting that the interstock acts as a trap for those toxic factors produced in B that elicit the incompatibility response in A. Incompatibility that is overcome by the use of an interstock has been termed localized incompatibility. Translocated incompatibility is the term used for incompatibility which cannot be overcome by an interstock.

Cellular necrosis is the most dramatic (and perhaps most widespread) incompatibility response. Incompatibility due to necrosis of callus cells at the graft interface differs from other mechanisms of incompatibility in that one or both tissues actively reject the other graft partner.

Biochemical studies. To date, only the incompatible graft between pear and quince has been characterized biochemically. This graft combination is often compatible, but in certain cases (for example, at elevated temperatures) it becomes incompatible. A cyanogenic glycoside from quince moves across the graft interface, where it is broken down by glycosidase in the pear tissue to liberate cyanide. While lower temperatures do not elicit the production of toxic levels of the cyanogenic glycoside in quince, the cyanide liberated at higher temperatures leads to the death of cells at the graft interface. Cellular necrosis has been observed in many incompatible grafts, but the cytological events associated with this type of incompatibility have been elucidated only in the incompatible heterograft between *Sedum telephoides* and *Solanum pennellii*.

Cytological studies. Early events of the incompatible heterograft between *Sedum* and *Solanum* are similar to those of a compatible graft. Adherence of the stock and scion occurs by 12 hours after grafting and is correlated with cell wall deposition and callus proliferation. Thus, incompatibility in this system is related neither to callus proliferation nor to stock-scion adhesion. Unlike a compatible graft, however, a dramatic woundlike response ensues at the graft interface within 2–3 days. A progressive cellular senescence extends several cells away from the necrotic layer and is characterized by cytoplasmic release of the hydrolytic enzyme acid phosphatase, cytoplasmic vesiculation, degeneration of the cell membrane and organelles, and finally cellular death and collapse. As a result of this lethal cellular senescence, the necrotic layer thickens and is never ruptured throughout the development of the incompatibility response (Fig. 3).

The reactions of the stock and scion differ by 5–7 days after grafting. *Sedum* cells bordering the graft interface continue to due and collapse. By contrast,

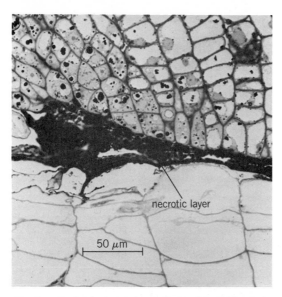

Fig. 3.   Graft interface between *Sedum telephoides* (bottom) and *Solanum pennellii* (top) at 8 days after grafting. *Sedum* cells adjacent to the necrotic layer are in various stages of collapse (arrows). (*From R. Moore and D. B. Walker, Studies of vegetative compatibility-incompatibility in higher plants, II: A structural study of an incompatible heterograft between Sedum telephoides (Crassulaceae) and Solanum pennellii (Solanaceae), Amer. J. Bot., 68:831–842, 1981*)

Solanum cells stabilize and apparently no longer detect incompatibility. Thus, only senescing *Sedum* cells contribute to the necrotic layer that separates the graft partners. However, the cumulative effect is that no vascular redifferentiation occurs across the graft interface, which results in starvation, desiccation, and abscission of the *Sedum* scion by 4–6 weeks after grafting. Incompatibility between *Sedum* and *Solanum* appears to be due to a diffusible factor that is normally associated with *Solanum* and which facilitates the incompatibility response in bordering *Sedum* cells. The cellular site of action of the incompatigen appears to be the endomembrane system.

Cellular recognition. Cellular recognition is not involved in the initial adhesion of the stock and scion. However, the differentiation of superimposed cellular structures (plasmodesmata, sieve areas and plates, and so on) must depend upon some type of cellular recognition or communication, which therefore occurs during at least one stage of the development of a compatible graft. The coordinated differentiation of vascular tissue between the graft partners may also require some type of cellular recognition.

The precise involvement of cellular recognition in incompatible plant grafts is more difficult to determine since there are several causes of graft incompatibility. While recognition may be operative in determining incompatibility in some graft combinations, it is not necessary to assume that a recognition phenomenon is operative in all incompatible grafts. For example, the incompatible graft between *Sedum* and *Solanum* can be satisfactorily explained without invoking cellular recognition.

For background information *see* GRAFTING OF PLANTS in the McGraw-Hill Encyclopedia of Science and Technology.

[RANDY MOORE]

Bibliography: A. Gur and A. Blum, The role of cyanogenic glycoside of the quince in the incompatibility between pear cultivars and quince rootstocks, *Hort. Res.*, 8:113–134, 1968; R. Moore and D. B. Walker, Studies of vegetative compatibility-incompatibility in higher plants, III: The involvement of acid phosphatase in the lethal cellular senescence associated with an incompatible heterograft, *Protoplasma*, 109:317–334, 1981; F. L. Stoddard and M. E. McCully, Effects of excision of stock and scion organs on the formation of the graft union in *Coleus*: A histological study, *Bot. Gaz.*, 141:401–412, 1980; M. M. Yeoman et al., Cellular interactions during graft formation in plants, a recognition phenomenon?, *Symp. Soc. Exp. Biol.*, 32:139–160, 1978.

Gravitation

Shortly after Isaac Newton proposed his theory of gravitation, he made an educated guess that the average density of the Earth was about 4½ times that of water. From this he estimated the value of the newtonian gravitational constant G. His estimate was confirmed roughly by Pierre Bouguer in Ecuador and by Nevil Mascalyne in Scotland, who measured the attraction of geological features. Their methods have not as yet lent themselves to precision measurement, since the extent of these geologic masses and their mass distributions are inaccurately measured and the systematic errors are poorly defined; yet a precision measurement of G using the sea as the large mass system has been proposed.

All recent precision determinations have been conducted in a laboratory setting, and precisely machined and measured masses and distances measured as accurately as laboratory techniques allow have been used. These laboratories all have used torsion pendula. A typical torsion pendulum consists of a cylindrical or dumbbell-shaped mass a few centimeters long, of 1–20 grams, which is suspended by a fiber of either tungsten or quartz and has a mirror attached to indicate the angular position of the torsion pendulum in the system. These torsion pendulums do not swing, but rather rotate about the vertical axis. Generally, this rotation is very slow, ranging from a few minutes to as much as an hour for one oscillation. However, in one experiment, which used a magnetic suspension system in place of the fiber, the small mass system weighed on the order of a kilogram and was about one-third of a meter long. This experiment used toroidal-shaped large masses which minimize the errors that result from their inhomogeneity and imprecise positioning and machining.

All recent measurements of G fall into one of three categories: the change-of-frequency method; the res-

Determinations of the newtonian gravitational constant *G*

Date	Investigators	Method*	$G \times 10^{-11}$ $N \cdot M^2/kg^2$
1942	P. R. Heyl and P. Chrzanowski	1	6.673 ± .003
1969	R. D. Rose, J. Beams, et al.	3	6.674 ± .004
1972	C. Pontakis	2	6.6714 ± .0006
1975	G. G. Luther, W. R. Towler, et al.	3	6.6699 ± .0014
1976	O. V. Karagioz et al.	1	6.668 ± .002
1981	G. G. Luther and W. R. Towler	1	6.6726 ± .0005

*Method 1 = change of frequency; method 2 = resonance; method 3 = servoed angular acceleration.

onance method; and the Beams servoed rotation method. The choice depends upon which systematic errors the experimenter thinks can be best overcome and which can be tolerated. The table lists recent determinations.

Change-of-frequency method. In this method, the frequency ω of a torsion pendulum in an evacuated chamber is measured as a function of the position of some large, precisely manufactured and positioned masses. The value of *G* is proportional to $(\Delta\omega)^2$, where $\Delta\omega$ is the change in frequency caused by moving the large masses, and *G* is calculable from the position of the large masses, their mass distribution, and the torsion pendulum. The advantages of this method are that during the experiment only a frequency is measured; the experiment is rather insensitive to the nonlinearities of the supporting fiber; it is rather insensitive to the damping caused by the fiber and the residual gas in the evacuated system; and it is insensitive to any drift in the ambient angular position of the torsion pendulum caused by changes in temperature.

The disadvantages of this method are its sensitivity to vibrations; its sensitivity to the changes in the gravity gradient; and its extreme sensitivity to any change in the torsion constant due to temperature changes or to molecules adsorbed onto or evaporated from the surface of the fiber. Figure 1 shows a typical arrangement of the elements in this method.

Resonance method. In this method, a torsion pendulum hangs in a vacuum. A pair of large masses are either rotated around or oscillated near the torsion pendulum in resonance with its frequency. The oscillation of the torsion pendulum is "pumped up" by the gravitational attraction between the large masses and the torsion pendulum, much like a child pumps up on a swing, and the increase of amplitude of the oscillation of the torsion pendulum is measured as a function of time. The calculation of *G* is made from this increase in amplitude.

This experimental method is quite sensitive to: the nonlinearities and dissipational effects of the fiber; vibrations which are inherent in the rotation of the large masses; and the coupling of the movement of the large masses to the torsion pendulum through vibrations and forces other than gravity. There is also the problem of synchronizing the phase and frequency of the oscillation of the large mass system with that of the torsion pendulum. In view of all these difficulties, it is not surprising that so few ex-

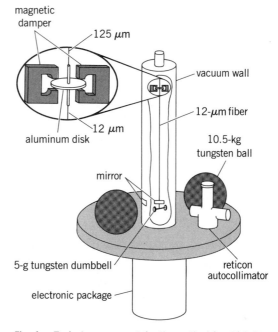

Fig. 1. Typical arrangement for the method in which the frequency of the torsion pendulum is changed by the attraction between it and the large spherical balls. (*After G. G. Luther and W. R. Towler, Redetermination of the newtonian gravitational constant G, Phys. Rev. Lett., 48:121–123, 1982*)

perimenters have attempted this method and that it has had only one recent successful application.

Beams servoed rotational method. In this method (Fig. 2), the large masses, the torsion pendulum, and a mechanism for determining their relative positions (an autocollimator) are all mounted on a rotating table, with the axis of the torsion pendulum oriented approximately 45° from the line joining the centers of the large masses. The attraction of the large masses causes the torsion pendulum to try to rotate so as to align itself with the large masses. The rotating table is accelerated in response to the signal from the autocollimator in order to maintain the relative positions of the two mass systems. The inertial reaction of the torsion pendulum to this angular acceleration is therefore equal to the gravitational attraction between the large masses and the torsion pendulum. Consequently, a measurement of the angular acceleration of the rotating table is equivalent to a measurement of *G*. The advantages of this method

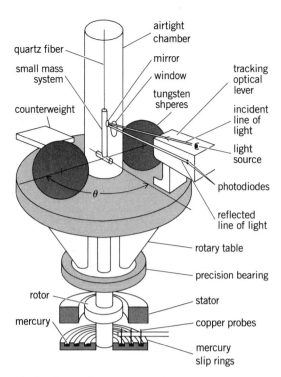

Fig. 2. Schematic arrangement of the Beams servoed rotational method. The angle θ is the angle between the axis of the small mass system and the line joining the centers of the large masses. (*After R. D. Rose et al., Determination of the gravitational constant G, Phys. Rev. Lett., 23:655–658, 1969*)

are that it is insensitive to the gravitational gradient and the damping constants and linearity of the fiber, since the torsion pendulum does not move with relation to its support. However, it is extremely sensitive to the drift in the null position of the fiber, caused by temperature changes and vibrations which will necessarily be generated in order to rotate the table.

Future measurements. Several experiments to determine the value of G are now being constructed, and these span several orders of magnitude in scale.

By far the largest of them is an experiment being constructed in the Great Grotto of Trieste. The beam for the torsion pendulum small masses will be about 1 m in length. The fiber may be almost 100 m long, and the large masses will be about 500 kg each.

At the small end of the scale is an experiment where the mass separation is to be as small as 3 mm. Of a more ordinary size, but also cleverly conceived, is an experiment where the support function of the torsion fiber is replaced by floating it on a pool of mercury, and the torsion constant of the fiber is replaced by electrostatic forces. One experiment under development does not use a torsion pendulum but incorporates a horizontal beam balance. This experiment also uses roughly doughnut-shaped attracting masses in order to reduce the dependence of the measurement to the mass distribution of these large masses.

For background information *see* GRAVITATION in the McGraw-Hill Encyclopedia of Science and Technology.

[GABRIEL G. LUTHER]

Bibliography: D. N. Langenberg and B. N. Taylor (eds.), *Precision Measurement and Fundamental Constants*, NBS (U.S.) Spec. Publ. 343, 1971; C. Pontakis, Determination of the gravitational constant by the resonance method, *C.R. Hebd. Sean. Acad. Sci. B (France)*, 274(7):437–440, 1972; J. H. Sanders and A. H. Wapstra (eds.), *Atomic Masses and Fundamental Constants 5*, 1976; B. N. Taylor and W. D. Phillips (eds.), *Precision Measurement and Fundamental Constants II*, NBS (U.S.) Spec. Publ. 617, in press.

Gyroscope

The fiber-optic gyroscope or fiber gyro is a fiber-optic interferometric sensor capable of measuring mechanical rotation in inertial space. During the past several years it has progressed from being unable to detect the Earth's rotation rate to being able to detect some three orders of magnitude below the Earth's rate, and it is now of substantial interest for practical applications. This progress has resulted from understanding and controlling the mechanisms of noise and drift. Also, these devices are now susceptible to being packaged in small, rugged form, including all-solid-state implementations.

Principles of fiber gyro operation. The fiber gyro, like the ring laser gyro, is based on an effect demonstrated by G. Sagnac in 1913 and studied theoretically and experimentally by a number of investigators in the years following. In a basic Sagnac system an optical beam from a source (Fig. 1a) is separated by a beam splitter (BS) into two beams, which travel in opposite directions around a closed planar path established by mirrors. These two beams, starting at the beam splitter, experience different total phase shifts if the path is rotated about an axis perpendicular to its plane, because the beam circulating in the same sense as the path rotation has a longer transit time back to the beam splitter than the beam circulating in the opposite sense. The phase difference $\Delta\phi_R$ between the two waves on emerging from the beam splitter is proportional to the area and the rotation rate of the path. Measurement of $\Delta\phi_R$, for example, by shifts in a fringe pattern produced by the two waves, gives a measure of the rotation rate. For paths of reasonable diameter (on the order of 1 m) the sensitivity is too small to be of practical interest.

In 1975 V. Vali and R. W. Shorthill proposed the use of optical fiber to form the closed optical path. A multiturn fiber sensing loop or coil multiplies $\Delta\phi_R$ by the number of turns, yielding measurable values for rotation rates suitable for applications. The basic circuit of a fiber interferometer is shown schematically in Fig. 1b. Beam splitter BS-1 performs the function of the beam splitter of Fig. 1a. Measurement of $\Delta\phi_R$ by the photodetector, and the functions of the remaining elements in the circuit, are described below. For a circular N-turn coil of area A, diameter D, and fiber length L, the phase difference $\Delta\phi_R$ is related to rotation rate Ω by the equation below, where c and λ are the free space velocity and wavelength of the light.

$$\Delta\phi_R = 8\pi\frac{NA}{c\lambda}\Omega = 2\pi\frac{LD}{c\lambda}\Omega$$

In contrast to conventional gyros, optical gyros have no rotating parts, which, generally speaking, is an asset. They are of interest for many similar applications, including inertial navigation, guidance and tracking, and applications based on direct determination of the Earth's rotation vector.

In application, fiber gyros would be mounted rigidly on a vehicle or object whose rotation is to be measured, and the only rotation involved would be that of the vehicle itself. When fiber gyros are used together with accelerometers, the path of a vehicle can be determined entirely from measurements made within the vehicle. Sensitive accelerometers for such purposes are also potentially realizable in fiber-optic form.

Closed-loop fiber interferometers. The interferometers of Fig. 1 are referred to as closed-loop interferometers. They make possible the measurement of very small phase differences between the counter-circulating waves. Because both waves travel the same path, their phases tend to be changed similarly by slowly varying environmental factors, such as temperature and vibration, and these changes tend to cancel when $\Delta\phi_R$ is measured. This leaves the system quiet and able to respond to small rotation rates. The full exploitation of this property requires special procedures. The most successful approach to date uses single-mode fiber, together with a high-quality polarizer and an additional beam splitter, BS-2 (Fig. 1b). The polarizer provides single-mode operation by selecting one of the two possible polarization states of the fiber. This, together with BS-2, ensures strict optical reciprocity. Further reductions in environmental noise are achieved by winding patterns which place symmetrically located points along the sensing loop in mutual proximity.

Bias. The output current of the square-law photodetector is shown in Fig. 2 as a function of the phase difference $\Delta\phi$ between two mutually coherent waves incident upon it, where $\Delta\phi$ can have various components, including that due to rotation ($\Delta\phi_R$). To make the system sensitive when at rest ($\Delta\phi_R = 0$) and linear (to first order) in $\Delta\phi_R$, it is necessary to bias it so that the output current is given by point B (Fig. 2) when $\Delta\phi_R = 0$. A static, nonreciprocal phase shift $\Delta\phi_B = \pi/2$ can be imparted to the counterpropagating waves before they enter the sensing loop or when they are within the loop by using a Faraday cell, or it can be produced by shifting the frequencies of the two waves by different amounts by using single-sideband optical modulators (such as Bragg cells). Stabilization of $\Delta\phi_B$ can be a difficult problem.

An important stable dynamic approach to biasing

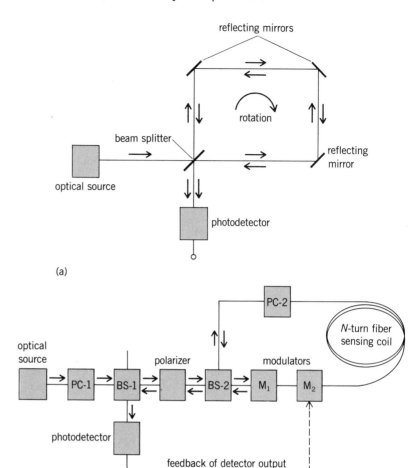

(a)

(b)

Fig. 1. Block diagrams of optical rotation sensors: (a) basic Sagnac interferometer; (b) fiber optic interferometer.

involves placing within the loop one or more phase modulators which will sinusoidally modulate the phases of the counterpropagating waves at frequency f_M. For a typical phase modulator located near one end of the loop (M_1 in Fig. 1b), the two waves pass through the phase modulator at different points on the modulation cycle because of the optical transit time around the loop (typically microseconds). The

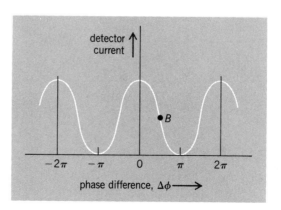

Fig. 2. Photodetector output current as a function of the phase difference between two incident optical waves.

result is that sinusoidal components are introduced into the detector current at frequency f_M and its harmonics. Measurement of the f_M component gives a value proportional to $\sin \Delta\phi_R$.

Sources of noise and drift. An important source of noise is light backscattered by irreducible inhomogeneities randomly distributed along the fiber (Rayleigh backscattering), which in single-mode fiber combines coherently with the primary waves at the detector, giving a phase error containing random, environmentally produced variations. A principal approach to lowering this noise level involves reduction of the temporal coherence of the source. Closed-loop interferometers can tolerate very short source coherence lengths because they have a very small optical path difference. Further important reduction of this noise is provided by the phase-modulation-biasing process described above.

Nonlinear characteristics of the fiber (Kerr effect) are a source of nonreciprocity, leading to a differential phase shift $\Delta\phi_K$ which constitutes an environmentally sensitive error. This $\Delta\phi_K$ can be canceled by proper modulation of the source or by the use of a source with a suitably broad spectral distribution (for example, superluminescent diodes and some lasers with multiple axial modes). Magnetic fields, such as the Earth's magnetic field, acting through the small Verdet constant of the fiber material, can also introduce nonreciprocal phase shifts, constituting environmentally sensitive random errors which can be suppressed by shielding the sensing loop.

Noise can arise from the laser source and the detector. Low insertion loss in the optical circuit between the source and detector is important for overriding the detector noise. Electronic noise is reduced by employing heterodyne detection (such as described for the modulation-biasing scheme), so that the signal may be amplified at an intermediate frequency which is above the range of f^{-1} noise.

Photon noise, the irreducible noise component at any given level of signal power P_D reaching the detector, gives rise to a signal-to-noise component proportional to $\sqrt{P_D}$. In the present state of development, the photon noise level produced by laser diode sources can be kept below other noise levels by holding the total optical insertion loss to low values (for example, 10–15 dB).

If standard fiber is used, which has low (but generally finite) spurious birefringence, the polarization will vary, resulting in signal fading due to power loss in the polarizer. Polarization controllers PC-1 and PC-2 (Fig. 1b) correct the polarization of light entering the polarizer—PC-2 being necessary only when the path between the laser and polarizer consists of a fiber. Polarization controllers can consist of piezoelectric elements which apply transverse stress to a section of the fiber to introduce controlled birefringence in response to an applied control voltage. Recently developed polarization-maintaining fiber may eliminate the need for polarization controllers.

Implementation. Fiber gyros can be implemented with miniature, ruggedly mounted bulk optics components, such as beam splitters, polarizers, and modulators, and with miniature lenses to focus the light into and out of the ends of the fiber sensing coil—all of which can fit inside the coil, whose diameter can be as small as a few inches (1 in. = 2.5 cm). An alternative all-solid-state implementation involves all-fiber components (Fig. 3a). A directional coupler (DC) is constructed by placing modified sections of fiber side by side such that energy transfer can take place between them, giving the equivalent of a beam splitter. The polarizer is formed by placing a birefringent crystal in contact with a modified section of fiber. A very promising all-solid-state approach involves the use of integrated optics components employing single-mode optical waveguide elements on planar substrates which can be mass produced inexpensively and with high precision. In the example of Fig. 3b waveguide Y junctions serve as beam splitters. The polarizer and modulator consist of deposited metal films. Further improvements in such components are needed to bring their performance up to the standards required for the most sensitive gyro applications.

For many applications in which the minimum rotation rate is greater than 0.1°/h some components

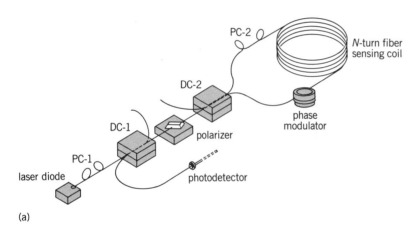

(a)

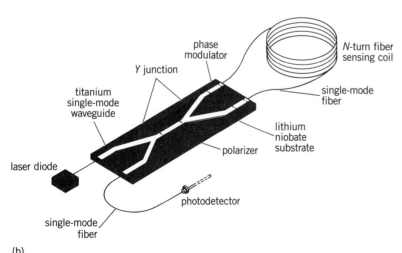

(b)

Fig. 3. Solid-state, single-mode fiber gyro diagrams (not to scale): (a) all-fiber form; (b) integrated optics form.

can be omitted, resulting in simpler, less expensive devices. However, other requirements may be paramount, such as fast response time, high dynamic range, high temperature capability, and high acceleration capability. Alternate approaches to fiber gyros which are under study or have been proposed include pulsed recirculating systems in which optical signals travel many times around the sensing loop to multiply $\Delta\phi_R$; systems using multimode fiber or unpolarized light which could reduce the parts count; and passive resonator systems in which $\Delta\phi_R$ shifts the measurable resonant frequency of a small, single-turn resonant sensing loop.

Status. Fiber-gyro sensitivity and stability have been progressively improved during the past several years. Using circular sensing coils with lengths up to 600 m and diameters of about 6 in. (15 cm), a minimum detectable rotation rate $\Omega_{min} = 0.01°/h$ has been reached, with integration time $T \cong 30$ s (Ω_{min} scales as $T^{-1/2}$). The minimum is $\Delta\phi_R \cong 10^{-7}$ rad. Extrapolated random drift is $0.001°/\sqrt{h}$. Drift values of $0.02°/h$ over a period of 12 h have been demonstrated under laboratory conditions. Within this performance range many applications can potentially be included, some permitting a scaling back of one characteristic or another, such as allowing larger Ω_{min} or longer integration time, which tend to simplify the design. Scale factor linearity and dynamic range can be achieved by adding a closed-loop servo system by feeding back (Fig. 1b) the detector output to elements in the sensing coil (for example, single-sideband frequency modulators) which keep the interferometer balanced at point B (Fig. 2), and by using feedback amplitude as a measure of rotation rate. Further development is needed on components for closed-loop operation of all-guided systems. Engineering development, environmental testing, and system evaluation of fiber gyros are in very early stages, and most of these activities lie in the future.

For background information *see* GYROSCOPE; INERTIAL GUIDANCE SYSTEM; INTEGRATED OPTICS; OPTICAL FIBERS in the McGraw-Hill Encyclopedia of Science and Technology.

[H. J. SHAW]

Bibliography: H. J. Arditty and S. Ezekiel (eds.), *Fiber Optic Rotation Sensors and Related Technologies*, Springer Series in Optical Sciences, vol. 32, 1982; R. A. Bergh, H. C. Lefevre, and H. J. Shaw, Overview of fiber optic rotation sensors, *IEEE J. Lightwave Technol.*, March 1983.

Hall effect

The quantum Hall effect is a remarkable phenomenon in which the combination of fundamental constants e^2/h (where e is the electron charge and h is Planck's constant) can be measured as the ratio of a current to a voltage in a solid-state device. It was discovered by Klaus von Klitzing and coworkers in 1980 during routine measurements of the electric properties of metal oxide semiconductor (MOS) transistors at low temperature and in the presence of

high magnetic fields. The effect is scientifically important for two reasons. First, together with an independent measurement of the speed of light c, it provides an extremely accurate theory-independent measurement of the fine-structure constant $\alpha = e^2/\hbar c$ (where $\hbar = h/2\pi$). This dimensionless number ($\cong 1/137$) is of central importance in quantum electrodynamics, the theory of the interaction of light with matter. Second, the existence of the effect has broad ramifications for understanding electrical conduction of solids in strong magnetic fields.

Two-dimensional electron motion. In order to understand the quantum Hall effect, it is first necessary to understand the two-dimensional nature of electron motion in the systems in which it occurs. An MOS transistor (Fig. 1a) consists of layers of metal and silicon separated by a thin layer of insulating

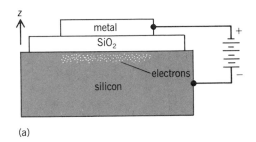

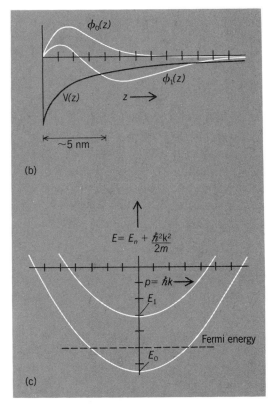

Fig. 1. Two-dimensional electron motion in an MOS transistor. (a) Cross section of transistor. (b) The potential $V(z)$ trapping electrons in wave functions $\phi_0(z)$, $\phi_1(z)$, . . . , at the Si-SiO$_2$ interface. (c) Plot of energy E versus momentum p in the interfacial plane for trapped electrons. All states below the Fermi level are occupied with electrons.

silicon dioxide (SiO_2). When a potential, called the gate voltage, is applied between the metal and the silicon, electrons are drawn out of impurity sites in the silicon, where they were immobile, to the Si-SiO_2 interface, where they are free to move in the x-y plane. The interface thus becomes a two-dimensional metal. The electrons are forbidden from moving in the z direction by quantum mechanical effects. Let the potential holding the electrons to the interface be designated $V(z)$. Then each electron drawn to the interface must reside in a state ψ of total energy E satisfying the Schrödinger equation in the presence of $V(z)$, namely, Eq. (1), where m is the electron mass. ψ is a solution of this equation when ψ has the form given in Eq. (2). The function $\phi_n(z)$ satisfies Eq. (3) and E is given by Eq. (4). The wave function ψ describes an electron trapped by $V(z)$ in a quantum state $\phi_n(z)$ [Fig. 1b] and moving

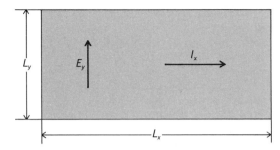

Fig. 2.   Currents and electric fields in a typical quantum Hall experiment.

$$-\frac{\hbar^2}{2m}\left(\frac{\partial^2\psi}{\partial x^2} + \frac{\partial^2\psi}{\partial y^2} + \frac{\partial^2\psi}{\partial z^2}\right) + V(z)\psi = E\psi \quad (1)$$

$$\psi = e^{ixk_x}e^{iyk_y}\phi_n(z) \quad (2)$$

$$-\frac{\hbar^2}{2m}\frac{\partial^2\phi}{\partial z^{2n}}(z) + V(z)\phi_n(z) = E_n\phi_n(z) \quad (3)$$

$$E = \frac{\hbar^2}{2m}(k_x^2 + k_y^2) + E_n \quad (4)$$

with a momentum $\overrightarrow{p} = \hbar\overrightarrow{k}$ in the x-y plane. The energy E of this electron is the sum of its negative trapping energy E_n and its translational kinetic energy in the plane (Fig. 1c). In a quantum Hall experiment, all the electrons trapped at the interface have the same z motion $\phi_0(z)$. It is thus customary to ignore this motion and to speak of the electrons as though they resided in a perfect two-dimensional metal.

Quantum Hall effect in ideal systems. The quantum Hall effect occurs when a two-dimensional metal is forced to carry electric current in the presence of a strong magnetic field. Imagine a magnetic field of strength H_0 piercing the surface of the two-dimensional metal. In the presence of this field, an electron with charge e moving at speed v experiences a force $(ev/c)H_0$ at right angles to its direction of motion, called the Lorentz force. (In this article, cgs gaussian units are used throughout.) In the absence of other forces, this Lorentz force will cause the electron to execute a circular orbit, with the orbit radius always matched to v, so that the electron completes an orbit every $T = 2\pi/\omega_c$ seconds, where $\omega_c = eH_0/mc$ is called the cyclotron frequency.

In the presence of other forces, specifically those due to an electric field of strength E_0, the electron also executes drift: it moves in a direction perpendicular to both the force and the magnetic field with a speed $v_D = c(E_0/H_0)$, called the drift velocity. It is important that the electron does not drift in the direction it is forced. If a rectangular sample with sides of length L_x and L_y is filled with electrons (Fig. 2), then forcing the electrons with an electric field

in the y direction E_y will cause them all to move in the x direction, giving rise to a current I_x. This is the only way the electrons can be caused to move in the x direction. If a current I_x is flowing, it follows that E_y must be present. E_y need not be caused by an external source, such as a battery, as it can be caused by the redistribution of electrons inside the sample. The magnitude of E_y depends on the number of electrons per unit area inside the sample, designated σ. In terms of σ, I_x and E_y are related by Eq. (5). The quantity in parenthesis can be seen to be the voltage drop across the sample.

$$I_x = ev_D\,\sigma\,L_y = \frac{ec\sigma}{H_0}(E_y L_y) \quad (5)$$

The simplest way to understand the quantum Hall effect is in terms of the quantization of σ. In a two-dimensional metal free of impurities, σ can have only the values given by Eq. (6), where N is an

$$\sigma = N\left(\frac{eH_0}{hc}\right) \quad (6)$$

integer. This is because quantum mechanics forces the cyclotron orbits to be certain sizes. When electrons are packed together so that their orbits touch, additional electrons cannot be added easily. A packed layer of electrons subjected to an electric field E_y carries a current I_x given by Eq. (7). The Hall resistance, the ratio of the y voltage drop to the x current, is thus given by Eq. (8). This is the quantum Hall effect.

$$I_x = N\frac{e^2}{h}(E_y L_y) \quad (7)$$

$$R = \frac{V_y}{I_x} = \frac{h}{Ne^2} \quad (8)$$

The quantization of σ can be described quite precisely. Let a Landau gauge vector potential, given by Eq. (9), be adopted for describing the magnetic field. In this gauge, the Schrödinger equation for the x-y motion of the electron is Eq. (10). Its solutions are given by Eq. (11), where a_0, given by Eq. (12), is a characteristic distance called the magnetic length, and where y_0 is related to k by Eq. (13). The energy of the state is independent of the quantum number k but dependent on N in the manner given by Eq. (14). The set of states with the same N is called the Nth Landau level. When the Fermi level is raised above $(N + 1/2) \hbar\omega_c$, all the states in the Nth Landau level become occupied at once. This results in a discontinuous jump in the number of electrons trapped at the interface. The number is easily calculated. k must be given by Eq. (15) for integer M, and y_0 must lie inside the sample and therefore must satisfy notation (16). The number of states in a Landau level is therefore given by Eq. (17). This number is also $\sigma (L_x L_y)$.

$$\vec{A} = H_0 y \hat{x} \tag{9}$$

$$-\frac{\hbar^2}{2m} \left(\frac{\partial^2 \psi}{\partial x^2} + \frac{2ieH_0}{\hbar c} y \frac{\partial \psi}{\partial x} - \left(\frac{eH_0}{\hbar c} \right)^2 y^2 \psi + \frac{\partial^2 \psi}{\partial y^2} \right)$$
$$= E\psi \tag{10}$$

$$\psi = e^{ikx} e^{1/2(y-y_0)^2/a_0^2} 2 \frac{\partial^N}{\partial y^N} e^{-(y-y_0)^2/a_0^2} \tag{11}$$

$$a_0 = \sqrt{\frac{\hbar c}{eH_0}} \tag{12}$$

$$y_0 = a_0^2 k \tag{13}$$

$$E = \left(N + \frac{1}{2} \right) \hbar\omega_c \tag{14}$$

$$k = M \frac{2\pi}{L_x} \tag{15}$$

$$-\frac{L_y}{2} < y_0 < \frac{L_y}{2} \tag{16}$$

$$\text{Number of states} = \frac{L_x L_y}{2\pi a_0^2} = \frac{eH_0}{hc} (L_x L_y) \tag{17}$$

Experimental results. In a typical quantum Hall experiment, both the quantization of the Hall conductance and the absence of an electric field in the direction of current flow are observed. Figure 3 shows the results of von Klitzing's original measurements, and Fig. 3a shows the geometry of the experiment. A magnetic field of 18 teslas points out of the paper, the temperature is 1.5 K, and the length of the sample is 400 μm. A current of 1 μA is forced to flow between the source and drain. Voltages U_{pp} and U_H are plotted versus gate voltage V_g or Fermi level, in Fig. 3b. At certain values of the Fermi level, U_{pp}, the voltage drop in the direction of current flow, drops essentially to zero, as though the sample were conducting current with no resistive loss. Resistive loss is greatest when the Fermi level lies in a Landau level, since resistive scattering is most efficient when empty and occupied states coexist at the same energy. It can also be seen from Fig. 3a that the lossless conduction is accompanied by plateauing of the Hall voltage U_H. Because it plateaus, U_H can be measured accurately.

The ratio of this voltage to 1 μA gives a resis-

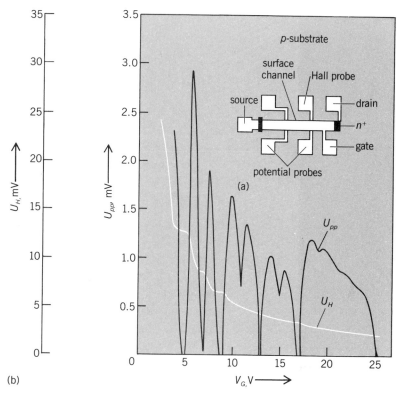

Fig. 3.   Original quantum Hall measurements. (a) Apparatus. (b) Results. U_{pp} is the voltage drop between the pads marked potential probes. U_H is the voltage drop between the right-hand potential probe and the Hall probe. V_g is the gate voltage, or Fermi level. (*After K. von Klitzing, G. Dorda, and M. Pepper, New method for high-accuracy determination of the fine-structure constant based on quantized Hall resistance, Phys. Rev. Lett., 45:494–497, 1980*)

tance, which is $1/N$ times 25,813 ohms, the value of h/e^2 in ordinary electrical units.

Quantum Hall effect in real systems. The extreme accuracy of the effect indicates that the explanation in terms of the quantization of σ is overly simplistic, for real systems contain imperfections which can cause the interface charge density not to be quantized. Figure 4 compares the density of states of an actual system to that of the idealized one discussed above. The density of states at energy E is just the number of states having energy E. In the absence of imperfections it consists of spikes at $(N + 1/2) \hbar\omega_c$. In the presence of imperfections it consists of a continuum with peaks at $(N + 1/2) \hbar\omega_c$, but with states at any energy. The interface charge density of such a system clearly does not change abruptly as the Fermi energy crosses $(N + 1/2) \hbar\omega_c$.

The states of a real system fall into two categories: extended states, those contiguous from one side of the sample to the other and capable of carrying electric current, and localized states, those not contiguous and not capable of carrying current. All the states in the ideal case are extended. In the real system a given region of the density of states is either completely localized or completely extended. Extended and localized regions are separated by a sharp boundary, called the mobility edge. The den-

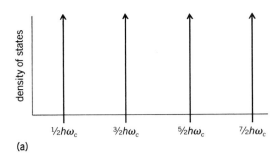

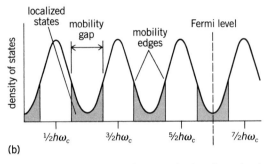

(a)

(b)

Fig. 4. Density of electronic states for two-dimensional metals in a strong magnetic field. (a) Ideal system, without imperfections. (b) Real system, with imperfections.

sity of states consists of extended regions centered at $(N + 1/2)\,\hbar\omega_c$ separated by localized regions, with sharp mobility edges delimiting the boundaries. The total number of states in an extended state band is sample-dependent. The quantum Hall effect occurs when the Fermi level lies in a region of localized states, as shown in Fig. 4b.

Systems with macroscopic imperfections. It is possible to understand why the Hall conductance is quantized, even though the carrier density is not, by imagining the imperfections to be caused by a

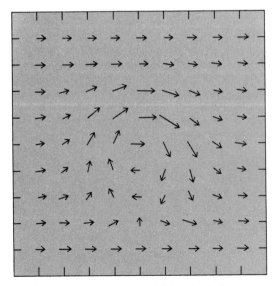

Fig. 5. Current density in the vicinity of a shallow imperfection in a quantum Hall experiment. Localized states contribute to the vortex current but not to the net flow from left to right.

slowly varying potential $\Delta V(x,y)$. All the states near local minima and maxima of ΔV are localized. However, since ΔV is macroscopic, a local version of the quantum Hall equation can be written, Eq. (18),

$$\vec{J} = \frac{Ne^2}{h}(\hat{z} \times \vec{E}) \qquad (18)$$

where $\vec{J}$ and $\vec{E}$ are the current density and electric field locally. This equation can be integrated across the sample, as in Eq. (19), to retrieve the original

$$I_x = \int J_x dy = \frac{Ne^2}{h}\int E_y dy = \frac{Ne^2}{h}V \qquad (19)$$

relation. Figure 5 shows the current density near a local minimum of ΔV. The vortex of current in this minimum is caused by the localized states. Since the vortex does not contribute to the net flow of current across the sample, whether or not these states are occupied with electrons is irrelevant. This is crudely why the Hall conductance is the same no matter where the Fermi level lies in a localized state band.

Systems with microscopic imperfections. That this is always the case, even when the imperfections are microscopic, was shown by R. B. Laughlin. His proof is based on a thought experiment in which the ribbon of two-dimensional metal is bent into a loop, as is shown in Fig. 6a. There is a current I flowing around this loop, a potential V between one edge and the other, and a magnetic field H_0 piercing the ribbon's surface. The Fermi level lies in a region of localized states, as is shown in Fig. 4b. Current is carried by the extended states below the Fermi level. There is no resistive loss at zero temperature because there are no current-carrying states at the Fermi level. The quantum Hall effect relates I to V. It is quite difficult to evaluate the current explicitly in the presence of imperfections, but these difficulties can be circumvented with a trick.

Suppose that a small amount of magnetic flux ϕ is slowly added through the center of the loop. According to Faraday's law of induction, the changing flux induces an electric field around the loop in the direction of the current satisfying Eq. (20). This elec-

$$\int_{\text{loop}} \vec{E} \cdot \vec{ds} = \frac{1}{c}\frac{\partial\phi}{\partial t} \qquad (20)$$

tric field does work on the moving electrons. Since these experience no resistive loss, energy is conserved, and thus the work is stored into the total energy U of the system as given by Eq. (21). Thus, the current can be related to the derivative of total energy with respect to flux, as in Eq. (22). Letting L denote the circumference of the loop, this flux derivative can be expressed as a derivative with respect to a uniform vector potential A pointing around the loop, as in Eq. (23).

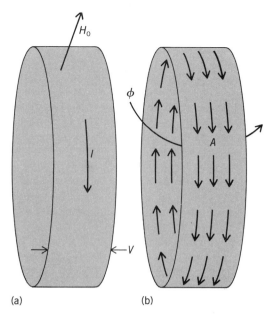

(a)   (b)

Fig. 6. Loop used in showing that the Hall conductance is quantized in systems with imperfections, showing (a) current I flowing around loop, Hall voltage V across loop, and magnetic field H_0 perpendicular to surface of loop; and (b) small amount of magnetic flux ϕ added through center of loop and corresponding vector potential A pointing around loop.

$$\frac{\partial U}{\partial t} = \frac{I}{c}\frac{\partial \phi}{\partial t} \qquad (21)$$

$$I = c\frac{\partial U}{\partial \phi} \qquad (22)$$

$$I = \frac{c}{L}\frac{\partial U}{\partial A} \qquad (23)$$

It is now easy to determine the effect of adding a small flux increment on the electrons in the sample. These electrons are described by a Schrödinger equation (24), where V is the potential caused by

$$\frac{1}{2m}\left(\frac{\hbar}{i}\vec{\nabla} - \frac{e}{c}\vec{A}\right)^2\psi + V\psi = E\psi \qquad (24)$$

the imperfections. If the system were not bent into a loop, then the addition of a vector potential increment $\Delta A\hat{x}$ would have no effect on it, since the solutions ψ' of the new Schrödinger equation (25) would be simply the solutions to the old one multiplied by a phase factor, as given by Eq. (26).

$$\frac{1}{2m}\left(\frac{\hbar}{i}\vec{\nabla} - \frac{e}{c}(\vec{A} + \Delta A x)\right)^2\psi' + V\psi'$$
$$= E\psi' \qquad (25)$$

$$\psi' = \psi e^{i(ex/\hbar c)\Delta A} \qquad (26)$$

This property of the Schrödinger equation and its solutions is called gauge invariance. Gauge invari-

ance is simply a formal statement that vector potentials describing the same magnetic field are physically equivalent. The vector potential increment in the loop is not a gauge transformation, since it corresponds to a physical change of the magnetic field threading the loop. Since this magnetic field is zero where the electrons actually are, its presence is manifested subtly in the equations. If one attempts, in the case of the loop, to multiply ψ by a phase factor, as in Eq. (26), with x denoting the coordinate around the loop, then, since going a distance L takes one around the loop to the starting point, Eq. (27) must be satisfied.

$$e^{i(eL/\hbar c)\Delta A} = 1 \qquad (27)$$

The new and old wave functions cannot generally be related to one another by a gauge transformation unless ΔA satisfies Eq. (27). However, if ψ is localized, then it becomes zero somewhere around the loop, and Eq. (27) does not apply. This leads to an important principle: The addition of any ΔA looks like a gauge transformation to a localized state. This is important because it demonstrates simply that localized states carry no current. In order for a state to carry current, its energy must change with the addition of ΔA. Gauge transformations leave the energy of a state invariant. If ψ is not localized, then the addition of ΔA does not correspond to a gauge transformation. Such a state responds to the addition of ΔA in a different manner. For example, if there are no impurities, one can see from Eq. (9) that the addition of ΔA simply redefines the origin of y. Therefore ψ responds by moving a distance $-(\Delta A/H_0)$ in the y direction. This changes U because of the presence of an electric field in the y direction. The contributions to ΔU from each electron can be summed to calculate I. However, a simpler way is to increase ΔA adiabatically until Eq. (27) is satisfied. When this occurs, a total flux through the loop of $\Delta \phi = hc/e$ has been added. This is the condition for being able to perform a gauge transformation. Thus the new states must be just the old ones, up to a phase factor. Accordingly, each state ψ has simply moved over to occupy the position of its neighbor. The net result is the transfer of exactly one electron per occupied Landau level from one edge of the ribbon to the other. The current is therefore given by Eq. (28).

$$I = c\frac{\Delta U}{\Delta \psi} = c\frac{(NeV)}{hc/e} = \frac{Ne^2}{h}V \qquad (28)$$

However, this reasoning also works in the presence of imperfections. The system is physically identical before and after the addition of $\Delta \phi$. Also, as in the ideal case, there is an energy gap across which the electrons cannot be excited if the addition of $\Delta \phi$ is adiabatic. This is the mobility gap, the region between the highest occupied and the lowest unoccupied delocalized state. The mobility gap functions as a real gap because the states in it are not mixed with one another or with extended states with the addition of $\Delta \phi$. Since electrons cannot be excited across this gap, the only possible excitations which can result are the charge-transfer variety discussed in the ideal case. Thus Eq. (28) is always true, for some ineger N, whenever the Fermi level lies in a mobility gap.

Understanding of the quantum Hall effect may be summarized in the following way. The quantization of the Hall conductance is a direct consequence of gauge invariance and the existence of a mobility gap at the Fermi level. The quantum of Hall conductance is related, not to a quantum of surface charge density, for there is no such quantum, but to the quantum of electric charge itself.

For background information *see* BAND THEORY OF SOLIDS; DE HAAS–VAN ALPHEN EFFECT; ELECTROMAGNETIC INDUCTION; HALL EFFECT; NONRELATIVISTIC QUANTUM THEORY; PARTICLE ACCELERATOR; POTENTIALS in the McGraw-Hill Encyclopedia of Science and Technology.

[ROBERT B. LAUGHLIN]

Bibliography: R. B. Laughlin, Quantized Hall conductivity in two dimensions, *Phys. Rev.*, B23:5632–5633, 1981; F. Stern, Self-consistent results for *n*-type silicon inversion layers, *Phys. Rev.*, B5:4891–4899, 1972; K. von Klitzing, G. Dorda, and M. Pepper, New method for high-accuracy determination of the fine-structure constant based on quantized Hall resistance, *Phys. Rev. Lett.*, 45:494–497, 1980.

Herpes

The herpesviruses are ubiquitous among the viral agents of human diseases. During recent years genital infections caused by the herpes simplex viruses have reached epidemic proportions, and it is estimated that more than 20 million are affected in the United States alone. Because of spontaneous recurrences of the infection, its effects on newborns, its relationship to cancer of the cervix, and the difficulty of controlling the infections, genital herpes is one of the most important sexually transmitted diseases.

Virology. The Herpesviridae family consists of structurally similar enveloped, icosahedral, linear double-stranded DNA viruses which average 180–200 nanometers in diameter and 80–100 million daltons in molecular weight. The viral complex consists of elemental materials arranged in concentric layers. The core consists of DNA coiled in the form of a doughnut, with proteins arranged as a bar bell passing through the hole. The core is surrounded by additional protein layers collectively known as the nucleocapsid. This in turn is surrounded by two outer elements, the tegument (a granular zone composed of globular proteins) and an encompassing envelope. While deficient in viral transcriptase, the host cell nucleus must serve as the exclusive source for transcription of viral DNA, thus necessitating a nuclear phase of replication.

Following viral RNA synthesis, migration into the cell cytoplasm occurs, and the protein synthesis in the host cell is taken over. The newly constructed viral proteins are then transported back into the nucleus of the host cell, where viral nucleocapsids are assembled. Completion of replication occurs with acquisition of the viral envelope from the inner nuclear membrane of the host cell, followed by transfer of the viral particles into the cytoplasm, where they leave the host cell.

The herpesviruses that frequently produce disease in humans include herpes simplex and herpes zoster. Herpes simplex virus type I (HSV-I) most commonly produces infection above the waist, and is usually the cause of fever blisters, or cold sores. Herpes simplex virus type II (HSV-II) is the most common cause of infections below the waist, including 80–90% of genital herpes infections. Although not related to genital infections, three other members of the herpes group of viruses commonly produce disease in humans: the varicella-zoster virus (VZV), the cytomegalovirus (CMV), and the Epstein-Barr virus (EBV). In children, VZV causes varicella (chickenpox), while in previously sensitized adults the same virus produces herpes zoster (shingles). These diseases are not sexually transmitted, even though they may be transferred by direct contact. CMV, which is sexually transmitted, may affect both newborns and adults. Although minimally symptomatic in most healthy adults, a progressive form of pneumonitis may sometimes be fatal in the immuno-compromised individual. EBV, which is transmitted by close or intimate contact, is the cause of infectious mononucleosis.

Latency. Specific characteristics of the herpesviruses that have eluded research and clinical investigations are their neurotropic affinities and their unique ability to remain latent within certain nerve tissues to produce recurring infections. Typically, HSV-I may be found latent in nerve tissues of the cervical or trigeminal ganglia, while HSV-II are usually found in the sacral ganglia which innervate the perineal and genital areas. During these inactive or latency phases, little is known about the metabolic activities of the viruses, and their presence can be detected only by tissue culture techniques. Exactly how latency becomes established remains unanswered, and which cellular or immune factors control latency is also not known. The questions of whether the inactive virus is "switched on" (static theory) or whether the local suppressive mechanisms periodically fluctuate (dynamic theory) also remain unanswered.

Immunology. The immune factors associated with herpesvirus infections are not completely understood. Intercellular infection causes the fusion of adjacent cells, allowing the virus to travel freely between the fused cells and impeding the circulation of humoral immune components when infected host cells aggregate with uninfected cells. Only after viral particles are released from the host cells do immune responses develop (although levels of immunoglobulins are unrelated to the severity of infection), and viral-induced cell surface antigens cause specific antibodies to be activated. It is surmised that host cell–mediated immune responses play an important role in controlling the herpesvirus infections. Clinically, individuals most at risk are those having defects in their cellular immune responses, while those with subnormal levels of immunoglobulins are somewhat less at risk for severe infections. The role of immune responses, cellular immunity, reactivity to viral antigens, response regulators, and T-cell functions are currently under investigation.

Epidemiology. The incidence of genital herpesvirus infections varies in different population groups. However, during the past two decades there has been a tremendous increase in the numbers of affected individuals. Unlike other sexually transmitted diseases, such as syphilis and gonorrhea, which are reportable, precise statistics relating to genital herpesvirus infections are unknown. Collected data from major research institutes and from the Centers for Disease Control estimate that 300,000 to 500,000 new cases occur annually, and that perhaps 9–35% of the population in the United States (more than 20 million individuals) have been exposed to the genital herpesviruses. Of adults in higher socioeconomic groups, 30–50% have antibodies to HSV, compared to 80–100% in lower socioeconomic groups. The prevalence of HSV antibodies increases with age. A significant uptrend in genital herpes has been related to changing sexual practices and increased numbers of sexual partners, especially in upper-middle-class groups. Overall, 1–2% of all women attending gynecology clinics have cytologic evidence of HSV infection.

Clinical features. Genital herpesvirus infections may be divided into two distinct clinical categories: primary infections and recurring infections. The primary infection is immunologically distinct, demonstrating no prior antibodies. Following infectious contact, symptoms may appear within 2–21 days, the average being 5–7 days. The average duration of lesions ranges from 15 to 20 days, while viral shedding may persist for 8–10 days. With those persons who become symptomatic (between 50–75% remain asymptomatic), compared to those with recurring infections, the primary infection tends to be more severe, to last longer, to include systemic symptoms, and to produce more pain; and it usually involves the genitals bilaterally.

Subsequent to primary infection, measurable specific antibody levels may be detected 3 weeks after onset of disease (see table). Initial clinical transmission of the herpesviruses most often occurs by means of intimate contact, causing viral penetration into mucous membranes or through damaged epidermal surfaces, although autoinoculation from the person's own infected skin may lead to inadvertent genital infection. Oral-genital sexual practices produce genital transmission involving HSV-I from active cold sores or fever blisters. The role of fomites (toilet seats, towels, clothing, and so on) in causing genital infections appears to be unimportant. Although clinical variability exists, the characteristic herpes lesions, once past the initial blister or vesicle stage,

Clinical differences between primary and recurrent genital herpesvirus infections

Signs or symptoms	Primary	Recurrent
Incubation time	5–7 days	12–48 hours
Duration of lesions	15–20 days	3–10 days
Duration of viral shedding	8–10 days	3–5 days
Duration of pain	8–12 days	3–7 days
Local pain	Moderate to severe	Mild to slight
Flulike symptoms (viremia)	Common	Rare
Inguinal adenopathy	Common	Unusual
Number of lesions	Multiple	Few or singular
Size of lesions	Variable; 2–15 mm	Small; 1–5 mm
Location of lesions	Bilateral	Usually unilateral
Specific antibody titer	Absent	Present

appear as small superficial ulcerations, superimposed upon an erythematous base.

Recurrent genital herpesvirus infection occurs in 50–60% of infected individuals following primary infection. The recurrent lesions are often inconspicuous and difficult to identify. Local symptoms (prodrome) may be experienced by the individual prior to an outbreak. Descriptions include itching, burning, tingling, throbbing, sensitivity, or a neuralgia-like pain. The genital lesions are vesicoulcerative and usually develop unilaterally in small localized patches. Vesicles rupture rapidly on mucosal surfaces, while on epidermal skin surfaces they persist longer, followed by crusting over until complete healing occurs (see illustration).

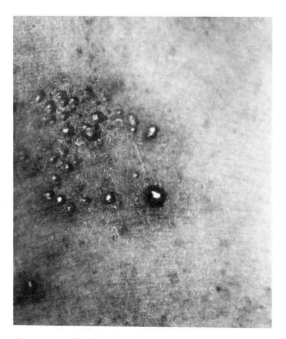

Recurrent genital herpesvirus infection of the upper inner buttocks. Grouped vesicles typify cutaneous lesions which rupture and then crust over until healing is complete.

Certain precipitating factors may place a person at greater risk for recurring infections. Herpesvirus reactivation may be associated with emotional or metabolic disturbances, premenstrual changes, systemic diseases, fever, sunlight, and local genital infection or trauma.

Diagnosis. The most specific diagnostic test is viral isolation, utilizing tissue culture techniques. Though less accurate, cytological tests such as the Papanicolaou smear or the Tzanck preparation may confirm the diagnosis. On the sole basis of the clinical pattern, the experienced practitioner may be highly accurate in diagnosing the infection. Routine antibody titers are of limited value in establishing the diagnosis due to the immunologic cross-reactivity between the herpesviruses.

Importance during pregnancy. Women at risk for genital herpes during pregnancy, although asymptomatic, are tested frequently during the last month to prevent inadvertent delivery through an infected birth canal. In the presence of active herpesviruses, the risk of infection to the neonate is approximately 50%. Where neonatal herpes infection occurs, the majority of newborns ultimately develop disseminated or central nervous system disease, and the mortality rate approaches 50%. Of those surviving, approximately 80% may be permanently neurologically damaged. If required, appropriately timed abdominal delivery can prevent neonatal virus transmission and infection.

Relationship to cancer. The herpesviruses are known to be mutagenic agents in tissue cultures, and those women experiencing genital herpesvirus infections have a five to eight times higher risk for developing cervical abnormalities. Prevention of dysplastic changes is best accomplished by recommending cytologic cervical smears twice yearly. Recent evidence has also demonstrated a similar association of genital herpes with cancer of the vulva.

Management and therapy. There exists no guaranteed medical cure for genital herpesvirus infections. Recently developed antiviral compounds, such as acyclovir, exert temporary control and lessen the duration of clinical infections; however, they may not prevent future recurrences. The other available antiviral agents are ineffective for genital herpesvirus infections. Because of viral metabolic and replication factors, other forms of therapy, including vaccines, immunopotentiators, vitamin and dietary therapies, various locally applied agents, and diverse and varied home remedies, have been unsuccessful. Management consists basically of genital hygiene, prevention or therapy of associated genital infections, and treatment of discomfort or pain. Negative emotional attitudes may be minimalized by professional support groups or individual stress-reduction techniques.

For background information *see* ANIMAL VIRUS; CELLULAR IMMUNOLOGY; VENEREAL DISEASE; VIRUS INFECTION, LATENT, PERSISTENT, SLOW in the McGraw-Hill Encyclopedia of Science and Technology.

[JEROME M. EDER]

Bibliography: L. Corey, The diagnosis and treatment of genital herpes, *J.A.M.A.*, 248:1041–1049, 1982; H. L. Gardner and R. H. Kaufman, *Benign Diseases of the Vulva and Vagina*, 1981; J. H. Grossman, Herpes simplex virus infections, *Clin. Obstet. Gynecol.*, 25:555–561, 1982; A. J. Nahmias, W. R. Dowdle, and R. F. Schinazi, *The Human Herpesviruses: An Interdisciplinary Perspective*, 1980.

Human-factors engineering

Each optical orbit, commonly known as the eyeball, has movements which are controlled by three complementary muscle pairs: the medial rectus and lateral rectus for horizontal movement, the superior rectus and inferior rectus for vertical movement, and the inferior oblique and superior oblique for torsional movements. These muscles form the greatest musculature per weight of any movable organ of the body. As a consequence, the eyes are capable of moving over great latitudes with high velocity—up to 900°/s.

Eye movements generally function to (1) bring information from peripheral visual areas to the center of vision, at the fovea, for more in-depth examination (saccades); (2) hold the visual scene at the fovea when the scene or viewer is in motion (pursuit); (3) serve a combination of the following/holding, reflexing/compensating sequences (nystagmus); and (4) maintain clear vision with approaching or receding targets (vergence). There are also fine tremors whose purpose is not yet well understood, and data which indicate that if there is no eye movement, the visual scene disappears.

Saccades. By far the most studies eye movements are saccades. These occur about three or four times each second, or up to 230,000 times each waking day. Each saccade lasts about 20–30 milliseconds with an intermovement or fixation time of 200–400 ms. Generally, the larger the saccade, the greater the velocity, although most saccades span less than 15° of the visual angle and infrequently reach velocities of about 400°/s. Saccades are considered ballistic in that once executed, they must reach the target before any correction can be made. Saccadic extent, that is, distance traveled, frequency, and associated fixation durations seem variable and flexible, depending on scenic cognitive aspects, such as meaning and meaningfulness, and perceptual aspects such as size, shape, and density of targets. Studies have shown that when a subject reads, the primary eye movements are the saccades. Research has offered a plethora of new findings regarding the contribution of the reader, the author, and the text to understanding.

It is not likely that successive saccades occur in less than 150 ms, since it takes about 50 ms to program the next saccade during a fixation; 20–30 ms to execute it; 50 ms to regain clear acuity; and a minimum of about 50 ms to acquire the new visual scene for assessment, interpretation, and integration with previous scenes. Regularity of text, positioned in successive horizontal arrays, has contributed to interest in reading dynamics, since regularity allows for more precise specification of content than is possible in the seemingly more haphazardly arrayed scenes and pictures.

Collecting eye-movement data can be a formidable task. During the 1930s and 1940s, the most commonly used methods consisted of following traces of corneally reflected light patterns on movie film or intricately measuring blips and dashes on strip-chart recordings. Recording the eye movements during reading for 4 or 5 min could mean weeks and months of detailed plotting and measuring. With the advent of digital computers, the same assessment can now either be online or is completed minutes after the task. Since the mid-1970s, the analysis of eye movements during reading has taken two separate paths: microanalysis and macroanalysis.

Microanalysis. Research efforts using microanalysis are generally aimed at the precise specification and designation of where a saccade lands during text reading. Recording systems are generally accurate to within 1 to 2 minutes of arc of visual angle, which corresponds to the ability of telling where the reader is looking on the letter H, in normal text at a distance of 14–17 cm. Researchers involved in the microanalysis of reading are interested in which words and parts of words are fixated between saccades, in the precise distance between saccades, and in the number of letters to words to the left or right of fixation which determine saccade length and fixation duration. More recent research efforts seem to indicate that little, if any, word identification or semantic information is available beyond a few letter spaces to the left or right of fixation. Both identification of words and semantic analysis seem to be performed during direct fixation at the fovea. Other examples of saccadic specifications are: fixations are more likely to be focused on the centers of words rather than on their ends; fixations are not likely to land on short words unless a short word is to the left of a longer word; and fixations are not likely to fall in regions between sentences.

Other microanalysis data indicate that fixations are longer in duration on low-frequency and technical words, and also when the text to the right of the fixation is blocked by a mask of some sort. The first fixation on a line is longer than average, while the last fixation on the line is generally shorter than average. Saccades generally tend to bring the eye close to a convenient viewing position in words of five or six letters, so that the entire word may be viewed. In large words like hippopotamus more than one fixation would probably be necessary for readers to process the entire word, and would certainly be necessary for beginning readers. Saccadic extent increases when easy text is read, or when a long word lies to the right of a fixated word.

One of the more commonly used instruments for the study of the microanalysis of eye movements is an eye tracker (Fig. 1). Generally, stimuli are presented on a cathode-ray-tube display in either short sentences or in longer, single sentences. Because of

Fig. 1. Subject with head positioned in eye tracker viewing an acuity target. (*Courtesy of Hewitt Crane, Stanford Research Institute International, Menlo Park, CA*)

the precision required, the subject's head is immobilized by a bite board, a chin rest, and a head rest, generally providing an unnatural position for viewing, especially for reading. The instrument, however, is used to determine only where and for how long the eyes land on particular regions of the text. These types of eye-tracker systems combined with high-speed digital computers are capable of monitoring eye movements with a resolution of milliseconds in time, and then accumulating clusters of millisecond samples, in which the eye fixates on a given area in a process called a fixation.

Macroanalysis. Macroanalysis research into text reading tends to be less concerned about which letter is being viewed at a particular time; it concentrates rather on the pattern of fixations and words fixated. Macroanalyses address more global questions, such as do text variations, constraints on typeface, or the level of text difficulty change about fixation and saccadic characteristics. Traditionally, reader flexibility has been demonstrated by the changes in eye-movement dynamics which occur when a subject reads a novel for pleasure as opposed to a chemistry or physics textbook. Where reading material becomes difficult, fixation duration increases and saccades themselves become much more closely spaced. As text becomes easier, the fixation duration decreases to about the average (240 ms) and the extent of saccades increases up to eight letter spaces. More recently, however, reader flexibility within text has been defined. Dramatic changes in eye movements

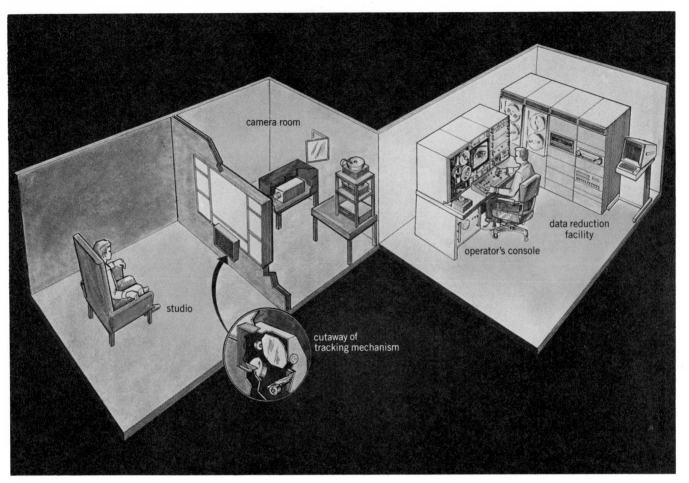

Fig. 2. Schematic of oculometer showing subject studio, camera projection room, and data reduction facility at the U.S. Army's Human Engineering Laboratory, Aberdeen Proving Ground, MD. (*From R. H. Lambert, R. A. Monty, and R. J. Hall, High speed data processing and unobtrusive monitoring of eye movements, Behav. Res. Meth. Instrum., 6:525–530, 1974*)

were found within text or paragraphs, meaning that readers are capable of varying eye-movement dynamics in order to identify sections of text, such as topical sentences and phrases, or those propositions which are of more importance than other sections. When readers identify sections of the text that they believe the author considers to be important, eye-movement dynamics change dramatically. If new and unfamiliar information is read, again the saccades occur quite close together and fixation durations lengthen. If readers are familiar with the new information, but also consider it to be important, saccadic extent as well as fixation duration may lengthen. Reader flexibility seems to be useful in the prediction of good and poor readers; poor readers are much less flexible. They tend to maintain the same saccade and fixation characteristics, irrespective of the text, part of the text, or value placed on the text.

Most macroanalysis recording systems are video-based, and 16 frames per second can be recorded. Their accuracy level is about 0.5–1° (corresponding to three to five letters), which is much lower than the microanalysis systems. However, they have much greater capabilities for variation in types and amounts of materials that can be presented to the subject.

For example, full texts of 2000 to 3000 words are not uncommon in macroanalysis recording sessions. Large texts are preferable because they allow the assessment of perceptual features, such as typeface change and line width, which are likely to lead to changes in central or foveal viewing as well as peripheral viewing. For example, the change to italics causes a change in the eye movement dynamics, and greater line width seems to enhance reading speed, probably because successive saccades give overlapping or multiple views of text segments, which aids recognition and understanding. The use of larger texts also allows the assessment of context effects that occur between paragraphs or within paragraphs, for example, when the reader takes extra time on a topic sentence but reads supportive, consistent subordinate paragraphs very rapidly; and the assessment of individual differences between two readers who exhibit very different eye movement characteristics but have a good understanding of the text. In addition, there is little or no constraint placed on the subject, and so when the reader chooses to halt in order to analyze text, macrosystems can identify the location where the reader left off.

An oculometer (Fig. 2) is a macroanalysis recording

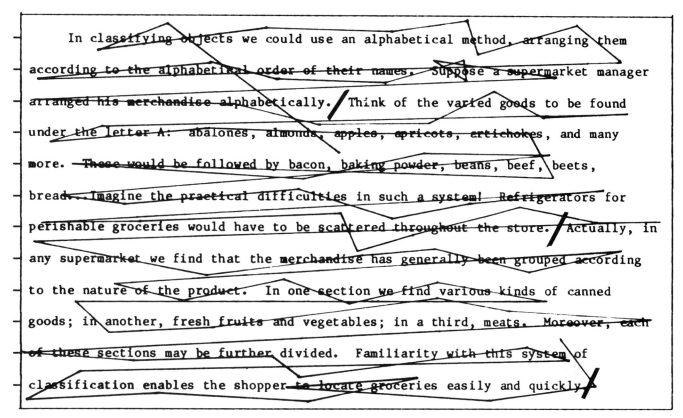

Fig. 3. Recorded eye-movement scan pattern superposed on text. (*From W. L. Shebilske and D. F. Fisher, Eye movements reveal components of flexible reading strategies: Thir-* *tieth Yearbook of the National Reading Conference, in M. L. Kamil, ed., Directions in Reading: Research and Instruction, 1981)*

system. The video camera is placed beneath the screen, and the subject is seated comfortably approximately 2 m (6 ft) from the screen. The screen is used to present any type of visual information, such as text, TV filmage, or pictures, and there are no physical restraints placed on the subjects. In fact, it is not likely that the subjects are aware that their eyes are being monitored.

While macroanalysis systems are less accurate in defining the precise location of a fixation than microanalysis systems, each system has its own limitations and set of questions. A typical macroanalysis scan pattern is shown in Fig. 3. Each discontinuity or angle found in the horizontal line represents a fixation. Here, the fixation scan pattern has been superposed on the text. The dark lines within the text are added later, and separate portions of text are designated as more or less meaningful, that is, conveying more important or less important information. Analysis indicates that readers spontaneously changed eye-movement characteristics to match the meaningfulness of the material displayed.

Theoretical issues. All micro- and macrolevel analyses of texts seem to verify one of four possible purposes for eye-movement sequencing. First, there is the possibility that eye movements are necessary to maintain vision; that is, without eye movements, receptors fail to respond, as though they fatigue, and scenes and pictures fade from view. Second,

eye movements and their fixations might well be re sponsible for perception; that is, they seem to exhibit a psychomotor program that is directly related to a particular picture or scene that is viewed. Third, eye movements and their fixations may well be directed by perception; that is, because fixations seem to fall in locations containing high levels of information, such as contours, objects, and shadings, it is believed that some overall semantic content directs the eye-movement pattern. Fourth, the eye-movement sequence seems to provide a means for encoding, storing, and subsequently reconstructing the successive retinal images of the detail found in a scene, text, or picture; that is, individuals provide idiosyncratic eye-movement patterns as a reflection of their own viewing and interpretation of the scene. There is also the possibility that at one time or another all four of these alternatives are valid.

For background information *see* EYE (VERTEBRATE); HUMAN-FACTORS ENGINEERING; VISION in the Mc-Graw-Hill Encyclopedia of Science and Technology.

[DENNIS F. FISHER]

Bibliography: D. F. Fisher, R. A. Monty, and J. W. Senders (eds.), *Eye Movements: Cognition and Visual Perception*, 1981; K. Rayner (ed.), *Eye Movements in Reading: Perceptual and Language Processes*, 1982; E. A. Taylor, *Controlled Reading*, 1937; A. L. Yarbus, *Eye Movements and Vision*, 1967.

Human genetics

Studies of the possible genetic consequences of exposure to ionizing radiation from atomic bombs were initiated in 1947 and have continued uninterruptedly since that time. They were prompted by the demonstrations more than a half century ago of the Nobel laureate H. J. Muller that ionizing radiation increases the frequency of transmissible changes in the genetic material, termed mutations. Muller showed that most of these mutations are harmful to the offspring of the exposed persons. In the 35 years since the inception of these studies, further insights into the origin of new mutations have been gained. Four different approaches have been developed to measure radiation-related genetic damage among the offspring of the survivors of the atomic bombings: clinical assessments of health, mortality surveillance, biochemical studies, and cytological evaluations.

Clinical assessments of health. The first steps toward a continuous surveillance of the children born in Hiroshima and Nagasaki subsequent to the atomic bombings were taken in 1946, and a full-scale program was initiated in 1948. The latter rested on a provision within the rationing system then existing in Japan, which entitled pregnant women, who registered their pregnancies after the fifth lunar month, to have access to supplementary rations. The vast majority of eligible women reported their pregnancies. This made it possible to identify in these cities most, if not all, of the pregnancies with a gestation period of at least 20 weeks, and to examine the outcomes of these pregnancies. These clinical observations were supported by an autopsy program which sought to study as many stillborn infants or infants that died in the first few days of life as possible. A second examination studied 20% of surviving infants 8–10 months after their birth.

The indicators of possible genetic effects which could be gleaned from such an examination program—all, of course, potentially influenced by a variety of extraneous factors—were sex, birth weight, viability at birth, presence of gross malformations, occurrence of premature death, and physical development at the age of 8–10 months. From 1948 through 1953, as part of this clinical program, 70,082 pregnancy outcomes were studied. The clinical findings are summarized in illustration *a*, which includes pregnancies which terminated in a child who had a major congenital defect (illustration *b*), who was stillborn (illustration *c*), or who died during the first week of life (illustration *d*). The vertical axes express the percentage of untoward outcomes based on the total number of pregnancies, and the horizontal axes are the radiation exposures received by the child's father and/or mother, expressed in rads. To facilitate this presentation the individual exposures have been grouped into four categories, namely, less than 1 rad, 1–9 rads, 10–99 rads, and 100 rads or more. Thus, there are only 16 possible combinations of parental exposure, and the outcomes associated with each of these alternatives are placed, in the illustration, at the point which constitutes the average pa-

rental exposure within a particular category. The outcomes of the "100 rads or more" group are placed on the "100 rads" lines. Clearly, the percentage of untoward outcome does not increase or decrease in a simple manner as exposure varies.

Etiologically, not every major congenital defect, stillbirth, or premature death is attributable to transmitted or new genetic factors; nongenetic causes loom large as well. Thus, these various pregnancy outcomes are indicative of genetic damage only insofar as they are genetic in origin, and this proportion is a matter of conjecture. Genetic theory does predict an increase in these untoward events proportional to the radiation dose received by the parents because of the induction of mutations with deleterious effects. After concomitant variables which are known to influence the outcome of pregnancies, such as parental ages and inbreeding, were taken into account, it was found that the events pictured in the illustration increase in frequency with increasing parental dose, although not in a statistically significant way. However, there is no real reason to doubt that mutations were produced in Hiroshima and Nagasaki, and these data can be used to estimate how large an increase in mutations could have arisen and yet have gone undetected statistically. It has been shown that this increase could not have been so large as to double the frequency of the untoward pregnancies.

Mortality surveillance. Originally the cohort of offspring of survivor parents in whom the exposure of both was known, and of comparable nonexposed parents, whose survival status is under continuous scrutiny, consisted of 52,621 live, single births occurring between May 1946 and December 1958 in Hiroshima and Nagasaki. Recently 22,984 births occurring from January 1959 through December 1980 have been added, of which 11,196 are from parents of known exposure status. It should be noted that most of these latter parents were too young at the time of exposure to have had children when the original cohort was defined. Deaths among these individuals are ascertained through periodic examination of the household censuses, the *koseki*, required by Japanese law. A copy of the death schedule is obtained on all individuals who have died since their *koseki* was last perused.

Among these 63,817 individuals of whom the exposures of both parents are known, there have been 3786 deaths (3552 in the original cohort and 234 in the extension). Cumulative mortality in this cohort does not increase significantly with parental exposure, although it does increase. Given the frequently cited relationship of mutation (somatic as well as germinal) to cancer, it is interesting to note that only 72 of the 3552 deaths in the original cohort were ascribed to malignancies. The most common of these latter causes of death was leukemia; indeed, 35 of the 72 deaths were attributed to this specific cancer. However, there is as yet no unequivocal relationship of exposure to deaths due to leukemia.

Chromosomal abnormalities. Efforts to assess radiation-related chromosomal damage in the survivors and their offspring began as early as 1948, but these early attempts were not successful. Reliable information had to await the development in the late 1950s of the use of short-term cultures of the white cells in peripheral blood and other new methods. These advances have made the establishment of the frequency of chromosomal aberrations possible on a scope not practicable with earlier methods. Accordingly, on the basis of a pilot study conducted in 1967, an investigation of the children of exposed parents was initiated in 1968, the subjects being drawn from the cohorts established for the mortality surveillance program just mentioned. Since the age of the youngest children in this study was 13 and of the oldest 34 in 1980, the survey will not yield adequate data on the frequency of chromosomal abnormalities associated with increased mortality rates. The data on sex-chromosome abnormalities and on those rearrangements of the autosomal chromosomes which lead to no loss or gain in genetic material, the so-called balanced structural rearrangements, should, however, be relatively unbiased.

Recently it has been reported that the frequency of sex-chromosome aneuploids (individuals with the incorrect number of sex chromosomes) among the children of parents exposed beyond 2400 m (7870 ft) from the hypocenter (the spot immediately beneath the burst location) is 13/5058 or 0.00257. The frequency among children of parents exposed within 2000 m (6560 ft) is 16/5762 or 0.00278.

Biochemical studies. Electrophoretic techniques for the identification of abnormal protein molecules have made possible a new approach to an assessment of the genetic effects of atomic bomb exposure. This approach, like the cytogenetic one, is free of many of the ambiguities inherent in the study of population characteristics whose genetic bases are unclear, and as a result has been vigorously advocated. It is argued that a mutation at a locus which controls a given enzyme may result either in a change in the structure of that enzyme, possibly reflected in the net charge on the enzyme molecule and hence recognizable electrophoretically, or in a change which alters the catalytic efficiency of the enzyme, or in the amount of enzyme produced. In the latter event, a deficiency in the enzyme can arise. Mutations of this variety are often referred to as null mutants or deficiency variants.

The subjects being studied are drawn from the cohorts identified for the mortality study previously described. Each individual is examined for rare electrophoretic variants of 28 proteins of the blood plasma and erythrocytes. Since 1979, a subset of the children has been examined further for deficiency variants of 10 of the erythrocytic enzymes. A rare electrophoretic variant is defined in this context as one with a frequency of less than 2% in the population, and an enzyme deficiency or low-activity variant as one which results in an activity level three standard deviations below the mean for the enzyme in question (or less than 66% of normal activity). When either variant is encountered, its occurrence is first verified, and then blood samples from both parents are examined for the presence of similar variants. If the variant is

not also found in one or both parents, and a discrepancy between putative and biological parentage is improbable, then it presumably represents a mutation.

Recently it has been estimated that information on the equivalent of 419,660 locus tests on children born to parents exposed within 2000 m of the hypocenter are now available. Two probable mutations have been observed: a slow-migrating variant of the enzyme glutamate pyruvate transaminase, and a slow-migrating variant of phosphoglucomutase-2. One probable mutation has been seen in the equivalent of 282,848 tests on children whose parents were either so far removed from the hypocenter or so effectively shielded that they in essence received no ionizing radiation, and one mutation where the parents received some irradiation but in an amount substantially less than the average for those parents exposed within 2000 m. As yet, the data on enzyme deficiencies or low-activity variants are too preliminary to provide much insight into the mutational process.

Doubling dose. A convenient and frequently used measure in the assessment of radiation risk is the doubling dose. This is the amount of radiation needed to double the natural incidence of a genetic or somatic event, in this case a mutation. This amount is difficult to ascertain, for estimates of the spontaneous or natural rate of mutation in humans are still quite poor. Generally indirect estimates must be used, and these have many weaknesses, including: the uncertain overall representativeness of the array of mutation-rate estimates currently available; the uncertain number of genetic loci which may actually be involved in a particular disease on which an estimate is based; and the proportion of such events as major congenital malformations which are genetic in origin. Nonetheless, it is still possible to estimate the doubling dose, although such estimates may incorporate large errors. For these estimates, however, exposures must be expressed quantitatively. The units commonly used in this quantification are the rad (the energy imparted to matter by ionizing radiation per unit mass of material, for example, the skin or the gonads) and the rem (a comparative measure which represents the quantity of a particular radiation, for example, neutrons, that is equivalent in biologic damage to 1 rad). Since the radiation spectrum released by an atomic bomb includes many different energetic particles as well as gamma rays, exposures are most conveniently represented in terms of the rem, but this requires at least a measure of the biologic effectiveness of neutrons to produce mutations as compared to gamma rays. Experimental evidence suggests this latter proportion to be about 1 to 5; that is, 1 rad of absorbed neutron energy is as effective in producing mutations as 5 rads of gamma rays.

Although the biochemical data are still too sparse to provide a meaningful estimate of the doubling dose, based on the foregoing considerations and the Hiroshima and Nagasaki experience, a doubling dose can be estimated for untoward pregnancy outcomes, for mortality ascertained through the surveillance program, and for sex-chromosome abnormalities. These

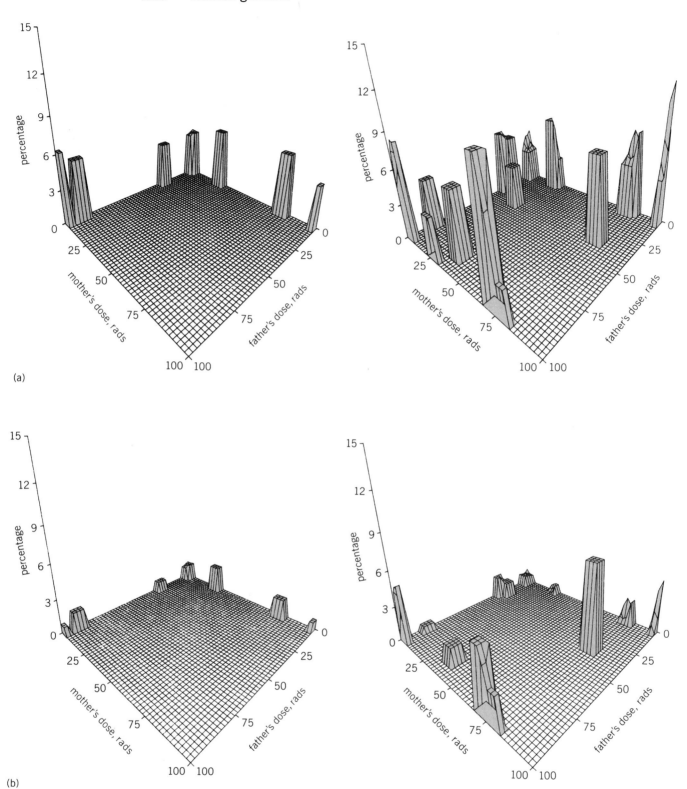

(a)

(b)

Genetic consequences of radiation exposure in Hiroshima or Nagasaki. (a) Summary of clinical findings—the untoward pregnancy outcome. (b) Pregnancies which terminated in congenital malformations, (c) in stillbirths, and (d) in neo-natal deaths (1–7 days after birth). In each set, left and right parts refer, respectively, to the experience when only one parent was present and when both parents were present.

estimates are, respectively, 69, 171, and 535 rems. A single estimate can be derived from these three by multiplying each by the inverse of its variance (a measure of uncertainty), and dividing the so-weighted sum of the estimates by the sum of the weights (the inverses). This gives rise to a value of 139 rems with a standard deviation of 157. To provide perspective to this estimate, it should be noted that the Committee

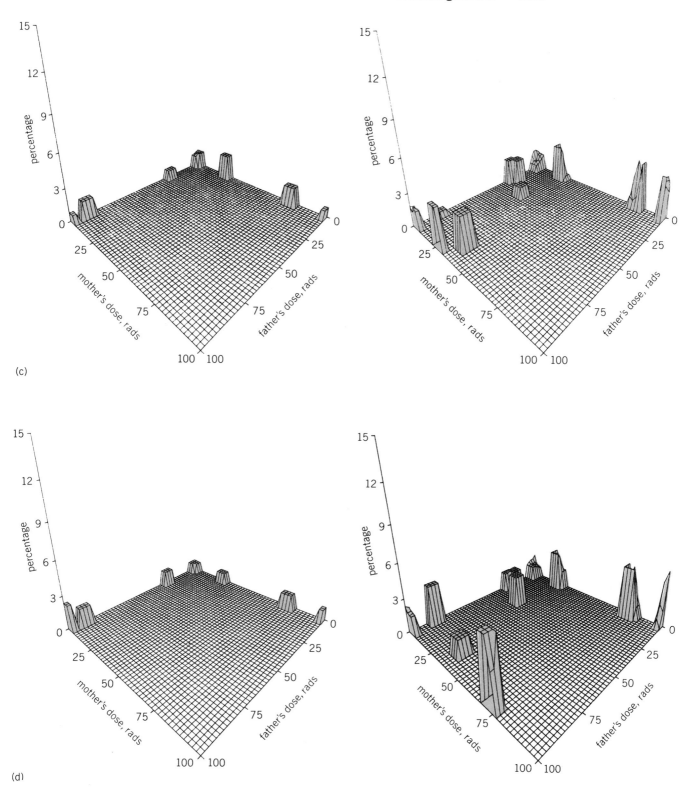

(c)

(d)

on the Biological Effects of Ionizing Radiation of the National Academy of Sciences of the United States has estimated that the genetically significant dose received by a typical individual in the United States in his or her first 30 years of life from natural and other types of radiation is approximately 3.5 rems. This value is higher, of course for those persons occupationally involved with radioactive materials or other sources of ionizing radiation.

The doubling dose previously cited falls near the center of the range of values which has been used by the United Nations Scientific Committee on the Effects of Atomic Radiation (UNSCEAR), or the Committee on the Biological Effects of Ionizing Radiation in their estimates of the hazards which confront humans through increasing exposure to ionizing radiations. Some of the doubling doses these agencies have used, for example, 50 rems, may be too low, however, and consequently the hazards somewhat overestimated. Nevertheless, given the present uncertainties, it seems wise to overestimate the impact of ionizing radiation on the public health rather than to underestimate it.

For background information *see* RADIATION BIOLOGY; RADIATION CYTOLOGY; RADIATION INJURY (BIOLOGY) in the McGraw-Hill Encyclopedia of Science and Technology. [WILLIAM J. SCHULL]

Bibliography: J. V. Neel and W. J. Schull, *The Effect of Exposure to the Atomic Bombs on Pregnancy Termination in Hiroshima and Nagasaki*, National Academy of Sciences–National Research Council, 1956; W. J. Schull et al., Hiroshima and Nagasaki: Three and a half decades of genetic screening, in T. Sugimura, S. Kondo, and H. Takebe (eds.), *Environmental Mutagens and Carcinogens*, pp. 687–700, 1982; W. J. Schull, M. Otake, and J. V. Neel, Genetic effects of the atomic bomb: A reappraisal, *Science*, 213:1220–1227, 1981.

Immunology

Natural killer cells are a particular subpopulation of normal cells and have been found in most normal individuals of a wide range of mammalian and avian species. Typical natural killer cells have the morphology of large granular lymphocytes, being somewhat larger than typical lymphocytes. Their nuclei are indented, and the cytoplasm contains reddish-staining granules. These cells have the spontaneous ability to lyse a wide variety of tumor cells or cell lines derived from tumors. Natural killer cells are nonadherent and nonphagocytic cells which are clearly distinct from either typical macrophages or T cells, but share some markers and other properties with each of these other cell types. The derivation of natural killer cells remains unclear, but it seems likely that these cells are derived from an offshoot of the T-cell lineage development, or are derived from a separate lineage from bone-marrow precursor cells and merely share some characteristics with cells of other, better-known lineages. In the past few years, natural killer cells have attracted increasing attention, particularly because of their possibly important role in defense against cancer.

For a considerable period of time, T cells were considered the central effector cells for immune surveillance, particularly against cancer. However, it has become increasingly clear that T-cell–mediated immunity alone cannot account for resistance against development of tumors or against infection by various microbial agents. Exceptions to a central role for T cells have led to pessimism as to whether there is any type of immunological protection against cancer. However, other effector cells, particularly natural killer cells, may be important for immune surveillance and may account for those instances in which T cells do not appear to play a significant role.

There is considerable evidence indicating that natural killer cells may be important in defense against disease. They have been shown to possibly mediate a wide range of functions in the living organism, including: (1) resistance against natural killer–sensitive transplantable tumors; (2) resistance against metastatic spread of tumor cells; (3) immune surveillance against certain types of tumors; (4) early natural resistance against infections by certain microbial agents; (5) natural resistance against bone-marrow transplants; and (6) development of certain diseases, including graft-versus-host disease, atopic dermatitis, asthma, and some cases of aplastic anemia.

Resistance against tumor-cell lines. There is substantial evidence that natural killer cells play an important role in resistance against established cell lines of tumors, particularly those that show susceptibility to cytolysis in culture by natural killer cells. A major approach has been to look for correlations between resistance to the growth of the tumor cell lines inside the organism and the levels of natural killer activity in these organisms. In several different situations, a good correlation was observed. The various types of correlations between natural killer activity and resistance against tumors have been observed only in tumor lines with some susceptibility to lysis by natural killer cells. The growth of completely natural killer–resistant cell lines has not been affected by the levels of natural killer activity in the recipients. Moreover, in a study of two sublines of a mouse lung tumor, the metastatic subline was resistant to natural activity, whereas the nonmetastatic subline showed some susceptibility to lysis by natural killer cells.

Beige mice, with low natural killer activity associated with their recessive point mutation, have also provided a convenient model for examining the role of natural killer cells in resistance to growth of transplantable tumor cell lines. Several natural killer–susceptible syngeneic tumors have been shown to produce a higher incidence of tumors, and particularly of metastases, in beige rather than in normal, heterozygous littermates.

Another approach to the role of natural killer cells in the growth of transplantable tumors has been the attempt to transfer increased resistance by natural killer cell–enriched populations. Mixture of such cells with a natural killer–sensitive tumor resulted in reduced tumor incidence after transplantation. Systemic adoptive transfer of cells (by injection into the bloodstream), with the characteristics of natural killer cells, from normal or nude mice to mice inoculated with a transplantable leukemia was also found when given in conjunction with chemotherapy to increase protection against progressive tumor growth.

In an alternative approach that utilized information about selective markers on natural killer cells in mice, nude mice treated with antiasialo GM1 (an antiserum directed against a glycolipid on the surface of natural killer cells) had almost no detectable natural killer activity and showed increased susceptibility to transplantation of syngeneic, allogeneic, and human tumors.

To obtain more direct information about the role of natural killer cells in the direct and rapid elimination of tumor cells in living organisms, ^{125}I-iododeoxyuridine (^{125}IUdR) labeled tumor cells were injected intravenously and clearance from the lungs and other organs was measured. In mice with high natural killer activity, there was a greater clearance of radioactivity when measured 2–4 h after inoculation than there was in mice with low reactivity.

As further confirmation of the role of natural killer cells in resistance to growth of natural killer–susceptible transplantable tumors, transfer of natural killer cell–containing populations into mice with cyclophosphamide-induced depression of natural killer activity was shown to significantly restore both internal clearance and natural killer reactivity. The effectiveness of the transfer correlated with the levels of natural killer activity of donor cells in a variety of situations.

Similar results were obtained when radiolabeled cells were inoculated subcutaneously into the footpads of mice. Clearance correlated in several ways with the levels of natural killer activity in the recipients, and cells with the characteristics of natural killer cells were effective in increasing destruction of tumor cells when inoculated at the same site.

Although the results in some studies, especially those with intravenously inoculated radiolabeled tumor cells, suggested that natural killer cells may be particularly involved in resistance against hematogenous metastatic spread of tumors, the results of the footpad assay and the various demonstrations of natural killer–related differences in the outgrowth of subcutaneous tumors indicate that natural killer cells can also enter and be active at sites of local tumor growth. Thus, natural killer cells have the potential to be involved in the primary line of defense against both the local outgrowth and the metastatic spread of transplanted tumors. However, it appears that the effectiveness of this natural resistance mechanism is rather limited. Even with tumor cells that are highly susceptible to natural killer activity, development of progressively growing tumors can occur in animals with high natural killer activity.

Role in resistance against primary tumors. From the available evidence that was summarized above, it seems likely that natural cells play an important role in resistance to growth and metastatic spread of some tumor cell lines. In addition, there is some evidence that natural killer cells can also have a similar role in defense against growth and metastasis of primary tumors. The majority of spontaneous mammary tumors of C3H mice and of spontaneous lymphomas in AKR mice have been found to have detectable, albeit low, susceptibility to lysis by natural killer cells. Similarly, some human tumors have been significantly lysed by natural killer cells. Such lysis has been appreciably augmented, and clearly detected with a higher proportion of tumors, when the effector cells were pretreated with interferon. Of particular importance has been some evidence that

natural killer cells from autologous or syngeneic individuals react against primary tumor cells. Normal C3H mice have been found to be reactive against some syngeneic mammary tumors, and some cancer patients also have had detectable, interferon-augmentable natural killer activity against their autologous tumor cells.

Another line of evidence in support of the possibility for natural killer cells to interact with autologous primary tumor cells is the demonstration that natural killer cells can enter and accumulate at the site of tumor growth. Natural killer cells have been detected in small spontaneous mouse mammary carcinomas and in small primary mouse tumors induced by murine sarcoma virus. In contrast, natural killer activity has usually been undetectable in large tumors in mice or in clinical tumor specimens. This may be due, at least in part, to the presence of suppressor cells, which have been demonstrated in cell suspensions from some tumors.

Immune surveillance against tumors. Of primary interest is whether natural killer cells may be involved in immune surveillance against the initial development of spontaneous or carcinogen-induced tumors. There is circumstantial evidence consistent with, or suggestive of, a role for natural killer cells: (1) Patients with the genetically determined Chédiak-Higashi syndrome have a high risk of development of lymphoproliferative diseases. In recent detailed studies on several patients with this disease, all were found to have profound deficits in natural killer–cell and K-cell (the effector cells mediating antibody-dependent cell mediated cytotoxicity) activities, whereas a variety of other immune functions, including cytotoxicity against tumor cells by T cells, monocytes, and granulocytes, was essentially normal. (2) Similarly, beige mice, which have an analogous genetic defect, have a substantial, but incomplete, selective deficiency in natural killer activity. A small colony of aged beige mice have recently been reported to have a high incidence of lymphomas. (3) Another human genetic abnormality, X-lined lymphoproliferative disease, has been associated with a defect in the ability to control proliferation of B cells injected with Epstein-Barr virus (EBV). Recently, low natural killer activity has been found in such individuals, and this deficit has been linked to the pathogenesis of the disease. In support of this possibility, cells with the characteristics of natural killer cells have been found to inhibit the proliferation of autologous EBV-infected B cells in laboratory culture. (4) Patients on immunosuppressive therapy after kidney allotransplants also have a high risk of developing tumors, both reticuloendothelial tumors and a variety of carcinomas. Patients on such treatment regimens have recently been found to have very low natural killer activity, and this has been suggested as a contributing factor to the subsequent development of tumors. Each of these lines of evidence fits one of the major predictions of immune surveillance theory: that tumor development would be associated with, and in fact preceded by, depressed immunity.

Carcinogenic agents. A related prediction of immune surveillance theory is that carcinogenic agents will cause depressed immune function, thereby impairing the ability of the host to reject the transformed cells. This postulate has been examined by many investigators in regard to the possible role of mature T cells and humoral immunity, and conflicting results have been obtained. In contrast, the initial and still fragmentary data on this point in relation to natural killer cells are promising: (1) Urethane, which produces lung tumors in only some strains of mice, caused transient and marked depression of natural killer activity in a susceptible strain but not in resistant strains. Administration of normal bone-marrow cells, which can reconstitute natural killer activity, to urethane-treated mice reduced the subsequent development of lung tumors. Also, infection during the latent period with various viruses, each known to induce interferon and thereby augment natural killer activity, also reduced the incidence of lung tumors induced by urethane. (2) Carcinogenic doses of dimethylbenzanthracene also were found to produce depression of natural killer activity during the latent period. (3) Sublethal irradiation of mice has been found to cause considerable depression of natural killer activity. Of particular interest is the schedule of multiple low doses of irradiation of C57BL mice, highly effective in inducing leukemia in this strain, which was found to produce a substantial deficit in natural killer activity. The depressed natural killer activity can be restored by transfer of normal bone-marrow cells, a procedure which has been reported to interfere with radiation-induced leukemogenesis. (4) Natural killer activity also has been strongly inhibited by two different classes of potent tumor promoters, phorbol esters and teleocidin. All of these observations support the possibility that one of the requisites for tumor induction by carcinogenic agents may be interference with host defenses, including those mediated by natural killer cells.

Natural resistance against microbial infections. There has been increasing evidence that natural killer cells may promote resistance to microbial infections. Most of the studies are related to viruses, with several investigators showing that cells infected by a variety of viruses become considerably more sensitive to lysis by natural killer cells, and persistently virus-infected tumor cells grow poorly in nude mice, apparently as a result of interferon induction and reactivity by natural killer cells. Furthermore, resistance of living cells to infection by several types of viruses has been found to correlate with natural killer activity. This is considerable evidence for a natural killer–cell role in genetic resistance of mice to severe infection by herpes simplex virus type 1. It also seems likely that natural killer cells are involved in natural resistance to infection by mouse cytomegalovirus. Natural genetic resistance to another herpes virus, Marek's disease virus in chickens, may also be mediated by natural killer cells. Cells with the characteristics of natural killer cells were found to transfer resistance to susceptible new-born chicks. In contrast to the suggestions that natural killer cells offer protection against infection by several viruses, however, it should be noted that there is evidence against a role for natural killer cells in resistance against some other viruses.

Natural killer cells may also be involved in resistance against some other types of microbial infections. A correlation has been observed between natural killer activity and resistance of mice to the malarial parasite, *Babesia microti*. Beige mice were highly susceptible to infection, whereas heterozygous, normal mice were resistant. Natural killer cells may be involved in infection by *Trypanosoma cruzi*, and this parasite has been shown to be susceptible to cytotoxicity by natural killer cells in laboratory culture. Natural killer cells may contribute to the natural resistance of mice to infection by the fungus *Cryptococcus neoformans*. During the first two weeks after infection, nude mice were found to be more resistant to growth of the organisms than euthymic mice, and cells with the characteristics of natural killer cells were found to mediate inhibition of colony formation by the fungus.

For background information *see* CANCER (MEDICINE); IMMUNOLOGY; PHAGOCYTOSIS; TRANSPLANTATION BIOLOGY; TUMOR in the McGraw-Hill Encyclopedia of Science and Technology.

[RONALD B. HERBERMAN]

Bibliography: R. B. Herberman (ed.), *NK Cells and Other Natural Effector Cells*, 1982; R. B. Herberman and H. T. Holden, Natural cell-mediated immunity, *Adv. Cancer Res.*, 27:305–377, 1978; R. B. Herberman and J. R. Ortaldo, Natural killer cells: Their role in defenses against disease, *Science*, 214:24–30, 1981; G. Möller (ed.), Natural killer cells, *Immunol. Rev.*, 44:1–250, 1979.

Integrated circuits

Since the mid-1960s, integrated circuits have significantly improved in performance quality and have been widely accepted into general use. Rapid advances in integrated circuit fabrication have been responsible for computers that can perform 100 million operations per second; for inexpensive hand-held calculators; for digital watches; and for the recent boom in home computers. The success of integrated circuit technology is the result of the development of more complex circuits, operating at greater speeds yet being produced at decreasing cost. The number of transistors per circuit has increased annually by about a factor of 2, while the processing cost per circuit has remained stable. Reduction of device dimensions has spearheaded the advance in circuit integration. That reduction has been brought about by continual refinements in microlithography allied with semiconductor-processing techniques that preserve lithographic resolution.

Silicon MOSFET. The principal switching element in most very-large-scale integrated (VLSI) circuits is the silicon metal-oxide-semiconductor field-effect transistor (MOSFET). MOSFET fabrication will be used to illustrate the application of techniques discussed in this article, and the reduction of MOS-

FET dimensions will illuminate the general problems and approaches involved in producing smaller devices for VLSI.

A prototypical *n*-channel MOSFET (Fig. 2) consists of source and drain regions that are electrically isolated from each other in the absence of any charge (voltage) applied to the gate. The electrical isolation is a result of a reverse-biased diode at the interface between the drain area and the electrically different surrounding silicon region. The diodes are *pn* junctions and, for the *n*-channel device of Fig. 1, the source and drain regions are *n*-type (conduction by electrons) and the surrounding silicon is *p*-type (conduction by holes). The regions are formed by the introduction of the appropriate impurity or dopant atoms; this process will be discussed in more detail below.

The *n*-channel MOSFET behaves like an open switch when the gate voltage is zero. As increasing voltage is applied to the gate, surpassing a threshold value, a "sheet" of electrons is attracted to the region under the gate oxide to form a continuous conducting path between source and drain, setting the switch into an "on" state. The sheet of electrons is called an inversion layer, since the electrical character of the *p*-type silicon region has been locally inverted to be *n*-type, thus removing the barrier to current flow across the *pn* junction.

One measure of the speed of the device is the time required to charge the MOSFET gate capacitance. In most cases, this is largely determined by the gate width *L*, and therefore much concern has centered on shrinking this parameter. However, in order to take advantage of the gain in speed from a reduced *L*, the other dimensions must also be appropriately scaled; that is, the source and drain regions should be smaller in area and more shallow, the oxide must be thinner, and the accuracy of alignment of structures (such as gate, source, and drain) must be greater.

Fabrication masks. At the heart of VLSI fabrication is the integrated construction of all components of the circuit upon a single foundation, or wafer substrate, rather than separate construction and final assembly. The fabrication of the separate components of the integrated circuit is done successively, using masks or stencils created directly on and overlaying the substrate. These dictate which portions of the device are under construction at a particular point in the fabrication sequence. The stencils are formed by coating the substrate with a uniform thickness of a few micrometers of resist material, which is usually an organic polymer. An exposing source (such as optical light or x-rays) illuminates selected areas of the resist-covered substrate through an intervening master template for a particular pattern. Exposure of certain areas of the resist results in chemical changes that make them more or less soluble in a developer solution. The resist which remains after development is either a positive or negative image of the template (Fig. 2).

A good-quality resist stencil retains the original feature sizes throughout the subsequent pattern-transfer process. Generally, the more sharply defined the edges of the resist, the better the stencil serves its purpose. Examples of the uses of the resist stencil are: as an etch mask to cut windows in underlying films of insulating oxide, so that the resulting pattern of oxide regions determines which areas of the circuit are in electrical contact with each other and which are isolated; to define the source and drain regions, so that those areas of the substrate covered by the stencil are protected from being doped with impurity atoms; and to delineate the dimensions of the gate, serving as an etch mask for transfer of the gate pattern into the gate material. To first approximation, then, device dimensions are limited by the fineness of feature that can be produced in the resist mask itself. That resolution is determined by the lithography used in forming the pattern, categorized by the type of exposure source.

Optical lithography. At present, optical lithography is most commonly used for integrated circuit fabrication. The exposing radiation lies in the visible or ultraviolet range. The laws of optics limit resolution to approximately the wavelength of the exposing light. In practice, to achieve sharp profiles in conventional resist material, features must be considerably larger than radiation wavelength. Exploratory work indicates certain inorganic resists can extend the capability of optical lithography close to the limits of resolution given by physics alone (Fig. 3).

The exposure tool dominant in production today is the 1:1 projection printer which casts a full-size image of the template mask onto the entire wafer in a

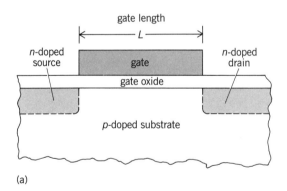

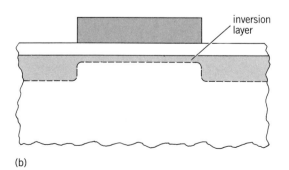

Fig. 1. An *n*-channel MOSFET. (*a*) Gate voltage is zero; MOSFET is off. (*b*) Gate voltage is greater than its threshold value; a conducting inversion layer is formed; MOSFET is on.

exposing radiation

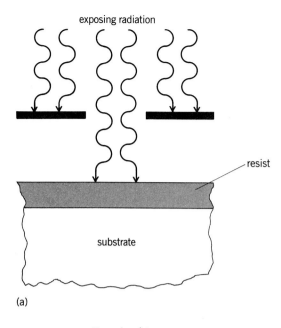

resist

substrate

(a)

patterned resist

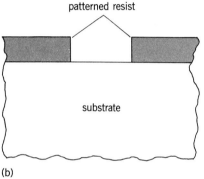

substrate

(b)

Fig. 2. Lithographic exposure of resist-covered substrate. The case shown is for positive resist, where the exposed regions are chemically removed. (a) Exposure. (b) Resist pattern after development.

single exposure. This method allows rapid processing of wafers (high throughput) but cannot produce the local focus corrections needed to compensate for distortions (for example, bowing) in the wafer itself. The alignment accuracy between one level and the next is approximately 0.5 μm, while the minimum achievable line width using projection printing is now about 1.25 μm.

Step-and-repeat projection tools can give better resolution and alignment accuracy than 1:1 projection printers. A segment of the wafer (whose size is much smaller than the wafer size) is exposed according to the desired pattern; the procedure is then repeated across the entire wafer. Because local refocusing and alignment can be done, resolution and alignment accuracy are improved. Often the original template mask pattern is reduced when projected onto the wafer surface. This has the practical advantage of shrinking mask defects that at full size might be fatal to the operation of the circuit, but are negligible when printed at one-fifth or one-tenth full size. The additional cost of using a step-and-repeat scheme

is due to the increased time needed to print an entire wafer step by step.

X-ray lithography. X-ray lithography is similar to optical lithography except that exposures are made with x-rays rather than with visible light. The shorter wavelengths of x-rays essentially eliminate the resolution-degrading diffraction effects found in lithography in the visible and ultraviolet regions, and high-contrast resist profiles can be produced. The areas of active concern in the development of this technology are: well-collimated, high-intensity x-ray sources; good-quality x-ray template masks; and adequate alignment capability of one level with respect to another. The electron synchrotron is an x-ray source under active consideration that can provide both the necessary intensity and collimation. Strongly collimated x-rays are emitted by high-energy electrons circulating in a closed-loop path under a strong magnetic field. Although x-ray lithography is not yet used in production, test circuits in which x-rays defined the MOSFET gates hold the record for switching speed in silicon. These devices transfer signals from their inputs to outputs in less than 20 picoseconds.

Electron-beam lithography. The resolution achievable by using electron beams for pattern generation is far better than that possible with visible light. Electron-beam lithography is currently the primary means of producing master templates for high-resolution optical lithography. In addition, it is used to write high-resolution patterns directly into resist on individual wafers, typically producing features as small as 0.25 μm. In electron-beam lithography the

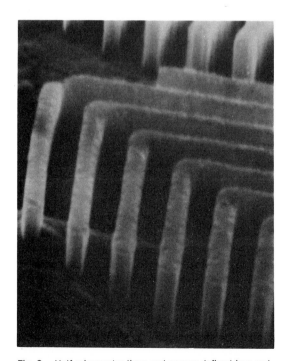

Fig. 3. Half-micrometer lines and spaces defined in a multilayer resist having an upper-layer inorganic resist (a germanium-selenium compound). Exposure was made by using a 1:1 projection printer.

Fig. 4. Ten-nanometer-wide gold-palladium lines formed by using an experimental high-resolution electron-beam lithographic technique. The lines, which are 40 nm apart, are about a hundred times finer than the smallest features used in production.

pattern information is stored in a computer memory and reproduced under computer control onto the wafer. The beam is serially directed from point to point across the wafer. Although direct writing can mean greater flexibility in pattern generation and more quickly implemented changes in circuit design, this serial exposure mode generally results in slow throughput. Therefore, the application of electron-beam lithography is usually limited to mask generation, or to hybrid schemes with optical lithography. Hybrid techniques use the more rapid throughput of optical printing to form coarser resolution features (such as the source and drain in Fig. 2), and the greater resolution and alignment accuracy of electron-beam lithography for the more critical features (such as the gate). In addition, general-purpose circuits may be customized with electron-beam-defined features connecting optically patterned circuit elements.

Electron-beam lithography also provides the means for studying the ultimate limits of miniaturization of electronic components. Electron beams can be focused to less than 1 nm, and in special circumstances they have been used to form patterns comparable to this fine spot size. However, the major complication in using electron-beam lithography is the effect of electrons backscattering from the substrate into the resist (electron backscattering). The backscattered electrons expose the resist in a larger area than was originally intended. The net effect of the backscatter is to limit the obtainable resolution. Laboratory demonstrations have shown that the effect of backscatter can be reduced to produce patterns with feature sizes as small as 10 nm (Fig. 4). These studies may point the way to achieving features of comparable resolution in a production environment.

Resist materials. A lithographic process depends crucially not only on the exposing tool but also on the medium exposed, that is, the resist. Much effort is devoted to synthesizing resist materials having desired properties, some of which are the fol-

lowing: the resist should be sensitive enough to a particular exposing radiation to allow shorter exposure times, hence faster throughput; it should produce high-resolution features and be durable enough for the pattern transfer process; and it must adhere well to the substrate and then be cleanly and completely removed when processing is complete.

Multilayer resist structures allow the combination of several desirable resist properties into one system. An upper resist layer is chosen to optimize exposure characteristics, while the lower-layer properties are designed to match wafer-processing requirements. For example, a particular lower-layer material may be chosen for its durability in etching processes, or its ability to withstand high temperatures. The pattern is exposed and developed in the upper layer and transferred by reactive-ion etching, discussed below, to the lower layer. Sometimes an intermediate layer is used to aid in this transfer and to better ensure fidelity to the original pattern. Thick lower layers also serve to planarize the wafer, allowing improved exposure resolution in the upper layer. In the course of wafer processing, a three-dimensional circuit is formed into and over the substrate surface. For high-resolution lithography, it is preferable that the wafer present a smooth, uniform surface to the exposure tool, so that a good focus may be maintained across the entire wafer. The thick lower layer of a multilayer resist can fill in the valleys of the substrate topography, so that the upper layer coats a flat surface. Multilayer resists are now being introduced into production.

Pattern transfer processes. Two important pattern transfer processes are etching and doping.

Etching. Wet chemical etching has been the mainstay of pattern transfer; however, the last few years have seen a large-scale acceptance of "dry" etching processes. Wet chemical etching is usually isotropic: etching takes place wherever the chemical solution is in contact with the substrate. Therefore, some material is removed in areas covered by the resist stencil as well as in resist-free areas, a situation called undercutting (Fig. 5). Reactive-ion etching, also known as reactive sputter etching, is a directed chemical etching process: reactive, or chemically active, ions are accelerated along electric field lines to meet the substrate perpendicular to its surface. The by-products of the chemical reaction are gaseous; they are pumped out of the reaction chamber and do not further perturb the etch-

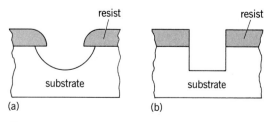

Fig. 5. Difference in profiles between (a) isotropic chemical etching and (b) anisotropic reactive-ion etching.

ing process. The directionality of the ions (hence lack of undercut) provides an extremely high-resolution pattern transfer. Reactive-ion etching has been used to form three-dimensional structures as narrow as 0.025 μm (Fig. 6).

Doping. Until the mid 1970s *n* and *p* regions in semiconductors were formed by placing wafers in a high-temperature gas containing the dopant atoms and allowing the atoms to diffuse through windows in an oxide or nitride on the wafer into the silicon. Diffusion doping is an isotropic process like wet etching, so that it is difficult to dope with high spatial resolution. Diffusion has now been largely replaced by a process called ion implantation. Dopant ions are accelerated through a high voltage so that they penetrate the silicon and become embedded a fraction of a micrometer below the surface. A resist stencil or silicon dioxide stencil defines those areas on the substrate that will receive an implantation dose. The ions are prevented from penetrating the silicon under the masked areas. By adjusting the implantation energy and ion species, sensitive control over doping profiles can be achieved in both the lateral and vertical dimensions. Spatial doping resolution of 0.1 μm is possible.

Ion implantation lends itself especially well to self-aligned doping. This technique can allow the elimination of one lithography level and provides critical fabrication control. In a MOSFET, the source and drain must be adjacent to the gate. If the source and drain are defined on a different level of lithography

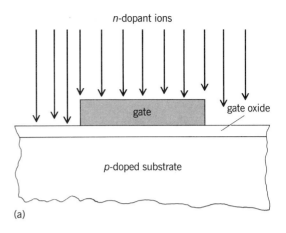

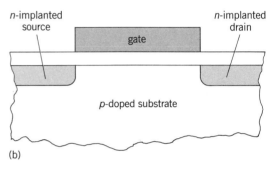

(a)

(b)

Fig. 7. Schematic of the self-aligned ion implantation scheme. (*a*) Gate serves as a mask for implantation. (*b*) Resulting device with the implanted source and drain areas.

than the gate, the gate must be patterned to overlap the source and drain. In this way, if there is an alignment error, there will still be continuity from source to drain through the inversion layer. However, the overlap introduces extra capacitance, reducing circuit speed. This problem can be avoided if the gate is first formed and then used as a mask for self-aligned ion implantation of the source and drain (Fig. 7). The dopant atoms penetrate the silicon substrate in the regions adjacent to the gate, but are stopped by the gate, so that the region under the gate is not implanted.

Prospects. Progress in circuit integration should continue unabated through the 1980s. Because of the high throughput achievable and the already advanced state of the technology, optical lithography will remain the dominant patterning method. With improved resists and advanced reduction printers 0.5-μm line widths should be attained in mass production in about 1990. Random-access memory (RAM) chips will be built that store 4 million bits of data. (By comparison 250,000-bit RAMs were just being introduced in 1982.) Finer line widths will also improve circuit speeds, perhaps by a factor of 2 in this decade.

Though revolutionary in its consequences, the continual increase in circuit density has proceeded along a fast-moving, but nevertheless evolutionary, track. A steady accumulation of small, individual improvements has allowed the present high level of circuit integration to be achieved. Larger-scale cir-

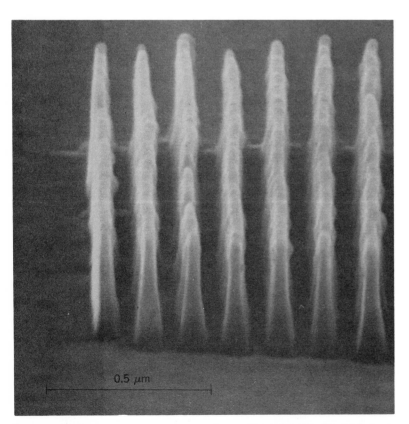

Fig. 6. Fine structures formed in silicon by using reactive-ion etching. The original patterns were generated using high-resolution electron-beam lithography.

cuit integration has relied greatly on the straightforward reduction of circuit elements as the commensurate lithographic and pattern-transfer tools have become available. For this reason, the discussion here has centered on high-resolution lithography. However, the limits of the materials, processes, and device designs currently in use are rapidly being encountered as device dimensions are reduced to less than 0.5 μm.

In fact, the technology that will provide device sizes below 0.5 μm will itself be limited by the practicalities of real processing constraints at one end and by the physics of device operation at the other. Mundane but severely challenging problems, such as the cleanliness of the wafer, its flatness, and its freedom from crystalline imperfections, will put a limit on the complexity and density of circuits for VLSI. On the other hand, a point will be reached where device performance will not smoothly scale with decreasing device dimensions. As MOSFET lateral lithographic dimensions are reduced, the oxide that forms the gate capacitor must also be made thinner. Oxides thinner than about 5 nm do not provide good electrical insulation. As the wires connecting circuit elements become narrower, their resistance increases. Thus it becomes difficult to send electrical pulses from one end of a chip to the other. Another problem that may limit further miniaturization is the statistical nature of the distribution of dopant atoms. Inversion layers may contain only 10,000 dopant atoms on average. The random fluctuation from one device to another may be hundreds of atoms, a value large enough to cause circuit failure.

Lithographic tools already exist for patterning features 100 times finer than current production line widths. The interrelated nature of integrated circuit technology dictates that simple reduction of individual features, or even individual devices alone, cannot be a successful means of achieving VLSI. Clever circuit designs combined with innovative processing techniques and clear understanding of materials problems must all contribute.

For background information *see* INTEGRATED CIRCUITS; ION IMPLANTATION in the McGraw-Hill Encyclopedia of Science and Technology.

[EVELYN L. HU; LAWRENCE D. JACKEL]

Bibliography: Joint special issue on very large scale integration, *IEEE Trans. Electr. Devices*, vol. ED-29, April 1982; Proceedings of the 16th Symposium on Electron, Ion and Photon Beam Technology, *J. Vac. Sci. Technol.*, vol. 19, November/December 1981; Special issue on high-resolution fabrication of electron devices, *IEEE Trans. Electr. Devices*, vol. ED-28, November 1981; E. S. Yang, *Fundamentals of Semiconductor Devices*, 1978.

Interstellar matter

The study of interstellar matter was dramatically changed in 1968 with the discovery of complex interstellar molecules. Since then, astronomers have recognized that in abundance the molecular component is comparable to the atomic form of interstellar gas in the Milky Way Galaxy and in other galaxies. The molecular gas dominates in large regions within these galaxies. Several classical subjects in astronomy, such as the birth and death cycles of stars and the structure and nucleosynthetic history of the Milky Way, have been greatly advanced by the study of these molecules. The new field of astrochemistry has been born, and new ideas on the chemical evolution of life on Earth have arisen.

The interstellar medium is now known to comprise several different physical regimes. At one extreme is very hot (10^6 K), tenuous (10^{-2} atom/cm^3) coronal gas which may exist only near the hottest stars. Intercloud gas is somewhat cooler (10^3–10^4 K) and denser (0.1–1 atom/cm^3). Next comes diffuse interstellar clouds (approximately 100 K and 0.1–100 atoms/cm^3). Finally, there are molecular clouds which are much colder (approximately 10 K) and denser (more than 10^2 molecules/cm^3) than the other regimes.

Molecular clouds range in size from approximately 1 parsec (3.28 light-years or 3.1×10^{13} km) to well over 100 pc, and in mass from perhaps 1 to 10^6 solar masses (1 solar mass = 2×10^{30} kg). There are countless small molecular clouds, but there are estimated to be about 4000 molecular clouds of 10^5 solar masses or greater in the Galaxy. These objects, called giant molecular clouds (GMCs), are the most massive entities in the Galaxy (see illustration); their gravitational forces materially affect the dynamics of the stellar populations of the Galaxy.

The total gaseous component of the interstellar medium constitutes about 10%, and the stars about 90%, of the total mass of the Galaxy (about 10^{12} solar masses). Dust makes up 1% of the gas mass, and its particle density is only 10^{-12} that of the gas density. However, the dust is almost entirely responsible for the extinction of starlight, and renders the molecular clouds completely opaque to visual and ultraviolet light.

Star formation. The cooler envelopes of the larger molecular clouds contain several warm, dense cores (several hundred kelvins or more, 10^6 molecules/cm^3 or more) which are sites of star formation. These hot cores are where most of the complex molecules reside (see table). While all stars form inside molecular clouds, the hottest, most massive stars form inside only the largest clouds, the GMCs. Once formed, these stars ionize the surrounding gas, and it is then blown away by stellar winds (see illustration). These agents compress the adjacent gas, possibly triggering the formation of successive generations of massive stars. A GMC is completely disrupted about 30 million years after it forms, perhaps reduced to diffuse atomic gas, perhaps broken into many small molecular clouds. Low-mass stars like the Sun form in both large and small molecular clouds.

Galactic structure. Radio studies of atomic hydrogen conducted 30 years ago showed that the Milky Way Galaxy is disk-shaped, with a radius of about 25,000 pc but a thickness of only 100 pc. By contrast, studies of the CO molecule (which traces the much more abundant but harder-to-observe H_2 molecules) have shown that the molecular gas forms a giant doughnut or annulus extending around the Galaxy between 4000 and 8000 pc from the center. The inner and outer edges of the doughnut are not sharp; there is still considerable molecular gas near the Sun, 10,000 pc from the center. There is also a huge excess of molecular gas in the central or nuclear region of the Galaxy. Certainly the Sun is at the outer reaches of the annulus, that is, the region of most active star formation in the Galaxy. From the Sun to the galactic center, the interstellar medium is mostly molecular; outward from the Sun, it is mostly atomic.

In the inner Galaxy, both atomic and molecular gases exist between and in the spiral arms, making the arms difficult to map. However, recent analyses of radial velocity versus spatial location of the brightest CO emission, and of certain types of OH emission, are beginning to reveal large-scale patterns of the GMCs which seem to be associated with the overall spiral structure of the Milky Way. The interarm regions are apparently populated by smaller molecular clouds. It appears that GMCs are formed from smaller clouds or from the atomic gas by the compression effects of the spiral density wave, a strong shock wave which accompanies spiral arms as they sweep through the interstellar medium as a result of rotation of the Galaxy.

Interstellar chemisty. The known interstellar molecules are largely organic (see table). The relative lack of ions and free radicals is a selection effect caused by lack of laboratory microwave spectra for many of them. Among the organics, long-chain species (cyanopolyynes) are dominant. Ring compounds are absent. There is no preference for the order of carbon bonding. N=O bonds are absent in all species except NO itself. The molecules mostly contain H, C, O, and N, reflecting the high cosmic abundances of these elements; S and Si are seen in fewer species, and Mg and Fe, which have cosmic abundances similar to S and Si, have not yet been observed. While about half the known species are familiar terrestrially, many others are truly exotic, nonterrestrial species previously unknown to chemists. Relative to H_2, the other molecules are just a trace, ranging in abundance from CO, at about 20 parts per million (2×10^{-5}) down to HC_9N and $HC_{11}N$, just detectable at a few parts per trillion (2×10^{-12}). Generally, the more complex the species, the less abundant.

Because of the very low temperatures and pressures of the interstellar medium, interstellar molecules cannot be formed from atoms by processes that normally occur on Earth. Astrochemists have focused on three different chemistries, of which gas-phase reactions involving molecular ions seem the most important. The simplest of all molecules, H_2, cannot be made in this way, however, but is believed to be formed by the catalytic reaction of two H atoms on interstellar dust grains. Once formed, H_2 is ionized by cosmic rays, and the resultant ion then reacts efficiently with other species to produce a large number of the observed molecular species. Catalytic surface reactions in the cold clouds can produce only H_2 (which alone is volatile enough to evaporate from the grain surface); in the hot cores of clouds, where stars form, these surface reactions may locally produce many other molecular species. A third type of interstellar chemistry is shock chemistry: massive stars produce expanding ionized re-

Interstellar molecules

Molecule	Chemical symbol	Year of discovery	Part of spectrum
Methylidyne	CH	1937	Visible
Cyanogen radical	CN	1940	Visible
Methylidyne ion	CH^+	1941	Visible
Hydroxyl radical	OH	1963	Radio
Ammonia	NH_3	1968	Radio
Water	H_2O	1968	Radio
Formaldehyde	H_2CO	1969	Radio
Carbon monoxide	CO	1970	Radio
Hydrogen cyanide	HCN	1970	Radio
Cyanoacetylene	HC_3N	1970	Radio
Hydrogen	H_2	1970	Ultraviolet
Methyl alcohol	CH_3OH	1970	Radio
Formic acid	HCOOH	1970	Radio
Formyl ion	HCO^+	1970	Radio
Formamide	$HCONH_2$	1971	Radio
Carbon monosulfide	CS	1971	Radio
Silicon monoxide	SiO	1971	Radio
Carbonyl sulfide	OCS	1971	Radio
Acetonitrile	CH_3CN	1971	Radio
Isocyanic acid	HNCO	1971	Radio
Methylacetylene	CH_3C_2H	1971	Radio
Acetaldehyde	CH_3CHO	1971	Radio
Thioformaldehyde	H_2CS	1971	Radio
Hydrogen isocyanide	HNC	1971	Radio
Hydrogen sulfide	H_2S	1972	Radio
Methanimine	H_2CNH	1972	Radio
Sulfur monoxide	SO	1973	Radio
Imidyl ion	N_2H^+	1974	Radio
Ethynyl radical	C_2H	1974	Radio
Methylamine	CH_3NH_2	1974	Radio
Dimethyl ether	$(CH_3)_2O$	1974	Radio
Ethyl alcohol	CH_3CH_2OH	1974	Radio
Sulfur dioxide	SO_2	1975	Radio
Silicon sulfide	SiS	1975	Radio
Vinyl cyanide	H_2CCHCN	1975	Radio
Methyl formate	$HCOOCH_3$	1975	Radio
Nitrogen sulfide	NS	1975	Radio
Cyanamide	NH_2CN	1975	Radio
Cyanodiacetylene	HC_5N	1976	Radio
Formyl radical	HCO	1976	Radio
Acetylene	C_2H_2	1976	Infrared
Cyanoethynyl radical	C_3N	1976	Radio
Ketene	H_2CCO	1976	Radio
Carbon	C_2	1977	Infrared
Cyanotriacetylene	HC_7N	1977	Radio
Ethyl cyanide	CH_3CH_2CN	1977	Radio
Cyanotetracetylene	HC_9N	1977	Radio
Nitric oxide	NO	1978	Radio
Methane ?	CH_4 ?	1978	Radio
Butadiynyl radical	C_4H	1978	Radio
Methyl mercaptan	CH_3SH	1979	Radio
Isothiocyanic acid	HNCS	1979	Radio
Thioformyl ion	HCS^+	1980	Radio
Cyanopentacetylene	$HC_{11}N$	1981	Radio

Horsehead Nebula in Orion (south of the star Zeta Orionis). The opaque cloud of molecules and dust (lower part of figure and beyond) is the Northern Orion molecular complex, comprising several tens of thousands of solar masses. The arc of emission nebulosity (ionized hydrogen) represents the edge of an expanding HII region progressing into the molecular cloud and producing a shock front at the edge of the Horsehead. (*From Kitt Peak National Observatory*)

gions of gas and later supernova remnants, both of which cause strong shock fronts to travel ahead of them through the interstellar medium. These shocks heat and compress the gas for a few hundred years, producing ideal conditions for the occurrence of many high-temperature chemical reactions. None of these chemistries proceed under even remotely equilibrium conditions, so the products depend upon the particular reaction pathways which are most efficient, and upon the input elemental abundances.

Ion-molecule reactions are the best understood of the three proposed chemistries, and appear to explain fairly well the simpler interstellar molecules—those containing up to four or five atoms and existing in the cooler molecular clouds lacking hot, dense core regions. The more complex species are generally observed in only a few hot cores, and their chemistry is very speculative. Surface reactions on grains which consist of silicates, graphite, or metal oxides have been proposed. Laboratory experiments featuring silicates and graphite fail to produce many of the observed interstellar species, and produce others that are not observed. Graphite rapidly disintegrates in the experiments. Metal oxides appear to require ultraviolet radiation to release molecules from their surfaces, radiation which rapidly dissociates the molecules. As for shock chemistry, it is not possible to simulate interstellar conditions. Model calculations are in their infancy, but they do offer hope of explaining many interstellar species, such as the nonvolatile ones. *See* SCATTERING EXPERIMENTS (ATOMS AND MOLECULES).

Molecules in the solar system. The protosolar nebula condensed from an interstellar molecular cloud. The most primitive remnants of this material exist in the Oort cloud of comets and in certain types of meteorites known as carbonaceous chondrites. Comets and possibly meteorites are the only objects whose chemistry would not have been completely modified from that of the original interstellar cloud by the high temperatures and densities that occurred in the protosolar nebula durings its condensation.

Graphite inclusions found in carbonaceous chondrites could conceivably have originated as graphite grains in the primordial interstellar cloud. It is unlikely that interstellar-type surface reactions occurred on these grains once inside the protosolar nebula, because under even slightly elevated temperatures the reactions would be of the Fischer-Tropsch type, known to produce molecular species unlike those of the interstellar medium. Perhaps instead the original grains were coated with mantles of organic polymers, formed and condensed on them in the interstellar medium and later released into the protosolar nebula. The problem with the hypothesis is that infrared studies of interstellar grains show no conclusive evidence for organic mantles. Also the amino acid glycine, abundant in carbonaceous chrondrites, is not found in the interstellar medium. Current evidence is thus against meteorites as carriers of primordial interstellar molecules.

Comets are more likely repositories of the original interstellar chemistry. The Oort cloud of comets probably formed when the outer parts of the protosolar nebula were still immersed in the interstellar cloud at low temperatures. Of the myriad of planetesimals that formed early on, the majority may have formed Uranus and Neptune over a period of 3×10^8 years, while the small subset whose orbits were perturbed remained as comets. It has been speculated that comets in turn deposited primordial organic material in the inner solar system, where most such material was earlier evaporated during the condensation phase. While this hypothesis is difficult to disprove for Earth, it appears to be contradicted by the lack of organic surface molecules on Mars.

For background information *see* GALAXY; INTERSTELLAR MATTER; SOLAR SYSTEM in the McGraw-Hill Encyclopedia of Science and Technology.

[B. E. TURNER]

Bibliography: B. H. Andrew (ed.), *Interstellar Molecules*, 1980; L. Blitz, Giant molecular-cloud complexes in the Galaxy, *Sci. Amer.*, 246(4):84–94, 1982; R. H. Gammon, Chemistry of interstellar space, *Chem. Eng. News*, 56(40):20–33, October 2, 1978; B. E. Turner, Interstellar molecules, *J. Mol. Evol.*, 15:79–101, 1980.

Ion implantation

Ion implantation is emerging as a viable process by which unusual properties of solid surfaces can be achieved. The treatment of tools, specifically cemented tungsten-carbide tools, is on the verge of commercial realization.

Ion implantation is a process by which ionized atoms of an element are accelerated by an electric potential toward the surface of a solid metal or ceramic material. Because of the high kinetic energy imparted to the ions, they displace atoms in the target material and embed themselves at a shallow depth below the target surface. In principle, therefore, any ion species can be embedded into the surface of a solid, regardless of the chemical affinity between the two. Ion implantation is a nonequilibrium process by which surface constitution of solids can be designed from the starting point of the desired properties, and is not bound, to a first approximation, by chemical equilibrium and thermodynamic constraints.

In an ion implantation machine, the ions to be implanted are generated in the source (Fig. 1). First the atoms have to be brought to the gaseous state by evaporation, sublimation, and so on. Implantation of ions already in the gaseous state, for example oxygen or nitrogen, is relatively simple. Ionization is accomplished by bombarding the atoms with energetic electrons obtained from a hot filament. The bombarding electron strips an electron from the atom, thus creating a charged atom, that is, an ion. The ions are extracted from the source by a potential, typically 20–30 keV; move through a series of focusing lenses; and then are given the final acceleration toward the target by the applied potential, typically 50–400 keV. It should be noted that the whole process has to be carried out in a vacuum of the order of 10^{-6} torr (10^{-4} pascal).

Ion penetration. Within the range of energies used for modifying solid surfaces, the interaction of the target surface with the bombarding ion can be described by independent, elastic two-body interaction, not unlike the interaction between hard billiard balls. The energy T transferred from the incident ions of mass M_1 moving toward the target atoms of mass M_2 with energy E is described by the equation below. This will cause the target atoms to scat-

$$T = \frac{4M_1M_2}{M_1^2 + M_2^2} E$$

ter, while the incident ions will slow down as a result of energy loss. Because the collision rate, and thus the rate of energy loss of the incident ions, are statistical in nature, not all ions will come to rest at the same place. Usually, a normal distribution (gaussian) curve describes the range of penetration of the incident ions (Fig. 2). The implantation (penetration) depth is very shallow; the peak concentration for typical conditions occurs at range values of a few hundred to about 2000 angstroms (a few tens to about 200 nanometers). Thus, surface modification by ion implantation refers to the "real" surface, a zone with a depth of a few hundred atomic layers.

Improvement of mechanical properties. A large number of experiments have been carried out to explore the modification of friction and static microhardness by ion implantation. Most of these experiments involved the implantation of nitrogen into iron, steel, and cemented tungsten-carbide tools and components. Reductions by as much as a factor of 10 of the wear rate of steels implanted with nitrogen have been reported. In addition, reductions by as much as a factor of 15 in the sliding wear of tool steels implanted with either boron, carbon, or nitrogen have been reported. These effects have been

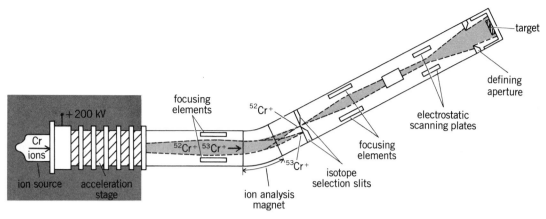

Fig. 1.   Schematic of an ion implantation system. (*After J. K. Hirvonen, ed., Treatise on Materials Science and Technology, vol. 18, p. 11, Academic Press, 1980*)

attributed to hardening caused by the implanted atoms. Significant reductions in wear in stainless steel implanted with nitrogen have been achieved by a hardening effect due to the formation of very small and dense populations of chromium nitrides just below the surface. Determination of this effect was based on transmission electron microscope studies.

Significant improvements in the useful lives of cemented tungsten-carbide tools implanted with nitrogen have been reported by a number of investigators. Wire-drawing dies for both ferrous and nonferrous materials last three to five times longer after implantation with nitrogen, and good improvements have been noted in other cemented tungsten-carbide cutting tools after nitrogen treatment. A variety of carbide tools implanted with 100-keV nitrogen have exhibited improvement by as much as a factor of 6 in the life of punch and die sets for index slotting of rotor laminations for electrical motors. In this case,

the implanted surface was selected to maintain the improvement after repeated sharpening steps. Similar improvements were observed with carbide drills, dies, wire-forming rolls, and steel-slitting knives.

There have been few explanations proposed to account for the rather significant modification of cemented carbide surfaces by ion implantation. One suggestion is that the implantation hardens the cobalt binder in the cemented carbide tools, thus delaying the disintegration of the material. Another interpretation is a toughening mechanism of the carbide grains based on microscopy data which showed a rather high dislocation density within the surface carbide grains following bombardment with nitrogen. The number of engineering components treated by ion implantation is rather limited at present. Experiments at the U.S. Naval Research Laboratory have been successful in implanting chromium into low alloy steel ball bearings. In the United Kingdom certain specialized valve components have been treated by ion implantation, resulting in large reductions in wear rate under operating conditions.

For background information *see* ION IMPLANTATION in the McGraw-Hill Encyclopedia of Science and Technology.

[RAM KOSSOWSKY]

Bibliography: G. Dearnaley et al., *Ion Implantation*, 1973; J. K. Hirvonen (ed.), *Ion Implantation in Materials Science and Technology*, 1980; S. T. Picraux and W. J. Choyke (eds.), *Metastable Materials Formation by Ion Implantation*, 1982; C. M. Preece and J. K. Hirvonen (eds.), *Ion Implantation Metallurgy: Conference Proceedings*, Metallurgical Society AIME, 1978.

Jupiter

Jupiter has four large satellites which are easily seen from the Earth through binoculars. They are named the galilean satellites after their discoverer and range in size from slightly smaller than the Earth's moon to slightly larger than the planet Mercury.

As observations of these satellites multiplied in

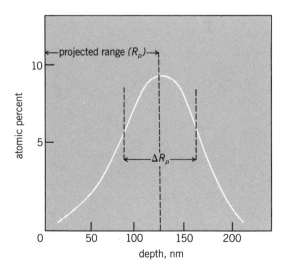

Fig. 2.   Distribution of nitrogen implanted into steel with an acceleration potential of 100 keV. The dose of nitrogen ions is $1.1 \times 10^{17}/cm^2$.

the last 30 years, it became clear that the innermost satellite, Io, was unlike any other body in the solar system. It was found to be anomalously bright in the red and infrared parts of the spectrum. Its surface showed no trace of water ice, which was abundant on its galilean companions. An observation of its radiation at an infrared wavelength of 10^{-3} cm, while it was eclipsed by Jupiter's shadow, showed it to be anomalously warm. Yet it was as cold as the other eclipsed satellites when observed during another eclipse at 2×10^{-3} cm. Sometimes it was brighter than normal at visible wavelengths for a few minutes after emerging from an eclipse—but not every time. Io was also found to have a tenuous atmosphere whereas its companions did not, and to be surrounded by an extensive cloud of sodium atoms. It was sometimes observed to brighten dramatically in the infrared only to return to normal in a few hours.

The intermittency of these strange phenomena made them difficult to interpret, but all became clear in March 1979 when L. A. Morabito discovered the giant plume of an active volcano on an image of Io made by the passing *Voyager 1* spacecraft. Eight more active plumes and numerous sulfurous flows recorded on *Voyager* photographs showed Io to be the most thermally active body known in the solar system. This discovery was preceded a few days earlier by the publication of a prediction that the interior of Io should be molten with only a thin surface crust, and that extensive volcanism was a likely consequence.

Gravitational and radioactive heating. Several lines of evidence point to the conclusion that rocky, terrestrial-type bodies were heated by the energy released during their initial accumulation from smaller solid pieces, and by radioactive elements, which release energy upon decay in the interior. This energy is sufficient to melt the planet or satellite, provided that the object is large enough that removal of thermal energy by convection and conduction to the surface (with subsequent radiation into space) would be at a lower rate than that of interior energy release. The Earth and Moon have been extensively melted and differentiated, with low-density minerals rising to the surface and, for the Earth at least, dense metallic iron and nickel settling to the core. After differentiation, their subsequent thermal history was a steady cooling as the radioactive elements became depleted. The Earth's nickel-iron core is still molten, but the much smaller Moon is nearly if not completely solid. Since Io is only slightly larger than the Moon and has about the same density, a similar initial composition and thermal history might be inferred, leading to the expectation that Io would be a cold, solid satellite with a surface scarred with old impact craters.

Mechanism of tidal heating. It has been suggested that another energy source could have affected Io's evolution. It is known that an otherwise spherical satellite is distorted into an egg shape by the gravitational field of its planet. The Earth is distorted by the Moon in this way, resulting in the fa-

miliar ocean tide. The solid Earth is also distorted, but less so because of its rigidity. A tidal bulge is also raised on the Moon because a volume of mass on the side closest to the Earth experiences a greater gravitational force from the Earth than does a volume at the center, which experiences a larger force from the Earth than does the far side. This differential force tends to stretch the Moon along the line joining it and the Earth. Io is similarly distorted in Jupiter's gravity field. If the tidal bulge changes its orientation on the satellite, or changes its magnitude in a periodic way, the satellite material is repeatedly flexed. Dissipation of the energy associated with repeated flexing causes the satellite to heat up, and this heating supplements the heat created by the decay of radioactive elements.

Decrease of tidal effects. If the satellite is rotating relative to its planet, the tidal bulge sweeps repeatedly around the satellite. Energy is dissipated in the interior, but at the same time the tidal bulge is displaced in the direction of rotation relative to the line joining planet and satellite. The planet attracts the tidal bulge on the near side more than the one on the far side, which results in a torque or twisting action on the satellite. This torque tries to realign the bulge to face the planet, thereby slowing the satellite's spin. Thus, the spin slows until the satellite keeps the same face toward the planet. The Earth's Moon is in such a state of rotation which is synchronous with its orbital motion, as are all the galilean satellites, including Io. In the synchronous state of rotation the tidal bulge is nearly fixed in the satellite. There is no flexing and hence no heating unless the satellite spin axis is not perpendicular to its orbit plane or the orbit is not circular. In the former case the planet appears alternately above and below the satellite's equatorial plane, and the tidal bulge on the satellite follows. In the latter case the satellite distance from the planet varies, causing the size of the tide to vary. In addition, the nonuniform motion in the orbit coupled with the uniform rotation causes the planet to oscillate in the satellite sky. All these factors lead to slight motions of the tidal bulge relative to the body of the satellite, and hence heat the satellite interior. However, as in the case for a relative rotation, the dissipative process decreases the parameter leading to the dissipation. The satellite eventually ends up rotating synchronously with its orbital motion, traveling in a circular orbit, and with its equator plane and orbit planes coincident. The maximum amount of energy that would be dissipated in attaining such a state could yield a mean temperature increase of only a few tens of kelvins for a satellite like Io. This is why the dissipation of tidal energy has not generally been considered important to the thermal histories of planets and satellites.

On the other hand, if the equator plane and orbit plane of a synchronously rotating satellite could be prevented from becoming coincident, or the orbit eccentricity could be sustained in spite of the dissipation of tidal energy, significant heating might occur. The Moon's equator plane remains inclined to its orbit plane by about 6° because the orbit plane keeps changing its orientation relative to the fixed stars owing to the gravitational effects of the Sun. Although some tidal dissipation is thereby sustained, it is too small to have contributed significantly to lunar thermal history. For Io, it is the orbit eccentricity which is maintained, and the consequences are quite different.

Maintenance of Io's eccentricity. A nonzero eccentricity for Io results from a special interaction with two of the other galilean satellites, Europa and Ganymede. The period of orbital revolution of Europa is nearly twice that of Io, and Ganymede's period is nearly twice that of Europa. The relationship of Io's motion to Europa's is more precisely represented by the relation $n_1 - 2n_2 + \omega_1 = 0$, where n_1 and n_2 are the mean orbital angular velocities of Io and Europa, respectively, and ω_1 is the angular velocity of the line joining the center of Jupiter with the point in Io's orbit closest to Jupiter, called the pericenter. A necessary condition in the maintenance of this orbital resonance is that conjunctions of Io and Europa (that is, both satellites on the same side of Jupiter in closest approach to each other) always occur when Io is at its pericenter. If a given conjunction does not quite satisfy this condition, restoring forces return the conjunction to the pericenter. The observed value of $n_1 - 2n_2 = 0°.739507/$ day, and the resonance condition requires $\omega_1 = -0°.739507/$day. This means that Io's elliptical orbit precesses slowly in a direction opposite the motion of Io in order to preserve the timing of the conjunctions to coincide with Io's passing through its orbit pericenter. This retrograde motion of the pericenter is forced by the interaction with Europa, and it is inversely proportional to Io's orbital eccentricity e_1. Hence, e_1 is forced to the value 0.0041, which adjusts ω_1 precisely to $-(n_1 - 2n_2)$. This means that Io is about 0.4% closer to Jupiter at pericenter than it is on the average. The resulting tidal dissipation in Io would tend to reduce e_1, but this would change ω_1 and thereby cause the conjunctions to occur when Io is slightly past the pericenter of its orbit. Instead the restoring accelerations keep ω_1 at the precise value given above and prevent the reduction of e_1. The existence of the orbital resonance sustains the value of e_1 in spite of tidal dissipation.

The energy converted to heat in Io must still come from the orbital motion, however, and that process would still decrease e_1, but now under the constraint of maintaining the resonance. Hence, ω_1 would become more negative, and n_1 would increase to compensate. Io would spiral closer to Jupiter, the orbital resonance would eventually be destroyed, and tidal heating of Io would cease. The process which prevents this involves the tide raised on Jupiter by Io. Just as in the case of a satellite in relative rotation, the high spin rate of Jupiter (period of about 10 hours) carries the tidal bulge ahead of the line to Io, with the result that Io exerts a torque on Jupiter which tends to slow it down. But Io experiences an equal and opposite torque which accelerates the satellite in its direction of motion. The tide raised on Jupiter causes Io to spiral away from the planet and more than restores the orbital energy lost in heating Io's interior. Thus the energy deposited as heat in Io ultimately comes from Jupiter's very large energy of rotation.

Effect of tidal heating on Io. The tide raised on Io by Jupiter is enormous—about 7 km high if Io were completely fluid. The variable part of this tide due to the relatively small eccentricity is about 100 m, compared to a 1-m mid-ocean tide on Earth, so substantial tidal dissipation in Io can be anticipated. In fact, if Io were made of rocks that behaved under strain like those on Earth, the tidal heating per unit volume in the center of Io would be about ten times that now produced by radioactive elements, if the latter had a concentration typical of the Earth or Moon. It seems likely, then, that the tenfold increase in the volumetric heating of the center of Io by tidal dissipation, supplemented by any radioactive decay, resulted in an initial melting of Io's center.

Once the core is melted, Io is less rigid and has greater deformation under the same distorting forces. The greater amplitude of the variable strain increases the heat deposited per unit volume in the remaining solid shell. In fact, the increase in the rate of heat deposition per unit volume in the shell overcompensates for the decrease in volume in which the dissipation is taking place (dissipation in the molten core is negligibly small), and the total rate of energy dissipation is higher. This in turn melts more core, allowing even larger deformations. A thermal runaway could result, which would produce a final state having a relatively thin solid crust through which the interior heat is mostly transported by volcanic conduits.

Whether or not this thermal runaway actually could occur depends on how efficiently the internal energy can be carried to the surface. The motion of material above the Earth's liquid core (manifested in continental drift), although typically only a few centimeters per year, is much more efficient in transporting heat energy from the Earth's interior than thermal conduction. It is likely that the heat generated in a body as large as Io is transported by the same process of solid-state convection. If heat transport by this mechanism is assumed, the heating rate necessary to sustain a given temperature difference across a layer of thickness L increases as L decreases approximately as $1/L$; but the rate of heating in such a layer on Io due to the tides increases even faster. The values of several parameters characterizing the material in Io are unknown, so the actual rate of energy transport cannot be calculated. But the fact that the rate of tidal energy deposition increases faster than that necessary to maintain a given temperature difference across the layer as the layer thins means that solid-state convection could not prevent the thermal runaway following an initial melting of Io's center.

Comparison of Io with the Moon, which (with much

less available energy) is at least nearly molten in the center, together with the high probability that the materials in the Moon and Io are sufficiently similar that convective properties are similar, showed that the interior of Io probably had been melted and initiated the thermal runaway. The end result would be an Io that was completely processed, that is, thoroughly differentiated. It would be cold over most of the surface, like the other galilean satellites, but hot enough to melt rocks at relatively shallow depths (perhaps 10 or 20 km). The extraordinary heat of tidal dissipation would have to escape through the movement of hot material through the thin crust, thus implying volcanic activity. Hence the prediction that *Voyager* images of Io may reveal evidence for a planetary structure and history dramatically different from any previously observed. Rarely is the coordination between theory and observation inherent in the scientific method exemplified in such a prompt and timely manner.

For background information *see* JUPITER; PLANETARY PHYSICS in the McGraw-Hill Encyclopedia of Science and Technology.

[STANTON J. PEALE; P. M. CASSEN; R. T. REYNOLDS]

Bibliography: M. H. Carr et al., Volcanic features of Io, *Nature*, 280:729–733, August 30, 1979; L. A. Morabito et al., Discovery of currently active extraterrestrial volcanism, *Science*, 204:972, June 1, 1979; S. J. Peale, P. M. Cassen, and R. T. Reynolds, Melting of Io by tidal dissipation, *Science*, 203:892–894, March 2, 1979; B. A. Smith et al., The Jupiter system through the eyes of Voyager 1, *Science*, 204:951–971, June 1, 1979.

Land navigation system

Self-contained navigational systems for land vehicles have been mechanized by using the vehicle odometer and a heading reference. These systems are independent of any external radio, radar, or beacon aids. A magnetic compass is used for the heading reference for those systems with less accurate requirements and also for the earlier developed systems. At present, the more accurate systems use one or more gyroscopes for determining heading, and the trend is to use inertial sensors operating in a strapped-down configuration.

Though similar to the inertial systems used for land surveying, land navigation systems are less expensive and less accurate—typical navigation performance is 0.1–0.5% of distance traveled. These systems have been used primarily in military combat vehicles to provide self-contained position-determination capability for tanks, self-propelled howitzers, observer vehicles, and mobile antiaircraft defense systems. For nonmilitary applications, there has been interest in providing this position-determining capability to emergency vehicles such as police, fire, and ambulance, and to ordinary vehicles operating in unmarked areas.

Besides present position of the vehicle, these systems also provide an output of vehicle heading, which

is useful for steering a predetermined course. In addition, some configurations provide vehicle pitch and roll.

System description. Vehicle navigation using an odometer and gyroscope is a form of dead reckoning. All the navigation measurements are made by instruments within the vehicle. The basic output of this type of system is distance traveled, and by proper computation this output is related to the map or navigation coordinate system.

Basically, the vehicle determines present position by resolving the odometer-derived velocity through a heading reference and integrating the resolved velocity components to obtain distance traveled in a coordinate system (Fig. 1a). The components of distance traveled are added to the coordinates that were used to initialize the system.

Grid coordinate systems such as UTM (universal transverse Mercator) are generally used. Because of the limited travel range of a vehicle, a flat earth approximation for navigation is usually sufficient. However, latitude-longitude or other angle-coordinate systems can be used with appropriate calculations.

The odometer provides a measure of vehicle distance traveled or, with computation, of velocity. Gyroscopic devices are used for the heading reference, and they can be mechanized to maintain the vehicle heading angle in the presence of vehicle motion.

Gyroscopes. Usually a directional gyro is used as the heading reference. This is a two-degree-of-freedom gyroscope, that is, the gyroscope can sense motion about two orthogonally disposed axes in the plane perpendicular to the spin axis. The gyroscope is oriented so that its spin axis is horizontal and pointed toward the angular reference for heading, usually true north. Once aligned, the gyroscope tends to maintain its orientation through all the maneuvering of the vehicle. The spin axis is decoupled from the vehicle by two gimbals. The outer, or azimuth, gimbal will decouple vehicle yaw motion from the gyroscope. The angular output from the gimbal is a measure of the heading of the vehicle (the angle between the gyro spin axis and the vehicle longitudinal axis). The inner, or level, gimbal is used to keep the gyro spin axis horizontal. A level sensor mounted along the gimbal, or a pendulous mass unbalance of the gimbal, is used to position the gimbal to keep the spin axis level. The setting into the instrument of the heading angle has been performed by using a magnetic compass reference, or by an external gyroscope. The setting into the instrument of the vehicle's position and other navigational parameters just prior to the vehicle's departure is covered by the term initialization.

Self-contained alignment. The desire for increased performance accuracy and self-contained operation has resulted in the elimination of the magnetic compass as the source of heading initialization and in the incorporation of the external reference within the navigation unit. This is accomplished by adding another gyroscope whose function is the initial alignment or north finding of the directional gyroscope, or by providing means for rotating the mechanical gimbals so that the directional gyro could be used for both the north-finding or alignment function and the directional gyro reference.

Gyrocompassing is used for the self-contained alignment. In gyrocompassing, a level or horizontal gyro axis is used to measure the horizontal component of the Earth's spin vector. The Earth's spin axis is oriented through the north and south poles. A level gyro input axis will measure the component of this earth rate spin vector that is along the axis. If the input axis is positioned directly east-west, it will not measure an earth rate component, as it is oriented perpendicular to the north-south earth spin axis. When it is in this east-west orientation, the gyro spin axis will be oriented north-south. So gyrocompassing is the angular positioning of the level gyro input axis either mechanically or computationally until it no longer senses an earth rate component.

A two-axis platform (Fig. 2) is used for heading reference. The platform functions in the same manner as the directional gyro previously described. In a directional gyro, the gimbals are inside the gyro, while with the platform, the gimbals are external. This configuration permits the construction of a more accurate gyro, as it does not have to be designed to provide the angular freedom that the interior gimbals require.

Strapdown configurations. In the preceding mechanization, the spin axis of the directional gyro was suspended from the vehicle by internal or external gimbals which provided the angular freedom to isolate the gyro from the vehicle motion. As part of a continuing effort to lower the cost and improve the reliability of inertial instrument configurations, a strepdown instrument configuration has been developed. With this approach, the gyroscope is mounted directly to the vehicle chassis; there are no gimbals. The output is vehicle angular rate rather than vehicle angles. The required navigation angles are obtained by integration and computation within a digital computer. The more expensive, less reliable mechanical gimbals are replaced by computational gimbals within the digital computer. This mechanization has become feasible with the development of high-speed digital processors and gyroscopes that have an accurate, wide dynamic range of rate measurements.

Systems have been configured in which the strapdown gyroscopes have been used only to develop the heading angle. These are analogous to the gimbal systems described previously. However, a further development of the strapdown configuration has included a complete inertial system that uses the odometer for velocity aiding (Fig. 3). This system uses three axes of accelerometers for linear motion measurements, and three axes of gyroscopes for angular motion measurements. A Kalman filter and error controller is provided to combine the velocity measurements made by the odometer and the iner-

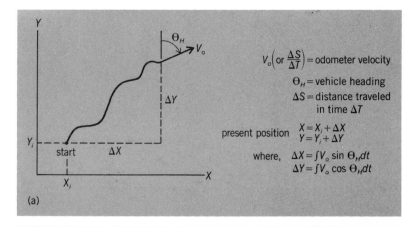

$$V_o \left(\text{or } \frac{\Delta S}{\Delta T}\right) = \text{odometer velocity}$$

$$\Theta_H = \text{vehicle heading}$$

$$\Delta S = \text{distance traveled in time } \Delta T$$

$$\text{present position } \begin{array}{l} X = X_i + \Delta X \\ Y = Y_i + \Delta Y \end{array}$$

$$\text{where, } \begin{array}{l} \Delta X = \int V_o \sin \Theta_H dt \\ \Delta Y = \int V_o \cos \Theta_H dt \end{array}$$

(a)

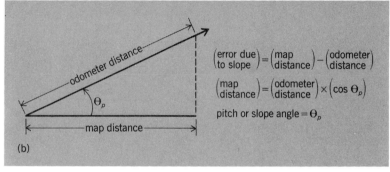

$$\begin{pmatrix}\text{error due} \\ \text{to slope}\end{pmatrix} = \begin{pmatrix}\text{map} \\ \text{distance}\end{pmatrix} - \begin{pmatrix}\text{odometer} \\ \text{distance}\end{pmatrix}$$

$$\begin{pmatrix}\text{map} \\ \text{distance}\end{pmatrix} = \begin{pmatrix}\text{odometer} \\ \text{distance}\end{pmatrix} \times \begin{pmatrix}\cos \Theta_p\end{pmatrix}$$

$$\text{pitch or slope angle} = \Theta_p$$

(b)

Fig. 1.   Summary of basic computations for determining present position. (a) Calculation of vehicle coordinates from odometer velocity and vehicle heading. (b) Correction for slope.

tial sensors in order to determine the optimal velocity to use in the navigation equations and to provide correction signals for the various system-error sources. By using a complete inertial system, the errors due to not resolving the velocity through the level angles are eliminated. Also, through the Kalman filter, the inertial system can be used to correct the odometer scale factor and angular boresight errors, and the odometer can be used to correct for inertial instrument biases and drifts.

Further system developments include the addition

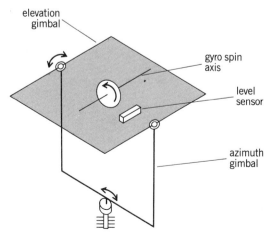

Fig. 2.   Typical two-axis plattorm used for heading reference.

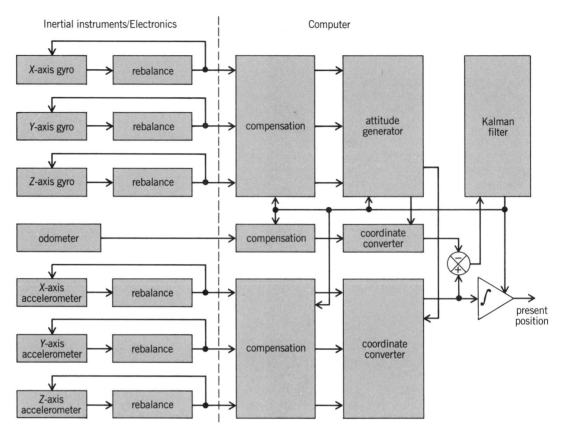

Fig. 3. Block diagram of strapdown land navigation system.

of position updates from external position references. These might be from the Global Positioning System (GPS), radio beacons, or survey points. The Kalman filter and error controller would incorporate these updates to further correct the various system-error sources.

Odometer. The vehicle velocity is measured by an odometer, a distance-registering instrument that measures the rotation of the vehicle wheels or of an appropriate transmission gear. This measurement is converted to an equivalent distance traveled by a scale factor; a typical scale factor is 1000 revolutions per mile of travel. Some of the devices used for this measurement are an incremental digital encoder; electromagnetic devices; and ac or dc tachometers. The devices have two output lines or other means to indicate whether the direction of travel is forward or reverse. Generally, the units are mounted to measure the rotation of an undriven or special fifth wheel (to minimize the slip of a driven wheel), or the rotation of a final drive gear in the vehicle transmission.

Determination of scale factor and boresight angle. For the conventional odometer navigator, the odometer scale factor is determined by driving the vehicle between two known points. The system measures the distance traveled between the points, and this is compared with the known distance. The scale factor is adjusted until the averaged measured distance corresponds to the known distance. In a similar

manner, the odometer angular boresight is determined by observing the cross-track distance error during the calibration runs and adjusting the boresight until the average cross-track error is zero.

When the complete inertial system mechanization is used, this calibration procedure is not required. The inertial system provides a moving reference which can be used for continuously monitoring the odometer velocity. The scale factor and boresight angle can be continuously computed and corrected. The errors discussed below are automatically compensated for with this mechanization. Errors in the use of the odometer result from slopes in the vehicle path, changes in tire or wheel diameter, change of slippage of the wheel, and inaccuracies in aligning inertial instruments.

The odometer measures the total distance traveled over the ground. In driving up and down slopes, the distance measured is greater than the horizontal or map distance traveled (Fig. 1b). The map distance traveled is equal to the cosine of the slope angle multiplied by the total distance traveled. The correction for this error is usually performed by estimating the error and adjusting the odometer scale factor as a function of the slope angle, or measuring the slope angle (in a complete inertial system) and resolving the odometer velocity through it to obtain the horizontal component for integration by the navigation equations. When the resolved components are available, the vertical velocity component can be integrated to obtain a measure of altitudes.

Scale factor variations result from changes in tire or wheel diameter due to wear, temperature, tire pressure, vehicle loading, and vehicle speed. Compensation, for the simple mechanization, is generally performed by manual adjustment of the scale factor. Scale factor errors from these sources are 0.5–1.0%.

Apparent scale factor errors result from change of slippage of the wheel. As a vehicle moves over a surface, there is some slippage of the wheel. The scale factor calibration accommodates the nominal slippage. If the terrain should change, such as from an asphalt road to a dirt road, or to a field, or the weather should change from sunshine to rain, ice, or snow, the wheel slippage will change. The changes are accommodated by estimating the amount of scale-factor change and adjusting it accordingly. Large values of wheel slippage can be detected by computer programming. By computing the apparent vehicle acceleration from the input odometer information, the odometer input can be rejected when the apparent acceleration is greater than the possible vehicle acceleration.

Angular inaccuracies in aligning the inertial instruments with respect to the odometer or longitudinal axis of the vehicle are termed boresighting errors. These errors may be caused by improper installation calibration or changes in the boresight angle over a period of time. This results in a position error proportional to distance traveled at right angles to the direction of travel (cross-track error); for example, a 1-milliradian angle is a 0.1% of distance traveled cross-track error.

System errors. There are two basic types of system inaccuracies; those that result in along-track errors and those that produce cross-track errors. The along-track errors can be equated to equivalent velocity scale-factor errors, and the cross-track errors can be equated to heading errors.

The previously mentioned odometer scale-factor errors will produce along-track errors. A 0.5% error in the velocity scale factor will result in a 0.5% error in distance traveled. If the error is constant, the position error is self-compensating over a closed course. That is, over a closed course, the maximum position error is at the maximum distance from the start-finish point. The terminal error will be zero.

The cross-track errors are related to inaccuracies in initial heading alignment and boresight determination, and drift in the azimuth gyro axis. A 2-milliradian error in the initial alignment or boresight angle will result in a 0.2% of distance traveled cross-track error. The position error resulting from these two sources over a closed course is similar to the previously mentioned scale-factor error. It is maximum at the greatest distance from the start-finish and is zero at the finish. The cross-track error due to azimuth gyro drift is proportional not only to the distance traveled but also to the time of traveling. This error does not close to zero on a closed course.

Another source of system inaccuracy is the error in the initial system position. The system is a dead-reckoning navigator, and any position errors that were introduced during the system initialization will be maintained.

For background information *see* ESTIMATION THEORY; GYROCOMPASS; GYROSCOPE; INERTIAL GUIDANCE SYSTEM in the McGraw-Hill Encyclopedia of Science and Technology.

[TOM J. RICKORDS]

Bibliography: B. Fleming, Honda onboard inertial navigation system, *IEEE Vehicular Technol. Soc. Newsl.*, 29(1):13–14, February 1982; T. Rickords and J. Yamamoto, A strapdown configured land navigation system, *Proceedings of the 37th Annual Meeting of the Institute of Navigation*, pp. 154–164, June 1981; H. Sorg (ed.), *Conference Proceeding of the Symposium on Gyro Technology*, German Institute of Navigation, Stuttgart, 1980; V. R. Vento, Singer's AN2000 land navigation system: A simplified approach to navigation of a terrestrially bound vehicle, *IEEE Position Location and Navigation Symposium Record*, pp. 267–272, December 1980.

Larva

The larva is a distinct developmental stage in the life cycle of many invertebrate and lower vertebrate animals. It succeeds the embyro, which develops within the egg or the parent, but differs markedly from the adult in appearance and life habits. The tadpole stage in the life cycle of the frog is a familiar example. Larvae metamorphose into immature juveniles, resembling the adult stage, through a remarkable alteration in form. The larval stage has a special significance in bottom-dwelling marine invertebrates, because in most species it is the free-swimming larva that maintains genetic connections among populations and enables the establishment of new populations when colonizing opportunities arise. The mode of larval development can be inferred from the earliest growth stages of mollusk shells, so evidence from both present-day and fossil species can contribute to an understanding of the role of larval biology in the biogeography and evolution of marine invertebrates.

Dispersal. Although the adult stages of most bottom-dwelling invertebrates are limited in their mobility, free-swimming larvae (Fig. 1) can be carried over great distances by ocean currents. For example, in the open ocean even a relatively slow current of 0.5 km/h (0.31 mi/h) would carry a larva about 350 km (217 mi) in a month. Closer to shore, larval stages can float past environments hostile to their bottom-dwelling adult stages; for example, rocky shore forms avoid intervening mud flats, and shallow-water species are carried across deep-ocean barriers. The swimming organs (usually rows of beating cilia that in many species also aid in feeding) serve mainly to control the larva's vertical position above the sea floor, enabling the larva to take advantage of differences in water flow at different depths or different times of day and to choose sites suitable for settlement and metamorphosis.

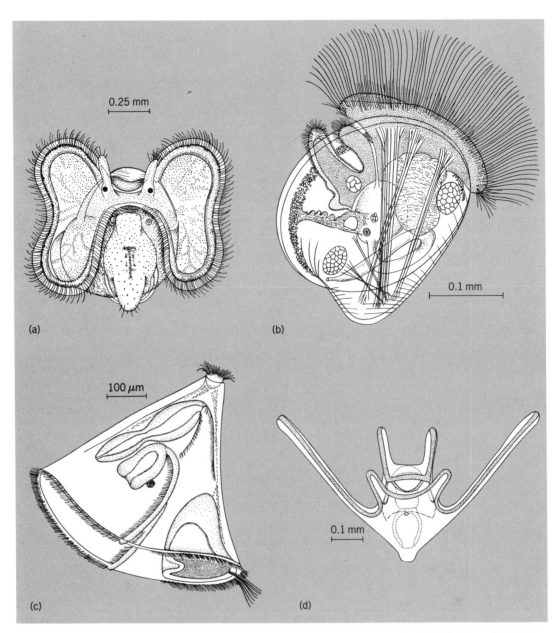

Fig. 1. Free-swimming, planktotrophic larvae of marine invertebrates. (a) Mud snail *Nassarius reticulatus*. (b) Oyster *Crassostrea virginica*. (c) Bryozoan *Membranipora* sp. (d) Brittle star *Ophiopholis aculeata*, with the stipple indicating ciliary bands. (*Part a after V. Fretter and A. Graham, British Prosobranch Molluscs, The Ray Society, 1962; b after P. S. Galtstoff, The American oyster Crassostrea virginica Gmelin, U.S. Dept. Interior Fish. Bull., no. 64, 1964; c after J. S. Ryland, Bryozoans, Hutchinson University Library, 1970; d after R. R. Strathmann, The feeding behavior of planktotrophic echinoderm larvae: Mechanisms, regulation, and rates of suspension-feeding, J. Exp. Mar. Biol. Ecol., 6:109–160, 1971*)

There are two basic modes of larval development, distinguished by whether or not the larva feeds on plankton; each mode has a characteristic dispersal capability. Planktotrophic larvae, those which feed on the plankton, hatch from small eggs with little yolk. Parents invest little energy reserve per offspring, and young are often produced in great numbers, for example, up to 70 million eggs per individual in a single spawning of the American oyster *Crassostrea virginica*. Predation, environmental stress, and failure to reach suitable settlement sites take an enormous toll on planktotrophic larvae, but so many are produced each year that only a small fraction of a percent needs to survive to perpetuate the species. The ability of the larva to feed during development not only reduces its energy cost to the parent, but allows a long free-swimming period in many species. A few species can delay settlement up to 6 months or more, permitting them to maintain populations on the continental shelves on both sides of the Atlantic.

In contrast, nonplanktotrophic larvae, those which

take little or no nourishment from the plankton, hatch from large, yolky eggs. These species produce fewer larvae than do planktotrophs, commonly hundreds to thousands per parent (for example, a thousand young per spawning in the slipper limpet *Crepidula convexa*), and the greater parental investment is accompanied by lower larval mortality rates. Nonplanktotrophic larvae have relatively brief freeswimming stages, usually no more than a few hours to a few days. Some species lack a free-swimming larva, and bottom-dwelling juveniles emerge from a brood chamber or egg capsule. This reproductive mode has been termed direct development, but for some groups this is a misnomer because the developing embryo still passes through a distinct larvalike stage that undergoes metamorphosis before hatching. The lack or brevity of a free-swimming stage greatly reduces larval dispersal capability in nonplanktotrophs.

Planktotrophs and nonplanktotrophs can be readily distinguished from each other in fossil bivalves and gastropods by the size and microornamentation of the earliest growth stage of the larval shell, which is often present on well-preserved fossils (Fig. 2). Most other groups of animals lack easily fossilized larval stages, but other clues are sometimes available, such as brood pouches in brachiopods and sea urchins. Such developmental indicators can be applied to any animal whose reproduction is not easily observed. For example, living deep-sea animals are difficult to culture in the laboratory, but a number of workers have been able to infer modes of development from egg size or larval shell dimensions. They found that patterns of reproduction in the deep sea are far more varied and complex than was originally believed.

Distribution. Larval types are not randomly distributed in the oceans. The most striking distribution pattern is latitudinal, with the great majority of species in shallow tropical waters being planktotrophic. Short-term advantages of the high fecundity and wide dispersal afforded by planktotrophic development include: avoidance of competition with parents or siblings; spreading the risk of mortality over a broad spectrum of environments or locations; greater opportunity to select favorable habitats; increased colonizing ability; and ability to exploit widespread but patchy habitats. The predominance of species with nonplanktotrophic development increases toward the poles and with depth. This trend has been attributed to the brevity of plankton blooms and the slowing of temperature-sensitive developmental rates at high latitudes; to the lower degree of habitat patchiness at depth (and thus less need for wide dispersal); and to the increased hazards with depth of transit between the sea floor and the plankton-rich surface waters.

Adult body size also constrains developmental modes. Below a critical body size, parents are apparently unable to produce enough planktotrophic offspring to survive the great mortality rate accompanying this mode of development. Consequently, while large-bodied species can follow either mode of development, very small-bodied species tend to be nonplanktotrophic. All of these constraining factors—body size, planktonic seasonality, habitat patchiness, and so on—interact and where mutually reinforcing, strengthen the general geographic and bathymetric trends in reproductive modes. Exceptions or even reversals in these trends occur, however, when conflicting requirements are present (for example, ephemeral habitat patches in the other-

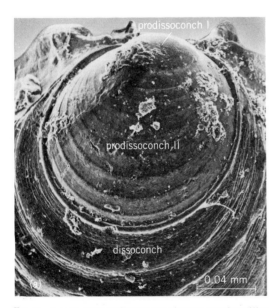 

Fig. 2. Larval shells of Cretaceous bivalves showing different modes of development. (*a*) Nonplanktotrophic *Vetericardiella crenalirata*, with large initial larval shell (prodissoconch I), and reduced or absent free-swimming larval stage (prodissoconch II). (*b*) Planktotrophic *Uddenia texana*, with small initial larval shell and well-developed free-swimming larval stage. (*After D. Jablonski and R. A. Lutz, Molluscan larval shell morphology: Ecological and paleontological applications, in D. C. Rhoads and R. A. Lutz, Skeletal Growth of Aquatic Organisms, Plenum, 1980*)

wise monotonous deep sea), or where rapid environmental changes have left relict forms. Specific interactions, and how they combine to give rise to the complex array of larval types in marine invertebrates, constitute an active field of research today.

Larval dispersal by ocean currents plays an important part in determining faunal similarities and differences among regions. Faunal differences can result not only from wide geographic separation, but also when currents flow in the wrong direction for the transport of larvae between two areas, or when currents range into inhospitably cold or warm latitudes before reaching suitable settlement sites. In addition, the two larval types respond to ocean barriers according to their different dispersal capabilities. Moderately isolated Pacific islands share planktotrophic species with other island groups, but nonplanktotrophic larvae are unable to cross the broad intervening distances, and consequently endemic species have evolved. On a geologic time scale, plate tectonics causes continental shelves to split and drift apart, and nonplanktotrophic groups become isolated and evolve along separate lines well before transoceanic distances exceed the dispersal powers of planktotrophic groups. Conversely, as continental plates approach and collide with one another, planktotrophic species will be shared before nonplanktotrophic species will.

Patterns of evolution. Differences in larval dispersal capabilities can strongly affect rates of evo-

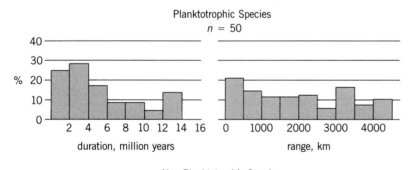

Planktotrophic Species
n = 50

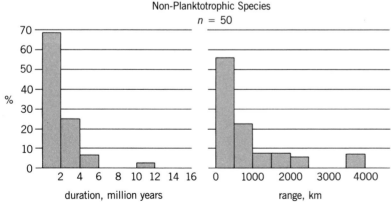

Non-Planktotrophic Species
n = 50

Fig. 3. Biogeographic and evolutionary effects of larval modes in Cretaceous gastropods. The sample of 50 planktotrophic species shows significantly greater geologic durations and geographic ranges than the sample of 50 nonplanktotrophic species. (*After D. Jablonski, Evolutionary rates and modes in Late Cretaceous gastropods: Role of larval ecology, Proceedings of the 3d North American Paleontology Convention, 1:257–262, 1982*)

lution as well. Because planktotrophic species are able to disperse over wide areas in a single generation, local catastrophes are unlikely to eliminate such species over their entire geographic range, and larvae from other, persistent populations can replenish regions in which a species has declined. These effects produce geologically long-lived species and lineages with low rates of extinction, as can be seen in Late Cretaceous gastropods (Fig. 3a). In contrast, species having nonplanktotrophic, low-dispersal larvae tend to have smaller geographic ranges, so that local catastrophes and random population fluctuations are more likely to result in total extinction. Accordingly, nonplanktotrophic species tend to be geologically short-lived and lineages have high extinction rates (Fig. 3b).

Dispersal capability also affects the rate at which lineages split and give rise to new species. Wide dispersal of planktotrophic larvae maintains the gene flow and thus suppresses divergence of scattered populations. On the other hand, in nonplanktotrophic lineages local extinctions or chance colonizations can create a patchwork of semi-isolated populations, each of which has the potential of adapting to local conditions (or otherwise losing its genetic continuity with other populations) and of giving rise to a new species. Nonplanktotrophic lineages, then, have high rates of species origination as well as high rates of extinction.

These differences in evolutionary rates at the species level can also affect large-scale patterns, such as the duration and diversity of entire evolutionary lineages. Groups with nonplanktotrophic larvae can diversify more rapidly than groups with planktotrophic larvae. Therefore, the expansion of a lineage may not be the result of some adaptive improvement over its competitiors, but simply a byproduct of a speciation-prone mode of development. However, the extinction resistance of planktotrophic lineages may be important in times of extreme environmental change, allowing these groups to survive while the more diverse, extinction-prone nonplanktotrophic lineages die out.

Finally, the distributional patterns in larval types outlined above play a part in shaping evolutionary patterns in time and space. Nearshore communities, because they are dominated by species having planktotrophic larvae, commonly exhibit low rates of species turnover. Offshore communities tend to have high rates of species turnover despite the greater (short-term) stability of the environment they inhabit, because more of the species have nonplanktotrophic larvae. Similarly, body size constraints require a lineage to shift to nonplanktotrophy if it evolves below its critical size threshold, and evolutionary rates will then change accordingly. One intriguing exception is found in the latitudinal gradient in larval types. Low latitudes are dominated by species with planktotrophic larvae, yet the species-rich tropics are generally regarded as having been the site of rapid evolution. Much progress has been made in recent years, but scientists still have much to learn about the links between larval biology and evolution in marine invertebrates.

For background information *see* GASTROPODA; INVERTEBRATE EMBRYOLOGY; MARINE ECOSYSTEM in the McGraw-Hill Encyclopedia of Science and Technology.   [DAVID JABLONSKI]

Bibliography: A. C. Giese and J. S. Pearse (eds.), *Reproduction of Marine Invertebrates*, vol. 1, 1974; T. A. Hansen, Influence of larval dispersal and geographic distribution on species longevity in neogastropods, *Paleobiology*, 6:193–207, 1980; D. Jablonski and R. A. Lutz, Larval ecology of marine benthic invertebrates: Paleobiologic implications, *Biol. Rev.*, in press; R. R. Strathmann, Why does a larva swim so long? *Paleobiology*, 5:373–376, 1980.

Laser

Lasers represent a technology for the production of highly directed intense radiation of a precisely defined wavelength, λ. Historically, laser devices have operated over a wide range of wavelengths extending from the millimeter region ($\lambda \cong 10^{-3}$ m) to the vacuum ultraviolet region ($\lambda \cong 10^{-7}$ m). During the last few years there has been a trend in the development of coherent sources of radiation in the high-frequency region of the spectrum, particularly in the range of wavelengths around 10^{-7} m.

Ultraviolet lasers. Among the significant advances is the refinement of laser sources in the ultraviolet. A recently discovered class of ultraviolet lasers, known as excimer lasers, exhibits an output in the ultraviolet range, high power, and energy-efficient operation—a combination of three desirable properties which was not available in any previously known laser system. In addition, for this class of lasers the fundamental parameter describing the quality of radiation, known as spectral brightness (the energy of the emitted laser pulse divided by its duration, spectral bandwidth, beam cross section, and divergence), can be increased essentially to the theoretical maximum. The experimentally demonstrated enhancement in spectral brightness is a factor of approximately 10^{10}. Overall, the discovery of excimer lasers, together with the subsequent technical improvements made in performance, represents a dramatic advance in the technology of sources of coherent radiation.

The efficiency characteristic of the excimer media is of special interest. The radiating species of this class of systems are mainly electronically excited diatomic molecules, typically consisting of a rare gas and a halogen atom, which are formed by a sequence of complex reactions in the gas phase at somewhat greater than normal atmospheric density. The necessary electronic excitation is provided by collisions of the constituent atoms with a current of electrons made to flow by external means through the material. Although the electron-atom collisions generate a rather unselective distribution of possible excited atomic configurations, the subsequent atom-atom interactions characteristic of the material create an unusually efficient and orderly flow of energy, and a very substantial specificity of excitation arises. In other words, excimer media can be regarded as an efficient energy funnel, directing practically all the broad, primary electronic excitation into the narrow band of excited states which participate directly in the laser action. Overall, the kinetic energy of electrons is used to produce a broad, unselective excitation of atoms; next, this energy is transferred first via atom-atom interactions to a few selectively excited states which serve as the upper laser level, and then via stimulated emission to the ultraviolet laser field, thereby leading to the efficient production of ultraviolet radiation. Several excimer laser systems have been developed involving ordered energy flow of this type.

Vacuum ultraviolet lasers. Until recently, coherent sources in the vacuum ultraviolet have been extremely limited in power, efficiency, and available wavelengths. This picture is now changing in direct response to the advances in ultraviolet laser technology. A basic impediment to the development of lasers in the vacuum ultraviolet region, as well as at shorter wavelengths, has been the general lack of a physical means for furnishing the required high rate of excitation in a sufficiently selective manner. Basically, a technology with high spectral brightness is needed to supply the energy density of excitation necessary to establish the conditions for amplification in the domain of short wavelengths. Since the power required to create the conditions for laser action at wavelength λ scales is $1/\lambda^4$, very severe demands are associated with short-wavelength generation. This limit can now be pushed to much smaller wavelength values by proper utilization of the efficient, high-spectral-brightness ultraviolet laser technology now available.

This short-wavelength capability has been realized in recent demonstrations of vacuum ultraviolet laser action in molecular hydrogen, the simplest and most abundant molecular material in the universe. Through the use of appropriate radiative coupling mechanisms to atomic and molecular material, long-wavelength radiation can be directly and efficiently converted into radiation at shorter wavelengths. The specific class of suitable couplings is described as nonlinear, since these mechanisms always involve the conversion of two or more long-wavelength quanta into the shorter-wavelength quantum, and therefore do not exhibit a simple linear dependence on the radiation field at the long wavelength. Since nonlinear couplings are, in general, very strongly favored by high-intensity electromagnetic fields, a premium value is placed on the ability to produce pulsed outputs of maximal power and intensity. In this context, ultraviolet laser technology is most efficacious when configured to produce output pulses with a duration in the picosecond (ps) range. Excimer lasers with pulse duration in the 10-ps range have recently been developed, and no major difficulties are expected for the development of excimer lasers emitting pulses of 1-ps duration.

Other technologies have been considered as alternatives for furnishing the excitation, including electron beams, electrical discharges, laser-produced plasmas, current pinch phenomena, nuclear reactors, and chemical explosives. Although many of these can develop very high power, the control of the energy is generally poor and the effective brightness is relatively low. This is not particularly surprising, since the use of coherent energy is generally expected to compete very favorably with that in incoherent form.

X-ray lasers. Although nonlinear scattering mechanisms have been used to generate coherent radiation in the extreme ultraviolet range at wavelengths shorter than 100 nm, no genuine x-ray laser was in operation in late 1982. That is, no amplification at x-ray wavelengths arising from the normal stimulated emission process associated with an inversion of population had been observed. Nevertheless, the developments discussed above clearly point to a significant change in this situation in the near future.

Applications. Ultraviolet excimer lasers can serve a broad spectrum of scientific and technical applications. A simple comparison is useful in illustrating the vitality of this field of activity. Normal matter is bound together by forces arising from the presence of valence electrons, which are normally subject to an effective electronic field strength of approximately 10^9 V/cm. On the other hand, the electric fields characteristic of the radiation that can be generated by the most recent ultraviolet laser technology can approach a value of 10^{10} V/cm, a considerably greater magnitude. In this circumstance, the electric field produced by the radiation, rather than the atomic field arising from the internal binding forces, is the dominant factor governing the electronic motion. This condition represents an abrupt reversal of the customary experimental situation and clearly indicates the potential of the ultraviolet laser technology for both basic material studies and the general control of the behavior of matter. Important applications to extensive areas of chemistry and the solid state are expanding rapidly.

The existence of a laser, or a comparably bright source of radiation, at x-ray wavelengths would open up an enormous range of applications in both pure and applied science. Basically, the laser devices would complement synchrotrons as sources of x-ray quanta and would permit the study of nonlinear optical processes in the x-ray region, an area of research which is accessible only with high-brightness laser sources. The applications of an x-ray laser can be placed into two classifications: those associated with fundamental physical measurements, which mainly represent the use of the x-ray laser as an instrument for detailed physical diagnostics; and those concerned with the actual processing of solid-state microstructures, such as photolithography.

In terms of output characteristics, an x-ray laser is anticipated to produce naturally the short-wavelength radiation in a pulse length on the scale of picoseconds or less. Such a source is ideally suited for detailed examinations of condensed matter, including biologically active materials. Examples of specific areas of study are microholography of live biological specimens, and absorption, reflection, and luminescence spectroscopy of solids for a wide range of materials, including metals, semiconductors, and insulators.

For background information *see* LASER in the McGraw-Hill Encyclopedia of Science and Technology.　　　[HANS EGGER; CHARLES K. RHODES]

Bibliography: C. K. Rhodes (ed.), *Excimer Lasers*, 1979; C. Kunz (ed.), *Synchrotron Radiation*, 1979.

Launch vehicle

The Ariane is a three-stage launch vehicle (see illustration) built for the European Space Agency (ESA) by a consortium of 36 European industrial companies. The design responsibility rests with the French space agency, the CNES (Centre National d'Études Spatiales). The launcher has a total height of 47.4 m and weighs 210 metric tons at liftoff. Its mass is 90% propellant; the structure and the usable payload account for only 9% and 1%, respectively of the launcher's total mass.

Components. The first stage weighs 13.32 tons empty and has a height of 18.4 m and a diameter of 3.8 m. It is equipped with four Viking V engines that develop a thrust of 245 tons at liftoff. Its inflight burn time is 146 s.

The 147.6 tons of propellants, unsymmetrical dimethylhydrazine (UDMH) and nitrogen tetroxide (N_2O_4), are contained in two identical steel tanks connected by a cylindrical skirt. The four Viking turbopump engines are mounted symmetrically on the thrust frame and can be swiveled in pairs about two orthogonal axes to provide three-axis control. Four tail fins, each 2 m^2 in surface area, improve aerodynamic stability.

Second stage. The second stage weighs 3.13 tons empty (without the interstage and jettisonable acceleration rockets) and has a height of 11.6 m and a diameter of 2.6 m. It is equipped with a single Viking IV engine that develops a thrust of 72 tons in vacuum for 136 s of flight. The engine is attached to the tapered thrust frame by a gimbal with two degrees of freedom for pitch and yaw control; roll control is effected by auxiliary jets fed by hot gas tapped from the same-stage gas generator. The fuel tank, of aluminum alloy with a common bulkhead, is pressurized with gaseous helium (3.5 bars, or 350 kilopascals) and contains 34.1 tons of propellants (UDMH and N_2O_4).

Before liftoff, during the wait on the launch pad, the second-stage tank is thermally protected from the heat exchange between the propellants and the external environment by cold-air-ventilated external blankets. These thermal blankets are jettisoned at liftoff (see illustration).

Third stage. The third stage is the first cryogenic fuel stage developed in Europe. It weighs 1.164 tons empty and is 9.08 m high and 2.6 m in diameter. It is equipped with an HM7 engine that develops a thrust of 6 tons in vacuum for 545 s of flight.

The fuel tank, which contains 8.23 tons of propellants (liquid hydrogen and liquid oxygen), is of aluminum alloy and, like the second stage, has a common bulkhead to separate the fuel and the oxidizer. The tank is clad with an external thermal protective layer to prevent heating of propellants. The hydrogen and oxygen tanks are pressurized in flight by gaseous hydrogen and helium, respectively. As in the second stage, the engine is attached to a tapered thrust frame, which is gimbal-mounted for pitch and yaw control; roll control is provided by auxiliary jets ejecting gaseous hydrogen.

Stage separation. Stage separation is effected by linear-shaped pyrotechnic charges located in the aft skirts of the second and third stages. The stages are moved apart by retrorockets mounted on the lower stage and by acceleration rockets attached to the upper stage. Separation of the first two stages is commanded by the on-board computer upon detection of first stage thrust tail-off (propellant depletion). Separation of the second and third stages is commanded by the on-board computer when the velocity increment due to the second-stage thrust has attained a predetermined value.

Equipment bay. The equipment bay houses the vehicle's electrical subsystem, including the on-board computer. All the electrical equipment necessary for carrying out the flight profile, (that is, equipment for sequencing, guidance, flight control, and tracking), the flight safety destruction subsystem, and the telemetry is in the vehicle equipment bay (VEB) mounted atop the third stage. The VEB is a donut-like aluminum structure 3.2 m in diameter and 1.15 m high with the top of the third-stage tank in the middle, and weighs 316 kg, including the various electronic black boxes. The VEB also supplies power to the payload, usually one or two communications satellites, as well as providing the attachment points for the large protective fairing.

Fairing. The metal fairing, which protects the payload during the ascent phase through the atmosphere, is jettisoned in flight during the second-stage burn at an altitude of approximately 120 km. The fairing, which weighs 826 kg and has a diameter of 3.2 m and a height of 8.65 m (external dimensions), is bulb-shaped to provide a diameter and useful volume compatible with satellites requiring Ariane capability. The bottom, or boat-tail, section of the fairing is made of radio-transparent material to allow communications with the payload. If two satellites are to be carried, the bottom satellite is placed inside an egg-shaped structure called the SYLDA [system to launch a double Ariane (payload), or système de lancement double Ariane]. The SYLDA is a 180-kg aluminum-honeycomb carbon-filament-covered structure that separates to allow the bottom satellite to be launched. The top satellite is carried on top of the SYLDA and is launched first. Both satellites, the "top rider" and SYLDA- enclosed

Liftoff of the Ariane launch vehicle. The falling rectangular objects are thermal blankets that protected the second-stage tank and have been jettisoned from the vehicle.

"bottom rider," are enclosed by the large external fairing.

Fairing ventilation on the ground is provided by a filtered airflow of 150 normalized m³/h (normalized implies standard temperature and pressure) within the service tower, and of 2350 normalized m³/h outside the tower, on the pad.

Follow-on development. The cost of a "kilogram in orbit" is becoming a major factor for customers and a key element in the competition between the various launch systems. The development of the *Ariane 1* launch vehicle began in 1973, and was completed in 1981. Accordingly, an uprating program is under way to reduce the launch costs. This economic consideration, together with a growing trend toward larger spacecraft in orbit, has led to proposals for a first phase of uprating aimed at dual launches of satellites and at reducing the cost per kilogram in orbit by about 20%. The launches which will be developed through this program are called *Ariane 2* and *Ariane 3*.

Ariane 2 and Ariane 3. The *Ariane 2* launcher will be able to place a payload of more than 2000 kg in geosynchronous transfer orbit. This improvement over the current version is obtained by an increase of the thrust of the Viking engines used in the first and second stages; a 20% increase in the mass of the third-stage cryogenic propellants, from 8 to 10 tons; an increase of about 4 s in the specific impulse of the third-stage engine; and finally, an increase in the volume beneath the fairing by modifying the forward conic section.

Lower launch cost may come about by recovering the first stage by means of a parachute system deployed during its reentry. Feasibility studies have been completed and have shown the economic benefit of this modification. Testing of this recovery system in flight is planned for early 1983. The first launch of *Ariane 2* is to take place toward the end of 1983.

The *Ariane 3* launcher, which will be available simultaneously with *Ariane 2*, is the same as the latter, with the addition to the first stage of two strap-on boosters, each containing 7 tons of solid propellant and delivering 70 tons of thrust. It will be able to place a payload of more than 2580 kg in transfer orbit, or two payloads of 1195 kg each in a dual launch (utilizing the internal support structure, SYLDA).

Ariane 4. Trends in space use will lead to the launching of even greater payloads after 1985, and studies have been completed for an even more powerful version of Ariane, the *Ariane 4*, which will be able to match a wider range of possible requirements.

Ariane 4 will make use of the development carried out under earlier programs. The characteristic feature of this launcher will be the existence of six different versions, with performances in transfer orbit of between 2000 and 4300 kg. All the *Ariane 4* vehicles will have a stretched first stage with a capacity of 220 tons of propellant, powered by four Viking engines operating at a chamber pressure of 58.5 bars (5.85 megapascals). They will have the same second stage and third stages as *Ariane 2* and *3*; an equipment bay similar to those of *Ariane 2* and *3*, subject to structural strengthening; and a new, larger 4-m-diameter fairing available in three versions: normal, lengthened, and dual-launch. The various versions of the *Ariane 4* launcher will differ according to the number of strap-on boosters: either two or four solid-propellant boosters derived from those of *Ariane 3* or two or four liquid-propellant boosters using the Viking engine and approximately 38 tons of propellant. There are also plans for a hybrid version with two dry-propellant boosters and two liquid ones and for another version without boosters. These *Ariane 4* configurations will ensure great flexibility and will enable the launcher's performance to be matched to the payload, thus keeping the occupancy rate high.

The development of *Ariane 4* was formally approved by ESA in January 1982. The program started in early 1982 and will be concluded by a test launch in the second half of 1985. *Ariane 4* will be able to launch satellites as large as the *Intelsat 6* telecommunication satellite whose launch is scheduled from 1986 onward.

Ariane launch site. The Ariane is launched from a modern complex in Kourou, in French Guiana, South America. This is the launch site nearest to the Equator and therefore allows maximum benefit from the Earth's rotation. A second launch pad, for the *Ariane 4*, called ELA-2, is planned to be operational in late 1984.

The reasons for proposing this new site are to provide redundancy for the present launch site, ELA-1; to increase operational flexibility by reducing the interval between launches to 1 month; and to enable launches of the improved versions of the vehicle beyond *Ariane 3*.

Commercial use of Ariane. Marketing and operational use of the Ariane launch vehicle has been entrusted to a new French company called Arianespace based in Evry, near Paris. Arianespace is unique in that it is a private company whose shareholders include the 36 firms that build the Ariane, 13 banks, and the CNES. To date, the launch service has been purchased by European, North American, and South American governments and by private companies, usually for the launch of telecommunication satellites.

For background information *see* LAUNCH COMPLEX; ROCKET ENGINE; SPACECRAFT STRUCTURE in the McGraw-Hill Encyclopedia of Science and Technology.

[RALPH W. JAEGER]

Bibliography: Arianespace, *Ariane 2-3 User's Manual*, issue no. 1, 1981; ESA Information Retrieval Service, *ESA Bulletin*, no. 15, 1978.

Lectins

Although body fluids of marine invertebrates, the seeds of leguminous plants, and the cell surfaces of certain bacteria represent diverse biological sites, they share an interesting property: in each of these locations can be found one or more types of compounds called lectins. Lectins are protein molecules with the ability to recognize particular sugars. When these sugars are found in association with certain types of cells, lectins can be used to distinguish these cells from others. Lectins are thus important tools in cell biology.

Erythrocytes are the cells most commonly used in studying lectins. In recent years it has become clear that a variety of structures and products of other cell types, including single-cell organisms, also react with these sugar-specific proteins (Fig. 1). These interactions have increased knowledge of microbial topography and the general structure of the complex sugars associated with them.

Lectin-microorganism interactions. Most studies of lectin-microorganism interactions have focused on

Gram-positive
bacterium

Gram-negative
bacterium

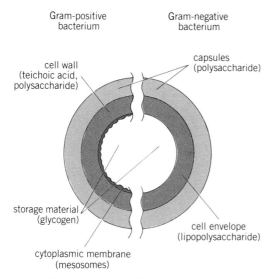

cell wall
(teichoic acid,
polysaccharide)

capsules
(polysaccharide)

storage material
(glycogen)

cell envelope
(lipopolysaccharide)

cytoplasmic membrane
(mesosomes)

Fig. 1. Schematic representation of gram-positive and gram-negative bacteria, illustrating the various sites to which lectins are known to bind.

the surface structures of these organisms. Capsular material is usually a polysaccharide, and may be composed of a single sugar such as glucose, or of a combination of two or more sugars. In the latter case, different lectins may react with the separate components of the polysaccharide.

The cell walls of gram-positive bacteria usually contain a complex phosphate polymer known as teichoic acid. If this molecule is glycosylated, the resulting complex sugar may be detected by lectins. Sugars often associated with teichoic acids are glucose, galactose, and their *N*-acetylated aminosugar counterparts. Additional lectin-binding sites in the form of polysaccharides are also found in the cell walls of certain gram-positive bacteria.

Lipopolysaccharides, found in most gram-negative bacteria, interact with many types of lectins. This is due to the wide variety of sugars found in lipopolysaccharide molecules from different microorganisms. By noting the reactivity of a particular lipopolysaccharide with lectins, the component sugars and their arrangement within the molecule can be determined to a certain degree. Besides bacteria, lectins have been shown to react with many fungi and yeasts, certain protozoa and algae, and even viruses.

Bacterial agglutination. Interactions between lectins and microorganisms can be demonstrated in a variety of ways. In most cases the assay systems are analogous to serological assays using antibodies. The simplest detection system is the agglutination test. Organisms bearing the particular receptor site to which a lectin will bind are aggregated in the presence of the lectin, assuming it is polyvalent. Bacterial agglutination can be observed with the unaided eye, the light microscope, or the electron microscope (Fig. 2). Because agglutination can occur nonspecifically, it is important to include proper

controls in such assays. Inhibition of the agglutination reaction by the addition of the corresponding free sugar is often used to confirm that the original aggregation was lectin-mediated. Direct binding of the lectin to microorganisms may be demonstrated by prior introduction of a label in the lectin. This label may be a fluorescent dye, an enzyme, a radiolabel, or an electron-opaque substance for visualization in the electron microscope.

Applications in microbiology. Lectins are widely applied as tools in microbiology. By interacting with a particular microorganism, a given lectin may indicate not only the presence of a particular sugar

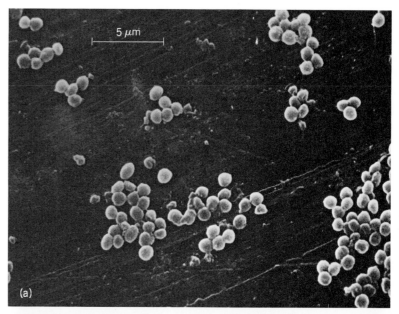

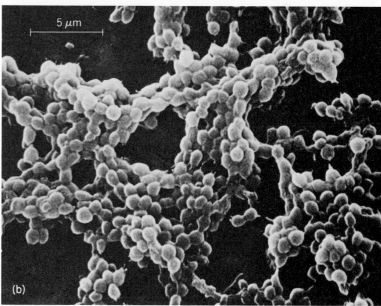

Fig. 2. Scanning electron micrographs of *Staphylococcus aureus* (*a*) in the absence of and (*b*) in the presence of a staphylococcus-binding lectin from the horseshoe crab, *Limulus polyphemus*. (*From K. J. Gilbride and T. G. Pistole, Isolation and characterization of a bacterial agglutinin in the serum of Limulus polyphemus, Prog. Clin. Biol. Res., 29:525–535, 1979*)

but also its anomeric specificity. For example, concanavalin A, which is a commonly studied lectin derived from the jack bean, will bind to α-glucosylated, but not β-glucosylated, teichoic acids. Lectins may also be used to recover structural components of bacteria in a purified form from whole-cell preparations. Teichoic acids, capsular polysaccharides, and lipopolysaccharides have been obtained in this manner. One particular form of this procedure, known as affinity chromatography, involves passing a mixture of these materials through a chromatography column in which the particular lectin has been immobilized. Nonreactive substances pass through the column, but lectin-binding material is retained. The latter can be eluted by adding the corresponding free sugar.

Because certain bacteria have rather unique surface structures, lectins may find increasing use in the clinical laboratory. A lectin found in the snail *Helix pomatia* will agglutinate streptococci belonging to Lancefield's group C, but no others. This reaction may be used to distinguish group C from the closely related group A streptococci, which are major human pathogens. Another lectin, wheat germ agglutinin, may be useful in screening tests for gonorrhea. Lectins have also been used to aid in the identification of pathogenic fungi in surgical and postmortem specimens. Plant-derived lectins have been shown to distinguish microorganisms with disease-producing potential from their avirulent counterparts.

Role of lectins in nature. Despite the many studies that have been carried out with lectins, their role in nature remains unclear. Lectinlike substances appear to be involved in species-specific recognition in sponges and other invertebrates. They may also participate in sugar transport within living systems. Some of the most interesting possibilities involve localized lectin-microbe interactions. Lectins found in the roots of certain legumes may mediate the symbiotic relationship between these plants and species of soil bacteria. Root nodules filled with species of *Rhizobium* are capable of fixing atmospheric nitrogen, and growing evidence suggests that each particular legume species has its corresponding *Rhizobium* symbiont.

An intriguing idea proposed by both scientists in plant research and those in animal research is that lectins may participate in host defense mechanisms. Although the lectins which have been chemically and physically characterized appear to have no obvious structural similarity to vertebrate immunoglobulins, it is possible that these lectins function like antibodies. The binding of potential pathogenic microorganisms by endogenous lectins may prevent their spread within the potential host. In addition, serum from certain invertebrate species, containing one or more lectins, can effect the selective killing of bacteria. Lectins may also function as natural opsonins, coating microorganisms and rendering them susceptible to destruction by circulating blood cells. These observations are based on assays in laboratory culture;

whether they can be extended to living animals is not known. While studies on this aspect of lectins have been confined until now to plants and invertebrate animals, it is possible that lectins from vertebrate species may contribute to their natural defense systems, complementing the antibody-mediated specific defenses found in these animals. It is also possible, of course, that lectin-microorganism interactions are mere laboratory artifacts and that the true role of these proteins is unrelated to their ability to bind to bacteria. Further studies are needed before these speculations can be substantiated.

Besides the interaction of microorganisms with plant- or animal-derived lectins, it appears that bacteria themselves possess surface structures with lectinlike properties. For a number of intestinal diseases due to bacteria, an initial binding of the microorganism to susceptible host tissue seems to be a prerequisite of overt infection. In some cases this adherence can be blocked with simple sugars, for example mannose, suggesting that lectins may be involved. Similar mechanisms of adherence have been found in organisms populating the oral cavity and in those causing urinary tract infections. The phenomenon is not an isolated one, although ascribing the microorganism–host cell interaction to lectins must await further characterizations of bacterial surface structures.

For background information *see* Cellular adhesion; Cellular affinity; Lectins; Nitrogen fixation in the McGraw-Hill Encyclopedia of Science and Technology.

[THOMAS PISTOLE]

Bibliography: I. J. Goldstein and C. E. Hayes, The lectins: Carbohydrate-binding proteins of plants and animals, *Adv. Carbohyd. Chem. Biochem.* 35:127–340, 1978; H. Lis and N. Sharon, Lectins in higher plants, in P. K. Stumpf and E. E. Conn (eds.), *The Biochemistry of Plants: A Comprehensive Treatise*, vol. 6: *Proteins and Nucleic Acids*, ed. by A. Marcus p. 371–447, 1981; T. G. Pistole, Interaction of bacteria and fungi with lectins and lectinlike substances, *Annu. Rev. Microbiol.* 35:85–112, 1981; R. W. Yeaton, Invertebrate lectins: II. Diversity of specificity, biological synthesis and function in recognition, *Develop. Compar. Immunol.* 5:535–545, 1981.

Leukemia

The role of viral agents in animal leukemia has been known for a long time. The demonstration of virus particles in humans affected with leukemia has been less successful. However, the discovery of an enzyme—reverse transcriptase—in T cells of leukemia and lymphoma patients strongly supports a viral etiology for certain forms of the disease.

Retroviruses in animal leukemias. RNA tumor viruses (oncornaviruses and retroviruses) have long been identified as etiological agents in the natural incidence of leukemias and lymphomas in a wide variety of animal systems. In 1908 V. Ellermann and O. Bang reported transmission of leukemia in chickens with cell-free filtrates. Ludwig Gross later showed the induction of leukemias in certain inbred strains of mice with murine leukemia virus. Similarly, feline leukemia virus was shown to induce leukemia in cats. Seroepidemiological surveys revealed that most, if not all, naturally occurring leukemia in cats is caused by feline leukemia virus. Other species where an etiological association between retroviruses and certain leukemias or lymphomas has been established include cattle, gibbons, some wild mice, and turkeys. Retroviruses are a class of RNA viruses, and they carry a special DNA polymerase called reverse transcriptase. Upon infecting a cell, this reverse transcriptase transcribes the viral genomic RNA into a complementary DNA. This DNA, called provirus DNA, is integrated into the cellular genome. When the infected cell divides, the provirus is replicated along with the host genome and is passed on to the progeny cell.

In the natural incidence of leukemias in animals, the viruses are transmitted horizontally by exogenous infection. In this respect these viruses as a class are unlike the endogenous retroviruses, contained in the germ line of certain species, which are transmitted vertically (genetically). Endogenous retroviruses are not usually expressed in RNA, protein, or extracellular virions, and do not seem to be pathogenic except in some inbred animals. Pathogenic retroviruses are grouped into transforming viruses and nontransforming viruses. The transforming viruses cause acute malignant transformation in animals and carry a transforming gene, variously described as src, onc, leuk, and so on. These genes are derived from normal host-cell DNA in the past, presumably by recombination events between viral and host DNA. Some results indicate that increased expression of one particular onc gene may lead to growth of particular cells, but not of another cell type. Therefore, study of transforming genes and their expression may elucidate one general type of carcinogenesis. The nontransforming retroviruses, those that carry no onc gene, usually do not transform cells in culture, and they cause leukemias and lymphomas in animals only after a long latent period. Most naturally occurring animal leukemia viruses belong to this category. Several of these "chronic" leukemia viruses are thought to induce leukemias by activating cellular onc genes as a result of provirus integration in the proximity of the cellular genes. Such integration may provide either a viral promoter or an enhancer sequence for the expression of these normally unexpressed host genes.

Evidence for retroviruses in human leukemia.
Although there have been a large number of reports suggesting the involvement of retroviruses in some human leukemias and lymphomas, until recently the presence of retroviruses in humans was not clearly demonstrated. Reverse transcriptase, resembling the enzyme from animal retroviruses, was partially purified from some human leukemic leukocytes. Further nucleic acid sequences hybridizing to retroviral RNA have been detected, specifically in a leukemic

twin where such sequences were absent in the normal, identical twin. These results suggested that exogenous infection by an unidentified retrovirus may be involved in some human leukemias. Yet, unequivocal isolation of a human retrovirus was not forthcoming. Recent evidence indicates that the difficulties in the past stemmed partly from the uncertainty of what cell and tissue sources should be examined, and partly from the paucity of available culture systems to successfully grow appropriate types of cells in culture. For instance, many viral-induced leukemias in animals are of T-cell origin, yet the ability to grow T cells in the laboratory has only recently been acquired. In 1976 the Laboratory of Tumor Cell Biology (National Cancer Institute), which has shown continued interest in the long-term growth and differentiation of human hematopoietic cells, discovered a T-cell growth factor in the conditioned media of lectin-stimulated human peripheral blood lymphocyte culture. As a result, for the first time several neoplastic, mature human T-cell lines were established and were systematically examined for expression of retroviruses. Cultures from the lymph node and peripheral blood lymphocytes of a patient with cutaneous T-cell lymphoma were found to express type-C retrovirus particles, identified by electron microscopy and by reverse transcriptase activity in the cell-free culture fluid. This reverse transcriptase activity had all the biochemical characteristics of a retroviral reverse transcriptase, and both the enzyme activity and the particles are banded in a sucrose gradient at a specific gravity of 1.16, which is typical for a type-C retrovirus. This virus, identified as human T-cell leukemia/lymphoma virus (HTLV), has now been characterized in detail. Subsequently, the same virus or very closely related viruses have been isolated in certain laboratories from several other cases of T-cell leukemias and lymphomas from around the world.

Human T-cell leukemia/lymphoma virus is a unique human retrovirus, substantially unrelated to all the known animal retroviruses. Its nucleic acid sequences are not present in normal human DNA. Even in the patient from whose cell cultures the virus was initially isolated, viral genetic sequences were present only in the neoplastic T cells, and were absent in the normal B cells. Therefore, HTLV is not an endogenous human virus transmitted through the germ line, but must be acquired by postzygotic infection. Patients with HTLV-related T-cell malignancies usually possess serum antibodies to HTLV proteins. A survey of thousands of human sera for anti-HTLV antibodies has shown some extremely interesting correlations. First, HTLV is usually correlated with aggressive forms of T-cell malignancies associated with visceral organ enlargement, sometimes hypercalcemia, and often with skin manifestations. The neoplastic T cells of patients often have convoluted nuclei and ultrastructurally resemble a small cell variant of Sezary syndrome. This typical disease is described by Japanese clinicians and pathologists as adult T-cell leukemia (ATL). The occurrence of this disease has been documented in geographic clusters in the southwestern parts of Japan, particularly on the islands of Kyushu and Shikoku. Sera of nearly all the Japanese ATL patients contain natural antibodies to HTLV proteins, including the core proteins, suggesting that these patients may have a replicating virus. In contrast, sera of normal donors from nonendemic areas have only a small incidence of antibodies to HTLV. As expected, the incidence (about 10%, but higher in some regions) of antibody-positive nonleukemic people in the ATL-endemic regions is higher than in the nonendemic regions, a reflection of their increased exposure to the virus in the endemic areas. A clinical entity very similar to the Japanese ATL is encountered among Caribbean blacks, and has been called T-cell lymphosarcoma cell leukemia (T-LCL). In an initial screening of sera of such patients, eight out of eight have been found to be antibody-positive, and about 3% of the normal individuals in the Caribbean basin were also found to have antibodies to HTLV, indicating that this is another endemic area for clusters of HTLV-associated T-cell leukemias. When T cells of Caribbean T-LCL patients were grown in culture, almost all of them expressed HTLV. Based on the results of extensive serological screening, some of the T-cell malignancies in the United States do not appear to be HTLV-related. However, by selecting patients who displayed similar clinical features as the Japanese ATL and the Caribbean T-LCL, several have been found in the United States and in other countries who have serum antibodies to HTLV or have HTLV in their neoplastic cells. In general, the HTLV-associated neoplasms in Japan, the Caribbean, and elsewhere represent a spectrum of clinicopathological forms, including cutaneous T-cell leukemias, diffuse large and mixed cell lymphoma of T cells, and ATL. The T-cell lymphomas of interest are of peripheral T-cell origin, and therefore the malignant cells are mature and express differentiated functions. Typical studies using monoclonal antibodies suggest that these T cells are of the helper-inducer phenotype (the subset of T-cells that stimulate antigen activated B-cells to become antibody-producing plasma cells).

A major feature of HTLV-associated leukemias and lymphomas is that the patients' close family contacts have a significantly higher incidence of serum antibodies than the normal population living in the same areas. While relatives living in the endemic areas may be exposed to the virus outside the family, and therefore exclusive intrafamily transmission cannot be ascertained in every case, such a mode of spread has to be a significant factor. In nonendemic areas, such as the United States, relatives of virus-positive patients form the only major group besides the patients themselves who express serum antibodies to HTLV.

In laboratory culture systems, human peripheral blood lymphocytes and umbilical cord blood lymphocytes can be infected by HTLV. These infective transmissions cause transformation of the recipient lymphocytes, resulting in the expression of several phenotypic markers characteristic of HTLV-positive leukemic T lymphocytes. As in most animal leukemias induced by chronic leukemia viruses, HTLV-positive leukemic T cells appear to be of clonal origin with respect to provirus integration sites. Since there is no evidence to indicate that HTLV carries a transforming gene, it is postulated that the integrated provirus may induce leukemia by activating some normal cellular genes. Such mechanisms are already known in some animal leukemias.

As implied earlier, antibodies to HTLV were not ubiquitous among all leukemias and lymphomas. A large number of childhood leukemias and lymphomas were screened during the course of the initial survey. They were all invariably negative, irrespective of whether they were treated or untreated, and independent of their clinical status at the time of obtaining serum sample. The patients included children with acute leukemias and lymphomas of B-cell, null-cell, and T-cell types, and children with miscellaneous solid tumors. A few antibody-positive non-ATL leukemias have been found, and some of these cases were known to be from the ATL-endemic areas. Fortuitous infection with HTLV cannot be excluded in these cases.

From the foregoing discussion it is clear that at least one subtype of T-cell leukemias and lymphomas is associated with an infectious retrovirus. There are, however, other types of human leukemias for which there is as yet no such evidence available. Leukemias are believed to be the end result of a block or aberration in the normal maturation process of the hematopoietic precursor cells. The cells may become arrested in particular differentiation states and never become terminally differentiated as they normally do, and they may retain proliferative capacity for abnormally long periods. This aberration may be brought about by any agent, ranging from environmental factors to viruses. Different factors may impact different target-cell types, resulting in different types of leukemias. It is very likely that not all leukemias are virally induced. However, the fact that no virus has been demonstrated in a majority of leukemias does not exclude the possibility that viruses are involved with these leukemias. Rather, the evidence of a viral etiology in at least one type of leukemia should be an incentive to explore further the possibility of additional viruses in other forms of human leukemia.

For background information *see* ANIMAL VIRUS; LEUKEMIA; LYMPHOMA in the McGraw-Hill Encyclopedia of Science and Technology.

[ROBERT C. GALLO; M. G. SARNGADHARAN]

Bibliography: V. Ellermann and O. Bang, Experimentelle Leukämie bei Hühnern, *Zentralbl. Bakteriol.*, 46:595–609, 1908; L. Gross, "Spontaneous" leukemia developing in C3H mice following inoculation, in infancy, with Ak-leukemic extracts, or Ak-embryos, *Proc. Soc. Exp. Biol. Med.*, 76:27–32, 1951; V. S. Kalyanaraman et al., Natural antibodies to the structural core protein (p24) of the human T-cell leukemia (lymphoma) retrovirus (HTLV) found in sera of leukemia patients in Japan, *Proc. Nat. Acad. Sci. USA*, 79:1653–1657, 1982; D. A. Morgan, F. W. Ruscetti, and R. C. Gallo, Selective *in vitro* growth of T-lymphocytes from normal human bone marrows, *Science*, 193:1007–1008, 1976; B. J. Poiesz et al., Detection and isolation of type-C retrovirus particles in fresh and cultured lymphocytes of a patient with cutaneous T-cell lymphoma, *Proc. Nat. Acad. Sci. USA*, 77:7415–7419, 1980.

Leukotrienes

Many of the observed responses in allergic reactions and inflammation are now thought to be due in part to a group of compounds called leukotrienes. Leukotrienes are a family of oxidized metabolites of certain polyunsaturated fatty acids, predominantly arachidonic acid, produced in specific cells as chemical mediators of the biologic response following stimulation. For example, leukotriene C_4 causes powerful, sustained contraction of bronchial smooth muscle as well as increased capillary permeability or edema. Leukotriene B_4 stimulates the motility of polymorphonuclear leukocytes (chemokinesis) and their movement toward the source of biosynthesis (chemotaxis), a phenomenon common in inflammation. The recognition of the existence of leukotrienes is a recent development in the area of arachidonic acid metabolism. Previously, oxidative metabolism of arachidonic acid to biologically active compounds was thought to be limited to prostaglandins, thromboxanes, and prostacyclin, which originate from a common endoperoxide intermediate. Leukotrienes arise from a different intermediate by way of a specific lipoxygenase route of oxidation. Recent scientific research on these compounds originated with the elucidation of the structure of the slow-reacting substance of anaphylaxis (SRS-A).

SRS-A. In 1940 two Australian physiologists performed experiments with the lung isolated from a guinea pig which has been made allergic to egg albumin. They found that when the lung was exposed to egg albumin, a substance was released from the lung tissue which made certain smooth muscles contract very slowly. They termed it slow-reacting substance of anaphylaxis (SRS-A). After the discovery of antihistamines, it was found that this SRS-A was not entirely due to histamine, which is also produced by the antigen-antibody reaction. During the period 1950–1969 experiments conducted in several laboratories suggested the importance of this substance. SRS-A biosynthesis was linked with the production of antibodies of the IgE class, which suggested the tissue mast cell as a site of biosynthesis (Fig. 1). SRS-A was found to profoundly contract human bronchial smooth muscle. The treatment of lung segments from asthmatic individuals with various pollens resulted in the production of SRS-A, and in prolonged contraction of the bronchial smooth muscle again upon exposure to pollens. These experiments, as well as others, suggested that SRS-A was an important chemical mediator of the clinical symptoms observed in asthma as well as other acute hypersensitivity reactions.

Leukotriene C_4. In spite of the potential importance of SRS-A, its chemical nature was not discovered until 1979. The reasons for this 40-year delay were the chemical instability of SRS-A and the fact that only miniscule amounts are synthesized in cells, typically picomole quantities in an observable allergic response. However, various alternative means of SRS-A production were found, including the stimulation of certain neoplastic (basophilic leukemia and mastocytoma) cells with a drug that stimulates cells by transporting calcium ions across cellular membranes. Advances in purification techniques permitted the isolation of a few micrograms of pure SRS-A. One striking physicochemical property of SRS-A was that it absorbed ultraviolet light

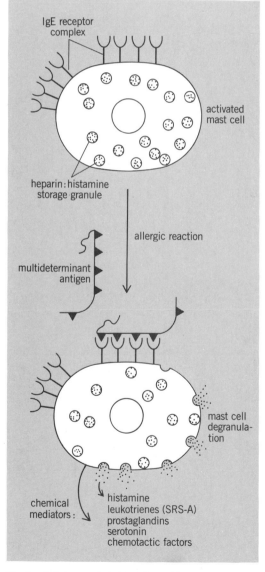

Fig. 1. Cellular events in the IgE-sensitized mast cell leading to the production of SRS-A (leukotriene C_4 and D_4) and other chemical mediators of allergy.

in a manner characteristic of molecules having a conjugated triene structural unit. It was found by using radiochemical tracer techniques that SRS-A was synthesized from arachidonic acid as well as from cysteine. Mass spectrometry was used to characterize in part the arachidonic acid backbone, and amino acid analysis revealed two additional components, glycine and glutamic acid, in the structure of SRS-A. This compound has been synthesized by several organic chemists and its structure verified by numerous chemical, physical, and biological means. The chemical name of SRS-A from the mastocytoma cell is 5(S)-hydroxy,6(R)-S-glutathionyl-7,9-*trans*; 11,14-*cis*-eicosatetraenoic acid (LTC$_4$). Other workers using different species of animals found another SRS-A with an identical structure, except that it lacks either the glutamic acid portion (LTD$_4$) or the glutamic and glycine residue (LTE$_4$). These compounds were named leukotrienes because of their common conjugated triene structure and the fact that white cells (leukocytes) could synthesize them from arachidonic acid.

Leukotriene B$_4$. Another metabolite of arachidonic acid made by leukocytes and containing the conjugated triene structure is leukotriene B$_4$ (LTB$_4$). Its chemical structure was deduced in a series of elegant experiments done prior to the leukotriene C$_4$ studies, and the characteristic absorption of ultraviolet light of leukotriene B$_4$ was an important clue to the final structure of leukotriene C$_4$. The potential biological importance of leukotriene B$_4$ was revealed well after its discovery. Important features of leukotriene B$_4$ are its ability to stimulate eosinophils and polymorphonuclear cells to move (chemokinesis and chemotaxis) and to aggregate. Furthermore, these cells then begin to release lysosomal enzymes which are very powerful in destroying bacteria. The inflammatory response includes the influx of cells like polymorphonuclear leukocytes and the ingestion (phagocytosis) of bacteria and release of lysosomal enzymes by these cells. Leukotriene B$_4$ may be a chemical mediator of these important events in inflammation.

Biosynthesis. The biochemical synthesis of leukotrienes is outlined in Fig. 2. Arachidonic acid is a plentiful fatty acid in mammalian tissue, but it occurs almost exclusively as an ester in phospholipids. In certain cells, like the mast cell, arachidonic acid can be liberated from these phospholipid stores by a specific stimulus. It is thought that the antigen binding to the IgE antibody embedded in the mast cell plasma membrane could be an appropriate initial stimulus. The free (unesterified) arachidonic acid is then available to be a substrate for many enzymatic reactions, including the complex cyclooxygenase system which produces thromboxanes, prostaglandins, and prostacyclin, as well as the 5-lipoxygenase system which produces leukotrienes. This latter pathway takes molecular oxygen and incorporates it into the arachidonic structure to yield the reactive 5-hydroperoxyeicosatetraenoic acid (5-HPETE), or dehydrates it to a reactive allylic epoxide which is also a conjugated triene, leukotriene A$_4$ (LTA$_4$). This is the common biochemical precursor for all leukotrienes. Enzymatic addition of water to leukotriene A$_4$ yields leukotriene B$_4$. Addition of the very common tripeptide glutathione (GSH) to leukotriene A$_4$ yields leukotriene C$_4$. Sequential peptide cleavages converts leukotriene C$_4$ to leukotriene D$_4$ and to leukotriene E$_4$.

Medical implications. The role of leukotrienes in allergy and inflammation is not well established and is the subject of a great deal of research activity at this time. It has been suggested for a number of decades that SRS-A is a chemical mediator in allergic asthma. Pharmacological studies have revealed that purified leukotriene C$_4$ and leukotriene D$_4$ are 1000 times more potent than histamine as a spasmogen in human bronchi. These leukotrienes also increase capillary permeability, have direct and indirect effects on the cardiovascular system, and can have direct effects on neuronal activity in the central nervous system. There is also recent evidence that prostaglandins and leukotrienes may work in concert to produce certain biological responses. Since most of the recognized properties of leukotriene C$_4$

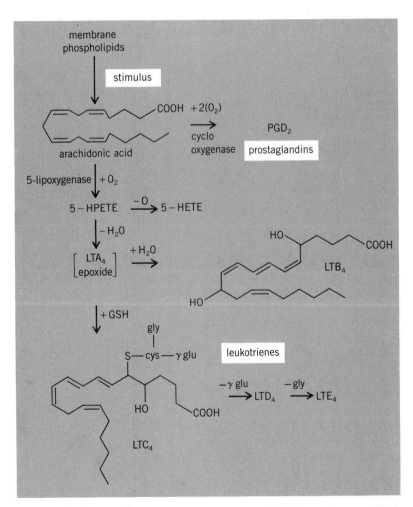

Fig. 2. Biosynthetic events involved in the production of leukotrienes from arachidonic acid.

and D_4 are rather harmful, the target of drug research is to find a substance that will prevent leukotriene biosynthesis. Steroids suppress inflammation in part by reducing the level of arachidonic acid for oxidative metabolism, that is, both leukotrienes and prostaglandins. Nonsteroidal anti-inflammatory drugs, such as aspirin and indomethacin, will prevent the production of prostaglandins, thromboxanes, and prostacyclin. However, they may increase the production of leukotriene metabolites of arachidonic acid. One important goal of current research is to find drugs that will specifically inhibit the lipoxygenase reaction, in particular the formation of leukotriene C_4 and D_4, with the anticipation that many adverse biological responses, such as severe allergic reaction, can be controlled. In any event, the discovery of leukotrienes has enabled researchers to expand their understanding of chemical mediators involved in allergy and inflammation and the importance of arachidonic acid biochemistry in related disease processes.

For background information *see* ANAPHYLAXIS; CELLULAR IMMUNOLOGY; INFLAMMATION in the McGraw-Hill Encyclopedia of Science and Technology.

[ROBERT C. MURPHY]

Bibliography: P. Borgeat and B. Samuelsson, Transformation of arachidonic acid by rabbit polymorphonuclear leukocytes: Formation of a novel dihydroxyeicosatetraenoic acid, *J. Biol. Chem.*, 254:2643–2646, 1979; A. W. Ford-Hutchinson et al., Leukotriene B, a potent chemokinetic and aggregating substance released from polymorphonuclear leukocytes, *Nature*, 286:264–265, 1980; R. A. Lewis and K. F. Austen, Mediation of local homeostasis and inflammation by leukotrienes and other mast cell-dependent compounds, *Nature*, 293:103–108, 1981; R. Murphy, S. Hammarstrom, and B. Samuelsson, Leukotriene C, a slow reacting substance (SRS) from mouse mastocytoma cells, *Proc. Nat. Acad. Sci. USA*, 76:4275–4279, 1979.

Life, origin of

All living systems have basically the same machinery for the translation of the genetic information stored in strands of nucleic acid into the amino acid sequence of proteins. A fundamental difficulty is that this machinery operates by means of proteins, the production of which requires many proteins in the first place. The crucial question is how such a translation device could have arisen. Experimental data are not available, but a solution to the problem, at least in principle, can be found. This may be done by considering a detailed model of a pathway that consists of a sequence of many small, physically and chemically plausible steps, the probability of each step being sufficiently large to permit its occurrence under model conditions with near certainty.

A major premise in the development of such a model is that a particular preexisting external spatial and temporal structure is essential for the formation of the living machinery. A machine is constructed by fitting its interacting parts together through externally directed action. On the primordial planet the role of the external operator is replaced by the enormous variety of environmental influences which offer appropriate conditions somewhere by chance. Important influences are periodic temperature changes at crucial minute sites, caused, for example, by the pattern of light and shadow associated with the daily motion of the Sun, and a structurally diversified environment exemplified by porous rock formations.

Self-organization of matter under the regime of a highly structured environment can obviously occur under appropriate conditions. The crucial problem of the origin of life—finding particular conditions that lead to a solution of the basic puzzles—should not be confused with the problems of the general conditions for the formation of structure in a structureless system.

Initial conditions and first steps. Amino acids and nucleotides (consisting of a sugar and a nucleobase) are the building blocks of proteins and nucleic acids, respectively, the essential components of living systems. These building blocks were presumably present on the primordial planet. Many researchers have been able to obtain amino acids, sugars, and nucleobases by using simulations of presumed conditions on the prebiotic Earth. It has also been demonstrated that these compounds can be made to yield nucleotides (Fig. 1a) and short nucleotide polymer strands on the one hand, and activated forms of amino acids on the other hand, under conditions believed to be realistic. Moreover, the enzyme-free polymerization of nucleotides on nucleic acid templates has been accomplished, with more than 90% of the nucleotides of the replicate strands being complementary to those on the template strands.

It may thus be assumed that suitable conditions exist within pores in rocks to favor the creation and later to drive the replication of short nucleotide strands in such a way that these strands, (+) strands, can act as templates for the polymerization of the complementary (−) strands (Fig. 1b). Pores in rocks above a critical size cannot be populated by such strands because these strands diffuse too quickly from the favorable region. Longer strands may arise by fusion of shorter strands (Fig. 1c), and be able to colonize larger rock pores. The availability of larger pores then provides a selective gradient toward longer strands, but there is a limit on strand length because of copying errors. The probability that replicable strands are error-free must be large enough to ensure their survival.

Macromolecule cooperation. The resulting impass to further evolution is broken, in the model, by the appearance of strands with sequences of nucleotides allowing intramolecular base pairing and aggregate formation (Fig. 1d). The resulting decrease in diffusion rate permits the invasion of still larger pores, and the aggregate formation serves as an error filter because of the fact that only correct

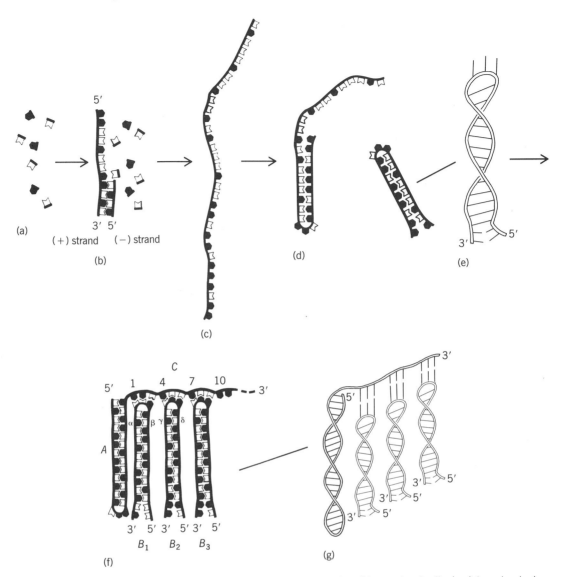

Fig. 1. Molecular model of a translation in the synthesis of proteins. (*a*) Spontaneous condensation of nucleotides; (*b*) replicable short strand obtained under particular conditions; (*c*) formation of longer strands allowing intramolecular base pairing and (*d*, *e*) aggregate formation; (*f*, *g*) particular aggregates.

copies interlock, while erroneous copies are not complementary in shape. In this manner erroneous copies are rejected during aggregation. The simplest conformation that permits such interlocking is that of a twisted hairpin (Fig. 1*e*). A great selective premium is associated with machinery that facilitates such aggregation and a hairpin *A* with open end *C* provides such machinery (Fig. 1*f*). Filament *C* acts as the collector strand, guiding the hairpins B_1, B_2, . . . , B_i one by one to the region where they are incorporated into the aggregate, while *A* acts as nucleation site by side-by-side interlocking with approaching molecule B_i, thus forming a stable complex in case of pairing of anticodon triplet and corresponding triplet on *C*. This picket fence–like arrangement may well be stabilized by polyvalent cations that form links between negatively charged phosphatidyl groups on the outside of neighboring hairpin strands.

Triplet reading as a by-product. The three nucleotides in the middle of the strand of unit B_i form the loop of the hairpin, while its ends, twisted into a base-paired double helix, form the hairpin legs. The nucleotides in the hairpin loop form hydrogen bonds to the complementary nucleotides contained in filament *C*. This imposes restrictions on the sequence of nucleobases of filament *C* and of the loop of hairpins B_i. Assuming the use of only two kinds of nucleobases (guanine G and cytosine C) and assuming that the sequence on the loop of the (+) strand (read from the 3′ end of the strand to the 5′ end) is CCG, the sequence on the loop of the (−) strand will be CGG (strand B_2 in Fig. 1). This is because the 3′–5′ direction of the strand obtained in the template-assisted polymerization is opposite to that of the template strand. Since α was pairing with δ during replication, δ is complementary to α; for a hairpin, γ is complementary to δ and therefore α is identical with γ.

At the positions of the collector strand that correspond to the first position of the triplet of the hairpin bending (at positions 1, 4, 7, . . .; Fig. 1c) there must therefore be G, and at the positions that correspond to the third position (at positions 3, 6, 9, . . .), there must be C. The middle positions can be occupied randomly by G or C. In the first case a (+) hairpin strand would be attached, in the second case a (−) strand. The bonding energies involved in base pairing are known to be insufficient at room temperature to establish stable bonding between two triplets of complementary bases G and C; therefore the hairpins will become firmly attached to strand C only if it reaches the region of growth of the aggregate, if its shape allows for a close enough fit to be incorporated in a stable manner, and if its triplet is complementary to the corresponding triplet on filament C.

Cellular envelopes. The open ends of the hairpin strands (Fig. 1g) do not affect the lateral steric fit and the bonding between adjacent subunits in the aggregate. In the model these open ends possess affinity toward activated amino acids and facilitate the formation of polypeptide bonds between amino acids attached to them. Polypeptides of hydrophobic amino acids, such as glycine and alanine, can form impediments in pore channels and coalesce into confined cellular envelopes. This has a great selective advantage, since regions with pores too large to curb strand diffusion can now be colonized. Again a multifaceted environment provides selective pressure because of the existence of a large variety of niches available for colonization by emerging life forms of increasing sophistication.

Emergence of translation machinery. If amino acids a_1 and a_2, activated by linkage with C and G nucleotides, respectively, existed prebiotically, a specific binding to the open ends of a (+) and (−) strand, respectively, could be obtained and (+) and (−) hairpins would serve as adapters for a_1 and a_2. This would be the beginning of an automatic correlation between amino acids and nucleotides in the anticodon triplets on the loops of the hairpins B_i. The collector strand C would be the precursor of the messenger strand and the hairpins B_i precursors of transfer RNA molecules.

Eventually a sequence along filament C arises, encoding a polypeptide acting as a primitive RNA replicase, that is, a molecule, that decreases replication errors sufficiently so that its information coding would not be lost during the number of generations required to fix it by selection. Such a replicase may, for example, be a suitable short polypeptide that fits into a notch of the double helix which exists during strand replication in the region where the new strand is formed, slipping along in the notch as the new strand is lengthened. The rate of evolution can then increase enormously. More diversified enzymes and increasingly intricate translation machinery evolve.

Molecular models of the specific binding of the amino acids at the open 3′ ends of the hairpins can be realized considerably better with L than with D amino acids. This result may explain the astonishing fact that all amino acids associated with the genetic machinery are of the L kind. The open 3′ ends with the amino acids have sufficient freedom of motion to allow the formation of peptide bonds in the aggregate (Fig. 2). These conclusions are based on the assumption that the original template for replication was made of nucleotides with D-ribose (or some

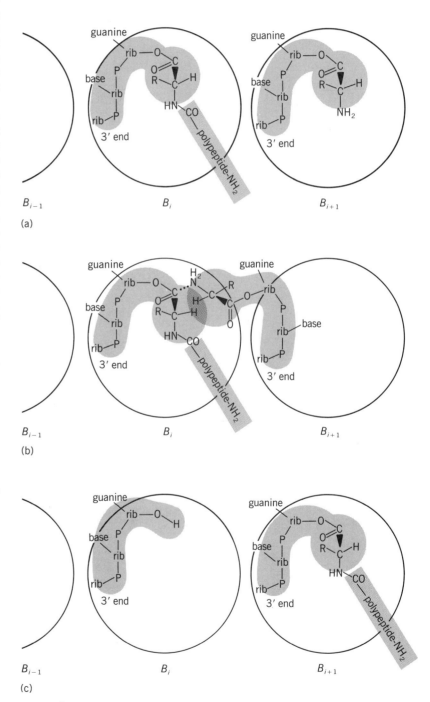

Fig. 2. Formation of peptide bond between polypeptidyl amino acid at the 3′ end of B_i and an amino acid at the 3′ end of B_{i+1}. The hairpins B_{i-1}, B_i, and B_{i+1} are viewed end on. (a) Polypeptide (i amino acids) at 3′ end of B_i. (b) 3′ end of B_{i+1} turned into bonding position of amino acids. (c) Polypeptide (i+1 amino acids) at 3′ end of B_{i+1}.

equivalent), which then induced an evolution based on D-ribose. If the original template had been made with L-ribose, which could have had occurred with the same a priori probability, then systems with D amino acids would have evolved according to the present view.

Supporting experimental evidence. The proposed specific activation of two different amino acids with C and G nucleotides is supported by simulating prebiotic conditions. It has been demonstrated by eluting a mixture of the amino acids glycine (Gly) and alanine (Ala) and the C and G nucleotides (GMP and CMP) with a diluted salt solution that glycine and GMP are eluted with the same speed while alanine and CMP are eluted with another, significantly different speed. This opens the possibility of in-place formation of GMP-Gly and CMP-Ala while the specimens CMP-Gly and GMP-Ala would not be formed.

The view that the collector strand is the primordial form of the carrier of genetic information given in the reading frame GNC (N = G or C) is supported by a recent sequence analysis of the DNA of viruses, prokaryotes, and eukaryotes. This research clearly revealed periodic correlations indicating that originally there existed the reading frame PuNPy (Pu = purine, like G;, Py = pyrimidine, like C; N = purine or pyrimidine). The difference in the deviations from the PuNPy reading frame in different organisms was found to be correlated with the time of their phylogenetic diversion, and from this the calculated time of last use of the PuNPy frame was found to be of the order of 3 billion years ago, a number which agrees well with fossil traces of early life. It can be well imagined that G and C were first used exclusively or predominantly and that later G was partially exchanged for another purine (A) because of its steric similarity, and C by another pyrimidine (U) leading to the PuNPy reading frame.

The model assumption that the transfer RNAs of all amino acids have a common ancestor is well supported by recent comparative studies. These studies indicate that these tRNA sequences have a common ancestor which consists predominantly of C and G, supporting the hypothesis that G and C were used first. The sequence is very similar to another recently obtained sequence of an archebacterial tRNA, which is a very old organism in the phylogenetic sense. If a strand with that sequence is assumed to be in a hairpin conformation, purine-pyrimidine base pairs can be found in 85% of all pairs. This supports the view that the ancestral adapters were in a hairpin conformation and that only a few pyrimidines later changed to purines or purines to pyrimidines. It was demonstrated recently in other work that an association of tRNA molecules is induced during the codon-anticodon binding. These findings may point to vestiges of the postulated early translation device.

For background information *see* DEOXYRIBONU-CLEIC ACID (DNA); LIFE, ORIGIN OF; NUCLEIC ACID; PROTEINS (EVOLUTION) in the McGraw-Hill Encyclopedia of Science and Technology.   [H. KUHN]

Bibliography: M. Eigen and R. Winkler, *Naturwissenschaften*, 68:217, 282, 1981; T. Inoue and L. E. Orgel, *J. Amer. Chem. Soc.*, 103:7666, 1981; M. W. Kilpatrick and R. T. Walker, *Nucleic Acids Res.*, 9:4387, 1981; H. Kuhn, *Angew. Chem. Int. Ed.*, 11:798, 1972; H. Kuhn and J. Waser, *Angew. Chem. Int. Ed.* 20:500, 1981; H. Kuhn and J. Waser, in W. Hoppe et al. (eds.), *Biophysik*, 1982; H. Kuhn and J. Waser, *Nature*, 298:585, 1982.; R. Lohrmann, P. K. Bridson, and L. E. Orgel, *Science*, 208:1464, 1980; D. Pörschke and D. Labuda, *Biochemistry*, 21:53, 1982; J. C. W. Shepherd, *J. Mol. Evol.*, 17:94, 1981; J. C. W. Shephard, *Proc. Nat. Acad. Sci. USA*, 78:1596, 1981.

Lignin

Lignin is a biochemical polymer found in most terrestrial plants, including ferns, horsetails, and large mosses. It provides much of the structural support that enables plants to grow upright against gravity and to withstand physical forces such as wind and snow, and it also functions as a barrier to plant pathogens. Lignin formation occurs outside the cell membrane in the plant cell walls. The lignin is deposited among existing cellulose fibers, forming chemical bonds with the cellulose or other cell wall polysaccharides. Lignin deposition begins near the completion of cell enlargement and significantly increases the rigidity of the plant cell wall. The rigid cell walls then provide the structural support for plants to grow upright and withstand physical forces.

Structure and distribution. The chemical structure of lignin is generally known, except for certain details. It is an amorphous, three-dimensional polymer composed of phenylpropanoid units. The estimated molecular weight is about 100,000 daltons. A portion of conifer lignin is shown in Fig. 1. The assembly of phenylpropanoid units which form lignin does not follow a repeatable pattern, and so the structure of lignin molecules is species-specific. Lignins are classified into one of three major groups: guaiacyl lignin, principally found in conifers (softwoods); guaiacyl-syringyl lignin, found in dicots, including the hardwoods; and guaiacyl-syringyl-*p*-hydroxyphenyl lignin, primarily found in monocots (for example, grasses). The principal building blocks of the three types of lignin are: in conifers, coniferyl alcohol; in dicots, coniferyl and sinapyl alcohol; and in monocots, coniferyl, sinapyl, and *p*-coumaryl alcohols. The proportion of phenylpropanoid units varies between different plant species within each lignin group.

After cellulose, lignin is the second most abundant biological material in the biosphere. The lignin content of higher plants ranges from about 5 to 40%. The lower amounts are found in low-growing plants, and the larger amounts in trees. In general, the lignin content of plants is related to plant mass, and therefore it is probably a function of gravitational forces. Furthermore, the lignin content within individual plant regions varies and is dependent upon the amount of mechanical loading; therefore more

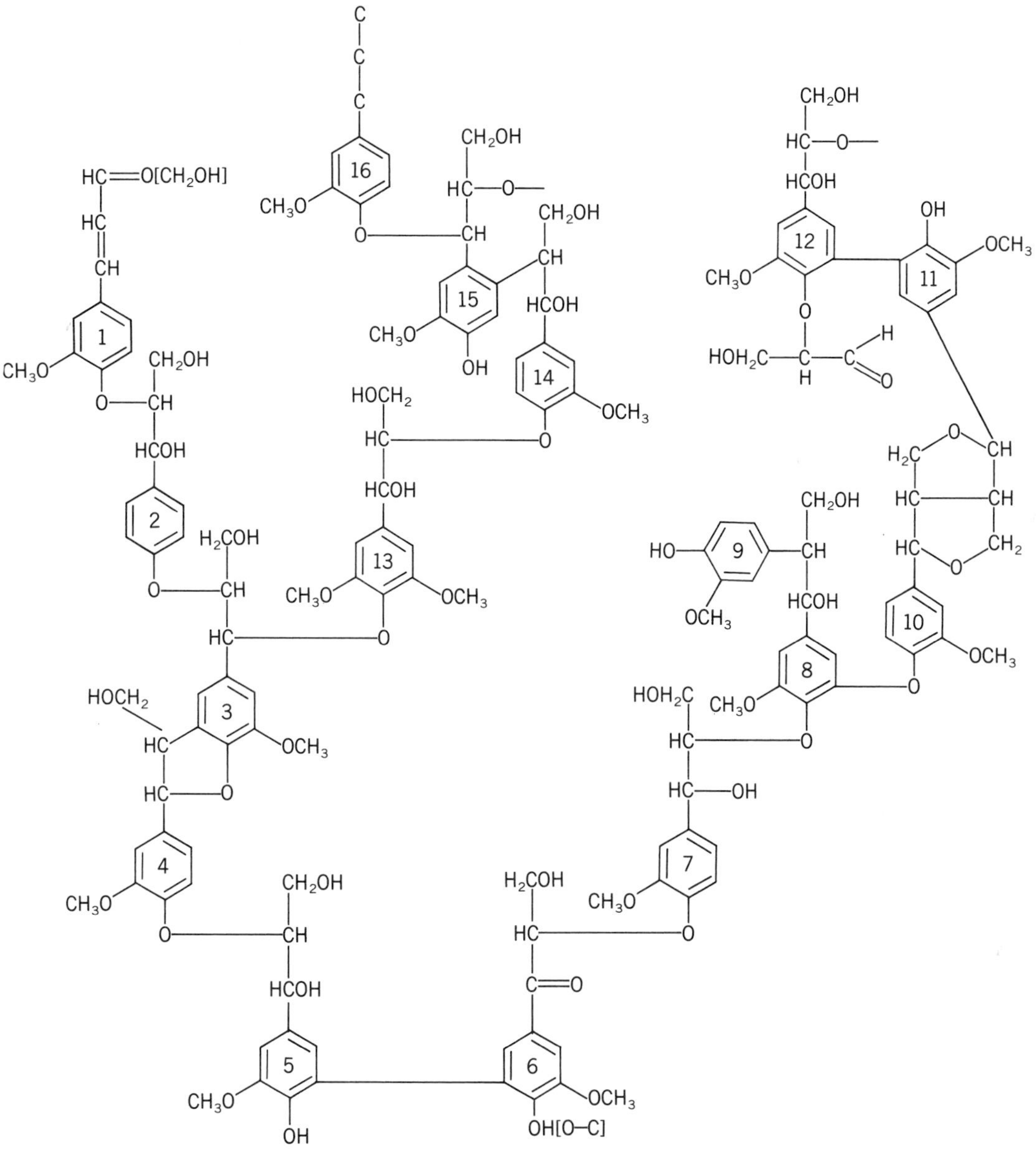

Fig. 1. Structure of conifer lignin (softwood). (*After E. Adler, Lignin chemistry-past, present and future, Wood Sci. Technol., 11:169–218, 1977*)

lignin is usually formed at the base of a plant than near the apex. Most cells within plants contain some lignin, and certain ones, such as xylem and sclerenchyma cells, are particularly lignin-enriched. The first cells to lignify in elongating tissue are xylem cells, which carry water and nutrients up the plant. The large lignin content found in trees results from crushing of the lignin-rich xylem cells to form a dense material. This lignin-enriched material provides the structural qualities of wood.

Synthesis. As with most plant constituents, the initial substrate for lignin synthesis is carbon dioxide. The respiratory products of the cells that provide the initial materials for lignin synthesis are phosphoenol pyruvate from glycolysis and erythrose-4-phosphate from the pentose phosphate pathway. These compounds are used in the shikimic acid pathway, which ultimately produces the aromatic

amino acids phenylalanine and tyrosine. Phenylalanine is directed into the phenylpropanoid biosynthetic pathway by the enzyme phenylalanine-ammonia-lyase (PAL). The product of this reaction, cinnamic acid, is chemically altered to form phenylpropanoid esters and primary alcohols (Fig. 2). The alcohols, or monoalcohol polymers, presumably are synthesized within the cell, probably near the cell membranes, and transported across the cell membrane into the cell walls. Polymerization of the phenylpropanoid units into lignin occurs in the cell wall and is catalyzed by the enzyme peroxidase.

Regulation of synthesis. The regulation of lignin synthesis is not well understood. Obviously with a very complex biosynthetic pathway there are several potential sites of regulation. Other biochemical reactions compete for the substrates and intermediates used for lignin biosynthesis. For example, phenyl-

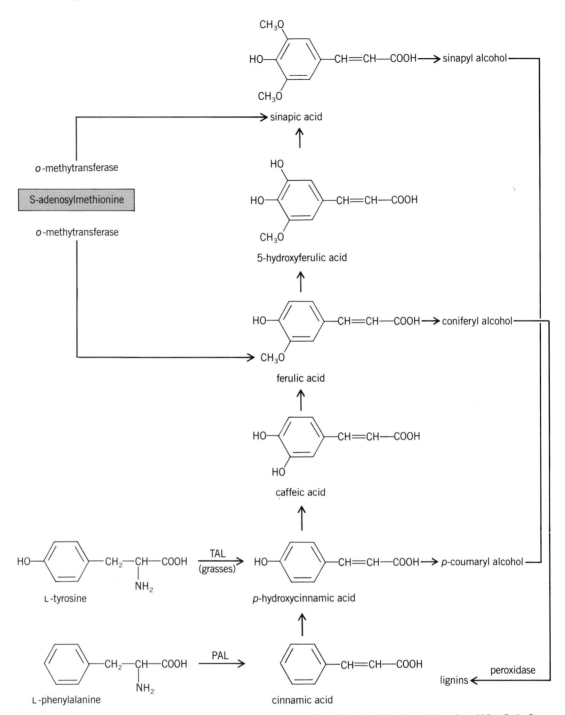

Fig. 2. Lignin biosynthetic pathway, beginning with the aromatic amino acids phenylalanine and tyrosine. (*After R. L. Crawford, Lignin Biodegradation and Transformation, 1981*)

alanine, the principal substrate for lignin biosynthesis, is also required for protein synthesis. Even after phenylalanine is directed into the phenylpropanoid pathway by PAL, it can be used for products other than lignin. Pathways producing cinnamate esters and flavonoids compete with the lignin biosynthetic pathway for phenylpropanoid units. Coumerate:CoA ligase (CoA is coenzyme A) has been suggested as an important regulatory enzyme in distributing cinnamate derivatives between the cinnamate ester, flavonoid, and lignin biosynthetic pathways. As with substrates, other potential users compete for the cellular energy needed to synthesize lignin and, therefore, the rate of lignin synthesis is affected. The plant hormones auxin, gibberellin, and ethylene are reported to affect lignin synthesis, but their specific mode of action is not well understood. One of the largest affectors of lignin synthesis is plant pathogens, which cause substantial increases in PAL and lignin synthesis in the region of attack. The stimulation of lignin synthesis, along with overall aromatic synthesis, is a defense mechanism by plants in an attempt to ward off attack by pathogens.

Several environmental factors affect lignin synthesis. Light is particularly critical for lignin biosynthesis. Young pine seedlings grown in the dark contain less than 10% of the lignin found in light-grown seedlings. The dark-grown seedlings, upon exposure to light, rapidly increase their rate of lignin synthesis. The new rate of synthesis is greatest at the highest light intensity. It is known that light is required for the induction of PAL activity in lignin synthesis. Light is also required for overall substrate and energy synthesis.

Lignin synthesis is also reduced if seedlings are grown at low oxygen levels. Cucumber and bean seedlings grown in an atmosphere of 5% oxygen contain about half the lignin of seedlings grown at 21% oxygen. Young pine seedlings grown at 5% oxygen also produce substantially less lignin than seedlings grown at 21% oxygen.

Exposure of seedlings to increased water stress and to gravity levels above 1 g are reported to increase the lignin content of plants. Young cucumber seedlings exposed to 5 days of water stress contain at least 50% more lignin than the watered controls. Young dicots exposed to reduced gravity by clinostatting (slow horizontal rotation) or flotation exhibit reduced lignification. Young mung bean, oat, and pine seedlings were grown for 8 days in space (near-zero gravity) on the flight STS-3 of the space shuttle in 1982. Preliminary data show a small reduction in the lignin content of whole pine stems and a larger reduction in the lignin content of the mung beans.

Degradation. Once synthesized, lignin is difficult to break down either biologically or chemically. Normal digestive processes are ineffective in hydrolyzing lignin, and it passes through animals, including humans, undigested. Some microorganisms, the best known of which are the white rot fungi, members of the class Ascomycetes, are capable of degrading lignin. Other members of the class Ascomycetes, fungi of the class Fungi Imperfecti, certain groups of bacteria, a number of eubacteria, and actinomycetes can also degrade lignin. Lignin is not easily hydrolyzed by chemicals, including hot sulfuric acid. It can be solubilized, however, by hot alkali and bisulfite. Industries have directed considerable research effort toward efficient ways of removing lignin from cellulose fibers, since it can be a hindrance in the production of certain useful products, such as paper. Lignin is an extremely valuable polymer, however, supporting higher plant growth against gravity and providing a source of building materials.

For background information *see* CELL WALLS (PLANT); SCLERENCHYMA; WOOD ANATOMY; XYLEM in the McGraw-Hill Encyclopedia of Science and Technology.

[JOE R. COWLES]

Bibliography: E. Adler, Lignin chemistry: Past, present and future, *Wood Sci. Technol.* 11:169–218, 1977; R. L. Crawford, *Lignin Biodegradation and Transformation*, 1981; W. J. Schubert, Lignin in L. P. Miller, *Phytochemistry*, vol. 3, pp. 132–153, 1973;

Liquid helium

Physicists are continually on the lookout for ideal systems in which a restricted number of phenomena may be studied free of complications. For those interested in the properties of two-dimensional electron gases, the charged surface of liquid helium is such an ideal system. Liquid helium, largely through its inert gas nature, has the property of negative electron affinity. That is, positive energy is required to insert an electron into bulk liquid from a vacuum. As such, electrons may be trapped near the vacuum-fluid interface of liquid helium by using suitable electric fields. The electronic motion normal to the surface is that of a particle bound in a potential well about 8 nm from the liquid, while motion along the liquid surface resembles that of an electron in free space.

Recently, experiments have revealed that two different types of crystallization occur for electrons on the helium surface. In the first case, as the electric field acting on a uniformly charged surface is raised beyond a critical value, an instability results, with the subsequent formation of a macroscopic hexagonal lattice (Fig. 1). This ordered structure reflects an equilibrium balance of fluid-mechanical and electrostatic forces. The second type of crystallization is of statistical-mechanical origin. As the temperature of the electron-helium system is lowered below a critical value, the electron gas undergoes a phase transition. A two-dimensional solid is formed in which individual electrons are located at the lattice points of a crystal with microscopic hexagonal order.

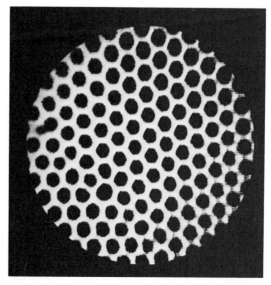

Fig. 1. Dimple crystal on the charged helium surface. Distance between dark regions is 2 mm.

Experimental configuration. Typically, the charged liquid helium surface lies in the gap between a pair of metallic plates across which electric fields up to 3000 V/cm may be applied. Metallic sidewalls are also included to confine electrons in a radial direction. By shining light through the liquid surface or by applying time-varying fields via electrodes beneath the helium layer, the behavior of the system may be studied. Experiments are typically performed at temperatures from 0.3 to 4.2 K, and with charge densities between 10^7 and 10^9 electrons/cm^2.

Charged surface instability. The response of the charged liquid surface to a small disturbance is determined by three forces. The first two, gravity and surface tension, resist deformations (since such changes cost energy) and thus provide the restoring force which is responsible for ordinary surface waves. The charge sheet adds, effectively, a negative restoring force. A surface deformation will result in local electric fields that act to increase the deformation, and thus tend to destabilize the surface. By increasing the externally imposed electric field acting on the charged surface, the destabilizing force due to the charge sheet can be made larger relative to the gravity–surface tension forces.

At a critical value of this electric field, the competing forces are in balance and a small disturbance of one particular wavelength will feel no net restoring force. Such a disturbance acts as a wave with zero frequency. Figure 2 is a plot of the frequency versus wave number for a charged helium surface, for several different values of applied electric field. As the electric field is increased to a critical value, E_c (typically of order 2500 V/cm), the frequency of waves of a particular wave number, k_c, drops to zero, indicating the onset of an instability. Both E_c and k_c depend upon gravitational acceleration, liquid density, and surface tension.

Dimple crystal. Once the critical value of electric field is reached, the surface rearranges itself into the crystalline pattern shown in Fig. 1. The picture is obtained by shining a parallel beam of light vertically through the liquid surface. Depressions in the surface refract the light, producing dark regions. The lattice constant is determined by the critical wave number, k_c, mentioned above, and is typically about 0.3 mm. The dark regions in Fig. 1 correspond to surface depressions about 10 μm deep, and contain about 10^6 electrons apiece; hence the name dimple crystal. Optical measurements of the dimple depth as a function of electric field indicate that, in fact, the electrons are localized in the center of each dimple, while the remainder of the helium surface is free of charge. To a good approximation, these islands interact with each other like point charges, whose lowest energy configuration is a lattice of hexagonal symmetry. Crystalline order is often disrupted by defects, just as it is in ordinary solids. Closer examination of the figure reveals a grain boundary separating two regions of perfect crystal.

The detailed mechanism by which the uniformly charged surface spontaneously orders at the critical field involves the interaction of surface waves. In a small-amplitude or linear approximation, surface waves have the frequency–wave number relation shown in Fig. 2, and do not interact with each other. However, above the critical field, a surface fluctuation of wave number k_c would grow without bound, and can no longer be considered small-amplitude. A more accurate picture must now incorporate the interaction of these finite amplitude waves of wave number k_c. This wave-wave coupling results in additional energy which tends to stabilize the charged surface in a new configuration. As it turns out, the most stable configuration allowed is one in which three waves, whose directions vary by 60° relative to one another, couple together, thus "freezing in" a surface distortion with hexagonal symmetry and lattice constant $2\pi/k_c$. It is this nonlinear wave-wave coupling that produces the dimple crystal. As the electric field is increased, electric charge concentrates in the center of each dimple, and the system more closely resembles the lattice of point charges discussed above.

One consequence of the wave-wave coupling picture is that each dimple should acquire a finite depth abruptly as the critical field is reached, rather than growing continuously. In more familiar terms, the transition should be first-order. Precise measurements of the crystal formation near E_c do indeed show the predicted abrupt change. First-order transitions generally exhibit hysteresis, which is behavior dependent upon the history of the system. In this case, the dimple crystal would be expected to remain stable after the electric field was lowered below the value, E_c, at which it was initially formed. Hysteresis has been observed, but it is difficult to accurately measure, requiring very high stability of the electric field and exceptional isolation of the liquid surface from mechanical vibration.

Coulomb crystallization. Nothing in the discussion so far has involved temperature in a crucial way,

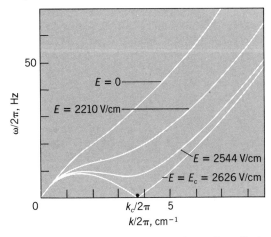

Fig. 2. Frequency versus wave number for small-amplitude waves on the charged surface, with applied electric field as a parameter. The dimple crystal forms when ω = 0, at E = E_c and k = k_c.

except insofar as it slightly changes parameters such as liquid density and surface tension. However, the temperature of the electron gas determines the mean kinetic energy per particle. The potential energy arises from the Coulomb forces of interaction between electrons, and will vary as $\sqrt{n}$ where n is the charge density in two dimensions. The thermodynamic properties of the electron gas are determined by the dimensionless ratio of potential energy to kinetic energy. As this ratio, called Γ, is increased, either by raising the charge density or by lowering the temperature, the potential energy eventually dominates the kinetic energy, and the electron system is expected to crystallize. In this case, individual electrons would be localized at the lattice points of a two-dimensional crystal, in contrast to the islands of charge in the dimple crystal. (If the electric field is smaller than E_c, the dimple crystal will not form, and the helium surface will remain perfectly flat).

Experiments in recent years have demonstrated that this Coulomb crystallization does indeed occur. For surface electron concentrations of $10^8/\text{cm}^2$, the phase transition takes place at 0.35 K. By varying both temperature and charge density, the phase diagram of the Coulomb crystal can be mapped out, as shown in Fig. 3. Over the range studied in this figure, the ratio Γ, at which crystallization occurs, is approximately 130, in excellent agreement with various computer simulation studies of the two-dimensional electron system.

Since individual point charges make only microscopic depressions in the helium surface, the Coulomb crystal cannot be imaged with light as is possible with the dimple crystal. However, by applying time-varying electric fields to the charged surface, an absorption of energy results which is a signature of the presence of a hexagonal electron crystal. Essentially, the electron sheet acts as a transducer to excite high-frequency waves on the helium surface. At frequencies such that the wavelength of these surface waves matches the electron crystal lattice

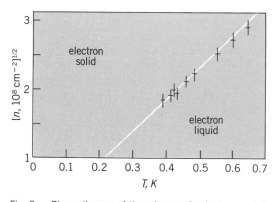

Fig. 3. Phase diagram of the microscopic electron crystal in the charge density–temperature plane. Charge density n is in units of 10^8 electrons/cm^2. Along the line, the ratio Γ is 137. (*After C. C. Grimes and G. Adams, Evidence for a liquid-to-crystal phase transition in a classical, two-dimensional sheet of electrons, Phys. Rev. Lett., 42:795–798, 1979*)

spacing, a resonance condition occurs, and a detectable amount of energy is absorbed.

The detailed behavior of the melting transition of the electron crystal is still under study. Theoretical work suggests that two-dimensional melting will be rather different from its three-dimensional counterpart. In particular, a partially ordered hexatic phase is predicted to lie intermediate between the electron crystal and the disordered electron liquid. To date, this intermediate phase has eluded experimenters, but its observation would be an important step in understanding the physics of two-dimensional ordering.

For background information *see* LIQUID HELIUM; PHASE TRANSITIONS in the McGraw-Hill Encyclopedia of Science and Technology.

[RUSSELL GIANNETTA]

Bibliography: R. Giannetta and H. Ikezi, Nonlinear deformation of the electron charged surface of liquid ^{4}He, *Phys. Rev. Lett.*, 47:849–852, 1981; C. C. Grimes and G. Adams, Evidence for a liquid-to-crystal phase transition in a classical, two-dimensional sheet of electrons, *Phys. Rev. Lett.*, 42:795–798, 1979; B. I. Halperin and D. R. Nelson, Theory of two-dimensional melting, *Phys. Rev. Lett.*, 41:121–124, 1978; M. Wanner and P. Leiderer, Charge-induced Ripplon softening and dimple crystallization at the interface of ^{3}He $-$ ^{4}He mixtures, *Phys. Rev. Lett.*, 42:315–317, 1979.

Marine ecosystem

The structural heterogeneity of habitats such as forests and fields, and the way this heterogeneity relates to the diversity and adaptations of terrestrial organisms are well-known phenomena. In contrast, the open sea appears essentially homogeneous, so that the diversity in the forms of animal life found there has long puzzled ecologists. This has given rise to several concepts including one that has been termed the paradox of the plankton. Simply stated, the argument runs that, since water seems to have very little structure as compared with land, there must be fewer ways to live as a planktonic organism than as a terrestrial one, and yet there is an unexpected diversity of planktonic species. With the increasing use of direct underwater observational techniques, such as scuba diving or deeper-diving submersibles, this paradox is shown to be due to earlier ecologists' limited imaginations. In fact, the open ocean is not nearly as poor in ecological niches as previously supposed. Gelatinous animals provide one type of structural heterogeneity for other pelagic organisms. Many different species have evolved to exploit gelatinous animals in highly diverse and often specific ways—as food, shelter, and breeding grounds. In the simplest cases, these large gelatinous organisms serve as substrates, allowing other organisms to live a pseudobenthic existence in the plankton. However, many animals are more specifically adapted to gelatinous hosts such as salps.

Salps. Salps are transparent, barrel-shaped animals, ranging in size from about 1 to 30 cm. They have evolved from sessile, filter-feeding ascidians (often called sea squirts). Salps are found primarily in the open ocean. They are so widely distributed, and often occur in such large numbers, that they must be considered as one of the major groups of organisms living on the Earth. Although they superficially resemble jellyfishes, salps are complex animals of the phylum Chordata, with a heart, stomach, muscle bands, and a well-developed nervous system. They feed by straining particulate material from the water with a mucous net, and they propel themselves through the water by rhythmic contractions of the circular body muscles, in a manner similar to the swimming of squids. Since salps are the size of fishes, rather than copepods, it is perhaps better to consider than as small nektonic organisms rather than as large planktonic animals.

A diverse assemblage of crustaceans, flatworms, protozoans, and fishes, all living with salps, have been found by scuba divers. Among the crustaceans are copepods (primarily species of *Sappharina*) and hyperiid amphipods (primarily species of *Vibilia*, *Lycaea*, *Brachyscelus*, *Parathemisto*, and *Oxycephalus*). The hyperiid amphipods are particularly diverse in their behavior on salps. Therefore, it comes as little surprise to find that the midwater fish, *Tetragonurus*, is also associated with salps in a highly specific way. This association is reflected in both its morphology and its behavior.

Tetragonurus relations. *Tetragonurus* is a genus of perciform fishes in the suborder Stromateoidei. It is the only genus in the family Tetragonuridae, and there are only three species: *atlanticus*, *pacificus*,

and *cuvieri*. They are found (albeit rarely) in all tropical and subtropical regions of the open sea. The young forms are found in the upper waters, and as the fishes grow they move deeper and deeper in the water column. All stromateoids appear to associate with large gelatinous animals or with floating objects. In general, the associations have been regarded as rather tenuous, and previous studies of *Tetragonurus* have listed salps, pyrosomes, ctenophores, and medusae as major sources of food. However, this fish has been found only in association with salps. Species of *Tetragonurus* observed in the field are always seen inside, or close to, salps (Fig. 1). Often, the fish darts into the body cavity of salps, apparently attempting to elude the divers. The fish is well adapted for living in cavities, since its body is elongate, which facilitates swimming backward and making tight turns. The dorsal and pelvic fins can be retracted into grooves, so that the fish can avoid getting stuck in the salp's body cavity when it moves backward.

Feedings habits of Tetragonurus. Although the young fish has never been seen eating salps in the field, they actively consume them in the laboratory. They refuse to eat other gelatinous animals, such as ctenophores and jellyfish. The fish do not usually eat the entire salp, but only the stomach, and sometimes the gill bar. Since the stomachs of salps are packed with food, and the salp does not appear to digest most of the food that it ingests, salp stomachs are a highly nutritious and convenient packet of food. The teeth of the fish appear to be specially modified for the consumption of salp stomachs. The upper teeth are widely spaced conical pegs which serve to hold the salp, while the lower teeth, which form a structure like a hacksaw blade, neatly slice out the stomach (Fig. 2). The fish rotates and thrashes about as it excises the stomach. Salps have often been found in the field with their stomachs missing; often the mutilated salps are still swimming. Preferential feeding on salp stomachs is not restricted to species of *Tetragonurus*, as unidentified filefish (Balistidae) have been seen doing the same thing.

After the stomach of the salp is eaten, the fish usually loses interest in the salp, but interest can be rekindled if a false stomach is placed in the still-swimming salp, regardless of where it is put in the salp. It is not known whether the color of the false stomach is important. In addition to salp stomachs, species of *Tetragonurus* have been seen eating the hyperiid amphipods and copepods that live on salps. Since these crustaceans are also highly specific, they provide useful tools for the study of feeding on gelatinous animals by fishes, as crustaceans are much less digestible than their hosts. These crustaceans have been used to study the feeding habits of specimens of *Tetragonurus* that live too deeply to be studied directly. Studies on the stomach contents of trawl-collected specimens of *Tetragonurus* indicate that they are highly specific predators on pelagic tunicates. Fragments of salps, small intact salps, pieces of pyrosome colonies, copepods, and hyperiid amphipods have been found in their stomachs.

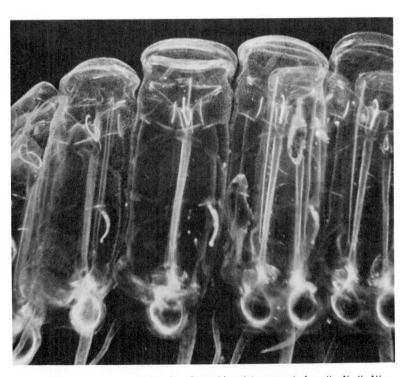

Fig. 1.   Two *Tetragonurus atlanticus* in a *Pegea bicaudata* aggregate from the North Atlantic. The fish are visible inside the second organism from the right. (*From J. Janssen and G. R. Harbison, Fish in salps: The association of squaretails (Tetragonurus spp.) with pelagic tunicates, J. Mar. Biol. Ass. U.K., 61:917–927, 1981*)

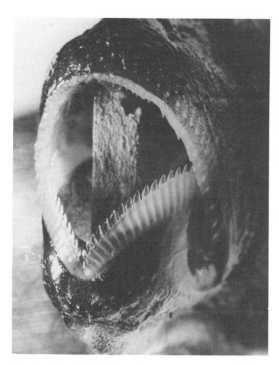

Fig. 2. Jaw of a 137-mm *Tetragonurus cuvieri*. (*From J. Janssen and G. R. Harbison, Fish in salps: The association of squaretails (Tetragonurus spp.) with pelagic tunicates, J. Mar. Biol. Ass. U.K., 61:917–927, 1981*)

There is thus no evidence, from these observations, that species of *Tetragonurus* eat other gelatinous animals.

Salps do appear to have some defenses against the fish that feed on their stomachs. The stomachs of some species are protected by a tough shield which is difficult for the sawlike teeth of *Tetragonurus* to cut. The aggregate stages of some salps are also arranged so that the stomach of one salp is protected by an adjacent salp. In these salps the ones at the ends of the colonies are most vulnerable and are eaten first. Tentaclelike protuberances on either side of the stomach of one salp species interfere with feeding by *Tetragonurus*. Salps also make it difficult for the fish to enter their body cavities by contracting their mouths and atrial siphons. This often traps the fish, which must then struggle to escape. It is likely that *Tetragonurus* is not the major predator on salps, but these defenses are probably effective against other predators on salp stomachs.

As the fish grow and move to deeper water, they become too big to enter even the largest species of salp. The literature and gut-content studies indicate that the larger fishes feed upon pyrosome colonies, and that they take refuge in them. Although individual pyrosomes (which are closely related to salps) are small, the colonies can be extremely large, up to several meters in length, with a large central cloaca that is big enough for large fish to enter.

Summary. The open sea is generally considered to be a region where food is in very short supply. Further, gelatinous animals are also considered to be a very poor food source, as their bodily tissues are mostly water. The fact that a genus of midwater fishes has evolved to exploit a single group of gelatinous animals in the open sea indicates that both of these speculations are not well founded, and that they are more the product of an anthropocentric view of the open sea as an environment than the result of fact. Much more information is needed on the life styles and food requirements of animals which live in the open ocean before a realistic picture of the life of a pelagic or planktonic organism can be obtained. As technology increases the ability to study oceanic animals directly, for longer periods, and at greater depths, many of the present conceptions (or better, misconceptions) about the nature of the open sea will probably be discarded.

For background information *see* MARINE ECOSYSTEM; TUNICATA in the McGraw-Hill Encyclopedia of Science and Technology.

[G. R. HARBISON; JOHN JANSSEN]

Bibliography: J. Janssen and G. R. Harbison, Fish in salps: The association of squaretails (*Tetragonurus* spp.) with pelagic tunicates, *J. Mar. Biol. Ass. U.K.*, 61:917–927, 1981.

Marine engineering

Today's oil shortages and price uncertainties have led shipowners and designers to investigate alternative fuel types. Coal, long a standard fuel for electric power generation, offers several advantages for ship propulsion. Coal burning uses existing, proven technology, and components that are similar or identical to most oil-burning marine propulsion equipment; and coal has a decided price advantage over current residual oils. These facts have provided the impetus for the renewed interest in coal as a ship's fuel.

Advantages of coal-burning ships. In today's economic environment the shipowner has become increasingly concerned with the cost of fuel. The cost of the traditional residual fuel oils has risen so markedly in the last several years that an average owner can expect about 35% of yearly operating expense to go for the purchase of fuel oil. Moreover, availability of fuel oils has become a question. The technology is known, and now there is an economic incentive, for refining basic petroleum further than in the past. This extended refining reduces the amount of residual oil produced, and the residual contains such concentrated quantities of sulfur and vanadium that its continued use as marine fuel becomes less acceptable. For ship propulsion, there are alternatives. Probably the most promising and most adaptable, in terms of technology, is coal. Coal reserves in the United States and Europe are substantial.

The National Coal Association estimates that the United States has 30–50% of the world's coal reserves. This natural resource is particularly free of foreign dependency and it is important for the United States economically, since stable coal prices can be predicted relative to the forecast escalation of oil prices. Based on recent studies of the cost advantages of coal over fuel oil, savings range between $1.5 and $4.0 million per ship per year with coal-fired plants. In addition, coal can be utilized as a propulsion fuel with shipboard components that are almost identical to those using fuel oil. Little new technology is required to introduce coal burning in a new-design vessel. In fact, this ready adaptability makes coal fuel attractive for conversion of existing steam-powered vessels.

These advantages notwithstanding, a detriment to using coal as a shipboard fuel is that bunker coal is not readily available in many United States ports. Until the demand increases, coal storage and the mechanisms for delivering the coal to the vessels will be available only at select ports—those normally handling and shipping coal as a cargo. This situation indicates that the initial use of coal as a ship fuel will be seen on those ships that normally trade at such ports, such as bulk carriers and colliers.

Ship design considerations. The major features of coal-fired vessels which differ from conventional oil-fueled vessels are provisions for coal storage (that is, bunkers) and processing equipment, and ash storage and removal equipment. Other propulsion-related equipment and auxiliaries, although of perhaps modified capacity, are virtually identical to such components currently found in today's oil-fired steamships. The illustration shows the relationship of these components.

Storage bunker design requirements. There are three factors influencing the design of coal storage bunkers.

First, the volume of coal that must be carried is significantly larger than the volume of fuel oil of equivalent total heating value. Coal has a heating value of approximately 12,000 Btu/lb (28 megajoules/kg), compared with 18,500 Btu/lb (43 MJ/kg) for oil. The cubic volume of coal varies with its lump size, but is typically 50 lb/ft^3 (800 kg/m^3) versus 54 lb/ft^3 (865 kg/m^3) for oil. Thus the space and weight requirements for coal storage bunkers are considerably greater than for oil tanks. On ships that are volume-limited, this factor will increase the ship size.

Second, coal imposes design constraints on the location and configuration of the storage bunkers. Coal must flow downward by gravity to a conveyor system or pump that moves the coal to the service bunker. Gravity feed requires the storage bunkers to be relatively high in the ship and to have steep sloping sides, typically about 70°. If the angle of the slope is much less than this, the sides of the bunkers may require special treatment to reduce friction and to allow the free movement of the coal. Regardless of the slope angle used, the space under the slopes is of limited use for auxiliary machinery or systems.

Third, coal imposes special safety considerations, because under certain conditions it may be subject to spontaneous combustion. Such factors as the moisture content, air flow or lack thereof, length of time stored in one position, elapsed time since the coal was mined and crushed, and methane content of the coal as mined must be evaluated. The hazard of spontaneous combustion does not normally present a problem, since bunker storage is usually less than 2 weeks; however, it does require that certain areas of the ship be treated as potential methane and coal dust areas. Electrical equipment must be explosion-proof. Provision should be made for drainage at the bottom of the coal bunkers, since coal could be received wet or in rainy weather. Current regulations also require that coal bunkers be isolated from heat sources and ventilated to prevent the buildup of gas.

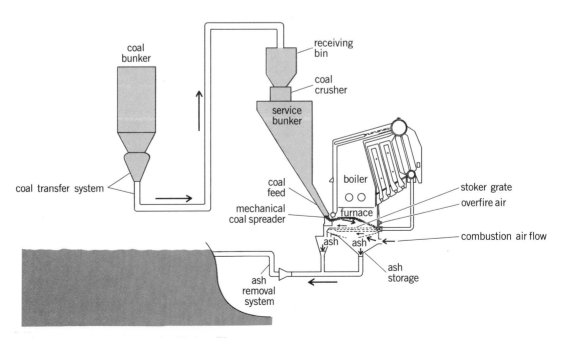

Components of coal-firing system for ship propulsion.

Coal-forwarding system. A coal-forwarding system moves coal from the storage bunkers to the service bunkers from which it is gravity-fed to the boilers. The coal-forwarding system can consist of either pneumatically actuated dense-phase pumps or mechanical conveyors. If a dense-phase system is used, special attention is required for system layout and arrangement. Due consideration must be given to the amount of compressed air required for the pump activation and for the coal transport system. Since pneumatic dense-phase pumps require about 50% more compressed air than is required for the usual ship functions, that is, control air and service air, the ship's air compressors must be sized accordingly. Regardless of whether a pneumatic or mechanical system is used, or whether a dedicated or backup air compressor is used, adequate space must be allocated for straight runs of piping or conveyor belts. From a safety and cleanliness point of view, the completely enclosed pneumatic transport systems are preferable for coal transfer aboard ship.

Coal crusher. In a stoker-fed boiler, to permit the stoker and boiler to operate at maximum efficiency, the coal fed to the service bunker must be of a lump size of about 1½ in. (38 mm), with no more than 50% of the coal being less than ¼ in. (6.4 mm). In pulverized plants, all of the coal must be less than ¼ in. If the coal supplied to the ship is of a larger lump size, it must be passed through a crusher or pulverizer. If the coal has not been screened for foreign material onshore, the foreign material must be trapped at some point in the ship processing procedure. The installation of screening equipment on the ship, although not mandatory, enables the ship to be independent of shoreside coal quality control.

Furnace design. From the crusher, the coal passes into the service bunker, then into a conical distributor directing the coal into the feeder for the boiler. In a stoker-feeder configuration, the stoker-feeder spreads the coal onto a grate which moves the burning coal through the furnace in approximately 3–5 min.

If the boiler is configured to burn pulverized coal rather than lump coal, the pulverized coal passes from the service hopper through high-pressure nozzles and directly into the furnace for combustion. For either combustion process, the resulting coal ash is collected into a hopper below the furnace.

Ash storage and removal. The ash hopper is located directly beneath the boiler and can be designed for short- or intermediate-term storage. If storage is required for periods of several days or longer, a long-term ash storage facility is necessary, external to the hopper. If sufficient space is available, dry ash can be accumulated and stored aboard ship indefinitely.

Ash removal can be accomplished in several ways. If long-term storage capability is provided, the ash can be off-loaded ashore during loading or unloading of cargo. Because of shoreside disposal problems in some areas, it is more likely that an ash removal system would be used to discharge the ash overboard while at sea. Current studies and regulations permit such disposal when not in coastal waters. For ash removal, either a vacuum eductor system or pneumatic dense-phase pump system can be used.

The volume required for ash storage depends upon the quantity of ash that must be stored and discharged overboard. Depending upon the coal, the quantity of ash can vary between 9 and 12% or more of the volume of the coal burned. For a ship of the 30,000-shaft-horsepower range (22.4 megawatts), the ash produced could amount to about 30 tons/day (27 metric tons). Consequently, the volume required for ash storage and removal systems must be taken into consideration in the arrangement of the lower-level machinery spaces.

Fly ash collection and reinjection. As the coal is forced into the boiler, small particles of coal known as fines are burned in suspension. A portion of the fines ash is carried in suspension by the furnace gases as fly ash. This fly ash collects in the boiler area after passing through the superheater, and in the dust collector located in the exhaust passage. The fly ash is then reinjected into the furnace for complete combustion.

Dust collectors are available with 90% efficiency ratings at boiler full-load conditions. At lesser loads, dust collectors require sectionalization to improve efficiency. An overfire air fan is provided to reinject the fly ash and assist the combustion process by providing air turbulence. In addition to the forced-draft fan and overfire fan, an induced-draft fan generates a negative pressure in the furnace and thus prevents coal dust or soot from entering the fireroom.

Propulsion redundancy considerations. To meet regulatory requirements for redundant propulsion capability, environmental protection, and smooth maneuvering, most shipowners currently prefer boilers that can be fired with coal at sea and then switched from coal to fuel oil for in-port operation. The use of fuel oil for docking maneuvers is advantageous because the oil-fed boiler turndown ratio is greater than that of a stoker-fired coal boiler. When a ship is maneuvering on coal, a steam-conditioning valve is employed to dump excess steam into the main condenser, thus preventing excessive steam buildup in the boiler. The main condenser must therefore be designed to accommodate the dump steam through a series of baffles and orifices. Oil-firing capability can also be used to ignite coal at plant startup.

Engine-room design considerations. The requirements for ash storage under the boiler, and the ash removal equipment fitted there, tend to elevate the boiler higher than a conventional oil-fired boiler. The size of the boiler and other considerations tend to force the boiler location forward within the machinery space. This feature of coal-fired boiler propulsion plants contributes to arrangement considerations that are unique for the ship designer. Locating the main boiler or boilers is probably the most difficult and critical aspect in the design of ships of this type. Other aspects of the ship's engine room are relatively unaffected by the choice of coal as a fuel.

Efficiency of coal firing. Coal is generally considered to be a lower grade of fuel than ship fuel oils. For example, Bunker C fuel oil nominally has a heat content of 18,500 Btu/lb (43 MJ/kg), while power-plant coal has a heat content of about 12,000 Btu/lb (28 MJ/kg). However, on a volumetric basis, the heat content of oil is about 1,017,500 Btu/ft^3 (37,900 MJ/m^3) versus 600,000 Btu/ft^3 (22,400 MJ/m^3) for coal. Since the size of a furnace is determined by the heat generation requirements, a coal-fired boiler must be significantly larger than an oil-fired boiler. Other considerations peculiar to firing coal may increase the size even more, depending on the firing mode. The larger a boiler is, for an equivalent heat rate, the less its efficiency will be from an overall system point of view, simply because the parasitic loads represent a larger percentage of the output. These efficiency losses, however, seldom represent much more than 3–4% of the total plant capacity. Efficiency of the boiler alone rivals that for a comparable oil-fired boiler of the same configuration, coal-fired boilers being in the order of 2–3% less efficient. These lower efficiencies, which are due to the boiler and the related systems, are more than compensated for by the price advantage of coal.

For background information *see* CRUSHING AND PULVERIZING; MARINE BOILER; SHIP POWERING AND STEERING in the McGraw-Hill Encyclopedia of Science and Technology.

[ROBERT B. GEARY]

Marine sediments

In deep water on the western sides of ocean basins, fast currents flow along the continental margins. Rapid flow through deep-ocean channels and areas of strong deep-sea tidal currents are also found in the middle and on the edges of ocean basins. These areas are characterized by distinctive sedimentary bedforms and deposits. They range in scale from crag-and-tail structures of about a centimeter in height and several centimeters in length to sediment drifts that are hundreds of meters thick and up to a thousand kilometers in length. The material stirred up by the fast currents in these regions yields a dilute layer of suspended sediment up to 1500 m thick, the nepheloid layer, responsible for much of the sediment transport and deposition in these areas.

Deep-sea currents. The major components of the deep-sea current system are driven by density. Cold dense water that is formed and sinks in polar regions then flows around the Antarctic continent and northward into the three major oceans along their western sides (Fig. 1). In the Pacific and Indian oceans this is the only source of fast-flowing bottom water. However, in the Atlantic there is a second source in the north which provides cold water that flows around the banks and ridges of the North Atlantic and eventually moves southward along the margin of the United States. These currents achieve speeds of a few tenths of a meter per second at many places along their path. Exceptionally, they may flow

for short periods at over a half a meter per second. These regions of fast flow, sediment resuspension, turbulence generation, and sinking of bottom waters are termed high-energy benthic boundary layers.

Not only the deep dense flows achieve high speeds; other currents are locally fast as well. Deep-sea tidal currents are found in most current-meter records, and in a few places they exceed 0.15 m/s (0.34 mi/h). It has been proposed that in the western North Atlantic the Gulf Stream system shows a deep return flow to the west and southwest. This return-flow system achieves high speeds along the margins of the Bermuda Rise, an extensive topographic feature, rising 1500 m (5000 ft) above the ocean bottom and largely composed of mud. Very probably there are other areas of the deep ocean affected by fast currents in systems related to surface circulation.

Finally, very fast currents can be found in narrow deep-sea passages where the flow of dense bottom water is constricted. These places, while not really extensive, are important because they allow a determination of the means and the direction in which fast water and sediment are being transferred from one basin to another.

Sediments and bedforms. The sediments deposited by swift ocean-bottom currents have been called contourites because the currents generally flow along contours (in contrast to turbidites, the deposits of turbidity currents that flow downslope across contours). It is impossible to generalize about the composition of contourites. They may be clays, silts, or sands of any mineralogy. Commonly they are a mixture of terrigenous siliciclastic and calcareous biogenic components of muddy texture with a small content of sand that is often composed of foraminiferal skeletons. An important feature of their texture is that they show lamination and cross-lamination, indicating a degree of current sorting and deposition that is usually absent in deposits of weak-current systems. Exceptionally, they may be composed entirely of sand and display well-developed cross-lamination formed by deposition on migrating ripples. These sandy contourites are much less common than muddy contourites of silt plus clay.

Although the muddy contourites may show little obvious winnowing or sorting, their magnetic fabric does reveal the effects of currents. The anisotropy of magnetic susceptibility shows the efficiency of alignment of sediment particles, and well aligned grains correlate well with fast currents. It has also been shown that there is an increase in the mean size of the silt fraction from about 11 μm to 13 μm on the continental margin southeast of New York, under faster currents.

The deep-ocean floor displays many surface sedimentary bedforms. Some of these originate through the activities of animals, but the main effect of currents is smoothing-off of the sediment surface and removal of most organic traces. After smoothing, faint lineations are seen, and with increasing current speed, crag-and-tail features that have developed behind some more resistant lumps (stones or worm tubes, perhaps) are found (Fig. 2). Mounds of bio-

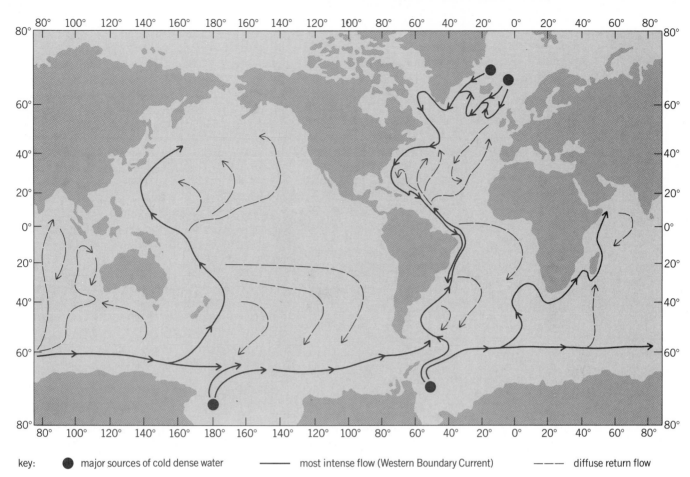

key: ● major sources of cold dense water ——— most intense flow (Western Boundary Current) − − − diffuse return flow

Fig. 1. A summary of the paths taken by bottom-water flow in the world ocean. (*After D. A. V. Stow and J. P. B. Lovell,* *Contourites: Their recognition in modern and ancient sediments, Earth Sci. Rev., 14:251–291, 1979*)

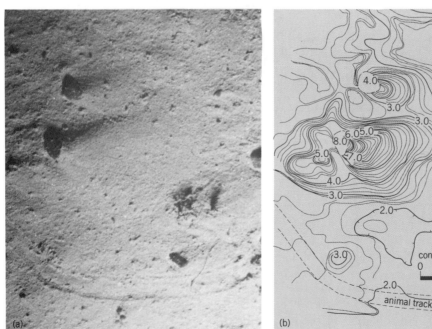

Fig. 2. Crag-and-tail structure. (*a*) Photograph taken by stereo camera. (*b*) Contour map with millimeter contour intervals of this bedform and the adjacent seabed with an an-imal trail. (*Courtesy of C. D. Hollister, Woods Hole Oceanographic Institution*)

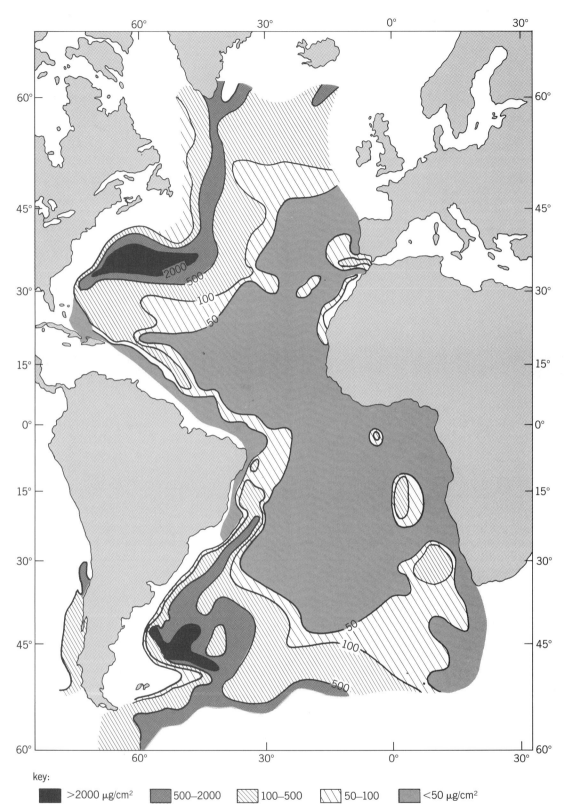

key:

| ■ >2000 µg/cm² | ■ 500–2000 | ⫽ 100–500 | ⫽ 50–100 | ■ <50 µg/cm² |

Fig. 3.   Distribution of suspended material originating from bottom resuspension in the bottom nepheloid layer in the Atlantic Ocean. (*After P. E. Biscaye and S. L. Eittreim, Suspended particulate loads and transports in the nepheloid layer of the abyssal Atlantic Ocean, Mar. Geol., 23:155–172, 1977*)

logical origin are abundant on the ocean bottom. Under fast-current systems these are smoothed and often develop tails of sediment. In some cases these tails have been shown to extend downcurrent as longitudinal triangular ripples. These are symmetrical ripples aligned parallel with the mean current, up to 10 cm (4 in.) high, 60 cm (24 in.) wide, over 10 m (33 ft) long, and spaced several meters apart.

Some authors think these features develop in response to helical circulation in the lower part of the oceanic boundary layer. One of the most striking features of abyssal topography is the long furrows parallel to the current. The majority of these are several kilometers long, spaced 50–150 m (160–500 ft) apart and less than 5 m (16 ft) deep. Although they occur in places where sediment is presently being deposited and are perhaps maintained by modern flows, they may initially have been formed by erosion. Some furrows are certainly eroded and are deep (20 m or 65 ft) and wide (50–150 m or 160–500 ft) with outcropping layers on the furrow walls. This type of feature has also been ascribed to laterally variable bottom stress in a boundary layer with helical circulation. One puzzle is why in one situation furrows are the most frequently observed bedform, whereas in another, longitudinal triangular ripples dominate.

Transverse ripples are also found under deep-current systems. When they are of noncohesive foraminiferal sands, the ripples are just like those found in streams and shallow seas. Where there are abundant winnowed foram skeletons and rapid currents, large dune bedforms are encountered, again resembling those found on the continental shelf or in deserts. Surprisingly, however, transverse ripples have been found in cohesive muds. This may indicate a degree of noncohesive behavior on the part of these muds, possibly due to movement of aggregates such as sand-sized fecal pellets.

Much larger bedforms are mud waves, which are symmetrical waves 5–50 m (16–160 ft) high spaced 1–5 km (0.6–3 mi) apart. Some of these are furrowed and most are skewed to the current direction. Mud waves in turn are most commonly located on contourite drifts, major accumulations of sediments several hundred meters thick and up to a thousand kilometers long, located under the path of the abyssal thermohaline current systems.

Sedimentary processes. The behavior of muddy sediments is not at all well understood, even in shallow seas and rivers, because of cohesion of the sediments that is not simply related to easily determined properties, such as grain size. The few laboratory studies that have been undertaken on deep-sea sediment suggest that current of about 0.15–0.2 m/s (0.34–0.46 mi/h) are necessary to erode it. This sediment on the ocean bottom is under the organic influences of mucous binding and pelletization, which are difficult to reproduce in the laboratory, so extrapolation from experiments is unwise.

When eroded by currents generally in excess of 0.15 m/s (0.34 mi/h), the finer components are put into suspension while sand-sized particles are rolled along the bottom. The suspended sediment is distributed turbulently throughout the approximately 50-m-thick (160-ft) bottom mixed layer to form a homogeneous region of the bottom nepheloid layer. The nepheloid layer becomes thicker by detachment of the more concentrated bottom mixed layers which override zones of denser bottom water. These detached bottom layers are thus injected at levels above

the bottom homogeneous layer. This can result eventually in nepheloid layers up to 1500 m (5000 ft) thick over deep abyssal plains.

The total amount of material suspended in the bottom nepheloid layer, coming from the ocean bottom, has been estimated by using light-scattering records from many points in the Atlantic (Fig. 3). There is a remarkable correspondence between the position of the maximum suspended sediment load and the axes of maximum cold bottom-current flow (Fig. 1). The broadening of the area in the North American Basin is due to the Gulf Stream return flow. While it has been suggested that there may be a similar effect in the Argentine Basin, it is now thought that there are three recirculating loops of Antarctic bottom water in this basin which may be responsible.

The major contourite drifts and their associated mud waves probably represent deposition from these nepheloid layers. Within the nepheloid layers the suspensions contain a large amount of fine material of small settling velocity, but this becomes aggregated into larger units that settle faster and are deposited. Nevertheless some of the finest material is probably swept away and deposited elsewhere in the oceans. Aggregation occurs both physicochemically, by particles colliding and sticking together, and biologically, by particles being incorporated into the biological systems (for example, the gut of a filter feeder) and packaged for disposal. The rate of deposition from suspension is partly dependent on both particle-settling velocity and current speed. Thus when the current slows down, the rate of deposition goes up as the aggregates are laid down. When the current speed goes up, some of the finer components are broken out from aggregates and resuspended, leaving a coarser deposit behind. In extreme cases this material is of sand-sized foraminiferal tests, and sandy contourites result.

For background information *see* DEPOSITIONAL SYSTEMS AND ENVIRONMENTS; MARINE SEDIMENTS; SEDIMENTATION (GEOLOGY) in the McGraw-Hill Encyclopedia of Science and Technology.

[I. N. McCAVE]

Bibliography: P. E. Biscaye and S. L. Eittreim, Suspended particulate loads and transports in the nepheloid layer of the abyssal Atlantic Ocean, *Mar. Geol.*, 23:155–172, 1977; B. C. Heezen and C. D. Hollister, *The Face of the Deep*, chap. 9, pp 335–421, 1971; I. N. McCave (ed.), *The Benthic Boundary Layer*, 1976; D. A. V. Stow and J. P. B. Lovell, Contourites: Their recognition in modern and ancient sediments, *Earth Sci. Rev.*, 14:251–291, 1979.

Mars

In 1971 the *Mariner 9* spacecraft returned the first pictures of channels and valleys on Mars. The probability that these features were formed by flowing water has now been established by studies of high-resolution photographs provided by the orbiting Viking spacecraft in 1976–1980. This discovery and its paleoclimatic implications are perhaps among the most important new developments in planetary geology.

Imaging. The channels and valleys of Mars are interpreted from images produced by the sensing devices on spacecraft. The interpretation of landforms on those images is accomplished by analogic reasoning used to reconstruct the complex interaction of processes responsible for the observed features. The highest-quality Viking orbiter images can resolve surface features as small as 10 m (33 ft). This is better than the resolution available for portions of the Earth. The landscape revealed by the Viking vidicon cameras indicates an abundance and variety of geomorphic processes, both past and present, on the Martian surface. The similarity between Martian and terrestrial landforms and the inferred nature of Martian surface processes lead to the conclusions below.

Environment. The Martian environment is best described as an exceedingly cold, dry desert. The atmosphere is thin, with a pressure ranging from about 700 to 1000 pascals (7 to 10 millibars), or only about 1% of the terrestrial value. It is composed of approximately 95% carbon dioxide, 3% nitrogen, 1.6% argon, and lesser amounts of oxygen, carbon monoxide, water, and noble gases other than

argon. Temperatures range from −103°C at the southern polar cap in winter to 20°C at mid-southern latitudes in summer.

Channel formation. Martian channels (Fig. 1) are immense features, as much as 100 km (60 mi) wide and 2000 km (1200 mi) in length. These features seem best explained by the action of great outbursts of water onto the Martian surface, perhaps with heavy loads of sediment and jams of ice. The channel floor gradients range from about 1 to 2.5 m/km (5 to 13 ft/mi). A suite of bedforms on the channel floors indicates that large-scale fluid flows were the primary agents of channel genesis. Most channels show evidence that the fluid flows emanated from complex collapse zones known as chaotic terrain. Such channels are termed outflow and probably formed by a headward extension of the collapse zones and concomitant erosion of downstream troughs by the channel-forming fluids. Morphological relationships in the channels show that the eroding fluid was characterized by a free upper surface, demonstrated by its ponding upstream of flow constrictions. It eroded scour holes, deposited barlike sediment accumulations, spilled over low divides, and shaped magnificent streamlined hills (Fig. 2), forms developed by minimum resistance in the turbulent erosive fluid.

Fig. 1.   Large channels eroded into the heavily cratered Martian surface. Maja Vallis is the prominent channel extending from lower left to upper right. Note the constricted central portion of Maja Vallis and the streamlining and grooves that developed when fluid flows spilled through this constriction and debouched into a plains region (upper right). Also note that many craters exhibit lobate ejecta flows that probably result from impacts into a subsurface layer rich in ground ice. The region shown in this mosaic of Viking images is approximately 500 km (300 mi) wide. (*NASA*)

Among cosmically abundant substances water seems best suited to satisfy these and other constraints on the nature of the primary agent of channel genesis.

The outflow channel geometry and bedforms require aqueous erosion on an immense scale, a scale achieved only in some terrestrial examples of catastrophic glacial outburst flooding. After an initial phase of immense flooding, the outflow channels experienced extensive modification of their floors and walls by processes that include cratering, ground-ice melting, eolian erosion and deposition, landsliding, debris flowage, and rilling. Even though polygenetic and highly modified features abound, the analogic reasoning process with terrestrial counterparts requires that water was a necessary ingredient in channel formation.

The physical character of the erosive water flows on Mars may have differed somewhat from common processes on Earth. The lower Martian gravity would have facilitated sediment entrainment, and some investigators postulate that the so-called Martian floods behaved more as debris-laden slurries than as less viscous clear water flows. If flows occurred when temperatures were as cold and the atmosphere as thin as today, the water would have frozen on its surface. As flow continued beneath the ice cover, great jams and drives of river ice would develop. Some ice accumulations themselves may also have flowed slowly, as occurs in terrestrial glaciers. These various models of channel formation are still being compared and evaluated.

Channel age. The ages of Martian channels and valleys are determined by crosscutting relationships and by the densities of impact craters on the terrain. Age interpretation is complicated by assumptions concerning cratering rates and the resurfacing of cratered areas. Given this limitation, most researchers conclude that channels and valleys on Mars are extremely ancient, many such features as much as 4 billion years old. Some channels, or at least the depositional mantles that cover their floors, may be less than 1 billion years old, relatively young by Martian standards.

Valley networks. The term valley network applies to the Martian trough systems, which appear to form by fluid flow but lack a suite of bedforms on their floors, unlike the outflow channels. The Martian valleys of greatest interest consist of interconnected, digitate networks that dissect extensive areas of heavily cratered uplands on the planet (Fig. 3). Some networks may have formed by surface runoff requiring the concentration of rainfall, but most show low drainage densities, theater-headed valleys, and short tributaries, which indicate that rainfall need not have been a direct cause of the patterns. Rather, the observed morphology is most likely the result of a headward sapping process. Groundwater, derived from an ice-rich permafrost, probably played a key role in the genesis of many valley networks.

Aqueous processes. On Mars the appearance of flowing surface water must be explained in terms of known present atmospheric conditions or postulated

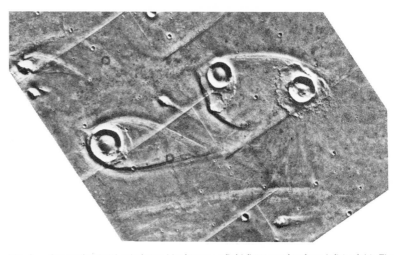

Fig. 2.   Streamlined uplands formed by immense fluid flows moving from left to right. The teardrop-shaped upland at left is about 40 km (25 mi) long. The raised crater rims acted as barriers to the flow. The ejecta material was eroded from the upstream sides of these craters, but it was preserved on the downstream sides. (*NASA*)

ancient conditions. For the valley networks a relatively slow release from a subsurface water system poses few difficulties. The biggest problem is to explain prolonged flows on the planet's surface under present conditions. Prolonged surface flow requires such assumptions as a warmer, denser, ancient atmosphere, ice-covered rivers, or freezing-point depressants. For the outflow channels, the required short-duration floods are possible even in the present Martian atmosphere, but a mechanism must be found to account for the immense quantities of water needed. The proposed release mechanisms include outbursts of melted ground ice from an ice-rich permafrost heated by volcanism, meteor impacts into the ice-rich permafrost, liquefaction of sensitive subsurface materials on Mars, and sudden release of immense aquifers of very high permeability confined by the ice-rich permafrost. Although an extensive planetwide permafrost system appears to have been involved, the precise release mechanism for fluids to form outflow channels remains highly speculative.

In addition to the channels, numerous other Martian landforms are consistent with aqueous processes and ice-rich permafrost. Many Martian craters are surrounded by layered debris that was probably emplaced by flow. This morphology contrasts markedly with that of lunar craters, generally believed to be surrounded by ejecta of ballistic origin. The peculiar morphology of Martian craters may result from the entrainment of water into the ejecta from a zone of Martian permafrost. Even the Martian volcanoes show evidence of phreatic (steam-explosive) phases in their early eruptive history. This would occur as magma erupted through water-saturated (or ice-rich) crustal materials. The eruptions produced extensive ash deposits, but the style of volcanism changed as the crust was depleted in water near the eruptive sites.

Thermokarst is the process of melting ground ice

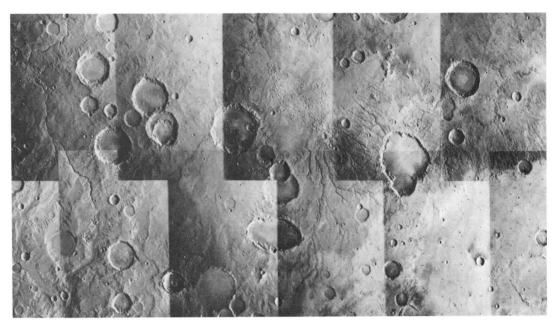

Fig. 3.   Valley network development in the heavily cratered terrain of Mars (latitude 25°S, longitude 26°W). The long valley, named Parana Vallis, extends approximately 200 km (120 mi) at the lower center of this mosaic of Viking images. Another valley, Samara Vallis, is at the left. (*NASA*)

to produce local collapse of the ground surface. The extent of thermokarst development depends on the ice content of the ground material and on the degree and rate of disruption of the thermal equilibrium in the permafrost. The process on Mars appears to have produced irregular depressions and scalloped scarps, chaotic terrain, and great scarps surrounded by mantles of debris. Some debris mantles experienced outward flow radially from upland massifs. Flow lines show that the debris was locally deflected around obstacles. The debris flowage probably occurred when debris or talus became charged with interstitial ice over a long time span. Eventually the debris flowed as a rock glacier.

Conclusion. The channels and valleys of Mars indicate an ancient atmospheric environment with higher temperatures and pressures than at present. If a groundwater flow system was involved in channel and network formation, the cessation of that system indicates that major changes occurred in the hydrologic cycle. The role of ground ice seems important, indicating that Mars possessed or retains a thick, ice-rich permafrost. Outbursts of water confined by ice, climatic warming, local volcanism, impact events, and scarp retreat may have all led to disruptions of the ice-water subsurface system. Extensive valley networks formed by sapping where slow seepage undermined resistant cap rocks. Some ancient networks may have formed by runoff of surface rainfall during an epoch of warmer climate and denser atmosphere. Differences in the scale and morphology of Martian features, in comparison to terrestrial ones, probably developed because of the immense spans of time available for landform development, and the apparent absence of rainfall and

related overland flow processes throughout most of Martian history. However, the abundance of channels, valleys, and related geomorphic features clearly indicates that Mars was either volatile-rich during its early history, especially the period of valley-network formation, or extremely effective at recycling its limited inventory of water. In either case, recent studies confirm that water has been a dominant agent in landscape development on Mars just as it has been on Earth.

For background information *see* Fluvial erosion landforms; Mars; Remote sensing in the Mc-Graw-Hill Encyclopedia of Science and Technology.

[Victor R. Baker]

Bibliography: V. R. Baker, *The Channels of Mars*, 1982; V. R. Baker, Erosional processes in channelized water flows on Mars, *J. Geophys. Res.*, 84:7985–7993, 1979; V. R. Baker, The geomorphology of Mars, *Prog. Phys. Geog.*, 5:473–513, 1981; V. R. Baker and R. C. Kochel, Martian channel morphology: Maja and Kasei Valles, *J. Geophys. Res.*, 84:7961–7983, 1979.

Mathematical ecology

Mathematical ecology originated more or less simultaneously with the work of A. J. Lotka and, independently, with that of V. Volterra in the 1920s. Lotka derived the mathematical techniques necessary for dealing with age-distributed populations and also represented two interacting populations as a set of coupled differential equations. At about the same time Volterra, using the same coupled differential equations, derived many of the results which were to be rediscovered much later. This approach received support from the famous experiments of the

Soviet biologist G. F. Gause, in which it was shown that the population behavior of several microorganisms corresponded closely to the behavior of Lotka's and Volterra's coupled differential equations. Subsequent to Gause's work mathematical ecology was relatively static; the competitive and predatory forms of the Lotka-Volterra equations and Lotka's actuarian life tables were found in virtually every ecology text, but little active research was devoted to the field.

The next major surge of activity was in the 1960s, largely stimulated by the work of R. MacArthur and R. Levins. More or less simultaneously, the new availability of high-speed computers and the development of systems analysis brought the promise of new ways of dealing with the complexity of ecological systems. These two lines of research gave rise to a dichotomy that seems to persist to the present day. On the one hand is the approach that derives largely from the work of MacArthur and Levins, in which mathematics is used more or less in a qualitative sense, applying approximate equations for the purpose of deducing general patterns in nature. On the other hand is the approach that derives from systems analysis, in which specific systems are modeled with a high degree of precision to produce accurate description and prediction. These two extremes may be conveniently viewed as existing along a continuum; the "strategic" approach sacrifices a great deal of precision for generality and conceptual economy, while the "tactical" approach places the emphasis on precision.

Single-species populations. Single-species populations are usually approached in either a continuous or a discrete fashion. The continuous approach employs a first-order differential equation, $dN/dt = f(N)$, where the state variable N is population size. In the simplest case, f is a linear function, which leads to the law of exponential growth. The exponential case is unrealistic since most populations do not grow without limit, and thus f is usually considered to be quadratic. This results in the popular logistic equation, usually written in the following form to emphasize the biological significance of the parameters:

$$\frac{dN}{dt} = rN\left(\frac{K - N}{K}\right)$$

Here r is the intrinsic rate, the rate at which the population would grow if it had no limitation, and K is the carrying capacity, the size to which the population will eventually tend. These two concepts, K and r, are the most important qualitative characteristics of a population in a given environment.

The discrete approach to single-species populations, in which individuals in the population are grouped according to age in age classes, extends back to the life-table work of Lotka but has its first true formalization in the so-called Leslie matrix and finally in the stage-distribution matrix. For a population with two age classes, the Leslie matrix takes the following form:

$$\begin{pmatrix} N_{1,\,t+1} \\ N_{2,\,t+1} \end{pmatrix} = \begin{pmatrix} m_1 & 0 \\ P_{ij} & m_2 \end{pmatrix} \times \begin{pmatrix} N_{1,\,t} \\ N_{2,\,t} \end{pmatrix}$$

where $N_{i,\,t}$ is the number of individuals in the ith age class at time t, m_i is the fecundity of the individuals in the ith age class, and P_{ij} is the probability of surviving from age class i to age class j. That approach, as well as most of those that followed, deals with the basic discrete equation, $N(t + 1) = f[N(t)]$. Again, if f is a simple linear function of $N(t)$, the model population is exponential. When f represents a matrix, and all elements are constant (that is, all age classes are linear), the population is exponential.

When f is nonlinear, population behavior is usually more complicated with discrete models than with continuous ones. Generally, as the intrinsic rate rises or the density-dependent feedback (decline of growth rate with increasing population density) becomes stranger, the population behavior goes from simple damped oscillations to simple permanent cycles; to complicated permanent cycles; to noncyclic, nonasymptotic, totally unpredictable behavior; to noncyclic, nonasymptotic, yet regular behavior. The behavior which is totally unpredictable is known as chaos, and the nonasymptotic noncyclic, yet regular behavior is called resolved chaos.

The existence of chaos and similarly complicated behavior patterns in simple models offers a remarkable challenge to population biologists. Some populations, even those with simple dynamical rules, exhibit behavior patterns too complex to describe even in principle.

The analysis of spatially distributed single-species populations has not received nearly as much attention as it probably deserves. Yet the static description of individual populations has become quite sophisticated. The pioneering work of Grieg-Smith and other plant ecologists has given way to the more sophisticated techniques, introduced by Morishima. In turn these techniques are being replaced by even more advanced techniques deriving from a wedding of the study of geographic variation with geography.

A final single-species topic which has received a great deal of attention among plant ecologists is the nondynamic question of the relationship between population yield and population density. It had been known for quite some time that population yield, when plotted against density, gave empirical results which fell into two qualitative categories: the yield either increased with density to an asymptote, or increased to a particular density and decreased as densities rose further. This relationship, as proposed by Bleasdale and Nelder, is represented as the reciprocal relationship:

$$\frac{1}{Y^\theta} = A + BD$$

Here A, B, and θ are constants, D is density, and Y is yield. When $\theta = 1$ the equation takes on the basic form of the Michaelis-Menton equation. Recently yield-density–based equations have been easily transformed into the discrete dynamic population growth equations.

Two-species populations. Treatment of two-species populations falls into three categories: competition (sometimes referred to as interference), predation (including parasitism), and mutualism (including symbiosis and commensalism).

The mathematical treatment of competition has generally continued along the lines laid out by Lotka and Volterra, usually beginning with a set of coupled differential equations, $dN_i/dt = f_i(N_i N_j)$ ($i = 1, 2, j = 1, 2, j \neq i$). Such simple equations continue to generate surprising results, although initially they are approached strictly from the point of view of neighborhood stability analysis, an approach which apparently hid some of their complexities.

A large body of empirical work has resulted from various interpretations of these simple equations. This empirical work has taken two directions. First, a great deal of effort has gone into attempting to apply the equilibrium form of the equations to natural communities. This work takes the form of measuring competition coefficients in nature, a procedure that was probably faulted from the start and is now seldom used. Along with this approach a small but popular body of literature associated with niche theory developed, ostensibly aimed at establishing a link between earlier notions of the ecological niche and more recent attempts at estimating competition coefficients in the field. The second direction was the attempt to test directly whether or not the simplest form of the equations fit experimental data. While all of these tests were flawed, the general conclusion seems to be that the simple form of the equations is useful only for simple (or small) organisms and possibly for qualitative insights, but the equations must be appropriately modified to be applied to more complicated life forms. Various specific modifications of the basic equations have thus been proposed.

Independently, and largely through the same group of workers associated with demography and yield-density relations in plant monocultures, a tradition of studying plant competition evolved from the early work of C. DeWitt. In this approach two populations are conceived as occurring as fractions of one another (from 0% of type A and 100% of type B to 100% of A and 0% of B). The dependent variable is either yield or fitness. DeWitt's approach has become standard for experimental plant competition studies.

Models for predation (including parasitism) also originated with the simple Lotka-Volterra equations. In this case the initial results of the simplest form of the equation showed that populations of predators and prey should oscillate with respect to one another, a phenomenon commonly held to occur in nature. The original permanent oscillations of the simple equations gave rise to damped oscillations and limit cycles as various complications were added. Most complications include, at least implicitly, the notion of density-dependent feedback, an inclusion that generates damping of the oscillations and frequently generates limit cycles.

Perhaps stemming from C. S. Holling's original work on the relationship between foraging pattern and predator-prey theory, a rather large literature has recently emerged associated with foraging theory. Much of this literature is in the form of optimal strategies—in an evolutionary sense—for the predator. Nevertheless, the nature of these foraging patterns is what dictates, if implicitly, the complicated models of predation and parasitism which are currently being generated.

A further noteworthy development in predator-prey theory is concerned with the dynamics of spatially distributed predator-prey systems. Dispersion is theoretically capable of stabilizing a locally unstable system or destabilizing a locally stable system, depending on the parameters.

Finally, mutualism has been relatively ignored until recently, when it was shown how the classical elementary differential equations could be cast in a mutualistic frame yet still generate qualitatively reasonable results. Several other mathematical approaches have been proffered, but the richness of competition and predation theory has yet to be reached.

Multiple-species assemblages. The mathematical treatment of communities (multiple-species assemblages) has a long history of curve fitting, in which various mathematical formulations were utilized to describe species frequencies. MacArthur first noted how this approach was related to the more dynamic approach using species interactions, a point later studied by R. May. This set the stage for further elaboration of dynamic approaches, in which rates of extinction and colonization are balanced, of the theory of island biogeography, and of another theory, the community matrix approach, in which the matrix of species interactions was considered to be the focus of analysis. Recent work has emphasized the subtle and somewhat surprising indirect effects that can arise when many populations are coupled in an interactive fashion. Spatial distribution is included first in a static form, mainly by plant biogeographers, and second in a dynamic form wherein rates of migration between patches in the environment are seen as major determinants of community structure.

Applications. The history of mathematical approaches to natural resource management is most extensive in fisheries research, in which the common-property models of Gordon, and their extensions by Ricker and Larkin, were summarized and expanded by Clark. These models are more specific forms of the general discrete models described earlier. More recently, classical dynamic equations have been modified to deal qualitatively with general questions of resource management, especially in common-property fisheries.

A most important application of predator-prey theory is in epidemiology. Standard mathematical techniques in epidemiology are based on earlier ecological models, especially life-table and projection-matrix approaches. Recently, epidemiological questions have been approached in a more qualitative sense through the use of modified classical ecological equations.

Application of ecological theory to agriculture has been extensive and has been concentrated mainly on the fields of biological control and agronomy. The large-scale systems approach is frequently advocated for specific cases of integrated pest management. Recently there have been several attempts to apply modified classical ecological theory to agriculture for the purpose of qualitative prediction.

By and large, as an applied discipline, mathematical ecology has yet to provide a great number of insights to practical problems. Exceptions exist, but in general the practical potential of mathematical ecology remains elusive.

For background information *see* ECOLOGICAL INTERACTIONS; POPULATION DYNAMICS in the McGraw-Hill Encyclopedia of Science and Technology.

[JOHN VANDERMEER]

Bibliography: R. May, *Theoretical Ecology: Principles and Applications*, 1976; E. C. Pielou, *An Introduction to Mathematical Ecology*, 1969; J. Vandermeer, *Elementary Mathematical Ecology*, 1981.

Membrane separation techniques

The concept of separating gas mixtures by using membranes has been recognized for over a century. Until recently, however, the practical utility of membrane separation has been limited to a few applications, such as uranium isotope separation by a microporous metallic membrane, hydrogen purification with a palladium alloy membrane, and separations of industrial gases (mostly hydrogen) by hollow-polyester-fiber membrane units.

The latest outburst of activity in commercial membrane gas separation is due to the development of asymmetric polymeric membranes. Such a polymeric membrane has a very thin and dense skin layer on one side and a porous structure for the rest. The leading industrial gas separator, the Monsanto Prism Separator, is a system of hollow polysulfone fibers coated with a thin layer of silicone rubber. It is used mainly to separate hydrogen in ammonia synthesis and in petrochemical and refinery processes. Another membrane system is the oxygen enricher, a small portable-type unit designed for medical applications. Other commercial units are made of spirally wound, flat sheets of cellulose acetate used primarily for the separation of carbon dioxide from methane.

In addition to these commercial developments, there have been a number of significant research advances during the last decade. Several theoretical models have been developed to explain transport mechanisms through membranes, including the free-volume theory, the dual-mode sorption model, the matrix model, and the asymmetric membrane model. The free-volume theory assumes the existence of holes or free volume in the membrane. Diffusion takes place when there is a hole of sufficient size adjacent to the diffusing molecule. The temperature and pressure dependence of diffusivity has been explained rather successfully on the basis of this theory. The same theory has been extended to the transport mechanisms of gas and liquid mixtures through polymeric membranes.

The dual-mode sorption approach models a glassy polymer with two distinctly different sorption modes. This assumption leads to a dual mobility model for the transport of gases in glassy polymers. Sorbed gases are transported by Henry's law mode as well as the Langmuir mode. The temperature and pressure effects on the permeability of gases through glassy polymers can be well explained by this model.

The matrix model assumes that the penetrant gas molecules alter the structure and dynamics of the polymer matrix and thereby influence the solubility and diffusivity. The concentration dependency of solubility, diffusivity, and permeability can be modeled naturally. The experimental data of carbon dioxide in polycarbonate show that the matrix and dual-mode models describe the transport phenomena equally well.

For an asymmetric membrane, the porous supporting layer prevents the mixing of local permeate fluxes, resulting in a cross-flow pattern irrespective of the feed and bulk permeate flow pattern. The model incorporates the permeate pressure drop and can be applied equally to hollow-fiber and spiral-wound modules.

Studies have also been made to improve the engineering of membrane processes. In order to increase the degree of enrichment, different schemes have been suggested, such as using a recycle stream, purging a part of the permeate, using two membranes in a cell, cascading and recycling in various connections, and using a revolutionary technique called the continuous membrane column. Another interesting development is the so-called facilitated transport membrane, which consists of a liquid membrane with a carrier-forming complexing agent. This is a chemical substance inside the membrane that shuttles back and forth between the two interfaces, carrying the permeant molecules by combining and dislodging.

Specifications. There are many polymeric membranes which can separate a given gas mixture. For a membrane to be commercially successful, it should have certain characteristics which will make the process energy-efficient, versatile, and material-conserving. To achieve this the membrane should: (1) possess a high selectivity, which guarantees the separation of the mixture into purified components; (2) have high permeabilities in order to produce large quantities of products; (3) make it possible to compact a large surface area into a relatively small volume; (4) have a long life without any deterioration, so that it does not have to be replaced frequently, and (5) be able to function normally under a high pressure difference.

Asymmetric membranes. All of these requirements are met by the asymmetric hollow-fiber membranes. The first asymmetric membrane, developed

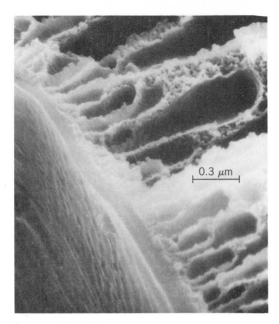

Fig. 1.   Cross-sectional view of a Monsanto hollow fiber. (*Monsanto Company*)

around 1960, was a thin-skinned cellulose acetate membrane for desalination by reverse osmosis. This type of membrane is made by incorporating a suitable water-soluble additive in the film-casting solution and leaching out the additive with water after casting. A very porous substrate structure is produced which is covered with a thin, dense layer acting as a permselective membrane with very high flux. By changing the amount of additive and the casting conditions, the size and amount of the pores can be controlled. Although this type of asymmetric membrane was highly satisfactory in reverse osmosis, it could not be used in gas-phase separations, primarily due to the presence of surface pores on the dried membrane. This difficulty was overcome by coating the surface with a very thin layer of highly permeable silicone rubber material.

The Monsanto Prism Separator membrane was produced by introducing the casting solution into a tube-in-orifice spinneret which is immersed in a water bath. Hollow fibers (Fig. 1) are spun from polysulfone with a thin dense skin on the surface, which is coated with silicone rubber to eliminate the surface pores. The separation factor of this composite membrane is primarily derived from the high selectivity of the substrate, polysulfone. The thin layer of highly permeable silicone rubber does not contribute appreciably to the overall separation factor. The same technique may be applied to produce different permselective membranes by changing both the substrate and coating materials. Effective substrates, such as polycarbonate, polyphenylene oxides, and styrene copolymers, can be used, while hydrocarbon rubbers or even polyethylene can replace silicone rubber as coating materials.

The Monsanto hollow-fiber membranes are very sturdy and durable. They can withstand high pressures up to 2000 psia (13.8 megapascals) and pressure differentials up to 1600 psia (11.0 MPa) at temperatures from 0 to 55°C. The fibers are assembled into a compact bundle which is housed in a steel shell resembling a shell-and-tube heat exchanger. Individual units can be hooked up in series or in parallel to boost the capacity and product purity (Fig. 2).

The main application of the Prism Separator has been to recover hydrogen from petrochemical process streams and ammonia purge streams. With feeds containing over 30% hydrogen, the units can provide hydrogen of 86 to 96%, or even up to 99%, purity. The Prism Separator system can also economically separate carbon dioxide from various gas streams. A possible application is in tertiary oil recovery by using carbon dioxide flooding with a membrane separator to recover carbon dioxide for reinjection while upgrading the methane for higher value.

Other membrane technologies. The membrane oxygen enricher is widely used by patients with respiratory impairment, especially for home therapy. This oxygen-enrichment system employs ultrathin supported silicone polycarbonate membranes used with flat plates in a parallel stack. The unit is capable of generating 40% oxygen continuously out of

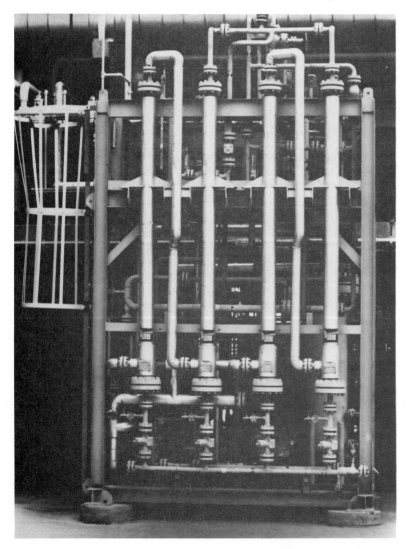

Fig. 2.   Prism Separator installation in a naphtha hydrotreater plant. (*Monsanto Company*)

room air. No additional humidification is needed since the oxygen-enriched air also contains sufficient water vapor. A vacuum, rather than pressure, is used to separate oxygen from nitrogen through the membrane. A typical patient receives 3 to 6 liters of 40% oxygen-enriched air per minute through the use of a nasal cannula.

Other commercial-scale membrane units are membrane separators that are made of spirally wound cellulose acetate membranes with appropriate spacers. Their primary application is in the separation of carbon dioxide from methane. The asymmetric cellulose acetate membranes are made by a solution casting technique similar to those made for reverse osmosis. Care is exercised in drying so as not to lose the asymmetric character or to cause collapse of the membrane.

The potential applications of these techniques seem very attractive. Membrane separation of hydrogen sulfide and carbon dioxide from natural gas would facilitate sour gas purification, since the permeabilities of hydrogen sulfide and carbon dioxide are much larger than that of methane through cellulose acetate. These membrane units can also separate hydrogen from gas streams coming off various petrochemical processes. Hydrogen is frequently found in the mixture of carbon monoxide, methane, and nitrogen. The cellulose acetate membrane offers an excellent selectivity for these gases. Another application may be found in coal gasification and synthetic fuel processes.

An additional application for cellulose acetate membrane units is in the oxygen enrichment of air. The separation factor ranges from 4 to 5.5, thus providing a reasonable separation. Since this is a rather low separation, applications would be for products requiring low oxygen concentrations in order to be cost-effective. Examples are secondary sewage treatment, wastewater treatment, high-temperature furnaces, and oxygen therapy.

Another promising membrane technology is the use of liquid membranes loaded with a carrier-forming complexing agent. An example is the facilitated transport of oxygen in liquid membrane. Both selectivity and flux are reported to exhibit tremendous increases even compared to those of silicone rubber.

Most gas permeators used in industrial applications operate in the form of a single-stage permeator. The degree of enrichment is limited by the selectivity of a given membrane. If a single permeator cannot attain the enrichment required, two or more permeators are cascaded. The disadvantages of cascades are the large membrane area, many interstage compressors, and cumbersome operation. The recently developed continuous membrane column (Fig. 3) greatly reduces these problems. The continuous membrane column is viewed as a cascade rather than as a single-stage permeator. Therefore, the degree of enrichment is unlimited as the compressor recycles the permeate stream continuously around the enriching section. The most- and least-permeable gases can be separated as highly concentrated products from a feed mixture of any composition.

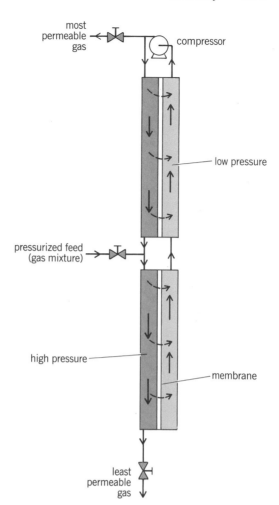

Fig. 3. Schematic diagram of a continuous membrane column. (*After S. T. Hwang and K. Kammermeyer, The continuous membrane column, AIChE. J., 26:558, 1980*)

Economic potential. The key to the success of membrane gas separation depends upon the economics. All of the conventional processes, such as cryogenic, adsorption, and electrolytic technologies, demand intensive energy and capital. As the cost of energy continues to rise and membrane fabrication techniques improve, the future of membrane processes looks ever brighter.

For background information *see* CHEMICAL SEPARATION TECHNIQUES in the McGraw-Hill Encyclopedia of Science and Technology.   [SUN-TAK HWANG]

Bibliography: J. M. S. Henis and M. K. Tripodi, A novel approach to gas separations using composite hollow fiber membranes, *Separation Sci. Technol.*, 15:1059, 1980; S. T. Hwang and K. Kammermeyer, *Membranes in Separations*, 1975; S. T. Hwang and J. M. Thorman, The continuous membrane column, *AIChE. J.*, 26:558, 1980; S. Loeb and S. Sourirajan, *Adv. Chem. Ser.*, 38:117, 1963.

Memory

There is a growing consensus among researchers in the area of memory and aging that memory does not deteriorate globally during normal aging. General knowledge, as indexed by performance on the vocabulary and information subtests of the Wechsler Adult Intelligence Scale or by questionnaires tapping knowledge of current events or historical facts, remains stable or improves with increasing age. Similarly, ability to engage in highly skilled activities, such as choosing a good move in a chess game, is not affected by age. Nevertheless, the ability to acquire new information appears to deteriorate with advancing age. This conclusion is based on work using a large number of experimental paradigms, including free and cued recall of lists of words, paired-associate learning, and retention of narratives and expository prose. Although age-related decrements in recognition memory are smaller than those found for recall, they are pervasive. Thus, the encoding of information for future use, as well as retrieval of new information from memory, appears to become less efficient in old age.

The age at which these changes occur is not easy to pinpoint. Most studies compare groups of students or recent college graduates with people aged 60 or older and do not include adults in their thirties or forties. Consequently, it is not clear whether memory declines, in the areas in which they occur, are gradual, or whether there are fairly abrupt changes. The most that can be said with confidence is that adults over 60 may be slower than those in their twenties to acquire new information in laboratory tasks devised by cognitive psychologists. There have been few studies of skill acquisition or on-the-job learning in real-life situations in young and old adults, and there is also a considerable range of learning abilities in all age groups. The findings described are thus not predictive of how any given individual, young or old, will perform in a situation which requires learning of new information; rather, they are generalizations based on group averages.

A large number of explanations have been proposed as to why old adults do not remember new information as readily as young adults. The major ones are discussed below.

Cohort differences. Virtually all studies of memory and aging employ cross-sectional rather than longitudinal designs, that is, different people are tested in each age group and the same people are rarely studied over a period of years. In cross-sectional studies there is always a risk that age will be confounded with other variables associated with cohorts (groups with the same year of birth, or generations) that contribute to memory differences. In particular there may be differences in educational level since, on the average, old adults have had fewer years of schooling. Since performance in many memory tasks is related to educational level this is a serious problem. According to hypotheses which make use of cohort differences to explain age-related changes in memory, the poorer performance of old adults is more apparent than real; there are no age-related differences in memory that cannot be accounted for by cohort differences. However, longitudinal studies in which the same individuals are tested at several-year intervals during adulthood also show a decline in memory performance in old age.

Environmental differences. The physical and social environment of many old adults is quite different from that of young adults. Old adults may suffer a loss of control over the events in their lives; experiences such as death of family members, inadequate financial resources, joblessness, illness, and, for old adults residing in institutions, loss of autonomy and control over daily routine may all contribute to feelings of "learned helplessness" or beliefs that their actions are no longer effective. This sense of loss of control may contribute to a decline in physical and psychological well-being and to clinical depression, all of which affect cognitive functioning. It has been shown that when residents in a nursing home were given an opportunity to increase control over their environment, their general mental alertness improved. When residents were rewarded for correctly answering questions about recent events, their performance on other memory tests improved. These results suggest that memory deficits in nursing home residents might be lessened by a more demanding environment over which they have some control.

This sense of loss of personal effectiveness may be exacerbated by the negative stereotyping of old adults, namely that they are incompetent and must be retired from productive work and looked after by others. Together with a lack of intellectual stimulation, this may contribute to old adults' debilitating anxiety about memory tests. In fact, their memory performance is improved when a memory test is presented as an activity or game rather than a test. These experiential factors may contribute to memory problems in old adults, but they cannot explain the age-related differences in memory found in highly educated, intellectually active, community-dwelling populations.

Differences in strategies. One popular explanation of aging deficits in memory is that old adults are less likely than young adults to spontaneously engage in certain helpful strategies while studying material or trying to retrieve it. This production deficiency hypothesis suggests that old adults' memory deficit can be eliminated if they are induced to use these strategies. There is evidence that old adults do not spontaneously use certain strategies that improve memory. In particular, old adults do not appear to rehearse word lists while learning them or to organize words by clustering in conceptual categories while recalling them. The latter result suggests that old adults may not generate retrieval cues such as category names to aid the recall of a list of words. Similarly, in learning pairs of words, old adults are less likely to use imagery or verbal mediators to link the words.

The reason why old adults do not use such mnemonic strategies is unknown. Perhaps it is caused by a lack of motivation induced by feelings of helplessness, or by an unfamiliarity with effective strategies in test situations. Young people in college or recently graduated may be more accustomed to memorizing large amounts of new information. Therefore they may be more attuned to the cognitive demands of laboratory memory tasks, and this would contribute to their superior performance. There may also be differences in the demands that daily life places on the use of memory skills.

In any case, a deficiency in producing strategies is not the cause of poorer memory among old adults. If it were, age differences would disappear when the experimenter imposed organization on the material or gave old adults instructions on effective strategies. However, age differences in recall remain under these conditions when, for example, lists are presented with items grouped temporally, or explicit instructions to use images or linking words are given for learning paired associates. When experimenters provide the subjects with retrieval cues at test, young adults still outperform old adults. Old adults may be unable to use these cues because cues are effective only when encoded during initial study with the material to be remembered. In an attempt to guarantee that cues are encoded with the material, they have been presented at the time of study as well as at recall, but age differences in performance remain. Finally, if, in general, old adults do not use effective strategies to aid in retrieving information they are trying to recall, and if this is the cause of their poorer recall, age-related differences should disappear on recognition tests. Recognition is believed to be less dependent on retrieval and memory search. However, as mentioned earlier, age-related differences in recognition have been found with a wide variety of materials.

Differences in semantic processing. Perhaps the most popular current approach to the study of memory decline in old age is derived from the levels-of-processing framework. In this view, incoming information can be encoded in different ways. For instance, a visually presented word can be encoded in terms of its appearance (for example, whether it is printed in upper- or lowercase type), in terms of its acoustic properties (for example, whether it rhymes with frog), or in terms of its meaning (for example, whether it is a member of the category fruit). Deeper or more semantic analysis produces a richer or more elaborated memory trace which is more resistant to forgetting. Old adults may not spontaneously encode items semantically, which would lead to poorer recall. According to this view, old people do not spontaneously process semantically, but they are able to do it when the task encourages them to do so. A number of studies have attempted to equate semantic processing in young and old people, but age differences in free recall were not eliminated. It has been argued that the residual difference is due to retrieval problems. However, the use of recognition tests, which minimize retrieval, do not invariably result in equivalent performance across ages. Thus there is no clear empirical support for the levels-of-processing explanation of age differences in memory.

The levels-of-processing explanation assumes that there are age differences in the semantic processing of words. However, as suggested by the fact that scores on vocabulary tests show little decline with age, there is no evidence for age-related changes in the processing of word meanings. Old and young adults also give similar responses on word association tests and have comparable facilitation when processing semantically related, rather than unrelated, words in lexical decision tasks. In fact, some studies have found no age differences in amount of semantic facilitation in lexical decision tasks, indicating that there are no age differences in semantic processing. The same studies also report that the same subjects show age differences in the level of recall. Memory problems in older people are therefore probably not due to difficulties in semantic encoding of individual words.

Capacity limitations. While there may not be age differences in the processing of single words, there are deficits in higher-level semantic processes, such as memory for text and ability to draw inferences from stated information. These appear to be the result of limitations of capacity in working memory, which involves the short-term storage and manipulation of information. One indication of such limitations is that the age difference in span on the Digits Backward subtest of the Wechsler Adult Intelligence Scale is typically greater than that on the Digits Forward subtest. The Digits Forward task involves passive storage of information, while the requirement to reverse the order of the digits in the Digits Backward task involves taxing mental operations. Differences in working memory capacity may contribute to problems in comprehension and memory for spoken and written discourse. Comprehension of discourse requires simultaneous processing of current inputs and integration of these with earlier portions of the communication. Old adults are less able than young adults to perform these operations simultaneously, especially in spoken discourse when the input is rapid and there is no opportunity to recover earlier portions of a message.

A related view is that the slowing of the central nervous system which occurs in old age results in a concomitant slowing of mental operations. It is well known that old adults do not perform as well on tasks demanding speed, such as the Digit Symbol subtest of the Wechsler Adult Intelligence Scale. They are also slower in tasks which require manipulation of symbols in working memory, for example, in mental rotation tasks and in short-term memory search. It is easy to see that a slowdown in the speed of mental operations could result in memory impairment when the presentation rate exceeds the processing rate and storage capacity is insufficient to preserve the incoming information. According to this view, age-related differences should be reduced or eliminated when the rate of input is slow or under the individual's control, but the evidence does not consistently support this hypothesis of cognitive slowing.

Automatic versus effortful processes. A view which incorporates elements from several of the hypotheses described above is that old adults may exhibit poorer performance than young adults in situations which require effort or capacity, that is, any task involving the use of mnemonic strategies or the conscious manipulation of information, but that there will be no age differences in tasks which are automatic, that is, do not require capacity or attention and are not susceptible to conscious control. For example, there is substantial evidence that access to word meaning occurs automatically, and indeed there are no age differences in tasks which involve lexical access. It has also been suggested that memory for certain types of information, such as the frequency with which events occur and their spatial or temporal characteristics, is automatic and that there should be no age differences in memory for these three classes of information. The available evidence supports this hypothesis for frequency and temporal information, but not for spatial information.

Conclusion. There is general agreement about which aspects of memory change with increasing age in adulthood and which remain stable. There is less agreement on the explanation of these changes. No single existing theory provides an adequate account of all the facts. The capacity limitation approach appears to be the most satisfactory in that it handles age-related declines in a number of cognitive tasks. However, this approach postulates, but does not explain why, there should be a capacity decline in old age.

For background information *see* MEMORY in the McGraw-Hill Encyclopedia of Science and Technology.   [LEAH LIGHT; DEBORAH BURKE]

Bibliography: Botwinick, J., *Aging and Behavior*, 2d ed., 1978; D. M. Burke and L. L. Light, Memory and aging: The role of retrieval processes, *Psychol. Bull.*, 90:513–546, 1981; N. Charness, Aging and skilled problem solving, *J. Exp. Psychol. General*, 110:21–38, 1981; G. Cohen, Language comprehension in old age, *Cog. Psychol.*, 11:412–429, 1979; F. I. M. Craik and E. Simon, Age differences in memory: The roles of attention and depth of processing, in L. W. Poon et al., (eds.), *New Directions in Memory and Aging: Proceedings of the George Talland Memorial Conference*, Hillsdale, NJ, 1980; L. Hasher and R. T. Zacks, Automatic and effortful processes in memory, *J. Exp. Psychol. General*, 108:356–388, 1979; D. V. Howard, M. P. McAndrews, and M. I. Lasaga, Semantic priming of lexical decisions in young and old adults, *J. Gerontol.*, 36:707–714, 1981; E. J. Langer et al., Environmental determinants of memory improvement in late adulthood, *J. Personality Soc. Psychol.*, 37:2003–2013, 1979; S. M. McCarty, I. C. Siegler, and P. E. Logue, Cross-sectional and longitudinal patterns of three Wechsler Memory Scale subtests, *J. Gerontol.*, 37:176–181, 1982; M. Perlmutter, What is memory aging the aging of?, *Dev. Psychol.*, 14:330–345, 1978; M. Perlmutter et al., Spatial and temporal memory in 20 and 60 year olds, *J. Gerontol.*, 36:59–65, 1981; K. W. Schaie, Quasi-experimental research designs in the psychology of aging, in J. E. Birren and K. W. Schaie (eds.), *Handbook of the Psychology of Aging*, 1977.

Meson

Recently, firm evidence for the existence of a new type of matter was found in a series of experiments carried out at the Cornell Electron Storage Ring (CESR) accelerator. In particular, a new stable particle, called the B meson, was found; it is more than five times heavier than the neutron or the proton and in a broad sense is as stable as the neutron. As background to the discovery, a series of experiments begun at the Fermilab accelerator in 1976 and continued at the DORIS storage ring at DESY (Deutsches Elektronen-Synchrotron Laboratory, Hamburg), and at CESR had led to the wide belief that a new quark should exist.

Quarks and their transmutations. Quarks are the fundamental building blocks of matter. Ordinary matter is composed of two kinds, or flavors, of quarks: the up quark, or u, and the down quark, or d. Protons and neutrons are made up of (uud) and (udd), respectively. Quarks of different flavor can change into each other only under the influence of the weak interaction, whenever energetically possible, and typically emit an electron and a neutrino. Thus a free neutron decays into a proton, an electron, and a neutrino with an energy release of 0.78 MeV (1.25×10^{-13} joule). The neutron decay is a result of the elementary process $d \to ue^-\bar{\nu}$. A free neutron has a lifetime of about 15 min. Neutrons in most nuclei are stable because their decay becomes energetically forbidden. When quarks were first invoked as the ultimate building blocks of matter, three quarks were in fact required to explain both the nucleons and the ordinary mesons, as well as the so-called strange particles, such as K mesons and hyperons. The third quark was naturally called the strange quark, or s. The s quark can transmute into the u quark through the weak interaction.

Decays mediated by the weak interaction speed up when the energy released in the reaction increases; more precisely, the lifetime decreases as the inverse of the fifth power of the energy release. Thus the Λ^0 hyperon, comprising uds, decays according to $\Lambda^0 \to pe^-\bar{\nu}$, just as the neutron, but with a lifetime of about 10^{-10} s. While these times might appear extremely small, they should be compared to the reaction times characteristic of the strong interactions, which for similar energy release are of the order of 10^{-23} s. It is in this context that the neutron, the pi mesons, the K meson, and the sigma and lambda hyperons are called stable particles. The recently discovered B meson is almost 40 times heavier than the pi meson and more than 5 times heavier than a hydrogen atom; yet it is stable, providing the first direct evidence for the existence of a very heavy quark, carrying a new flavor, called beauty or bottom, the b quark, or b. Since the b

quark can change into a u or c quark with a very large energy release, the lifetime of the B meson could be as short as 10^{-15} s. This value, short as it might appear, is still some 10^{11} times longer than the expected lifetime for an object as heavy as the B meson if it were just an excited state of ordinary matter. In this sense the B meson is referred to as a new stable particle.

Heavy quarks. In 1974 a very heavy meson of puzzlingly narrow width, equivalent to a lifetime of 10^{-20} s, too short for the weak interaction but 1000 times too long for the strong interaction, was discovered at the SLAC (Stanford Linear Accelerator Center) laboratory and at the Brookhaven National Laboratory. This new state, the J/ψ, was to become a cornerstone of the newly emerged theory of strong interactions, the so-called quantum chromodynamics (QCD). The J/ψ meson was interpreted as being composed of a new heavy quark, the charmed quark, or c, bound with an anticharmed quark, $\bar{c}$. While the circumstantial evidence was very strongly in favor of this explanation, the J/ψ itself carries no net charm. Shortly thereafter, a family of mesons named the D^0, $\bar{D}^0$, D^+, and D^- were discovered comprising $(c\bar{u})$, $(\bar{c}u)$, $(c\bar{d})$, and $(\bar{c}d)$, respectively. These mesons contain a new flavor and therefore can decay only through the weak interaction, for instance $c \rightarrow se^+\nu$ or $c \rightarrow de^+\nu$. These decays have been observed, and the D meson lifetime has been measured to be of the order of 10^{-13} s, a value appropriate for a weak interaction process.

While the c quark was in some sense required to explain phenomena observed in other fields of high-energy physics, the discovery of the b quark was a surprise. In 1976 an experiment at Fermilab reported the discovery of a very narrow state with a mass of approximately 10 GeV, observed in the mass spectrum of muon pairs produced in proton collisions against nuclei. The original experiment suggested that the observed structure could in fact be composed of two or possibly three very narrow lines, unresolved due to the limited experimental accu-

racy. The state discovered at Fermilab was named the upsilon, Υ. DORIS first resolved two states, the Υ and Υ'. In 1979, CESR proved the existence of the Υ'', and in early 1980 a fourth upsilon, the Υ''', was discovered.

The storage ring at Cornell has two experimental halls where two detectors are in operation simultaneously at all times. The two detectors are based on completely different approaches to the study of e^+e^- annihilations at high energy and to a large extent are complementary in their capabilities. Figure 1 shows the four Υ's as observed in the reaction $e^+e^- \rightarrow \Upsilon \rightarrow$ hadrons by groups operating both detectors at CESR. The three narrow peaks in the cross section are due to the Υ, Υ', Υ'', while the smaller and broader last peak is the Υ'''. The first three upsilons have masses of 9.4, 10, and 10.3 GeV and total widths of 40, 30, and 20 keV. Once again the existence of mesons with masses greater than ten times that of the proton and with such narrow widths can be explained in the framework of quantum chromodynamics by postulating the existence of the b quark of mass around 5 GeV.

B meson. The upsilons are bound states of a b and an anti-b ($\bar{b}$) quark and therefore carry exactly zero beauty. The ultimate proof of the existence of this new flavor requires the discovery of particles containing only one b quark bound to ordinary quarks. The lightest such system should be composed of a b and an anti-u or -d quark. The very small widths of the first three Υ's are explained by quantum chromodynamics as due to the annihilation of the $b\bar{b}$ pair, a very slow process in the scale of the standard strong interaction, but still extraordinarily fast compared to the weak interactions. In the continuing series of increasingly heavy Υ states, a point will be reached when the mass of such a Υ is larger than that of two $(b\bar{u})$ or $(b\bar{d})$ systems, called before their discovery B mesons. When this happens, the corresponding Υ can decay into a pair of B and anti-B ($\bar{B}$) mesons with the full speed of the strong interaction, limited only by the possibly small energy release. An Υ of width between 10 and 100 MeV is therefore typically expected. This is in fact the case for the Υ''', which was observed at CESR to have a width of about 20 MeV. The Υ''' is dubbed sometimes the B meson factory, since each Υ''' is supposed to decay into two B mesons.

This long chain of circumstantial evidence, however, is not yet a proof of the existence of the b quark. Only the telltale decay $b \rightarrow qe^-\bar{\nu}$, where q is some other quark, is the ultimate evidence which the two groups working at CESR sought and found. While it is reasonably simple to exhaustively study the weak decays of pi mesons ($u\bar{u}$, $u\bar{d}$, and so forth) and K mesons ($s\bar{u}$, $s\bar{d}$, . . .), it is much harder to do this for very heavy mesons. In the case of the B meson, for instance, the weak interaction can induce decays to final states consisting of more than eight charged light mesons plus four neutral ones. At the present level of the CESR experiments it is practically hopeless to attempt to prove that compli-

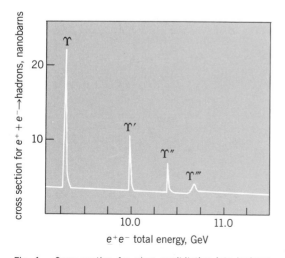

Fig. 1. Cross section for e^+e^- annihilation into hadrons versus total energy. 1 nanobarn = 10^{-37} m².

cated final states of this kind are indeed due to decays of B mesons. The basic decay $b \rightarrow e^-\bar{\nu}c$ or u (together with the similar one where the electron is replaced by a muon), however, is still expected to be a significant fraction of all decays, and it has a striking signature. The requirements of energy and momentum conservation in any reaction limit the energy of the emitted electron or muon to about one-half of the b mass, approximately 2.5 GeV. Moreover, the properties of the weak interaction are such that this maximum value is statistically favored. Finally, it is extremely improbable that the reaction $e^+e^- \rightarrow$ hadrons should result in the production of electrons of energies much larger than about 1 GeV, and this fact can be directly checked.

Measurements of the yield of electrons and muons of energies greater then 1 GeV, just below the Υ''' peak, at the Υ''' peak, and above the Υ''', show a striking increase in the production of high-energy electrons and muons at the fourth upsilon. The results of these measurements are shown in Fig. 2 in the form of the combined high-energy electron and muon yield observed by the two experiments. For comparison, the cross section for hadron production is also shown. While the latter shows an increase of less than 30%, the production of high-energy electrons and muons jumps by about a factor of 3 at the Υ''' peak. The large enhancement in electron and muon yield is a clear signal of the B meson. All the electron and muon production observed away from the Υ''' peak is well explained in terms of known processes, mostly weak decays of lighter mesons produced with high energy. The results reported so far are based on a total sample of approximately 10,000 produced Υ''' or 20,000 B mesons. From the observed electron and muon excess, it follows that approximately one out of every four B mesons decays into an electron or a muon. If the B meson were to decay by the strong interaction, only one electron decay for 10^{11} B mesons studied could be observed. Preliminary results on the shape of the energy spectrum of the electrons from B meson decays have also been obtained at CESR, suggesting that the dominant beta decay mode of the B meson

is $B \rightarrow De^-\bar{\nu}$, where the D meson is, as mentioned before, the charmed and much lighter counterpart of the B meson. This last result indicates that the weak interaction favors the decay of the b quark into a c quark rather than a u quark, in agreement with the present theoretical understanding of these interactions.

Studies involving the B meson have barely begun and will continue for several years with improved detectors at CESR and a rebuilt DORIS accelerator. The discovery of the first b-flavored particle, the B meson, not only proves the existence of the b quark, but also provides experimental and theoretical high-energy physics with new challenges and the means to confront them.

For background information *see* ELEMENTARY PARTICLE; LEPTON; MESON; QUARKS; UPSILON PARTICLES; WEAK NUCLEAR INTERACTIONS in the McGraw-Hill Encyclopedia of Science and Technology.

[PAOLO FRANZINI]

Bibliography: P. Franzini and J. Lee-Franzini, Upsilon physics at CESR, *Phys. Rep.*, 81(3):239–291, January 1982; G. 't Hooft, Gauge theories of the forces between elementary particles, *Sci. Amer.*, 242(6):104–138, June 1980; L. Lederman, The upsilon particle, *Sci. Amer.*, 239(4):72–80, 1978.

Metallurgical engineering

Microbial (microbiological) fouling can be defined as the process of microorganisms or their products attaching to or accumulating on a surface. This process may be accompanied by the growth of some of these organisms. In microbiological corrosion, microorganisms initiate or otherwise become a factor in the corrosion process. This article discusses the fouling and corrosion of metals.

Fouling processes. Surfaces submerged in an aqueous medium tend to adsorb compounds, especially organics, from the medium. As a result, the concentration of these organic substances on the surface increases relative to that in the bulk fluid. Since these organic materials are the primary source of nutrients for the microorganisms and are usually in very short supply, the microbes will move to those surfaces which provide a richer food supply. They attach by means of appendages (flagella or pili); ultimately some of them "cement" themselves into place on the surface by forming microcolonies, and secrete sticky exopolysaccharides (slime) which form a film, the glycocalyx, around the organisms.

Successive stages in the biological fouling process may involve the trapping of other types of microorganisms, debris, inorganic materials, and eventually macroorganisms, for example, barnacles. This buildup of fouling material has several serious consequences, including reduction of the heat transfer across metal tubes or pipes used in heat exchangers, reduction of the flow rate of liquids in or over these tubes, and in some cases complete plugging of the tubes. These factors lead to much lower operating efficiencies and higher operating costs, and may lead to extensive repair and replacement of pipes (Fig. 1). Finally, the presence of these fouling ma-

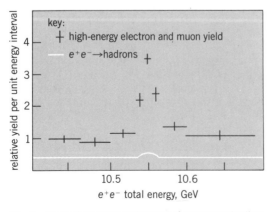

Fig. 2. Yield of electrons and muons of energy greater than 1 GeV in e^+e^- annihilations into hadrons versus energy around the mass of the Υ'''. The hadronic cross section (solid line), greatly scaled down, is also shown for comparison.

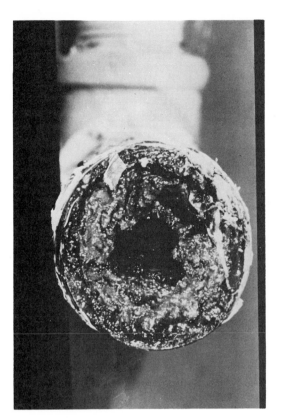

Fig. 1.   Cross-sectional view of a carbon steel water pipe showing biological fouling material and associated accumulation of iron, manganese, corrosion products, and debris. Pipe was in service for only 3 years. (*Photograph by W. K. Link, DuPont Company*)

terials can initiate or otherwise influence corrosion of the underlying metals (Fig. 2).

Microorganisms. The microorganisms involved in the fouling and corrosion processes are bacteria, algae, and fungi.

Bacteria. As a group, bacteria can grow over wide ranges of temperature, pH, and oxygen concentration (including the total absence of oxygen), and can use many different organic compounds as food sources. Some species can use inorganic materials such as nitrate (NO_3^-) and sulfate (SO_4^{2-}) ions or carbon dioxide as terminal electron acceptors for respiration under anaerobic conditions. Under such conditions they produce nitrite ions (NO_2^-), which are toxic and mutagenic; elemental nitrogen (N_2); hydrogen sulfide (H_2S), which is toxic and corrosive; and methane (CH_4). Other organisms produce inorganic compounds such as sulfuric acid, and organics such as amino and other acids, and vitamins. The latter are important as nutrients but potentially harmful to metals.

Among the microorganisms involved, microbes such as *Legionella pneumophila* and *Pseudomonas aeruginosa* are also harbored by and disseminated from such fouling materials. Growth of such organisms in water pipes has been implicated as a source of outbreaks of Legionnaires' disease on several occasions.

Algae. These organisms are photosynthetic and therefore directly affect the fouling process under conditions where light is available. They require very little for growth: light, carbon dioxide, water, and a few inorganic or organic nutrients. Most important to algae in many aquatic systems are nitrogen and, especially, phosphorus nutrient sources. Algae also indirectly affect slime production, as they are a primary source of nutrients for bacteria and fungi.

Fungi. Molds and yeasts are a varied group which can form large masses of filaments capable of trapping debris and greatly accelerating the fouling process. They use many organic compounds and are almost all strict aerobes, that is, they require oxygen for respiration. They also produce various acids and other substances that participate in corrosion processes. In addition, fungi are capable of degrading wood, a common component of some cooling systems.

Mode of action. The growth of organisms on surfaces leads to differences in the way that materials (for example, oxygen or chloride ions) diffuse to the surface of the underlying metal. This may lead to the formation of differential aeration cells (the areas under the biological material that are oxygen-rich or -poor as compared to adjacent areas) and differential-ion-concentration cells (for example, concentrations of chloride, sulfate, or ferrous ions within the

Fig. 2.   Weld seam in a stainless steel tank showing holes in the tank associated with biological material. The biological deposits in this case are small, but the corrosion is very severe. (*From G. Kobrin, Corrosion by microbiological organisms in natural waters, Materials Performance, 15(7):38–43, 1976*)

biological material). Both situations could lead to corrosion of metals.

There are also a number of other mechanisms which may be employed by microorganisms in the corrosion process. Limited diffusion of the oxygen or the ions may lead to inefficient operation of sacrificial anodes; the acids produced by many microbes can directly attack certain metals; or microbes may possibly consume atomic hydrogen at the cathodic site, leading to cathodic depolarization and thus accelerated corrosion of metals. It has been suggested that the formation of the amino acid cysteine or the enzyme hydrogenase is involved in certain corrosion processes. In addition, some *Pseudomonas* species have been reported to directly reduce ferric iron to ferrous iron. This is thought to accelerate the corrosion of iron by converting the various possible ferric compounds on the surface to ferrous forms which are removed because they are more soluble in aqueous solutions, leaving the surface reactive and corrosion-prone.

The production of H_2S by *Desulfovibrio desulfuricans* under anaerobic conditions, with the subsequent consumption of the iron in the steel to form ferrous sulfide and ferrous hydroxide, is a very important type of corrosion. This is probably also the most firmly established case for microbiologically related corrosion. Direct oxidation or reduction of iron, manganese, and perhaps chromium and nickel by bacteria may play an important role in the corrosion process. Microorganisms probably also shield the surface of the metal from the action of some corrosion inhibitors.

Control. Several metals and metal alloys, including brass, titanium, and stainless steel, are more resistant to some types of corrosion under certain conditions. Unfortunately they are not resistant to all types of corrosion or to fouling and may in fact be more prone to certain types of microbial fouling.

Recently there has been a sharp upturn in the number of cases of suspected microbiological corrosion at industrial sites, and therefore a heightened interest in learning more about the role of bacteria in the fouling and corrosion processes. Much current research is aimed at trying to characterize the nature of the biofilm community and its properties. Another area of continuing interest is in the development of mechanical, chemical, or biological methods of keeping metal surfaces clean. Thus far, no one method appears to be totally satisfactory in preventing biofilm from accumulating on metal surfaces.

For background information *see* CORROSION in the McGraw-Hill Encyclopedia of Science and Technology.

[DANIEL POPE]

Bibliography: J. W. Costerton, G. G. Geesey, and K. J. Cheng, How bacteria stick, *Sci. Amer.*, 238:86–95, 1978; R. Mitchell and P. H. Bensen, *Micro and Macro Fouling in the OTEC Programs: An Overview*, Argonne National Laboratory, 1980; D. H. Pope, R. J. Soracco, and E. W. Wilde, Studies on biologically induced corrosion in heat exchanger systems at the Savannah River plant, Aiken, S. C., *Mater. Perform.*, 1982; R. Tatnall, Case histories: Bacteria induced corrosion, *Mater. Perform.*, 20:41–48, 1981.

Micropaleontology

Micropaleontology—traditionally the study of microfossils for the solution of geologic problems—is now gaining importance as a tool in evolutionary biology and paleobiology. Microfossils include tiny remains representing all the major groups of organisms from bacteria to vertebrates. No logical basis for the organization of micropaleontology exists other than its usefulness in solving problems in other fields. At least 16 major groups of microfossils are now commonly studied (see table) by micropaleontologists.

The study of microfossils has clarified the history of the Earth's oceans (paleoceanography) and climates (paleoclimatology) of the last 150 million years; defined paleobiogeographic patterns; provided detailed zonation and correlation of sedimentary rocks of the Earth's crust (biostratigraphy); aided in understanding the way evolution proceeds; indicated causes of massive extinctions in the Earth's biota; and provided evidence for long-distance tectonic transport of large pieces of the Earth's crust.

Geologic applications. Microfossils are used for the determination and refining of stratigraphy, both for the oil industry and in basic research. Additionally, in the constant search for better stratigraphic resolution and correlation, the systematic relationships among microfossils have been more carefully determined. For example, drilling of deep-sea sediments as old as the Jurassic by the Deep Sea Drilling Project necessitated increasingly detailed biostratigraphic information based on groups of microfossils not previously utilized. The stratigraphic occurrences of planktonic foraminifera, radiolarians, diatoms, calcareous nannoplankton, silicoflagellates, and ebridians have been described in detail. By using these microfossils, time divisions as small as 20,000 years in the Quaternary and 100,000 years in the Cenozoic and Mesozoic can be distinguished and correlated over vast regions of the Earth. The same groups of microfossils are proving to be powerful tools for age dating and correlation in oil-bearing sedimentary basins. On land, radiolarians, acritarchs, dinoflagellates, and benthic foraminifera provide as sound a basis for age dating and correlation of Jurassic and older rocks as benthic foraminifera have long done in younger sedimentary rocks.

Because accurate correlation is now possible, especially when the biostratigraphy of microfossils is coordinated with the paleomagnetic and radiometric time scales, paleoceanographic and paleoclimatologic inferences can be made based on the distribution of microfossils at certain times. Events marked by changes in the microfossil distributional patterns and species composition are: the circulation patterns during the opening of the Atlantic Ocean (150 mil-

Some important microfossils

Microfossil	Size	Composition	Age range
Bacteria	<10 μm	Organic	Precambrian–Recent
Blue-green algae	<50 μm	Organic	Precambrian–Recent
Dinoflagellates	5–150 μm	Organic	Silurian–Recent
Acritarchs	<100 μm	Organic	Precambrian–Recent
Ebridians	<100 μm	SiO_2	Paleocene–Recent
Diatoms	<2 mm	SiO_2	Jurassic–Recent
Silicoflagellates	<100 μm	SiO_2	Cretaceous–Recent
Calcareous nannoplankton	<50 μm	$CaCO_3$	Pennsylvanian–Recent
Foraminifera	0.01–100 mm	$CaCO_3$, foreign materials, organic	Cambrian–Recent
Radiolarians	0.03–1.5 mm	SiO_2	Cambrian–Recent
Tintinnids	<300 μm	Organic, foreign materials, $CaCO_3$	Ordovician–Recent
Pteropods	0.1–5mm	$CaCO_3$	Cretaceous–Recent
Ostracods	0.1–5mm	$CaCO_3$	Cambrian–Recent
Conodonts	0.1–3mm	Apatite	Cambrian–Triassic
Chitinozoans	75–700 μm	Chitin	Ordovician–Devonian
Fish parts (ichthyoliths)	0.05–2mm	Apatite	Silurian–Recent

lion years ago to the present); the closing of the Panamanian land bridge (7–2 million years ago); the drying and flooding of the Mediterranean Sea (6.5 million years ago); the changes in bottom waters of the oceans during the last ice ages; and the shrinking of the Pacific Ocean over the last 200 million years. Paleoclimatic inferences have been made on the basis of these changing oceanographic patterns and the distribution and the stable isotopic composition of microfossils.

The abundances of various oxygen and carbon isotopes can be measured in the $CaCO_3$ (calcium carbonate) skeletons of microfossils, and these measurements are now very powerful tools in paleoceanographic and paleoclimatic reconstructions. These isotopes are assumed to be incorporated in the microfossils in equilibrium with seawater. Since seawater composition is controlled by temperature, the amount of ice on land, and other factors that can be determined for particular times and places, these microfossils are highly sensitive indicators of changing oceanic and continental environments.

Paleotemperature curves for the Earth's oceans for the last 150 million years show that the oceans have generally become considerably cooler. The high-latitude surface- and deep-water temperatures have decreased from values of 15–20°C to present-day temperatures of 0–4°C. The record shows that this decline was not uniformly gradual, but was marked by shorter periods of warming and cooling. At the end of the Cretaceous and continuing into the earliest Tertiary, the oceans warmed about 5°C. Throughout the early Tertiary, the microfossil isotopic record shows a general cooling from 15°C to less than 10°C, followed by a particularly rapid decrease of about 5°C at the end of the Eocene, about 36 million years ago. This event seems to mark the onset of Antarctic glaciation and the formation of sea ice around Antarctica. Temperatures remained near 5°C in high latitudes until the latest Miocene (14 million years ago), when Antarctica developed ice shelves and an extensive ice cap, and sea temperatures declined to present values. The deep-sea water masses also changed in parallel, and likewise affected the organisms living there.

From these same data, paleoclimates have been inferred for the Earth's surface, particularly during the last ice ages. The relative abundance of carbon isotopes seems to be related to productivity in the oceans when the microfossil skeletons were secreted. Data accumulated to date indicate that productivity declined especially at the boundary between the Cretaceous and the Tertiary, and during the Oligocene. The reasons for these major changes are related to the circulation of the oceans, but the causes of these changes remain uncertain. Possibly they include tectonic rearrangements of continental masses, mountain building, extraterrestrial impacts, and variations in the solar energy received on the Earth.

Paleoenvironmental reconstructions are made by using mainly benthic microfossils, such as foraminifera and ostracods, because these groups are abundant and were ecologically sensitive to environmental changes. They not only have confirmed the larger-scale changes noted above but have also provided evidence for local changes in paleoenvironment, such as depth changes in particular basins, regressions and transgressions of the sea, development and cessation of anoxic conditions, upwelling, and local circulation and water-mass patterns.

Microfossils have been found in ancient rocks of western North America and elsewhere around the rim of the Pacific Ocean that indicate tropical, low-latitude environments. This information, in combination with paleomagnetic and field geologic studies, shows that rocks have been transported on pieces of the Earth's crust through plate tectonic action for many thousands of miles from the central Pacific Ocean to their present locations. In western North America alone, over 45 large tectonic blocks have been identified.

Paleobiologic studies. The Earth is about 4.6 billion years old. The earliest known life is some 3.5 billion years old and is represented by microfossils, mostly bacteria and blue-green algae. Eukaryotic microfossils appeared about 1.5 billion years ago. These microorganisms dominated the Earth until megascopic animals first appeared about 0.7 billion years ago.

Evidence from the microfossil record shows that at the end of the Cretaceous and the end of the Eocene, planktonic and reef-associated benthonic microfossils underwent major, sudden mass extinctions. The total species diversity was decreased as much as 80–90% in some groups within a time period too short to be resolved geologically (less than 1000 years). The mass extinctions occur at precisely the same stratigraphic intervals where abnormally large amounts of iridium and related elements are found. These facts suggest that extraterrestrial catastrophes occurred at those times that deposited the iridium and killed off not only the microbiota but many larger groups as well (dinosaurs, reef corals, bivalves, ammonites, and others). Although evidence is still accumulating, it seems that asteroids impacted the Earth at those times, causing widespread ecological changes. The mechanisms linking the impacts of asteroids with extinctions are still not clear, but one hypothesis is that an Earth-encircling dust cloud thrown up by the impact cut off sunlight for a period of several weeks, causing photosynthesizing microplankton and other plants to die off. This would cause the collapse of oceanic food chains, resulting in the extinction of organisms dependent on the primary producers. Other, more specific hypotheses have been proposed because the extinctions were selective. Planktonic species with complex morphologies were eliminated, leaving the simple forms; and certain ecologic zones (reefs, open ocean) were affected, but not others (deep sea, continental slope).

Evolution. Different models for the mode of evolutionary change have been proposed. Some workers suggest that evolution proceeds with a series of geologically rapid speciation events followed by long periods of little or no change (punctuated equilibrium), while others claim that it proceeds in a gradual, slow fashion (gradualism, or phyletic evolution). Because microfossils are so abundant and have such a complete geological record, they have been used to test which of these two hypotheses may be correct. In Neogene radiolarians and planktonic foraminifera, evolutionary change occurred gradually and continually, thus supporting earlier observations of gradual evolution in a Permian foraminifer. *See* EVOLUTION.

Summary. Micropaleontology has continued to serve geology, particularly through age dating, detailed correlations, and paleoenvironmental interpretations of sedimentary rocks across the globe. In the last decade, microfossil studies have provided a basis for the new discipline of paleoceanography and new methods of interpretation in paleoclimatology. Microfossils indicate that gradual changes are perhaps more common in biologic evolution than other recent ideas suggest. Finally, microfossils are increasingly studied to understand the living habits of the ancient organisms themselves.

For background information *see* FOSSIL; MICROPALEONTOLOGY; PALEONTOLOGY in the McGraw-Hill Encyclopedia of Science and Technology.

[JERE H. LIPPS]

Bibliography: B. U. Haq and A. Boersma, *Introduction to Marine Micropaleontology*, 1978; J. P. Kennett, *Marine Geology*, 1982; J. H. Lipps, What, if anything, is micropaleontology? *Paleobiology*, 7:167–199, 1981.

Mineralogy

The visual appearance of rocks and minerals depends on two interrelated properties, color and luster. Color depends on the way that minerals absorb light, and luster depends on the way that minerals reflect light. In addition, dispersion, scattering, interference, and diffraction influence the way that light moves through mineral crystals. Absorption and reflection are related to the crystal and electronic structure of minerals. A new understanding of mineral color has come about as the principles of solid-state physics have been applied to mineralogy. The table lists such sources of color in minerals. Dispersion, scattering, interference, and diffraction are described by classical optics and relate more to the microstructure of minerals than to the atomic arrangement.

The intensity of light passing through an absorbing mineral grain falls off exponentially with distance. Absorption is described by the Lambert-Beer law [Eq. (1)], where I is the intensity of light trans-

$$I/I_0 = e^{-\alpha t} \tag{1}$$

mitted through a mineral slab and I_0 is the incident intensity. The absorptivity $\alpha(\lambda)$ [in units of reciprocal length] gives the absorption spectrum. Bands of strong absorption in the spectrum relate to the physics of the absorption process; intermediate windows of low absorption are responsible for the color.

The luster of minerals depends on the refractive index. Diamond, with a refractive index of 2.41, has a brilliant or adamantine luster, and fluorite, with a refractive index of 1.43, has a dull and vitreous luster. Luster also depends on absorption and reflectance. Pyrite, with very high absorption and reflection, has a metallic luster although it is colored. Most silicate minerals are weakly absorbing and also weakly reflecting and have a vitreous luster. The refractive index of an absorbing material, N^*, is a complex number, expression (2), where the real part, n, is the usual refractive index, and the imaginary part, k, is the extinction coefficient.

Sources of color in minerals

Physical mechanism	Classes of minerals	Examples	Theoretical model
Plasma edge	Metals	Gold, copper	Band theory
Absorption edge	Semiconductors	Cinnabar, proustite	Band theory
Band structure	Metals and semiconductors	Pyrite, bornite, arsenopyrite, galena	Band theory
Transition metal ions	Insulators	Olivine, emerald	Crystal field theory
Ligand-to-metal electron transfer	Insulators	Crocoite	Molecular orbital theory
Metal-to-metal electron transfer			
Short-range	Insulators	Blue kyanite, vivianite	?
Long-range	High-resistivity semiconductors	Magnetite	Band theory
Charges trapped in defects	Wide-gap semiconductors	Amethyst, colored fluorite	Band theory

The relation between the dimensionless k and the α of the Lambert-Beer law is given by Eq. (3). The reflectance R from mineral surfaces depends on both the real and imaginary parts of the refractive index [Eq. (4)]. If k is small, the minerals have a vitreous

$$N^* = n + ik \qquad (2)$$

$$\alpha = 4\pi k/\lambda \qquad (3)$$

$$R = I_{\text{(reflected)}}/I_{\text{(incident)}} = \frac{(n-1)^2 + k^2}{(n+1)^2 + k^2} \qquad (4)$$

luster that may be brilliant or dull, depending on whether n is large or small. If k is large, the reflectance approaches unity, and the minerals have a metallic luster.

Optical effects. Many of the apparent colors of minerals arise from optical effects. Dispersion is the change in refractive index with wavelength. It describes the ability of a mineral grain to act as a prism and is responsible for the "fire" seen in cut gems and in some mineral fragments. Scattering distinguishes translucent from transparent materials. Light passes straight through a slab of transparent mineral but is scattered in all directions in a translucent one. Scattering is caused by voids, fluid inclusions, cleavage cracks, grain boundaries, and particles with a refractive index different from the bulk material. The most efficient scattering occurs when the scattering centers are the same size as the wavelength of light. Moonstone, stars, eyes, and related features are scattering phenomena.

Interference can produce colors in minerals within thin surface layers because, for a given layer thickness, some wavelengths interfere constructively and others destructively. The patterns of color are similar to those formed by thin oil films floating on water. Interference is responsible for the play of color, called *iridescence*, seen on labradorite, iridescent chalcopyrite, and certain volcanic glasses.

Diffraction is the separation of light into its component wavelengths. It is a rare but spectacular source of color in minerals such as fire opal. The scanning electron microscope reveals that opal consists of spheres of hydrated silica. The diameter of the spheres is comparable to the wavelength of light. If the spheres are of uniform size and are regularly packed, they act as a three-dimensional diffraction grating and are responsible for the fire. If the spheres are of nonuniform size or are irregularly packed, they merely act as scattering centers, and the result is a dull, milky-white opal with no fire.

Delocalized electrons. Minerals may be classified into insulators, metals, and semiconductors.

Insulators. In insulators the electrons are all bound, and these minerals, such as quartz, are usually transparent to visible light. As the wavelength of light becomes shorter and shorter, its energy becomes higher and higher. Eventually, an energy will be reached that is capable of pulling the electrons from their bound states and putting them in a band of empty states where they are delocalized and free to move under the influence of an electromagnetic field. These states are shared by the entire crystal. The low-energy bound states are called the valence band, the empty set of excited, delocalized states is called the conduction band, and the forbidden energy region between them is called the band gap. The wavelength corresponding to the band gap is called the absorption edge. At wavelengths longer than the absorption edge, the crystal is transparent ($\alpha \cong 0.05$ cm^{-1}); at wavelengths shorter than the absorption edge, the crystal becomes nearly opaque ($\alpha = 10^3$–10^5 cm^{-1}). In transparent crystals the absorption edge lies in the ultraviolet, but in certain strongly colored minerals the absorption edge lies in the visible region of the spectrum. The absorption edge of cinnabar is at 630 nm. Thus blue light and green light are strongly absorbed, but red light is transmitted, giving cinnabar its characteristic deep red color. The absorption edge of orpiment, a deep yellow mineral, lies at shorter wavelengths so that only blue light is cut off (Fig. 1).

Metals. In metals the conduction band overlaps the valence band so that empty delocalized states are continuously available to the electrons. When metals are excited by light, the electrons simply move in phase with the field. Metals are nearly opaque ($\alpha = 10^5$–10^6 cm^{-1}) but are usually highly reflective. As the wavelength becomes shorter, there is a point at which the free electrons in the metal can no

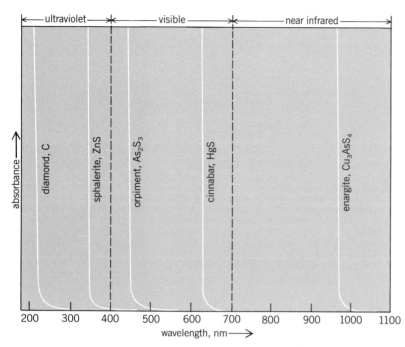

Fig. 1. Some representative absorption-edge spectra for transparent, colored, and metallic-luster minerals.

longer remain in phase with the oscillating field. This is the plasma edge, and it is marked by a pronounced dip in the reflectivity. The plasma edges of most metals occur in the ultraviolet, and so the visible color is a nondescript white or gray. A few metals, notably gold and copper among minerals, have plasma edges in the blue-to-green regions of the spectrum. Gold reflects predominantly yellow light, and copper reflects red to red-orange light.

Semiconductors. In semiconductor minerals the absorption edge lies in the infrared region. The energy of visible light is sufficient to excite electrons into the conduction band. To visible light, semiconductor minerals behave like metals and have a metallic luster. The visible spectrum is very narrow compared with the width of the conduction bands, and therefore many semiconductor minerals, such as galena and arsenopyrite, are a colorless gray. Others, such as pyrite and chalcopyrite, have maximum reflectivities in the middle of the visible spectrum and thus have bright metallic-yellow and yellow-orange colors. A few minerals, such as bornite, have maximum reflectivity in the blue range and exhibit brilliant metallic blue and blue-violet colors. The complex reflectivity spectra responsible for the colors are produced by corresponding details in the electronic band states.

Transition and rare-earth elements. Ions of the transition or iron group, Fe^{2+}, Cr^{3+}, Mn^{2+}, Ni^{2+}, and so on, have in common a partially filled d-electron shell. Absorption of light rearranges the d-electrons within tightly bound states. These absorption processes are weak, and minerals colored by transition metal ions are still partially transparent. The energy levels are controlled by the arrangement of ligands that coordinate the transition metal ion. A

coordination octahedron of water molecules produces the blue of copper solutions, the deep green of chromium solutions, and the yellow-green of nickel solutions. Similar colors are produced when the ions are substituted into minerals. However, in minerals the size and geometry of the site is fixed, and the crystal field sensed by the ions varies between different mineral structures. Factors that influence the crystal field are the coordination number of the cation, the deviations from regular octahedral symmetry, and the cation-to-anion distances. Green emerald and red ruby are both colored by chromium and have similar spectra. Chromium enters ruby as a substitute for the smaller Al^{3+} ion. This increases the crystal field and shifts the spectrum to produce a red color (Fig. 2).

The rare-earth and actinide ions contain partially filled f-shells which also produce absorption. The f–f transitions are very weak and produce only pastel colors even in pure compounds.

The behavior of f- and d-electrons is described by crystal field theory, which provides good predictions of the spectra and color to be expected from ions in sites of known geometry.

Electron transfer. Molecular anions such as chromate and vanadate are strongly colored. The absorption of light excites an electron from the p-orbitals

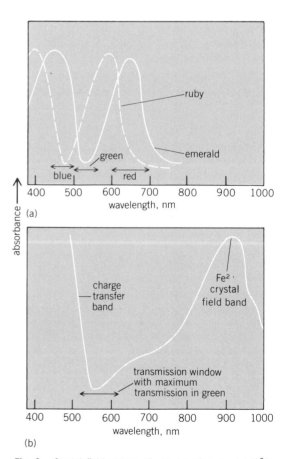

(a)

(b)

Fig. 2. Crystal field spectra of colored minerals. (a) Cr^{3+} crystal field bands of emerald and ruby. (b) Absorption spectrum for olivine, $(Mg,Fe)_2SiO_4$.

of the oxygen ligands to the empty d-orbitals of the metal ($\alpha = 10^2-10^4$ cm^{-1}). Ligand-to-metal charge transfer also occurs in the coordination polyhedra of the other transition metals and is responsible for a very intense absorption band in the ultraviolet. The tail of the charge-transfer band often extends into the visible. The characteristic blue-green color of many ferrous iron–containing minerals is due to the window formed between the tail of the charge-transfer absorption in the blue-violet and the tail of the Fe^{2+} crystal field band which occurs in the near-infrared. Molecular orbital theory gives a rather precise description of ligand-to-metal charge transfer.

When metals are present in more than one valence state, the absorption of light can excite direct metal-to-metal electron transfer, the valence states being reversed in the process. Couples that commonly produce color in minerals are $Fe^{2+} \rightleftarrows Fe^{3+}$ and $Fe^{2+} + Ti^{4+} \rightleftarrows Fe^{3+} + Ti^{3+}$. The latter is responsible for the blue color of some kyanites and of sapphire. The ferrous-ferric exchange results in an absorption band near 700 nm in the near-infrared, producing deep green in jade and violet in vivianite. Maximum intervalence electron-transfer absorption occurs when the light is polarized parallel to the line joining the two metal ions, so the colors are strongly pleochroic. The absorptivity is about ten times that of crystal field transitions. Intervalence electron transfer produces deeper colors than transition metals alone, but the absorption is sufficiently weak that grains are transparent and of vitreous luster.

Deep-green and blue colors are produced by the interactions of metal-metal pairs. When more metal atoms—in chains, sheets, or clusters—interact, the absorption becomes much more intense, and the colors produced are browns and blacks. Magnetite, pyrolusite, biotite, and many other deeply colored oxide and silicate minerals owe their color to long-range intervalence electron transfer. These minerals are often high-resistivity semiconductors as well, since the exchanged electrons are sufficiently mobile to act as carriers.

Otherwise colorless minerals may be colored by defects introduced by impurities or by radiation damage. The energy levels arise from the way in which the defects or impurities modify the band structure of the host crystal. The color of blue diamond is caused by part-per-million quantities of nitrogen substituting for carbon. The pink of amethyst is related to iron impurities in quartz. The blue of sodalite is due to halogen substitution. Minerals that are associated with radioactive minerals are dosed with radiation over long periods of time. Radiation-induced defects color smoky quartz. Some radiation damage centers are linked to impurities and are stable; others can be bleached by exposure to strong light.

The energy levels for absorption are produced by electrons or holes trapped near the impurities, in anion vacancies (F-centers), or in cation vacancies (V-centers). The theory of defects and their influence on the electrical and optical properties of crys-

tals is well developed because the same principles apply to semiconductors.

For background information *see* ABSORPTION OF ELECTROMAGNETIC RADIATION; COLOR; CRYSTAL DEFECTS; MINERALOGY in the McGraw-Hill Encyclopedia of Science and Technology.

[WILLIAM B. WHITE]

Bibliography: R. G. Burns, *Mineralogical Applications of Crystal Field Theory*, 1970; A. S. Marfunin, *Physics of Minerals and Inorganic Materials*, 1979; A. S. Marfunin, *Spectroscopy, Luminescence and Radiation Centers in Minerals*, 1979; K. Nassau, The origins of color in minerals, *Amer. Mineral.*, 63:219–229, 1978.

Molecular biology

Since 1975, plant molecular biology has mushroomed into a major area of research. The field includes studies of the structure and expression of genes from nuclei, chloroplasts, and mitochondria of the plant cell, and of the genes of viruses, plasmids (extrachromosomal DNA), and symbiotic bacteria associated with plants. Recombinant DNA techniques are being applied in plant studies with two main goals: first, to understand the basic mechanisms of gene expression, and second, to manipulate genes for agricultural improvements.

Nuclear genes. Nuclei of flowering plant cells contain between 2×10^8 and 8×10^{10} base pairs of DNA, depending on the species. The DNA is organized into chromosomes and is arranged so that sequences of about 1000 base pairs of single-copy DNA (sequences that are present once or a few times per genome) are interspersed with sequences that are repeated dozens to thousands of times. A few sequences may be clustered together on the chromosome. Genes (sequences of DNA that are transcribed into RNA and then usually translated into proteins) are mostly single-copy DNA. The RNA sequences present on polysomes, and therefore being translated into proteins, have been studied in different organs of the tobacco plant (root, leaf, stem, petal, ovary, and anther) to estimate what fraction of the DNA in a plant nucleus actually codes for proteins. It was found that about 25,000 different messenger RNAs of average size (1200 bases) are on polysomes in each organ, with 8000 of these sequences shared in common between all organs. From this analysis it is estimated that at least 60,000 genes are expressed in an adult tobacco plant. This accounts for 10% of the single-copy DNA and less than 2% of total DNA. Therefore, more than 90% of the DNA does not appear to code for protein and is of unknown function.

Since the advent of recombinant DNA technology in the late 1970s, one can study the structure of an individual gene or DNA sequence—to isolate and amplify 1000 base pairs out of 10^8 base pairs! So far, a handful of nuclear genes have been cloned and their base sequences determined: the genes for soybean actin, a protein involved in intracellular movements; soybean leghemoglobin, an oxygen-binding protein from nitrogen-fixing root nodules;

phaseollin, the storage protein from bean embryos; zein, the corn endosperm storage protein; and corn alcohol dehydrogenase, an inducible enzyme. It has also been learned that many plant genes contain introns (regions of DNA within genes that are transcribed into RNA but then processed out before the protein is made) and that the sequences of DNA-bordering introns are the same in plant and animals genes. Signals for RNA polymerase binding also seem to be conserved. Researchers hope that by studying the sequences of individual genes they will be able to decipher the signals genes use to be expressed at particular times and places in the plant.

Chloroplast and mitochondrial genes. In plants there are two organelles other than the nucleus which contain DNA: the chloroplast, site of photosynthesis, and the mitochondrion, involved in respiratory energy production. In both organelles more than 50% of the proteins involved in their function are encoded in the nucleus rather than in their own genomes. In fact, there are some proteins (for example, ribulose-bisphosphate carboxylase, the enzyme that captures CO_2 in photosynthesis) where one subunit is coded by the nucleus and another subunit is coded by the chloroplast or mitochondrion. "Cross talk" must exist between organelles, but little is presently known about such intraorganelle communication.

A flowering-plant chloroplast genome consists of a circular molecule of about 1.5×10^5 base pairs. In general, chloroplast genes do not contain introns. Many of the genes have sequence homologies with genes from bacteria, and there are many indications that both chloroplast and mitochondrial genomes are more similar to prokaryotic genomes than to the eukaryotic nucleus with which they share cytoplasm. The chloroplast codes for its own ribosomal RNAs, transfer RNAs, and many of the proteins involved in photosynthesis; these genes have been cloned in bacteria and are being sequenced. Although the chloroplast genome is thousands of times smaller than the nuclear genome, there may be several circular molecules per chloroplast and many chloroplasts in each cell that contains chloroplasts. In some leaves, chloroplast DNA comprises 30% of total cellular DNA.

Less is known about plant mitochondria. There is some debate about the size and form of the mitochondrial genome, although the average size is probably more than ten times larger than the genome of animal mitochondria, and both circular and linear forms may exist. It codes for transfer RNAs, ribosomal RNAs, and 18 to 20 proteins. In most higher plant cells, mitochondrial DNA contributes less than 1% of the total cellular DNA. The mitochondrial genome is currently under intense investigation because it has been implicated in the susceptibility of corn to the fungus, *Helminthosporium maydis*, that causes southern corn leaf blight. In 1970 an epidemic caused over $1 billion in losses in the United States. It appears that a mutation in the mitochondrial genome that causes cytoplasmic male sterility, a trait used in production of hybrid

corn seed, also increases plant susceptibility to the fungal toxin.

Genetic engineering. Plant molecular biology is still in its infancy, with genes being described in terms of location, sequence of bases, and function. In order to study more sophisticated questions of gene regulation, researchers need to be able to remove a gene, alter its sequence, and determine how the alteration affects expression. This is necessary both for basic studies and for genetic engineering. Some possible applications hold interest for investors as well as scientists: disease and pest resistance, stress tolerance (particularly to drought and salt), increase in photosynthetic efficiency, transfer of nitrogen-fixation to different species, better amino acid composition of storage proteins, and so on.

Many steps are involved in genetic engineering. The gene has to be identified, isolated, amplified, and altered; all of these steps are already possible by using proven recombinant DNA techniques. The gene must then be reintroduced into the plant in such a way that it will be maintained through cell division, be expressed, and be passed on through meiosis to progeny. Little is known about the processes involved in these steps, although it is an active area of research. Since many plants can be regenerated from individual protoplasts (cells without walls) from tissue cultures or from plant organs, a promising approach is to introduce foreign DNA into protoplasts by using vehicle (or vector) DNA which contains information for autonomous replication, integration, or both, into a plant chromosome. The goal is to insert the gene of interest into the vector in such a way that it will be carried along.

Success in the area of experimental genetic engineering has come mainly from studies of the crown gall bacterium, *Agrobacterium tumefaciens*. These bacteria, which contain the Ti plasmid (a tumor-inducing plasmid, which besides oncogenicity, also encodes opine biosynthesis and catabolism, and other functions), have the remarkable ability to transfer a piece of DNA from the Ti plasmid (T-DNA region of the Ti plasmid) into the nuclear DNA of a plant they have infected. The T-DNA is covalently linked into the plant DNA where it causes the plant cells to proliferate into a tumor and to make opines, compounds that only the bacteria can eat. Crown gall formation is the first known example of naturally occurring DNA transfer between members of different kingdoms. In one study, foreign genes (from fungi, animals, bacteria, and plants) were substituted for a gene on the T-DNA; plants were injected with the altered plasmid; and it was found that the new gene was integrated into the plant nuclear DNA along with T-DNA. In some cases the new gene was also expressed, but at a low level.

This experiment is a very promising first step, but there are some problems other than low expression. For one thing, the transferred T-DNA plus new gene causes a tumor rather than a normal, nontumorous plant that can be propagated. Recently the regeneration of normal plants from tumor tissue has succeeded, so perhaps the problem will be overcome.

Also, *A. tumefaciens* infects only dicotyledonous plants, which excludes the major grass crops, such as corn, rice, wheat, and barley. As yet, not enough is known about the causes of host specificity to alter this trait.

A few plant viruses are also being studied as possible vectors. However, most plant viruses contain RNA as their genetic material, which is not as easily manipulated for recombinant DNA as double-stranded DNA (most of the enzymes used do not recognize the RNA). Cauliflower mosaic virus is one of the few double-stranded DNA plant viruses. The basic molecular biology of this virus is not yet understood sufficiently to use it as a vector.

It may also be possible to use pieces of DNA from plant nuclear DNA itself as vehicles. Transposable elements (transposons) are pieces of DNA that can "jump out" of one place in the chromosome and then "jump back in" at another place. Transposable elements have been demonstrated by standard genetic techniques in corn. The possibility of cloning a transposable element, inserting a gene into it, and then having it reintegrate in the chromosome is under investigation now.

Nitrogen fixation. One of the aims of genetic engineering is to transfer nitrogen fixation to plants (such as wheat or corn) other than legumes (such as peas or beans), thus making expensive nitrogen fertilizers unnecessary. Symbiotic nitrogen fixation involves a complex interaction between bacteria of the *Rhizobium* genus and legume roots and requires many genes in both the bacteria and plant. Most work is centered on finding all of these genes by mutational analysis, and then the cloning and sequencing of the genes. The genes involved in nitrogen fixation have recently been transferred to *A. tumefaciens* from *Rhizobium*, and when this *A. tumefaciens* infected clover, rudimentary nodules formed. They did not fix nitrogen, however.

Plant molecular biology is moving rapidly as more scientists come into the field, and it can be expected that there will be both basic and practical results in the near future.

For background information *see* DEOXYRIBONUCLEIC ACID (DNA); GENE; GENETIC ENGINEERING; NITROGEN FIXATION; RECOMBINATION (GENETICS); RIBONUCLEIC ACID (RNA) in the McGraw-Hill Encyclopedia of Science and Technology.

[MARTHA CROUCH]

Bibliography: W. J. Brill, Agricultural microbiology, *Sci. Amer.*, 245(3):199–213, 1981; J. L. Fox, Plant molecular biology beginning to flourish, *Chem. Eng. News*, pp. 33–44, June 22, 1981; S. H. Howell, Plant molecular vehicles: Potential vectors for introducing foreign DNA into plants, *Annu. Rev. Plant Physiol.*, 33:609–650, 1982; A. Marcus (ed.), *The Biochemistry of Plants*, vol. 6: *Proteins and Nucleic Acids*, 1981.

Molecular chirality

The recent discovery of chiral amino acids in a meteor that landed in Australia has quickened interest in the physical basis for the origins of molecular chirality. It is very clear that biological processes can operate only on one set of enantiomeric molecules. If such processes involved racemic mixtures of amino acids, carbohydrates, and other chiral molecules, the information-processing problem in biosynthesis would be virtually insurmountable. The scientific interest in this problem centers on why R-glyceraldehyde (I) and S-alanine (II) are basic building materials instead of their enantiomeric, or mirror-image, counterparts.

(I)   (II)

Chiral biomolecules. There are two viable theories relating to the origin of the biological chiral molecules. One school of thought contends that the selection of R-glyceraldehyde, or the first chiral biomolecule in the prebiotic universe, was due to stochastic processes. The most frequently mentioned process is the crystallization of one enantiomer from a solution containing a racemic mixture of molecules. This spontaneous resolution process is known to occur in the laboratory, and if it is done a sufficient number of times, the number of R-crystallites and S-crystallites will be the same. The discovery of chiral amino acids in a meteorite with the same chirality as the amino acids on Earth makes the stochastic theories seem improbable, since the same chance event would have to have occurred in the same way at distant points in space and time.

The second school of thought concerning the origins of biological chirality contends that absolute asymmetric synthesis occurred in the prebiotic universe. Absolute asymmetric synthesis is the synthesis of dissymmetric molecules without the intervention of dissymmetric catalysts, reagents, solvents, or separation systems. Absolute asymmetric synthesis can occur only in the presence of a chiral physical field which can interact at the molecular or atomic level with prochiral chemical reactants. Prochiral chemical reactions involve achiral reagents which would be converted into a racemic mixture in the absence of a chiral perturbing influence.

Chiral physical fields. A general theorem was recently published which states that physical fields which can induce chiral motion in molecules, molecular fragments, or subatomic particles can in principle cause asymmetric synthesis. This theorem is at first surprising in that it is well known that the hamiltonians which describe molecular structure are invariant on reflection in either space or time. Physical fields which are described by polar vectors, such as an electric field, will be reversed on reflection in space but not on reflection in time. Physical fields that are described by axial vectors, such as magnetic fields, are not reversed by the parity operation which reflects a right-handed coordinate system through a point in space to a left-handed coordinate system, but these fields are reversed on reflection in time. If a chiral physical field is constructed by using parallel polar and axial fields, the field chirality would be reversed on reflection in either space or time, but the field would be unchanged on reflection in both space and time. Thus, a chiral molecule and its enantiomer would be in the same physical state and have the same energy in a chiral field structure. The possibility for absolute asymmetric synthesis depends upon the fact that prochiral chemical reactions which produce stable chiral or racemic products are thermodynamically irreversible and are therefore asymmetric in time.

There are three naturally occurring chiral physical fields. These fields are respectively due to: (1) the weak force, which is important in nuclear structure; (2) parallel electric and magnetic fields which occur in circularly polarized light and macroscopic field structures in which the polar electric vector and the axial magnetic vector are parallel; and (3) gravitation. The general theory of relativity states that gravitation is due to the distribution of matter in the universe. The distribution of matter in the universe is indeed chiral, since the universe, or even the solar system for that matter, cannot be superimposed upon its mirror image. The weak chirality of gravity can be substantially amplified in local environments, for example, the motion of air in tornadoes and related fluid dynamic phenomena.

It is interesting to note that the chirality of gravity, which was evidently apparent to Pasteur, has escaped the notice of a number of physicists who work in this area. The Schwarzschild solution to the equations of general relativity, which is a source for the idea of black holes, deals with an achiral model for gravity. The model is a simple sphere that is not perturbed by any other matter in its universe.

Asymmetric synthesis. The possibility of absolute asymmetric synthesis caused by chiral physical fields

is dependent on the induction of chiral motion at the level of elementary particles, molecular fragments, or transition states. There may be many circumstances in which chiral motion can be induced at the particle or molecular level. Four of these circumstances have received experimental attention, namely: (1) interaction of molecules with circularly polarized light; (2) interaction of molecules with longitudinally polarized electrons (electrons which have their electron-spin axial vector parallel to the polar vector which describes their linear motion in space); (3) interaction of prochiral transition states with parallel electric and magnetic fields; and (4) interaction of these transition states with a strongly chiral gravitational field. Absolute asymmetric synthesis and asymmetric photodestruction with circularly polarized light have been well documented over the last 30 years. The use of longitudinally polarized electrons, which are emitted in beta-decay reactions for asymmetric destruction of racemic mixtures, has been the subject of a number of experiments. Some have demonstrated enantiomeric excesses of the order of 1% for asymmetric destruction of solid amino acid racemates by longitudinally polarized electrons. This particular area of research has been plagued by poor reproducibility between laboratories. There are three experiments described in the literature on the use of magnetic fields to induce an enantiomeric excess in prochiral chemical reactions. Reproducibility in these experiments is very poor even for replication within one laboratory.

Gravitational fields. In a series of recent experiments a chiral combination of the Earth's gravitational field and a centrifugal gravitational field was used to shift the product distribution in a prochiral chemical reaction. These experiments involved a prochiral chemical reaction in which the epoxidation of isophorone (as in the reaction below) was carried out in a partially filled 30-mm tube which was spun in either vertical or horizontal attitudes at rates of roughly 10,000 rpm.

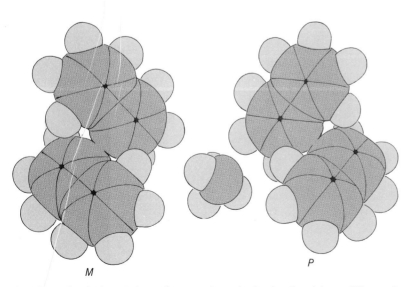

All of the experiments started with 7.6 g of isophorone. When the reaction was conducted with the tubes spinning clockwise in a vertical orientation, the average optical rotation of five experiments was $+10.9 \pm 0.3$ millidegrees with an average rotational rate of 9140 rpm. The average optical rotation for the corresponding counterclockwise spinning experiment was -2.3 ± 0.8 mdeg, with an average rotational rate of 10,660 rpm. When the reaction was conducted with the tubes spinning in a horizontal attitude, the average optical rotation for five experiments was -0.4 ± 0.5 mdeg, with an average rotational rate of 11,760 rpm. By using the Student t distribution, it is possible to reject the null hy-

P- and *M*-chiral 1-binaphthyl enantiomers and a molecule of methanol (space-filling models).

pothesis for both the clockwise and counterclockwise spinning experiments at a significance of 0.05 (one-sided test). If an artifact were responsible for the observed optical rotations, it would have to have been very consistent in the series of experiments. The optical rotations in these experiments were measured in a double-blind manner, so it seems unlikely that an artifact in the measurement of optical rotations could have been responsible for the observed results.

Chiral motion. The theorem concerning the possibility of absolute asymmetric synthesis very clearly requires chiral motion of prochiral transition states for elementary particles in thermodynamically irreversible chemical reactions in order for a consistent enantiomeric excess to be observed. A fluid contained within a spinning tube that is vibrating in an air turbine and is partially filled so that while spinning the center of the tube contains gases rather than liquids, has chiral motion that is both microscopic and macroscopic. The macroscopic motion of the liquid would certainly not couple well enough to molecular motion to cause absolute asymmetric synthesis due to the chiral motion of reactants. Microscopic helical motion in the liquid is known to occur as the result of a phenomenon called Ekman sucking. Ekman sucking, a boundary layer phenomenon, causes a pumping of fluid on the bottom of a tube from the interior to the exterior as the tube rotates. This results from the viscous drag at the exterior of the tube being higher than that at the interior because of the higher velocity of the exterior. The microvortices in the Ekman layers at the boundary of the fluid would have radii approaching colloidal dimensions, and so coupling of the motion of the fluid with the motion of molecules could be significantly above part-per-million levels. The enantiomeric excess observed in the experiments was at approximately part-per-million levels. Unfortunately, exact calculations of the effect of fluid flow in the Ekman layer on molecular motion are not yet possible. This is a major impediment to estimating the enantiomeric excess to be generated by conducting a prochiral reaction in a chirally flowing fluid.

For absolute asymmetric synthesis to occur in a chiral set of gravitational fields, it is necessary for molecular motion in the field to be chiral so that encounters of intermediates to form a transition state of one chirality are more frequent than encounters to form transition states of the opposite chirality. The illustration shows the molecular shapes of two enantiomeric 1-binaphthyl molecules and a low-molecular-weight solvent, methanol. It should be clear from the illustration that the two enantiomeric binaphthyl molecules would behave as propellers of opposite configuration in a chiral flowing environment. The probabilities of encounters to form one of the two enantiomers should be higher for chiral motion in one sense. Specifically, the formation of the P-enantiomer (the enantiomer which is a right-handed propeller) of the binaphthyl should be favored in a right-handed vortex when starting from a naphtha-

lenelike precursor. This assumes that for molecular interactions in fluids, molecules have three-dimensional shapes. This assumption is backed by a host of experiments which include very specific enantiomeric recognition in solution.

Tetrahedral dissymmetry can be directly related to helical dissymmetry. This ensures that the synthesis of tetrahedrally dissymmetric molecules, such as isophorone oxide, will show enantiomeric selectivity when the reaction occurs in chirally moving fluids.

Implications. Research in the general area of absolute asymmetric synthesis is continuing at a vigorous pace in a number of laboratories throughout the world. Since gravitation may have been considerably stronger in the early universe, the possibility of gravitational chirality influencing molecular chirality is certain to receive more attention in the future.

For background information *see* GRAVITATION; OPTICAL ACTIVITY; PROCHIRALITY; STEREOCHEMISTRY in the McGraw-Hill Encyclopedia of Science and Technology.

[RALPH C. DOUGHERTY]

Bibliography: W. A. Bonner et al., Asymmetric degradation of DL-leucine with longitudinally polarized electrons, *Nature*, 288:419–421, 1975; P. G. deGennes, Sur l'impossibilité de certaines synthèses asymétriques, *C. R. Acad. Sci. Paris*, 270:891–898, 1970; R. C. Dougherty, Chemical geometrodynamics: Gravitational fields can influence the course of prochiral chemical reactions, *J. Amer. Chem. Soc.*, 102:380–381, 1980; D. Edwards et al., Asymmetric synthesis in a confined vortex: Gravitational fields can cause asymmetric syntheses, *J. Amer. Chem. Soc.*, 102:381–382, 1980.

Molecular genetics

Geneticists have focused their attentions on *Drosophila melanogaster* for many years. Its short life cycle and the small number of chromosomes offer considerable cytological advantages. In many tissues chromatids do not separate after replication, but become laterally aggregated to form giant chromosomes visibly banded in a highly specific way. This has meant that chromosome inversions, duplications, translations, and deficiencies could be mapped not only by genetic means, but also by direct observation of the polytene chromosomes. In recent years molecular biologists have followed this well-trodden path, and have begun to analyze the organization of DNA in *D. melanogaster* chromosomes. Several types of mobile DNA elements have been found which have no fixed abode and so impact a considerable degree of fluidity to the genome. Neither is the genome constant throughout development, but certain genes undergo tissue-specific amplification at stages when their products are needed in abundance.

Cloning DNA. DNA segments can be cloned from a heterogeneous mixture of chromosomal fragments by first joining them to either a plasmid or bacteriophage DNA molecule from the bacterium *Escherichia coli*. These recombinant DNA molecules constructed in laboratory culture can then be introduced into *E. coli* cells under conditions in which a single bacterial cell receives a single recombinant molecule. Then by picking single colonies selected by a marker function on a plasmid vector, or single plaques which represent areas in which *E. coli* cells have been killed by replicating recombinant bacteriophage, it is possible to cultivate clones of recombinant DNA molecules. The very self-complementary nature of the DNA molecule then permits these cloned segments of DNA to be identified. For example, cloned genes for ribosomal RNA (rRNA) would anneal with radiolabeled rRNA if the DNA was denatured. Fortunately it is possible to transfer colonies or plaques onto nitrocellulose disks in order to denature the DNA within *E. coli* or phage and immobilize it on the disk in the position of the original colony or plaque. The result is viable colonies or plaques on a petri dish on the one hand and a replica disk holding immobilized DNA on the other. This disk can be incubated with a suitable ^{32}P-radiolabeled probe, and then those colonies or plaques containing DNA which anneal with the probe can be identified on the replica disk by autoradiography.

Transposable elements. A similar technique makes it possible to determine the location of any cloned segment of *D. melanogaster* DNA on the giant chromosomes. The DNA within chromosomes immobilized on glass slides is denatured and allowed to anneal with cloned ^{3}H-labeled DNA. The site of annealing is again localized by autoradiography. Figure 1 shows one such example in which the cloned transposable element (transposon) *copia* is annealed to salivary gland chromosomes. *Copia* elements are present at many sites, which are not necessarily identical in all flies. This can be readily seen in the micrograph where the maternal and paternal chromosome homologs are not fully synapsed. The *copia* family represents 1 of perhaps 30 families of transposable elements in the *D. melanogaster* genome which share common structural features. The terminal segments of the elements are direct sequence repeats and are flanked at each of the chromosomal sites which they occupy by a duplication of a few nucleotide pairs, present as a single sequence in the uninterrupted chromosome. Such duplications are found alongside a variety of transposable elements from both prokaryotes and eukaryotes. It has recently been demonstrated that there are also circular extrachromosomal copies of *copia* elements in cell nuclei. These could be intermediates in the process of transposition, but much remains to be done in order to clarify their exact role. The *white apricot* (w^{a}) mutant allele at the X-linked *white* (w^{+}) locus is one example of a mutation caused by the insertion of a *copia* element. This can be seen in stocks in which the w^{a} allele and a closely linked gene, *roughest* (rst^{+}), are located at a number of alternative positions as a consequence of being part of a large transposable element (TE). In these stocks one chromosomal site to which *copia* sequences hybridize always correlates with the cytogenetic position of the TE. The *copia* element is not responsible for the transposition of the TE, but it just happens to be a transposon within a larger chromosomal segment (approximately 10^{5} nucleotides), itself capable of transposition (Fig. 2).

This association of *copia* with the w^{a} allele has allowed the cloning of DNA from the *w* locus. A collection of cloned chromosomal DNA segments containing *copia* were first selected from the DNA of flies carrying w^{a}. Roughly speaking, 1 in 50 of such clones should contain the *w* locus, but such a clone would be difficult to identify since it would hybridize to 50 chromosomal sites. The solution was to hybridize the *copia* clones to the chromosomes of a

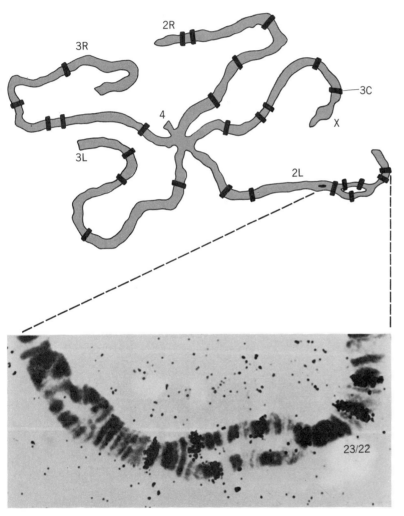

Fig. 1.   Hybridization of *copia* to salivary gland chromosomes. The four constituent chromosomes are labeled (L and R refer to the left and right arms). Each major chromosome arm is divided into 20 regions, which in turn are subdivided into regions with alphabetic designations. The *white* locus is within region 3C on the X chromosome. This, together with other sites of the *copia* element, is shown by black bars. The micrograph shows part of 2L in the vicinity of regions 22 and 23 in which the chromosome homologs are not fully paired. The diameter of the chromosome is about 2 μm. Silver grains deposited by autoradiography show that *copia* sequences are not present at identical sites in the paternal and maternal homologs. (*From E. Strobel, P. Dunsmuir, and G. M. Rubin, Polymorphisms in the chromosomal locations of elements of the 412, copia and 297 dispersed repeated gene families in Drosophila, Cell: 17, 429–439, 1979*)

closely related *Drosophila* species in which there are only two copies of *copia* on the chromosomal arms. In this way it becomes feasible to identify within pools of cloned DNAs those which hybridize to the *w* locus in addition to the *copia* sites. An alternative approach makes use of stocks in which *w* is transposed to other genomic sites as part of the TE. In one such stock the TE is located on the third chromosome and has been brought very close to a previously cloned heat-inducible gene by a deficiency of part of the intervening chromosomal segment. It is then possible to radiolabel sequences from the site of the heat-inducible gene at 87AC and use these as a probe for overlapping chromosomal segments in a library of cloned fragments. By selecting successively overlapping segments of chromosomal DNA it is possible to "walk" from one chromosomal site to another, and to monitor the progress by hybridization of the DNA segments to wild-type chromosomes. Chromosomal segments from the first steps of the walk hybridize to the third chromosome at the location 87AC, but when the TE is reached, the DNA is then seen to hybridize at region 3C of the X chromosome, the site of the *w* locus. At the boundaries of the TE there are copies of another set of transposable sequences, known as FB elements since they terminate in self-complementary inverted-sequence repeats which form "fold-back" structures following denaturation and reannealing. These elements are capable of independent transposition but also may be able to mobilize large chromosomal segments (Fig. 2).

Mutations of the *w* locus generated by hybrid-dysgenesis have been characterized and show that each mutation is a consequence of the insertion of a transposable sequence, the P element. Hybrid dysgenesis occurs when male flies of P strains, which carry the transposable P elements in their chromosomes, are crossed with M-strain females, which lack the elements. P-element transposition is presumably repressed by the gene products of the elements themselves, but when P-strain sperm enter M-strain eggs then this repression is relieved and transposition can occur, resulting in insertional mutations or other chromosomal rearrangements. The phenomenon can be mimicked by injecting cloned P elements into M-strain embryos. There are also indications that it is possible to coinject a P element, into which another *D. melanogaster* gene has been inserted, together with a helper P element to introduce both elements in a stable manner into the genome. This raises the possibility of transforming *D. melanogaster* with cloned genes mutated in laboratory culture, a protocol which should greatly extend understanding of gene regulation.

Gene amplification in development.

The genome is not stable throughout development, and the capacity of the fly to produce polyploid cells seems to provide a mechanism for the disproportionate replication of certain genes. It has been known for some time, for example, that the genes for rRNA (rDNA) are underreplicated in polytene chromosomes. Only

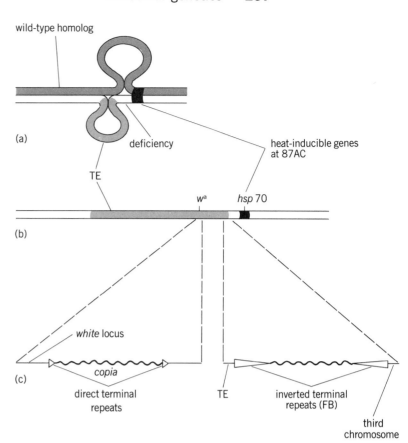

Fig. 2.   Transposable sequences associated with the transposable element (TE). (*a*) A wild-type chromosome is shown paired with its homolog carrying the TE and deficient in the sequences between the TE and the genes for a heat-inducible protein (hsp70). This is a diagrammatic representation of the salivary gland chromosomes of such a heterozygote. (*b*) A segment of linear DNA of the third chromosome having the TE, a 100-kilobase (kb) segment of X-chromosomal DNA marked with the *white apricot* allele (*w*ª). (*c*) Transposable DNA segments associated with the white locus and the ends of the TE. *Copia* is a 5-kb element with a 0.3-kb direct terminal repeat. The 1.5-kb inverted terminal repeats of the FB element flank a 4.5-kb sequence.

subsets of rDNA units from one of the two nucleoli undergo replication during polytenization. Cytological observations made some 25 years ago indicated that gene amplification occurred in the DNA puffs in polytene chromosomes of the sciarid family of dipteran flies. These indications have been recently verified by experiments with cloned sequences from these regions. It is now clear that a similar phenomenon occurs for the genes which encode the proteins of the chorion, the shell which encases flies' eggs. These genes are located in two clusters, one on the X chromosome which can undergo 16-fold amplification and the other on the second chromosome which can undergo 60-fold amplification. This occurs during oogenesis in the polyploid chromosomes of the follicle cells which produce the chorion. The amplification is not confined to the immediate vicinity of the genes, but spreads as a gradient involving 50 kilobase (kb) or so of the chromosomal DNA on either side of the genes. A small chromosomal inversion involving the X-linked cluster reduces the level of amplification, but the sequences brought next to the chorion genes, which normally do not amplify, now do so. These observations suggest a model for amplification in which there are specialized subsets of replication origins, capable of repeatedly reinitiating replication, thereby generating a geometric series of diverging replication forks. An understanding of these processes not only will illuminate many aspects of the structure of polytene chromosomes, but should have considerable implications in understanding the control of the replication of eukaryotic chromosomes.

For background information *see* GENETIC ENGINEERING; MOLECULAR BIOLOGY; RECOMBINATION (GENETICS) in the McGraw-Hill Encyclopedia of Science and Technology.   [DAVID M. GLOVER]

Bibliography: P. M. Bingham, R. Levis, and G. Rubin, Cloning of DNA sequences from the *white* locus of *D. melanogaster* by a novel and general method, *Cell*, 25:693–704, 1981; A. J. Flavell and D. Ish-Horowicz, Extrachromosomal circular copies of the eukaryotic transposable element *copia* in cultured *Drosophila* cells, *Nature*, 292:591–595, 1981; M. L. Goldberg, R. Paro, and W. J. Gehring, Molecular cloning of the *white* locus region of *Drosophila melanogaster* using a large transposable element, *EMBO J.*, 1:93–98, 1982; A. C. Spradling, The organisation and amplification of two chromosomal domains containing *Drosophila* chorion genes, *Cell*, 27:193–201, 1981.

Mollusca

Most mollusks with external shells have virtually no protection except their shells. During the last 500 million years, predators of mollusks have improved their ability to attack, while mollusks have become more difficult to subdue. Apart from defensive mechanisms, such as burrowing, now adopted by many bivalves, the ability of shelled animals to resist an attack depends on two features: the size and shape of the shell (its build), and the mechanical characteristics of the shell material. These two features interact, of course, but they can be discussed separately.

Shell material. Mollusk shell material is composed almost entirely of calcium carbonate in the crystalline form of aragonite or calcite, with a little organic material mainly in the form of protein. The microscopic arrangement of the material falls into a number of rather different structural types that are easily recognizable under the scanning electron microscope (see table): nacre, prisms, crossed-lamellar, foliated, and homogeneous. To some extent certain mechanical properties can be assumed from the structure. Nacre, with its brick wall arrangement, is well suited to prevent cracks from traveling across the sheets, because the crack is continually being blunted as it reaches a new layer. This crack-stopping ability is enhanced because of the energy needed to shear the protein matrix. Similarly, the plywood arrangement of a crossed-lamellar structure also impedes cracks, although it is not as effective as nacre. On the other hand, the homogeneous

structure has a low tensile strength as a consequence of its very low organic content and of its rubblelike makeup and therefore cannot impede the travel of a crack.

For most mollusks, the shell stiffness is probably more than adequate. Tensile strength (as opposed to compressive strength) is usually most important in resisting attack. The ability of shell material to stop cracks gives a measure of how well the material resists impact. Shells differ also in their resistance to chemical attack and abrasion, such as in the assaults they may suffer from predatory gastropods.

Two conclusions emerge from studying the shell characteristics (as shown in the table) concerning mechanical properties of shell material. First, it is possible, as in nacre, to achieve respectable defensive properties from a very unpromising basic material—calcium carbonate. Second, there is a great diversity of properties among shell types found in living mollusks. Paleontological studies have shown that the strong material, nacre, evolved first, and that the tendency in molluscan evolution has been to replace nacre with materials of lesser mechanical strength, such as homogeneous and foliated structures. This paradoxical finding implies that for some molluscs there are advantages in using these poorer materials. It is probable that the speed of construction is one such advantage. Some snails are limited in their overall growth rate by the rate at which they can secrete shell material. Because of its structure nacre is difficult to produce, and it cannot grow more than a few micrometers a day in thickness. Prisms, however, have been observed to grow at a rate of several centimeters a day. Therefore, for sedentary animals, it may be advantageous to produce the needed strength by laying down a thick shell of weak material quickly, rather than a thinner shell of stronger material more slowly. It also seems likely that some materials require more energy for their production that others.

Build of the shell. There are many devices that can be used to prevent mollusk shells from being successfully attacked by predators; some of these are inappropriate for gastropods or bivalves, but most are common to both. Most obvious is the thickness of the shell in relation to its overall size. Increased protection is associated with certain disadvantages. In particular, a smaller, but safer animal cannot produce as many, or as large, gametes as a larger one can.

There are other antipredator devices. A snail having a globular, as opposed to a more pointed, shell is not easy for a predator, such as a crab or a fish, to crush because it is more difficult to grip the globular form. Spines or ridges on the outer surface also make the shell less accessible to the predator. Additionally, for gastropods, it is advantageous to have a shell with a narrow, elongated mouth (aperture) rather than a rounded one, because this makes it more difficult to peel the lip back. Many antipredator devices have corresponding disadvantages to their bearers, such as weight, time and energy required

Characteristics of shells

Structure	Crystal type	Arrangement	Young's modulus (stiffness), gigapascals	Tensile strength, megapascals	Crack stopping	Resistance to attack	
						Chemical	Abrasive
Nacre (mother of pearl)	Aragonite	0.5-μm-thick sheets, each made of flat tablets; sheets and tablets separated by thin protein sheets	30–60	40–120	Good	Fair	Good
Prisms	Aragonite or calcite	Columns, roughly hexagonal, 20 μm to 1 mm across, separated by a protein layer up to 20 μm thick	15–40	60	Fair	Good	Good
Crossed-lamellar	Aragonite	Like plywood; each ply about 20 μm across; alternate layers at right angles; little organic matrix	40–80	10–90	Fair	Fair	Good
Foliated	Calcite	Crystalline needles joined side to side in a vague plywood structure; very little organic matrix	30–60	30–45	Poor	Poor	Poor
Homogeneous	Aragonite	A fine-grade rubble; very little organic matrix	60	5–30	Negligible	Fair	Good

for construction, and restriction of locomotion.

Balance of forces. When selection acts in two opposite directions, one can look for alterations in the balance of selective forces to see how the organisms respond. The strength of the shells of two species of gastropod, dogwhelks and winkles, from different but nearby sites and with great differences in the intensity of crab predation, were examined. The dogwhelks exposed to intense predation were relatively thick-walled, with a smaller body size for a given external shell size, and were stronger weight for weight, than those exposed to less predation. The winkles from the same places showed no such differences. However, these winkles have planktonic larvae, and so genetic differentiation between nearby populations is difficult. But since the dogwhelks produce nonpelagic young, there is little gene flow, and interpopulation differences can appear.

One of the differences between tropical and temperate sea mollusks is that the shells of tropical forms tend to be thick, spiny, slit-mouthed, and so on, thereby serving better as antipredator devices. A problem in interpreting such observations is that different species are being compared in different regions, and the differences may be related to taxon-

omy rather than to ecology. This has been partially overcome by comparing species within the gastropod family Thaididae. The comparison showed tropical species to be almost three times stronger weight for weight. The higher strength was not the consequence of stronger materials, but of thicker shell walls. There is good corresponding evidence that the predators of mollusks in the tropics tend to be more specialized and effective than those of temperate regions. It seems that the development of attack and defense devices has proceeded further in the tropics than elsewhere. A possible reason is that, because calcium carbonate is less soluble in warm water, less energy and time are required to build a shell in the tropics. Therefore, the selective balance swings toward thicker shells in the tropics.

One way to study this is by looking at the problem of shell strength from the point of view of the predator. By observing boring predation by naticid gastropods, it is possible to develop a model that predicts which, among a number of prey species, should be preferred, and what size of prey should be attacked. The model takes into account the difficulty of boring through shells of different thickness, and shows that the predators behave appropriately according to their own size and anatomical limitations—assuming that what is being selected for is a maximization of energy gained per unit amount of energy expended. Many mollusks can, if they survive long enough, grow so large that their habitual predators cannot subdue them anymore, or, if they can, take so long to do so that they learn that it is not worthwhile for the amount of food obtained.

For background information *see* BIVALVIA; GASTROPODA; MOLLUSCA in the McGraw-Hill Encyclopedia of Science and Technology.

[JOHN CURREY]

Bibliography: J. D. Currey, Mechanical properties of mollusc shell, in *Symposia of the Society for Experimental Biology*, no. 34, pp. 75–97, 1980; J. D. Currey and R. N. Hughes, Strength of the dogwhelk *Nucella lapillus* and the winkle *Littorina littorea* from different habitats, *J. Anim. Ecol.*, 51:47–56, 1982; J. A. Kitchell et al., Prey selection by naticid gastropods: Experimental tests and application to the fossil record, *Paleobiol.*, 74:533–552, 1981; G. J. Vermeij, *Biogeography and Adaptation: Patterns of Marine Life*, 1978.

Musical instruments

Strings, wood, and air are the basic elements of most stringed instruments, with the strings acting as generators and the wood and air as amplifiers of the sound. The acoustics of this seemingly simple system is actually rather complicated and not yet completely understood. However, with the coordinated application of highly specialized, sophisticated technology in a wide range of disciplines, considerable progress is being made.

Resonator construction. The vibrating string by itself produces almost no audible sound, for it does not set up enough vibrations in the surrounding air for the sounds to reach the listener's ears. Thus some kind of amplification is needed. In the centuries since the mouth bow was discovered (it is still used in folk music today), many forms of resonators have been developed to make audible the vibrations of plucked and bowed strings—for example, skins stretched over hollow logs, gourds, bamboo stems, and even tin cans, creating nearly enclosed resonating air columns with thin vibrating walls. Eventually thin-wooded boxes with various shaped openings were developed, which were the forerunners of the highly sophisticated plucked string instruments, such as lutes, guitars, and mandolins and of the viols and the violins of the bowed string families.

In essence, the bowed and plucked string instruments consist of a set of strings mounted on a wooden box with an almost enclosed air space. The many resonances of the body and the enclosed air serve to amplify the string vibrations, which are communicated to the body through the bridge as well as through the two end fastenings of the strings.

In a fine instrument these resonances must be suitably spaced over the range desired—a condition with which luthiers have been concerned for centuries. Traditionally, spruce of various kinds has been used in the top soundboards of all stringed instruments because of its high stiffness, low density, and long bell-like ring when struck (long sound decay time). The plucked string instruments, such as the guitar, have flat soundboards 2–3 mm ($\frac{1}{12}$–$\frac{1}{8}$ in.) thick, while the violin family instruments have arched tops and backs carved from solid pieces and thinned to 2–6 mm ($\frac{1}{12}$–$\frac{1}{4}$ in.). For the back and sides of guitars, woods with a long decay time, such as

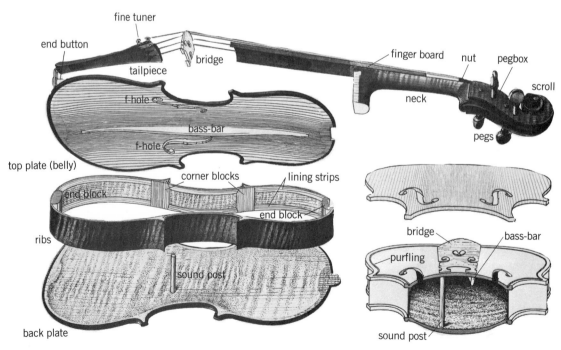

Fig. 1. Exploded view of the violin, showing the position of the sound post. (*From C. M. Hutchins, The physics of violins, Sci. Amer., 207(5):78–93, November 1962, copyright © 1962 by Scientific American, Inc.*)

rosewood and cypress, are preferred; while curly or flamed maple with a shorter decay time is preferred for the back and sides of violins.

These thin top soundboards of spruce are supported by narrow strips of spruce glued to their undersides. In guitars there may be as many as ten of these braces set in various traditional patterns. In viols and violins there is only one brace, the bass-bar, glued lengthwise along the top under the string of lowest tuning, and approximately under one bridge foot. Under the other bridge foot there is a small rod of spruce, fitted snugly, but not glued, between top and back, known as the soundpost (Fig. 1).

In the violin the combination of arched top, bass-bar, and soundpost supports 7–9 kg (15–20 lb) weight from the downward force through the bridge from string tension, which is about 25 kg (55 lb) weight. The soundpost not only serves to support the thin wood of the top plate, but also imposes a nearly stationary nodal point on the top and back at the points of contact. Although the soundpost transmits some vibrations between top and back over the lower range of each instrument, it functions primarily as the fulcrum of a rocking lever, partially immobilizing one foot of the bridge while increasing the motion of the other foot over the bass-bar. This unsymmetrical action is critical in causing the important volume change that enables the body to radiate sound effectively (like a zero-order radiator), particularly in the lower range where the wavelengths of sound in air are larger than the instrument body.

The two elegantly shaped sound holes of the violin provide a flexible platform for this bridge motion and allow the air to move in and out as the instrument body expands and contracts. It has been estimated that in the violin the air moves in and out through the f-hole on the bass-bar side at about 15 ft/s (4.6 m/s).

Even in the ingenious vibrating system of the violin, brought to near perfection by the luthiers of the seventeenth and eighteenth centuries, only about 1% of the energy from the bow finally radiates as sound power into the surrounding air. Thus every adjustment to increase the effectiveness of the violin's vibrating system is critically important—a fact well known to master violin makers.

String motion. Under the bow the string seems to move in a lens-shaped curve extending from bridge to nut. However, this is an optical illusion, for the string actually takes the form of a sharply bent straight line with a kink, or discontinuity, known as the Helmholtz corner, which travels from bridge to nut and back at the frequency of the bowed tone (Fig. 2a). For example, at 440 Hz the kink makes 440 trips per second from bridge to nut and back.

The waveform of the force of the bowed string at the bridge is sawtooth with the waves traveling in one direction under an up-bow and in the opposite direction under a down-bow (Fig. 2b). When the string is plucked, the pull of the finger creates a kink that divides the string into two straight sections. On release the two kinks travel in opposite directions, one toward the bridge and one toward the nut (Fig.

2c). These are the same as the modes of motion of the bowed string. Since they are both present at the same time, the wave shape of force at the bridge is rectangular, with its width depending on the point of plucking. Plucking close to the bridge gives a narrow rectangular wave with many higher partials. Plucking at the center produces a square wave with fewer partials (Fig. 2d), a phenomenon well known

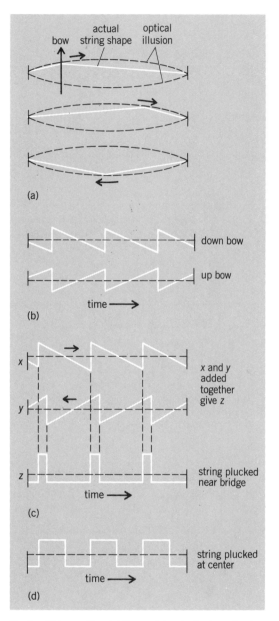

Fig. 2. String motion. (a) Shapes taken by the bowed string at three discrete points in time as the kink, or Helmholtz corner, created by the bow travels to the fixed end of the string and back, once in every vibration, causing the optical illusion of a lens-shaped curve. (b) Sawtooth waveforms of force in the bowed string produced by alternate sticking and release of the rosined string by the rosined bow hair, showing reversal of waveform with down- and up-bowing. (c) Waveforms, x and y, traveling in opposite directions, produced by plucking string near bridge, and waveform z that results from adding them together; the sharp rectangular waveform of z contains many higher partials. (d) Square waveform produced by plucking string at center, with fewer high partials.

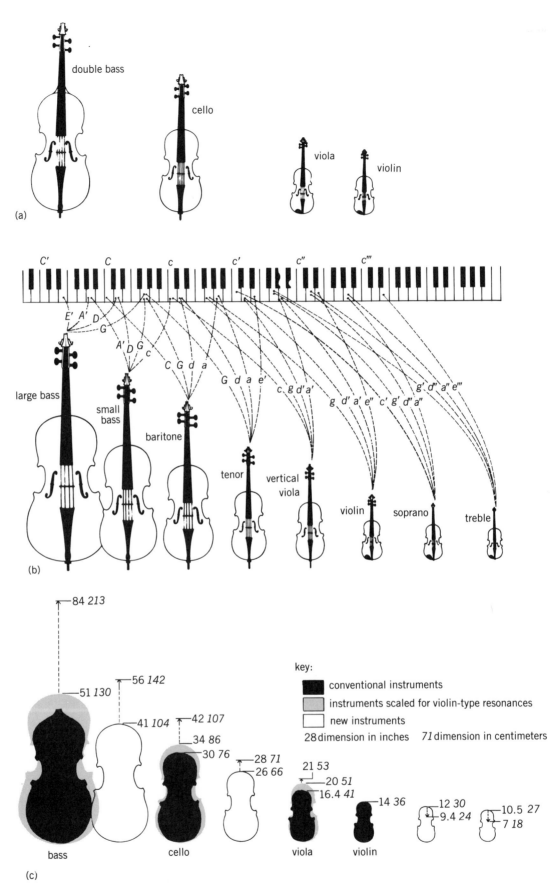

Fig. 3. Acoustically matched set of instruments of the violin family, compared with conventional instruments. (a) Conventional instruments. (b) Acoustically matched instruments with ranges indicated on piano keyboard. (c) Size comparison of conventional and acoustically matched instruments. Broken lines show theoretical body length. (From the New Grove Dictionary of Music and Musicians, Grove's Dictionaries of Music, Inc., 1980)

to guitar players. Another important difference between the bowed and plucked string is that with the bowed string the phenomenon is periodic, so that overtones are kept in a strictly harmonic relationship to the fundamental. With the plucked string any stiffness makes the higher overtones, or partials, sharper than integral multiples of the fundamental.

Study of vibrational characteristics. To be understood fully, a stringed instrument, and particularly the violin, must be studied in context. Not only the physical characteristics of the instrument itself, but the acoustics of the room, the style of music being played, the player, and the background of the listener are involved in any final evaluation.

Scientific investigations of the vibrational characteristics of the violin have been done since the early 1800s. This has not been a steady development, but has occurred in spurts of activity associated with technological innovations, particularly the development of increasingly sensitive measuring equipment.

Twentieth-century electronic and optical technologies have made possible many studies of the vibrational characteristics of the violin, including its spherical sound field radiation and some of the bending vibrations in the body which produce the radiation. Visualization of these body vibrations, which are on the order of a few micrometers in amplitude, has been done with laser light by using a technique known as hologram interferometry.

Since the mid-twentieth century, eight instruments of the violin family, one at each half-octave from the tuning of the double bass to an octave above the violin, have been developed through a combination of mathematics, acoustical theory and testing, and skilled violin making. Designed to project the resonance characteristics of the violin into seven other tone ranges, these instruments (Fig. 3) provide consistent string tone quality over the musical range and bring to modern focus a concept described by M. Praetorius in 1619. [The theoretical body lengths in Fig. 3c are based on the violin's length of 14 in. (36 cm). For example, the tenor, which is one octave below the violin, has a theoretical body length exactly twice that of the violin, 28 in. (71 cm). However, the actual instrument is constructed with a somewhat smaller body length, without loss of tone quality.] These same construction techniques are now being applied to the making of conventional violins, violas, and cellos.

In an effort to understand more of the complicated acoustics involved in stringed instruments, research in fields as widely separated as psychoacoustics, musical composition and performance, materials research, vibration analysis, and violin making is now being coordinated and applied to the violin and guitar families. Through creative thinking and interchange among such disciplines, significant advances are being realized in the art, science, and construction of stringed instruments.

For background information *see* HOLOGRAPHY; MUSICAL INSTRUMENTS; SOUND in the McGraw-Hill Encyclopedia of Science and Technology.

[CARLEEN MALEY HUTCHINS]

Bibliography: C. M. Hutchins (ed.), Pt. I, Violin family components, 1975, Pt. II, Violin family functions, 1976, *Benchmark Papers in Acoustics*; C. M. Hutchins, The Acoustics of violin plates, *Sci. Amer.*, 245(4):170–186, October 1981; C. M. Hutchins, The physics of violins, *Sci. Amer.*, 207(5):78–93, November 1962; Catgut Acoustical Society, Inc., *The Violin Octet*, 1981.

Nobel prizes

Nine recipients of the Nobel prizes for 1982 were announced by the Swedish Royal Academy.

Medicine or physiology. Half of this prize was awarded to Sune Bergström and Bengt Samuelsson of the Karolinska Institute of Sweden for their elucidation of the chemical structure of prostaglandin, a biologically active substance which affects smooth muscle, uterine, and blood vessel contractions. The English pharmacologist John R. Vane, of the Wellcome Research Laboratories, also received recognition for his contribution to prostaglandin research, revealing the anti-inflammatory effect of aspirin as due to its ability to inhibit prostaglandin production, and for his discovery of prostacyclin, a highly unstable prostaglandin capable of preventing platelet aggregation.

Physics. Kenneth G. Wilson of Cornell University was honored for his work with renormalization group techniques applied to the critical phenomena of metal-liquid and water-steam changes. By applying renormalization techniques to materials at the point when they become magnetized, Wilson was able to analyze changes involving billions of particles, which were previously thought to be too numerous to deal with in a controlled manner.

Chemistry. Aaron Klug, an associate of the Medical Research Council Laboratory of Molecular Biology in Cambridge, England, was recognized for his work with the structure of viruses and nucleosomes, and his discovery of the principle of quasi-equivalence in the subunit packing of protein particles in viruses. The South African–born scientist was also honored for his contribution to the development of crystallographic electron microscopy.

Economics. George Stigler was awarded this prize for his research on monetary regulation. A member of the conservative "Chicago School" of economics—which supports free markets, monetarism, and minimal government supervision—Stigler was cited for his work with the economics of information, in which he demonstrated that information should be treated as a market commodity.

Literature. Colombian Gabriel García Marquez, author of *One Hundred Years of Solitude* and noted political activist, was selected for his epic writings, which combine tragedy, comedy, and fantastic imagery to champion the oppressed peoples of Latin America and to reveal the nature of political struggle.

Peace. The peace prize was shared by sociologist Alva Myrdal and diplomat Alfonso García Robles. Myrdal, head of Sweden's delegation to the disarmament talks in Geneva, was cited for her pacifist efforts and her revealing book *The Game of Disarmament*. Robles was the coauthor of the Nuclear Nonproliferation Treaty and a sponsor of the Treaty of Tlatelolco (1967), which, if adopted, would make Latin America the largest inhabited nuclear-free zone in the world.

Nuclear molecule

The existence of nuclear states possessing either special symmetries or unusual shape or cluster configurations is often signaled by the observation of

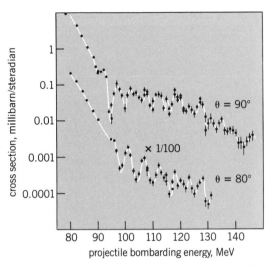

Fig. 1. Cross section for elastic scattering of ^{28}Si + ^{28}Si at two scattering angles θ plotted as a function of projectile bombarding energy. The points are spaced by 1 MeV. 1 millibarn = 10^{-31} m^2.

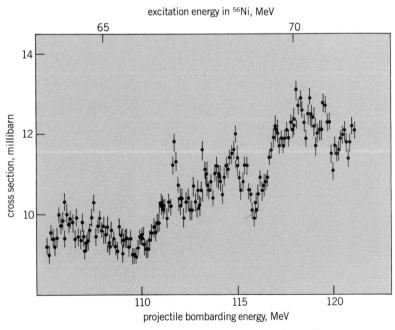

Fig. 2. Cross section for elastic and inelastic scattering of ^{28}Si + ^{28}Si averaged over scattering angle plotted versus projectile bombarding energy. The points are spaced by 100 keV. 1 millibarn = 10^{-31} m^2.

narrow, and therefore relatively long-lived, states at high excitation energies in the nucleus. In 1960, unexpectedly narrow (approximately 100 keV width) resonances were found in the energy dependence of cross sections for ^{12}C + ^{12}C nuclear interactions. These resonances correspond to states in the compound nucleus (^{24}Mg) at excitation energies of approximately 20 MeV. The existence of these narrow states, together with features of their decay, led to an interpretation in terms of extremely deformed states of ^{24}Mg, possessing some of the characteristics of a moleculelike state of the nucleus.

For many years, the occurrence of this phenomenon was thought to be restricted to nuclei close to ^{24}Mg and confined to a relatively small region of excitation energy. There is now, however, abundant evidence for a much wider occurrence of nuclear molecular phenomena. In particular, a recent series of measurements has demonstrated the existence of similar modes of excitation in the nucleus ^{56}Ni at excitation energies as high as 70 MeV. The significance of these new observations lies in the fact that, whereas the resonance states in ^{24}Mg have quite low angular momentum (2–4 $\hbar$, where $\hbar$ is Planck's constant divided by 2π), the ^{56}Ni resonances have spins which have been measured to be as high as 42 $\hbar$. This places them among the highest-spin nuclear excitations yet observed directly and also indicates the existence of a new class of nuclear excitation, namely, a high-spin shape-isomeric state. The understanding of these high-spin resonances will severely test current models of nuclear structure at high spin and large deformation.

Experiments and results. A beam of ^{28}Si nuclei accelerated by a tandem Van de Graaff accelerator was used to bombard a thin foil of ^{28}Si. The scattered particles were detected and identified, and the probability or cross section for various nuclear reactions was measured as a function of both scattering angle and projectile bombarding energy. The reactions of primary interest in this work were elastic scattering, a process in which the projectile nucleus scatters from a target nucleus without being excited from its ground state, and inelastic scattering, in which either or both the target and the projectile are internally excited to higher quantum states.

The variation of the elastic scattering cross section with ^{28}Si bombarding energy is shown for two different scattering angles in Fig. 1. Instead of the smooth variation expected on the basis of conventional theories of the elastic scattering process, the data fluctuate dramatically with energy, suggesting the occurrence of resonances in the scattering process. The data shown in Fig. 1 were measured only for every MeV of bombarding energy and yet seem to show structure that is at least as narrow as the spacing of the data points. This observation led to a further experiment in which the cross sections for elastic and inelastic scattering were measured in much smaller energy steps, 100 keV. The results of these measurements are shown in Fig. 2 where the total cross section for elastic and inelastic scattering, averaged over detection angle, is plotted as a function of ^{28}Si bombarding energy. These data reveal a much richer structure than was apparent in Fig. 1. Several peaks with widths as narrow as 200 keV appear quite prominently, indicating the existence of long-lived states (mean life on the order of 10^{-20} s) of the compound system ^{56}Ni, at excitation energies in the vicinity of 65–70 MeV—a result unexpected on the basis of conventional theories of nuclear structure.

Given the observation of narrow resonances, it is of considerable interest to attempt to determine their angular momentum or spin, which is one of the most important quantities characterizing nuclear quantum states. This, in principle, can be accomplished by careful measurement of the angular dependence of the elastic scattering cross section as a function of projectile energy. There are, however, several practical problems which make this particularly difficult in the present case. These were overcome by the use of a new technique which exploits simple kinematic features of the reaction, rather than direct measurement of the scattering angles, and thus allows the precise determination of the elastic scattering cross section at many angles simultaneously. One example of the results of these measurements is shown in Fig. 3, taken at 118-MeV bombarding en-

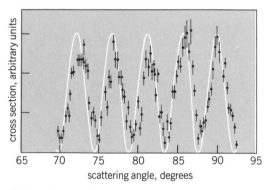

Fig. 3. Angular variation of the elastic scattering cross section for ^{28}Si + ^{28}Si at a bombarding energy of 118 MeV. The curve is the theoretical prediction for a resonance of spin 40 $\hbar$.

ergy. The curve shown together with the data is the theoretically expected behavior for elastic scattering proceeding via a resonance with a spin of 40 $\hbar$. Other spin values lead to predicted angular distributions which do not agree as well with the data.

The spins of the resonances observed at other energies have also been measured by using this technique, and the variation of resonance spin with bombarding energy is consistent with that expected for two ^{28}Si nuclei rotating around each other in a moleculelike configuration.

Interpretation of results. The observation of such narrow, high-spin resonances was a completely unexpected result, and the fact that such narrow states exist at very high excitation energies implies an unusual underlying structure for them. In analogy with

the previously known low-spin resonances in ^{24}Mg, they may be described as nuclear molecular states. In this description, there are two nuclei touching and rotating about an axis joining their centers. Such a configuration gives rise to a spectrum of states with energies E and spins I related by the expression

$$E = E_0 + \frac{\hbar^2}{2\mathscr{I}} I(I + 1)$$

where E_0 is a constant and $\mathscr{I}$ is a moment of inertia corresponding to the rotation of the two touching nuclei. Within this model, additional complexity is introduced into the spectrum by allowing one or both of the participating nuclei to be in an excited state, thus providing several different molecular states with the same spin.

The present data are certainly not at variance with this macroscopic picture, although some recent theoretical work has suggested an approach which may provide a better insight into the microscopic nuclear structure underlying the observed resonances. In this approach, the motions of single nucleons within the nucleus are calculated and the effects of changes in shape and rotation of the nucleus investigated. It has been found that for large deformations and high rotational frequencies new shell gaps appear, analogous to those responsible for the exceptional stability of certain "magic" nuclei. These new shell gaps imply the possible existence of relatively stable states of ^{56}Ni with large deformation and high spin—in the vicinity of 40 $\hbar$. The possibility of associating the observed high-spin resonances with the predicted high-spin shape isomers is an exciting one. Further experiments are in progress to test this possibility. In any case, the present observations indicate a new kind of nuclear behavior which, when explained, will lead to a deeper understanding of the structure of the nucleus.

For background information *see* MAGIC NUMBERS; NUCLEAR MOLECULE; NUCLEAR STRUCTURE in the McGraw-Hill Encyclopedia of Science and Technology.

[RUSSELL BETTS]

Bibliography: R. R. Betts, B. B. Back, and B. G. Glagola, Intermediate structure resonances in ^{56}Ni, *Phys. Rev. Lett.*, 45:23–26, 1981; R. R. Betts, S. B. DiCenzo, and J. F. Petersen, Energy dependent structure in ^{28}Si + ^{28}Si scattering, *Phys. Rev. Lett.*, 43:253–256, 1979; R. R. Betts, S. B. DiCenzo, and J. F. Petersen, High spin resonances in ^{28}Si + ^{28}Si scattering, *Phys. Lett.*, 100B:117–120, 1981; D. A. Bromley, J. A. Kuehner, and E. Almquist, Resonant elastic scattering of ^{12}C by carbon, *Phys. Rev. Lett.*, 4:365–367, 1960.

Nuclear physics

The atomic nucleus is held together by a short-range force that has been the subject of a half century of investigation. A subatomic meson, called the pion, is the principal carrier of that force. Only in recent years has the role of pions in the substance of nuclei, "nuclear matter," been studied in detail.

Pions and nuclear force. The fact that atomic nuclei are small and that their constituent neutrons and protons must be held together by a strong, short-range force was realized early in the history of nuclear physics. Such a force was then unprecedented in nature, since the two major forces, gravitational and electromagnetic, had an inverse-square dependence on distance. H. Yukawa postulated in the 1930s that such a force may be carried by the exchange of a tightly bound particle with finite mass—a zero-mass particle gives the inverse-square dependence while a finite mass gives an exponential one. From the range of the force, Yukawa predicted the approximate mass of this particle.

This particle was subsequently found, and it was called the pi meson, later mostly abbreviated to pion, designated by π. It is now known that even though pion exchange is the principal component of the nuclear force, other mesons also play a role.

The nuclear force comes about from the exchange of pions, and in the quantum-mechanical sense, even free protons and neutrons are surrounded by a virtual pion cloud. When protons from accelerators bombard targets, they interact rather violently with the nuclei of target atoms, and real, free pions are generated in the process. Pions come in three different forms, with a positive, negative, or neutral charge (π^+, π^-, or π^0). None of them are stable, but the charged pions live long enough (about 10^{-8} s) to allow physicists to form them into beams and study the properties of their interactions with other targets. In the interaction of pions with protons and neutrons many features have shown up, but the first and strongest is the delta (Δ) resonance, which has an important role in nuclei.

The force between nucleons immersed in the nuclear matter, within a nucleus, is different from the force between free nucleons. This is somewhat analogous to the way in which an electric field in a medium is modified by the dielectric properties of that medium. But the quantitative aspects of the modified force inside the nucleus are more complicated than those for a dielectric and are not understood completely. Because the delta resonance is the lowest in energy of the pion-nucleon resonances, it must play a crucial role in the effective force.

Pions in nuclei. In the last decade several powerful accelerators were built around the world: LAMPF at Los Alamos (United States), SIN in Zurich, and TRIUMF in Vancouver (British Columbia). These accelerators do not produce the very highest energies that are required in high-energy physics for studies of new elementary particles; instead they have intense beams of protons, at energies between 500 and 800 MeV, which produce large yields of pions when bombarding nuclei of other atoms. With a suitable arrangement of magnets, these pions are collected into beams of 10^7–10^8 pions per second so that they can be used in further experiments.

The pions produced by these "pion factories" are of the energy at which their interaction with target nucleons is dominated by the delta resonance. A pion

interacting on a free nucleon forms a delta which decays rapidly (about 10^{-23} s) by reemitting the pion. But if the nucleon is embedded in nuclear matter, then as soon as the pion is emitted it will recombine with another nucleon to form a delta. This may be treated in terms of a delta excitation (or a delta-hole excitation, since the nucleon which becomes a delta leaves a nucleon hole in the sea of nucleons). If this excitation can reach the surface, it may eventually decay by reemitting a pion outward. However, even within nuclear matter other options are also available, in which the delta excitation loses energy and the pion disappears, in some sense losing itself in the bound pion cloud and giving up its original energy, including its rest mass, to heating up the nucleus. This absorption process cannot happen with a free nucleon, and its understanding is perhaps the simplest path available to comprehending the role of real pions within a nucleus.

The simplest nucleus is the deuteron, a heavy hydrogen nucleus consisting of a proton and a neutron. If a pion is absorbed in a deuteron, the energy of the pion must be shared between the two nucleons. This process was known for about 25 years and was thought to be the mechanism for pion absorption in all nuclei—in other words, a pion always becomes absorbed between a pair of nucleons. Recent work has shown that this is not the case: the energy of the pion does not heat up the whole nucleus, nor does it just heat up a pair; instead it seems to be shared between about four or five constituent nucleons. At first, it was thought that the larger number of nucleons could be the consequence of the nucleons getting in each other's way. That this was not the case was demonstrated by comparing experiments with pions and experiments with high-energy photon beams. The probability of a pion interacting is so large that any pion incident on a nucleus will combine with the first nucleon it encounters to form a delta resonance, very near the outer surface skin of the nucleus. A photon of the right energy can also produce a delta from a free nucleon, but with a much lower probability. In fact, most of those photons will penetrate straight through a nucleus with no interaction. Clearly, a delta resonance may be produced by photons anywhere in the nucleus, in contrast to pions, which can only form it on the surface. Effects due to nucleons scattering from each other would have to be worse for deltas produced deep inside a nucleus than for those on the surface. Yet comparing experiments with photons and pions showed no significant differences—indicating that such secondary scatterings were not very important.

Quark structure. Nucleons and pions are some of the "elementary" particles that were classified with beautiful simplicity in the quark model. In this model the proton and neutron each consist of three quarks, while the pion is a quark-antiquark pair. In the delta resonance, the antiquark in the pion annihilates one of the quarks in the nucleon, and a new three-quark object is formed in which all three quark spins are aligned. This is unlike the nucleon, where the spin of one quark is opposite to that of the other two. The delta resonance is therefore an object that is as simple, or elementary, as a proton or a neutron—the pion facilitates the rearrangement of the three quarks.

To understand nuclear matter, the quark nature of its constituents must certainly be studied. For the most part, the quarks within the nucleus are thought of as being contained in separate "bags" of three quarks as nucleons, or in pairs as pions. But on some level the properties of nuclear matter as quark matter are likely to show up; this has been the subject of considerable thought and effort in nuclear physics.

In recent experiments physicists are beginning to explore the course of quark recouplings in nuclear matter. Pion propagation in the nucleus is a delta excitation or quark recoupling, and study has begun on how this excitation behaves and becomes damped. This is likely to lead to a better understanding of the interface between the quark degrees of freedom and the traditional nucleonic degrees of freedom within the nucleus.

For background information *see* MESON; NUCLEAR STRUCTURE; QUANTUM FIELD THEORY; QUARKS; STRONG NUCLEAR INTERACTIONS in the McGraw-Hill Encyclopedia of Science and Technology.

[JOHN P. SCHIFFER]

Oil and gas, offshore

Satellite subsea well systems (SSWS) are methods used to produce oil from the ocean floor. They differ from other techniques (drilling platforms and drill ships) in that the wellhead and Christmas tree, which are required to drill and produce the well, are located on the ocean floor rather than on a platform. Platforms are the customary way to support drilling and producing facilities, and they are normally located immediately over the oil and gas reservoir. Platforms can be used to drill wells at an angle (Fig. 1) to reach the outer extremities of the reservoir. Sometimes it is not possible to drill wells at an angle great enough to reach the outer limits of the reservoir; in these instances, SSWS are used. These systems differ from the platform well in that the horizontal flow lines and control lines are located on the sea floor. They have the potential to reach reservoirs in water depths to 2000 ft (610 m). However, they are complex systems, and a continuing development effort is needed to improve their ability to produce oil at lower operational costs.

Types of SSWS. There are three types of SSWS: single, cluster, and template. Cluster and template systems have several wells located within 7–10 ft (2–3 m) of each other, while single satellite wells are located at large distances from each other, sometimes as much as 3 mi (4.8 km). The wellhead, which is common to each type of system, is the device used to control the pressure of the well, and consists of a casing head and a tubing head. These support the pipes, called casing and tubing, that are run into the hole. During the drilling of a well several strings of casing are run into the well

and cemented. Cementing is a drilling process that provides a seal between the casing and the borehole to prevent unwanted intrusion of fluids into or out of the wellbore. The tubing head is similar in function to a casing head and supports a smaller string of pipe inside the casing. It seals off the casing annulus (the void space between the inside of the casing and the outside of the tubing) and provides a connection for a Christmas tree at the ocean floor. A Christmas tree is a series of high-pressure valves and fittings that control and direct the flow of oil and gas from the well. Its shape resembles a tree branching out above the wellhead. There are three basic types of subsea Christmas trees. Wet trees are exposed to the marine environment, dry trees are enclosed in a housing under atmospheric pressure which allows maintenance work to be done in a "shirt-sleeve" environment, and caisson trees are located below the ocean floor.

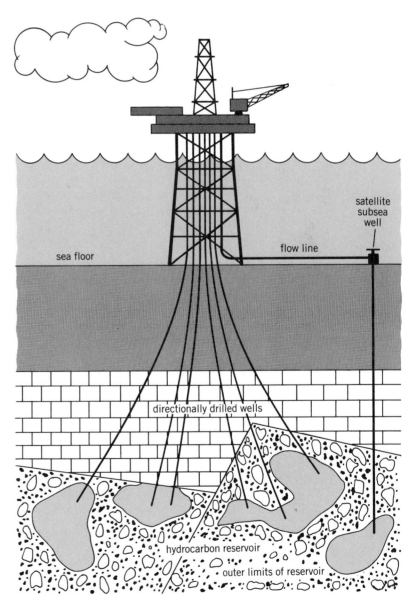

Fig. 1. Drilling platform and single satellite subsea well.

The valves on the trees must be controlled from a remote surface location. This is accomplished by one or a combination of the following control schemes: (1) direct hydraulics—each subsea function, that is, the opening or closing of a subsea Christmas tree valve, is controlled by a hydraulic pressure from the surface; (2) sequential hydraulics—a subsea controller contains pressure-sensitive pilots that activate at different pressure levels; a single line is used to transmit pressure from the surface; (3) electrohydraulics—this system comprises a number of electrical conductors in the form of an electrical "cable"; each conductor has a distinct function that will activate subsea hydraulic pilots when energized from the surface; (4) multiplex control—this system uses a single electrical conductor over which many electrical signals can be sent to a complex subsea controller; a subsea controller commands the required function. The sequential, electrohydraulic, and multiplex controls require a sea-floor module, while direct-hydraulic schemes require extra hydraulic lines from the well to the surface location.

Within the tubing string, about 100 ft (30 m) below the ocean-floor Christmas tree, is located a surface-controlled subsurface safety valve (SCSSV). Its purpose is to stop the well from flowing in case of an emergency, for example, an anchor dropping on the Christmas tree.

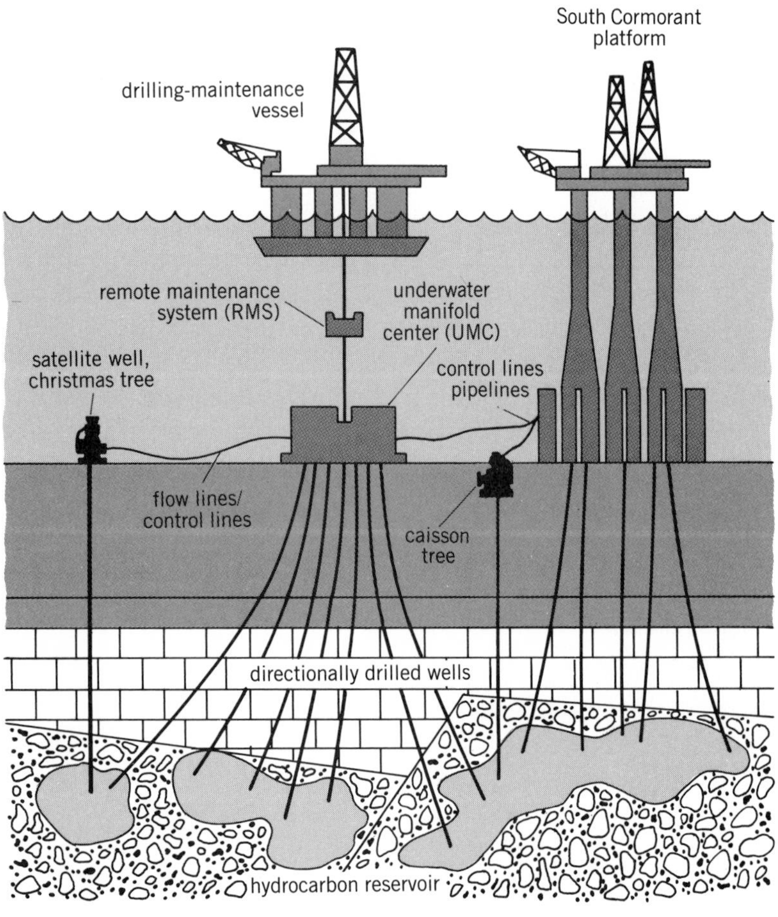

Historical development of SSWS. The use of SSWS has been gradual since the first installation, in 1943, at Lake Erie in 35 ft (10.7 m) of water. Between 1978 and 1982, 35 systems were installed off the coast of Brazil, one in 650 ft (200 m) of water. Since 1943 there have been a total of 230 installations, of which 30% have been abandoned for various reasons. To date, most installations have been single satellite subsea well systems, using wet trees. The first complex system was installed in 1982 at the Cormorant Field in the North Sea. Here, an underwater manifold center (UMC) concept (Fig. 2) was used, including both single and template systems. With the UMC, nine wells can be drilled at an angle from the template, and nine single wells can be connected to the template. The Cormorant UMC has a production capability of 50,000 barrels of oil per day (bpd), or 7.9 million liters.

Production capabilities of SSWS. Satellite subsea well systems are used to produce oil and gas, and also to inject water back into the formation to maintain pressure in the reservoir, thus increasing oil recovery. Of the installed SSWS, 72% have been used for producing oil, 20% for gas, and 8% for water injection. A single-satellite well using tubing of 5½-in. (14-cm) outside diameter has produced 20,000 bpd (3.16 million liters). The capacity of an oil well to deliver oil is a determining factor in selecting the size of tubing. If the tubing is too small, it hinders the flow of oil from the reservoir. However, there is a trend to use two strings of smaller tubing because it allows the servicing of downhole equipment. Two strings of 3½-in.-outside-diameter (8.9-cm) tubing will allow about 15,000 bpd (2.37 million liters) to flow with no erosion of the walls of the steel tubing.

Downhole equipment, such as the surface-controlled subsurface safety valve, is required to produce an oil well. This valve may need replacement within a 1- to 1½-year period, and this can be accomplished by using a "through the flow line" servicing system. Using two strings of tubing, tools are "pumped" into and out of the well from a remote surface location. These tools will fasten onto the SCSSV, and by reversing the pumping action the SCSSV is retrieved, and another one can be pumped into the well. Such a system is incorporated into the UMC at the Cormorant Field.

Servicing ocean-floor equipment. As illustrated in Fig. 2, a drilling-maintenance vessel is required to service the complex UMC. This system has a remote maintenance system that can be lowered to the UMC. It consists of a mechanical robot that lands on a rail, located around the UMC, which directs the robot to the well that requires servicing. Complete Christmas trees can be exchanged and other maintenance functions can be accomplished. This system does not require human intervention such as by divers. Generally all of the other systems which have been installed require divers to assist in the replace-

Fig. 2. Underwater manifold center, connected to satellite wells. (*After Shell's underwater manifold ready for launching, Petrol. Eng. Int., February 1982*)

ment of Christmas tree valves or subsea controllers.

Today, diving technology allows a diver to work for about 3 h at a depth of 1400 ft (427 m). However, the diver has to stay in a pressure chamber for 12 h prior to work at that depth. After performing the 3 h of work on the sea bottom, the diver requires 10 days to decompress. Special diving suits have been developed so that, by using a 1-atm breathing system, the diver can penetrate to a depth of 3000 ft (914 m). These diving suits show great promise; however, ocean-floor equipment must be designed to allow for the limited movement that is associated with their diving apparatus.

Flow line and control lines. An element of SSWS which has taken years to develop is a reliable flow-line and control-line system. Although the operational characteristics have been clearly defined, only recently has this technology been applicable. Satellite wells are often located 3 mi (4.8 km) from the control point and require several lines to connect them to the platform. Early installations suffered damage by fishing gear and other objects dropping on them. To overcome this problem, attempts were made to bury the lines, an operation which further damaged them. A reliable burial system uses flexible pipe installed from a specially built ship which lays the pipe on the sea floor, in the same manner as transatlantic cables are laid, and buries it. The flexible pipe combines steel and thermoplastic layers in a structure resembling a heavy rubber hose. Several North Sea single satellite wells have been connected to platforms by using this method.

A second method is the buoyant offbottom tow and connection method, which has been used to connect several wells to platforms in the North Sea. The lines are constructed onshore, towed to the field by two tugboats, lowered to the ocean floor, connected to the wellhead, and then buried. No subsea controller is used with this system; the many lines which are required to produce the satellite wells are bundled into a pipe that serves as a buoyancy chamber (Fig. 3). This system is designed so that when it is lowered to the ocean floor it comes to rest at a controlled height above the floor. This is accomplished by short ends of chain, attached to the pipe, which come to rest on the sea floor, giving the pipe buoyancy. The ends of the flow-line bundle are then connected and the outer housing of the pipe is flooded, causing the bundle to sink. The line is buried without damage to the internal pipes, using conventional bury barge methods. It is expected that this technology can be extended to water depths of 3000 ft (914 m).

Future development. Perhaps the greatest economic factor in favor of single SSWS or compared to the UMC concept is that no loss of production results when the template wells are worked over, an operation to increase or restore production. Adjacent template wells must be shut during a workover process. Single SSWS have been installed which do not use subsea controllers. The ocean-floor Christmas tree has been eliminated in favor of caissons,

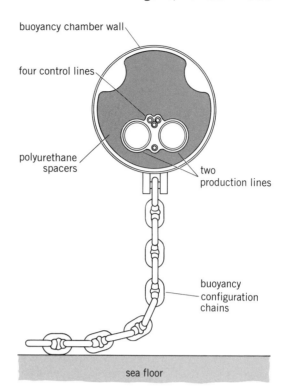

Fig. 3. Flow-line and control-line arrangement. The chain, once it touches the ocean floor, gives the pipe buoyancy. (*After A. W. Morton, New concepts used in satellite subsea well systems, J. Petrol. Technol., March 1982*)

which can be covered with concrete protective housings. The development of reliable SCSSV will prolong the service life of satellite subsea wells and reduce their operating costs. These developments should allow economical exploitation of oil reservoirs in water depths of 3000 ft (914 m).

For background information *see* OIL AND GAS, OFFSHORE in the McGraw-Hill Encyclopedia of Science and Technology.

[ARTHUR MORTON]

Bibliography: D. Booth (ed.), *Subsea*, vol. 2, pp. 109–150, 1982; *A Deepwater Oil and Gas Technology Assessment*, NEW England River Basins Commission (NERBC)—Resource Planning and Analysis (RPA) Program, pp. 121–177, April 1981; A. W. Morton, New concepts used in satellite subsea well systems, *J. Petrol. Technol.* pp. 477–481, March 1982; Shell's underwater manifold ready for launching, *Petrol. Eng. Int.* pp. 82–90, February 1982.

Oncogene, human

Oncogenes are dominant genetic elements whose expression within a normal cell leads to neoplastic transformation of the cell. Until recently, the only known oncogenes were those present in animal tumor viruses. In the case of retroviruses it has been well established that their oncogenes (designated v-*onc*) were generated by transduction of certain subsets of cellular sequences (designated c-*onc* or protooncogenes) that, upon recombination with viral ge-

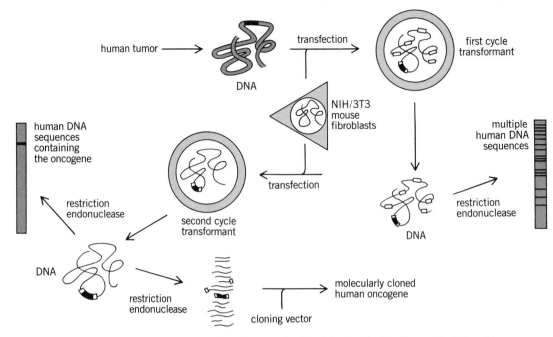

Fig. 1. Schematic representation of the experimental approach utilized for the identification and isolation of human oncogenes.

Summary of human oncogenes

Type of tumor	Cell line	Relationship with retroviral *onc* genes
Carcinomas		
Bladder	T24	*has, bas*
	EJ	*has, bas*
	A1698	*kis*
Breast	MCF-7	—
Colon	SW 480	*kis*
	SK-CO-1	*kis*
	A2233	*kis*
	Tumor sp. 1665	—
	Tumor sp. 2033	—
Gall bladder	A1604	*kis*
Liver	SK-HEP-1	—
Lung	SK-Lu-1	*kis*
	Calu-1	*kis*
	A2182	*kis*
	LX-1	*kis*
	A427	*kis*
	Tumor sp. 1615	*kis*
Pancreas	Tumor sp. 1089	*kis*
Sarcomas		
Fibrosarcoma	HT-1080	—
Rhabdomyosarcoma	Tumor sp. 1085	*kis*
Leukemias/lymphomas		
Promyelocytic leukemias	HL-60	—
Pre-B lymphocyte neoplasia	207	—
	697	—
	Tumor sp. CB1	—
	Tumor sp. CB2	—
B-cell lymphomas	GM 2132	—
	GM 1500	—
T-cell lymphomas	Tumor sp. T10	—
	Tumor sp.*	—
Other tumors		
Neuroblastoma	SK-N-SH	—
Astrocytoma	SW-l088	—
Chemically transformed cells	MNNG-HOS	—

*Tumor derived from a patient with Sézary syndrome.

netic information, acquired the ability to induce malignant transformation. About 30 types of acute transforming retroviruses have been isolated, some of which contain the same or closely related *onc* genes. These findings have suggested that vertebrates contain only a limited number of cellular genes that possess the potential to become transforming genes.

Over the last few years, an independent approach has been utilized to identify cellular oncogenes in tumor cells. The development of gene transfer techniques has permitted the demonstration that the malignant phenotype can be transmitted from tumor to normal cells via transfection with discrete fragments of DNA, thus providing direct evidence for the presence of dominant oncogenes in tumor cells. This experimental approach was used to investigate whether human tumor cells may also contain oncogenes in an effort to understand the molecular events involved in human carcinogenesis (Fig. 1). Studies have demonstrated that a significant fraction of the human tumor cell lines examined contain dominant transforming genes (see table). Human oncogenes have also been detected in unmanipulated tumors, thus eliminating the possibility that such oncogenes may have been generated during the establishment or continuous passage of human tumor cells in laboratory cultures.

Identification of human oncogenes. The demonstration that the serial transmission of the malignant phenotype of human tumor cells was mediated by an oncogene required the detection of human DNA sequences in the genome of the transfected cells (cells that have uptaken chemically pure DNA). This can be accomplished by molecular hybridization techniques which utilize a probe specific for human repetitive sequences (that is, the *Alu* family), providing that the recipient cells are of nonprimate origin, such as the widely utilized NIH/3T3 mouse cell line. Biochemical detection of human DNA fragments that cosegregate with the malignant phenotype permitted the identification of those sequences encompassing human oncogenes. As a consequence, these oncogenes can be preliminarily characterized and subsequently isolated by molecular cloning techniques.

Characterization of human transforming genes. As indicated in the table, at least one representative sample from each of the major types of human cancers has been shown to contain a dominant transforming gene. In the case of lymphoid tumors, it has been demonstrated that different oncogenes are activated in different types of tumors, such as pre-B, mature-B, intermediate T- or mature T-cell lymphomas. Yet, the same oncogene is activated in independent tumors of the same differentiated cell type.

Similar results have been found in tumors of epithelial origin, although without the high degree of specificity exhibited by lymphoid tumors. Of 17 human carcinoma cell lines or solid tumors in which oncogenes have been partially characterized, 12 (70%) contained the same oncogene, whereas 2 others possessed highly related transforming sequences. Carcinoma oncogenes do not exhibit specificity regarding the organ of origin. For instance, the same transforming gene has been detected in carcinomas of the bladder, colon, lung, and pancreas.

Oncogenes have also been identified in tumors of mesenchymal origin, although with less frequency. Whereas the oncogene present in HT-1080 fibrosarcoma cells appears to be unique, that identified in a solid embryonic rhabdomyosarcoma is highly related to the oncogenes widely found in carcinomas. Human tumor cells thus contain a series of different oncogenes which exhibit a limited specificity for the type of tumor in which they have been identified. However, tumors of different differentiation lineage may sometimes contain the same oncogene. These results indicate that the number of different human oncogenes is limited and may not exceed a few dozen.

Human and retroviral transforming genes. As indicated above, vertebrates contain a subset of genes that can acquire malignant properties upon recombination with retroviral sequences. Thus, it was reasoned that overlapping may exist between this group of c-*onc* genes and the oncogenes identified in human tumors. Intensive research in the last decade has made it possible to isolate the genomes of representative retroviruses by molecular cloning techniques. Probes specific for retroviral *onc* sequences were hybridized to the genomes of cells transformed by human oncogenes in order to detect possible homologies between these two independently identified groups of transforming genes. None of the oncogenes present in lymphoid tumors appeared to be related to known retroviral *onc* genes. Similar results were found with oncogenes detected in breast

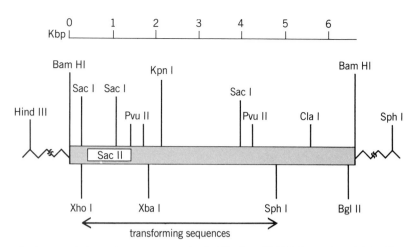

Fig. 2.   Restriction endonuclease map of a 6.6 kbp Bam HI DNA fragment containing the T24 human bladder carcinoma oncogene.

carcinoma (MCF-7) and fibrosarcoma (HT-1080) cells. In contrast, the oncogene present in all six lung carcinomas, as well as in certain carcinomas of the gall bladder (A1604), pancreas (1189), colon (SW 480, SK-CO-1, A2233), and urinary bladder (A1698) was found to be highly related to *kis*, the transforming gene of the Kirsten strain of murine sarcoma virus (MSV). Similarly, two other bladder carcinoma cell lines (T24 and EJ) contain oncogenes related to v-*has* and v-*bas*, the transforming genes of the Harvey and BALB strains of MSV. Findings that certain human oncogenes are related to acute transforming retroviruses have justified the extensive research efforts in which these viruses were utilized to study the mechanisms of malignant transformation as an approach to understanding human carcinogenesis.

Molecular cloning of bladder carcinoma oncogene. To date, attempts have been made to isolate human oncogenes by molecular cloning techniques. So far (1982), only one oncogene, present in T24 and EJ bladder carcinoma cells, has been isolated in its biologically active form (Fig. 2). When 1 microgram of this molecularly cloned oncogene is introduced more than 10,000 foci of transformed cells result. Biochemical characterization of this bladder carcinoma oncogene has indicated that it is small, less than 4.6 kilo base pairs (kbp). The relationship between the T24/EJ bladder carcinoma oncogene and Harvey- and BALB-MSV transforming genes has made it possible to characterize the transcriptional and translational products of this human oncogene. The T24 oncogene is transcribed as a 1.2-kbp mRNA that contains the retrovirus-related sequences. Translation of this mRNA molecule yields a protein of 21,000 daltons (p21) antigenically related to the gene products of the Harvey- and BALB-MSVs. When this human oncogene was compared by heteroduplex analysis (a technique in which two different DNA or RNA molecules are annealed to each other, homologous and nonhomologous regions being detected by electron microscopy) and restriction enzyme analysis with a molecular clone of its normal human homolog, no significant differences were detected. These findings suggest that rather subtle genetic alterations led to the activation of this oncogene in T24 and EJ human bladder carcinoma cells.

Nucleotide sequence analysis of the T24/EJ oncogene and its normal human homolog has revealed that they differ in the simplest possible way. A single nucleotide difference results in the change of the twelfth amino acid residue of the p21 protein coded for by these genes. Whereas glycine is present in the protooncogene product, valine was found in the p21 protein of the bladder carcinoma oncogene. Interestingly, a similar structural alteration—substitution of glycine by another amino acid residue at position 12 of the p21 protein—has been observed in the gene products of the Harvey, BALB, and Kirsten strains of MSV. Therefore specific structural changes within these p21 proteins profoundly alter their normal function—as yet, to be determined—triggering the pathways that lead to malignant transformation.

Implications and perspectives. The presence of oncogenes in human tumors suggests that the development of at least certain human neoplasias may be mediated by the activation of dominant transforming genes. However, there is no direct evidence linking these human oncogenes with human cancers as yet. It is difficult to understand how a gene, which is by itself capable of immediately inducing the malignant transformation of normal cells under experimental laboratory conditions, can be involved in the development of human cancers, a process that in most cases takes more than a decade.

There are several hypotheses that attempt to explain this apparent discrepancy. It is possible that activation of oncogenes is a common event that in most occurrences goes undetected because the transformed cell cannot grow, either because of the body's defense mechanisms (for example, immune response) or because of the limitations imposed by the cellular environment that surrounds it (extracellular factors, accessibility to nutrients, and so on). Overcoming these barriers by the activated cell may require complex secondary changes (for example, antigenic modulation and adaptation to the environment) that may account for the slow development of most human malignancies. Alternatively, it is possible that the subtle genetic alterations that lead to the activation of transforming genes may represent a late, irreversible step in oncogenesis. This hypothesis is supported by the observations that certain human cancers undergo preneoplastic changes. In fact, the irreversible change that commits a preneoplastic cell into the neoplastic pathways may not necessarily be the activation of oncogenes. This would explain why such transforming genes have only been detected in a fraction of human tumors.

A limitation in the interpretation of the biological significance of human oncogenes derives from the fact that such genes have only been shown to transform continuous cell lines in laboratory cultures. Some

of these cells, including the widely used NIH/3T3 mouse cell line, are heteroploid and occasionally undergo spontaneous transformation. It has been demonstrated that some human oncogenes are highly related to the *onc* genes of the Harvey, Kirsten, and BALB strains of MSV. Each of these three retroviruses is capable of inducing a variety of malignancies in animals. Thus, it seems likely that the human homolog of these retroviral *onc* genes, when it becomes activated as an oncogene, would also possess the capacity to transform normal cells. However, direct demonstration of this postulate will be required before the full malignant capabilities of human oncogenes can be established.

The above parallelism between the transforming properties of human oncogenes and those of retroviruses is only one of the few insights that the relationship between these two types of cellular oncogenes will provide to the understanding of human carcinogenesis. In fact, the extensive investigations on acute transforming retroviruses have led to considerable knowledge concerning their genomes and translational products, as well as their interaction with putative cellular targets. Researchers should now be able to apply this information toward the understanding of the mechanisms by which human oncogenes may induce malignant transformation.

Summary. Oncogenes have been identified in a significant fraction of human tumors. At least one representative sample of each of the major forms of human cancer has been shown to contain a dominant transforming gene. It appears that different oncogenes are activated in different types of tumors; yet the same oncogene is preferentially activated in tumors of the same differentiated cell type. This specificity has been best illustrated in lymphoid tumors. Similar results have also been obtained with tumors of epithelial origin, in particular lung carcinomas. To date, about ten different oncogenes have been identified. At least two of them, found mostly in carcinomas, are highly related to the transforming genes of Harvey, Kirsten, and BALB strains of murine sarcoma virus.

An oncogene present in T24 and EJ bladder carcinoma cell lines has been isolated by molecular cloning techniques. This oncogene was shown to be biologically active in transformation assays in laboratory culture, and did not undergo major genetic rearrangements. In fact, studies comparing the T24/EJ oncogene with its normal human homolog have demonstrated that these genes differ in the simplest possible way: a difference of a single deoxynucleotide that results in the substitution of the twelfth amino acid residue in the p21 protein coded for by this gene. Rapid progress has been made over the past 2 years on the detection, isolation, and characterization of human oncogenes. However, more information needs to be gathered before it can be unequivocally established that these dominant transforming genes are involved in the development of human neoplasias.

For background information *see* CANCER (MEDICINE); NEOPLASIA; ONCOLOGY; TUMOR; VIRUS in the

McGraw-Hill Encyclopedia of Science and Technology.

[MARIANO BARBACID]

Bibliography: J. M. Bishop, Oncogenes, *Sci. Amer.*, 246:80–93, 1982; G. M. Cooper, Cellular transforming genes, *Science*, 217:801–806, 1982; E. P. Reddy et al., A point mutation is responsible for the acquisition of transforming properties by the T24 human bladder carcinoma oncogene, *Nature*, 300:143–152, 1982; C. J. Tabin et al., Mechanism of activation of a human oncogene, *Nature*, 300:143–149, 1982; R. A. Weinberg, Fewer and fewer oncogenes, *Cell*, 30:3–4, 1982.

Oncology

Somatic cell hybridization has been a powerful tool in the study of gene mapping and control of gene expression of cells in culture. Investigators have utilized this technique in attempts to unravel the mechanisms that control the expression of transformation and tumorigenicity. Much of this work has been done by using intraspecific rodent cell hybrids or interspecific human-rodent cell hybrids. Although many studies have indicated that tumorigenicity is suppressed when a normal cell is fused with a malignant one, others show that tumorigenicity is expressed dominantly in such hybrids. The major reason for the differences in results is that both types of hybrid cells are often chromosomally unstable. In addition, rapid loss of chromosome material from the hybrid cells makes the analysis and interpretation of chromosomal control of neoplastic expression a difficult task.

Suppression of tumorigenicity. Recently, an intraspecific human hybrid cell system that eliminates this problem was developed. When malignant and normal human cells are fused, the resulting hybrids behave as transformed cells in culture but are nontumorigenic when inoculated into congenitally athymic nude mice (mice which lack a thymus and are unable to reject allogeneic or xenogeneic tissue grafts or tumors). The suppression of tumorigenicity is complete and, because of the chromosome stability of the hybrid cells, is stable. Rare tumorigenic segregants arise after prolonged periods in culture and express the same degree of tumorigenic potential as the malignant parental cell. Chromosome analysis of paired combinations of nontumorigenic and tumorigenic segregant HeLa/fibroblast hybrids has shown that the loss of a single copy of two specific chromosomes, namely 11 and 14, is associated with reexpression of tumorigenicity. Recent studies with multiparental combinations, for example, two malignant parental cells fused with one normal parental cell, or two normal parental cells fused with one malignant parental cell, indicate that control of expression of both transformed and tumorigenic phenotypes is governed by gene dosage effects rather than by simple dominant or recessive genetic traits.

Transformation versus tumorigenicity. Results obtained to date strongly suggest that the progression of a normal cell to a frankly neoplastic cell is a multistep process that is probably controlled by multigenic events. It appears certain that the transformed and tumorigenic phenotypes in these cells are under separate genetic control. Comparative analysis of paired nontumorigenic and tumorigenic segregant hybrid cells has proved to be an extremely powerful tool in analyzing those phenotypic traits that are specifically correlated with tumorigenic expression and therefore almost certainly represent tumor-associated markers. Reports on tumor-associated

Properties of parental and hybrid human cells in culture

Phenotype in culture	Parental cells		Nontumorigenic HeLa/fibroblast hybrids	Tumorigenic HeLa/ fibroblast segregants
	HeLa	Fibroblast		
Morphology	Epithelial	Fibroblastic	Intermediate	Epithelial*
Density-dependent inhibition of growth	No	Yes	No	No
Requirement for serum growth factors	Reduced	High	Reduced	Reduced
Lectin agglutination	+ + +	Minimal	Extensive	Extensive
Anchorage-independent growth in soft agar and methyl cellulose	Yes	No	Yes	Yes
Fibronectin expression	None	High	Reduced (short branched filaments)	Reduced (unbranched* stitch pattern)
Cytoskeleton				
microtubules	Organized	Organized	Organized	Organized
microfilaments	Poorly organized	Organized	Organized	Poorly organized*
Placental alkaline phosphatase	High	Low	High	High
Ganglioside analysis	Simple	Complex	Relatively complex	Relatively complex
Human chorionic gonadotrophin synthesis	Present	Absent	Absent	Present
75-kilodalton membrane phosphoprotein	Present	Absent	Absent	Present

*Reversible properties; addition of dexamethasone or sodium butyrate induces a phenotypic shift to that of the nontumorigenic hybrid cells.

markers often involve a comparison between neoplastic and normal cells, but do not take into account that in the multistep progression to cancer many of these markers may be expressed on transformed, nontumorigenic preneoplastic cells.

A large number of phenotypic traits considered to be tumor-specific or tumor-associated markers have been examined in this human hybrid cell system. The results (see table) indicate that many of these phenotypic traits are expressed in both nontumorigenic hybrids and their tumorigenic segregants. Thus the expression of these "tumor-specific" markers may be necessary, but is not sufficient, for full tumorigenic expression. Those phenotypic traits that distinguish tumorigenic HeLa/fibroblast hybrids from their nontumorigenic hybrid counterparts are morphology, fibronectin (a major cell-surface glycoprotein) distribution, and microfilament organization. All three traits are completely reversible. Addition of low concentrations of dexamethasone or sodium butyrate to tumorigenic hybrid cells results in a shift to the phenotype which corresponds to the nontumorigenic hybrids. Two tumor-associated markers have, however, been identified in HeLa/fibroblast hybrids. They are the occurrence of a 75-kilodalton phosphoprotein in the membranes of parental HeLa cells and tumorigenic segregant hybrid cells, and the production of the α subunit of human chorionic gonadotropin in these cells. The functional significance of these compounds is not yet understood. Neither phenotypic trait is expressed in normal parental cells or nontumorigenic hybrids.

Differentiation as a control mechanism. An extensive analysis of the growth behavior of the nontumorigenic and tumorigenic-segregant HeLa/fibroblast hybrid cells has shown that the hybrids behave almost identically in culture. However, dramatic differences are found when the cells are inoculated into congenitally athymic nude mice. Both cell types divide rapidly for the first 3–4 days in the animal. The tumorigenic hybrid cells continue to proliferate and form progressive, undifferentiated carcinomas (see illustration *a*). A dramatic cessation of mitotic activity is seen with the nontumorigenic hybrid cells, and they remain as a viable, nondividing tissue for many weeks. There is no evidence of rejection by the host. In experiments with hybrids formed between HeLa and different normal parental cells, for example, fibroblasts or keratinocytes, the nontumorigenic cells cease proliferating and then differentiate (see illustration *b*). Thus, the controlling element that the normal parental cell contributes to the hybrid in attaining the nontumorigenic phenotype is the reacquisition of the ability to differentiate in response to appropriate regulatory signals. Interestingly, the hybrid cells differentiate into the cell types of the normal parental component, for example, HeLa/fibroblast hybrids differentiate as fibroblasts and HeLa/keratinocytes differentiate as keratinocytes. Thus, the differentiated, nontumorigenic hybrid cell population in the intact animal takes on the phenotypic signature of the normal parental cell,

irrespective of the phenotype of the malignant parental cell.

Based upon observations of this intraspecies human hybrid cell system a model has been developed that may describe many of the observations. In this model, normal cells in culture respond to autoregulatory signals; stop dividing when confluency is reached; and, particularly in the case of keratinocytes, terminally differentiate. These same normal cells also cease dividing rather abruptly and differentiate when grafted in nude athymic mice and other immunoincompetent animals. Nontumorigenic hybrids, on the other hand, do not appear to exhibit division control, nor do they differentiate in culture. However, the cells maintain the capacity to respond to growth regulation and differentiation signals when inoculated into the host animal. The key difference between the nontumorigenic hybrids and their tumorigenic segregants is that the latter cells, as well as the malignant parental cells, appear not to re-

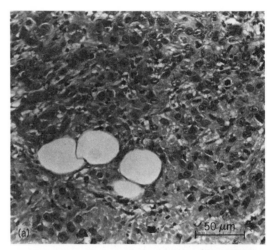

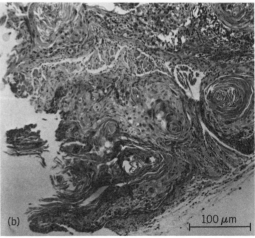

Histological appearance of 5-week nodules of HeLa/human keratinocyte hybrid cells inoculated subcutaneously into congenitally athymic nude mice. (*a*) A tumorigenic segregant; there is no evidence of differentiation, and the tumor has the appearance of an actively proliferating, undifferentiated carcinoma. (*b*) A nontumorigenic clone showing extensive differentiation, including horny pearls and keratin whorls.

spond to growth regulation or differentiation signals either in culture or in the intact animals, and therefore form progressive tumors.

Future plans for this research program include the development of techniques whereby individual chromosomes, and eventually DNA sequences, are transferred directly into recipient cells rather than by whole cell fusions. This will allow the identification of those specific chromosomes and genes involved in the control of neoplastic expression in human cells.

For background information *see* HUMAN GENETICS; ONCOLOGY; SOMATIC CELL GENETICS; TUMOR in the McGraw-Hill Encyclopedia of Science and Technology.

[ERIC STANBRIDGE]

Bibliography: C. J. Der and E. J. Stanbridge, A tumor-specific membrane phosphoprotein marker in human cell hybrids, *Cell*, 26:429–439, 1981; E. J. Stanbridge, Suppression of malignancy in human cells, *Nature*, 260:17–20, 1976; E. J. Stanbridge et al., Human cell hybrids: Analysis of transformation and tumorigenicity, *Science*, 215:252–259, 1982; E. J. Stanbridge et al., Specific chromosome loss associated with the expression of tumorigenicity in human cell hybrids, *Somatic Cell Genet.*, 7:699–712, 1981.

Optical telescope

The art of building large optical telescopes has advanced in several discrete steps over the past century, and now appears to be at another threshold. In the nineteenth century refracting telescopes held center stage, advanced by the art of making large (1-m) lenses of satisfactory quality. These lenses were favored over speculum mirrors by high light efficiency and insensitivity to warping—two problems encountered with reflectors. G. E. Hale pioneered a new generation of telescopes at Mount Wilson Observatory at the turn of the century using silvered glass mirrors and new methods of support to minimize warping. The largest of this class, the 100-in. (2.54-m), was put into operation in 1920.

The Palomar 200-in. (5.08-m) set the pattern for the next decades of telescopes through sophisticated engineering advances for large precision structures and through the use of a glass of low thermal expansion coefficient, Pyrex. Fused silica mirrors were deemed to be even better, but attempts to make them large enough failed.

In the 1950s ultralow-expansion glassy materials became available as a spin-off of defense activities, and a whole new generation of compact, 3- to 5-m telescopes benefitted from adding this new mirror material to 200-in.-type construction. In the meantime, radio telescope engineers had advanced the art of making altazimuth mountings, confronting the problem of nonlinear drive rates by precision shaft-angle encoders and digital computers. The last of the old generation of conventional telescopes was the Soviet 6-m (236-in.) telescope. It had an altazimuth design, but it was based on a solid Pyrex-type mir-

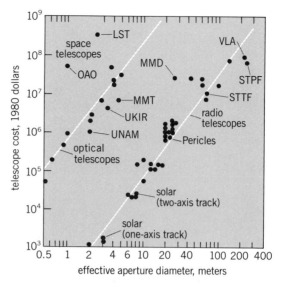

Fig. 1. Relationship between telescope aperture diameter and telescope cost, excluding building and dome, for space, optical, and radio telescopes and for solar-concentrating collectors. (*After A. B. Meinel, cost relationships for non-conventional telescope structural configurations, J. Opt. Soc. Amer. 72:14–20, 1982*)

ror with a rather slow focal ratio of *f*/4.

The real barrier to larger telescopes has been cost. Only space telescopes have thus far crossed the $100 million barrier, an acceptable cost in space programs because of the high cost of all space activities. Ground-based astronomy has had to remain below the $25-million level, but now it is hoped that telescopes in the 10- to 20-m class can be built at acceptable cost. The cost barrier to very large conventional designs is apparent from Fig. 1. Costs rise at about the 2.6 power of mirror diameter; thus a 20-m telescope would cost on the order of $1 billion. It is thus clear that new concepts must be developed, breaking the traditional conservatism of astronomers for following established approaches. Ideas are now given serious consideration if they promise to meet increasingly strict image-quality specifications and a cost below the curve for optical telescopes in Fig. 1.

Multiple mirror telescope. The multiple mirror telescope (MMT) is one of the new generation of telescopes that cost less than conventional telescopes. It combines the beams from six complete conventional telescopes, all carried on a single altazimuth mounting. The result is a very short, low-mass telescope (Fig. 2). The mass of the mirrors is small compared to that of a single mirror of equal light-collecting power. The thin-walled eggcrate fused-silica mirrors are a spin-off from space technology. The mass of the mounting holding the mirrors together is thereby also decreased, resulting in a low-mass telescope. The cost of telescopes and other large structures is approximately proportional to their mass, so lower mass can mean cost savings. The cost of a telescope building is about equal to the cost of the telescope, so a short telescope like the multiple

Fig. 2. Multiple mirror telescope of the Whipple Observatory, Mt. Hopkins, AZ.

mirror telescope means a small building and further cost savings. The resulting project cost lies well below the cost trend for conventional telescopes.

Segmented mirror telescope. Another approach to lowered telescope costs is to make a single large mirror out of many smaller segments, as proposed for a 10-m telescope by University of California astronomers. The smaller mirror mass that results from use of 1.5-m segments combined with space-frame construction technology leads to a low-mass telescope, but of greater overall length than the multiple mirror telescope. The segmented mirror telescope (SMT) has a distinct advantage in that beam-combiner mirrors are eliminated. The challenge of the segmented mirror telescope is twofold: how to make precision off-axis segments of parabolic or hyperbolic shape; and how to align the many segments to act as a single monolithic mirror. A bend-and-polish technique has been demonstrated whereby a spherical mirror assumes the desired aspheric shape after the stress of bending has been relaxed. Advances in sensor and actuator technology are necessary and appear to have been successful in meeting the strict specifications desired by the astronomers.

Active mirror technology. The principal problem affecting image quality has been to mount a large mirror on a passive support system and have a satisfactory mirror for all occasions. The excellence of a very large telescope depends on achieving smaller images than those obtained with existing large telescopes, but passive technology cannot meet this demand. Some gain could be made by finding a better site or by controlling local seeing disturbances, but the ability to tune the mirror figure appears to be essential. A few especially skilled astronomers have deliberately tuned passive mirrors to help specific observations for a short time span. Modern sensor and actuator developments offer the possibility of automatically tuning a mirror to yield perfect performance upon demand. Such active mirrors have been demonstrated in the space program over the past decade, but the advances have yet to be felt in astronomy.

There are various degrees of mirror control. The proposed 7.6-m University of Texas telescope approaches this problem from the conservative viewpoint of making the mirror as rigid as feasible, considering mirror thickness and mass, and then tuning it automatically to remove low-order harmonic deformations as determined by analyzing a star image. The 10-m University of California telescope uses rigid segments in a fully active assemblage. The multiple mirror telescope uses rigid mirrors which are semi-actively aligned, because manual alignment combined with an exceptionally stiff mounting permits the mirrors to remain in proper alignment long enough for most critical observations. Upgrading the multiple mirror telescope to incorporate new alignment technology is continuing. No astronomical telescope has yet used the ultimate degree of figure control that results from using a membrane type of mirror with little intrinsic stiffness.

New mirror technology. The emphasis over the past two decades has been on the use of ultralow-thermal-expansion materials like fused silica, ULE (ultralow expansion) silica, and glass ceramics like Zerodur and Cervit. Telescopes of up to 4 m have used solid mirrors of these types. The low expansion ensures that the mirror will keep its correct figure in spite of temperature gradients in the mirror. Large solid mirrors of these materials are expensive. Fused silica and ULE silica can be cut to shape and fused together into very light structures; but this can be done only at considerable cost, in part due to the high processing temperatures involved. Glass ceramics can be machined into lightweight mirrors, but also at considerable cost.

For very large telescope mirrors, the solid approach with zero-expansion materials meets a new problem. The mirror figure remains satisfactory, but the mass of the mirror constitutes a thermal disturbance to the air it contacts, adversely affecting the image quality produced by the telescope, an effect termed internal seeing. For small mirrors, the thermal response time is small enough that the mirror reaches thermal equilibrium promptly. A large mirror, like the 40-ton Soviet 6-m disk, never reaches equilibrium during the night.

A current development at the University of Arizona is to cast lightweight ribbed mirrors of Pyrex glass. The 200-in. disk currently in use there is ribbed Pyrex, but with 4-in. (10-cm) ribs. The new lightweight mirrors have ribs of less than 1 in. (2.5 cm) thickness. This approach combines the low cost and low processing temperature of Pyrex with a quick thermal response time. The rationale is that if the mirror reaches thermal equilibrium internally and with respect to the surrounding air the thermal expansion coefficient is of no importance. The same philosophy applies to both slab and lightweight aluminum mirrors. The high conductivity of aluminum more than offsets its relatively large expansion coefficient. Astronomers at the European Southern Observatory are planning to use an 8-m aluminum mirror in a new telescope. The University of Texas is planning to use a 7.6-m lightweight Pyrex mirror.

Meridian transit telescope. A new concept for very large, low-cost telescopes is that of the meridian transit telescope pioneered by the Steward Observatory. In this case the telescope does not track the stars: only the receiver or the secondary of the telescope moves to track the object for a limited time while it is near the meridian. This means that the telescope needs only to be pointed at the desired declination and fixed. The telescope structure, then, is simple and the building does not need to rotate. The mode of observation, however, is dramatically changed. One can observe a given object only for a limited time, but digital data can be readily stored and added over as many nights as required. The observational advantage is that time sharing allows many

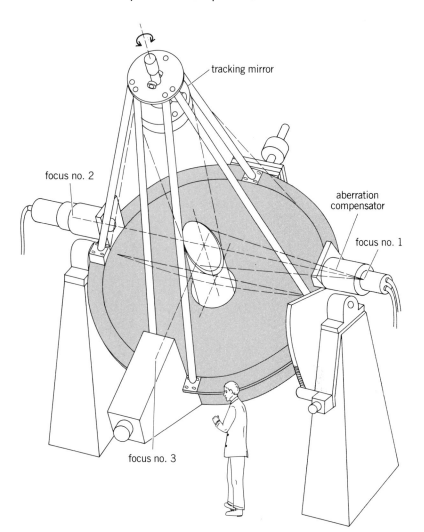

Fig. 3.   A 3.5-m extended-range transit telescope.

astronomers to be accommodated each night as long as a preprogrammed schedule is accepted, as it is with all the space telescopes today.

Feeding the light from a moving target to the desired instruments can be accomplished in two ways. The first is to use quartz fiber light pipes. These instruments can be completely off the telescope. The second is to track the object by moving the secondary mirror of a Cassegrain configuration. One concept for a 3.5-m transit telescope using the tracking secondary is shown in Fig. 3, illustrating the extreme simplicity of such a telescope.

Spherical bowl telescope. The ultimate in fixed-mirror telescopes is the spherical bowl of the Arecibo (Puerto Rico) radio telescope type or a meridian section of such a bowl. Both have much spherical aberration to be corrected, but relatively small two-mirror relay combinations appear to be able to provide a reasonable field of view. Multiple pickups could be provided along the meridian to observe several fields simultaneously in the limited transit mode. No large spherical primary telescopes of this type have been made so far, but cost-effectiveness arguments may render the meridian strip option attractive for future construction. The time-sharing mode could accommodate more astronomers in a given time than any other type of telescope.

For background information *see* ADAPTIVE OPTICS; OPTICAL TELESCOPE in the McGraw-Hill Encyclopedia of Science and Technology.

[ADEN MEINEL; MARJORIE MEINEL]

Bibliography: A. Hewitt, *Optical and Infrared Telescopes for the 1990s*, Kitt Peak Conference, Tucson, 1980; T. S. Mast and J. Nelson, Figure control for a fully segmented telescope mirror, *Appl. Opt.*, 14:2631–2641, 1982.

Organic reaction mechanism

Although single-electron transfer is not an unusual pathway for inorganic reactions, particularly those involving transition metal compounds, organic sub-

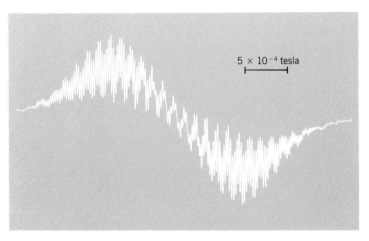

Fig. 1.   Electron paramagnetic resonance spectrum of the reaction of AlH_3 with dimesityl ketone (0.07 *M*) in tetrahydrofuran (THF); coupling constant $g = 2.0056$.

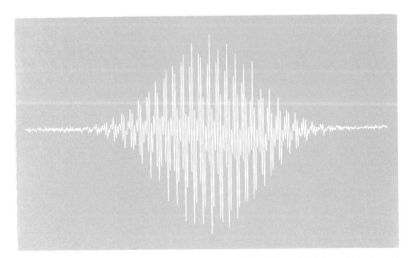

Fig. 2.   Electron paramagnetic resonance spectrum of the trityl radical intermediate formed in the reactions of trityl halides with metal hydrides in THF at room temperature. (*After E. C. Ashby et al., Spectroscopic evidence for the reduction of alkyl halides by metal hydrides via a single electron transfer mechanism, Tetrahedron Lett., 22(38):3729–3732, 1981*)

stitution reactions have been believed to proceed predominantly by polar pathways, as a result of either nucleophilic unimolecular substitution (S_N1) or bimolecular substitution (S_N2) reactions and by radical pathways as a result of homolytic reactions. Until recently, single-electron transfer was considered by organic chemists to represent a rather unusual reaction pathway followed by only a few rare reactions, and even then under some rather specific circumstances.

In the past few years, however, this understanding has changed dramatically because of the efforts of several physical organic chemists who have used modern mechanistic techniques to establish the credibility of single-electron-transfer reactions. It has been demonstrated on a firm mechanistic basis that certain nucleophiles, particularly carbanions, can act as single-electron donors toward substrates that are good electron acceptors. In reactions (1)–(3), the carbanion (I) is shown to react with a good electron acceptor, a nitro-substituted tertiary alkyl halide (II), to form the substitution product (V). The pathway is designated as the free-radical mechanism ($S_{RN}1$). It has been demonstrated that nucleophiles other than carbanions (alkoxides, thioalkoxides, and so on) will react with substrates other than *p*-nitrophenyl-substituted alkyl halides (for example, *p*-cyanophenyl and *p*-trifluoromethylphenyl will also react) to form the normal substitution products. Such studies have established a sound mechanistic basis for the $S_{RN}1$ mechanism.

Studies employing electron paramagnetic resonance (EPR) spectroscopy have established the existence of radical species in a large number of reactions, suggesting a larger scope to the concept of single-electron transfer.

There has been research involving substitution at aromatic carbon. Reactions (4)–(6) represent the reaction of bromobenzene (PhBr) with CH_2CN^- and the proposed mechanism of reaction. It can be observed that this mechanism is essentially the same as the $S_{RN}1$ mechanism, although such a mechanistic pathway was not expected for aromatic halide substitution. Some of the best-known and most fundamental organic reactions, previously thought to be polar in nature (S_N1 or S_N2), have now been shown to proceed by a single-electron-transfer pathway. For example, the reaction of a Grignard reagent ($R'MgX$) with a ketone [reaction (7)], which was thought to be polar in nature, has recently been shown to proceed by single-electron transfer [reaction (8)] when the ketone is aromatic.

Methods used to establish the radical nature of a single-electron-transfer pathway involve the use of probes, radical traps, kinetics, and direct observation of the radicals.

Probes. A radical probe involves a substrate or reagent that will cyclize or isomerize during the reaction, indicating that a radical is an intermediate [reaction (9)]. In this example the cyclic product appears because hexenyl radicals cyclize [pathway k_1, reaction (10*a*)] at $k = 10^{-5} s^{-1}$ and compete with

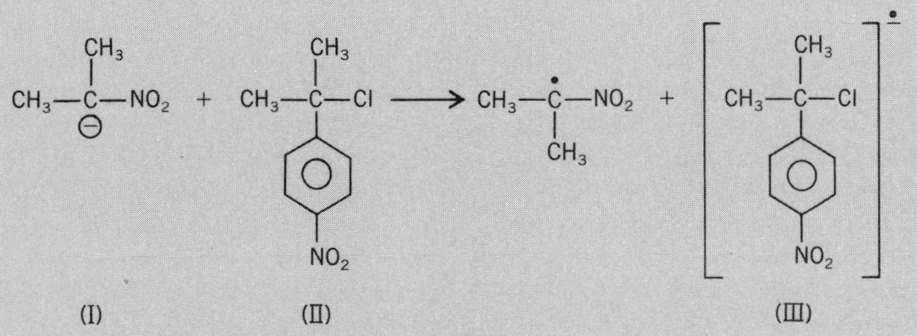

$$CH_3-\overset{CH_3}{\underset{\ominus}{C}}-NO_2 \;+\; CH_3-\overset{CH_3}{\underset{\underset{NO_2}{}}{C}}-Cl \;\longrightarrow\; CH_3-\overset{CH_3}{\underset{CH_3}{\overset{\bullet}{C}}}-NO_2 \;+\; \left[\overset{CH_3}{\underset{\underset{NO_2}{}}{C}}-Cl\right]^{\bullet \ -} \tag{1}$$

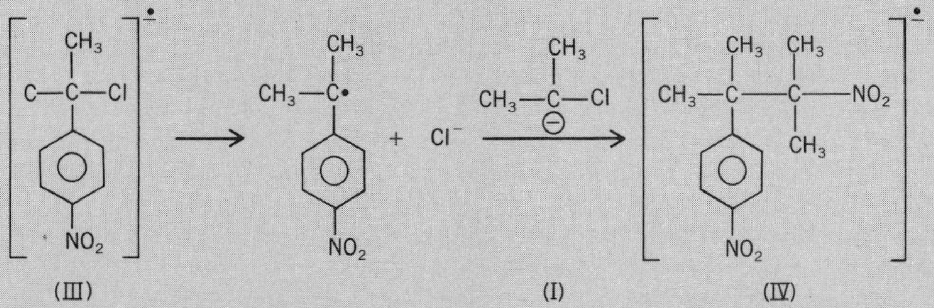

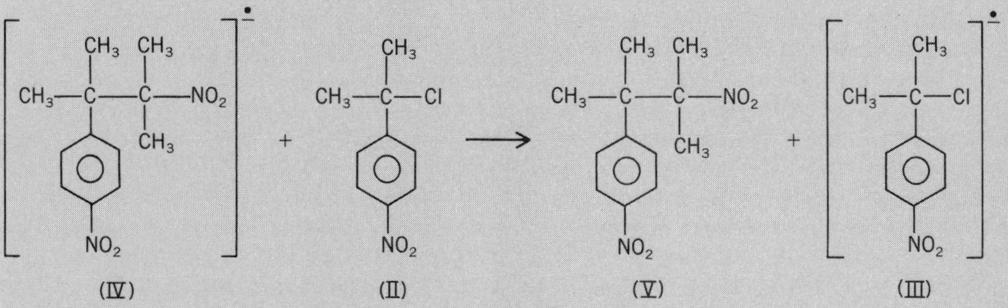

(I) (II) (III)

$$\left[\overset{CH_3}{\underset{\underset{NO_2}{}}{C}}-Cl\right]^{\bullet \ -} \longrightarrow CH_3-\overset{CH_3}{C}\bullet \;+\; Cl^- \quad CH_3-\overset{CH_3}{\underset{\ominus}{C}}-Cl \longrightarrow \left[\overset{CH_3}{\underset{CH_3}{C}}-\overset{CH_3}{\underset{CH_3}{C}}-NO_2\right]^{\bullet \ -} \tag{2}$$

(III) (I) (IV)

$$\left[CH_3-\overset{CH_3}{C}-\overset{CH_3}{\underset{CH_3}{C}}-NO_2\right]^{\bullet \ -} + CH_3-\overset{CH_3}{C}-Cl \longrightarrow CH_3-\overset{CH_3}{C}-\overset{CH_3}{\underset{CH_3}{C}}-NO_2 + \left[\overset{CH_3}{C}-Cl\right]^{\bullet \ -} \tag{3}$$

(IV) (II) (V) (III)

$$PhBr^{\bullet \ -} \longrightarrow Ph\bullet + Br^- \quad (4) \qquad Ph\bullet + CH_2CN^- \longrightarrow PhCH_2CN^{\bullet \ -} \quad (5)$$

$$PhCH_2CN^{\bullet \ -} + PhBr \longrightarrow PhCH_2CN + PhBr^{\bullet \ -} \quad (6)$$

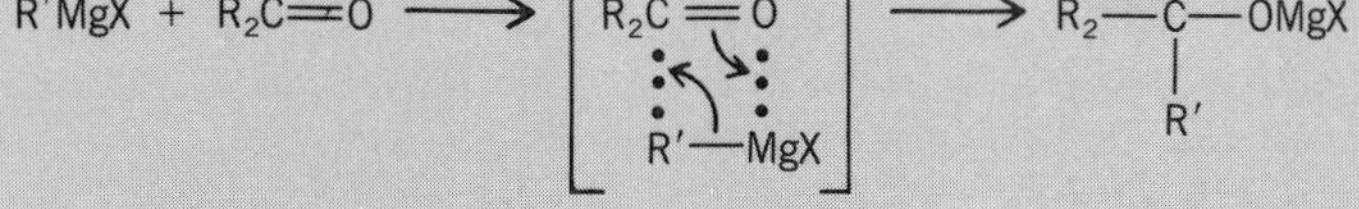

$$R'MgX + R_2C{=}O \longrightarrow \left[\begin{array}{c} R_2C{=}O \\ \\ R'{-}MgX \end{array}\right] \longrightarrow R_2-\overset{}{\underset{R'}{C}}-OMgX \tag{7}$$

$$R'MgX + Ar_2C{=}O \longrightarrow [R'MgX]^{\bullet +}\ [Ar_2C{=}O]^{\bullet -} \longrightarrow Ar_2-\overset{}{\underset{R'}{C}}-OMgX \tag{8}$$

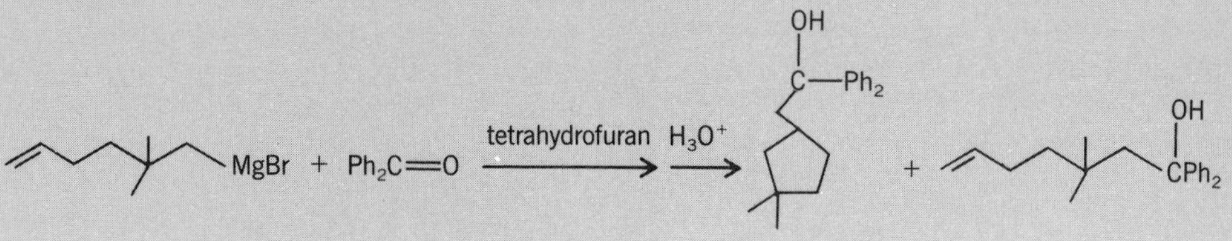

$$(10a)$$

$$(10b)$$

$$(11)$$

$$(12)$$

$$(13)$$

$$(14)$$

the subsequent coupling step [pathway k_2, reaction (10b)]. Another use of the same probe, is shown in reaction (11). Here the observation of a cyclic hydrocarbon shows that the reduction of the alkyl halide by lithium aluminum hydride ($LiAlH_4$) proceeded via a radical intermediate.

Radical traps. A radical trap is a compound that will accept an electron from a radical intermediate, or couple with it, and thereby slow down or stop the radical chain process. In reaction (12) there is a 65% yield in 75 s; however, if 10 mol % of *p*-dinitrobenzene is present, no reaction takes place during the same time period. *p*-Dinitrobenzene has a low reduction potential and therefore will accept an electron from a radical-anion intermediate very rapidly, thus short-circuiting the radical chain process. Other traps are diphenylhydrazines, dibutyl nitroxyl radical, galvinoxyl, and so on.

Kinetics. Kinetic studies have been used to establish the radical nature of reactions. For example, in the reaction of AlH_3 with dimesityl ketone, an electron paramagnetic resonance signal was observed (Fig. 1). The rate of disappearance of the paramagnetic intermediate giving this signal is equal to the rate of appearance of the product, a result consistent with the mechanism shown in reaction (13). This reaction is not believed to be a radical chain process, since its rate is not influenced by light or by a radical trap.

Direct observation of radicals. Electron paramagnetic resonance spectroscopy can be used to detect intermediates, such as radical anions or free radicals, in reactions involving electron transfer. This, of course, is true only when the radicals or radical anions have sufficient stability that their rate of subsequent reaction does not exceed their rate of formation.

When Ph_3CBr was allowed to react with $LiAlH_4$, the stable trityl radical was observed (Fig. 2), which slowly reacted to form Ph_3CH [reaction (14)]. As another example, in reaction (13), previously described, dimesityl ketyl is stable, and hence its rate of formation and disappearance could be observed by both visible and electron paramagnetic resonance spectroscopy.

For background information *see* FREE RADICAL; ORGANIC REACTION MECHANISM; REACTIVE INTERMEDIATES in the McGraw-Hill Encyclopedia of Science and Technology.

[EUGENE C. ASHBY]

Bibliography: E. C. Ashby, A. B. Goel, and R. N. DePriest, Evidence for an electron transfer mechanism in the reduction of ketones by main group metal hydrides, *J. Amer. Chem. Soc.*, 102:7779–7780, 1980; J. F. Bunnett, Aromatic substitution by the SRN1 mechanism, *Acct. Chem. Res.*, 2:413–420, 1978; N. Kornblum, Substitution reactions which proceed via radical anion intermediates, *Angew. Chem. Int. Ed.*, 14:734–745, 1975; G. A. Russell, E. G. Janzen, and E. T. Strom, Electron transfer process, I. The scope of the reaction between carbanions or nitranions and unsaturated electron acceptors, *J. Amer. Chem. Soc.*, 86:1807–1814, 1964.

Osteichthyes

In recent years, improved techniques for preparing fossil fishes for study, together with application of the principles of cladistic analysis have led to fundamental changes in the understanding of the interrelationships of Osteichthyes (the vast majority of living and fossil fishes), and to a reassessment of the derivation of tetrapods from Osteichthyes.

New preparation techniques. Historically, when fossil fishes were described, little or no effort was made to remove the matrix encasing a specimen. Limited morphological information could be obtained, and often important structures for the study of living osteichthyans were not visible. Consequently, systematically significant comparisons of living and fossil forms were precluded. In the 1950s, initially at the British Museum (Natural History) and later in many other museums and universities, technicians began, with much patience, to utilize dilute acid (mainly acetic acid) to dissolve the matrix away in order to reveal underlying bone. Exposed bone must be carefully covered with resins to strengthen it and to protect it from the acid. Sometimes a specimen is embedded partially in plastic to protect it, and also to keep the numerous small bones in place when the enclosing matrix is removed with acid. This technique has made it possible to see all the preserved parts, including elements of the splanchnocranium, and particularly the gill support system, which can provide useful data for tracing relationships. Well-prepared fossil specimens show as much osteological detail as do the skeletons of living fishes.

As a consequence of this technique, older collections have been opened to intensive reexamination, and the search for additional fossils, including those from new localities, is being vigorously pursued. Other techniques permitting expanded study of fossil fishes include x-rays, reconstructions in wax or other suitable materials of serially sectioned, three-dimensionally preserved fossils, paleohistological comparisons, and the removal of matrix with air-propelled abrasives of powdered aluminum oxide or dolomite.

Cladistic analysis. New preparation techniques have revealed an array of structures in fossils, permitting detailed comparison with living fishes. In addition, the rigorous application of cladistic analysis has become widely used for assessing which characters are primitive (plesiomorphic) and which are derived (apomorphic), and how such differences can be utilized to establish a cladogram showing sister-group relationships based upon shared derived characters. Much of the development and application of cladistic analysis was achieved through the study of living, and to some extent fossil, osteichthyans.

Until recently, the Osteichthyes were typically divided into two major groups, the Actinopterygii, or ray-finned fishes, and the Sarcopterygii, or lobe-finned fishes. The Dipnoi, or lungfishes, have been treated as a third group of somewhat uncertain relationship. Intensive reexamination of living and fossil fishes has yielded a complex classification, which, with some modifications, is shown in the following outline. A subdivisional hierarchy of the sarcopterygians is not included.

Class Osteichthyes
  Subclass Actinopterygii
    Infraclass Cladistia (bichirs)
    Infraclass Actinopteri
    Series Chondrostei (paleoniscoids, sturgeons)
    Series Neopterygii
    Division Ginglymodi (gars)
    Division Halecostomi
    Subdivision Halecomorphi (bowfins, some
      Mesozoic groups)
    Subdivision Teleostei (Pachycormidae, Aspi-
      dorhynchidae, and other families)
    Supercohort Osteoglossomorpha (bony tongues)
    Supercohort Elopocephala
    Cohort Elopomorpha (tarpons, bonefish, eels)
    Cohort Clupeocephala
    Subcohort Clupeomorpha (herrings, etc.)
    Subcohort Euteleostei
    Division Ostariophysii (minnows, suckers, cat-
      fishes)
    Division Neognathi
    Subdivision Protacanthopterygii (salmon, pike,
      etc.)
    Subdivision Neoteleostei
    Sect. Stenopterygii (stomiatoids)
    Sect. Eurypterygii
  Subsect. Cyclosquamata (lancet fishes, many ex-
    tinct genera)
  Subsect. Ctenosquamata
  Sept Scopelomorpha (myctophids)
  Sept Acanthmorpha (ctenothrissids)
  Superorder Paracanthopterygii (polymixiids, cods
    etc.)
  Superorder Acanthopterygii
  Series Atherinomorpha (flying fishes, cyprino-
    donts)
  Series Percomorpha (beryciforms, perciforms etc.)
  Subclass Sarcopterygii
    Rhipidistia (osteolepiforms, porolepiforms)
    Actinistia (coelacanths)
    Dipnoi (lungfishes)

Changes in the classification. Several changes in this classification are noteworthy. The division of the Actinopterygii into three parts, the Chondrostei, Holostei, and Teleostei, is eliminated; these chronologically sequential groups had always been regarded as convenient assemblages for the placement of fishes in the absence of strong reasons to sort them according to evolutionary lineages. The Chondrostei produced a morphologically, and no doubt ecologically, diversified radiation, but relatively few modern studies of the group have been published. While the chondrosteans share some basic skeletal characteristics with the early sarcopterygians, there is no evidence to suggest that one group was derived from the other, nor is it possible to identify a group from which they both may have been derived.

More significant changes, resulting from cladistic analysis of fishes usually placed in the Holostei and Teleostei, include elimination of the Holostei; combination of Teleostei and many Holostei into a new group (although the designation was first used more than 50 years ago), the Neopterygii; and expansion of the teleosts by the transfer of several holosteans in order to establish monophyletic status of the teleosts based on the structure of the premaxillary and the presence of uroneurals. Many extinct families and genera are placed in taxa to which their relationship is uncertain, or are given other indeterminate status pending further study. Therefore, taxonomic positions are established only for groups and genera which have recently been studied in detail.

Among the teleosts, changes are more dramatic. More categories above the familial level are necessary to recognize phylogenetic relationships of the major groups. The principal characters on which this classification is based involve morphology of the bones supporting the tail and the development of specific components of the jaws, gill, and pharyngeal support apparatus.

Some studies of Actinistia (coelacanths) have led to proposals that coelacanths and chondrichthyans are sister groups derived from undiscovered fossil forms. This radical conclusion is based largely on soft anatomy and physiology, including pituitary gland structure and function, presence of a rectal gland, osmotic adaptations with urea and trimethylamine oxide retention, and basic cation excretion. But, if the similarities of soft tissue structure and function between the chondrichthyans and coelacanths are shared derived characters, then all of the similarities which coelacanths and osteichthyans, especially sarcopterygians, share would have had to be independently derived. It is unlikely that such complex features as the intercranial joint, tropibasic neurocranium, osteichthyan hyomandibular and symplectic, operculum and scales, as well as the swimbladder, separate nasal openings, and so forth, would have evolved twice. While coelacanths are closest to sarcopterygians in all skeletal features, they are not ancestral to the tetrapods.

The study of the sarcopterygians remains an area of intense interest because this group includes the ancestors of the tetrapods. The generally accepted view is that tetrapods are derived from osteolepiform rhipidistians similar to *Eusthenopteron* or its close relatives. However, Swedish paleontologists have suggested that some of the amphibians (particularly salamanders) may have been derived from dipnoans or porolepiform rhipidistians, while other tetrapods were derived from osteolepiforms.

Recent research. In 1981 D. E. Rosen, and coworkers proposed that lungfishes are the sister group of the tetrapods, the coelacanths the sister group of dipnoans and tetrapods combined, and *Eusthenopteron* a primitive sister group to the sarcopterygians. Porolepiforms may be somewhat closer to the sarcopterygians, but these authors believe that important characters of both porolepiforms and *Eusthenopteron*, which would permit comparison with coelacanths and lungfish, are not sufficiently well known. The several characters of *Eusthenopteron* which are commonly used to link them with the tetrapods are considered primitive for all osteichthyans and all gnathostomes, or convergent with gnathostomes.

The conclusion that lungfishes and tetrapods possess critical shared derived characters is based on the contention that recent lungfish possess choanae (internal narial openings), two primary joints in each paired appendage, a hyomandibular not involved in jaw suspension, and some 17 other characters. This position revives a view of tetrapod descent which has not been seriously raised in over 100 years, and may be criticized at several levels, including the use of some characters determinable only in living fishes and others not yet known in particular fossils. It has been demonstrated, however, that many features of *Eusthenopteron* used to suggest the ancestry of the tetrapods, particularly the limb, are not distinct specializations of *Eusthenopteron*. Additional work will be needed to assess each of their characters. It will become equally important to develop a better characterization of tetrapods to permit more extensive comparison with fishes.

For background information *see* OSTEICHTHYES; TETRAPODA in the McGraw-Hill Encyclopedia of Science and Technology.

[DAVID BARDACK]

Bibliography: S. Lovtrup, *The Phylogeny of Vertebrata*, 1977; J. E. McCosker and M. D. Lagios, The biology and physiology of the living coelacanth, *Occ. Pap. Calif. Acad. Sci.*, no. 134, 1979; C. Patterson and D. E. Rosen, Review of ichthyodectiform and other Mesozoic teleost fishes and the theory and practice of classifying fossils, *Bull. Amer. Mus. Nat. Hist.*, 158:81–172, 1977; D. E. Rosen, Interrelationships of higher euteleostean fishes, in H. Greenwood, R. Miles, and C. Patterson (eds.), *Interrelationships of Fishes*, pp. 397–513, 1973; D. E. Rosen et al., Lungfishes, tetrapods, paleontology and plesiomorphy, *Bull. Amer. Mus. Nat. Hist.*, 167:159–276, 1981.

Paleobotany

Paleobotany has matured during the last two decades into a science employing modern instrumental and conceptual techniques. A biologically oriented approach is currently in use, and fossils are seen as the remains of active biological systems. Some of the problems of paleobotany are studied by sophisticated chemical techniques used in the field called paleobiochemistry. A few, but by no means all, major areas of paleobotany and paleobiochemistry are considered below.

Biological approach. In this approach, paleobotanists are now becoming interested in processes as well as in structure. In addition, better means of examining fossil plant remains have come into use, such as transmission and scanning electron microscopy.

Precambrian fossils. One of the areas that has had the greatest impact on paleobotany during the past 20 years is the study of Precambrian fossils. Much progress has been made since the days when it was assumed that there were no living creatures on the Earth prior to the Cambrian Period. Many rich microfossil floras in Precambrian rocks are now known. The oldest visible remains are microorganisms reported in rocks 3.4–3.5 billion years old. Most frequently, they appear as unicells or filaments. Precambrian unicells and filaments about 0.9 billion years old were reported from Australia, and within some of the spherical unicellular forms a single dark spot was found. These spots are considered by some to represent nuclei, and in that case, these fossils would be the oldest eukaryotes. Other unicells have marks resembling the triradiate ridges of spores of land plants, and because such spores are typically meiotically derived, it is suggested that these Precambrian organisms had a sexual cycle. However, these nucleuslike spots and triradiate ridges should not be interpreted too literally because similar features were observed in blue-green algal colonies in "stale" cultures.

A considerable span of time elapsed between the first occurrence of photosynthetic organisms and the colonization of land by more highly specialized plant forms. Undoubtedly some of the algal forms lived on land surfaces, but the first record of land vascular plants (plants with a specialized conducting system) was that of the simple land plant *Cooksonia* from the Late Silurian Period. There are reports of possible earlier remains of land vascular plants, but none is accepted without some reservation. Land plants could have evolved earlier, but there is no indisputable evidence for this.

Devonian fossils. An area currently receiving considerable attention is the evolution of land vascular plants during the Devonian Period. Initially, all simple Devonian vascular plants had been placed in the order Psilophytales. This group became unwieldy and unnatural until two principal lines of evolution were recognized: one represented by the Rhyniophyta—naked, dichotomously branched plants with sporangia borne terminally on the axes; and the Zosterophyllophyta—with dichotomously branched axes and sporangia borne along the sides of the axes. The latter group led quite imperceptibly into the Lycophyta, with the Rhyniophyta most likely serving as the source of all other vascular plants through the transitional group, Trimerophytophyta. This group somewhat resembles the Rhyniophyta, but the plants are larger and unequally branched, with a distinction between principal axes and smaller laterals. The Trimerophytophyta also had terminal sporangia.

Late in the Devonian, a considerable diversity of plant groups with large treelike forms developed. One of the most important discoveries during relatively recent years is the Late Devonian *Archaeopteris*, thought for many years to have been a large fern with flattened fronds which bore sporangia. In rocks of similar age, remains of large trunks that have an internal structure like that of conifers (called *Callixylon*) were found. In 1960 it was demonstrated that the coniferlike trunks were actually axes of *Archaeopteris*, and that what appeared to be fronds were actually flattened branch systems. An understanding of *Archaeopteris* led to the realization that there existed a group of plants called progymnosperms that combined characters of gymnospermous seed plants with the reproductive system of simpler, fernlike plants. It is likely that the seed plants originated from the progymnosperms.

Seed plants. The origin of the seed habit represented an evolutionary milestone in the plant kingdom. Seeds apparently appeared quite early during the course of the development of vascular plants. The oldest seed, Devonian in age, was reported in 1977 and had a flattened shape. In 1968 a primitive type of seedlike body from the Upper Devonian was described. It had an integument that incompletely enclosed the megasporangium, with tentaclelike appendages forming a micropyle. While these seeds were in the form of carbonized compressions, similarly constructed seeds in a petrified state, showing considerably more detail, were described in 1981.

The last two decades witnessed an emphasis on ontogenetic studies of ancient plants. In 1961 it was determined that arborescent Carboniferous lycophytes (*Lepidodendron* and related genera) had a determinate growth pattern, with more distal branches becoming progressively smaller, with a smaller conducting strand, smaller leaves, and fewer rows of leaves. Arborescent arthrophytes (*Calamites* and related genera) during the Carboniferous Period also had a determinate growth pattern, with lateral branches bearing smaller leaves, and fewer rows of leaves, than the parent axis. The stele, also, was smaller in branches. According to some researchers, the branch system of *Archaeopteris* is also determinate.

Noteworthy advances have been made in studies of the pteridosperms, with some biologically interesting observations. The occurrence of a pollen tube from a grain in the pollen chamber of the Carboniferous seed-fern ovule *Callospermarion* was demonstrated in 1972. It was the first evidence of siphonogamy in the fossil record, and an indication that this method of transmittal of the sperm cells is an ancient one. In 1977 pollen-drop remains in *Callospermarion* were described, indicating a pollination mechanism similar to that of many extant gymnosperms.

Progress has been made with the enigmatic genus *Glossopteris*, probably the most important component of Permian Gondwana floras. It is now agreed by virtually all paleontologists that *Vertebraria* represents axes with a rootlike internal structure and that they were borne at the bases of trees with *Glossopteris* foliage. For many years, the seed-bearing structures of *Glossopteris* were incompletely understood because they occurred only as impression fossils. It now appears, on the basis of study of petrified material, that the most common type of reproductive structure is a modified leaf with ovules borne on the ventral surface. The leaf is rolled around them, forming an incompletely closed carpellike structure. Many of the previously known impression forms may now be interpreted as preservational states of this basic megasporophyllous structure.

Cycadophytes (cycadeoids and cycads) have been studied with new material. *Leptocycas*, a Late Triassic cycad, shows that the habit of early members of the Cycadales consisted of a slender stem which lacked a dense, compact armor and bore distantly spaced leaves. Early cycadeoids also apparently had a similar habit; the more familiar and later *Cycadeoidea* most likely represented a derived and atypical form.

The history of conifers has been extended back to the Late Carboniferous with the discovery of shoots referable to the genus *Lebachia*. Conifers with features almost identical to those of extant genera were abundant in the Mesozoic Era, with the pinelike *Compsostrobus* occurring in the Late Triassic. The earliest reported seed plant embryo was in a conifer seed from the Permian. Excellent araucarian remains, including cones and seedlings, have been described from the Jurassic of Patagonia.

Flowering plants. Probably some of the greatest progress in paleobotany has been made in the study of flowering plant origin and evolution. There are still conflicting ideas concerning the time of the appearance of angiosperms, with one school postulating a pre-Cretaceous (even as early as late Paleozoic) origin, while others evaluate pre-Cretaceous angiospermlike fossils and indicate that the first definite angiosperms appeared in Cretaceous rocks. During the Cretaceous, however, radiation was extremely rapid.

Older works were typically floristic surveys of angiosperm fossils, with identifications based on superficial leaf shape and venation. It has become evident, however, that most of the earlier identifications were in error, and with the introduction of more painstaking techniques and approaches, more accurate and biologically significant work has emerged. Useful guides for the study of fine detail of angiosperm leaves have been compiled, and the progressive elaboration through time of vein patterns in angiosperms is maintained. While leaves are the most frequently encountered angiosperm fossils, there are obvious shortcomings in relying only on leaves to understand angiosperm evolution. In recent years, research has been devoted to flowers and fruits that are necessary for understanding the course of angiosperm specialization. For example, Cretaceous reproductive axes and Eocene flowers are now carefully studied.

The field of reproductive biology of plants of past ages is now receiving considerable attention. In spite of the inherent difficulties of incomplete fossils and the sporadic occurrence of suitable material, some interesting developments have been recorded. Elegantly preserved Carboniferous pteridosperm pollen grains show stages in the development of the male gametophyte. Developmental studies of the Paleozoic pollen *Monoletes* relate it to adaptations in the reproductive process. Well-preserved reproductive structures of lycophytes present insight into reproduction in some of the arborescent members. The female gametophyte in the megaspore of *Lepidostrobus* was identified, and the coevolution of various pollinating insect vectors and different flower types was demonstrated.

[THEODORE DELEVORYAS]

Paleobiochemistry. Traditional paleobotanical studies have emphasized the use of anatomy and morphology in uncovering the taxonomic and evolutionary relationships among fossil plants. Frequently, however, both anatomy and morphology are so poorly preserved in the fossil record as to preclude the definite assignment of a particular fossil to a specific taxonomic group. The use of chemical profiles has been shown to provide information which in many cases is useful even for extremely fragmented fossil remains. By analogy with biochemical studies, which are used to understand the taxonomy and phylogeny of extant organisms, the study of fossil chemical profiles is called paleobiochemistry.

Many geophysical and geochemical factors produce drastic changes in the organic profiles of fossils. Pressure and heat ultimately convert plant tissue into coal. Chemical processes, such as decarboxylation, polymerization, and fractionation, lead to the loss of taxonomically distinctive compounds during the coalification process. Under such circumstances it is often impossible to determine if the fossils formed by a coalification process were even vascular plants. Although many chemical and physical processes can alter the composition of a plant during and after fossilization, the longevity of molecular species can be exceptional—not infrequently ranging to hundreds of millions of years. Serum albumin has been immunologically detected in frozen mammoth tissue 40,000 years old, repeating amino acid sequences have been reported in mollusk shells over 70 million years in age, and mono- and disaccharides have been isolated from plant tissues over 390 million years old. Provided that fossilization does not result in severe diagenesis, the paleobiochemist can often determine taxonomic relationships at the suprageneric level and sometimes at the species level.

Three areas in which paleobiochemisty had been useful in determining critical features of major evolutionary events are: (1) the transition onto land of aquatic, nonvascular plants sometime during the Silurian, (2) the diversification of the early land plants (Devonian), and (3) the initial radiation and subsequent evolution of the flowering plants (Cretaceous-Tertiary). Each of these phases in plant evolution can be placed within the context of the paleobiochemical data.

Aquatic-to-terrestrial transition. Since lignin formation is characteristic of all vascular plants, the presence of lignin may be used to chemically distinguish these plants from other, nonvascular plants, and indeed from all other organisms. The origin and occurrence of lignin biosynthesis therefore is of critical importance to the paleobiochemist, since it marks the advent of the evolution of tracheophyte from nontracheophyte ancestors (such as the algae and bryophytes). While vascular land plants have been demonstrated from the Late Silurian on the basis of existing tracheids, various cell types showing parallel features with that of tracheids have been re-

ported from Early Silurian rock formations. The structural features of these tracheidlike cells (banded tubes) suggest morphological parallels between the early land plants and subsequent vascular plants. The precise evolutionary implications of banded-tube cell types, however, remain obscure. Chemical examination of Early Silurian plant remains containing banded tubes indicates that upon oxidative degradation these remains liberate *p*-hydroxybenzaldehyde and vanillin. The isolation of these two chemical constituents, which are also liberated upon the degradation of lignin from true vascular plants, suggests that banded tubes are biochemically similar to, but not identical with, the vascular tissue of tracheophytes.

The biosynthetic capacity to form ligninlike substances from prevascular plants suggests that the biochemical evolutionary stages required for the formation of lignin occurred well before the appearance of the first true tracheid. Chemical data indicate that plants with transitional characteristics between nonvascular and vascular land plants existed well before the first definite vascular plants and provide some insights into the occurrence of the biochemical, functional, and morphological features that led to the tracheids. The chemical data, however, do not resolve the question of whether or not plants possessing banded-tube cell types were direct precursors of the tracheophytes or represent a collateral lineage of land plants that eventually became extinct in the Devonian. Similarly, it is not possible on the basis of the chemical data to determine if the banded-tube cell types were parts of plants that were antecedents to the mosses and the liverworts (nonvascular embryophytes). On the basis of the chemical characterization of living mosses and liverworts, none of the bryophytes to date has revealed the presence of lignin within their conducting cell types or in nonspecialized cell types in other portions of the vegetative thallus. Biochemical analyses have revealed that the conducting cell types, or hydroids, of mosses contain biochemical components (polyphenols) that are very similar to lignin. If the Silurian banded-tube cell types that contain a ligninlike compound are the remains of forms ancestral to the bryophytes, then during their subsequent evolution there must have been a reduction in, or loss of, the synthetic systems regulating lignin production. There is not a single example of secondary thickenings in the form of rings, helices, or reticulae in the hydroids of a moss. Therefore, the banded-tube cell types of the Early Silurian plants are structurally quite distinct from the hydroids of living mosses, but are biochemically and structurally similar to tracheids of living vascular plants. Continued paleobiochemical analyses of Silurian and Early Devonian plant fossils eventually may be able to resolve the evolutionary patterns and diversification of the early nonvascular and vascular land plants.

Early vascular plants. Recent advances in paleobotany have included the formulation of a morphological system of classification of the early vascular land plants. Currently, three large groups are recognized as including the bulk of the early vascular land plant fossils: (1) the rhyniophytes which are simple, often small plants whose branching systems bore sporangia at the ends of axes; (2) the zosterophyllophytes which bore globose kidney bean–shaped sporangia; and (3) the trimerophytes which were much larger plants and possessed complex branching patterns, consisting of a main axis that bore clusters of spindle-shaped sporangia at the tips of lateral axes. With the discovery of genera having apparently intermediate morphologies between two of any of these three groups (such as *Renalia*), the classification of the suprageneric groups of Devonian plants is recognized to be in a state of flux which is unlikely to end until substantially more fossil plants are discovered or until the criteria for group classification are more clearly established.

Paleobiochemical profiles have been shown to be particularly useful in resolving the suprageneric taxonomic relationships among Devonian fossil plants. This approach is analogous to contemporary phytochemistry and is particularly valuable in cases where structural features of fossils are either poorly preserved, absent, or intermediate between previously recognized suprageneric taxa.

The resolution of suprageneric groups of Early and Late Devonian land plants on the basis of biochemical profiles often involves a complex statistical analysis known as principal component analysis. This method is necessary since suprageneric groups are often distinguishable on the basis of correlated biochemical features, rather than on the presence or absence of individual chemical profiles. Biochemical studies of these early vascular plants indicate that each of the suprageneric groups classified by morphology and anatomy can also be recognized on the basis of chemical composition. The rhyniophytes are characterized by the presence of relatively simple low-molecular-weight hydrocarbons and a lignin chemistry that differs from that found in contemporary plants. The apparent absence of true lignin in the rhyniophytes may be taken as evidence either for the extreme diagenesis of this macromolecule during fossilization or for the lack of canalization of the lignin biosynthetic system until Middle Devonian times. Subsequent, younger suprageneric groups of early vascular land plants show the presence of lignin as well as consortia of additional secondary metabolites which distinguish the trimerophytes from the zosterophyllophytes, and in turn distinguish both of these two groups from the rhyniophytes. In addition, biochemical analyses have shown that the lycopods which gave rise to arborescent forms during the Carboniferous and to the present club mosses are chemically distinguishable from all of the previously mentioned suprageneric groups. By Upper Devonian times, all of the early vascular land plant groups show canalization of biochemical profiles which remain distinct for the tenure of those lineages through successive geologic periods. Of particular interest is the dichotomy in biochemical profiles that occurred among those plants which on the one hand evolved

microphylls (lycopods) and those plants that evolved megaphylls (ferns, progymnosperms). This dichotomy in biochemical composition is evident by Early Devonian times and through all subsequent phases of the evolution of these two groups, suggesting that these organisms are both biochemically and morphologically distinct from one another. The biochemical data, however, are not capable of resolving whether both of these two large groups of organisms were convergent at the early stages of their evolution, or whether they may in fact represent polyphyletic evolution of the vascular land plant organization.

Angiosperm paleobiochemistry. Perhaps the most remarkable studies of chemical preservation in plants are those dealing from flowering plant remains of Cretaceous and Tertiary age. Studies of fossil angiosperms are valuable because living counterparts can be identified that provide a biochemical framework for evaluating paleobiochemistry. The extent to which paleobiochemical profiles are preserved in fossil plants is reflected by the fact that lower Miocene angiosperm leaves from the Succor Creek Flora of Oregon contain pheophorbides (chlorophyll derivatives) and carotenoids (plant pigments which frequently give fruits and flowers their yellow-orange coloration). In addition to these two compounds, flavonoids were also isolated from these fossil leaves. The flavonoids are pigments that give flowers some of their characteristic yellow, orange, and purple coloration and are extremely sensitive to heat and pH. To date, seven angiosperm leaf taxa have been chemically characterized and directly related to morphologically similar modern taxa. On the basis of these data, the biochemical profiles of morphologically similar fossil and extant angiosperms that are thought to be phyletically related may be compared and direct chemotaxonomic relationships inferred. In turn, biogeographic relationships of early angiosperm groups may be tested. For example, the fossil maple species *Acer oregonianum* is morphologically similar to the extant species *A. macrophyllum* found in western America. The fossil *Acer* possesses a flavonoid profile consisting primarily of flavonols and compounds that spectrographically appear to be tannins. However, the tannins found in the fossil *Acer* occur consistently in a number of extant Asian species rather than in western American species of *Acer*. Biochemically, the fossil is most similar to the extant species found in eastern Asia, therefore suggesting a paleobiogeographic relationship during the evolution of these two species.

Biochemical data, in addition to morphological studies of leaves, seeds, and fruits, may provide a more complete determination of rates of evolutionary change. The genetic basis for the biosynthetic pathways responsible for various secondary plant metabolites, such as flavonoids, is now well established, as are the mutations necessary to effect changes in these pathways with the concomitant production of novel or slightly modified compounds. Examining living and fossil biochemical profiles within the same genus could give an indication of the extent of biosynthetic alteration accompanying the observable morphological changes documented in the fossil record and in the comparison of fossils to living taxonomic referable species. Similarly, morphological stasis may now be assessed in terms of biosynthetic, genetic, and morphological changes.

For background information *see* PALEOBIOCHEMISTRY; PALEOBOTANY; PALEONTOLOGY in the McGraw-Hill Encyclopedia of Science and Technology.

[KARL J. NIKLAS]

Bibliography: H. P. Banks, The early history of land plants, in E. T. Drake (ed.), *Evolution and Environment*, pp. 73–107, 1968; H. P. Banks, Occurrence of *Cooksonia*, the oldest vascular land plant macrofossil, in the Upper Silurian of New York State, *J. Indian Bot. Soc.*, Golden Jubilee Vol., 50A:227–235, 1973; W. H. Gillespie, G. W. Rothwell, and S. E. Scheckler, The earliest seeds, *Nature*, 293:462–464, 1981; K. J. Niklas, Paleophytochemistry: Implications concerning plant evolution, *Paleobiology*, 7:1–3, 1981; A. C. Scott, The earliest conifer, *Nature*, 251:707–708, 1974; A. C. Sigleo, Organic geochemistry of silicified wood, Petrified Forest, National Park, Arizona, *Geochim. Cosmochim. Acta*, 42:1397–1405, 1978.

Paleoclimatology

Knowledge of the Earth's paleoclimatic history, particularly prior to 1.8 million years ago, has been derived almost exclusively from the geologic record of sediments and organisms. In contrast, the study of modern climates is based both on observations and on models which are mathematical representations of the physical laws which govern atmospheric and oceanic circulation. These physical laws are usually simplified or approximated and then applied to a hypothetical climate system in which some boundary condition (for example, solar input, atmospheric composition, or geography) has been modified.

With the coevolution of plate tectonic reconstruction in the earth sciences, atmospheric and oceanic modeling in meteorology and physical oceanography, and high-speed computers, there is tremendous potential for a computer-modeling approach to paleoclimatology. In this case, the geologic record can be used to test a set of model predictions. There are a number of fundamental problems with this approach which can be collectively described as the paleoclimate problem. The nature of this problem requires a research approach based on so-called sensitivity experiments in order to identify the factors which show potential for explaining the Earth's paleoclimates and to consider the model response to specific factors which may influence climate, for example, changes in solar output. The success of this research approach is best illustrated by recent model simulations which focus on the problem of how past continental configurations may have modified the Earth's paleoclimates.

Paleoclimate problem. There are three aspects to the paleoclimate problem. First, a number of potentially important factors may have jointly or independently influenced past climates. These include solar output, the Earth's orbit, galactic clouds, volcanism, carbon dioxide, orography, sea level, and continental configurations. Unfortunately, for most of geologic time these factors are poorly specified, and so the causes of climatic changes that have occurred throughout earth history cannot yet be identified with any certainty. Second, the primary task in climate modeling is to replace the complex natural system with a simplified version which can be used to investigate climatic change. Because the natural system is so complex, it is unlikely that any model can include all the feedbacks and interactions necessary to predict climatic change accurately. In addition, these models have been verified only in comparison with the present climate. Third, the paleoclimatic record is incomplete and difficult to interpret unambiguously. Paleoclimates were not recorded directly, but must be interpreted from proxy indicators which may not possess a clear quantitative relationship to climate. The interpretation is also dependent on the accuracy of paleogeographic reconstructions and stratigraphic resolution, which may be limited to a few million years. Areal synchroneity of climatic events cannot be demonstrated within this time frame. Consequently, for much of Earth's history the climate can only be reconstructed within broad limits.

Because of these fundamental problems it is not yet possible to reproduce the climate of a past period with any confidence. There are too many limitations in models, in the data, and in the knowl-

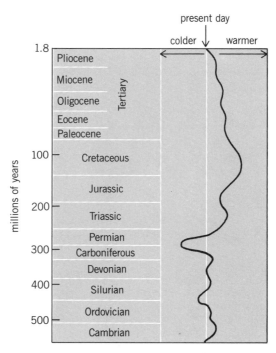

Fig. 1.   Schematic representation of climatic change over the last 570 million years. (*After L. A. Frakes, Climates Throughout Geologic Time, Elsevier, 1979*)

edge of the causes of past climates. Nevertheless, considerable insight may be gained through a modeling approach to paleoclimatology.

Modeling approach. Ideally the primary goal is to identify the specific forcing factors in a climate model which incorporates all the characteristics necessary to predict climatic change, and then to verify the simulation by comparison with the geologic record. Because of the paleoclimate problem, however, the actual approach is to perform sensitivity studies using climate models. In a sensitivity study the model is perturbed by making a change in a boundary condition or parameterization of a physical process, and the result is compared with an unperturbed or control simulation. By modifying one factor at a time, insight is gained into the model and into the mechanisms of climatic change which show the potential to explain the paleoclimatic record.

There are a number of ways to maximize the utility of this approach in the initial examination of a problem. There are a variety of climate models, each with different assumptions and limitations. A hierarchy of models from simple to complex should be applied in sensitivity studies. The largest reasonable changes in factors which may influence climate should be used in simulations. Large changes in forcing will produce the largest model response and will facilitate the interpretation of cause and effect. It will be easier to determine if the model response is similar to the geologic record if large contrasts from the present climate are examined. This approach has recently been applied to the study of the relationship between geography and climate.

Geography and climate. Since the development of the concepts of continental drift and plate tectonics, paleoclimatologists have considered paleogeog-

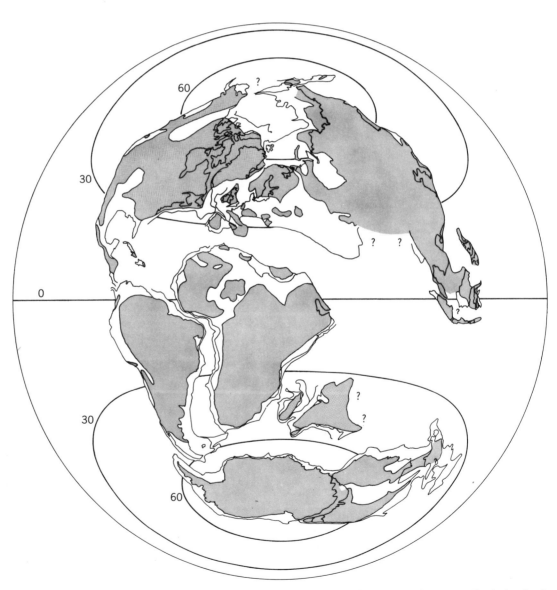

Fig. 2.   Cretaceous geography 100 million years ago, showing area of land above sea level (shaded areas) and 30° paleolatitude lines. Question marks denote areas of uncertain geography. (*Adapted from S. Thompson and E. Barron,* *Comparison of Cretaceous and present earth albedos: Implications for the causes of paleoclimates, J. Geol., 89 (2): 143–167, 1981*)

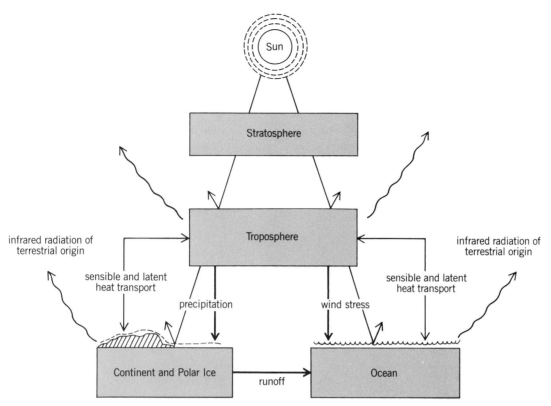

Fig. 3. Schematic of the processes included in highly resolved atmospheric general circulation models. (*After W. M.
Washington and D. L. Williamson, A description of the NCAR
global circulation models, Meth. Computat. Phys. 17:111–
171, 1977*)

raphy to be an important climatic forcing factor. In
particular, continental positions have been cited
frequently as an explanation of the contrast between
glacial and warm, equable ice-free climates (Fig.
1). The large changes in geography (continental positions, sea level, and orography) over the last 200
million years are a logical climatic forcing factor for
sensitivity studies. In particular, Cretaceous Period
geography (65–140 million years ago) contrasts
sharply with that of the present day (Fig. 2). The
Cretaceous also has the largest well-documented climatic contrast with the present climate. Cretaceous
temperatures can be defined within two broad limits:
a maximum case with mean polar temperatures near
15°C and an equator-to-pole temperature contrast of
only 17°C, and a minimum case with mean polar
temperatures near 0°C and an equator-to-pole temperature contrast of 26°C. The globally averaged
surface temperature of these two limits are 14 and
6°C higher, respectively, than at present.

A hierarchy of climate models has been used to
investigate the role of geography in explaining warm
paleoclimates. In a pioneering study, W. Donn and
D. Shaw used a thermodynamic atmospheric model
to examine the climate evolution over the last 200
million years. The calculated increase in surface
temperature 100 million years ago was approximately 2°C compared to a control. These simulations utilized an early version of continental reconstructions which assumed that the total land area of

the Earth was unchanging during the period under
study. A recently performed series of climate sensitivity studies used a planetary albedo model and an
energy balance climate model. These simulations
included experiments which considered the importance of geographic factors (continental positions,
sea level, vegetation) and climate feedbacks (clouds,
heat transport) in an attempt to explain Cretaceous
climates. One conclusion suggests that a 2 to 8°C
increase in globally averaged surface temperature
results from the geography and reasonable climate
feedbacks. However, most of this warming resulted
from feedbacks, not directly from the change in geography. Sensitivity experiments using the most
comprehensive model to date (the community climate model—a three-dimensional atmospheric general circulation model at the National Center for Atmospheric Research) with only a change in geography resulted in a 6°C planetary warming compared
to the present day. A major achievement of the model
simulations is that they have demonstrated quantitatively that geography is an important climatic variable. The estimates of the climatic importance of
Cretaceous geography are at the lower limit of the
warming as interpreted from the geologic record.
Therefore, there remain some significant problems:
Are the warm Cretaceous polar temperatures an incorrect interpretation of the record? Are other climatic forcing factors in addition to geography (such
as a change in atmospheric composition) required?

Or are the models inadequate, lacking some important feedback mechanism?

Other problems. There are also a number of important scientific issues in addition to the nature of the planetary warming or cooling. For example, previously paleoclimatologists thought that the equator-to-pole surface-temperature gradient controlled the nature of the atmospheric and oceanic circulation. During times of reduced gradient, such as the Cretaceous Period, the circulation would have been very weak and circulation features such as the easterlies and westerlies would have been displaced poleward. Experiments using a comprehensive general circulation model of the atmosphere (Fig. 3) question these long-held assumptions. These experiments with Cretaceous geography and two different equator-to-pole temperature characteristics revealed that geography modified the positions of major circulation features, but the experiment using a much reduced equator-to-pole surface-temperature gradient was not substantially different from the experiment with a larger surface-temperature gradient. In addition, the intensity of the circulation did not decrease with the decrease in the equator-to-pole surface-temperature gradient. The vertically averaged equator-to-pole temperature gradient, not the surface-temperature gradient, was the important factor.

These results illustrate the potential of a quantitative model approach to paleoclimatology. Considerable research on models, in data collection and interpretation, and in delineating the important climatic influences over geologic time remains to be completed before the climate of any geologic period can be confidently reproduced.

For background information *see* CLIMATIC CHANGE; CRETACEOUS; PALEOCLIMATOLOGY in the McGraw-Hill Encyclopedia of Science and Technology.

[ERIC J. BARRON]

Bibliography: E. J. Barron, S. L. Thompson, and S. H. Schneider, An ice-free Cretaceous? Results from climate model simulations, *Science*, 212:501–508, 1981; E. J. Barron and W. M. Washington, Cretaceous climate: A comparison of atmospheric simulations with the geologic record, *Palaeogeogr. Palaeoclim. Palaeoecol.*, in press; L. A. Frakes, *Climate Throughout Geologic Time*, 1979; S. L. Thompson and E. J. Barron, Comparison of Cretaceous and present earth albedos: Implications for the causes of paleoclimates, *J. Geol.*, 89(2):143–167, 1981.

Pareto's law

Pareto's law (sometimes called the 20-80 rule) describes the frequency distribution of an empirical relationship fitting the skewed concentration of variate-values pattern. The phenomenon wherein a small percentage of a population accounts for a large percentage of a particular characteristic of that population is an example of Pareto's law. When the data are plotted graphically, the result is called a maldistribution curve. To take a specific case, an analysis of a manufacturer's inventory might reveal that less than 15% of the component part items account for over 90% of the total annual usage value. Other analyses may show that a small percentage of suppliers account for most of the late or substandard deliveries; a few engineers get most of the patents; a few employees are responsible for most of the quality rejects, errors, accidents, absences, thefts, and so on.

The key to benefiting from Pareto's law is to recognize its existence in a given situation and then to handle the case accordingly. For example, if there is no Pareto's law effect in a population—for example, the number of assembly errors in a factory is equally spread among all the employees—then one course of action should be taken by the engineer; but if Pareto's law is present and 5–10% of the employees are making 85–90% of the errors, then a very different course of action should be taken. If this important distinction is not made, not only will the engineer's action be the wrong one for the problem, but more importantly the engineer's entire approach to engineering management will be wrong.

Good management requires the devotion of a major portion of the manager's time and energy to the solution of the major problems—with the less important problems delegated to subordinates, handled by an automatic system, dealt with through management by exception, or in some cases simply ignored. A manager who devotes too much time to unimportant things is in peril of failure. All things should not be considered equally when in fact they are not of equal impact on the total problem.

Pareto's law is the test and the answer. The cost of corrective action is high, and the payoff must exceed the expense. A Pareto's law analysis distinguishes the significant or vital few from the insignificant or trivial many. If it reveals that, for example, 2 out of 20 items (10%) account for 80% of the total problem, then by focusing attention on only that 10% of the population, up to 80% of the problem can be resolved.

The mathematics required to calculate and graph the curve of Pareto's law is simple arithmetic. It should be noted, however, that the calculations need not be done in all cases. It may suffice to merely make a rough approximation of a situation in order to determine whether or not Pareto's law is present and whether benefits may subsequently accrue. For example, a firm with 100 employees can easily review its personnel records and pick out the 5 worst chronic absenteeism cases. Suppose those 5 employees were absent for 120 of the 160 total days of work lost by all 100 employees. Pareto's law would apply, as 5% of the population accounts for 75% of the problem, while the remaining 95% of the employees account for only 25% of the total absences. The value of such a discovery would be the realization that absenteeism could be reduced by 75% by dealing with only 5 out of 100 employees.

[VINCENT M. ALTAMURO]

Bibliography: J. M. Juran, *Quality Control Handbook*, 3d ed., 1974; M. G. Kindall and W. R.

Buckland, *Dictionary of Statistical Terms*, 1960; M. O. Lorenz, *Methods of Measuring the Concentration of Wealth*, American Statistical Association Publication, vol. 9, pp. 209–219, 1904–1905.

Parity (quantum mechanics)

The modern formulation of the weak force was developed in the 1950s to describe nuclear beta decay, but it also predicted for the first time an extremely small, but possibly experimentally observable, weak force between protons. Evidence for this weak force started to accumulate from low-energy nuclear physics experiments. These experiments were unfortunately difficult to interpret because of the uncertainties associated with nuclear calculations. Physicists tried a new, sensitive experimental approach at low energies which was capable of observing this tiny internucleon force. They looked for the breakdown of mirror symmetry, or parity nonconservation, in a scattering experiment. With the advent of a high-energy, intense, polarized proton beam at Argonne National Laboratory in 1973 physicists became hopeful of observing these effects, which were thought to be larger at high energies. By using the technique of scattering an intense beam of polarized protons from a target, a team of scientists performed an experiment at high energy and observed a surprisingly large parity nonconserving force between nucleons, nearly ten times larger than that predicted by some estimates. This result was difficult to interpret since it was so entirely unexpected. The implications are numerous. New experiments are being undertaken to investigate this effect at different energies, and theorists are trying to understand where earlier calculations might have gone wrong. Scientists agree that this result will lead to better understanding of just how protons scatter at these energies.

Proton-nucleus scattering. One of the main goals of high-energy physics is to understand the structure of the proton and how it interacts with itself and other particles. A very successful experimental method used by physicists is to study how a beam of protons accelerated to high energy scatters from a target. The target appears to the incoming protons to be composed mostly of empty space sprinkled here and there with atoms. The atoms are themselves mostly empty space with a tiny nucleus, about 10^{-15} m in diameter, composed of protons and neutrons at the center, collectively called nucleons, surrounded by orbiting electrons. The ratio of the number of incident particles $n(i)$ impinging on the target to the number which pass undeflected $n(u)$ is given by Eq. (1), where R is called the transmission. This

$$R = n(u)/n(i) \tag{1}$$

quantity is related to the effective area of the target or total scattering cross section σ by Eq. (2), where ρ is the density of the target, and l is the length of

$$R = e^{-\rho \sigma l} \tag{2}$$

the target. The effective area is determined by how much the incident particles and the target nuclei interact. Protons scattering from nuclei interact via the four fundamental forces of nature in proportion to the relative strengths of the interactions or coupling constants. The nuclear or strong force has a coupling constant of order unity, the electromagnetic of order 10^{-2}, and the weak 10^{-12}. Gravity is so extremely weak it plays no observable part in high-energy physics interactions. In order to observe the weak force in the presence of the strong interaction, physicists employ a very unique property of the weak interaction: it violates mirror symmetry.

Nature of parity or mirror symmetry. Symmetry in nature provides phyicists with powerful tools to study properties of elementary particle interactions. One such fundamental symmetry is called parity or mirror reflection. For many years it was believed, and intuition suggests, that the laws of physics used to describe the results of an experiment were identical for the experiment and its mirror image. In 1956 T. D. Lee and C. N. Yang suggested that the force which causes heavy nuclei to undergo radioactive decay would not obey parity. C. S. Wu and collaborators conclusively showed parity to be broken in beta decay of an isotope of polarized cobalt nuclei. After this, many experiments proved that a unique property of the weak force, or, as is often said, the signature of the weak force, is parity nonconservation.

In order to test physical laws under reflection in space, the concept of left- and right-handedness of a proton must be introduced. Every proton has an intrinsic angular momentum S, called spin. To help understand this property, imagine the proton as a top spinning about its own axis. It can spin in two ways, clockwise or counterclockwise. The spin of the proton is a vector quantity called an axial vector and has a magnitude $h/4\pi$, where h is Planck's constant. Momentum denoted by p is another type of vector quantity, called a polar vector. A useful way of describing the spin of the proton in quantum mechanics is the helicity operator given by Eq. (3),

$$H = 4\pi S \cdot \hat{p}/h \tag{3}$$

where $\hat{p}$ is a vector of magnitude 1 in the direction of the proton's momentum. In the language of quantum mechanics, the expectation value of H is $+1$ for a proton with its spin pointed along its momentum direction, and -1 for a proton with its spin aligned opposite to the momentum direction. Helicity has the mathematical property that it changes from H to $-H$ under a reflection in space or equivalently a parity transformation. Therefore helicity is a useful quantity in describing an interaction, such as the weak force between nucleons, which does not conserve parity.

A proton of positive helicity is called right-handed, and one of negative helicity is left-handed. To produce a beam of left- and right-handed protons,

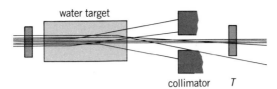

Fig. 1.   Main features of the transmission experiment.

physicists had to solve many technical problems.

Polarized beam. In 1973 polarized protons were accelerated for the first time to 6 GeV at the Argonne National Laboratory's Zero Gradient Synchrotron (ZGS). Production of this 6-GeV polarized proton beam was accomplished in several stages. First, a gas of hydrogen molecules was dissociated into neutral atoms and then passed through an inhomogeneous magnetic field. This technique causes those atoms whose electron spins point up (clockwise) to travel along the central axis of the magnet, while electrons with spins pointing down (counterclockwise) are steered to the poles. At the exit of the magnet all electrons have their spins pointing up, and the protons have their spins pointing both up and down. Then radio-frequency radiation is applied at a precise frequency to the atoms so as to flip the spin directions of the protons and electrons whose spins are antiparallel. Electrons are then stripped from the atoms, leaving only protons with spins pointing up. A Cockcroft-Walton column electrostatically accelerated this beam to 750,000 eV. It then entered a 50-MeV linear accelerator, and after acceleration the beam was injected into the main ring and accelerated to 6 GeV. The polarization P of the emitted beam is given by Eq. (4), where $N(\text{up})$ and

$$P = [N(\text{up}) - N(\text{down})]/[N(\text{up}) + N(\text{down})] \quad (4)$$

$N(\text{down})$ are the number of protons with their spins pointing up and down. The polarization of the beam from the ZGS, after all these stages of acceleration, was over 70%. In the final step the beam was delivered to the experimental area, and the spins rotated

such that the protons were in either the positive or negative helicity state, right- or left-handed, when they collided with the target.

Parity experiment. To test for the breakdown of mirror symmetry, a beam of right-handed protons was accelerated to 6 GeV and scattered from a water target. Next, the mirror-reflected experiment was performed. In practice this means firing a beam of left-handed protons onto the target. The effective area or total scattering cross section σ was measured for the left-handed and right-handed beams. A convenient description is given by the asymmetry A, defined by Eq. (5), where σ_+ and σ_- are the total

$$A = \frac{\sigma_+ - \sigma_-}{\sigma_+ + \sigma_-} \quad (5)$$

cross sections for right-handed and left-handed protons incident on the target. A nonzero value of the asymmetry A indicates the mirror symmetry is broken, or parity is violated. As stated earlier, the transmission R is the ratio of the number of incident particles to transmitted particles. The basic idea of the experiment is shown in Fig. 1. The incident beam intensity was measured by detector I. The T detector measured the intensity of the beam which passed through the water target and was not stopped by the collimator. In the actual experiment, I was really two different types of detectors, which gave independent measurements of the transmission. For simplicity only one type of detector, scintillation counters, will be discussed.

When a beam of charged particles traveling at close to the speed of light passes through material, energy is transferred to the atomic electrons of the material. In scintillating plastic this energy is quickly released in the form of electromagnetic radiation at a wavelength easily detectable by photomultiplier tubes. These vacuum tubes convert each quantum of light (the photon), after several multiplication stages, into many electrons. This electrical signal or voltage is them amplified, measured, and converted to a single number, representing the number of charged particles that pass through the scintillating material. Detector I in Fig. 1 was a scintillation counter which measured the intensity of the beam before it collided with the target, and detector T measured the beam intensity after the target. In this way it was possible to measure the transmission of both a left-handed and right-handed proton beam through the target and to then determine the asymmetry A. The magnitude of the asymmetry measured in this experiment was $A = (2.65 \pm 0.60) \times 10^{-6}$. This number can be understood from Fig. 2. The effective area of the target is larger for right-handed proton beams than for left-handed ones. This asymmetry is nearly ten times larger than expected from present-day knowledge of how protons and nuclei scatter at this energy. This result implies that the details of the collision process between the beam protons and the target nucleons are not understood.

The weak force is a short-range interaction, and

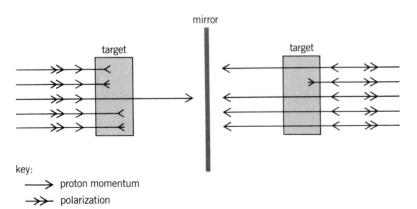

key:

→ proton momentum

⇉ polarization

Fig. 2.   Results of the experiment, which do not conserve parity because the measurement on the left can be distinguished from the mirror image shown at the right.

as the energy of the incident proton beam increases, the effects of the weak force are expected to grow. The experiment indicates that these violent collisions may be probing much shorter distances inside the proton than was expected and that the scattering is often taking place between the constituents of the proton, commonly called quarks. However, the collision process is very complex, and higher-energy beams tend to break up the proton before the quarks can collide violently. Of course, this is speculation, and further precision measurements combined with theoretical calculations will enable high-energy physicists to unfold this puzzling property of the weak force between nucleons, parity nonconservation.

For background information *see* FUNDAMENTAL INTERACTIONS: HELICITY (QUANTUM MECHANICS); PARITY (QUANTUM MECHANICS); QUARKS in the McGraw-Hill Encyclopedia of Science and Technology.

[NIGEL LOCKYER]

Bibliography: E. G. Adelberger, Nuclear parity mixing: An experimental review, in D. B. Cline and F. E. Mills (eds.), *Unification of Elementary Forces and Gauge Theories*, Ben Lee Memorial International Conference on Parity Nonconservation, Weak Neutral Currents, and Gauge Theories, Fermi National Accelerator Laboratory, October 20–22, 1977, p. 197–219; T. Khoe et al., *Particle Accel.*, 6:213–236, 1975; T. D. Lee and C. N. Yang, Question of parity conservation in weak interactions, *Phys. Rev.*, 104:254–258 1956; N. Lockyer et al., Parity nonconservation in proton-nucleus scattering at 6 GeV/*c*, *Phys. Rev. Lett.*, 45:1821–1824, December 8, 1980; C. S. Wu et al., Experimental test of parity conservation in beta decay, *Phys. Rev.*, 105:1413–1415, 1957.

Period doubling

Despite much study the weather remains largely unpredictable. A majority of the interesting examples of turbulent motion in fluids have defied insightful formulation that could enable reliable and accurate computation of this phenomenon. It was long thought that the difficulty in prediction of turbulent motion was the inextricably coupled motions of the vast number of particles which constitute the fluid. However, in 1963 the meteorologist E. Lorenz discovered that even with just three such coupled motions, truly erratic and unpredictable behavior occurred. Since that time, a branch of mathematics called dynamical system theory has begun to explain this phenomenon. In this study, an object called a strange attractor may account for the statistical behavior that precludes precise prediction. The systems studied are perfectly deterministic, so that one might imagine that, knowing the initial state of the system, complete prediction should result. The aspect of this general problem discussed here is the way that dynamical systems make a transition from predictable behavior to chaotic or unpredictable behavior. In the context of fluids, this is the problem of the onset of turbulence. However, the phenomenon is general, including problems as diverse as the onset of plasma instabilities and of cardiac fibrillation. The best understood general mechanism for the onset of chaos is called the period-doubling scenario and was discovered by M. J. Feigenbaum in 1975. This theory is especially interesting because even the precise quantitative results are independent of the particular system under consideration.

Discrete dynamics. While the systems that exhibit turbulent behavior vary continuously in time, it turns out that the phenomenon under study arises most naturally in the context of a discrete time. This means that the equations employed determine the state of system a finite time later given its present value, in contrast to differential equations which advance the time continuously. It turns out that the results from discrete time evolution carry over immediately to the continuous case. The prototypical example of a discrete system is given by Eq. (1).

$$x_{t+1} = 4\lambda x_t(1 - x_t) \tag{1}$$

$$(0 < \lambda \le 1,\ 0 \le x \le 1)$$

Here, λ is a parameter that is held fixed at some value while time advances (although it is essential to understand how the time evolution varies with different set values of λ), and x_t is the value of an observed variable at the integer-valued time t. The equation then determines the value of the observed variable at the next "legal" instant of time, $t + 1$. Clearly, if x_0 is specified, the equation determines x_1, from it x_2, and so forth. That is, the output of such an equation is a sequence of values x_1, x_2, . . . , x_t, . . ., denoted by $\{x_t\}$, where, for a fixed parameter value, a given initial state x_0 determines uniquely a sequence $\{x_t\}$. Of course, it is also important to know how this sequence varies with x_0.

Although Eq. (1) allows the easy computation of x_{t+1} from x_t, it is a very complicated matter to determine x_t from x_0 for large t. Indeed, x_1 is a quadratic function of x_0, so that x_2, also a quadratic function of x_1, is a quartic function of x_0. Generally, x_t is then a polynomial of x_0 of degree 2^t, which for large t is an immensely complicated function of x_0.

This has a serious consequence: while an error in specifying x_0 will propagate into an error of similar magnitude for x_1, the error in x_t can typically grow like 2^t—that is, exponentially—so that the ability to predict x_t from an initial x_0 quickly is eroded to ignorance no matter how small the error in x_0. The ability of completely deterministic systems to exhibit statistical behavior can always be tracked back to this "stretching" mechanism. In any case, it is plausible that Eq. (1) can serve as a model to understand these phenomena.

Stability and fixed points. Of course it is possible that the propagated errors will damp away rather than grow. This means that different initial states x_0 (that is, a fixed x_0 plus some error) will end up, asymptotically, following the same sequence of evolved states. In this case, the system can be even nicer than deterministic, in the sense that the system's initial state need not be known to determine its eventual evolution. This is an altogether usual occurrence in nonlinear systems [Eq. (1) is nonlinear because of the x_t^2 term] and accounts, for example, for the reproducible output of an oscillator independently of how it was turned on. This behavior, the reason for the reproducibility of familiar objects in the environment, is the phenomenon of stability, as opposed to the case of growth of propagated errors, which is instability. It will now be shown that, for suitable values of λ, Eq. (1) has both of these behaviors. It will then be possible to show how a system like Eq. (1) can undergo a transition from one type to the other.

The easiest way to show how Eq. (1) can result in the suppression of errors is to ask that the solution be so predictable that the state reproduces itself. That is, find an x_0 so that $x_1 = x_0$ (which then implies $x_2 = x_1 = x_0$ and generally $x_t = x_0$). This is easy: x_0 must satisfy Eq. (2), which implies Eqs. (3). The second possibility in Eqs. (3) is not inter-

$$x = 4\lambda x(1 - x) \tag{2}$$

$$x = 0 \text{ or } x = 1 - \frac{1}{4\lambda} \tag{3}$$

esting until $\lambda \geq \frac{1}{4}$, since otherwise $x < 0$. (For $0 < \lambda \leq 1$ every $0 < x_t \leq 1$ determines an x_{t+1} similarly bounded, which is then the interesting range of x's for this system.) Consider the first case, $x = 0$. States that reproduce themselves are called fixed points of the evolution. If the starting point is exactly at $x_0 = 0$, then $x_t = 0$. However, the object of this investigation is to learn how an error in x_0 propagates. Calling this error Δx_0, if it is sufficiently small, it follows from Eq. (1) that $\Delta x_1 \simeq 4 \lambda \Delta x_0$, which is again small. So $\Delta x_2 \simeq 4\lambda \Delta x_1 \simeq (4\lambda)^2 \Delta x_0$ and $\Delta x_t \simeq (4\lambda)^t \Delta x_0$, so long as Δx_t is sufficiently small. But, if $0 < \lambda < \frac{1}{4}$, $(4\lambda)^t$ decreases geometrically and Δx_t approaches 0 for large t. Accordingly, for $0 < \lambda < \frac{1}{4}$ and for small positive x_0, it follows that x_t approaches 0 (the fixed point) for large t. In fact almost all x_0 (the set of such x_0 has length 1) for this range of λ end up in the fixed point and prediction is altogether trivial.

More generally, if $x_{t+1} = f(x_t)$, then a sufficiently small error in x_t, Δx_t, results in $\Delta x_{t+1} \simeq f'(x_t)\Delta x_t$, where $f'(x)$ is the derivative of $f(x)$ with respect to x. If x^* is a fixed point [$x^* = f(x^*)$], then $\Delta x_{t+1} \simeq f'(x^*)\Delta x_t$ and $\Delta x_t \simeq [f'(x^*)]^t\Delta x_0$ if Δx_t remains sufficiently small. It is easy to see that the second fixed point of Eq. (1), $x^* = 1 - (1/4\lambda)$, has $|f'(x^*)| < 1$ for $\frac{1}{4} < \lambda < \frac{3}{4}$, $f'(x^*) = 1$ at $\lambda = \frac{1}{4}$, and $f'(x^*) = -1$ at $\lambda = \frac{3}{4}$. Again this stability is global in that almost all x_0 end up in x^* for $\frac{1}{4} < \lambda < \frac{3}{4}$, and prediction is again trivial. However, there are just two fixed points of Eq. (1) and neither is stable ($|f'| < 1$) for $\lambda > \frac{3}{4}$.

Chaotic dynamics. Before discussing what happens for $\lambda > \frac{3}{4}$, the behavior in the extreme case $\lambda = 1$ will be considered. As first noticed by S. Ulam and J. von Neumann, it is easy to determine the sequence $\{x_t\}$ in this case. Equation (1) becomes $x_{t+1} = 4x_t (1 - x_t)$, which under the substitution $x_t = \sin^2 \pi u_t$ becomes Eq. (4). This equation is satisfied by Eq. (5), whose right-hand side is just the fractional part of $2u_t$. Now the action of Eq. (5) is easiest to understand if u_t is given its binary expansion, Eq. (6). Then, multiplying by 2, the radix is shifted one position to the right and mod 1 truncates off the integer portion, giving Eq. (7). In fact,

$$\sin^2 \pi u_{t+1} = 4 \sin^2 \pi u_t(1 - \sin^2 \pi u_t) \tag{4}$$
$$= [2 \sin \pi u_t \cos \pi u_t]^2 = \sin^2 2\pi u_t$$

$$u_{t+1} = 2u_t \bmod 1 \tag{5}$$

$$u_t = 0.b_1 b_2 \ldots b_t \ldots \qquad b_i = 0,1 \tag{6}$$

$$u_{t+1} = 0.b_2 b_3 \ldots b_t \ldots \tag{7}$$

if $u_0 = 0.b_1 b_2 \ldots b_t \ldots$, then $u_t = 0.b_{t+1} \ldots$, and, by time t, the first t bits specifying u_0 have been completely lost. That is, the error in u doubles exactly each time, and the error in u_t is 2^t times the error in u_0, so that similarly the error in x_t grows as 2^t, and the possibility of worst exponential error growth has been realized. Indeed Eq. (5) generates a uniformly distributed pseudorandom sequence, since a typical u_0 is an irrational with a random sequence of b_n's. Since the leading bit of u_t is b_t, u_t is smaller or larger than $\frac{1}{2}$ as b_t is 0 or 1, which is to say that the sequence $\{u_t\}$ is randomly to the left or right of $\frac{1}{2}$. That is, it is a typical coin-toss Bernoulli sequence. Thus, at $\lambda = 1$, Eq. (1) generates chaotic dynamics. Evidently, somewhere between $\lambda = \frac{3}{4}$ and 1 the motion of Eq. (1) undergoes a transition to chaos. It does so through a cascade of successive doubling of the period of stable motions.

Orbits and attractors. Before discussing this process, it is necessary to introduce some terminology. The fixed point x^* is called an orbit of period 1; orbit refers to the succession of values x_t, and period 1 means that x^* recurs over one step. The fixed point is called an attractor if it is stable, so that values near it are "attracted" to it. A set of n points $x_1, \ldots, x_n$, with each recurring every n steps, is an orbit of period n. It is again an attractor if every n steps, a nearby value converges to it. An attractor of finite period always requires, for every period n, that the distances between nearby initial values diminish to 0 as the attractor is reached. An attractor of infinite period need not do so. In this case the distances can grow exponentially even between nearby points on the attractor. Such an object is a "strange" attractor, and produces statistical motions with specified distributions. As a final piece of terminology, if $x_{t+1} = f(x_t)$, then $x_{t+2} = f[f(x_t)] \equiv f^2 (x_t)$ and $x_{t+n} = f^n(x_t)$, where $f^{n+1} (x) = f(f^n(x))$ and f^n is called the nth iterate of f. f^1 is f itself, which is quadratic for Eq. (1), while f^t is the complicated polynomial of degree 2^t.

In this language, an orbit of period n is a set of n values $x_1, \ldots, x_n$, each of which is a fixed point of f^n: $f^n(x_i) = x_i$. It is an attractor if, for the identical reason as the argument concerning x^*, $|f^{n'}(x_i)| < 1$. (By the chain rule, this derivative is independent of i for each x_i of the periodic orbit.)

Doubling process. Now f^2 for the case of Eq. (1) is a more complicated function of x—explicitly it is given by Eq. (8), and so the equation $f^2(x) = x$ can

$$f^2(x) = f[4\lambda x(1 - x)] \qquad (8)$$
$$= 4\lambda\{4\lambda x(1 - x)[1 - 4\lambda x(1 - x)]\}$$

possess more real roots than does $f(x) = x$. Indeed, for $\lambda = \frac{3}{4}$, in addition to $x = 0$, it possesses $x^* = \frac{2}{3}$ as a triple root. For $\lambda > \frac{3}{4}$, two of its roots are just 0 and the old x^*, but the other two that were generated with x^* at $\lambda = \frac{3}{4}$ split apart from x^* and from each other. Since they are not fixed points of f, they are the points of a period 2 orbit. Evaluating the derivative of f^2 at these points shows that at $\lambda = \frac{3}{4} \equiv \Lambda_1$ the derivative is $+1$, that it decreases as λ increases, and that at $\lambda = \Lambda_2 < 1$ it becomes -1. That is, for $\Lambda_1 < \lambda < \Lambda_2$ a period 2 orbit is attracting (it is again global) and errors are decaying. The phenomenon in which a stable orbit throws off a new stable orbit as it becomes unstable at Λ_1 is called a bifurcation. In this case, where the new orbit has double the period of the old one, the process is called a period-doubling bifurcation.

The case for f^2 turns out to be the same as that for f. Between Λ_2 and $\Lambda_3 < 1$, f^2 has a stable orbit of period 2—that is, f has an attractor of period $2^2 = 4$. At Λ_3 it becomes unstable and throws off a stable period 2^3 orbit. Indeed, there is a sequence of parameter values, $\{\Lambda_n\}$ such that an orbit of period 2^n is stable for $\Lambda_n < \lambda < \Lambda_{n+1}$. These values are all smaller than 1 and converge to Λ_∞ which is also smaller than 1. That is, the system undergoes an infinite cascade of doubling bifurcations with the attractor at Λ_∞ being no longer finite in cardinality (that is, it is a Cantor set). This is the first point at which the system has nonperiodic behavior and in every sense is the transition point to chaotic behavior. *See* FRACTALS.

Another discrete system, given by Eq. (9), also

$$x_{t+1} = x_t \sin \pi x_t \qquad (9)$$

is a pseudorandom generator at $\lambda = 1$ and undergoes period doubling for another sequence of parameter values $\{\lambda_n\}$. Indeed the generic behavior of all systems whose evolution function is a one-parameter family like Eq. (1) have this identical behavior. (The more complicated behavior for $\Lambda_\infty < \lambda < 1$ is also the same.) Moreover, it is typical for systems of several first-order ordinary differential equations depending upon a parameter to exhibit these same period-doubling cascades; for such systems this can often be the first way in which the system comes to exhibit chaotic motions as its parameter is increased

from the laminar range in which the motion is simple. Such phenomena have been observed in geometrically or otherwise confined fluid systems.

Universal properties of doubling. What is remarkable, however, is that when an infinite period-doubling cascade occurs, near the corresponding Λ_∞ all the properties of the system are universal (the particular system does not matter) and are theoretically known. In particular, the theory has the following consequences:

(i) Λ_n converges to Λ_∞ geometrically and at a universal rate $\delta = 4.6692016 \ldots$ That is, Eq. (10) is valid.

$$\lim_{n \to \infty} \frac{\Lambda_{n+1} - \Lambda_n}{\Lambda_{n+2} - \Lambda_{n+1}} = \delta \qquad (10)$$

(ii) At Λ_n each element of an orbit of period 2^{n-1} throws off a pair of points whose separations increase to some value at Λ_∞. Calling this separation d_n for the pair at the maximum of f thrown off at Λ_n, there is a universal scaling factor, given by Eq. (11).

$$\lim_{n \to \infty} \frac{d_n}{d_{n+1}} = -\alpha \qquad (11)$$

$$\alpha = 2.502907875 \ldots$$

(iii) The very-long-time behavior at Λ_∞, when the expressions for f^t are too complex to evaluate, has a universal limit: the long-time fluctuations are universal and computable, and satisfy Eq. (12), where g can be obtained as a unique smooth solution of Eq. (13).

$$\lim_{n \to \infty} (-\alpha)^n f^{2n}(x/(-\alpha)^n) = g(x) \quad (\text{at } \Lambda_\infty) \qquad (12)$$

$$g(x) = -\alpha g(g(-x/\alpha)) \qquad (13)$$

(iv) For λ near Λ_∞ (either above or below) a limit like that of Eq. (12) again exists and is universal.

(v) In consequence of (iii) and (iv), the power spectrum for any coordinate of a system undergoing period doubling is universally self-similar and computable.

The results of (i), (ii), and (v)—which entail no free parameters—have been experimentally observed at the 10% level in real fluid experiments.

For background information *see* ALGEBRA; DIFFERENTIATION; FLUID-FLOW PROPERTIES; TRIGONOMETRY in the McGraw-Hill Encyclopedia of Science and Technology.

[MITCHELL J. FEIGENBAUM]

Bibliography: M. J. Feigenbaum, *Los Alamos Science*, vol. 1, no. 1, 1980; D. R. Hofstader, Strange attractors: Mathematical patterns delicately poised between order and chaos, *Sci. Amer.*, 245(5):22–43, November 1981; R. May, Simple mathematical models with complicated dynamics, *Nature*, 261:459–467, 1976.

Periodontal disease

Periodontal disease (periodontitis) is caused directly or indirectly by microorganisms residing in the sulci between the gingiva and the teeth. A less severe bacterial-induced inflammation involving only the gingiva, gingivitis, can occur suddenly, and in severe cases may be associated with pain, bleeding, and membrane formation. With local debridement (surgical removal of necrotic gingival tissue and foreign material) and improved oral hygiene the condition will subside; however, untreated cases may progress to a more severe stage of periodontal disease.

Microorganisms play a definite role in periodontal disease, but no specific single agent that is capable of invoking a positive disease response has been identified. As with caries, it appears that periodontitis is induced by an association of several microorganisms. The normal subgingival flora is a zoolike environment containing over 100 species of microorganisms, including the periodontitis-associated species. Some of the more prominent members are listed in the table. The normal flora interact with the pellicle on the tooth surface, and the subsequent aggregation of bacteria with salivary glycoproteins and extracellular bacterial polymers form dental plaque, which, if not removed, is mineralized, becoming calculus.

The clinical symptoms of periodontal disease are: gingival bleeding, foul breath, and loose teeth. If the condition is not treated, there will be progressive alveolar bone loss which may result in loosening and eventual loss of teeth (see illustration). Depending on the stage of the rapidly destructive, chronic, or juvenile periodontal disease, all or some of the clinical symptoms may be present. Currently, many patients with periodontal disease are treated by surgical procedures to remove subgingival calculus deposits and necrotic tissue, thus eliminating the periodontal pockets (infected sulci).

Microbial flora. The normal microbial flora in the oral cavity appear to include opportunistic pathogens; in a healthy individual the agents are not pathogenic, but in a stressed or unhealthy host they may produce disease. The microbiological population in periodontal disease brings about tissue and bone changes through a combination of bacterial products and actions, such as antigen production, enzymatic digestion, induced hypersensitivity, invasiveness, and toxin release. These bacterially induced changes in the oral cavity are influenced by age, diet, general health, genetics, hormones, and trauma in the host.

The microbiological population (see table) in periodontal disease varies somewhat, depending on the disease symptomology. Three distinct disease conditions can be labeled as chronic periodontitis, rapidly destructive periodontitis, and localized juvenile periodontitis, although several other forms of periodontal disease have been described clinically. In general the predominant organisms of all types of periodontal disease are anaerobes. Experimentally, a minimal periodontal infection can be induced in animals by using combinations of bacteria such as *Bacteroides*, a motile gram-negative anaerobic rod, and a facultative diphtheroid. However, it is apparent that many other oral cavity bacteria play a role in the periodontal disease process.

Microbial flora in chronic periodontitis consist for the most part of *Actinomyces* (*A. viscosus*, *A. israelii*, and *A. naeslundii*) and *Eubacterium* sp.; however, gram-negative microorganisms and spirochetes are also present in high numbers. In patients with acute gingivitis, *Actinomyces* spp. appear to become the predominant component of old plaques. *Actinomyces* spp. can be transmitted experimentally from one animal to another, and they have been isolated repeatedly from human patients with periodontal disease. Like most anaerobes, *Actinomyces* can be cultured in thioglycollate broth and on other media, and they may be identified by biochemical testing. For example, *A. israelii* and *A. naeslundii* do not produce catalase, but *A. viscosus* does. Serologically they can be grouped by fluorescent antibodies. Anaerobic conditions are required for the growth of *A. israelii*, but not necessarily for the facultative anaerobes *A. viscosus* and *A. naeslundii*.

In the rapidly destructive periodontal disease, large numbers of asaccharolytic, gram-negative microorganisms, such as *Fusobacterium* sp., *Bacteroides* sp., *Capnocytophaga* sp., *Eikenella corrodens*, and anaerobic vibrios, are present and are generally associated with the areas of greatest pocket depth. *Capnocytophaga* spp. are gram-negative, fusiform, gliding, anaerobic bacteria which, like the *Bacteroides* sp. and *Fusobacterium* sp., are found associated with localized juvenile periodontitis.

Localized juvenile periodontitis (periodontosis) microflora is generally composed of saccharolytic species of organisms, for example, *Fusobacterium*, *Actinobacillus*, *Capnocytophaga*, and *Eubacterium* species. With this condition there are little, if any,

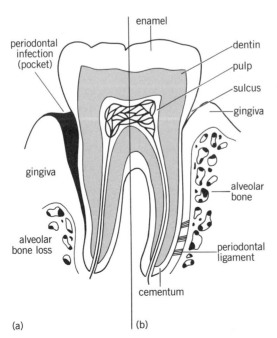

enamel

periodontal infection (pocket)

dentin

pulp

sulcus

gingiva

gingiva

alveolar bone

alveolar bone loss

periodontal ligament

cementum

(a)   (b)

Periodontal structures: (*a*) diseased and (*b*) healthy.

Some microorganisms of the healthy (normal) oral cavity and the microorganisms associated with various periodontal disease processes*

Normal oral cavity†	Chronic periodontitis‡	Rapidly destructive periodontitis§	Localized juvenile periodontitis (periodontosis)¶
Actinomyces spp.	*Actinomyces israelii*	Anaerobic spirochetes	*Actinobacillus* sp.
Anaerobic spirochetes	*Actinomyces*	Anaerobic vibrios	*Bacteroides* sp.
Anaerobic vibrios	*naeslundii*	*Bacteroides* sp.	*Capnocytophaga*
Bacteroides sp.	*Actinomyces viscosus*	*Campylobacter* sp.	sp.
Diphtheroids	Anaerobic spirochetes	*Capnocytophaga* sp.	*Eubacterium* sp.
Eikenella corrodens	*Eubacterium* sp.	*Eikenella corrodens*	*Fusobacterium* sp.
Eubacterium sp.	Gram-negative rods	*Fusobacterium* sp.	
Fusobacterium sp.	*Mycoplasma*	*Selenomonas* sp.	
Hemophilus sp.	*salivarium*		
Lactobacilli			
Mycoplasmas			
Neisseriae			
Pneumococci			
Rothia dentocariosa			
Staphylococci			
Streptococci			
(α-hemolytic and			
nonhemolytic)			
Veillonella sp.			
Yeasts			

*Microorganisms of the normal flora may also be recovered from periodontal lesions.
†Including spillover from upper respiratory tract (pharynx and trachea).
‡Most frequently observed periodontal disease process.
§Mostly asaccharolytic microorganisms.
¶Mostly saccharolytic microorganism.

calculus deposits on the teeth, and the bacteria are mostly associated with unattached portions of subgingival plaque. Two organisms associated with localized juvenile periodontitis, *Capnocytophaga* sp. and *Actinobacillus* sp., rapidly cause necrosis of the periodontal tissue in germ-free rats, which experimentally demonstrates the disease-producing potential of these organisms.

Other microbial flora. An organism not frequently considered in regard to periodontal disease is *Mycoplasma salivarium*; however, the incidence of this organism is high in periodontal disease patients and low or absent in those without disease. In addition, an increased *M. salivarium* antibody titer can be demonstrated in a patient with periodontal disease. This organism has a high affinity for attachment to tooth surfaces, and it can greatly reduce or completely destroy gingival adhesion, as well as produce a cytopathic effect in gingival cell monolayer cultures. Despite these conclusions, it is unknown what role this mycoplasma plays in periodontal disease.

Fungal (molds and yeasts) and viral agents apparently play no part in periodontal disease. However, acute herpetic gingivostomatitis and certain viral respiratory infections affect the health of the gingiva, although this impairment probably is only temporary. There is the possibility that specific viruses are associated with periodontal disease, but none has been isolated.

An important contributing factor to periodontal disease is the body's response to bacterial polymers such as endotoxin and peptidoglycan. These toxic agents can have a destructive effect on gingival tissue and bone. For example, the large number of gram-negative organisms associated with the subgingival pockets would provide a readily available source of endotoxins which could induce inflammation.

Experimental infections. Periodontitis can be induced in hamsters by oral infusion of microorganisms, clearly demonstrating the significance of bacteria in these diseases. These laboratory animals, which have been infected by microorganisms from either human periodontal diseases or analogous nonhuman periodontal diseases, appear to have tissue changes similar to those of humans. In some cases streptococci will produce periodontal-like disease. Concerning experimental infections in guinea pigs, it has been noted that approximately equal masses of bacteria, collected from a healthy crevice or from a periodontal pocket, do not differ markedly in number, type, or virulence. It could be concluded that no one microorganism appears able to produce periodontal disease, but that it is a microbial group effort that is somewhat dependent on host factors.

Therapy. Initial treatment of gingivitis or periodontal disease involves removal of calculus and plaque from the tooth surfaces, especially from the gingiva-cemento-enamel junction. This procedure should be carried out in conjunction with improved oral hygiene including the use of dental floss, gingival stimulation, and proper brushing of teeth to remove debris from the sulci. If this fails to reduce the gingival inflammation and periodontal pocket depth, the usual procedure is surgical excision of the necrotic tissue. The purpose of necrotic tissue eradication is to allow establishment of a healthy periodontium at a more apical position on the tooth. Success of these surgical procedures depends, to a great extent, on the patient and the maintenance of good oral hygiene.

The use of broad-spectrum antibiotics, for example, tetracycline, may temporarily relieve the symptoms of gingival or periodontal disease. In these cases, antibiotics suppress some but not all of the microbial flora. Without improved oral hygiene and other dental procedures the condition will progress upon antibiotic withdrawal. However, the use of antibiotics has shown promise as a means of controlling juvenile periodontitis.

A tooth-brushing technique utilizing baking soda, hydrogen peroxide, and sodium chloride may prevent, and in some cases alleviate, periodontal disease. The hydrogen peroxide helps oxygenate the periodontal pockets, eliminating many bacteria, especially anaerobes. This procedure may eliminate the need for surgery by controlling the oral bacterial flora that contribute to gingival inflammation and bone loss.

For background information *see* OPPORTUNISTIC INFECTIONS; PERIODONTAL DISEASE in the McGraw-Hill Encyclopedia of Science and Technology.

[RONALD FLETCHER]

Bibliography: P. H. Keyes et al., The use of phase-contrast microscopy and chemotherapy in the diagnosis and treatment of periodontal lesions: An initial report, *Quintessence Int.*, 1:50–55, 1978, W. A. Nolte, *Oral Microbiology*, 1982; G. I. Roth and R. Calmes, *Oral Biology*, 1981; J. M. Tanzer, Microbiology of periodontal diseases, *International Conference on Research Biology*: *Periodontal Research*, Chicago, pp. 153–191, 1977.

Photoelectrolysis

Photoelectrolysis, here narrowly defined as the photochemical cleavage of water into molecular products, has aroused a great deal of interest as a possible means of utilizing solar energy as a renewable alternative energy source. During the past 5 years, numerous experimental schemes have emerged for hydrogen harvesting via photocleavage of water. Although none of them is yet economically feasible, laboratory experiments have demonstrated that such processes, with reasonable yields, are possible. In fact, more than 10% quantum yields for hydrogen generation at the excitation wavelength of some sensitizers have been reported. These results show similar efficiencies to many of the better photovoltaic devices.

Water is one of the cheapest, most abundant, and chemically simplest raw materials on Earth, making it a prime target as a source for photochemical production of molecular hydrogen. Hydrogen is a particularly attractive fuel source not only because it has a high thermal content, but because it is a basic starting material and is ecologically clean. Most efforts have focused on splitting water into hydrogen and oxygen, and so far this direction has been the most promising. However, other substrates for photoelectrolysis, such as CO_2, N_2, or H_2S in combination with water, are currently being explored. A few examples of photoelectrolytic reactions that will store energy in chemical products are given in Table 1, along with the free energy required to bring about these processes; these thermodynamic requirements are translated in the last column into the threshold wavelength for a photon which might drive such a reaction. Since all photochemical redox reactions proceed by single-electron transfer steps, the threshold energy is calculated for each electron transferred in the reaction. Kinetic considerations and the need to overcome activation-energy barriers would require more energetic photons (by approximately 0.4 eV per electron transferred) than those listed in Table 1. Many more photoelectrolytic reactions have been observed to occur with highly energetic photons of the ultraviolet region. However, for practical applications, the visible and red spectral regions are

Table 1. Thermodynamic parameters for possible photoelectrolytic reactions*

Reaction	$\Delta H°$, kcal/mole	$\Delta G°$, kcal/mole	n†	$\Delta E°$,‡ V	λ_{th},¶ nm
$H_2O \rightarrow H_{2(g)} + \frac{1}{2}O_{2(g)}$	68.3	56.6	2	1.23	1008
$CO_{2(g)} \rightarrow CO_{(g)} + \frac{1}{2}O_{2(g)}$	67.6	61.4	2	1.33	931
$H_2O + CO_{2(g)} \rightarrow HCO_2H + \frac{1}{2}O_{2(g)}$	64.5	68.3	2	1.48	837
$\frac{3}{2}H_2O + \frac{1}{2}N_{2(g)} \rightarrow NH_{3(g)} + \frac{3}{4}O_{2(g)}$	91.4	81.1	3	1.17	1058
$H_2O + CO_{2(g)} \rightarrow HCHO_{(g)} + O_{2(g)}$	134.4	124.6	4	1.35	918
$2H_2O + N_{2(g)} \rightarrow N_2H_4 + O_{2(g)}$	148.7	181.3	4	1.97	630
$2H_2O + CO_{2(g)} \rightarrow CH_3OH + \frac{3}{2}O_{2(g)}$	173.6	167.9	6	1.21	1022
$3H_2O + CO_{2(g)} \rightarrow CH_{4(g)} + O_{2(g)}$	212.5	195.3	8	1.06	1171
$3H_2O + 2CO_{2(g)} \rightarrow C_2H_5OH + 3O_{2(g)}$	336.8	318.4	12	1.15	1078

*Unless otherwise specified, thermodynamic data is given for the liquid. All data are for 298 K. $\Delta H°$ and $\Delta G°$ are the standard enthalpy and the standard free energy of the reaction respectively.

†n is the number of electrons transferred in the electrochemical reaction.

‡By definition, $\Delta E°$ is the free energy of the reaction per one electron.

¶λ_{th} is the wavelength of the least energetic photon necessary to fulfill the thermodynamic requirements, assuming each photon promotes the transfer of only one electron.

of primary interest. All the reactions listed in Table 1 could be driven by sunlight.

Cyclic photoelectrolysis. A conceptual scheme for cyclic photoelectrolysis of water into H_2 and O_2 is given in Fig. 1. Since water does not absorb light at wavelengths greater than 200 nanometers, any photoelectrolytic system will have to contain a sensitizer, namely a light-absorbing component (S in Fig. 1). Upon absorption of a photon, the sensitizer is promoted to a highly reactive excited state with a short lifetime.

Another additive to the system, the electron acceptor relay (R), has to rapidly quench the excited sensitizer in an electron-transfer reaction. This electron-relay system is necessary in order to circumvent direct one-electron reduction of water, a process of very high activation energy which results in the extremely reactive hydrogen atom. As shown by Table 1, all photoelectrolytic processes require more than one-electron transfer from one atom or molecule to another. On the other hand, all photochemical reactions proceed by elementary steps of single-electron transfers. This mismatch in the number of electrons transferred is compensated through the use of redox catalysts. Usually two catalysts are used, one for the reduction of water, the other for its oxidation. The reduced relay reacts with the reduction catalyst again by an electron-transfer reaction to recover the relay. The reduced catalyst, which can accumulate a large number of electrons on its surface, then reduces water to H_2. The oxidized sensitizer similarly reacts with the oxidation catalyst to accept an electron from the catalyst (the process is commonly termed hole injection to the catalyst) to restore the starting sensitizer. Thus the system is cycling both the sensitizer and relay components; the only component used up is water (and photons).

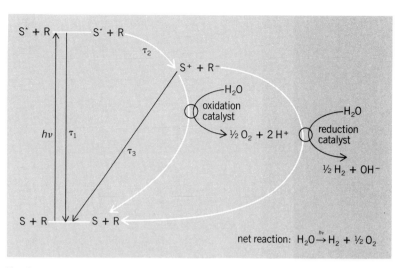

Fig. 1. Schematic diagram for cyclic photoelectrolysis of water into molecular hydrogen and oxygen. In this diagram S = sensitizer; R = electron-acceptor relay; $h\nu$ = excitation by light; S* = excited state of the sensitizer; τ_1, τ_2, τ_3 = lifetime of the excited state, of the electron-transfer quenching reaction, and of the back-electron transfer reaction, respectively; S^+, R^- = redox products (oxidized sensitizer and reduced relay) of the electron-transfer quenching reaction.

Several actual systems have been recently tested, and they have been proved to photochemically cleave water in a sustained cyclic process. All of them are based on the concepts embodied in Fig. 1, yet a variety of modifications exist. In some systems several electron relays are introduced during the reduction cycle. Others use electron relays in the oxidation cycle. Most systems utilize heterogeneous catalysts; others operate with homogeneous catalysts. An example of a successful operating system would include the inorganic complex $Ru(bpy)_3^{2+}$ as a sensitizer, methyl viologen (MV^{2+}) as an electron accep-

$$Ru(bpy)_3^{2+} \xrightarrow{h\nu} Ru(bpy)_3^{2+*} \qquad \text{excitation} \qquad (1)$$

$$Ru(bpy)_3^{2+*} + MV^{2+} \longrightarrow Ru(bpy)_3^{3+} + MV^+ \qquad \text{electron transfer} \qquad (2)$$

$$2MV^+ + 2H_2O \xrightarrow{\text{(Pt) catalyst}} 2MV^{2+} + H_2 + 2OH^- \qquad \text{catalytic water} \atop \text{reduction} \qquad (3)$$

$$2Ru(bpy)_3^{3+} + H_2O \xrightarrow{\text{(RuO}_2\text{) catalyst}} 2Ru(bpy)_3^{2+} + \tfrac{1}{2}O_2 + 2H^+ \qquad \text{catalytic water} \atop \text{oxidation} \qquad (4)$$

Net reaction: $2H_2O \xrightarrow[\text{(4 photons)}]{h\nu} 2H_2 + O_2 \qquad (5)$

Here

bpy (2.2'-bipyridine) =

$MV^{2+} = CH_3 - ^+N \quad N^+ - CH_3$

tor relay, colloidal platinum as a reduction catalyst, and colloidal RuO_2 as an oxidation catalyst. The sequence is given in reactions (1)–(5).

Many competing wasteful reactions can occur in such a system, and directing the whole ensemble to react in the desired direction poses a variety of problems which have been overcome only recently. A few of these competing reactions are shown (6)–(10). Thus in the cycle shown in Fig. 1, the photo-

$$Ru(bpy)_3^{2+*} \xrightarrow{\tau_1} Ru(bpy)_3^{2+} \quad \text{deexcitation, fluorescence, etc.} \quad (6)$$

$$Ru(bpy)_3^{3+} + MV^+ \rightarrow Ru(bpy)_3^{2+} + MV^{2+} \quad \text{back reaction} \quad (7)$$

$$2MV^+ \rightarrow MV + MV^{2+} \quad \text{disproportionation} \quad (8)$$

$$MV^+ + O_2 \rightarrow MV^{2+} + O_2^- \quad \text{products interference} \quad (9)$$

$$2MV^+ + H_2 \xrightarrow{\text{(Pt) catalyst}} 2MVH \quad \text{hydrogenation} \quad (10)$$

electron-transfer reaction (with lifetime τ_2) has to compete with relaxation of the excited sensitizer to its ground state (of lifetime τ_1, usually in the range of 10^{-7} to 10^{-3}). The hydrogen- and oxygen-production reactions will compete with the dark back reaction (with lifetime τ_3). Radical-radical reactions may also occur, and the catalytic reaction will have to be both rapid and specific (if other catalytic processes, such as hydrogenation, are to be avoided).

Back reaction. The most pressing problem which diminishes the yields of water photocleaving systems is that of the back reaction, with a lifetime τ_3 (Fig. 1). Any photochemical process that attempts to store energy will produce highly reactive products. These products, particularly in the initial stages of the cycle, will have a large driving force to back-react in a nonphotochemical reaction, which is the reverse of the initial electron-transfer reaction. Occurrence of the back reaction results in dissipation of the stored energy into wasteful heating of the medium. It is essential, therefore, to inhibit this back reaction and to allow enough time for the productive cycle to continue in the desired direction of fuel generation. A variety of chemical arrangements have been designed to achieve this goal, and some of them are illustrated in Fig. 2. All these arrangements basically utilize a local high electrostatic charge density of head groups at an interface between a water-hydrocarbon micro-subphase, and sometimes differences in solubilities (that is, hydrophobic interactions) between the reactants and the redox products in the various subphases. In all the examples illustrated in Fig. 2, a positively charged sensitizer is held close to the interface. An uncharged quencher in the main phase (usually H_2O) is free to approach the interface. Following excitation of S, electron transfer occurs between S^{+*} and R to yield the redox products S^{2+} and R^-. Since the interface contains a large number of negatively charged head groups, the product R^- is prevented from approaching the interface, while the product S^{2+} is

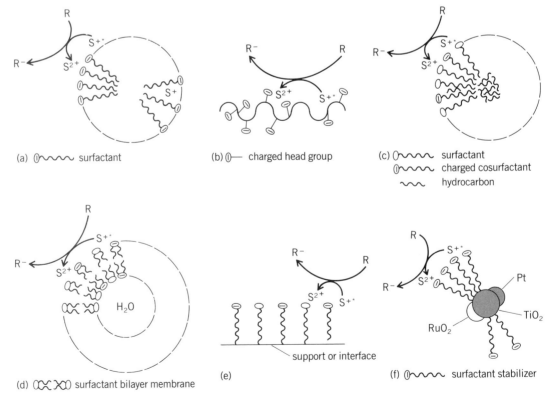

(a) ⊙〰 surfactant

(b) ⊙— charged head group

(c) ⊙〰 surfactant
⊙〰 charged cosurfactant
〰 hydrocarbon

(d) ⊙✕ ✕⊙ surfactant bilayer membrane

(e) support or interface

(f) ⊙〰 surfactant stabilizer

Fig. 2. Schematic illustration of colloidal assemblies employed to enhance charge separation and inhibit the back reaction in photoelectrolysis of water. (a) Micelles. (b) Poly-electrolyte. (c) Microemulsion. (d) Vesicle. (e) Monolayer. (f) Functionalized catalyst: TiO_2 carrier and Pt and RuO_2 catalysts.

Table 2. Specific compounds classified according to their function in photocleavage of water

Sensitizers	Acceptor relays	Donor relays	Catalysts
Inorganic complexes $Ru(bpy)_3^{2+}$, $Os(bpy)_3^{2+}$, and derivatives	*Inorganic* V^{3+}, Cr^{3+}, Eu^{3+}, U^{4+}, and their complexes (e.g., salicylates) $Rh(bpy)_3^{3+}$, $Co(bpy)_3^{2+}$, $Co(Me_6[14]diene\ N_4)^{2+}$	Ce^{3+}, Mn^{2+}, $Fe(bpy)_3^{2+}$ $Mo(CN)_8^{4-}$ $W(CN)_8^{4-}$ Hydroquinones Oxidized sensitizers	Pt, Au, Ag, PtO_2, RuO_2, IrO_2, TiO_2, Hydrogenase
Dyes Acridine dyes Proflavin Methylene blue Thionine Phthalocyanines Cyanine dyes	*Organic* Methyl viologen and its derivatives (4,4'-bipyridinium and 2,2'-bipyridinium salts)		
Biochemicals Porphyrins Chlorophyll Chloroplasts	*Biochemical* Cytochromes Quinones Ferredoxin		
Semiconductors TiO_2, $SrTiO_3$, CdS, ZnP, GaP			

strongly held at the interface. The rate of the back reaction of R^- and S^{2+} may thus be inhibited by several orders of magnitude. The systems that provide such microheterogeneous environments include micelles, microemulsions, polyelectrolytes, vesicles, and monolayers spread either on glass supports or at the water-air interface. Again, various modifications of this theme have been tested, the most ingenious being the use of

$$CH_3-(CH_2-)_{12}-^+N\bigcirc O\bigcirc O\ N^+-CH_3$$

as an electron relay in the presence of positive micelles. While this nonsymmetric viologen salt is highly water-soluble, its reduced cation is very water-insoluble and will therefore be solubilized in the positively charged micelles. However, effecting charge separation by stabilizing the redox products in such microenvironments necessarily reduces the free energy stored in the products.

Photoelectrolytic components. Some of the specific components of the various systems that have been studied so far are listed in Table 2. Various permutations of the candidate components have been tried with a wide range of effectiveness. In an effort to simplify the complete photoelectrolytic device, it is possible that one component will provide several of the functions simultaneously. Thus, for example, $Ru(bpy)_3^{2+}$ is often used as the photosensitizer, while its oxidized product, $Ru(bpy)_3^{3+}$, often functions as an acceptor relay. Generally speaking, the organic materials are less desirable because of their tendency to participate in a variety of destructive side reactions. On the other hand, some of the biochemical components, probably due to the sophistication achieved by natural selection, are extremely stable and highly specific. At present, however, the inorganic compounds seem to be the most promising for human-designed photoelectrolytic systems. Among

these, the semiconductor materials provide for the highest conversion efficiencies achieved so far.

Semiconductor systems. These are probably the most promising systems and the most intensively studied devices. Semiconductors could provide several of the functions required in a water-splitting system. Thus, the semiconductor material could be used as the photosensitizer, ejecting an electron to a relay acceptor upon band-gap irradiation. It could, however, also be used as the catalyst or as the catalyst carrier (for example, by codeposition of Pt and RuO_2 on the semiconductor material). The most attractive feature of the semiconductor, however, is its inherent capability to impose charge separation (that is, inhibit the back reaction) when brought into contact with an electrolyte solution. This concept is illustrated in Fig. 3 for two arrangements, a photoelectrochemical cell and a colloidal semiconductor particle. The potential of the conduction band at the semiconductor-liquid interface surface has to assume the redox potential of the electrolyte solution, while the conduction-band potential in the semiconductor bulk remains unchanged; a depletion region is thus created in the conduction band near the interface. Additional electrons that may be promoted to the conduction band by band-gap irradiation would then slide downhill into the bulk material. However, irradiation creates holes in the valence band which will climb uphill and be ejected into the electrolyte. This counterflow of electrons and holes provides the charge separation so essential for water-splitting processes. If the electrons are now directed toward catalytic H_2 production, either through an external circuit or by Pt deposition on the semiconductor particle while the holes are similarly directed toward catalytic O_2 evolution, the task of water cleavage is achieved.

The most serious problem with these arrangements is that of photodegradation of the material

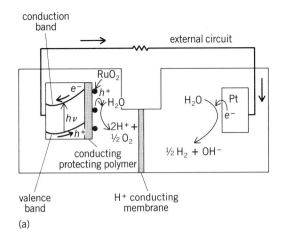

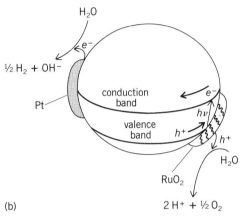

Fig. 3. Schematic representation of (a) photoelectrochemical semiconductor cell and (b) colloidal semiconductor particle for cyclic photoelectrolysis of water to H_2 and O_2.

(usually dissolution of the cation component of the semiconductor). However, this problem is presently being overcome by various surface modifications. For example, binding a conducting polymer to the surface of the semiconductor and then depositing the RuO_2 catalyst on the polymer will reduce the rate of photodegradation by many orders of magnitude.

For background information *see* PHOTOCHEMISTRY; SOLAR ENERGY in the McGraw-Hill Encyclopedia of Science and Technology. [DAN MEISEL]

Bibliography: J. R. Bolton and D. O. Hall, Photochemical conversion and storage of solar energy, *Annu. Rev. Energy*, 4:353–401, 1979; J. Kiwi, K. Kalyanasundaram, and M. Grätzel, Visible light induced cleavage of water into hydrogen and oxygen in colloidal and microheterogeneous systems, *Structure and Bonding*, 49:37–126, 1982; K. I. Zamaraev and V. N. Parmon, Potential Methods and perspectives of solar energy conversion via photocatalytic processes, *Catal. Rev. Sci. Eng.*, 22:261–324, 1980.

Photoreactivating enzyme

Photoreactivation was discovered in 1949 by Albert Kelner. He noticed, while studying the lethal effects of ultraviolet (uv) irradiation on bacteria, that the survival level of bacteria exposed to uv light could

be increased by exposure to visible light. In other words, damage caused by light in the uv range was apparently repaired by light in the visible range. Later investigations showed that this repair process involved a single enzyme which was consequently designated photoreactivating enzyme. Photoreactivating enzyme, originally thought to be an unusual repair enzyme of bacteria, has since been found to occur frequently among higher plants and animals. In recent years, the discovery of human photoreactivating enzyme has raised some interesting questions about the fundamental utility of this enzyme, its interaction with other repair processes, and its relationship to cancer in humans.

The damage that photoreactivating enzyme repairs is highly specific. DNA contains four bases: cytosine and thymine are pyrimidine bases, while adenine and guanine are purine bases. Along both of the strands that constitute the double-helix conformation of DNA, the bases are arranged in a specific sequence which codes the genetic information of the cell. At certain positions along the sequence, two pyrimidine bases will be adjacent. When this occurs, the potential for an uv-induced dimer exists. Ultraviolet light causes covalent bonding to occur between the ring structures of the bases, with the formation of cyclobutyl dimers, as shown in the following reaction. This constitutes genetic damage be-

cause the code can no longer be correctly read for transcription of genes, nor can complementary bases be correctly assigned during DNA replication. The expression of the damage will vary depending upon the particular gene sequences damaged, as well as the total amount of damage incurred. It is these cyclobutyl pyrimidine dimers which photoreactivating enzyme repairs.

Escherichia coli. As has been the case with most aspects of DNA biochemistry, *Escherichia coli* was the first organism used to study photoreactivation. Much of the knowledge of photoreactivating enzyme comes from *E. coli*, which contains concentrations of 5–10 molecules of the enzyme per cell. Isolated *E. coli* photoreactivating enzyme has a protein component with a molecular weight of about 35,000 which is glycosylated with mannose, galactose, glucose, and *n*-acetylglucosamine sugars. There is also an RNA cofactor of about 15 bases. This cofactor may be involved in the alignment of the enzyme along the damaged DNA strand. The enzyme has a notable lack of absorbance in the range usually associated with photoreactivation (300–450 nanometers). This was originally an enigma to researchers. However, recent work has shown that the required absorbance occurs only when the enzyme is associated with the dimer substrate. Separate solutions of photoreactivating enzyme, irradiated DNA, and unirradiated DNA fail to show the required absorbance. Only a combined solution of photoreactivating enzyme and irradiated DNA shows an absorbance peak in the required range.

The sequence of events for the photoreactivating repair process is believed to be relatively simple. A photoreactivating enzyme molecule recognizes a dimer and binds to it. The dimer-enzyme complex is stable and remains intact until a photon of visible light is absorbed. This causes lysis of the dimer and release of the enzyme, which is then free to bind to another dimer. There are several significant characteristics of this process. Photoreactivation does not involve incision into the DNA backbone or replacement of bases and, therefore, is not subject to mispairing of bases or to nuclease activity. Also, unlike other enzyme repair processes which require ATP, photoreactivation uses visible light as an energy source and requires no metabolic energy.

In addition to *E. coli*, photoreactivating enzyme has been isolated from a wide variety of species and cell types. The enzyme is non-species-specific in the sense that photoreactivating enzyme from one species will act on the dimers in the DNA of another species. Also, the different photoreactivating enzymes share the following: molecular weights in the 35,000–40,000 range; one subunit; an associated cofactor; and similar pH and ionic strength optima. A comparison of *E. coli* and *Cattleya aurantiaca* photoreactivating enzymes shows them to have nearly identical amino acid contents. Furthermore, antibody to *E. coli* photoreactivating enzyme displays some cross reactivity with human photoreactivating enzyme, suggesting an evolutionary conservation of some amino acid sequences.

Humans. Though similar in most respects, there are some species differences among photoreactivating enzymes. A good example is human photoreactivating enzyme. It was first discovered in leukocytes and later found in other body tissues, but for many years it was thought to be absent from humans. This oversight was probably due to a combination of three factors. First, human photoreactivating enzyme has a much lower optimum ionic strength than *E. coli* photoreactivating enzyme. If it is assayed under conditions suitable for *E. coli* photoreactivating enzyme, its activity is reduced to 10–20%. Second, the action spectrum of human photoreactivating enzyme (that is, the relative efficiencies of different wavelengths in producing photoreactivation) has a peak at a much longer wavelength then *E. coli* photoreactivating enzyme. Human photoreactivating enzyme isolation assays performed by using the shorter-wavelength yellow safelights rather than red safelights may have been affected. A third factor which has confounded ex-

perimentation is that the photoreactivating enzyme levels in cultured mammalian cells can be affected by the medium used. For example, human fibroblasts will synthesize photoreactivating enzyme when grown in Dulbecco's modified medium, but will not when grown on Eagle's minimum essential medium.

The discovery of photoreactivating enzyme in humans and other placental mammals confirmed the ubiquitous nature of this repair enzyme. Its widespread occurrence suggests that it has an essential and unique function in cellular DNA metabolism. Yet the pyrimidine dimers which it repairs can also be repaired by at least two other mechanisms (excision repair and postreplication repair). This leads to the question of why evolution has preserved what appears to be a redundant repair mechanism.

Experimentation using *E. coli* mutants deficient in different repair mechanisms has shown that photoreactivating ability increases cellular resistance to uv-induced death beyond that which can be achieved by other repair mechanisms alone. It is thought that the explanation for this might be related to the proximity of dimers along the DNA molecule. Repair mechanisms that involve excision of bases require an intact template in order to replace correctly complementary bases. However, if dimers in both strands of DNA are so close that excised regions overlap, there is no complementary strand which can be used as an error-free template. One function of photoreactivating enzyme may be to reduce the number of dimers which must be replaced by excision repair. In this way the probability of overlapping excised regions is reduced, and the probability of cell survival is consequently increased.

Another fundamental question still unanswered is why some tissues have photoreactivating enzyme at all. Certainly, during evolution all organisms have had to develop repair mechanisms to deal with damage caused by solar radiation. However, uv light is nonpenetrating, and damage due to uv-induced dimer formation is restricted to surface cells directly exposed. The fact that photoreactivating enzyme can be found in deep tissues never exposed to uv light can be explained by the possibility that dimers are formed in ways other than through uv exposure. However, photoreactivating enzyme requires visible light, which is also nonpenetrating, for activity. Photoreactivating enzyme may have some activities in addition to dimer removal, but they have never been demonstrated.

Possible role in carcinogenesis. One way in which scientists try to determine the function of a particular enzyme is to study mutant strains of cells defective in its production. As already mentioned, mutant *E. coli* defective in photoreactivating enzyme synthesis are less able to survive uv-induced damage than wild strains having a complete complement of repair enzymes. Now the question arises whether photoreactivating enzyme serves any protective function in humans. This is more difficult to answer. Carcinogenesis is of main concern in human somatic DNA damage, and protection against skin cancer would seem to be the probable function of photo-reactivating enzyme. However, there is a lack of mammalian cell lines available that are specifically mutant in photoreactivating enzyme synthesis. Cells from patients with xeroderma pigmentosum, a disease typified by a highly increased susceptibility to solar-induced skin cancer, show a multitude of repair defects, including variably suppressed photoreactivating enzyme levels. Furthermore, some apparently normal individuals have been found to have suppressed photoreactivating enzyme levels or deficient excision repair. These confusing cases may best be explained in terms of some type of summation effect for repair mechanisms, in which the weighted sum of the individual repair processes determines the normal or diseased state of an individual. However, under this scheme it is very difficult to analyze the relative importance of any one component. For this reason the role of photoreactivating enzyme in carcinogenesis is yet to be determined.

Much is now known about the specifics of photo-reactivating enzyme activity, but the most fundamental questions are still unanswered. The wide distribution of photoreactivating enzymes among species and tissues suggests its important role in DNA metabolism, but its significance in higher organisms remains to be demonstrated. It is evident, however, that the study of photoreactivating enzyme is intimately tied to other repair mechanisms. Further insight will come not by looking at isolated enzymatic steps, but by looking at DNA repair functions as a whole.

For background information *see* NUCLEIC ACID; NUCLEIC ACIDS, PHOTOCHEMISTRY OF in the McGraw-Hill Encyclopedia of Science and Technology.

[TIMOTHY J. JORGENSEN]

Bibliography: T. J. Jorgensen, Photoreactivating enzyme: A light activated repair enzyme of microbes and man, *BioScience*, 31:671–674, 1981; B. M. Sutherland, Photoreactivation, *BioScience*, 31:439–444, 1981; B. M. Sutherland, Photoreactivation in mammalian cells, *Int. Rev. of Cytol.*, *Suppl.* 8, pp. 301–334, 1978.

Photosynthesis

In the first electron paramagnetic resonance studies of photosynthesis in 1956, B. Commoner and colleagues observed the light-induced formation of paramagnetic electrons. Their work pioneered the application of a variety of magnetic resonance techniques in photosynthesis, including electron paramagnetic resonance, optically detected magnetic resonance, and reaction-yield-detected magnetic resonance. Electron spin-spin interactions can be determined exclusively by magnetic resonance techniques to provide unique structural and mechanistic information, such as the distance over which the initial electron transfer of photosynthesis takes place. Without distance measurements, the mechanism of primary photosynthesis is largely speculative. Consequently, for over 25 years the primary events have been probed by electron paramagnetic resonance in order to understand more fully the mechanism of the

initial charge separation in photosynthesis.

Unfortunately, electron paramagnetic resonance of the first step of photosynthesis has proved impossible because the crucial events last only 20 nanoseconds, much too short a time for detection by electron paramagnetic resonance. Instead, the magnetic resonance of the primary state has been observed by the technique of reaction-yield-detected magnetic resonance. In this method, optical absorption spectroscopy is used to monitor, as a function of applied magnetic field and x-band (9×10^9 Hz) microwaves, the yield of product formed after or during annihilation of charge separation. Microwaves in resonance with the radical energy-level differences alter the yield of reaction products.

Primary bacterial photosynthesis. In photosynthesis, charge separation occurs within the pigment system located in a protein matrix, known as a reaction center or a phototrap, where optical spectroscopy has demonstrated that the primary event occurs in about 5 picoseconds to form the initial charge separation state, P^F. For this experiment, an adequate description of P^F is a radical pair composed of a primary donor cation with one electron spin and a primary acceptor anion with another electron spin. The primary donor is a dimeric form of bacteriochlorophyll known as a special pair, and the primary acceptor is bacteriopheophytin. Two electrons are involved in this characterization of the donor-acceptor radical pair; the electron hole is located mostly on the primary donor, and the anion electron mostly on the acceptor. A radical pair can be characterized by four states, corresponding to the four combinations of spin angular momentum for singlet-triplet eigenstates of two electrons (Fig. 1), one $^1P^F$

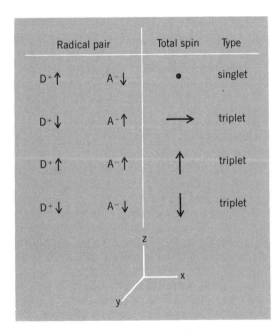

Fig. 1. Singlet-versus-triplet donor-cation acceptor-anion radical pairs. The arrows represent electron spin vectors. The singlet spin vectors add to give a null vector. For one of the triplet radical pairs the two spins add to give one spin perpendicular to the *z* direction in the coordinate system.

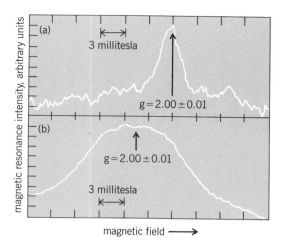

Fig. 2. Comparison of reaction-yield-detected magnetic resonance spectra of bacterial reaction centers with (a) ubiquinone removed and (b) chemically reduced ubiquinone. The symbol g represents the electron g value, which for a given frequency of radiation determines at what magnetic field resonance occurs. The additional electron of the chemically reduced ubiquinone results in a so-called environmental broadening of the magnetic resonance spectrum of P^F. (After M. K. Bowman et al., Magnetic spectroscopy of the primary state, P^F, of bacterial photosynthesis, Proc. Nat. Acad. Sci. USA, 78(6):3305–3307, June 1981)

and three $^3P^F$ states, respectively. The P^F state not designated as singlet or triplet refers to all four possible states of P^F combined.

In the normal course of photosynthesis, P^F lasts for only about 200 ps—the time necessary for the anion electron to travel from the primary acceptor bacteriopheophytin anion to the secondary acceptor Fe-ubiquinone complex. However, by eliminating secondary electron transport in specially prepared bacterial reaction centers the lifetime of P^F can be lengthened a hundredfold. For example, in reaction centers where the Fe-ubiquinone is chemically removed the second electron-transfer reaction cannot occur. Consequently, at room temperature the P^F state still forms in approximately 5 ps but now can remain for about 20 ns, the time during which charge separation self-annihilation occurs. These blocked reaction centers with longer P^F lifetimes are necessary for the reaction-yield-detected magnetic resonance experiments.

Spectra of primary events. The magnetic resonance spectra (Fig. 2) of the P^F state of the primary events in bacterial photosynthesis in blocked reaction center proteins are obtained with a reaction-yield-detected magnetic resonance spectrometer (Fig. 3). An optical cell containing reaction centers is placed in a standard optical-transmission electron paramagnetic resonance microwave cavity in an electron paramagnetic resonance electromagnet. A weak, continuous optical beam (detecting beam A) measures the optical absorption spectrum of the sample. An intense, pulsed laser beam (actinic beam B) excites the bacterial reaction centers while high-power microwaves (magnetic resonance beam C) are present. The actinic laser pulse B, with a duration of 5–15

ns, excites the primary donor, which ejects an electron to the primary acceptor to form P^F. The electrons behave analogously to pointing gyroscopes. When gyroscopes move from one location to another, their mutual spinning directions do not change even though their energy may. Likewise, when an electron moves from an unoxidized donor to form a reduced acceptor, the direction of spin is not changed. Energy changes alone are not sufficient to alter the direction in which an electron spin points; instead electron spins are flipped by magnetic forces. Prior to the transfer the two electrons in the donor point in opposite directions; that is, the spins are paired as singlet ground state. Immediately after the jump the spins are still paired due to the gyroscopiclike effect of the angular momentum of the electrons involved in donor-acceptor formation. Thus, initially P^F is in the singlet state, that is, $^1P^F$. The electrons in a small fraction of radical pairs flip and point in uncorrelated directions because of internal magnetic forces such as electron-nuclear spin-spin interactions. As a consequence, a small fraction of the radical pair electrons becomes partially correlated as triplet $^3P^F$. For the charge annihilation reaction the gyroscopic effect is again in operation such that $^1P^F$ reacts to ground-state singlet donor, keeping the electrons paired, whereas $^3P^F$ reacts to form triplet-excited donor (called $^3P^R$) in which the electron spins are unpaired. $^3P^R$ differs from $^3P^F$ mainly in that both electrons are on the donor, whereas in $^3P^F$ one electron is on the donor and one electron is on the acceptor. Also $^3P^R$ has a lifetime of about 100 microseconds and an optical absorption spectrum distinctly different from excited singlet donor, ground-state donor, or P^F.

For interpretation of the reaction-yield-detected magnetic resonance experiments, $^3P^F$ annihilates

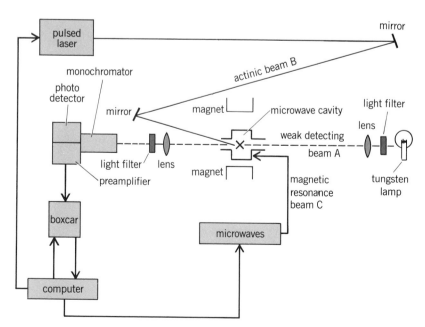

Fig. 3. Schematic of a computerized reaction-yield-detected magnetic resonance spectrometer.

about ten times faster than $^1P^F$. Approximately 10% of P^F correlates as $^3P^F$ and annihilates in about 2 ns to give metastable triplet donor $^3P^R$; approximately 90% of P^F correlates as $^1P^F$ and annihilates within 20 ns to give ground-state singlet donor. Thus, from 20 ns to approximately 100 μs after the laser pulse B, all $^1P^F$ and $^3P^F$ have vanished; instead mostly ground-state donor (about 90%) and a small amount of metastable $^3P^R$ (about 10%) are present, as is readily revealed by optical absorption spectroscopy. However, if the microwaves are resonant, then the separated electrons can be flipped by the magnetic field of the microwaves prior to annihilation. As a consequence, the microwaves catalyze formation of more $^3P^F$. Instead of the 10% formed in the absence of microwaves, a maximum of about 10.8% $^3P^F$ forms when optimal microwaves are present. This increased amount of $^3P^F$ quickly annihilated to produce a higher yield of $^3P^R$ detected by the weak monitoring beam A. Since the microwaves are effective only when in resonance, the magnetic resonance spectrum of P^F is obtained by the increase in metastable triplet donor $^3P^R$ yield as revealed by the detecting optical beam A while scanning the magnetic field.

Implications. The reaction-yield-detected magnetic resonance of the primary radical pair P^F of photosynthesis requires unusually high microwave powers (about 500 watts) compared to electron paramagnetic resonance (milliwatts). This high microwave power for such a small magnetic resonance signal suggests certain interpretations. The triplet form of the radical pair $^3P^F$ is very short-lived (about 1.8 ns) compared to the optically measured lifetime (15 ns) for all P^F states. The observed reaction-yield-detected magnetic resonance line width is about 25 (25×10^{-4} weber). Assuming lifetime broadening as the dominant source of magnetic resonance line width, a limit of about 1.8 ns is placed on the lifetime of $^3P^F$ in accordance with the uncertainty principle. Paramagnetic species with about 1.8-ns lifetimes do not require such high microwave powers to observe magnetic resonance. Thus, an additional source of low magnetic resonance intensity exists. Namely, P^F is predominantly nonparamagnetic singlet $^1P^F$. Electron spins in pure singlet states are not magnetic-resonance-active. The greater the spin-spin interactions between donor and acceptor, the more the radical pair states approach the pure singlet and pure triplet eigenstates. As P^F becomes a pure singlet and triplet-spin eigenstate, the more difficult is the process of singlet-to-triplet conversion, either by microwaves or by the intrinsic electron-nuclear forces that give rise to the approximate 10% yield for metastable triplet donor P^R in the absence of microwaves. According to the reaction-yield-detected magnetic resonance data, P^F must be dominated by $^1P^F$ for about 20 ns. However, a small amount of non-singlet-character $^3P^F$ develops in time such that reaction-yield-detected magnetic resonance gives weak magnetic resonance signals. That mostly $^1P^F$ and very few $^3P^F$ states exist in photo-

synthesis is one major consequence of these experiments, since ordinary optical experiments cannot distinguish between $^1P^F$ and $^3P^F$.

Some unique information is provided by these reaction-yield-detected magnetic resonance experiments.

(1) For radical pairs to be predominantly singlet or triplet in nature, an electron-electron spin-spin interaction must be large compared to the electron-nucleus interactions (that is, the local magnetic environment of the donor-acceptor electrons). Previous theoretical analysis of other experiments has shown that large isotropic spin-spin exchange interactions can be ruled out. The only remaining source for interaction between the electron spins of the radical members of P^F is the anisotropic electron-electron magnetic dipolar term. In order to explain this large electron-electron dipole interaction, the initial electron-transfer distance must involve bacteriochlorophyll-like molecules with a large area of almost van der Waals contact between primary donor and primary acceptor macrocyles. (2) The proposed singlet radical pair mechanism of bacterial photosynthesis is firmly established and does not involve singlet fission or triplet-triplet fusion. (3) Reaction-yield-detected magnetic resonance provides definitive criteria for the laboratory verification of biomimetic charge-separation devices based on photosynthesis. (4) Reaction-yield-detected magnetic resonance offers the opportunity for a new magnetic-resonance probe of photochemical and photosynthetic charge-transfer reactions that occur on the nanosecond-to-subnanosecond time scale. Thus, green plant photosynthesis should be investigated by reaction-yield-detected magnetic resonance.

For background information *see* BACTERIAL PHOTOSYNTHESIS; ELECTRON PARAMAGNETIC RESONANCE (EPR) SPECTROSCOPY; PHOTOSYNTHESIS in the McGraw-Hill Encyclopedia of Science and Technology.

[JAMES R. NORRIS, JR.]

Bibliography: M. K. Bowman et al., *Proc. Nat. Acad. Sci. USA*, 78:3305–3307, 1981; B. Commoner, J. J. Heise, and J. Townsend, *Proc. Nat. Acad. Sci. USA*, 42:710–718, 1956; E. L. Frankevich and A. I. Pristupa, Magnetic resonance of excited complexes with charge transfer revealed by fluorescence at room temperature, *Pis̀ma Zh. Eksp. Teor. Fiz.*, 24(7):397–400, 1976; J. Schmidt and I. Solomon, Modulation de la photoconductivité dans le silicium à basse température par résonance magnétique électronique des impuretés profondes, *C. R. Acad. Sci.*, Paris, 2366:169–172, 1966.

Plant pathology

Important areas of research in plant pathology concern the mechanisms by which pathogens are transmitted to plants, and the organisms which cause plant disease. This article discusses the role of aphids as vectors of plant viruses and the role of algae as plant pathogens.

Aphid transmission of plant viruses. Of all the different ways plant viruses are spread in nature,

Examples of plant viruses transmitted by aphids

Virus group	Virus	Relationship	Vector
Potyvirus	Potato virus Y	Nonpersistent	*Myzus persicae*
Carlavirus	Carnation latent virus	Nonpersistent	*M. persicae*
Cucumovirus	Cucumber mosaic virus	Nonpersistent	*M. persicae*
Alfalfa mosaic virus	Alfalfa mosaic virus	Nonpersistent	*M. persicae*
Caulimovirus	Cauliflower mosaic virus	Nonpersistent	*M. persicae*
Closterovirus	Beet yellows virus	Semipersistent	*Aphis fabae*
Pea enation mosaic virus	Pea enation mosaic virus	Persistent	*Acyrthosiphon pisum*
Plant rhabdovirus	Lettuce necrotic yellows virus	Persistent	*Hyperomyzus lactucae*
Luteovirus	Barley yellow dwarf virus	Persistent	*Rhopalosiphum padi*

transmission by insect vectors is clearly the most important. Many kinds of insects transmit plant viruses, but aphids are the principal vectors, with about 190 species transmitting one or more of about 170 different plant viruses. These numbers represent only about 5% of the known aphid species, but almost 50% of the described plant viruses. Nearly every crop plant is susceptible to one or more of the viruses transmitted by aphids, and in some instances the viral diseases limit crop production. Aphids transmit plant viruses effectively because their piercing-sucking mouthparts are efficient structures for the transfer of virus from the cells of one plant into those of another. Moreover, the tremendous reproductive ability of aphids, usually parthenogenetic during the growing season, increases the chances for spread of viruses not only within a field, but also over a wide area when the insects are carried by wind. A unique feature of plant virus transmission by aphids is the different ways virus is transmitted. Much current research is involved with understanding the diverse aphid-virus interactions, uncovering mechanisms that control them, and applying this knowledge to reduce crop damage.

Nonpersistent and persistent transmission. Aphids transmit plant viruses in two very distinct ways. One of the aphid-virus relationships is called nonpersistent (other designations for this relationship are stylet-borne and noncirculative), and the other is called persistent (also circulative, internal, or biological). Since the basis for the relationship is the virus, not the vector, viruses are often called nonpersistent or persistent (see table). The same aphid can transmit both kinds of viruses, simultaneously under some conditions. A major difference between the two relationships, and the origin of the descriptive designations, is the length of time that vectors continue to transmit virus after they acquire it.

An aphid can transmit nonpersistent viruses for only minutes or hours, whereas it can transmit persistent viruses for many days, often for as long as it lives. Another critical difference is that when aphids molt they lose the ability to transmit previously acquired nonpersistent viruses but not persistent ones. A brief summary of the transmission process illustrates other differences. Aphids, especially if they have been starved, acquire nonpersistent viruses often within seconds during probes in the epidermis

of infected plants; they acquire persistent viruses after hours or days of feeding in the phloem tissue. Aphids can transmit nonpersistent viruses as soon as they leave an infected plant; a delay, or latent period, may occur before aphids can transmit persistent viruses. Nonpersistent viruses are transmitted to a plant within seconds or minutes as the aphid probes in or between epidermal cells; aphids need several hours to transmit persistent viruses because their transmission is associated with feeding. Nonpersistent viruses often cause mosaic symptoms that can be recognized by clear leaf patterns. Persistent viruses usually cause diseases in which symptoms, such as dwarfing, are more subtle and often overlooked. Nonpersistent viruses, which infect many plant tissues, can easily be transmitted mechanically from plant to plant; most persistent ones cannot be transmitted mechanically, apparently because only the deeper-lying phloem tissue of plants is susceptible.

Some plant viruses, such as beet yellows virus, are intermediate between nonpersistent and persistent ones, and are often called semipersistent. For semipersistent viruses, feeding times are usually hours or days, rather than seconds or minutes as with nonpersistent viruses. Since viruliferous aphids do not retain the ability to transmit semipersistent viruses after a molt, most workers consider semipersistent relationships to be based on the same mechanisms as nonpersistent ones.

Interactions in the field. Understanding the diseases caused by aphid-transmitted plant viruses is often difficult because of the many variables that affect interactions among the three biological systems involved: plant, aphid, and virus. The complete picture in nature often includes interaction of the aphid vector with two plants in addition to the one that becomes diseased (Fig. 1). The weather can alter any one of the biological systems and can change interactions among them. Differences between virus-vector relationships also can be important. With many nonpersistent viruses, for example, the most serious disease outbreaks often occur when the source of virus (plant 2 in Fig. 1) is near the crop. Infected weeds around margins of fields frequently serve as the virus source. For persistent systems, however, the source of virus may be located some distance from the crop. Outbreaks of barley yellow dwarf in

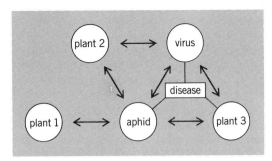

Fig. 1. Interactions in aphid transmission of a virus from one plant to another. Plant 1 represents the host on which the aphid develops and reproduces. Plant 2 (in some cases the same as plant 1) represents the source plant from which the aphid acquires virus that is transmitted to plant 3, the plant that becomes diseased. Disease occurs only when interaction among the aphid vector, the virus, and plant 3 results in infection of the plant. (*After W. F. Rochow, Vectors and variations, Proceedings of the 3d International Symposium on Virus Diseases of Ornamental Plants, International Society for Horticultural Science, The Hague, Netherlands, 1974*)

oats in Iowa have even been associated with aphids carried in low-level jet winds from Oklahoma. Control of disease by using insecticides is usually ineffective with nonpersistent viruses because virus is transmitted to plants during probing by aphids before the insecticide has time to act. Insecticides have been somewhat more useful in control of persistent viruses, especially in reducing secondary spread when the source of virus is within a field. In contrast, oil sprays seem to be more useful in reducing spread of nonpersistent viruses than of persistent ones.

The virus-vector relationship can also affect distribution of infected plants within a field. Plants infected with nonpersistent viruses are often most numerous along field margins; disease incidence may be high in rows adjacent to hedges or weeds and progressively lower with increased distance from the virus source. Distribution of persistent viruses is of-

ten associated with infection centers throughout a field or with infected plants randomly scattered over an entire field.

Large populations of aphids also cause much direct feeding damage to crop plants. High populations are not necessary for aphids to cause indirect damage by transmitting viruses, a distinction often overlooked. For aphids acting as vectors, movement is more important than numbers.

Transmission mechanisms. An early designation for the nonpersistent virus-vector relationship was "mechanical." At first, the process seemed to involve mere transfer of virus from contaminated aphid stylets. It was soon learned, however, that virus transmission was complex: a degree of specificity occurs between aphid species and plant viruses. In fact, the most infectious plant virus of all, tobacco mosaic virus, is not transmitted by aphids! Despite the efforts of many research workers over nearly 50 years, the nonpersistent transmission process is still only partially understood. For some years, nonpersistent viruses were referred to as stylet-borne because research suggested that transmissible virus was confined to the terminal portion of aphid stylets. A more current explanation is based on an ingestion-egestion hypothesis. Virus that is ingested as aphids probe infected hosts is thought to be egested, or regurgitated, from the foregut as aphids later probe the plant that becomes infected. Two of the problems encountered in pinpointing the mechanism for transmission of nonpersistent viruses are the short period of time, a few seconds, during which transmission occurs and the difficulty of manipulating vectors during study without altering their normal probing behavior.

Two separate mechanisms are involved in transmission of persistent viruses. The propagative mechanism includes replication of virus within the aphid; plant rhabdoviruses multiply in their aphid vectors as well as in the plants they infect. This is the same mechanism for insect transmission of viruses that infect animals. In the other mechanism (nonpropagative), viruses circulate through the aphid vector, but there is no evidence for replication. Luteoviruses, for example, are acquired from phloem tissue as aphids feed, and move through the vector (Fig. 2). Virus ingested along with food travels up the food canal of the stylet, through the esophagus, and into the stomach. Some virus then passes through the hindgut and is voided with honeydew. Other virus passes through the stomach wall, circulates in the hemolymph, enters the salivary gland, and is injected into plant tissue along with saliva when the aphid feeds. Passage of virus through the gut wall into the hemocoel is not a selective barrier for luteoviruses. Viruses that an aphid is unable to transmit can be detected by making tests of hemolymph collected from the nonvector. The major barrier in these circulative systems seems to be entrance of virus into salivary glands (Fig. 2).

A current working hypothesis is that circulative virus transmission by aphids is regulated by interactions between virus capsid protein and specific re-

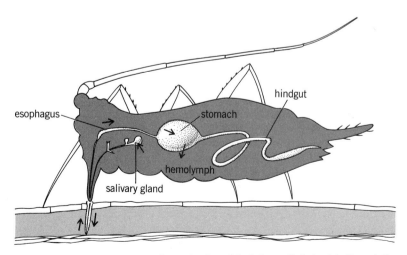

Fig. 2. Diagram of an aphid showing route of persistent viruses that circulate through the vector. (*After L. R. Nault, Aphid feeding and plant virus diseases, Ohio Rep., 53(2):24–25, 1968*)

ceptor sites in accessory salivary glands of the vector. Thus, selectivity of the interaction between virus protein and receptor sites of the salivary glands determines which virus an aphid can transmit and accounts for the marked specificity that occurs in luteovirus-aphid relationships. The concept of specific receptor sites on the glands also explains the persistent nature of the circulative virus-vector relationship. If the number of salivary gland virus-receptor sites is limited,· the flow of virus through the salivary system could be restricted, thus conserving virus circulating in the hemocoel and prolonging the time an aphid remains viruliferous.

Dependent virus transmission. Dependent virus transmission by aphids is an unusual feature of aphid-virus relationships. In dependent virus transmission, aphids transmit one virus, called the dependent virus, only in the presence of a second virus, called the helper virus. Examples of this phenomenon are known for nonpersistent, semipersistent, and persistent viruses. Two separate mechanisms explain dependent virus transmission. For nonpersistent and semipersistent systems, dependent virus transmissions occur not only following aphid probing on doubly infected plants, but also following sequential probing first on plants infected by the helper virus and then on plants infected by the dependent one. Thus *Myzus persicae* can transmit potato aucuba mosaic virus only in the presence of potato virus Y, not because potato virus Y itself is needed, but because of the proteinlike helper component produced in the plant infected by potator virus Y. Several workers have suggested that such a helper component might act as a connecting bridge between the aphid and the virus particle. This concept also suggests a basic mechanism that could be an important part of nonpersistent transmission.

In contrast, dependent virus transmission of persistent viruses occurs only when aphids feed on a doubly infected plant, apparently because interactions during simultaneous replication of the two viruses provide the basis for the phenomenon. This mechanism has been studied mostly with two isolates of barley yellow dwarf virus (designated MAV and RPV), which can be differentiated because they are serologically distinct. The aphid *Rhopalosiphum padi* does not regularly transmit the MAV isolate from oats infected only with MAV, but it regularly transmits MAV, together with RPV, from doubly infected plants. During simultaneous replication of RPV and MAV, some virus particles are formed that contain the nucleic acid of MAV encapsidated or coated with RPV protein (Fig. 3). The transcapsidated virus particles (also called genomic masking) function in *R. padi* like RPV because of the protein capsid, move through the vector, and are injected into the plant. In phloem cells of plants, such transcapsidated particles function like MAV because it is the nucleic acid of a virus particle that controls infection and subsequent virus replication. Such heterologous encapsidation is widespread among animal and bacterial viruses, but little is known about this

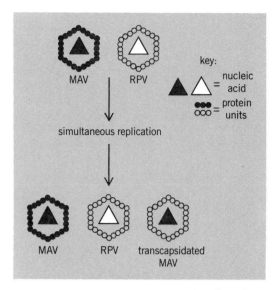

Fig. 3. Diagram of combinations of nucleic acids and protein capsids for two barley yellow dwarf luteoviruses that replicate simultaneously in a mixed infection.

phenomenon among plant viruses. Study of the mechanism for dependent virus transmission by aphids is a useful approach to understanding not only some of the unanswered questions of aphid transmission of plant viruses, but also how basic biological mechanisms operate in viral infections of different organisms.   [W. F. ROCHOW]

Plant parasitic algae. One branch of the science of plant pathology that has been relatively neglected for at least a century is the study of algae that damage crops and other plants, but a revival of interest in disease-causing algae is now taking place.

The first extensive list of plants infected by the most damaging of these algae was recently compiled for an area in the United States. Differences in disease resistance to this alga were demonstrated with a crop species for·the first time, which may lead to selection and breeding for host resistance in other susceptible crops. The alga involved is being grown in pure cultures in laboratories using artificial nutrient media. This is the first step required for the classic method of proving whether or not a suspected plant-damaging microorganism causes disease. Along with these fundamental discoveries, new algicidal sprays were selected from among the numerous fungicides being marketed for disease prevention.

Algae resemble bacteria and fungi with respect to their relationships with larger, more complicated terrestrial plants. Their association can range from living harmlessly near or upon plants to living on or within them in a mutually beneficial manner. Among those that cause plant disease are species which can live freely in the absence of plant hosts, if necessary, whereas others can exist only as parasites. Since most algal parasites contain chlorophyll for sugar production, they are less dependent upon their hosts then parasitic fungi and bacteria.

Algae that derive nourishment from vascular plants include eight green (*Chlorophyta*) and three blue-green (*Cyanophyta*) genera. One of the so-called green forms, *Synchytrium borreriae*, contains chlorophyll only in its resting spores; the vegetative, nonsexual stage is colorless and must therefore depend entirely upon its host for carbohydrates. Others, such as *Rhodochytrium* species, never contain chlorophyll and are classified as green algae because of their structure and reproductive processes. Two additional green forms, *Phyllobium* and *Stomatochroon*, appear to be harmless to the plants they inhabit. The effects of others on their hosts range from very slight injury to severe damage. The three blue-green species, *Anabaena cycadae*, *Nostoc punctiforme*, and *Chroococcus* species, kill host cells as they migrate from injuries in the host's protective epidermis to preferred sites in aerial roots of the cycad *Macrozamia communis*.

One of the less injurious green algae, *Chlorochytrium*, lives within a host's intercellular spaces and causes leaf nodules to develop in water weeds such as *Lemna trisulca* and *Hypnum* and *Sphagnum* species. *Chlorochytrium rubrum* stimulates the formation of leaf tubercles on leaves of terrestrial plants such as *Mentha aquatica* and *Peplis portulaca*. *Chlorochytrium limnanthemum* causes elevated, yellowish spots on leaves of *Limnanthemum indicum*. Another *Chlorochytrium* species is associated with galls in *Polygonum lapathifolium*. *Phyllosiphon* is a parasitic alga that causes gall production and yellowish spots and large dead patches in leaves of the Araceae family. *Phyllosiphon arisari* damages *Arisarum* species, and *P. deformans* and *P. philodendri* cause swellings in leaves of *Anchomanes difformis* and of the popular houseplant genus, *Philodendron*. Species of *Phytophysa* form galls on leaves of another houseplant genus, *Pilea*. Somewhat more injurious are the *Rhodochytrium* species, which contain no chlorophyll and are completely parasitic. They not only produce leaf and stem galls on species of *Ambrosia*, *Asclepias*, and *Spilanthis* but also cause considerable stunting of the roselle crop, *Hibiscus sabdariffa*.

Most injurious alga. Diseases caused by *Cephaleuros* species were first studied and reported during the last quarter of the nineteenth century, but relatively little research has been accomplished since that time. Although a resurgence of interest in this alga is a recent phenomenon, plant diseases caused by the genus are hardly a recent biological development. Forms very similar to it occur on fossilized leaves found in Tertiary strata in Germany, and its worldwide distribution and extensive host range mark it as an ancient, well-adapted plant parasite.

Cephaleuros is the most easily seen algal parasite because of its large thalli, and it undoubtedly ranks as the most injurious pathogen of many of its more than 500 hosts growing worldwide in the moist subtropics and tropics.

Hosts of Cephaleuros species. Susceptible plants occur in at least 53 plant families in Florida and include ferns, cycads, palms, and bamboo; a great variety of broadleaf plants that produce economically important fruits such as avocado, blackberry, blueberry, citrus, guava, and mango; and species such as thyme and cinnamon. Affected nut trees include cashew, macadamia, and pecan. A few of the many ornamental plants that are infected include trees such as acacia, banyan, bottlebrush, magnolia, maple, palms, silk oak, and sycamore. Among the numerous shrub hosts are azalea, bougainvillea, hibiscus, jasmine, juniper, oleander, photinia, privet, and wisteria. Gardenia, camellia, philodendron, ivies, and a great many other houseplants are also infected. In other tropical areas coffee, tea, and cocoa are severely damaged by this alga, particularly in India and Southeast Asia. Many hosts suffer damage to their leaves, branches, trunks, and fruits.

The vegetative stage of the parasite is most often seen on a host's upper leaf surfaces and consists of a rather circular, flat, green or gray thallus growing under the waxy leaf cuticle. It may be minute to a centimeter in diameter. The alga is often incorrectly called red rust because erect yellow to red filaments and fruiting structures arise from its surface during favorable weather. These reddish structures are all that can be seen on infected tree trunks and branches where thalli are not apparent, and they allow the parasite to spread from diseased to nearby healthy hosts by means of microscopic airborne spore cases. The cases settle onto wet surfaces of healthy plants, where they exude swimming spores that infect the new host. Fruit infection sometimes differs from the leaf and branch symptoms; dark indentations of the fruit surface with no obvious green thallus cause blemishes that diminish market value.

The dearth of research on these widespread and economically important plant parasites can be attributed to: their largely tropical distribution in areas where research facilities are rare; the fact that algal diseases have not caused famines; the relatively slow growth of the alga as compared to the more common plant-damaging fungi; and the apparent inability to grow the alga through its normal life cycle under laboratory conditions. In laboratory culture the alga is filamentous rather than disk-shaped and fruiting structures are abnormal; however, infectious swimming spores have been formed on artificial nutrient media.

Cephaleuros disease control. Algal disease severity and incidence is diminished by improving a host's general health, since debilitated plants are much more susceptible to disease than healthy ones. Reported causes of host debility are poor soil drainage, insufficient or excessive irrigation, malnutrition due to inadequate natural soil fertility or fertilization, or soil containing too much of one or more plant nutrients. Other diseases or insect infestations can also diminish host vigor enough to result in more severe disease. Early field studies of *Cephaleuros* diseases in agricultural crops were not devised to reveal epidemiological data suitable for statistical analysis. However, a recent study of guava plantings in Florida demonstrated for the first time differences in host susceptibility to algal disease. Disease control by application of fungicidal sprays has also been reported in Florida. Basic copper sulfate, Ferbam, Difolitan, and Daconil sprays provided some control, with Daconil being the most effective.

For background information *see* ALGAE; INSECTA; PLANT PATHOLOGY; PLANT VIRUSES AND VIROIDS in the McGraw-Hill Encyclopedia of Science and Technology. [ROBERT B. MARLATT]

Bibliography: K. F. Harris and K. Maramorosch (eds.), *Aphids as Virus Vectors*, 1977; K. F. Harris and K. Maramorosch (eds.), *Vectors of Plant Pathogens*, 1980; J. J. Joubert and F. H. J. Rijkenberg, Parasitic green algae, *Annu. Rev. Phytopathol.*, 9:45–64, 1971; J. J. Joubert and P. L. Steyn, Studies on the physiology of a parasitic green alga, *Cephaleuros* sp., *Phytopath. Z.*, 84:147–152, 1975; K. Maramorosch and K. F. Harris (eds.), *Plant Diseases and Vectors*, 1981; R. B. Marlatt, Susceptibility of *Psidium guajava* selections to injury by *Cephaleuros* sp., *Plant Dis.* 64:1010–1011, 1980; R. B. Marlatt and S. A. Alfieri, Jr., Hosts of a parasitic alga, *Cephaleuros* Kunze, in Florida, *Plant Disease*, 65:520–522, 1981; W. F. Rochow, Barley yellow dwarf virus: Phenotypic mixing and vector specificity, *Science*, 167:875–878, 1970.

Plant physiology

Silicon is the second most abundant element in the Earth's crust. When dissolved to form silicic acid [$Si(OH)_4$], it is taken up in sizable quantities by many types of organisms, such as sponges, radiolarians, diatoms, scouring rushes (*Equisetum* spp.), sedges, and grasses, and in more limited quantities by other organisms. In the grasses it can account for up to 20% of the dry weight of some organs. Shoots of scouring rushes collapse when grown in silicon-free nutrient solutions. Rice shoots also appear stunted and limp in the absence of silicon. However, some increase in yield and erect stature occur when silicon in the form of silica slag is added to rice paddies. In the diatoms, the cells will not divide to form new daughter cells unless silicon is in their external environment. Heavy deposition of silicon in plants may serve to ward off predators, including insects, pathogenic fungi, and herbivores. In grasses the presence of discrete silica bodies may help in directing light to the inner cells of the leaves, thus enhancing photosynthesis. Thus, for many organisms, silicon may be considered an essential element for normal growth and development.

Plants take up silicon from soil solutions, where monosilicic acid is available through dissolution of soil silicates (from rock and solid particles). Monosilicic acid enters the roots of plants along the same pathway as water and mineral nutrients: from root hairs to the cortex, into the endodermis and pericycle, and then into the xylem. Once in the xylem of the stem, it moves passively with the water aided by the transpiration stream. From the xylem it infil-

trates all types of cells and cell walls, eventually to be deposited as silica ($SiO_2 \cdot nH_2O$). Most of the monosilicic acid becomes concentrated in the epidermal system at the surface of these organs.

Biogenic silica. The conversion of soluble silicic acid to insoluble amorphous silica is not well understood. Silica-accumulating cells and regions may have special conditions that permit the nucleation and polymerization of silica from a supersaturated solution (150–300 ppm SiO_2). A more acidic pH (5–6) in this region is favorable for silica polymerization. It is also possible that some organic catalysts are available to bring about polymerization. The silica first forms as colloidal-sized particles under 5 nanometers in diameter which then aggregate into three-dimensional gel structures of short rods. Once deposited as silica gel, it is not further degraded or dissolved inside the living plant. However, when the plant dies, the more resistant silica bodies can survive in the soil for a long time as phytoliths and eventually enter the cycle through dissolution.

Deposition of silica in higher plants differs from that in diatoms. In the diatoms, silica is enclosed in small vesicles which fuse to form a layer of silica adjacent to the cell membrane, the silicalemma. Irrespective of the mechanism of deposition, silica in plants and other organisms is amorphous and noncrystalline. Biogenic silica is denoted chemically as $SiO_2 \cdot nH_2O$, with water of hydration varying to a considerable extent.

Biogenic silica can be studied by using a variety of microscopic techniques (Figs. 1–6). Treatment with strong acids (other than hydrofluoric acid) destroys almost the entire cell, except the resistant silica

Fig. 1. Scanning electron micrograph of living diatoms. An entire diatom (right) is seen beside two halves that make up a single diatom. Numerous epiphytic diatoms that grow on the surface are also seen.

and they can be used to distinguish various groups of silica-accumulating species from each other. An interesting feature of grasses is the difference that exists in silica deposition in the C_4 and C_3 grasses. The C_4 grasses, such as sugarcane, corn, and millet, use light energy more efficiently in food production. They differ from the C_3 grasses in making four-carbon acids in addition to three-carbon acids in photosynthesis. The C_4 grasses have 4.5 times more silica cells than the C_3 grasses. One possible reason for the greater efficiency of the C_4 grasses may be that these silica cells can act as "windows" to let more light penetrate to the inside of the leaf.

Even in silica-accumulating grasses, there are regions which contain very little or no silica. These include the growing regions at the bases of elongating internodes, called intercalary meristems; such regions later become silicified to strengthen the stem

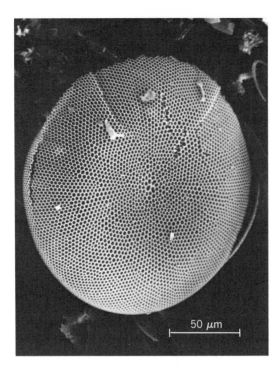

Fig. 2. Scanning electron micrograph of the wall of a living diatom showing intricate network of silica.

structure. Diatoms preserve their cell shape and intricate structure even after millions of years. Special staining techniques can also be used to localize the silica in plants (Fig. 5). When thin sections of a silica body are examined with a transmission electron microscope, silica appears as elongated beads linked to each other (Fig. 6).

Patterns of silica deposition. The patterns of silica deposition in plants are genetically determined,

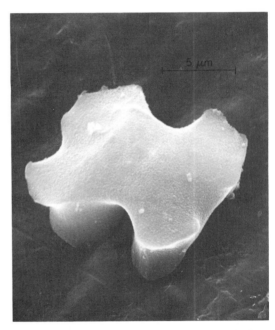

Fig. 4. Scanning electron micrograph of a silica body from a silica cell of the rice epidermis.

after growth ceases. Another important site is the swollen base of the leaf sheath that encloses the stem, called the leaf sheath pulvinus, which is the organ responsible for erecting the plant when the plant is horizontally prostrated by the actions of wind and water or animals. This region "recognizes" its relative position in the Earth's gravitational field. The lack of silica in this organ is of obvious significance when the tissues are stimulated to grow in order to erect the plant.

Mechanisms for silica deposition. Possibly there are two separate mechanisms by which silica is deposited in plants, one purely mechanical and the other involving metabolic reactions at the site of deposition.

Upside-down filter-cake model. There are plants in which silicon is carried in simple solution as silicic

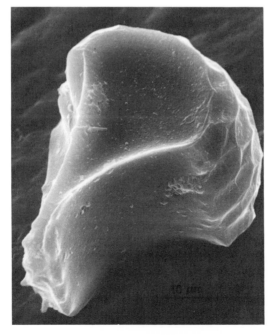

Fig. 3. Scanning electron micrograph of a silica body isolated from a bulliform cell of the rice leaf upper epidermis.

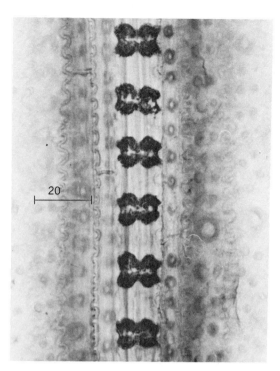

Fig. 5.   Silver-ammonium-chromate staining reaction for a row of silica cells in rice leaf epidermis.

acid, Si(OH)$_4$, in the xylem sap and into stems and leaves. Deposition is controlled by the pH value. Particles are formed by polymerization and deposited by ultrafiltration. Here, the shape of the silica mass is molded by the shape of the cavity next to the ultrafilter membrane. The following is a description of how this might occur.

Polymerization of silicic acid to the gel state can be divided into two stages: (1) the supersaturated solution of Si(OH)$_4$ first forms particle nuclei about 1–3 nm in diameter. This involves only a small part of the silicic acid. The monomer then polymerizes or adds to the surface of the particles, which grow to 3–5 nm. (2) At this point, the silica has a specific area of 500–1000 m^2/g. The particles then chain together to form a rigid network. During these further changes, the area is reduced by the degree to which the particles have grown together. It is predicted by the authors that most of the silica found inside plant tissues, when properly isolated, will be found to have a specific area of 400–500 m^2/g, regardless of the plant. The final area is likely to be determined by the pH and temperature.

Silicic acid is a very weak acid with a pK$_a$ of about 9.8. However, after particles are formed, the SiOH groups on the surface are much stronger, with a pK$_a$ of around 7.0. At this pH, the monomer is not ionized, but the surface of the polymer is ionized and consequently absorbs cations, such as Na$^+$, at the surface. As surface area is reduced by later changes forming the gel, Na$^+$ is liberated, resulting in an increase in pH. Thus, the pH of the region where silica is being deposited is higher than elsewhere.

It may be assumed that the concentration of sili-

cic acid in the sap moving up the xylem is higher than the solubility of amorphous silica, and may consist of a saturated solution of 100–150 ppm Si(OH)$_4$, along with the complex ions of silicon chelated with a tropolone derivative, (thpl)$_3$Si$^+$. The idea, then, is that when the pH rises above 6.8 or 7.0, the complex decomposes and more Si(OH)$_4$ is liberated, providing the supersaturation necessary for polymerization. The liberated Si(OH)$_4$ polymerizes to a low-weight polymer or to hydrated silica particles, 1–3 nm in diameter, in suspension (when particles this small are in suspension, it is the same as being in colloidal suspension, like a low-molecular-weight protein). In order for polymerization to take place, the SiO$_2$ concentration must be at least 100–300 ppm. P. Kaufman and R. K. Iler have actually measured the SiO$_2$ concentration in xylem sap collected from *Equisetum* internodes; sap can be collected with a Pasteur pipette from the "cups" formed by the leaf bracts after severing the young internode at the intercalary meristem locus. Using the molybdate blue method for silicon determination, the SiO$_2$ in xylem sap from mature shoots of *E. hyemale* var. *affine* was found to be 250 ± 25 ppm. Thus, in *Equisetum* shoots, the conditions for polymerization are satisfied. Rice plants grown in nutrient solutions containing 10–100 ppm SiO$_2$ had SiO$_2$ concentrations in the xylem sap of shoots varying between 150 and 400 ppm 21 h after adding SiO$_2$ to the nutrient solution; even after 37 h, the SiO$_2$ concentration in the xylem sap was between 120 and 650 ppm. The SiO$_2$ concentration varies in direct proportion to the amount of SiO$_2$ added to the nutrient solution. Thus, under these experimental treatments, the conditions for SiO$_2$ polymerization in rice shoots are also satisfied.

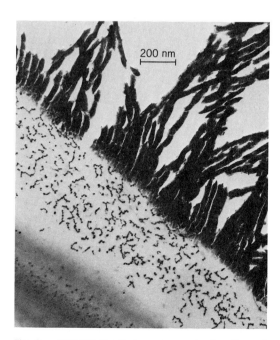

Fig. 6.   Transmission electron micrograph of a silica cell from the epidermis of oat stem showing amorphous silica bodies growing out from the cell wall as elongated beads.

The silica deposition process can now be visualized as follows in shoots of grasses and *Equisetum* (Fig. 7). In the intercalary meristem locus at the base of each internode, the monosilicic acid in solution moves upward with the transpiration stream. No polymerization occurs in the intercalary meristem locus while it is generating new cells for the elongating internode because of the low pH conditions. The same applies to the pulvini loci at both internodes and sheath bases of panicoid and festucoid grasses. Above the intercalary meristem at the upper end of the internode, the tissues collectively act as an ultrafilter (like a very-fine-pored membrane) and remove any silica polymers but still pass the monomer, $Si(OH)_4$, at the equilibrium concentration. Thus, an "upside-down filter cake" builds up in the upper portion of the internode, first in the epidermal tissue, and later in the internal tissues.

The older the internode, the thicker (or longer) the zone that is rich in silica. The deposited silica can, of course, be interspersed with organic polymers, such as the cellulose in cell walls, or it can fill channels such as intercellular spaces, but it is doubtful whether it could penetrate cell membranes. Thus, in silica cells of grasses, when the cell membranes break down during differentiation, the membrane barrier is removed, and monosilicic acid can polymerize in the lumens of these cells.

This is a somewhat mechanical concept of the silicic acid polymerization process and may be oversimplified, but so far as the authors know, none of the facts rule out this general model. Further testing must await results on measurement of internal pH at different levels of developing internodes and leaves, including pulvinus and intercalary meristem loci, and measurement of $Si(OH)_4$ concentrations in the xylem sap of rapidly developing and mature internodes of different grasses and *Equisetum* species.

Metabolic model. It is difficult to believe that in locations where sap movement does not occur, as in the tops of nettle hairs, any kind of ultrafiltration mechanism could be involved in molding the shape or pattern of the silica deposit. In this case, there must be some kind of membrane surface to which silica particles are attracted by ionic forces to form a layer of silica, which then grows thicker as monomeric silicic acid is deposited from supersaturated solution. Alternatively, the cell may produce cationic molecules that coat the first layer of particles so that the surface will attract another layer of colloidal particles. There are examples of such mechanisms in studies of the adsorption (or deposition) of organic polymer on the underside of monomolecular films at the surface of water.

The key difference in such a system, compared to that of a mechanical system, is that there is no evaporative process (transpiration) to concentrate $Si(OH)_4$ to a supersaturated state. Thus, some other mechanism must be involved. Possibly it is the formation of soluble chelates of silicon which become concentrated by physiological processes. These are in stable solution until the pH is changed, or the chelating agent is destroyed by enzymatic processes, thus liberating silicon in a very highly localized area. This involves metabolic control of the exact sites of deposition. There must be highly localized events that can lead to formation of silica in specific and characteristic forms, such as in silicified trichomes (hairs).

Both mechanisms could, of course, operate in the same plant at different locations. But in some plants, where the silica appears as casts of intercellular spaces, only the first mechanism need be involved. This is likely to be seen in plants grown at less than 100% humidity.

For background information *see* PHOTOSYNTHESIS; PLANT MINERAL NUTRITION; PLANT PHYSIOLOGY in the McGraw-Hill Encyclopedia of Science and Technology.

[P. B. KAUFMAN; P. DAYANANDAN;
C. I. FRANKLIN]

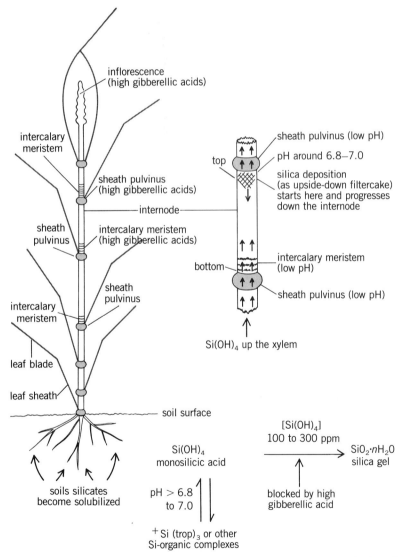

Fig. 7.   Diagrams illustrating the upside-down filter-cake model to explain mechanism of silica deposition in shoots of vascular plants which accumulate large amounts of silica. The grass shoot (highly schematic) is used as the example in the diagram. (*After P. B. Kaufman et al., Silica in shoots of higher plants, in T. Simpson and B. E. Volcani, eds., Silicon and Siliceous Structures in Biological Systems, 1981.*)

Bibliography: P. Dayanandan et al., Structure of gravity-sensitive sheath and internodal pulvini in grass shoots, *Amer. J. Bot.* 64:1189–1199, 1977; R. K. Iler, *The Chemistry of Silica*, 1979; J. D. Jones, P. B. Kaufman, and W. L. Rigot, Method for determining silicon in plant materials by neutron activation analysis, *J. Radioanalyt. Chem.*, 50(1,2):261–275, 1979; P. B. Kaufman et al., Silica in developing epidermal cells of *Avena* internodes: Electron microprobe analysis, *Science*, 166:1015–1017, 1969; P. B. Kaufman et al., Silica in shoots of higher plants, in T. Simpson and B. E. Volcani, (eds), *Silicon and Siliceous Structures in Biological Systems*, 1981; J. C. Lewin and B. E. F. Reimann, Silicon and plant growth, *Annu. Rev. Plant Physiol.*, 20:289–304, 1969; Y. Takeoka, P. B. Kaufman, and O. Matsumura, Comparative microscopy of idioblasts in lemma epidermis of some C$_3$ and C$_4$ grasses (Poaceae) using SUMP method, *Phytomorphology*, 29(3,4):330–337, 1979.

Pogonophora

Riftia pachyptila, the so-called giant tubeworm of the geothermal vents of the Galápagos Rift, was first collected in 1977. At certain of the vents it is the dominant member of the unique community of animals that is restricted to these unusual marine environments. Subsequent collections of *Riftia* were made on the East Pacific Rise at 13°N and at 21°N, as well as in the Guaymas Basin in the Gulf of California. All of these are sites of active tectonic plate movements involving either sea-floor spreading or faulting. At a depth of about 2500 m (8200 ft), the water temperature away from the vents is approximately 2°C (36°F). An appreciable amount of dissolved oxygen is present, but there is no hydrogen sulfide. In contrast, the temperature of the water issuing from the geothermal vents is up to 23°C (75°F); it contains no dissolved oxygen, and there is a rather high level of hydrogen sulfide, a substance usually considered to be toxic to animals.

Morphology. The largest specimen of *Riftia* is 1.5 m (60 in.) long and 37 mm (1.5 in.) in diameter, an amazing size for an animal which, like other members of the phylum Pogonophora, lacks a mouth, gut, and anus (Fig. 1). The most striking feature of the living animal is the bright red plume that extends from its white tube. The color of the plume is due to the hemoglobin of the blood passing through the blood vessels of the branchial tentacles (also called gill filaments) that form the plume. The largest specimen of *Riftia* may have as many as 224,000 individual tentacles.

The tubes are translucent to opaque; are closed at the basal end; and are affixed to a solid substratum, such as bare rock, other tubes, or mollusk shells. The tubes are quite tough but flexible, and are composed of many thin layers of a protein-chitin mixture secreted by the worm.

The most anterior of the four body regions of *Riftia* is the plume, a mass of gill filaments supported centrally by an axial structure, the obturaculum. The

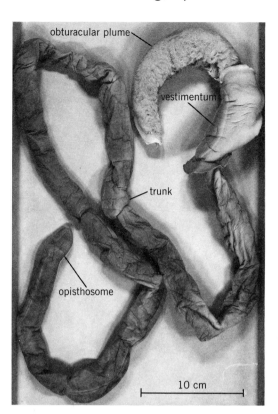

Fig. 1. Overall view of *Riftia pachyptila*. (*From M. L. Jones, Riftia pachyptila, new genus, new species, the vestimentiferan worm from the Galápagos Rift geothermal vents (Pogonophora), Proc. Biol. Soc. Wash., 93(4):1295–1313, 1981*)

second region, the vestimentum, is collarlike and has lateral extensions that meet over the dorsal surface to form a space into which the genital ducts open. Many glands that secrete tube material open on the outer surface of the vestimentum. The third region, the trunk, is relatively undifferentiated, although some tube-secreting glands are present; however, they are fewer in number than those in the vestimentum. The fourth and most posterior region, the opisthosome, is segmented because of the presence of internal septa that form as many as 100 separated spaces. The more anterior of these segments each bears thousands of minute, toothed bristles (setae) that allow this entire region to act as a anchor when the worm contracts to withdraw into its tube.

The obturaculum, which supports the plume, is a paired structure consisting of an internal gelatin-like material bounded by several layers of muscle. Along its length are transverse rows of filamentous branchiae (gill filaments), which are fused basally to form transverse lamellae (Fig. 2). Each branchia is provided with vessels that carry blood to the branchial tip and back. These vessels are connected along their length by intraepithelial capillary loops. The plume, as a whole, forms an organ of exquisite design for the exchange of gases or the uptake of compounds of low molecular weight.

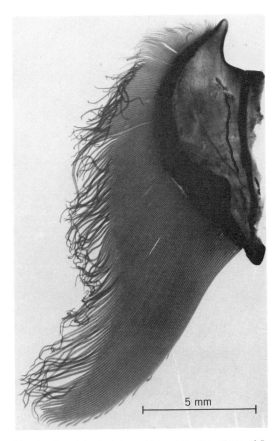

Fig. 2. Complete single branchial lamella from plume of *R. pachyptila*, obturaculum to the right. (*From M. L. Jones, Riftia pachyptila, new genus, new species, the vestimentiferan worm from the Galápagos Rift geothermal vents (Pogonophora), Proc. Biol. Soc. Wash., 93(4):1295–1313, 1981*)

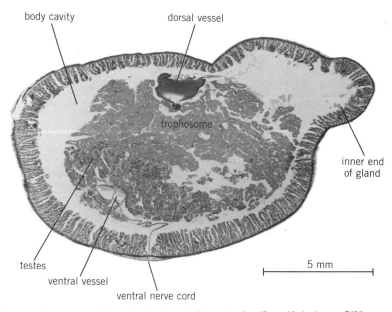

Fig. 3. Transverse section of trunk of male *R. pachyptila*. (*From M. L. Jones, Riftia pachyptila, new genus, new species, the vestimentiferan worm from the Galápagos Rift geothermal vents (Pogonophora), Proc. Biol. Soc. Wash., 93(4):1295–1313, 1981*)

The anterior region of the vestimentum is the site of the brain, from which arises the ventral nerve cord. Appearing at first as a pair of united cords, the pair separates and each segment diverges posteriorly to surround a ventral ciliated field that is underlain by a system of blood sinuses. Toward the posterior part of the vestimentum, the ventral nerve cords come together once again and remain united throughout the length of the trunk and opisthosome (Fig. 3). No ganglia have been observed along the length of the ventral nerve cord. The vestimentum is a solid structure and is composed of muscles, connective tissue, and tube-forming glands. There are no major body cavities. The dorsoposterior region of the vestimentum bears an anterior extension of the spacious cavity of the trunk. The male and female genital apertures are located on the dorsum of the vestimentum.

The extensive single pair of cavities of the trunk are lined with a thick layer of longitudinal muscles and a thin layer of circular muscles, and are dotted with the inner ends of glands. The dorsal blood vessel, probably carrying blood toward the plume, and the ventral vessel, probably carrying blood toward the opisthosome, traverse the length of the trunk. Their only branches lead to and from a central solid mass of variable development, the trophosome, and to and from the male or female gonads. The trophosome is invested with blood sinuses and is penetrated by small blood vessels and lacunae; it has no cellular organization apart from the vascular elements.

Symbiotic bacteria. The overall granular appearance of the trophosome at lower magnification can be recognized, at higher magnifications, as symbiotic bacteria, present in vast numbers. Because of the enzymes present in these bacteria, it has been suggested that the manufacture of organic carbon molecules takes place here. The energy for this bacterial processing is derived from the oxidation of sulfide carried from the outside environment by the blood of *Riftia* to the trophosomal bacteria. It has been demonstrated that the ratio of carbon-13 to carbon-12 in organic materials produced by the bacteria is the same as that found in the muscle and connective tissue of the vestimentum. This suggests that the carbon, fixed by the trophosomal bacteria, can be incorporated into the general tissue of *Riftia*. Thus, the bacteria or their metabolites serve as internal and nonphotosynthetic sources of "food" for these mouthless, gutless worms. The structure of the plume, with its prospective ability of taking up small organic molecules from the surrounding seawater, suggests a second method of obtaining nutrients.

Reproduction. The sexes of *Riftia* are separate, the only external difference being the presence of a pair of ciliated furrows leading forward from the male genital apertures; females have no such furrows. Unfertilized eggs are about 78 μm in diameter and sperm are about 80 μm in length, including tails. The body of spermatozoa are helical and give the sperm a corkscrew shape. No fertilized eggs or subsequent developmental stages have been observed. The smallest known *Riftia* is 750 μm in length; it had already formed its own tube on an adult tube and was essentially a small adult, with a plume, vestimentum, trunk, and opisthosome.

Other organisms in vent ecosystems. *Riftia pachyptila* is a member of a community of animals restricted to active geothermal vents. Other animals of the ecosystem include a crab, *Bythograea thermydron*; a shrimp, *Alvinocaris lusca*; a clam, *Calyptogena magnifica*; a leech, *Bathybdella sawyeri*; an undescribed mussel; and a number of undescribed limpets, shrimp, polychaetous annelids, and copepods. There have been some observations of *Bythograea* and *Alvinocaris* feeding on the plume of *Riftia*, and there are a number of museum specimens of *Riftia* with injured plumes, although the specimens had survived until collection.

Taxonomy. Along with other pogonophorans, *R. pachyptila* appears to be most closely related to the annelids. The primary reason for this conclusion lies in the structure of the opisthosome, the most posterior body region. Septa, similar to those occurring in worms of the phylum Annelida, delimit up to 100 segments, and most of these segments bear setae that have a basic structure and denticulation similar to that of the polychaetous annelids.

In addition to lacking a mouth, gut, and anus, the phylum Pogonophora is characterized by the presence of four regions: a tentaculate anterior region; a region that functions in tube formation; a long, relatively unspecialized trunk, which contains the gonads and symbiotic bacteria; and a posterior, segmented region. With the discovery off the coast of southern California of what was considered to be an aberrant pogonophoran in 1969, a separate branch of classification was suggested for this new species. This involved a new class (Afrenulata: lacking tube-forming ridges in the second body region), a new order (Vestimentifera: anterior plume with numerous fused tentacles or branchiae around a central axial support, and the second region collarlike with lateral folds meeting at the midline), a new family (Lamellibrachiidae), a new genus (*Lamellibrachia*), and a new species (*L. barhami*). In 1975 a second species (*L. luymesi*) from off the coast of Guyana in the Atlantic was described, differing from the first mainly in the relative length of the second region. Then, in 1981, a third species of aberrant pogonophoran was described from collection sites on the Galápagos Rift and from the East Pacific Rise at 21°N. On the basis of the structure of the plume, the number of excretory duct openings, and the general shape of the animal and its tube, a new family (Riftiidae) was established, with a new genus (*Riftia*) and a new species (*R. pachyptila*). Further, in consideration of the many aberrant and unique morphological characters of *Lamellibrachia* and *Riftia*, while recognizing that there is a definite, basic relationship to all other pogonophorans, it was decided that the two

genera should each be set apart in its own subphylum. Thus, the phylum Pogonophora is now divided into two subphyla: Perviata (those pogonophorans with a first body region lacking a central axial structure that serves as an operculum and closes off the tube opening) and Obturata (the two genera bearing an obturaculum).

For background information *see* ATHECANEPHRIA; POGONOPHORA; THECANEPHRIA in the McGraw-Hill Encyclopedia of Science and Technology.

[MEREDITH L. JONES]

Bibliography: C. M. Cavanaugh et al., Procaryotic cells in the hydrothermal vent tube worm *Riftia pachyptila* Jones: Possible chemoautotrophic symbionts, *Science*, 213(4505):340–342, 1981; H. Felbeck, Chemoautotrophic potential of the hydrothermal vent tube worm, *Riftia pachyptila* Jones (Vestimentifera), *Science*, 213(4505):336–338, 1981; M. L. Jones, *Riftia pachyptila* Jones: Observations on the vestimentiferan worm from the Galápagos Rift, *Science*, 213(4505):333–336, 1981; G. H. Rau, Hydrothermal vent clam and tube worm $^{13}C/^{12}C$: Further evidence of nonphotosynthetic food sources, *Science*, 213(4505):338–340, 1981.

Pollution

The use and disposal of chemicals has become an issue of major importance for environmental health in modern industrial society. It is estimated that more than 60,000 chemicals are now used in industry. About 5000 of these are produced in quantities greater than 400,000 kg annually, and 300–500 new chemicals enter production each year.

Possibilities for environmental pollution and for damage to human health exist in every phase of production, transportation, use, and ultimate disposal of chemicals. A broad range of laws have been enacted over the past decade to control such exposures and to protect public health and the environment. The problem is highly complex because of: the diverse properties of different chemicals, the variety of possible exposure pathways, and the difficulties in assessing the effects on human health.

Chemical properties. Toxic or potentially toxic chemicals are present throughout the entire spectrum of organic and inorganic compounds. The hazardousness of any one chemical is a matter of: its inherent toxicity; its capacity to interact with other chemicals to produce further toxic materials either in the environment or in the process of biologic modification (metabolism); and its ability to persist in environmental media or in tissue. Unfortunately, present understanding of toxic potential for most chemicals is severely limited, especially with respect to possible chemical synergy. The degree of inherent toxicity ranges from compounds with high initial toxicity but low biologic half-life, to materials with relatively low biologic activity but great stability in tissue and in the environment. Of particular concern are compounds such as lead and polyhalogenated biphenyls (for example, polychlorinated biphenyl, or PCB), or certain organic pesticide residues that are widely dispersed in large quantities, have at least moderate toxic potential, and are highly stable and persistent in the environment and in biological systems.

Exposure pathways. Biological systems, human or otherwise, are exposed to chemicals directly by contact with polluted air, water, or soil or indirectly through chains of food production. Exposures can occur accidentally or through unforeseen consequences of chemical use or waste-chemical disposal practices. Transportation accidents, particularly train derailments, are frequently cited as episodes of chemical exposure. Food contamination is an ever-present possibility which requires constant attention to safeguarding procedures for producing and handling food. A striking example of food contamination was the massive introduction in 1973 of the flame-retardant chemical polybrominated biphenyl (PBB) into dairy products in Michigan. The accident occurred through a mixup of PBB with cattle feed supplement. The potential toxicity of the chemical, together with its great biological and environmental durability, has made this particular pollution episode a major and continuing public health problem.

Although legal safeguards now address most conditions for occupational handling and consumer use of chemical products (for example, pesticide practices and the use of lead in paint and gasoline) and for disposal of industrial waste by-products, unforeseen exposures continue to occur, whether from lapses in current practices and gaps in protective rules or from exposures to persistent chemicals introduced into the environment before safeguards were enacted or considered necessary. Community exposure to organic pesticide by-products at the Love Canal dump site in Niagara Falls in the 1970s illustrates this problem as it relates to industrial chemical waste disposal. Erosion of the clay cap seal at this chemical waste site allowed rainwater to percolate waste chemicals through porous topsoil to yards and basements of nearby homes. Human contact with toxic chemicals in this setting occurred through direct contact on the unrestricted site and through chemicals dispersed in the air, soil, and surface water.

A different mode of human exposure to toxic industrial chemicals is illustrated by the presence of the pesticide DDT and its chemical residues in the small town of Triana, Alabama. A DDT manufacturing plant near that town disposed of chemical waste by discharging it into an adjacent stream. DDT consumed by fish in the stream was in turn consumed by local people, for whom such fish was a major source of food. Through this chain of biological exposure, exceptionally high levels of DDE (dichlorodiphenyldichloroethylene, a DDT congener) developed in local residents. While the health consequences of this episode are not yet clear, the toxicity of DDT compounds for laboratory animals, coupled with the persistence of such chemicals (like PCB and PBB) in fatty tissue, makes the problem a prime public health concern.

A particular issue concerning exposure pathways and the establishment of secure disposal practices for waste chemicals is protection of groundwater sources. While a toxic-waste site may pose health risks through direct exposures by air, soil, and surface water, the potential for chemical seepage through soil into underlying aquifers poses a greater danger. Once contaminated, these aquifers could affect more distant populations who may depend upon the aquifers as sources of drinking water. Toxic-waste sites not only should be located apart from areas of human activity, where potential direct contacts might occur, but should also be designed to prevent seepage of toxic material into ground-water. Construction of disposal sites should therefore be undertaken in locations where dense clay soils can block chemical migration and where groundwater pathways are not in close proximity. After construction of such a dump site, regular monitoring for possible surface-water contamination and groundwater seepage becomes, of course, a standard prerequisite.

Human health effects. Assessment of human health effects in relation to toxic chemical exposures is a complicated matter in which three aspects are of central importance: dose of chemical exposure, latency or interval between exposure and disease production, and nonspecificity of disease patterns. While it may be, as with ionizing radiation, that no dose level of chemical exposure may be without biological effect, severity of response is clearly a matter of dose. For most chemicals, immediate health effects (skin rashes, for example) result only from relatively high exposures. Dose levels usually seen with non-occupational toxic-waste exposures are considerably lower than those associated with acute illness. Instead, medical concern is for delayed health effects, such as cancer or chronic pulmonary disease, or for illness in population segments where susceptibility may be heightened, for example, the effects in young children, or in developing fetuses. In the absence of acute illness, attention also focuses on possible subclinical abnormalities, such as disturbances in liver enzyme levels or in immunologic or hormonal function, which might foreshadow clinical disease.

If an illness such as cancer or chronic pulmonary or cardiovascular disease is caused by toxic-waste exposure, in all likelihood it entails considerable passage of time (15–30 years in the case of most cancers) from the start of such exposure. Linking illnesses to their biological causes is obviously difficult under such conditions, particularly when the expected frequency of such conditions is relatively low. Long latency and low frequency are unfortunately expected features of cancer and other chronic illness for which there are possible toxic chemical causes. These two features of such diseases make it particularly difficult to make scientific determinations of cause.

A further difficulty is the fact that, aside from acute sickness arising soon after high exposure, almost no disease process can be identified as clinically specific for a particular chemical exposure. While rare exceptions exist (mesothelioma following exposure to asbestos, for example), this means that no particular case of illness can be attributed with certainty to any particular chemical cause. To establish such links usually requires complex and sustained epidemiologic investigation. Such studies require relatively large populations in which to record observations (especially if the diseases to be studied are rare), and they must take into account the possible competing effects of other disease causes, chemical or otherwise.

Thus far, in situations such as the Love Canal dump site, DDT in Triana, and PBB in Michigan, no clear links have been established between chemical exposures and particular human illness. Because of the known liver toxicity of such compounds as DDT and PBB, particular studies have been made regarding liver function in exposed populations in the latter two settings. At Love Canal, where diverse organic chemicals were involved, very few with capacity for persistence in tissue, studies linking specific health measures to specific tissue levels of chemicals have not been possible. Instead, attention has focused on the reproductive history of women living near the canal in an effort to define a subpopulation at particular high risk. Thus far, no clear evidence of reproductive abnormality (increased frequencies of spontaneous abortions, congenital malformations, or low birth weight) has been found, either because the subpopulation was too small to statistically detect such an effect or because no increase in abnormality was caused by exposure to Love Canal chemicals at the dose levels present. Studies such as these suggest that the investigation of toxic-waste health effects will become easier only when improved means have been developed to link specific illnesses to specific causes or to identify particular high-risk populations more precisely.

For background information *see* ENVIRONMENTAL PROTECTION; MUTAGENS AND CARCINOGENS; TOXICOLOGY in the McGraw-Hill Encyclopedia of Science and Technology.

[CLARK W. HEATH, JR.]

Bibliography: D. T. Janerich et al., Cancer incidence in the Love Canal area, *Science*, 212:1404–1407, 1981; K. Kreiss et al., Cross-sectional study of a community with exceptional exposure to DDT, *J. Amer. Med. Ass.*, 245:1926–1930, 1981; M. S. Wolff et al., Human tissue burdens of halogenated aromatic chemicals in Michigan, *J. Amer. Med. Ass.*, 247:2112–2116, 1982.

Powder metallurgy

Powder metallurgy is a process for forming precision metal articles directly from metal powders to net shape (or nearly net shape) of the desired parts. In conventional powder metallurgy, metal parts are formed by mechanically compacting the powder in a die at high pressure and by subsequently sintering the "green" compact to improve its physical properties. Recently, an extension of conventional techniques termed plastically formed powder metallurgy has become available. In this process, extremely fine metal powder (typically less than 10 μm) is blended with plastic materials, and the mixture is then molded into green parts by conventional plastic-molding techniques, for example, injection molding. The plastic binder is removed from the green parts, which are then sintered to achieve the desired physical and chemical properties (see illustration).

Sintering. Sintering is the heating of the green article to a high temperature (but below the melting temperature of the article) in a controlled atmosphere to develop its final properties. Pure, finely divided crystalline solid particles undergo sintering whenever the atomic mobility is sufficient to bring about internal mass transport under the influence of surface tension. Shrinkage, and hence densification, of a population of particles in intimate contact can be observed when D/r^2 (the ratio of volume diffusivity to the square of the particle radius) is about 10^{-6} s^{-1} in ionic solids, such as oxide ceramics, and 10^{-4} s^{-1} in metals. The theoretical treatment of various sintering mechanisms, for example, diffusion, liquid phase, and evaporation-condensation, has shown that sintering rates increase as the particle diameter decreases. Generally the relationship between sintering rate and particle diameter is greater than inverse linear.

A powder compact is composed of individual particles in intimate contact, having 20–60% porosity. To maximize many desired material properties, such as strength, ductility, permeability, and conductivity, it is desirable to eliminate as much of the initial porosity as possible. In other applications, it may be desirable to control the porosity at some level, for example, in filters.

The driving force for densification is the free-energy change that occurs during sintering by the elimination of solid-vapor interfaces. This usually takes place with the coincidental formation of new, lower-energy solid-solid interfaces. The net energy decrease that occurs on sintering a 1-μm-size powder corresponds to a free-energy decrease of about 10 cal/g (41.868 joules/kg). On a microscopic scale, material transfer is affected by the pressure difference and changes in free energy across a curved surface. If the particle size, and consequently the radius of curvature, is small, these effects may be of substantial magnitude. Hence, when the finest possible particles as the building blocks in a powder compact are used, the highest densities will be obtained. Thus, the smaller the particles that can be effectively utilized, the more favorable the sintering dynamics.

Metal dusts. By convention, the term dust is used to indicate a population of particles less than 10 μm in diameter, and the term powder is used to indi-

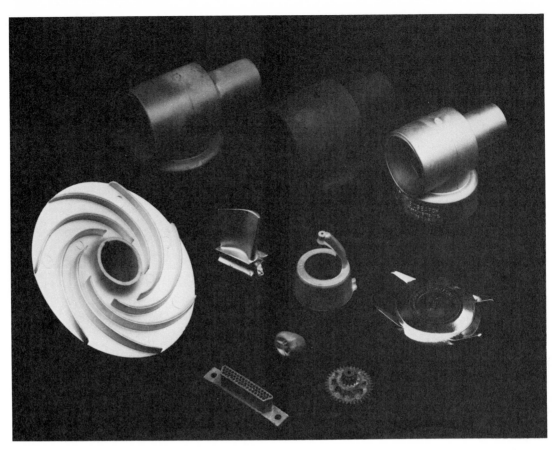

Plastically formed powder metallurgy parts. Note the difference between the as-molded, debinderized, and sintered parts. (*Multimaterial Molding Inc.*)

cate a population of particles greater than 10 μm in diameter. However, dusts, and especially metal dusts, have many characteristics that make their effective utilization difficult. Their high free surface energy and chemical reactivity with atmospheric constituents have indeed given metal dusts a bad reputation in industry. Metal dusts are pyrophoric and are relatively easily dispersed in air, causing dust explosion hazards. Airborne metals have also been shown, or at least suspected, to be carcinogens at worst and irritants at best.

Conventional powder-metal press-and-sinter methods are not particularly effective in producing parts from small particles due to the difficulties involved in the mechanical compacting of metal dusts. For example, the sintering of a low-alloy steel in which the initial particle size is 3 μm will result in the same final density of about 96% of theoretical density under the same sintering conditions for green parts with an initial porosity of 35, 45, or 55%. This indicates that unless the initial density of the green part is uniform, the sintered part will be seriously warped. Green parts that have been formed by compacting will generally display variations in the densities at different sections of the compacted piece. This effect is primarily due to wall friction in the die and is further amplified by the internal pressure angle of the compacting dust. For example, the green density of a compacted cylinder will be greater at the ends than in the middle, resulting in an hourglass shape in the sintered part.

The initial low density of metal dusts is another difficulty encountered in compacting dusts, causing large compaction volume-compression ratios. The resulting entrapped air will frequently cause cracks in the green part as the small pore size prevents rapid reduction of its pressure after compaction.

Plastically formed powder metallurgy. An effective method of utilizing metal dusts is the combination of the metal dust with a thermoplastic binder so that the pore volume is occupied by a thermoplastic material rather than atmospheric air. This method has several advantages: (1) By coating the dust particles with a thermoplastic, the particles are effectively removed from contact with the atmosphere. This greatly inhibits the chemical reactivity of the metal dust and virtually eliminates its pyrophoric nature (2) The thermoplastic effectively binds the particles into a uniform mass, eliminating airborne particles. Dust explosions and airborne-particle toxicity problems are eliminated. (3) The pore volume is occupied by an incompressible material; hence, the green part density is established by controlled, measurable mixture ratios of thermoplastic and metal dust. The green part density is obtained in a controlled mixing process rather than in an uncontrolled compaction process. (4) The mixture of the thermoplastic and metal dust will, under correct conditions, form a thermoplastisol feedstock that can be formed into complex geometries by conventional plastic-forming techniques. A green part of complex geometry can be injection-molded, vacuum-formed, and so on, at relatively low pressures and temperatures.

Before sintering can be accomplished, the thermoplastic binder must be removed from the pore volume. In the debinderizing process the binder is converted to a fluid (liquid, gas, or a combination of both) in stages so that the binder can flow from the green body to an external region that is maintained at a lower physical-chemical potential than the interior of the body. This can be accomplished by a variety of methods, but in order to be successful, the particle-particle tensile forces cannot be exceeded by any of the removal forces, or cracking, deformation, and bloating of the part will occur. This requirement places narrow limits on debinderizing which makes a much closer process control necessary, as compared to the conventional press-and-sinter processing. The debinderized parts are then sintered to produce the final dense part. Substantial shrinkage, on the order of 18%, takes place as the part densifies.

Plastically formed P/M parts that are produced from metal dusts have greatly increased geometrical capability (undercuts are easily made). The material properties are more similar to these of forged materials, for example the elongation of mild steel parts is superior to 30%. The densification forces here are not dependent upon high mechanical pressure as in conventional power metallurgy, but rather are a property of the dust.

The maximum sintered density is generally limited by the sintering atmosphere. In a typical body based upon metal dust, the sintering process results in a closed pore matrix with a pore density of about 10%. Gases that are not able to escape from the closed pores by diffusion and solution will be compressed as the sintering progresses. Inert gases, such as nitrogen or argon, will be compressed to about 10 atm (150 psi or 10^6 pascals) at a residual porosity of 1%. While the gas pressure is increasing, the negative radius of curvature of the pores becomes small. Thus, the negative pressure produced by the sintering forces is increased approximately proportionally to $1/r$, whereas the gas pressure increases proportionally to $1/r^3$. To achieve full-density materials, sintering under vacuum or special atmospheric conditions is employed. For example, injection-molded tungsten can be sintered to a density of 100% ($\rho \simeq 19.3$) under vacuum.

Plastically formed power metallurgy, the relatively new technique, is finding wide application in many areas. Parts that are produced by plastically formed power metallurgy are used in typewriters, computers, paper-handling machines, medical and dental instruments, tools, valves, and aerospace hardware.

For background information *see* POWDER METALLURGY in the McGraw-Hill Encyclopedia of Science and Technology.

[RAYMOND E. WIECH, JR.]

Bibliography: L. D. Peck, Metal injection molding comes out of the laboratory, *Machine and Tool Blue Book*, pp. 80–91, October 1981; R. E. Wiech and I. J. Weisenberg, *Superproductivity through Metal Injection Molding*, SAE Tech. Pap. Ser. 810240, 1981.

Printing

Computer-aided layout is an automated method of drawing lines and geometrical shapes with speed and precision which are superior to manual techniques. Drafter and design engineers have been utilizing this technology, referred to as computer-aided drafting and design (CADD) systems, for the past 5–7 years. In graphic arts applications, computer-aided layout eliminates much of the tedious hand drawing and cutting operations of mechanical art and film preparation for printing plates.

The new tools of the trade are used by printers to perform the image assembly (stripping) function of producing shaping masks for halftone and screen tint images, operations that normally consume 30–50% of total image assembly time on complex work such as catalogs and annual reports. The more complicated masking functions, those requiring dropouts (blocked-out areas for halftones) and undercuts to mortise the edges of one image to another, are easily produced with a computer-aided layout system. Once the necessary masks are produced, the technician completes the assembly by laying in (stripping) the halftone films and screen tints, using the masks for image cropping and final positioning of image elements for platemaking. Publication printers find these systems especially useful for correcting improperly prepared artboards.

Several advantages make computer-aided layout attractive to the printing industry: (1) The system has potential for improving the quality and turnaround time of film preparation. (2) Computer-aided drafting machines are now available, as is software designed to do specific line art and film preparation functions related to publication printing. (3) The hardware has proved to be successful in other fields related to printing, for example, cartography and electronics (printed circuit boards). (4) Modular design systems allow installation of equipment without major disruption of normal production procedures and work flow. (5) Computer-aided layout is a human-engineered system in which a technician can communicate with the system without training in computer programming.

All computer-aided layout systems currently on the market have several features in common: an input

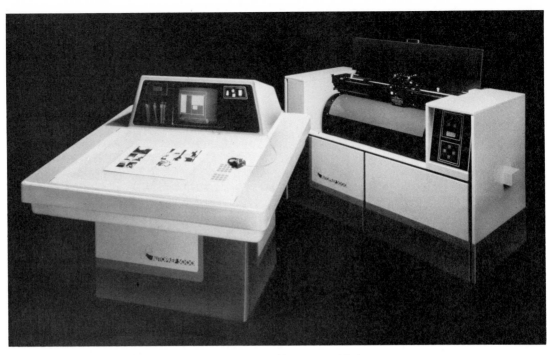

Fig. 1.   AutoPrep 5000 automated film system. (*Gerber Scientific Instrument Co.*)

station (an interactive image digitizer table with conversational input mode and disk storage); an output unit (a high-speed plotter to produce scribed, photographic, or masking film intermediates to augment the image assembly function); accuracy to ±0.004 in. (±0.102 mm) over a 48-in. (1.22 m) plotting area with excellent line edge sharpness and resolution; and equipment classified as modular (line graphics system) as opposed to an integrated image processing system that combines text, line art, and color graphics, performs color adjustment functions, and imposes multiple pages into final order for the press plate.

Gerber AutoPrep 5000. The Gerber AutoPrep 5000 is an automated film-preparation system consisting

Fig. 2.   DataPrep component of the AutoPrep 5000 system. (*Gerber Scientific Instrument Co.*)

of two modular components (Fig. 1): A DataPrep Station (Fig. 2) where the job is created, and an AutoPlot Station where photographic film masks are produced. The DataPrep Station consists of a 24 × 36 in. (61 × 91 cm) backlit digitizing table; a video display that provides scaled relationships of all line elements (except type); a puck-type digitizing cursor (detector for x-y coordinates); a datapad menu (input device for communication with the terminal); dual floppy disks; and a minicomputer. Figure 3 shows the high-precision film masks it produces. The AutoPlot Station features a drum-style, high-speed photoplotter which "draws" on photographic film over a 38 × 48 in. (97 × 122 cm) work surface with a precisely positioned beam of light at speeds to 21 in./s (53 cm/s). The AutoPlot Station also includes an integral vacuum power supply and plotter control unit. Register pins can be installed on the film vacuum drum to the printer's specifications.

Dainippon Cadograph System. The Cadograph System is an automatic drafting system for doing image layout and mask-cutting functions. Like the Gerber AutoPrep 5000, the Cadograph has an input digitizer that enables a skilled operator to manipulate images through a conversational input mode using a menu tablet, character display, and graphic display (video display terminal, or VDT) input-checking mode (Fig 4*a*). The output section of the Cadograph is the automatic drafting machine (Fig 4*b*). This is a flatbed "table" with a drawing head to create lines and geometric shapes according to data processed by the operator via the computer. Four different kinds of drawing heads are available: a pen head, cutter head, scribe head, and photo head. The drafting table has vacuum control that secures paper, scribe film, masking film, and lithographic film to it. The cutter head is selected for cutting masks for halftones and other image shapes from peelable masking film. The photo head (tungsten lamp) can draw line weights of 0.05–12.5 mm on unexposed lithographic film. The input digitizer can accommodate an original layout size of up to 24 × 33½ in. (610 × 860 mm), and the output section can produce shaping masks of up to 25 × 36 in. (640 × 920 mm).

Shukosha Cad-Ace System. The Cad-Ace System performs functions similar to the Cadograph and AutoPrep 5000 except that the Cad-Ace is not capable of photographic film output. The Cad-Ace is a computer-aided drafting system that cuts halftone masks from peelable masking film and creates graphic forms, illustrated maps, and business forms. The input digitizer table can accommodate an original layout size of up to 20½ × 30⁵⁄₁₆ in. (520 × 770 mm), and the output drafting table (plotter) can produce shaping masks of up to 23⁹⁄₁₆ × 33 in. (600 × 840 mm).

Advantages. Publication printers and film preparation houses benefit from computer-aided layout in several ways: the systems help relieve the production bottlenecks of conventional hand methods of preparing film assemblies for platemaking; they provide film output for intermediate four-color image assemblies that are vital to film preparation houses and publication printers for platemaking and for updating of pages for publication reprints; the systems offer printers new levels of mechanical accuracy and production efficiency in assembling two-, four-, six-, and eight-page imposed film flats; the systems are especially effective for augmenting assembly of complex magazine and catalog pages, which typically consist of multiple-color halftone images, screen tint backgrounds, and facing-page crossover images (single pictorial image that spans two pages); and finally, the computer-aided layout systems are being

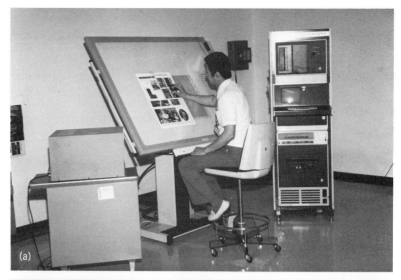

Fig. 4. Cadograph automatic drafting system. (*a*) Programming section. (*b*) Drafting section. (*Dainippon Screen Manufacturing Co., Ltd.*)

Fig. 3. Computer-produced film masks with all marks and windows in position, ready for assembly. (*Gerber Scientific Instrument Co.*)

accepted by technicians as new tools of the trade to enhance the quality and improve the productivity of prepress operations.

For background information *see* PRINTING in the McGraw-Hill Encyclopedia of Science and Technology.   [JEROME COZART]

Protein

Protein molecules have traditionally been described in static terms. The binding of a substrate to an enzyme, for example, has been compared to the linking of two pieces of a jigsaw puzzle. Recent studies have shown, however, that proteins are rather flexible molecules. Not only can proteins be deformed in response to applied forces, but the normal thermal energy in a protein causes its atoms to undergo sizable fluctuations around their average positions at all times. The resulting thermal motions range from irregular elastic deformations of the whole protein driven by collisions with solvent molecules, to chaotic librations of interior groups driven by collisions with neighboring atoms in the protein.

The flexibility and internal mobility of proteins are essential to much of their biological function. Hemoglobin and myoglobin provide good examples. These proteins serve as ferries for molecular oxygen. When the average structures of these proteins were determined by x-ray diffraction studies, however, no pathway was apparent by which oxygen could move between the outside of the protein and the binding site in the protein interior. The densely packed atoms of the protein appeared to present an impenetrable wall to the oxygen. The binding and release of oxygen in these proteins thus depends on the occurrence of structural fluctuations that open channels through this wall. This article describes recent studies that have clarified the nature of this and other important dynamical events in proteins.

Much of the new understanding of protein dynamics is due to theoretical studies. Of particular importance are molecular dynamics simulation studies. In these, the forces acting on the thousands of atoms in a protein are calculated; the forces are used in turn to compute the protein atom motions by solving Newton's equations of motion. The complexity of calculations is such that the motions can be simulated only for short intervals of time (about 10^{-10} s), even when fast computers are used. During these intervals, however, one can view the dynamics at the level of atomic detail. Specialized simulation methods have been developed to study certain slower motions. Increasingly detailed information about protein dynamics is also being developed by a variety of experimental methods, especially x-ray diffraction, nuclear magnetic resonance, and photophysical methods.

Local motions. The molecular dynamics method has been used to simulate the thermal motion in a variety of proteins. These studies show that at room temperature the typical magnitudes of the atomic displacements from their average positions are on the order of 50 picometers for atoms in the protein interior. That is, the amplitude of atomic motion is comparable to the sizes of the atoms themselves. Even larger fluctuations occur at the surface of the protein, where the atoms are not packed so tightly together. The illustration shows the atomic displacement magnitudes for a leucine residue in the protein cytochrome c. The axes of the ellipsoids represent the magnitudes of atomic displacement in each direction. The long axes of a number of ellipsoids point in the same direction; this shows that some of the motion in the protein has a collective character in which groups of atoms move together.

Examination of the time dependence of the atomic motion reveals a wide range of dynamic behavior. Many of the atomic displacements develop in large part over times on the order of 10^{-12} s. This corresponds to local motion of a bonded group of atoms relative to the groups that surround it. Collisions with these neighboring groups produce frictional effects

that slow the motion. This action is analogous to the viscous damping that slows the motion of a solute molecule in a liquid.

Modified simulation methods have been developed to study local processes that occur infrequently because of their large energy requirements. Such activated processes include the chemical reaction steps in enzymes. Application of these methods to study local structural changes involving side-chain isomerizations shows that these can be initiated by transient packing defects that give the side chain room to get a "running start." Frictional effects due to collisions can slow the rates of such structural changes considerably.

Large-scale motions. Motions that involve the collective displacement of many atoms tend to be intrinsically slow. This is due in part to the larger effective masses and frictional forces associated with larger groups of atoms. Some collective motions are apparent in molecular dynamics simulations (for example, collective displacement such as that in the illustration); these develop over 10^{-11} to 10^{-10} s. Slower motions must be studied by other methods, however.

A type of collective motion that is of particular biological interest is the relative displacement of two

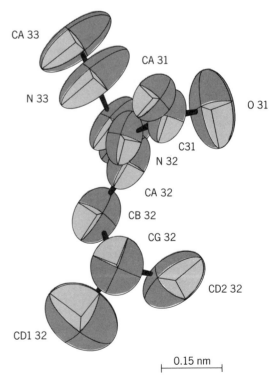

Ellipsoids representing the calculated magnitudes of the atomic position fluctuations for a leucine residue inside the protein cytochrome c. The first set of characters (for example, CD1) is the atom name (standard nomenclature for proteins), and the second set of characters (31, 32, or 33) is the number of the amino acid residue in the protein sequence. (*After J. A. McCammon and M. Karplus, The dynamical picture of protein structure, Acct. Chem. Res., 16, in press*)

or more globular domains that are linked by thin, hingelike regions in a protein. For the enzyme lysozyme, such motions may regulate substrate access to an active site that is located in the cleft between two globular domains. Such motions are apparently involved in the ligand-binding activities of a large number of other proteins, including enzymes, transport proteins, and antibody molecules. The character of such motions can be analyzed by using computer models to calculate the effective energy of bending the hinge and by estimating the effect of solvent molecule collisions on the protein deformation. For the L-arabinose-binding protein, the energy balance appears to be such that the open cleft of the protein is induced to close upon binding of an arabinose molecule within the cleft.

Experimental studies. A number of experimental methods have recently been developed that can provide information on the mobility of particular groups in protein molecules. X-ray diffraction, which has for many years been the key source of information on protein structure, can now be used to estimate the magnitudes of atomic fluctuations in protein crystals. The amplitudes of atomic motion determined in a recent x-ray diffraction study of cytochrome *c* were found to conform to those predicted in an earlier molecular dynamics calculation. Diffraction studies on myoglobin and several enzymes have pointed out regions of high mobility that may be involved in ligand-binding processes. Low-angle x-ray scattering is a related method that has been used to demonstrate the occurrence of large-scale hinge-bending motions in solution.

Detailed information on the actual time dependence of local motions in proteins has been obtained by nuclear magnetic resonance methods. These methods are sensitive to the rates of both short-time-scale motions and infrequent, activated motions. A variety of more specialized techniques have also been used to probe the time dependence of particular motions in proteins. An excellent example is the use of laser spectroscopic methods to monitor the rate at which oxygen and other ligands move through the myoglobin molecule to the internal binding site. These studies have shown that several energy barriers must be traversed during this motion, and that these barriers vary in time because of the motion of the protein atoms.

Protein function. The internal mobility of proteins is essential to their function. As has been mentioned, the motion within myoglobin creates transient channels that allow oxygen binding to occur. Moreover, the channels appear to discriminate against the binding of carbon monoxide and other harmful ligands. The conformational changes induced by ligand-binding in the L-arabinose-binding protein and other related proteins act as signals to stimulate such biological functions as the movement of bacteria toward sources of nourishment. Similar hinge-bending motions in enzymes act to bring together the groups necessary to catalyze chemical reactions of substrate molecules.

Dozens of other examples that connect protein function to protein dynamics could easily be cited. The recent work summarized here is providing the concepts necessary to understand the dynamics of protein function at the atomic level. The theoretical methods that are being developed will facilitate the interpretation of experimental data on protein function and will ultimately allow calculation of the rates of biochemical processes from first principles.

For background information *see* LASER SPECTROSCOPY; NUCLEAR MAGNETIC RESONANCE (NMR); PROTEIN; X-RAY DIFFRACTION in the McGraw-Hill Encyclopedia of Science and Technology.

[J. ANDREW McCAMMON; BORYEU MAO]

Bibliography: P. G. Debrunner and H. Frauenfelder, The dynamics of proteins, *Annu. Rev. Phys. Chem.*, 33, in press; F. R. N. Gurd and T. M. Rothgeb, Motions in proteins, *Adv. Protein Chem.*, 33:73–165, 1979; M. Karplus and J. A. McCammon, The internal dynamics of globular proteins, *CRC Crit. Rev. Biochem.*, 9:293–349, 1981; J. A. McCammon and M. Karplus, The dynamical picture of protein structure, *Acct. Chem. Res.*, 16, in press.

Psychology

Recently there has been renewed interest in the effects of the genes on human behavior. Partly this has been a result of the so-called IQ controversy, the dispute about the role genes play in determining differences in intellectual performance among individuals, and—more critically—among different racial and social class groups. The renewed interest in genes also reflects the emergence of sociobiology, the view that human nature in its social aspects is heavily influenced by its biological origins.

Psychologists share this renewed interest in behavioral genetics with zoologists, sociologists, economists, and philosophers. Nature-nurture controversies have had a long history in psychology; the nurture (environmentalist) side has been dominant during most of the last 50 years. During this period the study of learning and its effects was psychology's main theme. The current thrust of behavioral genetics is not so much to dispute the importance of learning in human behavior as it is to argue that in order to assess properly the effects of learning on behavior, one must first estimate and subtract out the effects of the genes.

Genes and personality. A good example of genetic studies' important role in psychological theory may be found in the area of personality development. Several behavioral-genetic methods permit the rough division of the population variation of a measured human trait into three portions: a portion associated with the genes, a portion associated with environmental factors common to members of a family, and a portion associated with environmental factors that differ among family members. An a priori argument for the importance of each of these three categories can be made. The genes should play a role, since it is possible to breed nonhuman mammals for temperamental and emotional characteristics, and it would be somewhat surprising if humans were completely unlike their biological relatives in this respect. Shared family factors ought to be important, because parents' child-rearing attitudes and practices tend to have a fair degree of consistency over time and are generally believed to have implications for their children's personalities. Finally, every member of a family has a history of experiences that is to some degree unique, and few would doubt that this history could have a lasting effect in giving each personality its special and distinctive character.

Recent behavioral-genetic studies suggest that only two of these three categories of influences—the genes and idiosyncratic experience—account for much variation in measured personality traits among people. Influences in the second group, environmental factors common to family members, seem to carry very little weight in explaining personality.

Three kinds of behavioral genetic studies have led to this conclusion: studies comparing the resemblance of identical and fraternal twins, studies of the resemblance of members of adoptive and biological families, and studies of identical twins reared apart in separate families. Some typical results from these studies are given in Table 1 for two personality traits: extroversion-introversion and neurotic tendencies or emotional instability. The studies in the table come from different Western countries and use slightly different tests, but the results are still reasonably consistent. Comparable data for non-Western populations (for example, Asian, African) are scarce or nonexistent, so that one cannot yet confidently generalize to all humans. The resemblances in the table are given as correlation coefficients which express average positive resemblance on a scale from 0.00 to 1.00. Here 0.00 represents the expected resemblance of individuals paired at random, and 1.00

represents perfect resemblance on the trait.

There is evidence in Table 1 that genes are important. Identical twins, who share all of their genetic variation, are more alike than fraternal twins, who share, on the average, only about half of theirs. And biological children resemble their parents more than adopted children do.

That the environment is important, is shown by the fact that the personalities of identical twins, who have identical genes, correalate only about 0.50. If their personalities were entirely determined by their genes, the correlation should be 1.00. Furthermore, a biological child gets half of his or her genes from a given parent, but average parent-child personality resemblances in the table are in the 0.10 to 0.20 range, instead of 0.50.

However, the environment that has these substantial effects on personality is apparently *not* the environment that family members share. In Table 1, parents and adopted children show practically no resemblance in personality (correlations of 0.01 and 0.02). Identical twins subjected to the same family environment are not more alike than identical twins reared in separate families. (If anything, the difference is in the other direction, but given the small sample of identical twins reared apart, a difference of this size could easily have occurred by chance.) Same-sex fraternal twins do not show more than half the resemblance of identical twins, as one would expect them to do if shared family environmental factors were contributing heavily to personality resemblances.

In short, both genes and environment appear to have substantial effects on the development of personality, but the aspects of the environment which are important are those which differ among family members, not those which are shared. It is this interesting and unexpected information which behavioral-genetic studies provide to psychologists interested in personality.

Genes and intelligence. Average IQ correlations are given in Table 2 for the same relationships as in Table 1. There is evidence in Table 2 for the importance of genetic factors in intelligence. Identical twins are more similar than fraternal twins, and biological children resemble their parents more than adopted children do. There is also evidence for the effect of the environment, but for intelligence, unlike personality, shared family environmental factors appear to be important. Parents and adopted children show an appreciable degree of resemblance in IQ (0.19). Identical twins reared together are more alike than identical twins reared in separate families (0.86 versus 0.72). Same-sex fraternal pairs show considerably more than half the degree of resemblance of identical twins. Also (though not shown in the table) fraternal twins have a higher correlation in IQ than ordinary siblings do (0.60 versus 0.47 in this survey), which suggests a greater sharing of environmental factors by twins, since the average degree of genetic resemblance of fraternal twins and siblings is the same.

Table 1. Personality resemblances of pairs of individuals with different levels of shared genes and environments*

Groups	Correlation	
	Extraversion-introversion	Neurotic tendency
Identical twins reared together (4987 adult pairs)	0.51	0.50
Same-sex fraternal pairs reared together (7790 adult pairs)	0.21	0.23
Parent and adopted child (362 pairs, late adolescent children)	0.01	0.02
Parent and biological child (510 pairs, late adolescent children)	0.12	0.18
Identical twins reared apart (65 pairs, mostly adult)	0.59	0.57

*From T. J. Bouchard, Jr., *Address to American Psychological Association*, Los Angeles, Aug. 25, 1981; B. Floderus-Myrhed et al., *Behav. Genet.*, 10:153–162, 1980; S. Scarr et al., *J. Pers. Soc. Psychol.*, 40:855–898, 1981; J. Shields, *Monozygotic Twins*, 1962.

Table 2. Intelligence resemblances of pairs of individuals with different levels of shared genes and environments*

Groups	Average of IQ correlations
Identical twins reared together (4672 pairs in 34 samples)	0.86
Same-sex fraternal pairs reared together (3670 pairs in 29 samples)	0.62
Parent and adopted child (1397 pairs in 6 samples)	0.19
Parent and biological child (8433 pairs in 32 samples)	0.42
Identical twins apart (65 pairs in 3 samples)	0.72

*From T. J. Bouchard, Jr., and M. McGue, *Science*, 212:1055–1059, 1981.

Unique, unshared experiences, though apparently not nearly as important for intelligence as for personality, still seem to play a role. Identical twins, while more highly correlated for IQ than for personality, still do not have a correlation of 1.00.

Thus for both intelligence and personality, behavioral-genetic studies point to broad classes of significant variables. In both domains, genes are important, as are environmental factors, in explaining differences among individuals, but family-to-family environmental differences are of considerable importance for IQ, whereas they seem to have almost no detectable effect on personality measures.

Genes and psychopathology. Much behavioral-genetic research has been directed toward syndromes of disturbed behavior known as psychosis, neurosis, and psychopathy. In the case of the major psychoses, schizophrenia and the manic-depressive disorders, twin and family studies during the last several decades have produced considerable evidence that genetic factors play a major role. As with normal personality, environmental factors also appear to play a substantial role, and again these mostly do not appear to be shared family factors, but idiosyncratic ones. Perhaps they may be unique histories of experience, but they also might be biochemical accidents. Evidence regarding the neuroses is somewhat less satisfactory. On the whole there is a suggestion that some genetic factors are involved, but less than with the psychoses, again with only minor shared family effects. Dispositions toward the psychopathic disorders, including criminality and alcoholism, appear to be at least partly, but not completely, genetically based. Information regarding the relative importance of familial and nonfamilial environmental factors in the development of these pathologies is inconclusive.

In studying a condition that affects a relatively small fraction of the population, the examination of the pedigrees of individual families containing affected persons is potentially highly informative. Such families permit the testing of specific hypotheses about a possible genetic basis for the predisposition. Such studies, for example, have identified a fairly large number of distinct genetic and chromosomal anom-alies that have mild or profound mental retardation as one of their consequences. Similar studies have long been pursued in the area of mental and emotional disturbance. Many intriguing suggestions have been raised, but the amount of information that has stood the test of time has so far been rather small. Since genes affect behavior via their biochemical effect on the body, and since a number of psychological conditions can be produced, aggravated, or attenuated by the administration of drugs, the triple hybrid area of psychopharmacogenetics is currently arousing considerable interest among researchers in psychopathology. Successes here could lead to dramatic progress in understanding the elusive relations between psyche and soma that have puzzled and challenged thinkers since the days of the ancient Greeks.

For background information *see* BEHAVIOR AND HEREDITY; INTELLIGENCE in the McGraw-Hill Encyclopedia of Science and Technology.

[JOHN C. LOEHLIN]

Bibliography: N. D. Henderson, Human behavior genetics, *Annu. Rev. Psychol*, 33:403–440, 1982; D. R. Peterson (ed.), Boulder symposium on behavioral pharmacogenetics, *Behav. Genet.*, 12:1–121, 1982; R. Plomin, J. C. DeFries, and G. E. Mc-Clearn, *Behavioral Genetics: A Primer*, 1980; L. Willerman, *The Psychology of Individual and Group Differences*, 1979.

Pyrometer

Pyrometry—more correctly called infrared thermometry or noncontact thermometry—has experienced great improvements in the state of the art over the last decade. These improvements have included the development of a microprocessor-controlled instrument, sighting systems to take the guesswork out of infrared surface-temperature measurements, and a precision instrument for use in the low-temperature range. Also, the capability of measuring the temperature of very tiny objects has been realized. Such unique features as measurement of temperature differential, data averaging, and sky radiation detection have been developed to meet the needs and demands of various users.

The current developmental work has led to applications of infrared thermometry to the study of the natural environment. These applications include the measurement of human skin temperature, temperatures of animals, and surface temperatures of plant life.

Computerized infrared thermometer. The computerized infrared thermometer has been introduced. A microprocessor provides five temperature calculations at once: it can measure surface temperatures, compute average temperatures, store maximum and minimum temperatures, and calculate the difference between them. This instrument also has a wide temperature range which increases its attractiveness. It measures temperatures from -20 to $2000°F$ and -30 to $1100°C$, or from 750 to $5400°F$ and 400 to $3000°C$.

Target sighting. One of the prior pitfalls of infrared thermometry was that the operator had no way of knowing exactly where a temperature measurement was being made. The object being measured must completely fill the field of view of the infrared thermometer or an erroneous reading will result. Various manufacturers have addressed this sighting problem.

For instance, various companies offer the option of a scope mounted on top of the instrument for sighting of the target. The cross hairs can be used to identify the center of the target, as long as the scope is not out of alignment. One company offers a laser sighting system, but the laser shows the center of the field of view. The major problem with these systems is that the breadth of the field of view is not known. It is most important that the entire field of view be identified in order to ensure correct readings.

This problem has been addressed and solved with an intraoptical light sighting system (Fig. 1). A visible, pulsating light is directed through the infrared optics, thereby sharing the same path as the field of view of the optics. This light gives a visual indication of the exact field of view, not just the center point. By using this light sighting system to determine where a measurement is being taken, an operator can scan individual parts on a printed circuit board, or components as small as a single resistor. The working distance is such that the operator can see the light on the target and the liquid crystal display simultaneously. Thus, the guesswork about where the instrument is being aimed and what surface is being measured is eliminated. Figure 1 provides an indication of the focus of the light sighting system.

Low-temperature measurement. Infrared thermometers have been used for some time in industry to monitor the temperatures of high-temperature processes. They have been used to measure: the temperature of fragile items that cannot be measured by any other means; extremely hot objects that would prove to be hazardous with other forms of temperature measurement; steam traps; and so forth. However, until recently there were no instruments available for making precision measurements for low-temperature applications.

The infrared emission signal levels are minute in the near-ambient temperature range and are zero when the object being measured and the instrument are at the same temperature. Developing instruments that would measure temperatures of objects which are near or at ambient temperature has proved the most difficult problem to be dealt with in infrared thermometry.

Recently solutions to this problem have been found. Low-temperature instruments now can operate in the range of -30 to $200°F$ or -30 to $100°C$, with a precision of $0.1°F$ or $0.1°C$ and accuracy of $\pm 0.5°F$ or $\pm 0.5°C$. These instruments are the only infrared thermometers in the world that can respond in a fraction of a second with $0.1°F$ readout. Other instruments will read out with this precision, but the response time of the instrument must be lengthened.

The major advantage of this breakthrough in technology is that it opens up the science of infrared thermometry to meaningful measurements of the human body, animals, and plant life. It provides reliable and stable instruments for making measurements in the natural environment.

Natural environment measurements. To further enhance the capabilities of making temperature measurements in the natural environment, a number of unique features have been developed. These include measurement of temperature differential, the method of data averaging, and sky radiation detection.

Agronomists have found that as crops become stressed for water plant temperature is elevated. By measuring the difference between the ambient air temperature and the absolute temperature of the crop canopy, they can optimize irrigation scheduling. A temperature differential feature has been developed to simplify this process (Fig. 2). The operator pulls the trigger on the infrared thermometer and the temperature differential registers immediately on the liquid crystal display of the instrument. The precision thermistor which senses the ambient air tem-

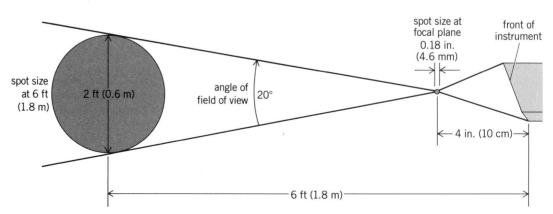

Fig. 1. Standard focus with intraoptical light sighting system.

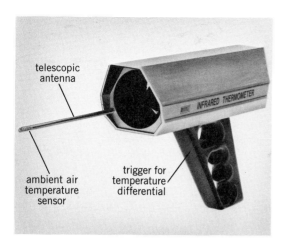

Fig. 2. Infrared thermometer with differential temperature capabilities. (*Everest Interscience, Tustin, CA*)

perature is housed in the end of a telescopic antenna which extends from the optical housing to remove the sensor from the heating or cooling effects of the operator's hand or the instrument itself. Also, the thermistor is covered with a gold coating to make certain that unwanted radiation is reflected.

Another feature developed for natural environment temperature measurements is data averaging. Wind gusts and varying temperature fields may cause difficulty in taking temperature measurements outdoors. To correct for this, an instrument's response time can be slowed down by flipping a toggle switch from "fast" to "data averaging," and the instrument will then average the data over time.

The latest development for these natural environment applications is a sky radiation detector. A window is cut in the top of the case of an infrared thermometer, and a second detector is positioned to measure the sky radiation. Circuitry to correct for errors caused by this unwanted sky radiation is added in order to improve the accuracy of the instrument.

These improvements are just the beginning in infrared thermometry. With the problems of precision, stability, and reliability solved, research continues with regard to sighting and the measurement of extremely minute objects. The applications for infrared thermometry will increase as people in all fields learn more about this noncontact, noninvasive method of measuring surface temperatures. Infrared thermometry is now fast, simple, safe, and precise.

For background information *see* PYROMETER in the McGraw-Hill Encyclopedia of Science and Technology. [CHARLES E. EVEREST]

Bibliography: C. E. Everest, Aiming takes the guesswork out of infrared thermometry, *Ind. Res. Dev.*, 24:138–139, 1982; C. E. Everest, Infrared thermometry, *Meas. Control*, 89:111–113, 1981; J. E. Galbraith, Applying and selecting portable IR thermometers, *Instr. Control Sys.*, 53:49–52, 1980; P. J. Panter, Jr., et al., Remote detection of biological stresses and plants with IR thermometer, *Science*, 205(4406):585–587, 1979.

Quantum solids and liquids

The interface between a quantum solid and its liquid phase is a physical system which merits considerable scientific interest. Experimental and theoretical studies of this system represent the intersection of a number of frontiers of condensed-matter research, including phase transition and critical phenomena, effects of system dimensionality, effects peculiar to quantum solids and liquids, and surface/interface physics (crystal equilibrium, growth, and epitaxy). Critical phenomena are dramatic effects, and the influence of dimensionality on them is correspondingly dramatic. Early theories indicated that systems with a dimensionality of less than three could have no finite-temperature phase transitions into states of long-range order. Subsequently, M. Kosterlitz and D. Thouless showed that for two dimensions one may introduce the concept of topological long-range order, thereby allowing observable phase transitions. The critical properties of these transitions are quite different from those of higher-dimensional systems. Since the development and experimental verification of the Kosterlitz-Thouless theory there has naturally been considerable interest in finding and studying new physical systems to which the theory might apply. The quantum solid/liquid interface is one of the newest and most promising of such systems. Research indicates that the interface critical phenomenon, called the roughening transition, should belong to the universality class of Kosterlitz-Thouless type two-dimensional phase transitions. The interface has additional features resulting from its quantum nature, and its study shows potential in furthering the understanding of surface physics.

Physics of solid/fluid interface. The general problem of solid/fluid equilibrium was first considered by J. Willard Gibbs, who carefully defined a surface free energy $F = E - TS$ (involving a surface energy E and entropy S at a temperature T) so that the configuration of the interface is one which minimizes F. A schematic representation of an interface configuration is shown in illustration a. Atoms which are part of the solid surface are drawn as terraces (flat areas where the solid-surface atoms have a relative maximum number of lateral neighbors), ledges (a terrace edge where an atom is missing one lateral neighbor), kinks (sites in a ledge where an atom is missing two neighbors), and clusters (consisting of one or more atoms which have few lateral neighbors). At low temperatures, where the TS contribution to F is small, the free energy is minimized by decreasing the surface energy E. Since the surface energy is lowered when a surface atom forms an attractive bond with a neighbor, F is minimized by having a maximum number of neighbors in the surface configuration. The best surface configurations would be large terraces corresponding to the high symmetry planes of the solid's crystal structure; these may then form the sharp facets of a macroscopic solid crystal.

At high temperatures, the effects of the surface entropy term TS must be considered. Configurations

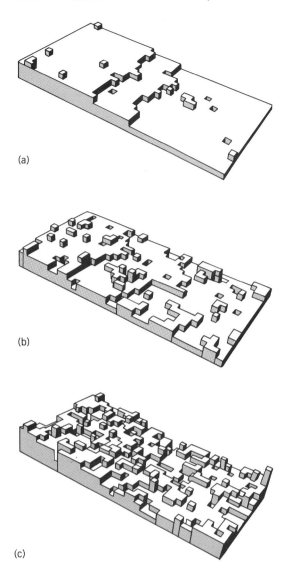

(a)

(b)

(c)

Computer simulation of the structure of a solid/fluid interface. (a) Low temperature; (b) temperature slightly below the roughening transition; and (c) temperature slightly above the roughening transition. (*After K. A. Jackson, The present state of the theory of crystal growth from the melt, J. Crys. Growth, 24/25; 130–136, 1974.*

formed by promoting an atom out of a terrace (or ledge) site into a single surface-atom state increase the entropy because there are many equivalent sites from which the atom can be removed and replaced. Since the entropy term appears in the free-energy expression with a minus sign, promoting atoms out of terraces and so forth decreases the free energy. As more atoms are promoted, clusters may form; atoms may then be promoted out of these clusters onto the cluster surface to form new clusters, and so on. Such a progression to higher-entropy configurations and rougher surfaces is depicted in illustration *b* and *c*. However, promoting atoms out of terraces breaks attractive bonds and increases the surface energy. Minimizing the free energy at finite temperatures involves a competition between the surface energy and entropy.

At some temperature the competition will be balanced, and the system will have no basis for preferring the maintenance of bonds or the promotion of clusters. Any small external perturbation can drive the system into a preferred state; this behavior corresponds to a singularity in some macroscopic susceptibility. The equilibrium state of the interface will be dominated by large-scale fluctuations, and the width of the interface (the distance from the lowest cluster to the highest cluster) will diverge to infinity as the interface area becomes large. These effects are characteristic of the critical behavior of the interface. Below the critical temperature the solid surface will be flat or sharp, while above it the surface will be rough or rounded; this is the roughening transition. The illustration is a computer simulation of the transition in a finite system. In illustration *a* and *b*, where the surface is slightly below the critical temperature, one can see distinct terraces and ledges, indicating a sharp surface. For the surface in illustration *c*, which is slightly above the critical temperature, such features are no longer discernible.

Theory. The first attempt to obtain detailed quantitative information about the solid surface was made by W. Burton, N. Cabrera, and F. Frank in 1949. They used a model that is now believed to be invalid; however, they made the first prediction of the roughening transition at the solid/fluid interface. With various approximations and a reasonable model for the interatomic forces, in 1976 S. Chui and J. Weeks showed that the roughening transition belonged to the universality class of Kosterlitz-Thouless-type two-dimensional transitions. The roughening transition is two-dimensional in nature, even though the solid/fluid system must be three-dimensional in order to allow for multilayer activity at the interface.

One of the most important aspects of the solid surface, in relation to the roughening transition, other surface equilibrium properties, and practical applications, is its behavior away from equilibrium, that is, when the solid is growing by the accretion of atoms onto its surface. The crystal growth process is very sensitive to the nature of the surface and in particular to the availability of bonding sites, the most favorable of which are kinks. If the solid surface is sharp, there are few kink sites, and the crystal growth is severely limited by the necessity of starting a new solid layer by the formation of clusters of a critical size (the two-dimensional analog of liquid droplet formation in a supersaturated vapor). If the solid surface is rough, there are many kink sites, and the crystal growth will be continuous and may occur at a high rate. Although classical treatments are adequate for describing crystal growth on low-temperature sharp surfaces and high-temperature rough surfaces, the temperature region near the roughening transition presents a formidable problem in surface kinetics. Studies of the roughening transition are relevant not only to the field of critical phenomena, but also to the field of surface physics and to practical applications in bulk crystal growth and thin-film autoepitaxy. In fact, it was these latter fields which stimulated and guided developments in the understanding of surface roughening.

Observation of roughening transition. The experimental observation of the roughening transition did not occur until 30 years after its prediction by Burton, Cabrera, and Frank. Different classical solids were observed to have either sharp or rough surfaces depending on whether the solid/fluid coexistence temperature was below or above the calculated roughening temperature. The coexistence temperature of the classical systems could be varied by changing the external pressure, but over the range of experimentally accessible pressures, the coexistence temperature could not be varied enough to cross the roughening temperature and thereby permit the observation of the transition. Even if the roughening temperature were crossed, the presence of minute quantities of an impurity at the interface could have destroyed the roughening behavior.

It was not until a quantum solid/fluid system (solid ^{4}He and superfluid He II) was studied that evidence for the roughening transition was found. Quantum effects in the helium system eliminate both the drawbacks of the classical system. As the temperature is lowered in the quantum system, the entropy may decrease because of an increase in order not only in position space in the solid, but also in momentum space in the fluid. Thus the two phases may have comparable entropies and can remain in coexistence over a wide temperature range. Because of the quantum zero-point motion, the coexistence temperature may extend to low values (absolute zero for helium) so that impurities may be precipitated out before the quantum solid/fluid interface is formed. By using a high-purity solid ^{4}He/He-II system, it was possible to make the first observations of the roughening transition. Two transitions at different temperatures were detected by visually observing the rounding of two different crystallographic facets. Quantitative measurements using holographic interferometry were made on one of the facets, and a sharp transition was indicated. The observed transition temperature agreed well with theoretical estimates.

Surface waves on helium crystals. The observation of roughening in the helium system did not occur earlier because of interference from the roughness of other crystallographic faces of solid helium at typical experimental temperatures; the behavior of these surfaces obscured the presence of the sharp facets and their roughening transitions. Prior to the observation of the roughening transition it was believed that solid helium might remain rough even down to absolute zero. To explain this, A. Andreev and A. Parshin proposed a model for the helium surface based on quantum-mechanical properties of the ledges and kinks at the interface. They showed that under certain conditions the interaction of nonlocalized zero-point defects could cause the ledge energy to vanish even at zero temperature, with the result that the surface would always be rough. Although their assumptions must be invalid for the sharp facets of helium below their respective roughening tempera-

tures, some of the predictions of Andreev and Parshin should be correct for the rough helium surfaces.

A significant prediction was the appearance of surface or capillary waves at the quantum solid/fluid interface which can propagate at bulk sound frequencies. These interface waves do not involve an elliptical motion of the surface mass as in a classical surface wave, but involve an effective motion produced by rapid freezing and melting at the interface. The high freezing and melting rates necessary to obtain weakly damped wave propagation are unique to the rough quantum solid/superfluid interface, possibly because of the quantum nature of the surface roughness and definitely as a result of the quantum nature of the fluid.

As discussed above, a rough classical surface may have a high growth rate (relative to a sharp surface), but in a classical solid/fluid system the surface growth rate is severely limited by the transport processes required for the removal of the latent heat of solidification. In the solid ^{4}He/He-II system the latent heat can be rapidly diluted by the appearance of the superfluid component which, by its quantum nature, carries no entropy. With the absence of transport (relaxation) loss mechanisms, the transfer of mass between the solid and fluid at the interface can occur as a propagating rather than as a diffusive wave. Shortly after the freezing-melting wave phenomenon was predicted, it was observed experimentally. The dispersion relation predicted for the wave was verified at acoustic frequencies as high as 2 kilohertz, and striking motion pictures of the undulating solid surface, with amplitudes as large as 2 mm, were made.

For background information *see* CRITICAL PHENOMENA; CRYSTAL GROWTH; LIQUID HELIUM; PHASE TRANSITIONS; QUANTUM SOLIDS in the McGraw-Hill Encyclopedia of Science and Technology.

[JULIAN MAYNARD]

Bibliography: S. T. Chui and J. D. Weeks, Phase transition in the Coulomb gas and the interfacial roughening transition, *Phys. Rev.*, B14:4978–4982, 1976; K. O. Keshishev, Ya. A. Parshin, and A. V. Babkin, Crystallization waves in ^{4}He, *Zh. Eksp. Teor. Fiz.*, 80:716, 1981; H. J. Leamy, G. H. Gilmer, and K. A. Jackson, Statistical thermodynamics of clean surfaces, in J. B. Blakeley (ed.), *Surface Physics of Materials*, pp. 121–188, 1975.

Quasars

Quasistellar radio sources, or quasars, are the most luminous objects known in the universe. Although they are located in the farthest reaches of space, billions of light-years away, many quasars are such powerful sources of radio emission that they can be easily detected by even modestly sized radio telescopes. Often the radio emission comes from giant clouds of radiating plasma hundreds of thousands of light-years in size; but the source of this tremendous energy content appears to lie in a small volume of space unresolved by ordinary optical or radio telescopes. When examined with high-resolution radio telescopes, these tiny radio nuclei show a bright core with a jet feature extending away from the core by some 10 to 100 light-years and pointing toward the giant radio cloud, which may be 1 million or more light-years away. Often the jet is seen to move away from the central core with a velocity typically five to ten times the speed of light. This is called superluminal motion and is widely interpreted by astronomers as the result of repeated cataclysmic activity in the central core of quasars, which results in the ejection of a sharply focused relativistic plasma with a remarkably high velocity. These plasmas are called beams or jets.

Noncosmological red shifts. When superluminal motion was first discovered more than a decade ago, the apparent conflict with accepted laws of physics was used as an argument in favor of noncosmological red shifts. By allowing quasars to be closer than indicated by their red shift and by reference to the Hubble law which relates red shift and distance, it was argued that the observed large angular motions would then correspond to linear velocities less than the speed of light. In the past few years, however, observations made with new electronic detectors on large optical telescopes have shown that many quasars appear to lie in galaxies which have essentially the same red shift as the quasar, and this has removed most lingering doubts as to the cosmological nature of quasar red shifts. Moreover, superluminal motion is observed not only in quasars but in at least one radio galaxy, 3C 120, whose distance is not in question.

It has also been speculated that it may not be possible to interpret the observed superluminal motion within the framework of the conventional physics and cosmology, but most astrophysicists do not consider such an extreme interpretation to be necessary.

Fixed component models. One early idea that received considerable attention was that the radiating clouds do not move at all, but that properly synchronized variations in the intensity of many stationary clouds could give the illusion of motion, much in the same manner as the lights appear to move on a theater marquee. But then contractions as well as separations would be expected, and this is contrary to observation.

Relativistic beaming. It is more likely that the observed superluminal motion is an illusion. If the true velocity of the radiating plasma is very close to the speed of light, then moving plasma nearly catches up with its own radiation. When viewed by a distant observer, located approximately along the direction of motion, this causes an apparent time contraction in the observer's frame and can result in an apparent transverse velocity that greatly exceeds the speed of light.

An example of this phenomenon is illustrated in Fig. 1. Suppose a cloud is moving with a velocity of 0.99 c (where c is the speed of light) at an angle of 8° from the line of sight, and after 10 years sends out a signal. In 10 years the cloud will have moved a distance of 9.9 light-years. Because the cloud has moved a distance of 9.9 × cos 8° = 9.8 light-years

along the line of sight, this signal is seen by a distant observer only (10 − 9.8) = 0.2 year after the signal which was radiated 10 years earlier. During this 10-year period the cloud has moved a distance of 10 × sin 8° = 1.4 light-years perpendicular to the line of sight. The apparent velocity is then 1.4 divided by 0.2, or seven times the speed of light.

In general, if the true velocity of the radiating material is v, then the apparent transverse velocity, v_{app}, in the plane of the sky is given by the equation below; θ is the angle between the motion and the

$$v_{app} = \frac{v \sin \theta}{1 - \beta \cos \theta}$$

line of sight and $\beta = v/c$. The maximum value of v_{app} is $\beta\gamma c$ where $\gamma = (1 - \beta^2)^{-1/2}$ and occurs at an angle $\theta \sim \sin^{-1}(1/\gamma)$. In the above example, where the true space velocity is about 99% the speed of light ($\beta = 0.99$, $\gamma = 7$), the maximum apparent velocity is about 7 c and occurs when the motion is oriented at an angle of 8° to the line of sight. As a further consequence of the rapid motion, the radiation is focused in a narrow cone along the line of sight. This causes an enhancement of the apparent intensity when viewed along the direction of motion by a factor of approximately γ^3, so the superluminal sources will appear to be very bright to a properly located observer.

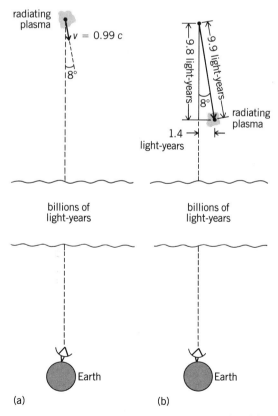

Fig. 1. Example of illusion of superluminal motion. (a) A cloud of radiating plasma, moving at a velocity of 0.99 c, at an angle of 8° from line of sight to Earth. (b) Same cloud 10 years later; the cloud has moved 9.9 light-years, which can be resolved into components of 9.8 light-years along line of sight and 1.4 light-years perpendicular to line of sight.

It has been known for some years that most quasars are not radio sources, and it has been suggested that because of the relativistic beaming, only those quasars which have their beams pointing roughly in the direction of the Earth appear as strong radio sources. All other quasars appear to be radio-quiet because their radio beams are pointed in other directions.

Superluminal motion has been observed in seven radio sources. Five of these are quasars, 3C 179, 3C 273, 3C 279, 3C 345, and NRAO 140, and one, 3C 120, is a radio galaxy. The other is BL Lacerta, an intermediate type of object with many properties similar to quasars, but without the bright emission lines characteristic of quasars.

Figure 2 shows the components of the quasar 3C 273. In the 3-year interval from July 1977 to July 1980, the component separation increased from 62 to 87 light-years, corresponding to an apparent velocity of about eight times the speed of light. The speed record, however, is held by the more distant quasar, 3C 279, whose radio components were clocked to be moving at nearly 50 times the speed of light. Such a high velocity is difficult to understand in the framework of the relativistic beaming picture, as it requires a true velocity greater than 99.98% the speed of light and precise orientation of the motion to within 1° of the line of sight. The probability that this would occur by chance is very small. More complex models are being investigated in an attempt to obtain a better understanding of the phenomena.

Very-long-baseline interferometers. The angular size of superluminal radio sources is typically less than 0″01 arc-second, and highly specialized techniques are required to study the individual features and measure their angular separation. Conventional parabolic dish-type radio telescopes have a resolution at best of only about 1 arc-minute, comparable to the resolution of the human eye. By using two or more dishes as elements of an interferometer, or array, it is possible to synthesize apertures as large as 10–20 mi (16–32 km) in diameter to obtain radio images with a resolution of better than 1 arc-second, roughly as good as the best pictures obtained with large optical telescopes. This is still not adequate to resolve the very small features found in the cores of quasars, however. Because of these small dimensions, interferometer baselines of thousands of miles are needed to study the compact radio clouds identified with the tiny quasar energy source. With such long baselines, direct electrical connection between the elements is impractical. Instead, high-speed tape recorders are used to record the data received at each telescope element, and later the tapes are transported to a central facility where they are correlated in a large special-purpose digital computer. Synchronization of the recordings to an accuracy of better than 1 microsecond is achieved by using hydrogen maser atomic clocks.

This technique of tape-recording interferometry is known as very-long-baseline interferometry (VLBI) and has been used to link up radio telescopes

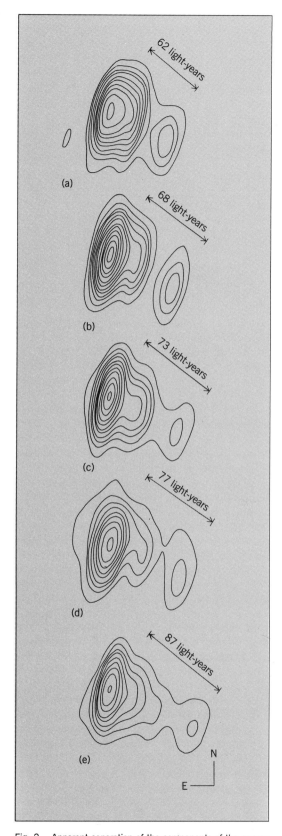

Fig. 2. Apparent separation of the components of the quasar 3C 273 at five successvie times, as observed by very-long-baseline interferometry using radio telescopes in the United States and Europe. Separation of components in light-years (ly) is indicated. (a) July 1977. (b) March 1978. (c) June 1979. (d) December 1979. (e) July 1980. (After T. J. Pearson et al., Superluminal expansion of quasar 3C 273, Nature, 290:365–368, 1981)

throughout the world to observe the superluminal radio sources and thus to better understand the energy source of quasars.

For background information *see* RADIO SOURCES (ASTRONOMY); RADIO TELESCOPE in the McGraw-Hill Encyclopedia of Science and Technology.

[K. I. KELLERMANN]

Bibliography: M. H. Cohen and S. Unwin Superluminal radio sources, in D. S. Heeschen and C. M. Wade (eds.), *IAV Symposium: Extragalactic Radio Sources*, pp. 345–354, 1982. K. I. Kellermann and I. I. K. Pauliny-Toth, Compact radio sources, *Annu. Rev. Astron. Astrophys.*, 19:373–410, 1981.

Radar

Recent developments in radar technology include the incorporation in ground surveillance radars of advanced detection and data-processing technologies; and the development of a simple height-finding radar system that will aid in evaluation of the navigation accuracy of aircraft.

Ground surveillance radar. A ground surveillance radar (GSR) is a system capable of detecting objects on the ground from points on the ground. Traditionally, radars of this type have been used for military or paramilitary applications. The targets of interest are primarily moving vehicles and troops. GSRs have maximum ranges of a few kilometers to 50 kilometers, but are often operationally limited to less than that by their line-of-sight operation. The equipment is either transportable by humans or by small vehicles. Recent developments have been the incorporation of advanced detection techniques and data-processing/display technologies previously found only in larger radar units. The primary benefits of these improvements are: greatly simplified operator tasks, more accurate and consistent performance, and the potential for automation.

Operational GSRs. The primary function of GSRs is the detection of ground traffic, ranging from individuals walking with a speed of about 1 mi/h (0.5 m/s) to convoys of vehicles with speeds of more than 30 mi/h (13 m/s). This is accomplished through various forms of moving-target indicator (MTI) techniques to achieve discrimination between the Doppler-shifted echoes from the desired targets and those from the stationary ground environment. In the current operational equipment, the detection and determination of target location (range and bearing) is accomplished manually by the operator. Many GSRs are also equipped with an auxiliary audio channel through which the operator can listen to an aural tone related to the Doppler frequencies of the target, and which can be used by a trained operator to provide a coarse identification of the target type (wheeled or track vehicle, a person walking, and so forth).

The U.S. Army's PPS-5 radar, shown in Fig. 1 in its jeep-mounted configuration, and the PPS-15, shown in Fig. 2, are typical of the performance and size of operational GSRs. Other versions range from hand-held to tank-mounted models. To achieve the desired compactness of equipment, GSRs usually operate at the upper end of the microwave frequency band, around 10 or 16 GHz. Such radars have been found to be useful in detecting other types of moving targets, such as helicopters, and stationary objects with rotating parts, such as fans or scanning antennas. Those with sufficient sensitivity have been found by the artillery to be useful in detecting and locating the impact point of its own shells. As these explode on the ground, large chunks of matter are kicked up with sufficient velocity and radar cross section to be detected in the MTI mode. This information, when relayed to the battery, can be used for fire correction.

Operational GSRs tend to be primitive as com-

Fig. 2.   Tripod-mounted PPS-15 radar with operator. (*General Dynamics, Electronics Division*)

Fig. 1.   Jeep-mounted PPS-5 radar with operator. (*AIL Division, Eaton Corporation*)

pared with other classes of radars, in the extent and nature of operator involvement in the interpretation of radar signals. This places an extensive burden on the individual's skills and training. In addition, since the operator is usually not the ultimate user of the information, target data must still be relayed. This overall process is a bottleneck with respect to the amount and speed with which GSR-observed information is available where it is needed.

Advanced GSRs. Recently, engineers at MIT Lincoln Laboratory have demonstrated the kind of advanced capability that can be expected from the next generation of operational GSRs. While their work was done in the context of demonstrating the tactical benefits of centrally integrated radar networks, this was made possible only by considerable advances in individual radar technology. These new radars feature radio-frequency hardware which provides high spectral purity of signals and utilization of advanced coherent Doppler processing of returned signals. This permits computer-automated discrimination between moving targets and undesired ground echoes, as well as their reliable detection, with low false-alarm rate even in a heavy-clutter environment, and the accurate determination of their location. This information, inherently in digital form, is then directly relayed to a data integration center, requiring no human operator invervention, and is available at the remote center within only a few seconds after its occurrence. Two types of radars have been demonstrated. One is an extensively modified PPS-5 with mechanical scanning. This feature provided coverage of a sector, typically 90° with a 5-s revisit time in the normal track-while-scan surveillance mode. In the shellburst-detection mode, the antenna is made to scan over a sector of a few degrees around the expected impact point with a 1-s revisit time (a burst typically lasts only a few seconds). Voice-grade very-high-frequency radios, interfaced to the radar by a full duplex digital modem operating at a rate of 2400 bits per second, provide the data link through which the radar can be remotely controlled, and the digitized target data are sent back to the integration center. Another radar features an electronically scanned cylindrical array with 360° coverage. Its beam agility makes possible the simultaneous and independent operation of the track-while-scan and shellburst-detection modes. In addition, this information remoting can be accomplished by timesharing the radar to establish a microwave data link.

The ultimate direct benefits provided by these advanced GSRs are evidenced by the remarkably clean and easy-to-interpret display of the information available to the user, as exemplified in Fig. 3. This takes the form of a plan-position indicator (PPI). This display is designed so that an operator can select an area of interest. The roads shown are produced from prestored map information. Targets are shown by symbols and are located in a standard coordinate system, in this case the universal transverse Mercator (UTM) system. Each target is assigned a numerical identification tag by the computer for track-

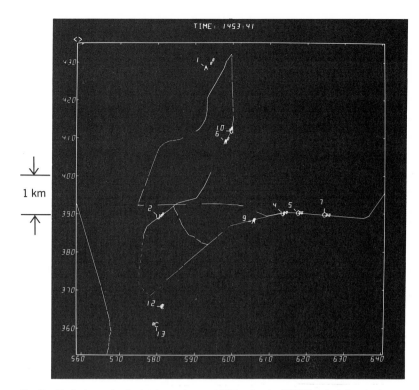

Fig. 3. Typical remote display from advanced GSR network. (*MIT Lincoln Laboratory*)

ing purposes. Also, the console operator can display the target-type identity, if determined, and the target speed. Past positions (small circles) can also be displayed, and the screen is updated every 5 s. Different symbologies are used to denote whether the target is being tracked simultaneously by several of the radars in the network (symbolized U), in which case all data are merged into one report, or if it is seen by only one radar, in which case it is given the symbol assigned to that radar (A, B, and so forth). Typically these GSRs, when associated with commensurate data processing, offer the user real-time target information in a format which requires practically no further manipulations.

In addition to military battlefield applications, GSRs are also used for security purposes in facility surveillance. As recently demonstrated, they can provide 24-h all-weather operational capability from an unattended location.

[JEAN-CLAUDE SUREAU]

Radar for navigation height finding. Separation standards are used for maintaining the safe and effective operation of aircraft in an air-traffic control environment. As aircraft operations increase in number, current separation standards must be reviewed. The possibility of reducing conventional separation standards on the basis of scientific data has been discussed. An effective way of evaluating these standards would be the use of mathematical methods employing a collision-risk model, for which data on the navigational accuracy of aircraft would be required as one of the parameters.

In general, three-dimensional radars are used for

height finding, but these are complicated and expensive. For many years the data on navigation accuracy in the vertical dimension were not collected because data collection was too expensive and time-consuming. In order to survey height-keeping errors, the Electronic Navigation Research Institute has developed a simple height-finding radar system called the navigation accuracy measurement system (NAMS).

There are various kinds of height-finding radars, such as the early "nodding" height finder and the V-beam, defocus, and phase-difference systems. Unlike those radars, NAMS uses a height estimation technique, called a curve-fitting method, which is a parameter estimation technique based on the least-squares method and the method of successive approximation. Thanks to this technique, NAMS has an advantage in its comparatively low cost and simple hardware structure, which is almost the same as that of a marine radar.

NAMS is designed to measure the height of a specified target (an aircraft in level and straight flight) whose altitude is between 20,000 and 41,000 ft (6100 and 12,500 m) with about 100 ft (30 m) of accuracy when installed under an airway. The system consists of an antenna whose radiation pattern is a fan beam, a transceiver, a digitizer, and a minicomputer. The radar antenna rotates about a horizontal axis and scans aircraft flying overhead. The height of an aircraft is estimated by processing the data obtained over several scans, using the minicomputer.

Principle of height estimation. NAMS is essentially a pulse radar with a fan-beam antenna which is installed under a trunk airway. The radar antenna produces a fan beam whose plane is parallel to the center line of the airway under observation. The antenna rotates continuously at 18 revolutions per minute about a horizontal axis parallel to the airway, so that the fan beam moves parallel to itself and provides information on the range, depression angle, and data-acquisition time for the target at each scan (Fig. 4). This information is not sufficient to determine three-dimensionally the instantaneous position of the aircraft. One more parameter is necessary.

To overcome this difficulty it is proposed that the data from several scans be combined, assuming that the aircraft maintains a constant height, speed, and heading while being scanned. If the dimensions of the fan beam are 20° × 1°, an aircraft flying at 480 knots (250 m/s) at an altitude of 33,000 ft (10,000 m) will be scanned over a period of about 15 s. With a beam width of 1°, an antenna rotation period of 3.3 s, and a pulse repetition frequency of 1500 pulses per second, about 14 pulses may be expected from the aircraft. From these reflected pulses a representative datum, corresponding to the center of the beam, can be extracted from the information.

While an aircraft traverses the fan beam, NAMS stores time-series data on the range, depression angle, and data-acquisition time for the aircraft. By

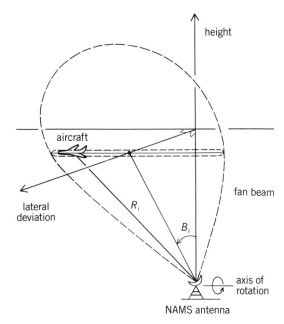

Fig. 4.   Geometrical relationship between an aircraft and the rotating fan beam in NAMS. R_i = range; B_i = depression angle.

using these observed data, NAMS estimates the height of the aircraft on the basis of a statistical parameter estimation technique. If the aircraft is in level and straight flight while it is scanned by the fan beam, the following relation among the observed data may be satisfied.

$$R_i^2 \cong (C_1 t_i + C_2)^2 + C_3^2/\cos^2 B_i$$
$$(i = 1, 2, \ldots, n)$$

Here C_1, C_2, and C_3 are coefficients, C_3 being the cruising height of the aircraft. R_i, t_i, and B_i are the range, data-acquisition time, and depression angle, respectively. The subscript i denotes the data derived from the ith scan. From this equation, C_3 can be determined by a curve-fitting method based on the least-squares method. The height of the aircraft is one of the coefficients of a mathematical equation which satisfies the theoretical relationship among the observed data. Therefore, if an estimate is made of the coefficients of the equation which give the best fit to the theoretical relation, then estimates of several parameters, such as the height and speed of the aircraft, can be obtained.

Figure 5 shows an example of the curve-fitting method. If there is no observational error, the range may vary on the hyperbolic curve described in Fig. 5. In fact, the observational data will be distributed around the theoretical curve because of the effect of measurement error. In this case, observational data are available for determining a curve which gives the best fit to the theoretical relation. This is equivalent to estimating the parameters included in the equation. A statistical estimation technique called the linearized least-squares method is used for determining the curve. Although this procedure is very complicated, it can easily be carried out within several seconds on a minicomputer.

System configuration. NAMS has been developed and gradually improved since 1978. The system so far includes the subsystems shown in Fig. 6. The antenna is a rotating-bar antenna whose rotation period is 3.3 s. The length of bar-antenna and beam dimensions are 2.26 m and 20° × 1°, respectively. The transceiver provides transmission pulses, with a repetition frequency of 1500 pulses per second, to the antenna and receives the returned signal from targets. The frequency and output power of the transmission pulse are 9.4 GHz and 50 kW, respectively. The digitizer provides the data required for height estimation, that is, B_i, R_i, and t_i. The digitized signals are derived from analog signals transferred from the transceiver. In this process several kinds of signal-processing techniques, such as sweep correlation, altitude gating, and analog CFAR (constant false-alarm rate) receiver, are used for improving target-detection capability. Data from the digitizer are processed by a minicomputer to estimate the height. This minicomputer can track four aircraft simultaneously. NOVA-01 (with a memory size of 16,000 words) is used as the processor. When the data from a given aircraft are completely processed, a typewriter prints the calculated results automatically. The temperature and atmospheric pressure at the radar site are measured by meteorological sensors set up in a shelter. These data are used to estimate the pressure altitude of aircraft. A meteorological data processor digitizes the analog signals obtained by sensors and provides digital signals to the minicomputer.

The antenna and the transceiver of NAMS are similar to those of conventional marine radars, but the direction of rotation of the antenna is different.

Height estimation accuracy. The accuracy, in general, depends upon observational errors, number of scans, and the geometrical situation of the aircraft. Observational errors in the measurement of range, depression angle, and time relate to the sensors and the methods of detection and processing.

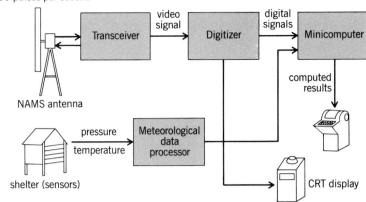

Fig. 6. System configuration of NAMS.

The number of scans obtained when the aircraft is within the range of the rotating beam depends on the speed, altitude, and size of aircraft; the beam width; and the antenna rotation period. The geometrical situation of the aircraft with respect to the radar antenna also affects the various errors.

The height measurement accuracy of NAMS is a complex function of these factors, and a computer simulation would be useful in estimating the accuracy. According to the evaluation of the accuracy of NAMS, it may be about 100 ft (30 m) for the cases under consideration.

For background information *see* RADAR in the McGraw-Hill Encyclopedia of Science and Technology. [SAKAE NAGAOKA]

Bibliography: *Jane's Infantry Weapons: 1980*, 1980; *Jane's Weapons Systems: 1980–81*, 11th ed., 1981; G. H. Knittel et al., The Netted Radar Demonstration at Fort Sill, Oklahoma, in *IEEE EASCON 1981 Conference Record*, pp. 78–88; S. Nagaoka et al., A simple radar for navigation accuracy measurements, *J. Nav.*, 34:462–469, 1981; S. Nagaoka et al., Radar estimation of the height of a cruising aircraft, *J. Nav.*, 32:352–356, 1979; J.-C. Sureau, An Advanced Ground Surveillance Radar, in *IEEE EASCON 1981 Conference Record*, pp. 71–78.

Radiation biology

Radiation biology is the study of effects of radiation on living systems. Radiation includes ionizing radiations (x-rays, electrons and heavy particles, neutrons, protons, and heavy ions) as well as ultraviolet rays. The process of ionization releases electrons from neutral atoms in the molecules, thereby making them reactive. Ultraviolet rays cannot produce ionization.

Radiation biology as a discipline started soon after the discovery of x-rays by W. K. Roentgen in 1895. The medical applications of x-rays were soon recognized when they were found to produce inflammatory changes such as dermatitis. This led to re-

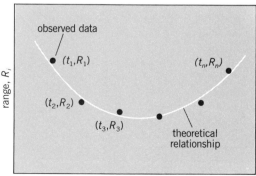

Fig. 5. Relations between theoretical curve and observed data (dots) from NAMS.

search to develop methods for treating cancer by radiation (radiotherapy). Research was further stimulated during and after World War II to study the hazards relating to the newly developed atomic bomb. This research was focused on the effects of ionizing radiation after total body exposure of mammals. In addition, radiation is used as a tool to study the complex biochemical processes required for cell maintenance and function. During the 1970s, research received a further boost with the development of cancer centers in major cities in the United States. A brief review of radiation biology related to cancer treatment, including historical development and future prospects, is presented in this article.

Radiation therapy, along with surgery, is the major form of cancer management. Nearly half of all cancer patients receive radiotherapy. One-third to one-half of those patients receiving radiotherapy have their life spans extended 5 years or longer. Radiobiological research and clinical investigations have the potential for further incremental improvements. However, major improvements can be expected only with the acquisition of a basic understanding of what makes a cell cancerous.

Radiation sensitivity of cells. Chemical agents that selectively kill cancer cells have been sought for many years. Unfortunately, the success with antibiotics in treating bacterial infections cannot be paralleled in cancer treatment. The large evolutionary difference between the chemistry of bacterial and mammalian cells makes it possible to find substances that are toxic to bacterial cells but not to mammalian cells. The lack of such differences in the chemistry underlying cancer cells and normal cells makes it difficult to selectively destroy the cancer cells.

A single, large dose of radiation that produced pronounced erythema of the skin was used in early treatments of cancer with the hope that such a dose would sterilize all cancer cells. When this treatment was found to be ineffective, multiple treatments of radiation with smaller doses were used. The radiobiological rationale for such multiple treatments came from the classical experiments by C. Regaud during the late 1880s and early 1900s. He used sterilization of rams' testicles without severe damage to the skin as a model for cancer treatment. He found that in an early stage of sperm development the spermatogonia were most sensitive, and in a latter stage the spermatids were least sensitive. He also demonstrated that cells in mitosis were the most sensitive. While multiple doses of x-rays sterilized rams' testicles, the single large dose failed to sterilize. These experiments suggested that fractionated doses could produce more damage to a tumor than to the surrounding normal tissue. It also showed that cells could recover from radiation damage in the time interval between fractions and suggested that normal tissues recover more efficiently than tumor cells.

Currently, a typical cancer radiotherapy regimen consists of daily treatments of about 200 rads (2 grays) each, 5 days a week for 4 to 6 weeks; this has remained nearly constant for the past 50 years. Since the primary function of radiotherapy is to sterilize cancer cells without severely damaging the normal tissues within and around the tumor volume, several techniques have been developed over the past 30 years that permit quantitative assessment of radiation damage at the cellular level. The techniques have utilized cell cultures of normal and malignant cells and have proved helpful in understanding some basic mechanisms in radiotherapy. At present, there is a rapidly growing body of cellular knowledge that could affect the strategy of radiotherapy in the near future.

Significant progress has been made during the past 30 years in understanding the radiation tolerance of normal tissues and in using penetrating high-energy x-rays or gamma rays from ^{60}Co (cobalt-60) to minimize the radiation doses to the normal tissues. Radiation effects on normal tissues depend on their proliferation characteristics. Acute radiation effects are seen in rapidly proliferating normal tissues such as the skin, intestinal tissues, and bone marrow cells, and also in rapidly proliferating tumors. Radiation effects on relatively slowly proliferating or nonproliferating tissues, such as connective tissue, cartilage, spinal cord, and lungs, are delayed, and are known as late effects. The tolerance of these relatively nonproliferating normal tissues is often a limiting factor in radiotherapy.

The table shows the spectacular improvements in survivorship resulting from improvements in radiotherapy. Despite these developments, local failures (that is, inability to sterilize all cancer cells) at the primary site are still common, and they account for approximately one-third of cancer deaths after treatment with radiotherapy.

Results of radiotherapy in 1955 and 1970, showing improvement in survivorship

Site or type of cancer	5-year survival, %	
	Kilovoltage (1955)	Megavoltage (1970)
Prostate	15	60
Bladder	5	35
Hodgkin's	35	60
Nasopharynx	30	60
Tonsil	30	50
Brain	20	30

Radiation, like surgery, is primarily intended to treat the localized primary tumor. Chemotherapy, while not so effective in controlling bulky primary tumors, is more effective on metastases. Hence, with improved methods of chemotherapy to control metastasis, local control of the tumor becomes even more important. Nuclear particles (neutrons, pions, protons, and heavy ions) and chemicals that either sensitize tumor cells to radiation or protect normal tissues from radiation are being experimentally investigated to improve local control of tumors. Evaluation of a new method requires 10 years or longer to yield convincing clinical evidence; thus, progress tends to be very slow. Experimental radiation biology in whole animal and cellular systems is playing an important role in evaluating new possibilities that might improve clinical treatments.

Particles. The rationale for using nuclear particles in radiotherapy is to increase the tumor cell killing without exceeding the tolerance of normal tissues. In principle this can be attained in two ways: improving the dose localization and increasing the density of ionization. Neutrons are densely ionizing radiation which offer no improvement in dose localization compared to the currently used megavoltage x-rays. The ionizing density of protons is similar to x-rays. However, proton beams have the best dose-localization characteristics because of the ionization properties of heavy charged particles. The other heavy charged particles of interest, such as pions and heavy ions, are densely ionizing, with better dose-localization characteristics compared with x-rays but slightly inferior dose localization compared with protons.

Dose localization. Dose localization of protons compared with gamma rays is shown in the illustration. The improvement in amount of normal tissue spared outside treatment volume for protons in contrast to ^{60}Co gamma rays is comparable to the normal-tissue-sparing improvements of ^{60}Co gamma rays in contrast to kilovoltage x-rays. Normal-tissue tolerance depends not only on dose but also on the volume exposed to the radiation. Hence, reduction of the volume of normal tissue exposed to the radiation may make it possible to increase the dose to the tumor without exceeding normal-tissue tolerance.

It is necessary to establish whether further im-

provements in dose localization could lead to improvements in radiotherapy by either increasing local tumor control or decreasing the morbidity of treatment, or both. The ongoing clinical investigations with protons should answer this important question. Since proton beams are radiobiologically similar to x-rays, clinical x-ray experience can be used directly to predict the advantage of improved precision in dose delivery. The experience so far indicates excellent results for the treatment of choroidal melanoma (a tumor in the back of the eye), preserving good eye function. The proton treatment may turn out to be the choice for malignancies that are close to critical structures such as the central nervous system.

Oxygen effect. Many tumors appear to have a small proportion of hypoxic cells (cells deficient in oxygen) because of inadequate blood supply within the tumor mass. Hypoxic cells are more resistant to x-rays by a factor of about 3 compared with cells in an oxygen-rich environment. This is known as the oxygen effect. Experimental tumor studies indicate that the presence of even a small proportion of hypoxic but viable cells in a tumor requires a considerable increase in dose for tumor control. Normal tissues in the radiation field may not tolerate such an increase in dose. The relevance of the oxygen effect in radiotherapy was recognized as early as 1930. However, it has recently been shown that in certain types of animal tumors, an increasing proportion of hypoxic cells become oxygenated during fractionated radiation exposures. Hypoxic cells that become oxygenated during treatment are not so resistant to subsequent fractions of radiotherapy. Therefore, it may be possible to overcome the oxygen effect with x-rays for many less advanced stages of cancer. However, reoxygenation may not take place in advanced tumors and in certain histological types. There is now considerable research effort directed toward synthesizing chemical compounds known as hypoxic cell sensitizers that selectively sensitize hypoxic cells to radiation. Densely ionizing particles such as neutrons, heavy ions, and pions considerably reduce the radiation sensitivity differences between hypoxic cells and oxygenated cells; this is one of the rationales for considering these particles in radiotherapy.

High density of ionization. It is also necessary to establish whether the dense-ionization characteristics of particles have any advantages over x-rays for the treatment of specific tumors. Since the dose-localization characteristics of neutrons are comparable with ^{60}Co gamma rays, the ongoing clinical trials of neutrons should answer this question. Current experimental and clinical results indicate that neutrons are very effective on tumor cells, and that they reduce the variability of x-ray response from tumor to tumor considerably. This is an advantage, but an equally important consideration is the sparing of normal tissues. Unfortunately, neutrons are also found to be very effective in damaging slowly proliferating, vital normal tissues. Neutrons, while having

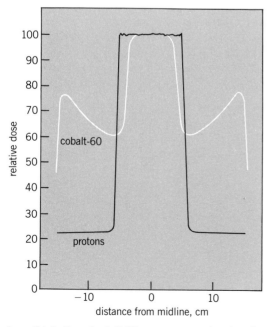

Dose distribution of cobalt-60 gamma rays and protons for irradiation of the uterine cervix. (*After A. M. Koehler and W. M. Preston, Protons in radiation therapy, Radiology, 104:191-195, 1972*)

greater effectiveness on tumors, appear to lack the advantage of x-rays in sparing late effects in normal tissues. A large number of centers in the world are investigating the potential role of neutrons in radiotherapy. There is a suggestion that neutrons may be effective for the treatment of slowly proliferating tumors. Because of the ability of normal tissues to recover from late effects of x-rays, doses lower than 200 rads (2 grays) per fraction of x-rays applied in two or three fractions per day may be more effective for treating rapidly growing tumors.

For tumors where neutron beams are effective, pion and heavy-ion beams may offer additional advantages of dose localization.

Hyperthermia. The application of heat to the treatment of tumors was proposed nearly 100 years ago upon observation that tumors regressed in patients with high fever. There has been a recent revival of interest in the use of hyperthermia for tumor treatment. Hyperthermia is very effective in inactivating cells that are normally resistant to x-rays (such as hypoxic cells and cells in resistant phases of the cell cycle). The technical problems in delivering heat optimally to deep-seated tumors are not yet resolved. However, considerable progress has been made, and about 1500 patients have been treated with hyperthermia and radiation, or hyperthermia alone. Hyperthermia may have potential applications in tumor treatment either alone or in combination with the other current methods of treatment: surgery, radiation, and chemotherapy.

For background information see RADIATION BIOLOGY; RADIOLOGY in the McGraw-Hill Encyclopedia of Science and Technology.

[M. R. RAJU]

Bibliography: J. F. Fowler, *Nuclear Particles in Cancer Treatment*, 1981; E. J. Hall, *Radiobiology for the Radiologist*, 2d ed., 1978; E. L. Travis, *Primer of Medical Radiobiology*, 2d ed., in preparation.

Reproduction (plant)

The vast majority of gymnosperms are wind pollinated, although there are observations indicating that insects may pollinate at least some species of cycads and Gnetales. In comparison with entomophilous plants, wind-pollinated plants produce large amounts of pollen with respect to the number of ovules produced, for example, approximately 1,000,000 pollen grains per ovule in *Pinus*, as opposed to 1000 pollen grains per ovule in insect-pollinated angiosperms. Pollen grains of many wind-pollinated plants are structurally adapted such that their settling velocities are much reduced, thereby extending their dispersal range. The buoyancy of some relatively large conifer pollen grains is increased by the presence of one or more air sacs, attached to the main body of the grain, which increase the volume and surface area of the grain without excessively increasing its weight. The efficiency of pollen transport is increased by the fact that most gymnosperm pollen grains lack an oily coating, which if present would result in the clumping of grains.

Wind tunnel experiments conducted on various gymnosperms indicate that the morphology of ovule-bearing axes often enhances the probability of pollen reaching the micropyles of ovules. Therefore, in addition to pollen grain features that facilitate efficient dispersal, gymnosperms have proficient pollen-trapping devices that often function on the basis of aerodynamically sophisticated principles.

Pollination in early seed plants. The seed habit arose presumably from a heterosporous antecedent condition by the development of a megasporangium containing a sterile layer of tissue called the integument. The earliest known seeds are from the Upper Devonian and possess integuments that are deeply lobed along their distal margin. Carboniferous seeds show a variety of morphologies that have led some to speculate that the integument arose from the progressive fusion or syngenesis of sterile axes that subtended an originally naked megasporangium. This hypothetically primitive condition is exemplified by the Mississippian *Genomosperma kidstoni*, where the integument is composed of eight separate elements fused only at their extreme base. In the smaller ovule *G. latens*, the fusion of the elements is greater, and they are adducted over the apex of the nucellus to form a rudimentary micropyle. More extensive fusion is seen in the integument of *Eurystoma angulare*, and fusion is complete in the integument of *Stamnostoma huttonense*. Although the Mississippian ovules are usually arranged in a phylogenetic sequence based on the relative degree of integumentary lobe fusions, it is worth noting that all these fossils are of approximately identical age, and more primitive than the earlier Devonian ovule *Archaeosperma arnoldii*.

Wind tunnel analyses of scale models of early seed fossils indicate that with the progressive fusion of integumentary lobes and their adduction over the megasporangium, there is a progressive canalization of turbulent airflow over the nucellar apex. On the basis of aerodynamics, the probability of wind-borne pollen (or prepollen) settling on a structure increases as turbulent airflow is focused on a particular region. Thus, the morphological changes of the hypothetical phylogenetic sequence which led to the integument may have facultatively increased the probability of pollination (Fig. 1).

Among the various ovule taxa described from Carboniferous strata, there are a number of mechanisms for receiving pollen into the megasporangium. The earliest of these in the geologic record is an elaboration of the nucellar apex into a trumpet- or cup-shaped funnel (the salpinx or lagenostome). As integument closure evolved, there appears to have been a concomitant reduction in the size and complexity of the nucellar apex. It has been speculated that in the earliest seed plants, pollen was directed to the pollen chamber by the salpinx, while in later forms, the fusion of the integumentary lobes (and subsequently the micropyle) took over this function. By early Pennsylvanian time, several groups of seed plants were producing resinous pollination droplets. The pollination droplet is prevalent in extant cycadophytes and coniferophytes, and serves to draw pollen into the pollen chamber to produce fertilization.

Pollination in cycads. Wind pollination appears to be the dominant mode in cycads. Observations indicating that insects pollinate some species reveal that entomophily either is accidental or is an irregular pattern in cycads. Pollen-eating beetles are known to visit the microsporangiate cones of *Encephalartos* and *Zamia integrifolia*, and small bees have been observed collecting pollen of *Macrozamia tridentata*. Since all cycads are dioecious, insect activity around pollen-bearing plants does not constitute pollination. However, beetles have been observed visiting pollen cones of *Encephalartos* (attracted presumably by smell) and also depositing eggs on ovule-bearing cones. Insect pollination mechanisms in cycads appear to be associated with semidestructive patterns of behavior, either by the collection and subsequent ingestion of pollen or by the deposition and subsequent herbivory of eggs and larvae.

The pollen of cycads is light and dry, and is produced in large quantities per cone; however, wind pollination is conspicuously inefficient. Seeds from an ovulate cone 4–5 m from a pollen cone may show 15–20 pollen tubes, while those 100 m away may show only 2 or 3. Ovulate cones 200 m from a pollen source show 0–3 seeds per cone.

Wind pollination experiments with receptive *Encephalartos* ovule cones and freshly released pollen reveal that at airflow velocities of 3–5 m/s most pollen grains accumulate on the distal portions of the cone (that is, about 35% of all the pollen found on the cone is concentrated on the upper 2–3 cm of the cone axis). With the application of water to simulate rain, pollen grains are observed to wash down the sides of megasporophylls, near or into micropyles. Similar experiments with the ovulate cone of *Dioon* reveal that pollen collects preferentially on the adaxial surfaces of sterile scales subtending the fertile (megasporophyll) portion of the cone. It is possible that beetles, foraging the cone for pollen, may enter the cone proximally and cause pollen to be carried into the recesses of the cone near the ovules. Examination of ovulate cones of *Zamia* indicates that these structures are aerodynamically inefficient. Pollen entering upwind jet streams are scattered over the entire surface of the cone, although there does appear to be a slight preferential settling of pollen on the downwind surfaces of the cone.

Wind tunnel and field observations suggest that pollination in some species of cycads may involve wind, wind and water, or wind and insect transport of pollen. Under any circumstances, however, it may be concluded that anemophily in cycads is an extremely inefficient mechanism.

Conifer wind pollination. The typical pine ovulate cone consists of an axis bearing spirally arranged scale-bract complexes. Each scale-bract complex is the developmental equivalent of a highly reduced and structurally modified branch, called a scale, subtended from and in part fused to a modified leaf or bract. Two ovules are borne on the adaxial surface of most scales, with their micropylar ends directed

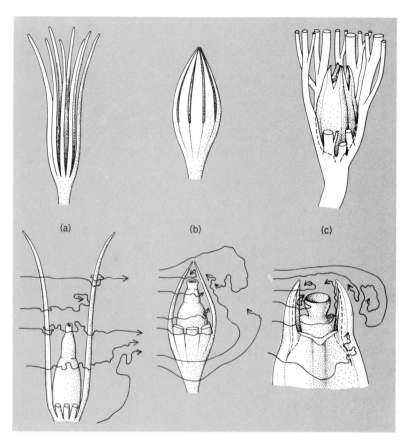

Fig. 1. Wind tunnel analyses of scale models of early Paleozoic seed fossils. (a) *Genomosperma kidstoni*; (b) *G. latens*; (c) *Eurystoma angulare*. The airflow-disturbance patterns show increased nonlaminar flow around the salpinx or micropyle from *a* to *c*. (*Courtesy of K. J. Niklas*)

toward the cone axis. The number and shape of scale-bract complexes, as well as the geometry of their insertion on the cone axis, vary from one species to another, thereby establishing morphological criteria for separating taxa. Ovulate cones which are receptive to pollination are typically small (0.5–2.0 cm in length), unsclerified, and borne upright, either singly or in small clusters, at the tips of branches. The ovules of many conifer species produce a slightly viscous exudate from the micropyle, which results from lysogeny of nucellar tissues.

When placed in a wind tunnel, models of ovulate cones (scaled to size) generate distinctive airflow patterns around and between scale-bract complexes. The airflow characteristics of an ovulate cone appear to be distinctive for each conifer species examined, but generally take on the configuration of three discrete patterns: airflow passing through or around the cone is seen to be deflected backward toward the downwind surfaces of the cone in a pattern conforming to a Kármán vortex street; airflow is canalized into a cyclonic vortex passing among scale-bract-complex orthostichies and parastichies, spiraling around the cone axis; and overlapping scale-bract complexes appear to generate doldrumlike eddies, where airflow enters along the adaxial surface of a bract and curls along the cone axis toward ovule micropyles, and then flows along the adaxial surface of the subtending

Fig. 2. Airflow-disturbance patterns created by an ovulate cone of *Pinus*. As shown in the enlargement at right, doldrumlike eddy of airflow (open arrow) can trap a reentrained pollen grain (black dot) and redirect it toward adaxial, proximal ovules (solid arrow). (*After K. J. Niklas, Airflow patterns around some early seed plant ovules and cupules: Implications concerning efficiency in wind pollination, Amer. J. Bot., 68:635–650, 1981*)

scale (Fig. 2). All three patterns of airflow appear to be generated by the geometry of scale-bract morphology and their insertion into the cone axis. Collectively, these patterns of airflow influence the behavior of wind-borne pollen grains as they approach and circulate around the cone.

The characteristics of wind-borne pollen grains are seen to conform in large measure to the airflow patterns generated by the ovulate cone. Pollen grains passing by the ovulate cone are swept into the downwind Kármán vortex street, and are often redirected in their movement such that they collide with the adaxial surfaces of downwind scale-bract complexes. Pollen grains trapped in a cyclonic vortex of airflow pass along successive scale-bract complexes and, when their velocities are sufficiently reduced, settle on the surfaces of scales. Pollen grains entering the doldrumlike eddies between overlapping scale-bract complexes either come into direct contact with micropyles or settle on the adaxial surfaces of scales to which ovules are attached. In the latter case, pollen grains have been observed to roll along the adaxial surfaces toward the distal edge of the scales and to then become reentrained into the airflow pattern around the cone axis. Often pollen grains reenter the airflow pattern of doldrumlike eddies and are directed back again toward micropyles.

For background information *see* POLLINATION; REPRODUCTION (PLANT) in the McGraw-Hill Encyclopedia of Science and Technology.

[KARL J. NIKLAS]

Bibliography: K. J. Niklas, Simulated wind pollination and airflow around ovules of some early seed plants, *Science*, 211:275–277, 1981; K. J. Niklas, Airflow patterns around some early seed plant ovules and cupules: Implications concerning efficiency in wind pollination. *Amer. J. Bot.*, 68:635–650, 1981; K. J. Niklas, Simulated and empiric wind pollination patterns of conifer ovulate cones. *Proc. Nat. Acad. Sci. USA*, 79:510–514, 1982.

Satellite (astronomy)

The total surface area of Jupiter's four large galilean satellites is comparable to the total explored areas of Mercury, Mars, the Moon, and Earth. These satellites are truly planetary in scale, ranging in size between the Moon and Mercury. In 1979 the two Voyager spacecraft returned a wealth of new information about these bodies and the processes by which they evolved. Contrary to expectations, these objects display a tremendous diversity in geologic style. Callisto, an icy world whose surface is saturated with ancient impact scars, has evidently remained unchanged for billions of years. Giant ring structures, analogous to the rings caused by tossing a rock into a pond, record cosmic-scale impacts on its early soft and fluidlike crust. The Voyager missions discovered that alien tectonic processes have operated on the crusts of Ganymede and Europa. Finally, *Voyager 1* showed Io to be the most volcanically active surface in the solar system.

Formation of galilean satellites. The galilean satellites probably formed in roughly their current orbits while Jupiter itself was forming. This system resembles, in miniature, the entire solar system, and many of the same physical processes controlled the evolution of the two systems. For example, Jupiter, like the Sun, has been losing heat since formation; even today Jupiter radiates more energy than it receives from the Sun. Jupiter's early period of high luminosity dramatically affected the final composition of the satellites; the Sun's radiation similarly controlled the composition of the planets. Low-density materials which condense at low temperature, like water ice, could not condense near Jupiter or near the Sun. Consequently the bodies that formed close in, for example, Mercury and Io, incorporated only the denser materials that condense at higher temperatures. These materials included metals, their oxides and sulfides, and silicates. As a result, the galilean satellites of Jupiter and the planets of the solar system show a common trend, that of decreasing density away from the primary or central body.

Voyager instrumentation. Although both Voyager spacecraft carried a broad array of scientific instrumentation, much of the new information about the surfaces of the galilean satellites came from the telescopic television systems. Two other experiments on the spacecraft made important contributions to understanding the satellites. The spacecraft radios, used to transmit information back to Earth, were used to track precisely the motion of the spacecraft. These data were then used to measure the gravitational pull of the satellites as the spacecraft neared them, and thereby provided accurate estimates of their masses. A second type of instrument, IRIS (infrared interferometric spectrometer), provided valuable data on surface temperatures and composition, as well as some information about the chemistry of Io's volcanic plumes.

Each of the spacecraft carries two television cameras, narrow-angle and wide-angle, that acquire images through a variety of color filters. From a range of 100,000 km, objects about 2 km in diameter can be resolved with the narrow-angle camera. The wide-angle camera is used when objects are so close that they exceed the narrow-angle field of view but global-scale images are needed.

Callisto. Farthest from Jupiter of the major satellites, Callisto is an object about 4850 km in diameter, slightly smaller than the planet Mercury. Its density, near 1.8 g/cm^3, suggests a composition of roughly half rocky material and half water ice. Most likely, the heavier rocky materials have sunk into its deep interior, leaving a thick water or water-ice mantle. Voyager images show Callisto to be very heavily cratered; impact craters nearly saturate its surface. Apparently Callisto's surface has changed very little in the last 4 billion years since gases and debris in the solar system coalesced to form the planets and left their surfaces battered by impact. Images of Callisto's limb show very little topography; evidently even the topography of the craters largely disappeared as the rims collapsed and the icy material flowed back into the crater floors. Callisto shows other evidence that its crust was soft and pliable during its earliest history. Giant concentric ring systems surround the largest and oldest impact scars on its surface. One of these systems, more than 3000 km in diameter, consists of about 15 rings (Fig. 1). The best model for the formation of the ringed structures is that the energy of mammoth-scale impacts was sufficient to penetrate to depths where the crust was fluid or soft and could flow easily. The surface then rebounded, perhaps coupled with enormous, concentric ripples, creating the ring systems. The slightly lower frequency of impact scars on the rings themselves shows that they formed late during the intense bombardment.

Ganymede. Callisto and Ganymede, the next regular satellite inward toward Jupiter, have very similar bulk properties. Ganymede is only slightly larger

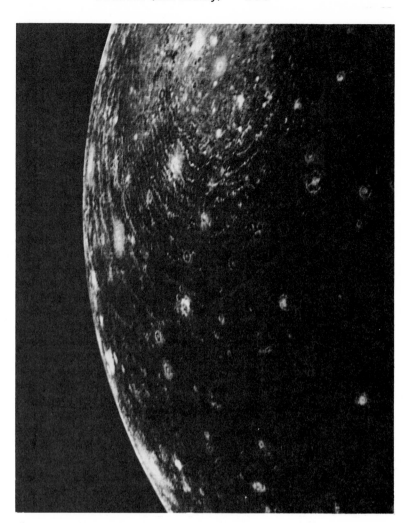

Fig. 1. Giant ringed structure on the surface of Callisto.

(diameter about 5275 km), nearly the same diameter as Mercury, and is slightly denser (about 1.9 g/cm^3) than Callisto. Like Callisto, the best model for Ganymede's composition is about half water or water ice, forming a thick mantle over a central rocky ball about the size of the Earth's Moon.

Voyager 1 images showed that Ganymede has undergone a far more complex geologic history than its neighbor, Callisto. Although parts of Ganymede's surface do closely resemble the intensely cratered, ancient dark terrain found on Callisto, Ganymede's terrain has been broken up and largely destroyed. Polygonal remnants of the old terrain, varying widely in size and shape, are separated by strips of younger terrain, termed grooved terrain, that has parallel grooves and ridges which are sometimes so regularly spaced that they appear as if made with a rake (Fig. 2). The ridges are a few kilometers in width and spacing, and they may extend for several hundred kilometers in length.

Evidently Ganymede underwent a phase of extensive global tectonism following its Callisto-like stage. The ubiquitous grooved terrain probably formed as internal pressures broke the surface, producing complex assemblages of parallel and steep (perhaps vertical) faults along the edges of the surviving poly-

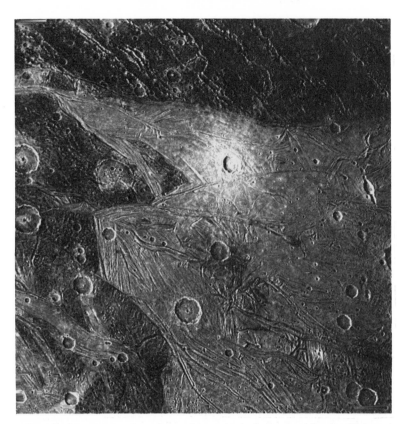

Fig. 2. Grooved and cratered terrains on Ganymede. Ancient cratered terrain with jagged furrows is visible along the top of the image. Swaths of grooved terrain dominate the rest of the view.

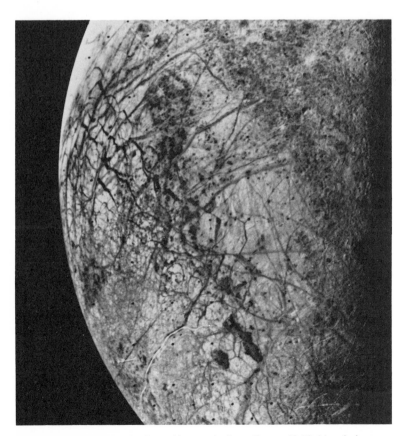

Fig. 3. Complex patterns of stripes, ridges, and pits on Europa. Mottled terrain is prevalent in the upper right of the image along the terminator. Curious ridges, ubiquitous in the bright plains, can be seen in the lower right along the terminator.

gons of the older, dark terrain. Fluids, perhaps water-ice slurries, were then intruded along these faults and were extruded onto the surface. In this senario, the fluids then froze and expanded, producing the grooved swaths that now frame the old cratered polygons.

Even the grooved terrain on Ganymede is ancient by terrestrial standards. The abundance of impact craters on even the youngest of Ganymede's terrains is so high that the grooved terrain must date back to Ganymede's early history. Apparently Ganymede's period of global tectonism was brief, leaving the icy world dormant for eons to follow.

Europa. Europa, the next major satellite in toward Jupiter, is about 3125 km in diameter and has a density of about 3.0 g/cm³. It is slightly smaller and less dense than the Moon. Measurements of Europa's near-infrared reflection spectrum from the Earth show strong absorption bands due to water ice. Europa's extraordinarily high albedo (near 70%) and flat reflection spectrum suggest further that the ice is relatively clean in contrast to that exposed at the surfaces of Ganymede and Callisto. Europa's density is lower than would be expected for a purely rocky planet, suggesting a composition of about 10% water. Prior to the Voyager encounters, a major question was whether or not most of the water had migrated to the surface. Europa could have a frozen ocean 50–100 km thick, and this ocean might still be liquid at depth.

Although *Voyager 1* glimpsed Europa only from great range, the images provided evidence of yet another unique, alien tectonic environment. The low-resolution images showed that enormous stripes, slightly darker than the bright background, laced the globe. They appeared typically 50–100 km wide; some ran more than half way around the satellite. The diffuse bright spots that had indicated abundant impact scars on Callisto and Ganymede from similar distance were not evident on Europa.

Voyager 2 flew past Europa at a range of about 200,000 km; features as small as 4 km were clearly visible. Two terrains were evident: bright plains with a dense network of intersecting dark stripes; and dark, mottled regions consisting of a dense population of small dark patches (Fig. 3). Along the terminator where grazing illumination accentuated what little topography exists on Europa, *Voyager 2* images showed that the dark, mottled terrains have a dense, almost spongelike population of shallow pits and subdued hills a few kilometers across. In this terminator region the bright plains showed a complex of intersecting narrow ridges that resemble strands wrapped on a ball of twine. Some of the ridges occur in the center of the dark stripes; others have bizarre geometric patterns resembling the trajectory of a bouncing ball. Very few craters were evident in the Voyager images; only three or four small craters, a few kilometers in diameter, were detectable. Two large circular brown smudges may be the scars of very old craters, long since collapsed in the icy crust.

The stripes and ridges indicate that Europa's crust

has expanded. The expansion may not be as great, however, as the area of the stripes might suggest. Many of the stripes may have formed when dark slurries flowed into graben (linear valleys formed as long wedges that break off and drop down into the crust). In some cases, however, pieces of the crust have actually rafted apart. In these areas the crust opened like a pair of scissors as one side rotated away from the other.

Various processes may have contributed to Europa's expansion. The addition of fluid from the interior beneath the frozen shell could cause local expansion. Freezing of water itself produces an expansion of about 10%, or about a 7% increase in area. Finally, changes in the mineralogy of the solid interior, forming hydrated minerals, have also been suggested as a cause for expansion.

Without better data on the rate of crater formation, the age of Europa's present surface cannot be determined, but it is now known that the planet remained geologically active long after the period of intense bombardment. It is possible that the current crust is only a few tens of millions of years old, a tiny fraction of geologic history, and Europa's crust continues to be episodically renewed.

Io. Closest to Jupiter of the four large galilean satellites, Io is about 3640 km in diameter, slightly larger than Europa and nearly the same diameter as the Moon. Unlike the other three major satellites, Io's reflection spectrum shows no evidence of water ice; the spectrum indicates its surface is extraordinarily dry. With nearly the same density (3.5 g/cm^3) as the Moon, Io was expected to be a lunarlike, rocky planet. Most thought it would probably show an ancient crust, perhaps with evidence of early volcanism whose activity had long since ceased. Earth-based observations had shown Io to be strongly colored, a bright yellow-orange object with barely resolvable darker poles. The likeliest substances to produce its rich color were sulfur and sulfur compounds. Elemental sulfur occurs in many forms, or allotropes, the colors of which can be white, yellow, orange, red, brown, or black. Other telescopic measurements showed that sodium, potassium, and sulfur ions were concentrated in a torus encasing Io's orbit. Somehow, these ions were being lost from Io's surface and swept away by Jupiter's magnetic field.

In the first few images acquired by *Voyager 1*, in which surface features could be seen, dark circular spots 50–100 km in diameter emerged, some surrounded with bright rings. It first appeared that Io's surface was, in fact, riddled with ancient impact scars, as most had expected. As *Voyager 1* neared, higher-resolution images began to indicate the reverse. The large circular spots with bright rings appeared as if painted on the surface; they did not have the topographic forms of impact craters. Additionally, smaller craters that should be very abundant and easily visible were totally absent. Not a single impact crater was detectable down to the limit of the highest resolution images (0.5 km). Evidently Io's surface was extremely young and geologically

very active; an upper limit of about a million years was placed on the surface age.

A prediction was published before the *Voyager 1* results were received that Io might be volcanically active. It was suggested that Io is locked in a gravitational struggle between Jupiter and two of the other large satellites, Ganymede and Europa. Whereas the gravitational pull of these satellites, mainly Europa, acts to distort Io's orbit out of circular form, Jupiter's gravitational field forces Io's orbit to remain nearly circular. The result is that Io's crust is flexed continually, and enormous tides migrating across its surface frictionally heat the crust. The authors of the idea concluded that Io has a thin rocky crust 10 or 20 km thick lying over a molten interior and is probably volcanically active. *See* JUPITER.

Voyager 1 made its closest approaches to Jupiter and Io on March 5, 1979. The *Voyager 1* data showed that the prediction of volcanic activity was accurate. The images showed numerous volcanic landforms: multicolored volcanic flows of red, orange, brown, and black material; and calderas, or volcanic collapse pits, some more that 50 km in diameter and some with immense radiating patterns of volcanic flows.

Even on the Earth, which is extremely active volcanically, the chances of observing a volcanic eruption are very small, particularly in an area as small as that viewed by *Voyager 1* on Io. Nonetheless, a few days after the encounter, *Voyager 1* acquired an image looking back at Io in which the first active volcano in the solar system beyond the Earth was discovered. In this image an enormous umbrella-shaped volcanic plume stood over Io's limb. Material was rising nearly 300 km above the surface and falling back into a ring nearly 1500 km in diameter. A detailed search of the rest of the imaging data revealed at least eight active plumes; with more recent analysis, the number has grown to 11.

When *Voyager 2* arrived 4 months later, most of the reobserved volcanic plumes were still active. The largest, the one originally discovered by Voyager 1, had ceased. The eruption velocities of Io's volcanic plumes are between 0.5 and 1.0 km/s, in the range of the velocity of the bullet fired from a high-powered rifle. The material remains in ballistic flight for up to 15 min.

At the conclusion of the encounter, most of the key elements explaining Io's volcanism had been identified: the energy was supplied by tidal pumping of the rocky crust; sulfur, which melts at temperatures that would be encountered at relatively shallow depths in Io's crust, was likely an abundant component of the crust. The IRIS experiment detected one other important ingredient. Sulfur dioxide gas was found to be a major component in one of Io's volcanic plumes. At about the same time, sulfur dioxide frost was discovered in Io's reflection spectrum when it was measured from the Earth. Sulfur dioxide gas, liquid, and frost are apparently abundant in Io's upper crust and explain many of the white deposits and rings around volcanic plumes.

In addition to the active plumes, in some regions, particularly near the south pole where the best images of highest resolution were acquired, white, snowlike deposits indicate that gases or fluids leak out of numerous fractures and faults in Io's crust (Fig. 4). Currently the best model to explain these deposits is that liquid sulfur dioxide migrates toward the surface along faults, and as it is exposed to the near-vacuum atmosphere at the surface, it explosively changes to a cloud of vapor and ice crystals which fall back to the surface like snow.

Two possible mechanisms have emerged to drive Io's volcanic plumes. Both are similar to the process that drives explosive volcanoes on the Earth. In the Earth's case, molten lava comes in contact with water or water is dissolved deep in the magma source. When the mixture reaches the surface, it explodes into an expanding cloud of rock debris and steam. In Io's case, two situations may well occur: molten silicate may come in contact with sulfur, or molten sulfur may come in contact with liquid sulfur dioxide. In either case, as the mixture rises to the sur-

face, it expands, exploding into the near vacuum. Io's volcanic landforms may be the composite deposits of both silicate and sulfur volcanic flows. On the one hand, the range of colors of the flows on Io's surface match those expected for sulfur flows, erupted at different temperatures and quenched at Io's extremely cold surface temperatures (125 K). On the other hand, many of the landforms with steep slopes and high relief probably could not be composed solely of sulfur, as they would collapse under their own weight. Silicate volcanic rocks are probably their primary source of strength. Perhaps on Io, silicate volcanism occurs beneath surface layers composed predominantly of sulfur and sulfur dioxide, analogous to submarine volcanism on the Earth. In some areas, islands akin to seamounts may build up from silicate lavas, eventually protruding above the sulfur–sulfur dioxide zone.

Geologic activity. To a large extent the Voyager findings were far from expected. Few anticipated that substantial geological activity could have taken place in cold, icy worlds like Europe and Ganymede, or that the small lunarlike body, Io, could conceivably be volcanically active. The basic premise of most scientists had been that the heat needed to drive a planet's internal activity came primarily from two sources: accretional heat as the material collided and aggregated, and radiogenic heat from radioactive decay. Further, it was reasoned that the smaller a planet, the faster it would lose its heat. All of these considerations led to the prediction that there would be too little heat too easily lost for these small objects to have been geologically active. The Voyager missions led to the discovery that there are many unpredicted processes, compositions, and conditions by which dynamic geologic evolution can occur.

For background information *see* JUPITER; PLANETARY PHYSICS; SPACE PROBE in the McGraw-Hill Encyclopedia of Science and Technology.

[LAURENCE A. SODERBLOM]

Bibliography: David Morrison (ed.), *Satellites of Jupiter*, 1982; B. A. Smith et al., The Galilean satellites and Jupiter: *Voyager 2* imaging science results, *Science*, 206:927–950, 1979; B. A. Smith et al., The Jupiter system through the eyes of *Voyager 1*, *Science*, 204:951–972, 1979; L. A. Soderblom, The Galilean moons of Jupiter, *Sci. Amer.*, 242(1):88–100, January 1980.

Scattering experiments (atoms and molecules)

During the collision of a charged atomic or molecular species X^+ with a neutral atomic or molecular species Y, an electron may be transferred from Y to X^+ so that after the collision the ion X^+ has become neutral and the neutral particle Y has become ionized. If the ion is initially multiply charged, more than one electron may be transferred during the collision. The rearrangement of the electron distribution is described as a charge transfer. Charge transfer is an important process in partly ionized gases or

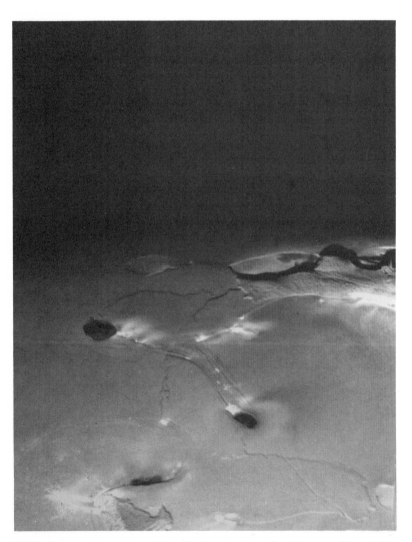

Fig. 4. High-resolution *Voyager 1* view of Io's south polar region. Numerous diffuse, bright features along fault scarps may be due to sulfur dioxide leaking out onto the surface and forming ice clouds and snowlike deposits.

plasmas, where it may exert a substantial influence over the distribution of the stages of ionization. Because charge transfer involves a rearrangement of the electron distributions, it shares some of the characteristics of a chemical reaction, and thus provides useful insight into the nature of reaction mechanisms.

Nonradiative charge transfer. At slow velocities the charge-transfer process can be envisaged as a transition from an initial molecular state of XY^+ formed by the approach of X^+ and Y, to a different final state of XY^+ which separates at large distances into X and Y^+. The molecule XY^+ is a temporary association which exists for the duration of the collision. The efficiencies of charge-transfer processes depend in a detailed way on the characteristic of the potential energy surfaces of XY^+ from which the forces determining the motion of the colliding particles are derived. The illustration is a schematic diagram of the potential energy curves of a diatomic molecule XY^+ created by the approach of an atomic ion X^+ and a neutral atom Y, shown as functions of the distance R between the nuclei of X and Y. If the potential energy curve of the initial state X^+–Y approaches the potential energy curve of the final state X–Y^+ over a range of internuclear distances (near R^* in the illustration), a transition may occur from

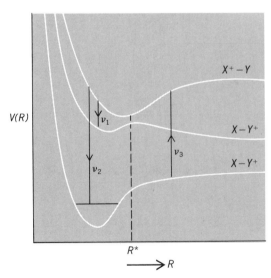

Curves showing potential energy $V(R)$ in various configurations which can result from the approach of an atomic ion X^+ and a neutral atom Y, as functions of the distance R between X and Y.

one curve to the other. An ion X^+ approaching a neutral atom Y along the uppermost potentail curve in the illustration may then cross over to the middle curve at distances near R^* and separate out to X and Y^+. If the potential energy curves remain far apart at all distances R, the probability of crossing from one curve to the other is negligible. If they approach very closely, the transition occurring on the inward path of the collision is followed by the

reverse transition to the original state on the outward path. In both cases, charge transfer is improbable. If the curves approach to some favorable energy separation, charge transfer is rapid. The probabilities of crossing depend upon the relative velocity of the colliding species, and measurements of the variation with velocity of charge transfer can be analyzed to obtain empirical constructions of potential energy surfaces and transition probabilities.

Radiative charge transfer. Collision-induced transition is not the only process by which charge transfer occurs. Because a quasimolecule is created during the collision, photons may be emitted in spontaneous radiative transitions from the initial state to some charge-transfer state, which may be different from that reached by the direct process. In the illustration, such a process is the emission of a photon ν_1 at any distance R, in a transition from the uppermost curve to the middle curve, again producing X and Y^+. The emission spectrum associated with radiative charge transfer, reaction (1), offers a

$$X^+ + Y \rightarrow X + Y^+ + h\nu \qquad (1)$$

powerful diagnostic probe of the internal dynamics of the rearrangement scattering taking place on a time scale on the order of 10^{-13} s. The emission intensity is weak. By dividing the mean size (5×10^{-8} cm) of the collision pair by the typical thermal velocity (10^5 cm s^{-1}), an estimated collision time of 5×10^{-13} s is obtained. The mean time for the spontaneous emission of a photon by an excited molecule is typically 10^{-8} s and may be much more, so that a photon is emitted during charge transfer in at most 1 in every 20,000 collisions. Because the radiation is produced in a transition from one scattering state to another, the emission does not consist of discrete lines but appears as a continuum. Often, though, the continuum emission will have structural features which reflect particular aspects of the scattering process.

The radiation emitted in the conversion of helium ions to neutral helium atoms in a charge-transfer reaction in neon gas, reaction (2), has been observed.

$$He^+ + Ne \rightarrow He + Ne^+ + h\nu \qquad (2)$$

In this experiment, pulses of low-energy helium ions were introduced into a drift cell containing neon gas, and a small electric field was applied. Radiation was observed downstream with a grating spectrometer of high optical speed and a photomultiplier detector. Signals were accumulated for periods exceeding 1 h at the low spectral resolution of 1.2 nm, and weak emission was recorded in the region between 410 and 414 nm—wavelengths that corresponded to the charge-transfer process in which the electron is captured into the ground state of helium.

Laser-induced charge transfer. In the reverse process to radiative charge transfer, reaction (3), a

$$X + Y^+ + h\nu \rightarrow X^+ + Y \qquad (3)$$

photon is absorbed during the scattering of Y^+ in a collision with X. In the illustration, such a process is the absorption of photon ν_3 in a transition to the uppermost curve which separates into X^+ and Y. In the absence of the radiation, the process is endothermic and charge transfer cannot occur. Because the collision time is short, an intense radiation field is necessary if charge-transfer products sufficient for detection are to be created.

In a series of experiments, two lasers were used to induce the charge transfer of calcium ions in collisions with strontium atoms, and the reaction probabilities were measured as functions of the applied laser frequency and power density. One laser produced a population of calcium ions, and the second drove the charge transfer to the excited level of the strontium ions. The strontium ions were observed through the fluorescence radiation they emitted.

Ions are a convenient high-capacity storage system, and laser-induced charge-transfer processes can be used in the construction of high-energy lasers. They are a potential means for obtaining selective control over the end products of chemical reactions.

Radiative association. In the collision of two atomic particles X^+ and Y during which a photon is emitted, the transition from the initial scattering state may terminate in a discrete bound level of the stable molecule XY^+, the excess momentum being carried away by the photon. In the illustration, such a process is the emission of photon ν_2 in a transition from the uppermost curve to the lowest one. Evidence was obtained in experiments on helium ions in neon of the formation of the stable molecule $HeNe^+$. In low-density environments where the simultaneous collision of three particles is a rare event, radiative association, reaction (4), is a major path to the formation of compound systems.

$$X^+ + Y \rightarrow XY^+ + h\nu \qquad (4)$$

If the colliding species X^+ and Y are themselves compound systems, they have many internal energy modes, and the collision time may be extended considerably by the temporary formation of a collision complex. The system energy is distributed among the large number of degrees of freedom, and some time elapses before it is concentrated again into a dissociation channel. The formation of complex species by radiative association is then a probable event, particularly at low velocities.

In interstellar molecular clouds, where the temperatures are usually less than 70 K and often less than 10 K, and gas-particle densities are less than 10^6 cm^{-3}, radiative associations are believed to be critical steps in the formation of the complex organic molecules which have been detected in interstellar space through their emission and absorption in the millimeter region of the electromagnetic spectrum.

For background information *see* INTERSTELLAR MATTER; MOLECULAR STRUCTURE AND SPECTRA; SCATTERING EXPERIMENTS (ATOMS AND MOLECULES) in the McGraw-Hill Encyclopedia of Science and Technology.

[ALEXANDER DALGARNO]

Bibliography: D. R. Bates, Theory of radiative association in interstellar clouds, *Astrophys. J.*, in press; S. E. Harris et al., Laser induced collisional energy transfer, in D. Kleppner and F. M. Pipkin (eds.), *Atomic Physics*, vol. 7, 1981; R. Johnsen, Spectroscopic observations of the radiative charge transfer and association of helium ions with neon atoms at thermal energy, *Phys. Rev. A*, in press.

Sea anemone

In 1973, while working with the sea anemone *Anthopleura elegantissima* in which asexually produced clonal assemblages occur, Lisbeth Francis showed that contact between individuals of different clones induces elaborate agonistic behavior but that no such response occurs upon clonemate contact. Since this inaugural study a broad interest has developed in the problem of self-recognition in coelenterates.

It has long been known that the discharge of nematocytes (the stinging cells that release toxic substances and are employed in food capture, defense, agonistic behavior, adhesion, and so on) cannot under normal conditions be caused by mechanical stimulation alone; chemical agents must be present. The most effective agents in nature appear to be lipids strongly linked to proteins. Sea anemones respond by nematocyte discharge upon contact with a host of animal species, and even with other closely related anemones. In some manner the mechanism controlling this discharge can distinguish between self and nonself, and is thus in direct relation with "recognition."

Nematocyte thresholds. Nematocytes have generally been considered independent effectors, that is, effectors whose thresholds are not markedly altered by influences originating within the body of the animal. A number of studies have shown that this concept has to be altered to a certain extent. It has been known for a long time that in *Hydra* the threshold of discharge to prey is much lower in starved animals than in well-fed ones. The existence of neuronematocyte junctions has been demonstrated in this animal, and also in the tentacles of the hydrozoan medusa, *Gonionemus*, providing at least circumstantial evidence that nematocytes may be affected by neuronal activity. Under certain circumstances the threshold of discharge can change with a rapidity commensurate with that produced by nervous conduction. Immediately after swimming is initiated in the anemone *Stomphia coccinea* by contact with a predatory starfish, the threshold of discharge into the starfish, which on first contact is very low, is dramatically raised. For a brief period of time before the anemone resettles, no discharge to the starfish can be observed.

That precise recognition and discharge may be closely related was indicated by the work of D. Davenport, D. M. Ross, and L. Sutton, who investigated thresholds of discharge in the anemone *Calliactis parasitica*, which lives on hermit crab shells. If this anemone settles on a glass plate and its tentacles come in contact with a shell, more specifically, with the shell's protective layer, the periostracum, it will first attach itself to the shell by nematocyte discharge, disengage its pedal disk, and then go through a behavioral sequence resulting in the assumption of its normal position on the shell. Discharge thresholds were found to be significantly lower when animals were on glass plates rather than on shells. Davenport and coworkers concluded that absence of contact between pedal disk and periostracum adaptively lowers the threshold of tentacular discharge so that when contact between tentacles and periostracum occurs, adhesion is easily accomplished. However, they could not draw conclusions concerning thresholds of single nematocytes, or even of fields of nematocytes, because they measured thresholds of discharge by comparing adhesion levels in similar samples of single tentacles ($N = 100$) in two groups of animals held under differing conditions. It is now known that when shell climbing is initiated in *C. parasitica*, electrical activity may be recorded in two conducting systems, one of which is ectodermal and may very well modify discharge. Perhaps it is unnecessary to postulate that there is any preexisting difference in discharge threshold for anemones on different substrates. Threshold lowering may be produced by the activity of the ectodermal system triggered by tentacular sensors upon first contact with the periostracum. Both models demand recognition of the periostracum, whether it be at the pedal disk or at the tentacles.

Consequently, many nematocyte discharge systems can no longer be considered to be independent of stimuli transmitted through the animal's internal system. The effects of agents in changing thresholds may in some cases be direct and may consist of no more than the activation of a sensor which is part of the nematocyte, perhaps the cnidocil. However, in others they may be the result of activity of extraneous sensors and conducting systems. In either case, recognition of agents must occur.

Francis demonstrated that *A. elegantissima* behaves very differently upon contact with a nonclonemate than it does upon contact with a clonemate. Anemones are covered with mucous coats, and when two anemones come into contact these coats are brought together. Investigation of the mucous antigenicity of a number of anemone species has given no indication of differences within species but has indicated marked differences between species. The suggestion has therefore been made that mucus may be a factor in recognition. S. Ertman and Davenport tested this suggestion using the level of nematocyte discharge as a criterion to investigate the possible role of mucus in the recognition of self from nonself. Here the level of discharge was measured by tests ($N = 100$) of single, randomly selected tentacles on ten animals for adhesion to mucus on glass rods. Using *A. elegantissima*, these researchers showed that isolated mucus from both clonemates and nonclonemates failed to elicit discharge of tentacular nematocytes. Further, while direct contact between clonemates failed to elicit discharge, similar contact between nonclonemates effected significant dis-

charge. It was shown that isolated mucus from a number of xenogeneic anemones elicits discharge, and the reciprocal tests were positive. Most interesting, however, was the fact that while mucus from the solitary congeneric *A. xanthogrammica* failed to elicit discharge in *A. elegantissima*, the latter's mucus produced massive discharge in the former.

Recognition mechanisms. Active and passive models may be used to explain these results. The passive model assumes that if an agent is applied to a tentacle and no discharge occurs, then either no effector substance is in the agent, or, if one is, it is simply ineffective in lowering the threshold of the specific discharge system being tested. According to this model, the effector substance in the mucus of *A. elegantissima* that lowers the threshold of discharge in other zoantharians simply has no effect on the animal's own system.

The active model assumes that zoantharian mucus contains an inhibitory agent which "protects" the animal's own discharge system against the effector substances it contains. Although this model assumes a high level of species specificity, it appears more tenable than the passive one, and agrees with the suggestion based on work with *Hydra* that discharge effector substances are enzyme substrates which react with strategically placed enzymes on the nematocyte, perhaps in the cnidocil, to effect discharge. It has been found that various organic phosphates augment nematocyte discharge and that discharge to the phosphates is suppressed by known enzymatic inhibitors. Such a system explains the diversity in specificity observed in many experiments to date. A study by Roger Lubbock is relevant in this regard. During agonistic behavior in *A. elegantissima*, heavily armed acrorhagi are employed. Lubbock compared the specificity of the acrorhagial nematocyte system with that of the tentacular system. He showed that neither group responds to excised syngeneic (clonemate) tissue but both respond to allogeneic (nonclonemate) tissue. Tentacular nematocytes respond to a broad range of tissues from diverse taxa, while acrorhagial ones are far more specific and respond only to certain allogeneic and xenogeneic (nonconspecific) zoantharians.

As yet there is no real knowledge of how this recognition of self versus nonself is mediated. Also, there is no clear understanding of the physiology of the stimulus-to-discharge chain. Some scientists have hypothesized that antigen-antibody phenomena may be involved in discharge. However, minimum criteria for an immunological reaction have not been established for nematocyte discharge; one criterion being the demonstration of inducible memory or selectively altered reactivity on secondary contact. In the above study, Lubbock, using acrorhagial discharge as a criterion, was unable to show that repeated aggressions against one particular clone specifically enhanced the aggressive response to that clone.

For background information *see* COELENTERATA in the McGraw-Hill Encyclopedia of Science and Technology.   [DEMOREST DAVENPORT]

Bibliography: D. Davenport, D. Ross, and L. Sutton, The remote control of nematocyst discharge in the attachment of *Calliactis parasitica* to shells of hermit crabs, *Vie et Milieu*, 12:197–209, 1961; S. Ertman and D. Davenport, Tentacular nematocyte discharge and "self-recognition" in *Anthopleura elegantissima* Brandt, *Biol. Bull.*, 161:366–370, 1981; L. Francis, Intraspecific aggression and its effect on the distribution of *Anthopleura elegantissima* and some related anemones, *Biol. Bull.*, 144:73–92, 1973; R. Lubbock, Clone-specific cellular recognition in a sea anemone, *Proc. Nat. Acad. Sci. USA*, 77:6667–6669, 1980.

Sea water

Measurements of the rare-earth elements (REE) and the $^{143}Nd/^{144}Nd$ isotope ratio in marine deposits and sea water have shown them to be important tracers of element supply to the oceans and indicators of chemical fractionation and redox reactions in sea water.

Rare-earth chemistry. The chemistry of the rare-earth elements makes them especially suited to marine geochemical applications. Chemically they are an extremely coherent group, and their relative abundances can be used as a "fingerprint" to deduce their sources in sedimentary deposits and in solution. Because elements with even atomic numbers are geochemically more abundant than those with odd atomic numbers (the Oddo-Harkins rule), rare-earth-element concentrations are normalized to concentrations in a standard on an element-by-element basis, and the rare-earth-element enrichment ratios plotted as a function of atomic number, thereby producing smooth curves known as REE patterns. The most useful standard for chemical oceanography is average shale because it is representative of continental input to the oceans and therefore allows small differences in the REE patterns of marine samples to be readily identified.

The chemical coherence of the rare-earth-element group is not its only contribution to marine geochemistry. Within the group there are subtle differences in chemical behavior which provide added important information. First, the stability constants of rare-earth-element complexes vary in an ordered way, which means that partitioning of the rare-earth elements between phases may lead to fractionation of the light rare-earth elements (La–Eu) relative to the heavy rare-earth elements (Gd–Lu). Second, although rare-earth elements exist predominantly as $3+$ cations in the marine environment, there are important exceptions, namely cerium, which can exist as Ce^{4+}, and europium, which may be found as Eu^{2+}. Thus, these two elements may fractionate from the $3+$ cations of rare-earth elements as a

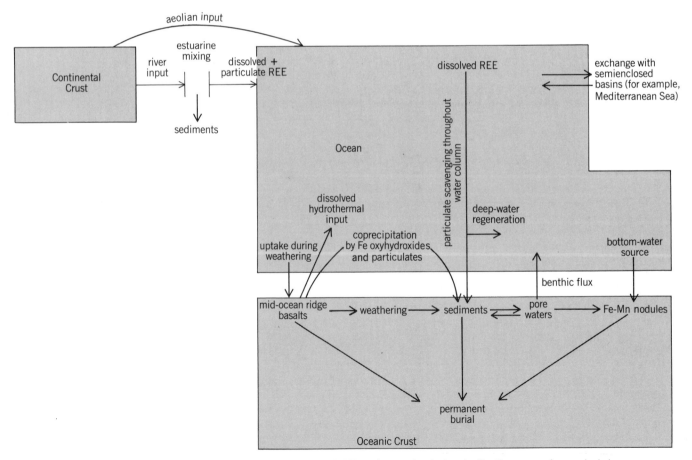

Fig. 1.   Diagram of the rare-earth-element (REE) marine geochemical cycle. The three reservoirs are shaded.

function of redox potential. Third, some of the rare-earth elements are weakly radioactive, particularly ^{147}Sm which decays to ^{143}Nd with a half-life of 1.06×10^{11} years. The fractionation of Sm and Nd between the crust and the mantle allows the $^{143}Nd/^{144}Nd$ ratio to be used as a further indicator of the source of rare-earth elements.

Marine geochemical cycle. Figure 1 is a diagram of the chemical behavior of the rare-earth elements in the marine geochemical cycle. The rare-earth-element cycle is still poorly understood, and several of the pathways illustrated are only just becoming resolved. As with all marine geochemical cycles it is necessary to consider both the internal cycle of the rare-earth elements within the oceans and also the relationship between rare-earth elements in the oceanic reservoir and in the oceanic (both the sediments and the basaltic rocks) and continental crusts.

Continental crust. There is evidence that the rare-earth elements are mobile during chemical weathering on land, but it has yet to be established what effect this has on river input of the rare-earth elements to the oceans. The shale-normalized REE patterns of river waters are reasonably flat, suggesting that little fractionation occurs (Fig. 2). The most important area where chemical continuity of rare-earth elements between continents and oceans may be disrupted is the estuarine environment where several other elements are actively removed from solution.

At present the only data on this aspect of rare-earth-element behavior are on the Gironde estuary in France. The flat REE pattern of the river water indicates that sedimentary rocks provide a major and possibly dominant source of rare-earth elements to the river, given that igneous rocks with quite different patterns outcrop in the drainage basin. About 98% of the rare-earth elements supplied to the oceans from the river are in particulate form (> 0.4 μm). Of the dissolved rare-earth elements entering the estuary, up to 80% are removed during the mixing of river water and sea water.

Another transport path of continentally derived rare-earth elements to the oceans is through the marine aerosol. Aeolian input is an important source of the rare-earth elements in North Atlantic surface water. This is suggested by an enrichment of rare-earth elements in surface waters compared to underlying waters and by the similarity in REE patterns of surface sea water and of Sahara dust which is known to be the source of the aerosol in this region (Fig. 2).

Oceans. Of the many important features of rare-earth-element marine chemistry, the most striking observation is the REE pattern for sea water showing a marked cerium depletion (a negative cerium anomaly). This is thought to be due to the oxidation of soluble Ce^{3+} from rivers to the more insoluble Ce^{4+} state. Also of note is the enrichment of heavy rare-earth elements relative to light rare-earth elements (Fig. 2).

Recent work has provided the first full vertical

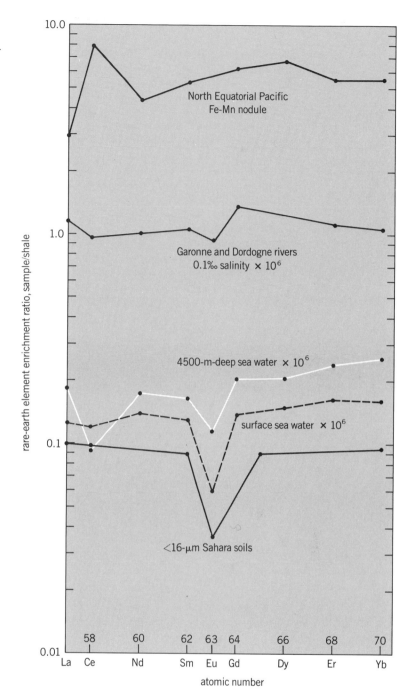

Fig. 2.   Shale-normalized REE patterns.

profiles of the rare-earth elements in the water column. The rare-earth-element concentrations are very low, in the range of about 70 picomolal (10^{-12} moles per kilogram) to less than 0.5 picomolal. Below the surface-water enrichment, rare-earth elements reach minimum concentrations but then increase in the deep waters. It appears that once dissolved in the water column they are rapidly scavenged from surface waters by particulates consisting of plankton fecal pellets and iron oxyhydroxide flocs incorporated on the outside of foraminifera and pteropod tests. The light rare-earth elements are preferentially removed from surface waters due to the higher stability of heavy rare-earth-element-soluble complexes. The residence time of the rare-earth elements in surface waters is short. At most it varies from approximately 6 years for the light species to 20 years for the heavy species.

After the rare-earth elements are scavenged from the water column, they are transported to the deep ocean and returned to solution. This is shown by the similarity of the deep-water REE pattern with that of surface water (except for cerium, which differs due to its existence in the quadrivalent state), which indicates that both have the same original source of rare earth elements. Although not yet proved, comparisons with copper (which shows a similar water-column profile in the Pacific) indicate that it is likely that the rare-earth elements are returned to the bottom waters by way of a benthic flux from the sediment pore waters. However, deep-water in-place regeneration cannot, as yet, be ruled out.

The use of the REE pattern as a water mass indicator is revealed by the detection of Mediterranean outflow water. Hydrographic signals (salinity and nutrients) show this water inflowing between 900 and 1500 m at a North Atlantic site. The REE pattern of this parcel of water is distinguished from that of the water above and below by a marked increase in the negative cerium anomaly and an enrichment of the heavy rare-earth elements.

The residence time of elements dissolved in the ocean is conventionally given by the equation here,

$$\tau_A = \frac{A}{dA/dt}$$

where τ_A is the residence time, A is the standing crop of dissolved element in the oceans, and dA/dt is the rate of supply of the dissolved element to the oceans.

When the residence time is calculated from the river input alone, values of the order of 2000 years are obtained. However, when the aeolian input is considered together with a possible benthic flux, the residence times drop to approximately 400 years. Thus, overall, the rare-earth elements may reside in the oceans for a few hundred years before being scavenged, but they are regenerated before being permanently buried.

Further support for the shorter overall residence times comes from the $^{143}Nd/^{144}Nd$ ratio (Fig. 3). Various studies have revealed that waters and hydrogenous sediments have Nd isotopic ratios that are distinctive for each ocean basin (Pacific samples having the highest values, Atlantic samples the lowest, with Indian Ocean samples having ratios that lie between these two end members), but that ^{143}Nd and ^{144}Nd are well mixed within an individual ocean. This indicates that the rare-earth elements have a shorter residence time than the mixing time for the world ocean (about 10^3 years) but longer or of the same order as the mixing time for individual ocean basins (about 10^2 years).

Comparisons of these marine $^{143}Nd/^{144}Nd$ ratios with

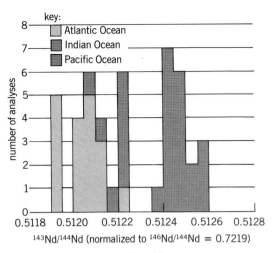

Fig. 3. Histogram of the $^{143}Nd/^{144}Nd$ ratios for the ocean basins.

Nd isotopic ratios of crustal rocks and ocean mantle source rocks reveal that at least 80% of the rare-earth elements supplied to the Atlantic Ocean and 50% of those supplied to the Pacific Ocean must have an original continental source. The remainder are thought to be derived from the alteration of island arcs and sea-floor basalts. This would account for the lower ratios of the Atlantic Ocean, since 60% of the world's fluvial waters drain into the Atlantic, and these rivers drain areas of ancient continent. In contrast, the Pacific Ocean receives relatively little continental runoff, and contains a more active and extensive mid-ocean ridge system.

Oceanic crust. As a first approximation, deep-sea sediments can be divided into two categories: lithogenous and hydrogenous. In general, lithogenous sediments are relatively inert in their effect on the rare-earth-element marine geochemical cycle. The hydrogenous sediments can be divided into biogenic and authigenic components. Biogenic sediments play relatively little part in the rare-earth-element budget (typical rare-earth-element concentrations in plankton tests being of the order of 1 ppm). Incorporation of soluble rare-earth elements into the sediments is largely controlled by coprecipitation with, and scavenging by, iron oxyhydroxide flocs. Once in the sediment, there is considerable evidence that the rare-earth elements become remobilized and enter the pore waters, although rare-earth-element concentrations in pore solutions have not yet been measured. Iron-rich coatings are also found on clays and biogenic material buried within the sediment. Rare-earth elements also appear to be strongly associated with phosphate. Concentrations of cerium of the order of 10,000 ppm have been found in fish teeth, which contain large amounts of biogenous apatite.

Ferromanganese nodules. These provide a major sink for rare-earth elements in the coeans. They are distinguished by having REE patterns which generally show a large positive cerium anomaly (Fig. 2).

This enrichment in cerium is a result of its oxidation to the more insoluble quadrivalent state. It appears that cerium is primarily incorporated into nodules directly from ocean water by coprecipitation with iron oxyhydroxide flocs. The 3+ cations of rare-earth elements exhibit more complex behavior which, together with the above mechanism, involves incorporation into the nodules from the pore waters through a phosphate-rich phase. Remobilization of rare-earth elements from the sediments and their eventual incorporation into nodules is supported by the close agreement of the $^{143}Nd/^{144}Nd$ ratios of ferromanganese nodules with their associated sediments.

Basalts. Low-temperature alteration of basalts appears to remove rare-earth elements from sea water. It has been found that the palagonitized rims of pillow basalts were enriched in the light rare-earth elements and with a negative cerium anomaly relative to the unweathered sections, believed to be due to uptake of rare-earth elements from bottom waters. Two alteration products of sea-floor basaltic debris are also important sinks for rare-earth elements. Phillipsite (a zeolite mineral) has an REE pattern very similar to that of sea water, indicating the probable source of its rare-earth elements. Although montmorillonite has an REE pattern similar to associated illitic clays, their absolute concentrations of rare-earth element are as much as three times as high as that of average shales. It has been suggested that this indicates extensive diagenetic addition of rare-earth elements.

Another way by which the basalts of the ocean crust may be important for the rare-earth-element balance of the oceans is through the discharge of submarine hydrothermal solutions at mid-ocean ridges and back-arc-spreading centers, but this is only now being investigated. This process has been shown to be responsible for an input of certain elements to sea water and a removal of others. Rare-earth-element analyses are not yet available for hydrothermal water samples, but recent work suggests that they are leached from the interiors of sea-floor basalts by circulating sea water. However, experimental studies suggest that the rare-earth elements are immobile at temperatures of 150–350°C even when the basalt is totally altered to clay.

Mid-ocean ridges. It has also been shown that processes associated with mid-ocean ridges actively remove dissolved rare-earth elements from the water column. The most important of these removal mechanisms occurs when iron released from hydrothermal vents forms oxyhydroxide flocs. These coprecipitate and scavenge the rare-earth elements from the bottom water and deposit a metalliferous sediment with a sea water–like REE pattern. The idea of a sea-water origin of the rare-earth elements in these sediments is supported by $^{143}Nd/^{144}Nd$ ratios, which are the same for the sediment and local bottom water.

For background information *see* RARE-EARTH ELEMENTS; SEA WATER in the McGraw-Hill Encyclopedia of Science and Technology.

[M. PALMER; H. ELDERFIELD]

Bibliography: H. Elderfield et al., Rare earth element geochemistry of oceanic ferromanganese nodules and associated sediments, *Geochim. Cosmochim. Acta*, 45:513–528, 1981; H. Elderfield and M. J. Greaves, The rare earth elements in seawater, *Nature*, 296:214–219, 1982; S. L. Goldstein and R. K. O'Nions, Nd and Sr isotopic relationships in pelagic lays and ferromanganese deposits, *Nature*, 292:324–327, 1981; D. J. Piepgras and G. J. Wasserburg, Neodymium isotopic variations in seawater, *Earth Planet. Sci. Lett.*, 50:128–138, 1980.

Seed germination

The understanding of seed germination on a molecular level has led to improvements in methods of seed treatment and seed testing. This article discusses recent research on the molecular aspects of seed germination and gives an overview of seed-testing methods.

Molecular aspects. A seed is a unit with an embryo and its storage material (endosperm) encased in a testa, or seed coat. The embryo consists of a radicle and a plumule (with one, two, or several cotyledons [seed leaves], depending on the plant group). Germination is a process that eventually leads to development of the embryo into a seedling. It is identified by protrusion of the radicle or shoot through the seed coat, and its onset represents growth processes. Thus, germination represents a number of sequential physiochemical steps which allow a desiccated quiescent seed to hydrate and to develop total metabolic activity leading to the formation of a seedling.

Viability. Genetic, environmental, and many internal factors (such as permeability, or hormones) influence seed viability and germinability. In general, viability is stable at low temperature, low humidity, high carbon dioxyde concentration, and other environmental conditions that are favorable to low metabolic activity. For most cereals, the relationship between viability period, p, temperature, t (in degrees Celsius), and percent moisture, m, follow Eq. (1), where K and C are constants.

$$\log p = K_v - C_1 m - C_2 t \qquad (1)$$

Partial loss of viability often implies that the overall metabolic processes that occur during germination are no longer fully functional. Such situations are caused by the development of dormancy factors, principally alterations in hormone balance. Although there is a genetic basis for longevity, it has been suggested that the ability of a seed to repair damaged DNA is the reason for the apparent stability of its viability.

The first process to occur during the onset of germination is water uptake by imbibition, a physical process determined by permeability factors and not by viability factors. Because the seed is filled with colloidal particles and hydrophilic polyelectrolytes, the imbibitional process generates imbibitional pressure and develops into an osmotic system. The imbibition of water by corn follows Eq. (2), where w is the water content, ft the water capacity, b the linear phase of water uptake, and k the permeability constant.

$$\frac{dw}{dt} = k(ft - w) + b \qquad (2)$$

Concomitant with imbibition, the cells establish, initially by fermentative and later by oxidative respiration, the production of energy in the form of adenosinetriphosphate (ATP) to sustain metabolic processes. In general, all seeds are naturally optimized to respire efficiently in air. High pressures (200 atm or 20 megapascals) and oxygen enrichment have no significant effect on germination.

Germination temperature. All seeds have an optimum germination temperature which, because germination has no single temperature coefficient, is described for a seed lot as the temperature range for the highest percent germination in the shortest time. Many factors, such as age, genetic makeup, composition of seed, and level of endogenous growth regulators, influence the optimum germination temperature. Temperature also affects the after ripening process and can impose secondary dormancy. The cellular mechanisms by which seeds perceive and respond to ambient temperature are not known. Alterations in membrane properties could be part of a temperature- sensing mechanism.

Sensitivity to light. Light-sensing mechanisms regulating seed germination are better understood than those for temperature. Imbibed seeds that are dried after exposure to light are known to retain the stimulatory effects. These light effects are mediated by the phytochrome system; however, many factors determine the equilibrium or actual ratio of the inactive form of phytochrome (P_R) to the active form (P_{FR}). Also, while it is known that irradiation with blue light (300–450 nanometers) as well as red light (660 nm) promotes seed germination, neither the precise mechanism of phytochrome action nor the relationship between the blue absorbing pigments and the phytochrome system in controlling seed germination is clearly understood. Explanations range from the existence of a hypothetical compound to alterations in membrane properties and specific gene expression. The interesting aspect is that light responses are modified by all internal and external factors (such as osmotic stress, oxygen enrichment, and growth regulators) that affect dormancy and germination processes.

Dormancy. The physiological state in which a viable seed fails to germinate under the most favorable environmental conditions is called dormancy. For mature seeds, which require a period of after ripening in storage or a period of chilling under moist conditions after harvest for germination to occur, this state is described as primary dormancy. The state in which seeds do not germinate because of some adverse storage condition (such as light and temperature, or low concentration of oxygen) is described as secondary dormancy. Many metabolic changes due to alterations in seed coat permeability, hormone levels, shift toward the inactive form of phytochrome, and other events are associated with the establishment of secondary dormancy. An interesting example is the dormancy of *Xanthium* (cocklebur) seeds. In the *Xanthium* fruit, the upper seed is very dormant and requires pure oxygen to germinate 100% at 21°C. In contrast, the lower seed is less dormant, requiring only 6% oxygen to germinate 100% at 21°C. One plausible explanation is that the seed coat is impermeable to oxygen and that probably some inhibitors have to be removed by oxidation. However, very low levels of ethylene will overcome this type of dormancy.

Hormonal mechanisms. It is generally acknowledged that cellular perception or cueing events are mediated by hormonal mechanisms. All seeds have the potential to synthesize, metabolize, and store gibberellins, cytokinins, auxins, and abscisic acid. Some seeds, for example, corn and rice, contain very high amounts (approximately 2500–4500 micrograms per gram) of bound forms (esters) of auxins (primarily the inositol ester of indole-s-acetic acid). Seeds produce many olefinic volatiles during imbibition, and auxins are known to enhance the production of ethylene. Treatment with gibberellic acid, cytokinins, and ethylene, singly and in combination, will overcome dormancy in most seeds. Gibberellic acid, and to some extent cytokinins, can overcome dormancy imposed by darkness, light, or abscisic acid. Treatment with gibberellic acid or cytokinins can also overcome osmotic and high-temperature stress at the onset of germination in different seed types.

Hormone levels change during the ripening, afterripening, and germination of a seed. For example, there are many reports of an increase in the level of gibberellic acid and cytokinins during dormancy breaking. It is not always clear whether the increase in hormone level is due to new synthesis, release from a previously bound form, or transport from another part of the seed. Equally confusing is the relationship between hormone metabolism and germination processes. For example, abscisic acid is rapidly metabolized to phaseic and dihydrophaseic acids, both believed to be inhibitors of germination; over 60 gibberellins are known to occur in higher plants; and many interconversions are known to occur in germinating seeds. Similarly, active cytokinins are derived from inactive forms, and cytokinin molecules such as zeatin are rapidly metabolized to mono-, di-, and triphosphates. The extent to which changes in the level of hormones are the cause and effects of the germination and dormancy processes remains obscure.

In considering the role of inhibitors and promotors, many studies have focused on the hormonal regulation of enzyme activity in seed germination. Cytokinins are known to induce the development of isocitric lyase and protease activity in squash cotyledons. In barley endosperm, gibberellic acid induces the synthesis of α-amylase and a few other hydrolases. Such responses are counteracted by abscisic acid. For example, abscisic acid will prevent the translation of carboxypeptidase and isocitratase messenger ribonucleic acid (mRNA) in cotton and of α-amylase mRNA in barley. However, the effects of abscisic acid are easily reversed by ethylene, which is not the primary inducer of enzyme activity.

Studies have been undertaken to determine the extent to which hormones control the transcription and translation steps of specific protein synthesis. By using modern methods of molecular biology, including recombinant DNA techniques, it was unequivocally demonstrated that gibberellic acid induces the synthesis of α-amylase mRNA in barley aleurone cells. Conceivably, hormones regulate genome activity in higher plants.

Several studies on the regulation of nucleic acid metabolism and specific enzyme activity have raised the question as to whether there are master enzymes that control the onset of metabolic processes in a quiescent seed. Other than the knowledge that when subjected to hydration, quiescent seeds develop functional polysomes, synthesize ATP, and initiate respiratory activity, little is known about the onset of metabolic activity.

A clear understanding of the onset of metabolic activity requires a knowledge of the influence of hydration in the control of specific gene expression in a desiccated seed with a quiescent genome. Without a clear conceptual knowledge of differentiation, development, and growth processes, a biochemical understanding of the molecular aspects of seed germination and dormancy will continue to remain vague.

[G. RAM CHANDRA; G. P. ALBAUGH]

Germination technology. Seed testing is an important step in ensuring maximum crop yield. Poor-quality seed is kept off the market through the seed-testing and labeling programs of each state or country. For most agronomic and vegetable seeds germination is completed in 24 to 96 hours from the start of imbibition. Seed technologists performing germination tests have extended the concept of germination period to include development of a seedling having a well-defined root and shoot. This would include the period of mobilization of the stored food reserves in the cotyledons or endosperm of the seed. The time frame for evaluation may extend from 1 to 4 weeks for most kinds of seeds and much longer (several months) for seeds showing a deep dormancy. Emphasis is placed on classifying germinated seeds as normal or abnormal, with only the normal seedlings included in the final germination percentage.

Seed-testing rules. Rules for performing and interpreting official seed germination tests have been developed by the International Seed Testing Association (ISTA) and the Association of Official Seed Analysts (AOSA). Covered in the rules are detailed procedures for sampling a seed lot for purity and germination tests, methods for germination and evaluation of the seedlings, and tolerances for statistical errors which may result from poor sampling or variability among replications in the germination test. Strict adherence to these rules ensures the quality of the seed being sold. All seed lots must bear a tag indicating the type of seed, purity (trueness to type or variety), and the germination percentage.

Purity analysis. To perform a purity analysis, a weighed working sample containing approximately 2500 seeds is examined seed by seed and separated into four categories: pure seed, other crop seed, inert matter, and weed seeds. The percentage of pure seed represents that fraction (by weight) of the sample which is true to type, or the variety which appears on the label. The second fraction is that proportion of other crop seed which is not true to type as printed on the label. The third fraction is classified as inert matter, which can include pieces of broken and damaged crop and weed seeds one-half or less than the original size and nonseed material such as nematode galls, fungal bodies, soil particles, chaff, stems, and pieces of bark. The final category includes weed seeds. Tolerance tables are used to determine when a sample fails to meet minimum purity standards for a particular crop seed. Exceeding the tolerance can result in that seed lot being barred from sale within a state or country.

Germination test. An official germination test consists of a minimum of 400 seeds tested in replications of 100 seeds or less. Because different seeds require different environments for optimum germination, the seed analyst must refer to tables in the rules which specify the germination temperature and substratum for each crop species. Special conditions, including light requirements or chemical treatments such as potassium nitrate for breaking dormancy, are also specified. Many seeds require alternating (day/night) temperatures, such as 30°C (86°F) during day and 20°C (68°F) at night for optimum germination. Finally, the number of days until first and final germination counts are made is given in the rules. If the replications vary too widely and exceed the tolerances allowed, the germination test must be repeated.

Classification of a seedling as normal means that it contains those essential structures (root and shoot) which will permit the seedling to become established and grow in the field under favorable conditions. Abnormal seedlings may have no root or shoot, may be deformed, or may lack essential food-storing tissues, such as cotyledons or endosperm. In the case of many vegetable crops, such as beans, lettuce, and tomatoes, where mechanical harvest is routine, uniformity of the crop in the field becomes extremely important. Abnormal seedlings may survive and grow in the field; however, they may mature too late and thus contribute nothing to the crop yield. Also, abnormal seedlings are weak and are more susceptible to diseases and adverse weather conditions. Thus, abnormal seedlings are not counted in the final germination percentage which appears on a seed label.

Special tests. Seed dormancy may occur within a

seed lot, requiring special tests to determine if the seeds are still alive. Breaking dormancy may require scarification (mechanical breaking of the seed coat), chemical treatment, or prechilling for extended periods of time at low temperatures under moist conditions (stratification). An alternative approach to breaking dormancy is the use of the chemical dye 2,3,5-triphenyltetrazolium chloride (TZ), which stains live seed tissue a pink to red color. Usually the seeds are cut in half to expose the embryo tissues, which are then soaked in the dye for periods of a few hours, depending on the type of seed. Each seed is evaluated for the degree of staining and the pattern of staining in the meristematic tissues. A trained analyst can then determine the percentage of viable seeds in the sample. TZ testing is accepted as a supplement to the official germination test.

Tests for vigor, although not incorporated into official seed-testing rules, have been used to separate seed lots into different vigor groupings. The AOSA has defined vigor as those seed properties which determine the potential for rapid, uniform emergence and development of normal seedlings under a wide range of field conditions. Thus, the vigor test attempts to reveal more information about a seed lot than the percentage of normal seedlings produced. There are many types of tests which measure such properties as growth rate (shoot or root elongation), enzyme activity, membrane integrity (leaching of nutrients), and response to suboptimum conditions (stress tests). Correlations between laboratory vigor test results and field emergence are made to determine the suitability of a particular test for each kind of seed.

For background information *see* DORMANCY; PHYTOCHROME; PLANT HORMONES; SEED; SEED GERMINATION in the McGraw-Hill Encyclopedia of Science and Technology.

[ERIC E. ROOS]

Bibliography: American Society for Horticultural Sciences, Seed quality: An overview of its relationship to horticulturalists and physiologists (Proceedings of a symposium), *HortScience*, 15(6):764–788, 1980; Association of Official Seed Analysts, Rules for testing seeds, *J. Seed Technol.*, 3(3):1–126, 1978; G. R. Chandra, Hormonal regulation of genome activity in higher plants, in N. Bhushan Mandava (ed.), *Plant Growth Substances*, ACS Symp. Ser. no. 111, pp. 246–261, 1979; L. O. Copeland, *Principles of Seed Science and Technology*, 1976; D. J. Osborne, Studies on DNA integrity and DNA repair in germinating embryos of rye (*Secale cereale*), *Israel J. Bot.*, 2(29):259–272, 1980/81; E. H. Roberts (ed.), *Viability of Seeds*, 1972; J. R. Thomson, *An Introduction to Seed Technology*, 1979.

Semiconductor memories

With the advances in semiconductor technology in the past few years, there have been attendant advances in the capability and applications of semiconductor-based computers and microprocessors. Such computers are currently used in a wide variety of applications which were not generally envisioned a few years ago, and the rapid growth of computing capability and applications to new areas is expected to continue.

Essentially all of these computers rely on semiconductor memories of various types. Semiconductor memories are devices for storing digital information and are used to store programs and data in digital computer systems. Different types of semiconductor memories are used to perform different storage functions. Such functions might include bulk data storage, program storage, and temporary or intermediate storage of programs and data. Virtually all of these memories are random access memories (RAM), in which any storage location can be addressed in the same amount of time. However, there are many different types of random access memories, including volatile writable and erasable, nonvolatile read-only memories (ROM), and nonvolatile electrically erasable programmable memories (EEPROM), among others. Volatile refers to those memories for which stored information is lost when the power is turned off, whereas nonvolatile memories retain the stored information in the absence of power. Although many memories are volatile, high-density nonvolatile memories which can be written electrically are desirable for many applications that require long-term data or program storage in the absence of power. Indeed, in the very-high-density memories envisioned in the future, it may be necessary to include nonvolatile portions in the memory to reduce the average power dissipation and associated heat load.

The two types of nonvolatile electrically writable and erasable memories currently used are the floating gate and the MNOS memory. In the floating gate memory, electrical charge is stored on a conducting layer (typically polycrystalline silicon) embedded in a thin silicon dioxide film on silicon. The acronym MNOS describes the metal–silicon nitride–silicon oxide–semiconductor structure (Fig. 1), discussed below.

MNOS structure. The basic charge storage element is illustrated in Fig. 1a. The MNOS structure consists of a metal contact (gate electrode) on thin (less than about 50 nm) silicon nitride film which has been deposited on a very thin (about 2–4 nm) silicon dioxide (SiO_2) film on silicon. The information is written by injecting charge from the silicon through the thin oxide into the silicon nitride by means of an electric field impressed between the metal and the silicon. The injected charge is stored in the silicon nitride near the silicon dioxide interface so that this MNOS structure behaves as a charged capacitor, and the thin oxide helps serve as a barrier to prevent the charge from returning to the silicon.

In a typical MNOS memory, the silicon nitride is incorporated into a field-effect transistor, as shown in Fig. 1b. Either positive or negative charge can be stored in the nitride, and the structure shown in

Fig. 1*b* is for positive charge storage (*p*-channel). As in the capacitor structure, information is written by injecting charge from the silicon into the nitride. For *p*-channel operation, the silicon is doped *n*-type by the introduction of electrically active impurities. The memory is read by application of a voltage bias between the source and drain, which are the two regions that have been doped *p*-type. If sufficient positive charge has been stored in the nitride to invert the charge population near the silicon–silicon dioxide interface, current will flow between the *p*-type regions, and the element is said to be written. Because of the nature of the charge traps and conduction mechanisms in silicon nitride, stored charge—and thus memory information—can be retained in the absence of an applied voltage. Furthermore, the properties of the silicon nitride can be tailored to permit preferential storage of either positive or negative charge. This feature allows the construction of semiconductor logic based on either electron (*n*-MNOS) or hole (*p*-MNOS) transport.

Retention time. The retention time is a measure of how long the injected charge is retained in the absence of an applied voltage. Since the stored charge can be determined by measuring the voltage that must be applied to overcome the internal electric field produced by this charge, performing such measurements as a function of the time yields the retention time. Retention time results for a typical nonvolatile MNOS memory element are shown in Fig. 2. The positive threshold voltage results when negative charge has been stored in the nitride, whereas the negative

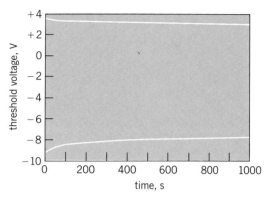

Fig. 2.   Typical decay of threshold voltage with time for *p*-channel MNOS.

voltage refers to storage of positive charge. The retention time can be estimated by extrapolating the decay curves to some predetermined threshold voltage, for example, 1 volt. The threshold voltage for both positive and negative charge storage is observed to decrease logarithmically with time, so that memories with retention times of years are predicted.

Radiation hardness. In addition to the nonvolatile character and versatility, MNOS memories are much more resistant to exposure to radiation than are conventional RAM and floating gate memories. This "radiation-hard" quality of MNOS makes it attractive for a variety of military, space, and satellite applications and has also served as an impetus for developing the MNOS technology. Radiation-hard MNOS microprocessors which will withstand more than 10^6 rads (10^4 grays) of gamma irradiation have been built.

Silicon nitride properties. The charge-storage mechanisms and radiation hardness can be further understood by examining the silicon nitride fabrication and properties. Silicon nitride for nonvolatile semiconductor memories is usually deposited from mixtures of nitrogen and silicon-containing gases (for example, NH_3 and SiH_4, or SiH_2Cl_2) by using chemical vapor deposition techniques with substrate temperatures between 700 and 900°C. Films deposited under these conditions are amorphous and typically contain between 3 and 9 atomic % hydrogen. The precise hydrogen concentration and chemical bonding configurations are determined by deposition parameters, such as deposition temperature, reactant gas composition, and reactant gas pressure, and the hydrogen has been found to play a central role in charge transport and trapping. An illustration of the differences in hydrogen concentration and bonding configurations obtained for silicon nitride deposited at atmospheric pressure and temperatures of 750 and 900°C is given in Fig. 3. The hydrogen depth distribution measured by nuclear reaction analysis techniques in the two films is shown in Fig. 3*a*. Increasing the deposition temperature decreases the hydrogen incorporation by about 25%; however, the chemical bonding configuration of the hydrogen

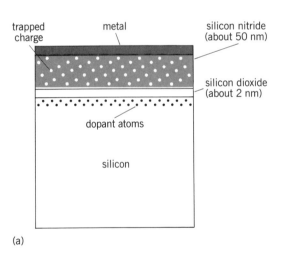

(a)

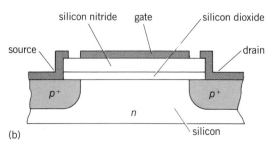

(b)

Fig. 1.   MNOS structure. (a) Basic charge-storage element in an MNOS memory. (b) Configuration of a *p*-channel MNOS.

changes significantly. The infrared absorption spectra (Fig. 3b) show hydrogen bonded to nitrogen (3350 cm^{-1}) and to silicon (2200 cm^{-1}) for the 750°C deposition, whereas hydrogen is bonded only to nitrogen for the 900°C deposition temperature.

From a wide variety of materials studies, such as those outlined above, and from correlation of the material composition and chemical bonding with the electrical behavior, the centers which trap positive charge are believed to be associated with hydrogen bonded to silicon (Si—H), whereas the negative charge is trapped at unsaturated or broken silicon bonds (Si—). The radiation tolerance of MNOS

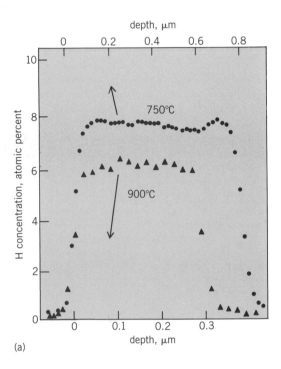

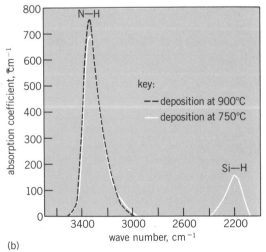

(a)

(b)

Figure 3. Comparison of hydrogen concentration and bonding configurations in silicon nitride films deposited at 750 and 900°C. (a) Hydrogen depth distribution measured by nuclear reaction analysis. (b) Infrared absorption measurements of the bonding of hydrogen to nitrogen (N—H) and silicon (Si—H).

memories is attributed to the large number of trapping sites produced throughout the nitride film in the course of fabrication, and to the fact that charges are strongly trapped at these sites.

While films with various operating characteristics can be fabricated by varying the deposition conditions to change the relative number of Si— and Si—H centers, the number of Si— or Si—H centers can be increased by postdeposition treatments. For example, annealing at temperatures above the deposition temperature will increase the number of Si— centers, whereas the number of Si—H centers can be increased by ion-implantation-induced transfer of hydrogen from N—H to Si—H bonds, or by postdeposition hydrogenation. All of these features combine to make MNOS a versatile memory material that is compatible with a variety of peripheral logic circuitries. As noted above, the ability to tailor the retention characteristics of silicon nitride permit either n-channel or p-channel operations and 8k p-MNOS and 16k n-MNOS memories have been fabricated (1k = 1024 bits).

For background information *see* AMORPHOUS SOLID; SEMICONDUCTOR MEMORIES; TRANSISTOR in the McGraw-Hill Encyclopedia of Science and Technology.

[P. S. PEERCY]

Bibliography: H. Nakayama and T. Enomoto, *Jap. J. App. Phys.*, 18:1773, 1979; P. S. Peercy et al., *J. Electr. Mater.*, 8:11, 1979; H. J. Stein, *J. Electr. Mater.*, 5:161, 1976; H. J. Stein, *Appl. Phys. Lett.*, 32:379, 1978; H. J. Stein, P. S. Peercy, and D. S. Ginley, in G. Lucovsky, S. T. Pantelides, and F. L. Galeener (eds.), *The Physics of MOS Insulators*, 1980; K. Uchiumi and T. Makimoto, *Electronics*, p. 154, February 24, 1981; Y. Yatsuda et al., *Jap. J. Appl. Phys.*, 19(suppl. 191):219, 1979.

Shale

The occurrence of minor elements (including uranium) in shales, and particularly in black shales, has been the subject of considerable interest in recent years, both scientifically and economically. Black (organic-rich) shales can be host to subeconomic uranium mineralization and economic base-metal mineralization, and in addition can provide a first-stage concentration of metals for subsequent higher-grade mineralization. These organic-rich shales, many of which are uraniferous, are also viewed as a potential fuel source.

Uranium concentration in shales. Shales form about 75% of all sediments and have a higher average uranium content than other sedimentary rocks. In discussing the occurrence of uranium in shales, these rocks can be adequately classified on the basis of their most obvious feature—color. The color is determined primarily by the amount of organic carbon present and by the Fe^{3+}/Fe^{2+} ratio. However, since this ratio depends on the oxidation state, which is a function of the amount of organic matter in the rock, the organic matter content ultimately controls the color of shales. Two color series have been distinguished within shales: a red–purple–greenish-gray series based on the Fe^{3+}/Fe^{2+} ratio and a greenish-gray–gray–black series based on carbon content.

Environmental variables that control the color of shales also affect the behavior of uranium, and therefore color (oxidation state) can be conveniently used in the genetic interpretation of the distribution of uranium in these rocks. A crucial factor in the geochemical behavior of uranium is its relative immobility in the tetravalent (reduced) state and high mobility in the hexavalent (oxidized) state in the form of the UO_2^{2+} cation or its complexes. However, it is important to realize that the mobility of uranium in oxidizing environments is controlled by the stability of uranium-bearing minerals. If the uranium is enclosed in resistate minerals (for example, zircon), it can be moved only if the enclosing grains are moved mechanically.

In view of the geochemical properties of uranium, it is not surprising that black shales, representing strongly reducing sedimentary environments, are usually characterized by high concentrations of this element. Red shales, representing the opposite extreme, that is, well-oxidized environments, contain relatively low concentrations of uranium. The great majority of shales (greenish-gray and gray, so-called common shales) lie between these two extremes in regard to their color and uranium content. The illustration shows the relationship of shale color to the percentage of organic carbon and the mole fraction of bivalent iron.

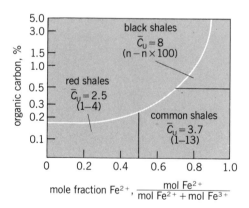

Uranium content in red, common (greenish-gray, olive gray, and gray), and black (dark gray, grayish-black, and black) shales. The symbol $\overline{C}_u$ denotes the average uranium concentration in each shale type; numbers in parentheses indicate the concentration range. (After P. E. Potter, J. B. Maynard, and W. A. Pryor, Sedimentology of Shale, p. 55, Springer-Verlag, 1980)

Black shales. The term black shale usually implies that the sediment was deposited in a marine or brackish-water environment in contrast to carbonaceous shale, which is usually deposited in a nonmarine environment. Nonmarine carbonaceous rocks are much less laterally extensive than black shales. Uranium in black shales is generally syngenetic (de-

posited simultaneously with the rock) and coextensive with a particular bed or beds, whereas in nonmarine rocks it is usually epigenetic (postdepositional) and locally distributed within a bed or several beds.

The most distinctive feature common to black and carbonaceous shales is their relatively high content of organic matter (normally black shales contain more than 2% organic carbon). Most black shales contain significantly more uranium than the average shale. The primary factors responsible for this phenomenon are: very slow sedimentation rates controlled largely by scarce input of detrital terrigenous material; abundant organic matter of terrestrial origin; the presence of an anoxic environment, and availability of uranium in solutions that come into contact with the organic matter.

The uranium content in black shales ranges from a few to a few hundred parts per million and averages about 8 ppm. Very few black shales contain more than 50 ppm of uranium. Two very large and well-known uraniferous black shales are the alum shales of Sweden and the Devonian-Mississippian Chattanooga shale and its stratigraphic equivalents in the United States. Reserves in the alum shale exceed 1 million short tons (907,000 metric tons) of uranium at ore grade of 300 ppm. The Chattanooga shale averages 60 ppm uranium.

It has been determined that about 3–4 ppm of the uranium in black shales is tied up in clastic material, mainly clays and resistate heavy minerals. The remainder of the uranium is clearly associated with either the organic or the phosphatic fractions of the shales. The proportion of uranium associated with these latter two fractions differs greatly. In the Chattanooga shale of Tennessee, organic matter is associated with more than 90% of the uranium, and the phosphate with only a few percent, whereas the phosphate-rich units of the Permian Phosphoria formation in Idaho display a reverse proportion. The uranium in phosphorites substitutes isomorphously for calcium in the crystal structure of apatite, the main constituent of phosphorites. It has been suggested that the enrichment of uranium in phosphorite compared to limestone, another calcium-bearing marine precipitate, is due to the adaptability of the apatite structure to coupled replacements, possibly SiO_4^{4-} to PO_3^{3-}.

Organic matter in shales has three sources: marine, terrestrial, and recycled. The syngenetic uranium content is related primarily to the terrestrial organic component of shales. This may account for the fact that although many black shales are characterized by a significant linear correlation between the organic carbon content and the uranium content, in some black shales this correlation is weak. It remains to be determined whether the association of terrestrial organic matter and uranium is due to scavenging of uranium by organic matter in rivers, the sea water, or the sediment-water interface. The exact process by which uranium is incorporated with the organic matter is unknown. It appears that organic carbon controls the enrichment of uranium by adsorption or chelation. Organic materials are strong adsorbents of uranium and may retain uranium as UO_2^{2+} either in adsorbed form, as organic compounds, or in reduced form, such as UO_2.

No uranium minerals have been identified in unweathered black shales. Disseminated uraninite and thucholite have been found in metamorphosed black shales, and secondary uranium minerals have been reported from weathered outcrops.

Common shales. Common (greenish-gray and gray) shales average 3.7 ppm of uranium with a range of 1–13 ppm, and have uranium values similar to the average uranium concentration in the continental crust. It has been suggested that this similarity may be due in many cases to rapid erosion and deposition that prevented significant leaching and fixation of uranium. However, even if these processes were relatively slow, uranium concentration would not be affected much by leaching if most of this element resides in the resistate fraction. Similarly, no significant fixation of uranium should be expected in common shales even at slow erosion and deposition rates. Clay minerals, a major component of shales, seem to be relatively unimportant as concentrators of uranium. Other minerals present in common shales are even less effective than clays.

Red shales. The color of red shales is due to pigmentation by ferric oxides. The shales are not red at the time of deposition, and the pigmentation originates from postdepositional destruction of iron-bearing minerals. In a recent study of the mobility and distribution of uranium in a geologically young (Pliocene-Holocene) redbed sequence displaying successive stages of development of the red pigmentation, it was determined that uranium concentration in 17 red shales averaged 3.3 ppm and ranged from 2.0 to 4.3 ppm. The highest concentration occurred in a sample with a relatively high content of amorphous ferric oxides, which are excellent uranyl sorbents.

The uranium content of forming red shales was found to be predominantly inherited from component detrital minerals. Reddening of shales, at least in the early stages of the process, does not promote major open-system migration of uranium. Rather, uranium is redistributed on an intergranular scale, moving from some of the detrital mineral hosts to secondary oxides of iron. About 30% of the uranium is associated with various iron oxide phases, and the remainder is tied up in insoluble detrital grains.

In geologically older red shales, the uranium content appears to be lower than in their younger equivalents. The uranium content in 17 red shale samples from the Permian of Oklahoma averages about 2.5 ppm and ranges from 1 to 4 ppm. This is in close agreement with values that had been reported for red shales in the late 1950s. It is not clear whether the difference in uranium concentration between geologically young and geologically old red shales is caused by differences in source composition or by uranium loss with time. Some loss of uranium could take place during aging of poorly crystalline iron ox-

ides, which are very effective uranyl sorbents, to well-crystallized secondary iron oxide (hematite). The removal of uranium excluded during the aging of iron oxides would be facilitated by the high mobility of the oxidized uranium species.

For background information *see* SEDIMENTARY ROCKS; SHALE; URANINITE; URANIUM in the McGraw-Hill Encyclopedia of Science and Technology.

[SALMAN BLOCH]

Bibliography: J. S. Leventhal, Pyrolysis gas chromatography–mass spectrometry to characterize organic matter and its relationship to uranium content of Appalachian Devonian black shales, *Geochim. Cosmochim. Acta*, 45:883–889, 1981; J. S. Leventhal and R. C. Kepferle, Geochemistry and geology of strategic metals and uranium in Devonian shales of the eastern interior United States, *Synthetic Fuels from Oil Shale II*, Institute of Gas Technology, pp. 73–96, 1982; P. E. Potter, J. B. Maynard, and W. A. Pryor, *Sedimentology of Shale*, 1980; R. A. Zielinski, S. Bloch, and T. R. Walker, The mobility and distribution of heavy metals during the formation of first cycle red beds, *Econ. Geol.*, publication pending.

Ship powering

Since the oil price rise, efforts have been made to increase the efficiency of propulsion of ships, in particular for very large bulk carriers such as very large crude carriers (VLCC) and ultralarge crude carriers (ULCC). Recently Japanese shipbuilding companies have carried out extensive research leading to the development of new devices which can be retrofitted to large tankers and other bulk carriers and can result in a substantial increase in propulsion efficiency. The cost of installing one of these devices is paid off after 3 to 5 years of operation.

Reaction fin. One of these units, called a reaction fin, was developed by Mitsubishi Shipbuilding Company in its towing tank in Nagasaki, Japan. The device (Fig. 1) is a multiple fin or stationary propeller installed ahead of the driving propeller. The fins are warped in such a way that a prewhirl of the water flowing into the propeller occurs in the opposite direction of propeller rotation. Through proper design the result is that little or no rotation of the race remains aft of the propeller.

Race rotation normally represents a substantial loss in propulsion energy which dissipates downstream. Elimination of the race rotation results in recovery of most of the lost energy. Efficiency improvements on the order of 4 to 8% in the fully loaded and ballast conditions, respectively, were found from model tests and in trial experiments on actual ships. Figure 2 shows the wake behind a ship before and after installation of the reaction fin. Figure 2a shows the typical boiling wake behind a single-screw ship. After installation of the reaction fin the wake is observed to be much flatter, as seen in Fig. 2b.

A special feature of this device lies in the fact that the installation can normally be made on an existing ship without change of the propeller. A small reduction in propeller revolutions per minute for the same speed is found. The fins can be built of steel forgings with a narrow ring welded to the outside ends of the fins for added strength. In general the operating characteristics of the vessel such as steering and maneuvering are little changed by fitting a reaction fin.

MIDP. Another very successful approach to improving propulsion efficiency with an add-on device, developed jointly by Exxon and Mitsui Engineering and Shipbuilding Company Ltd., Tokyo, Japan, is known as an MIDP (Mitsui integrated ducted propeller, Fig. 3). Full-scale trials of ships retrofitted with an MIDP have shown 5 to 8% improvement in efficiency.

In past years ducts have been fitted around propellers on all types of vessels. Tugboats and other craft with highly loaded propellers benefit substantially with ducted propellers or Kort nozzles. These installations involve special propellers completely surrounded by the duct, as opposed to normal open propellers with no duct.

To change a ship from an open propeller to a normal ducted propeller system would require a new propeller as well as a close-fitting nozzle built into the hull, and would be very expensive. The MIDP, however, has been developed to be retrofitted to existing ships, particularly very large, relatively slow-speed tankers. Normally no change in propeller is required. Since nozzle shape accelerates the flow into the propeller, a small increase in revolutions per minute for the same ship speed results. In most cases this small change presents no difficulties.

Fig. 1.   Reaction fin installed ahead of a ship propeller. (*Mitsubishi Heavy Industries*)

The conventional duct around the propeller requires close clearances between propeller blade tips and the inside of the nozzle for maximum efficiency. Thus the duct must be supported by the hull of the ship to be strong and rigid and not distort under operating conditions. The MIDP, on the other hand, is fitted in two halves ahead of the existing propeller. Therefore accuracy of the nozzle circularity is not highly critical.

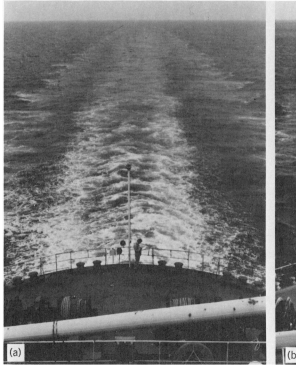

Fig. 2.   Wake behind a ship (a) before and (b) after installation of a reaction fin. (*Mitsubishi Heavy Industries*)

Fig. 3. Installation of an MIDP (Mitsui integrated ducted propeller). (*Mitsui Engineering and Shipbuilding Co., Ltd.*)

The MIDP is tapered from top to bottom for best efficiency. This device has been shown to have some other beneficial effects. Vibration and cavitation have been reduced. Since the accelerated flow produced by the duct impinges on the rudder, improvements in steering occur.

There are also new propeller types under development which are designed to improve efficiency over conventional propellers.

Stern end bulb. T. Inui, who conducted studies leading to the development of the bulbous bow for high-speed vessels with consequent reductions in horsepower, has (with the Kawasaki Heavy Industries, Ltd., Japan) also produced a new type of stern bulb. This has been shown to reduce drag by 3 to 6% on certain types of ships, particularly container ships (Fig. 4). Both bow and stern bulbs tend to reduce wave-making drag. However, the stern end

bulb (SEB) is designed to reduce eddy-making resistance as well as wave-making resistance. A properly designed bulb, of correct shape and in the right position, will smooth out the waves at the stern of the ship, resulting in substantial reduction in power required for the design speed.

The SEB aids course keeping, as the bulb acts like a fixed skeg and increases the ship's turning resistance. When using an autopilot, yawing angle is reduced and the rudder angle and frequency of rudder movement are also reduced. The SEB is fixed to the hull along the vessel's centerline at the transom and built with sufficient strength to withstand vertical loads (buoyancy and gravity) and forces through pitching, slamming, and yawing.

Three Japanese ferries were fitted with SEB in 1981, and further installations are under way. On trials the power savings measured were 5.6, 5.7, and 3.4%. It is estimated that a 36,000-horsepower (26.9 megawatts) ship could save about 1800 tons (1620 metric tons) of fuel per year assuming a 5% reduction in power. This savings alone will pay back the installation cost in 1 to 1½ years.

For background information *see* PROPELLER (MARINE CRAFT); SHIP POWERING AND STEERING in the McGraw-Hill Encyclopedia of Science and Technology.   [R. B. COUCH]

Bibliography: H. Narita et. al., Development and full scale experiences of a novel integrated ducted propeller, *Trans. Soc. Nav. Arch. Mar. Eng.*, vol. 89, November 1981.

Soil

This article discusses recent studies of the effect of soil aeration on chemical mobility, the characterization of salts in soils by using the techniques of scanning electron microscopy, and measurements of radionuclides in soils and tuff by means of x-ray and gamma-ray spectrometry.

Soil aeration and chemical mobility. Although the greatest part of the mass of the soil is relatively inactive, many soil constituents are mobile. These movements are important for such processes as plant uptake of nutrients, losses of nutrients through leaching, soil formation processes, and other processes that involve translocations within the soil or between the soil and other biosphere components. Some plant nutrients, such as nitrogen and potassium, move readily from the soil pore water or from adsorption sites on clay and organic matter to plant roots, where they are assimilated. In addition to plant uptake, other biological and chemical reactions result in the removal or precipitation of various soil constituents, creating concentration gradients that sustain the movement of these materials in the soil. Soil gases; mobile anions, such as nitrate, chloride, sulfate, and bicarbonate; and mobile cations, such as sodium, potassium, ammonium, and to a lesser degree, calcium and magnesium, respond rather rapidly to concentration gradients that develop in the soil.

Soil aeration plays an important role in the mobil-

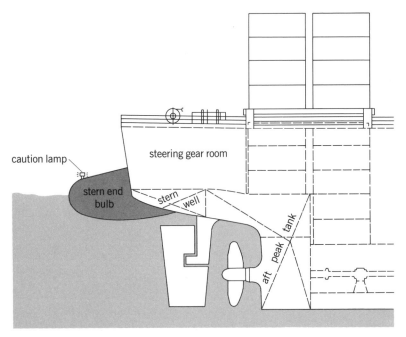

Fig. 4. Arrangement of stern end bulb as applied to a container-carrying vessel.

ity of soil constituents. As long as oxygen is supplied to a soil in adequate amounts, aerobic microbial processes are sustained, and there is little or no change in the form or mobility of the other soil constituents that can substitute for oxygen as alternate electron acceptors to maintain microbial respiration. However, the blockage of soil pores by water, or by severe compaction of the soil, reduces the diffusion rate of oxygen into the soil, and the resulting deficiency of oxygen causes the reduction of these alternate electron acceptors. Since some of these soil constituents have different mobilities under anaerobic as compared to aerobic conditions, the presence or absence of oxygen in the soil has pronounced effects.

Nitrogen. The first soil constituent to be affected by the disappearance of oxygen is nitrate, which is readily used by many microorganisms as an electron acceptor. Nitrate in an oxygen-deficient soil is rapidly converted to nitrogen gas (N_2) or to nitrous oxide (N_2O), both of which usually move in the gaseous phase. The absence of oxygen in the soil also results in the accumulation of ammonium (NH_4^+) because it cannot be oxidized to nitrate. Ammonium has a lower mobility than nitrate because it is a type of cation that is adsorbed onto the soil cation exchange complex and does not diffuse as rapidly as nonadsorbed anions, such as nitrate, which are free to move in the water phase.

Manganese and iron. Mobility of manganese and iron is especially affected by soil aeration. The oxidized forms of these elements are usually insoluble, whereas the reduced forms are slightly soluble. Many of these oxidized compounds replace oxygen as electron acceptors for microbial respiration and are thereby converted by reduction reactions from the immobile manganic (Mn^{4+}) and ferric (Fe^{3+}) forms to more soluble manganous (Mn^{2+}) and ferrous (Fe^{2+}) forms. The mobility of these cations is consequently greatly increased, and they move readily in the soil. A prerequisite for their movement is a concentration gradient but, as shown below, these gradients are readily established in flooded soils.

Sulfur. Another soil constituent affected by soil aeration is sulfur. Its stable inorganic form in aerated soils is sulfate (SO_4^{2-}) which is somewhat soluble and mobile. However, in the complete absence of oxygen, sulfate is reduced by anaerobic bacteria to sulfide (S^{2-}). Sulfide readily combines with ferrous iron (Fe^{2+}) to form insoluble FeS, and the mobility of the sulfur is consequently greatly reduced. Although the formation of insoluble FeS also decreases the solubility of iron, there is much more reducible iron than reducible sulfate in mineral soils. The result is that most of the reduced iron remains in the somewhat mobile Fe^{2+} form. (This is not always the case for high-organic-matter coastal systems, where much of the reduced iron can be immobilized as FeS.)

Flooded soils. In addition to restricting the supply of atmospheric oxygen, flooding has other effects on the mobility of some soil constituents. Complete filling of the pore space with water, instead of a mixture of air and water as in drained soils, increases the liquid cross-sectional area through which diffusion of dissolved constituents occurs. This speeds up movement of completely dissolved ions, such as nitrate and chloride, as well as the dissolved fractions of ammonium, manganous, and ferrous ions. There is a reverse effect on the movement of slightly soluble gases such as O_2, N_2, N_2O, and, under extremely reducing conditions, methane (CH_4).

Aerobic-anaerobic interfaces. Flooded soils and sediments are unique in having aerobic-anaerobic interfaces in which free oxygen exists in only one of the two adjacent zones. In most flooded soils there are two such interfaces. One occurs at the soil surface and consists of a thin layer of soil which is supplied with oxygen from the atmosphere by movement through the overlying oxygenated water column. The oxygen usually diffuses only a few millimeters into the soil before it is reduced microbially and chemically, a condition that results in an aerobic-anaerobic gradient extending over a few millimeters. The other major interface is in the oxygenated rhizosphere of wetland plants that have the capacity to transmit oxygen down to the roots and even to a thin layer of soil around the roots.

In both of these aerobic-anaerobic interfaces, reactions occur that affect both the form and the mobility of the various reduced and oxidized soil components discussed above. Reduced nitrogen, iron, manganese, and sulfur coming into contact with the aerobic portion of these interfaces are affected chemically or microbially, or both. Soluble manganous manganese and ferrous iron are oxidized into insoluble forms when they diffuse to the oxygenated zone of the interfaces. Here they may accumulate in amounts great enough to dominate the color of the aerobic surface layer and the root rhizosphere. This condition is most evident along the roots of wetland plants, where oxidized ferric iron is yellowish red. Ammonium diffusing to these oxygenated zones can be oxidized to nitrate, which in turn can undergo several reactions, including diffusion back into the anaerobic portion of the soil, where it is denitrified to nitrogen gas. Likewise, sulfide can be oxidized to elemental sulfur or to sulfate.

[WILLIAM H. PATRICK, JR.]

Scanning electron microscopy of salts. Scanning electron microscopy (SEM) has been widely used for studying the morphology of salts, particularly its crystal habits. As defined in soil science, salts are all inorganic compounds that are at least as soluble as gypsum. The crystallographic occurrence of salts in soils is determined by a number of factors, including ionic composition of the soil solution and soil temperature. In some soils, salts formed during the day are transformed to other kinds of salts at night. Even if the ionic composition is similar, one type of salt may form at the soil surface and another type in deeper horizons. For example, mirabilite ($Na_2SO_4 \cdot 10H_2O$) forms relatively deep in the soil, while thenardite (Na_2SO_4) occurs only at the soil surface. Crystal habits are determined largely by the microenvironment during crystallization, and researchers are currently evaluating this relationship.

Many of the salts occur in soils of the desertic or subdesertic areas of the world, including the deserts of the polar regions. The chemistry of the salts is generally well established, and much of the knowledge of their behavior in soil stems from geochemical studies of salt domes, evaporites, and so on.

Sulfate minerals. Gypsum, the most common sulfate mineral, is about 100 times less soluble than most other sulfates. In Aridisols, gypsum generally has a lenticular habit, and is occasionally fibrous or even massive. Gypsum in Sulfaquepts is characteristically tabular, as shown in Fig. 1. The other calcium sulfates—hemihydrate, or bassanite ($CaSO_4 \cdot \frac{1}{4}H_2O$), and anhydrite ($CaSO_4$)—are rare in soils. They are generally inherited with the sediment.

Many sulfate minerals occur as efflorescences on a crusted soil surface. Thenardite is the most common mineral among the sodium sulfates. It generally occurs as rosettelike aggregates. Other soil sulfate minerals are mirabilite ($Na_2SO_4 \cdot 10H_2O$), bloedite [$Na_2Mg(SO_4)_2 \cdot 4H_2O$], and hexahydrite ($MgSO_4 \cdot 6H_2O$). Scanning electron micrographs of these have been made. Celestite ($SrSO_4$), a rare mineral frequently associated with gypsum, generally occurs as short, stubby crystals with tapered ends. Authigenic barite ($BaSO_4$), which is less soluble than calcite, has also been reported in soils; the few micrographs that have been published show malformed crystals.

Halides. Of the halides, halite ($NaCl$) is the only mineral of any importance in soils. Halite crystallizes and dissolves very readily in the soil system. As a result, scanning electron microscope studies of air-dry soils may not illustrate its true habits.

In soils, the perfect cube—hexahedral form—is rare. The most frequent form is a diffuse, glassy aggregate coating the voids (Fig. 2). This glassy form permeates the void system and helps to cement the soil particles. Two specific habits occur when crystallization takes place from a brine formed by evaporation from ponds. The first is the cube. The second form may be described as a malformed cube,

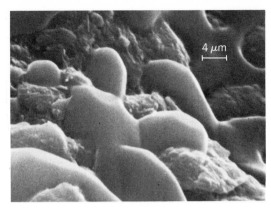

Fig. 2. Glassy aggregates of halite (NaCl) in a salt-affected soil.

where growth is partial and is confined to the corners or edges. Growth can take place preferentially only along one axis, forming a pseudotetrahedral form. A rare form is termed the hopper structure, where one face develops a series of depressions.

Soils with a high halite content develop a puffed surface about 1 cm (0.4 in.) thick. The puffing is caused by the growth of halite fibers, which are very brittle and collapse easily. The other two halides—sylvite (KCl) and carnallite ($KMgCl_3 \cdot 6H_2O$)—are rare in soils.

Carbonates. While they are not salts according to the definition used earlier, carbonates are so common in soils that they deserve mention. Calcite ($CaCO_3$) is the most common mineral, generally of rhombohedral form. Clay-sized and silt-sized calcites, which are the most common sizes in soils, are not always rhombohedral, however, since cementation during crystallization can result in irregular-shaped crystals.

One authigenic form of calcite frequently found in the lower horizons of some Alfisols has an acicular morphology and is sometimes referred to as lublinite. In the field, the void walls appear to be coated by mycelia of white calcite. Under the scanning electron microscope, they appear as a network of interlocking acicular crystals.

Other carbonates are rare but have been reported in soils. These include nahcolite ($NaHCO_3$), soda ($Na_2CO_3 \cdot 10H_2O$), trona ($Na_2CO_3 \cdot NaHCO_3 \cdot 2H_2O$), hydromagnesite [$Mg_5(CO_3)_4(OH)_2 \cdot 4H_2O$], thermonatrite ($Na_2CO_3 \cdot H_2O$), and huntite ($Mg_3CaCO_3$). Scanning electron microscope studies of these carbonates have been made but have not been published.

Other salts. Many other salts occur in nature but are rare in soils and when present indicate special soil conditions. A few of these, discovered in the Antarctic, are tachyhydrite ($CaCl_2 \cdot 2MgCl_2 \cdot 12H_2O$), antarcticite ($CaCl_2 \cdot 6H_2O$), darapskite [$Na_3(NO_3)(SO_4) \cdot H_2O$], soda niter ($NaNO_3$), and sodium iodate ($NaIO_3$).

Research goals. Scanning electron microscopy is a recent technique for evaluating salts, and much of the work has been done in the last decade. Current

Fig. 1. Rosettes of tabular crystals of gypsum in an acid sulfate soil.

Table 1. Comparison of two radionuclide assay techniques for the analysis of 100 soil samples

	Assay technique	
Radionuclide	Wet chemistry	Automated radionuclide assay system
Plutonium	Sample preparation Acid digestion Ion exchange Electroplating Alpha spectrometry	Sample preparation Counting by gamma spectrometry
Americium	Evaporation of column effluent Solvent extraction Ion exchange Precipitation Alpha spectrometry	Same as above
Total labor*	75 Worker-days	22 Worker-days

*The assumption is made in these estimates that only one technician is used to perform the assays.

research is directed toward characterizing morphology, establishing crystal habits, and describing growth phenomena in soils. It is anticipated that later research will determine the microenvironmental factors governing crystal habits and crystallization. The role of trace elements and other soil organic and inorganic properties must be evaluated. These studies will assist the soil scientist in clarifying some aspects of soil genesis and behavior and will aid in work on drainage, irrigation, and management of salt-affected soils.

[HARI ESWARAN]

Automated soil radioassay. An automated radionuclide assay system has been developed for performing soil radioassays. This method permits a more rapid and convenient determination than that provided by traditional wet chemistry techniques. By using L x-ray and gamma-ray spectrometry, the radionuclides ^{239}Pu, ^{240}Pu, and ^{241}Am can be analyzed simultaneously without prior separation.

Measurements of low-level radioactivity in soil samples are usually made with a wet-chemistry procedure in which the radionuclides are chemically separated from the soil, purified, and analyzed with spectrometric techniques. In this method, accuracy is dependent upon the quality of the chemical extractions. Careful analytical procedures involving wet chemistry are time-consuming and result in excessive personnel costs, but these problems are eliminated in automated radioassay. The only step that has to be completed by the technician before counting takes place is sample preparation. Table 1 shows a comparison of the time required by the two methods for analyzing 100 soil samples. The difference in technician time required will of course vary, depending on the size of the laboratory, but the difference shown in time consumed is significant. Once the wet chemical analysis has begun, the technician is locked into this regimen. However, with the automated system, once the samples are loaded and started, the technician is free to do other tasks as needed.

Assay system. The radionuclide assay system consists of a coaxial lithium-drifted germanium (GeLi) detector with a total active volume of 125 cm^3 and an intrinsic germanium (IG) detector with a total active volume of 14.7 cm^3. The intrinsic germanium detector is calibrated in the range of 0–200 keV, with 0.1387 ± 0.00006 keV per channel. The calibration range of the lithium-drifted germanium detector is 200–2000 keV, with 1.08601 ± 0.00000 keV per channel. Both detectors are interfaced to a multichannel analyzer and a minicomputer with an accompanying terminal console to supply hard copy.

After soil samples have been dried, approximately 25 g are transferred to a petri dish–like plastic container, which is 7.1 cm in diameter and 0.74 cm thick. The container is designed so that the lid, which is less than 1 mm thick, faces the intrinsic germanium detector, where the sample is assayed for low-energy (<200 keV) gamma-ray emitters. In contrast, the bottom of the container is about twice the thickness of the lid and is positioned facing the lithium-drifted germanium detector, where high-energy (>200 keV) gamma emissions are assayed. Twenty sample containers are positioned vertically in a lead-lined wheel. The programming in the minicomputer requires that each sample be arbitrarily counted within a time period of 1000 s (16.7 min) to 16,000 s (4.4 h). This time frame is dependent on the level of activity present and the time required to get <3% counting error. After subtraction of the room background contribution to each energy band, the number of counts in each energy band is determined and the concentration of the corresponding radionuclide is calculated, along with the measurement error. This information is then printed out on a hard-copy terminal, and the sample wheel advances to the next soil-sample dish.

Standards were made by using the base material, which was primarily tuff (volcanic rock). A radionuclide standard solution is added to each sample, bringing it to complete saturation. The amended samples are dried at 60°C for 48 h, quantitatively removed from the container, and homogenized. The samples are then analyzed for radionuclide content with both automated radionuclide assay and wet-chemistry techniques for [241]Am and plutonium.

The measurements of these samples by the radionuclide assay system are generally in good agreement with the wet-chemistry analyses. These analyses indicate less agreement between the two techniques for plutonium than for [241]Am. There is larger counting error associated with the plutonium assays than with the [241]Am determinations, and the total sample assayed for plutonium in the radionuclide assay is small compared to the wet-chemistry assay of the entire sample.

Detection limits. Radionuclide assays were performed on a set of over 850 tuff samples collected beneath a nuclear waste disposal site in an attempt to characterize the detection limits of this radionuclide assay system for plutonium and [241]Am (Fig. 3).

For plutonium, a detection limit of 30 picocuries per gram (pCi/g) was selected for routine assays, assuming a 16,000-s maximum count time. Figure 3a shows that the plutonium concentrations from 30 to 13,000 pCi/g exhibited counting errors of only 5–15%, whereas samples with concentrations <30 pCi/g had much higher counting errors.

Figure 3b shows that the [241]Am data for these samples exhibit a marked difference over the detection limits for plutonium. The majority of the samples had counting errors of <2%, and a detection limit of 0.8 pCi/g was chosen. Since the gamma-ray energy and intensity levels are higher for [241]Am than for the plutonium L x-rays, these differences are expected.

In order to determine reproducibility of the system, 85 samples were randomly selected from the 850 tuff samples that had already been assayed. Each sample was counted three times, and then the average [241]Am and plutonium concentrations were calculated. In addition, the coefficient of variation (standard deviation of the average concentration divided by the average radionuclide concentration times 100) for each sample was also calculated. The coefficients of variation for [241]Am assays ranged from 1.0 to 5.5%. In contrast, the coefficients of variation for plutonium ranged from 0.3 to 29%, although most were below 20%.

Applications. The automated radionuclide assay system promises to be a simple, fast, and cost-effective technique for routine soil analyses. It is suited for environmental monitoring, where a large number of samples must be screened quickly to determine if the radioactivity exceeds a given minimal level. The system could also be used in radioecological and waste management research, where concentrations of solid radionuclides usually demonstrate a high degree of variability and a large number of samples must be assayed before meaningful conclusions can be drawn.

[B. J. DRENNON]

Diffusion of ions in soils and soil clays. The rates of many important soil processes are determined, wholly or in part, by the rates at which ions or uncharged solutes will diffuse through the soil. Examples are the release of ions by weathering of minerals into positions where they readily equilibrate with the soil solution; the movement of plant nutrient ions to the absorbing surfaces of roots, often the rate-determining step in nutrient uptake; and the dissolution and spread of fertilizers or pesticides applied in solid form, thus determining their zone of influence.

Diffusion results from the random thermal motion of ions or molecules. The net flux of a substance

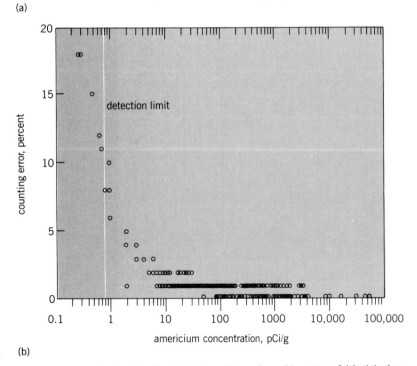

(a)

(b)

Fig. 3. Graphs showing detection limits in routine radionuclide assays of (a) plutonium and (b) americium.

(the amount crossing unit area in unit time) can be calculated by multiplying its diffusion coefficient by its concentration gradient in the whole soil. Values of the diffusion coefficient in soil systems of interest vary over an extremely wide range: from about 10^{-27} $m^2 \cdot s^{-1}$ for release of K^+ from an illitic clay to 10^{-9} $m^2 \cdot s^{-1}$ for ions like Cl^- and NO_3^- in a water-saturated soil. Knowing the diffusion coefficient D, the average distance x that a molecule will move in a given time t can be calculated from the formula $x = \sqrt{2Dt}$.

The release of ions by gradual weathering of minerals is characterized by diffusion coefficients of less than 10^{-24} $m^2 \cdot s^{-1}$. The release of K^+ from hydrous mica, fast enough to meet the needs of a crop, involves diffusion coefficients of the order of 10^{-19} $m^2 \cdot s^{-1}$. The exchange of cations between the interlayer positions in clays and the external solution, which is nearly complete in minutes, is governed by diffusion coefficients of the order of 10^{-13} $m^2 \cdot s^{-1}$. In moist soil, ions such as Cl^- and NO_3^-, which are usually in the soil solution only where they are mobile, have diffusion coefficients of about 10^{-10} $m^2 \cdot s^{-1}$; but cations, and anions, such as $H_2PO_4^-$, which are adsorbed by the soil solids and so spend much of their time immobile have correspondingly lower diffusion coefficients, 10^{-11} to 10^{-12} $m^2 \cdot s^{-1}$. In dry soil, near the wilting point, their mobility may be reduced by a factor of 10 to 100 from that in moist soil.

Ion diffusion coefficients in soil clays. The mobility of exchangeable cations in pure, fully hydrated clays such as kaolinites and montmorillonites relative to their mobility in aqueous solution is shown in Table 2. When the interlayer spacing is reduced by drying, the mobility of the ions is also greatly reduced. This is illustrated for Na^+ and Sr^{2+} in montmorillonites in Fig. 4.

The mobility of interlayer cations is closely related to the diffusion coefficient of the water adsorbed in interlayer positions. The coefficient may be accurately determined by neutron scattering spectroscopy, as illustrated in Fig. 5.

Diffusion of ions in soil. The diffusion coefficient

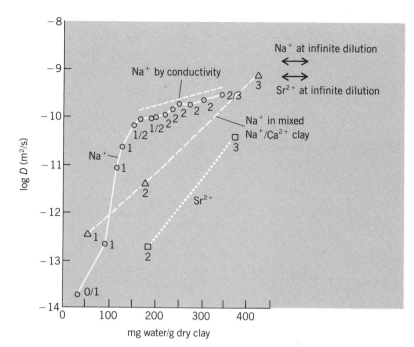

Fig. 4. Self-diffusion coefficients of Na^+ and Sr^{2+} in oriented montmorillonite at varying hydration. Integers on curves are the number of molecular thicknesses of water determined by c-axis spacing. (*After C. J. B. Mott, Cationic Mobility in Oriented Bentonite, Ph.D. thesis, Oxford University, 1967*)

of an ion in soils depends on: (1) the diffusion coefficient of the ion in free solution; (2) the fraction of the soil volume occupied by solution, which gives the cross section for diffusion; (3) an impedance factor, which reflects mainly the tortuosity of the diffusion pathway; (4) the buffer power of the soil for the ion; and (5) an extra term, which is zero when the exchangeable ions have no surface mobility, but represents their extra contribution to the diffusion coefficient if they are mobile.

The impedance factor falls steeply as the soil dries, as shown for a range of soils in Fig. 6. The diffusion coefficient may therefore be reduced a hundred fold over the field moisture range of water potential, -0.1 to -10 bar (-10 kilopascal to -1 megapascal).

Table 2. Mobility on ions in salt-free clay gels relative to solution at 25°C

Clay concentration	Ion							
	Li^+	Na^+	K^+	Rb^+	Cs^+	Ca^{2+}	Sr^{2+}	Ba^{2+}
			Montmorillonites					
0.1–10 g clay/100 g solution		0.37	—	—	—	0.13		
15–28 g clay/100 g solution		0.27	—	—	0.19	0.22		
4–6 g clay/100 g gel		0.37						
4.9–10.6 g clay/100 g gel		—	—	—	—	0.08		
3 g clay/100 ml solution		0.25	0.23	—	0.06	0.08	0.09	
60 g clay/100 g gel		0.13	0.06	0.02	0.01	0.08		0.04
			Kaolinites					
56 g clay/100 g gel	—	0.28						
70 g clay/100 g gel	0.19	0.18	0.08	—	0.03			
31–35 g clay/100 ml gel	—	0.14	—	—	—	—	0.05	

The buffer power is a measure of the proportion of time that an ion spends immobilized on soil particle surfaces rather than free in soil solution. The buffer power has an important role in determining the diffusion coefficient. For an ion, such as Cl^- or NO_3^-, which is not adsorbed the buffer power is less than 1. However, for an ion, such as $H_2PO_4^-$, which may be strongly adsorbed, the buffer power commonly lies in the range 100 to 1000. Consequently, the diffusion coefficient of phosphate is much lower than that of nitrate.

The extra term due to mobility on particle surfaces is difficult to estimate. Fortunately it has been found experimentally to be negligible except perhaps for Na^+. Exchangeable ions in real soils have much less surface mobility than they have in purified clays, possibly because their diffusion pathways are blocked by such clay decomposition products as polymeric $Al(OH)_3$.

When plant roots take up essential nutrient ions from the soil, the root may be pictured as growing into fresh soil zones from which nutrient ions have then to move to the absorbing surfaces of the root. For plants suffering from a nutrient deficiency, this rate of movement may be the step which limits the uptake rate. The ions move by diffusion and by the flow of soil solution that is induced by transpiration. Under deficiency conditions, diffusion is by far the more important process. Because phosphate has a low diffusion coefficient, roots take up nearly all their phosphate from distances of only about a millimeter from the root surface. Roots having long root hairs, or mycorrhizal hyphae can increase their phosphate uptake by developing these fresh absorbing surfaces in unexploited zones of soil. On the other hand, nitrate moves so readily through the soil that roots can take up all the nitrate within at least a centimeter. Consequently roots are usually in competition with each other for nitrate. Potassium is intermediate be-

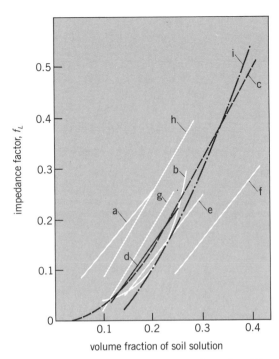

Fig. 6. Relation between the impedance factor f_L and the volume fraction of the soil solution, θ. (a) Wanbi sand (6% clay); (b) Urrbrae loam (19% clay); (c) sandy loam (24% clay); (d) Fort Collins loam (26% clay); (e) Apishapa silty clay loam (37% clay); (f) Pierre clay (53% clay); (g) sand (4% clay); (h) sandy loam (15% clay); (i) average of six silt loams. (*After P. H. Nye, Diffusion of ions and uncharged solutes in soils and soil clays, Adv. Agron., 31:225–272, 1979*)

tween phosphate and nitrate in these respects.

For background information *see* SCANNING ELECTRON MICROSCOPE; SOIL CHEMISTRY in the McGraw-Hill Encyclopedia of Science and Technology.

[P. H. NYE]

Bibliography: F. P. Brauer et al., Measurement of environmental ^{241}Am and the $Pu/^{241}Am$ ratio by photon spectrometry, *IEEE Trans. Nucl. Sci.*, NS-24(1):587–595, 1977; H. Eswaran and M. Carrera, Mineralogical zonation in salt crusts, *Proceedings of the Symposium on Salt-Affected Soils*, Karnal, India, pp. 20–28, 1980; H. Eswaran, G. Stoops, and A. Abtahi, SEM morphologies of halite (NaCl) in soils, *J. Microsc.*, 120:343–352, 1980; R. H. Howeler and D. R. Bouldin, The diffusion and consumption of oxygen in submerged soils, *Soil Science Society of America Proceedings*, 35:202–208, 1971; F. Moormann and H. Eswaran, A study of a paleosol from East Nigeria, *Pedologie*, 28:251–270, 1978; P. H. Nye, Diffusion of ions and uncharged solutes in soils and soil clays, *Adv. Agron.*, 31:225–272, 1979; P. H. Nye and P. B. Tinker, *Solute Movement in the Soil-Root System*, 1977; K. R. Reddy and W. H. Patrick, Jr., Effect of aeration on reactivity and mobility of soil constituents, *Soil Science Society of America Special Symposium*, in press; I. S. Sherman, M. G. Strauss, and R. H. Pehl, Measurement of trace radionuclides in soil by L x-

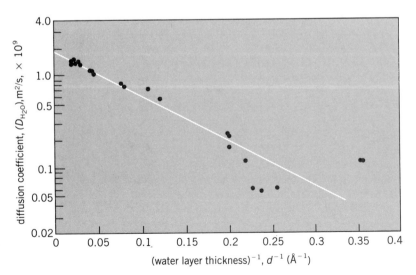

Fig. 5. Variation of the diffusion coefficient of water in montmorillonite and vermiculite with reciprocal of the interlayer spacing, *d*. 1 angstrom = 0.1 nanometer. (*After S. Olejnck and J. W. White, Thin layers of water in vermiculites and montmorillonites—modification of water diffusion, Nature (London) Phys. Sci., 236:15–16, 1972*)

ray spectrometry, *IEEE Trans. Nucl. Sci.*, NS-27(1):1–9, 1960; G. W. Snedecor and W. G. Cochran, *Statistical Methods*, 1967; G. Stoops, H. Eswaran, and A. Abtahi, Scanning electron microscopy of authigenic sulfate minerals in soils, in M. Delgado (ed.), *Proceedings of the 5th International Work Meeting on Soil Micromorphology*, Granada, Spain, pp. 1093–1110, 1978; M. G. Strauss et al., Considerations in measuring trace radionuclides in soil samples by L x-ray detection, *IEEE Trans. Nucl. Sci.*, NS-25(1):740–748, 1978.

Soil chemistry

Chemical reactions in the soil are controlled by a variety of complex mechanisms. This article summarizes recent research that has involved the degree of potassium availability to plants growing in high-potassium soils, the movement of heavy metals in soils, and the use of kinetics to study adsorption reactions in soils.

High-potassium soils. The concept of high-potassium soils is ambiguous. Often the term denotes a high total soil potassium content which is construed to be synonymous with a high availability of soil potassium to plants. A high soil potassium content, however, may provide a low availability of soil potassium to plants, and vice versa. This relationship is complex and dependent upon the resulting mineralogy and chemistry of the soil system.

Total soil potassium. The average potassium content of soils is 1.4% with values ranging from almost 0 to 8%. High total soil potassium content is generally greater than 3% and results from the presence of large quantities of potassium-bearing silicate minerals. The primary potassium minerals include the feldspars—microcline, orthoclase, sanidine, and so on; the micas—muscovite, biotite, glauconite, phlogopite, sericite, and so on; the feldspathoids—nepheline, kalsilite, leucite, and so on; and others such as the zeolite clinoptilolite. Generally these minerals are most abundant in the sand (diameter greater than 50 micrometers) and silt (50–2 μm) size fractions of soils. The secondary potassium minerals include illite, vermiculite, montmorillonite, and others such as noncrystalline allophane. These minerals generally dominate the clay (less than 2 μm) size fraction of soils.

The potassium present in minerals always neutralizes electrostatic deficiency of charge present within the mineral and never occurs within the lattice proper. These charge deficiencies result from the proxying of lower- for higher-charged cations within the crystal lattice. The chemistry of mineral potassium is a result of the characteristic properties of the site occupied by potassium within these various types of structures.

The potassium in feldspars occurs within interstices left between tetrahedral chains linked together to form a three-dimensional framework. This results in a very dense structure which requires complete decomposition for release of potassium. Since decomposition through dissolution is a surface phenomenon, most potassium occurring in soils within the feldspar structure occurs in the coarser size fractions and in younger, less weathered soil environments. Susceptibility of feldspars to weathering and release of potassium increases with increased presence of sodium and possibly iron within the structure, increased disorder of silica and aluminum within the structure, increased number of crystal twins, and increase in amount of lamellar perthitic structure.

The potassium in micas resides in a cavity in the interlayer position between layer silicate sheets in a largely covalent and partially ionic environment. Mica stability is largely controlled by the chemical composition of the octahedral sheet and bulk solution, and the structural cavity size and configuration.

The dioctahedral micas such as muscovite contain primarily aluminum in two out of three octahedral positions, whereas the trioctahedral micas such as biotite and phlogopite contain primarily iron or magnesium in all three of these positions. This chemical composition produces a dipole moment of the hydroxide vibration within the layer silicate cavity which is perpendicular to the basal plane for trioctahedral micas and inclined into the vacant octahedral site for dioctahedral micas. This configuration permits a greater repulsion of potassium by structural hydroxyls in trioctahedral micas, which results in a more rapid rate and quantity of potassium release into solutions. A solution composition of 10^{-5} to 10^{-4} mole per liter of potassium is capable of inhibiting dioctahedral mica decomposition, whereas 10^{-3} to 10^{-1} mole per liter of potassium is necessary for a similar effect with trioctahedral micas. Thus, the weathering rate of micas is also influenced by such factors as rainfall, soil percolation rates, plant uptake of potassium, and potassium fertilizer additions.

The mica structural cavity size and configuration is largely controlled by the dimension of the octahedral sheet, which increases as larger divalent cations replace smaller trivalent aluminum cations. The inherent oversized tetrahedral sheet rotates and tilts to obtain compatibility with the octahedral sheet, thus forming a ditrigonal cavity. This ditrigonal configuration produces six long and six short potassium-oxygen bonds, which results in increased bonding strength of potassium in the mica structure. Thus factors which decrease tetrahedral sheet rotation and tilting, such as larger cation size and greater abundance of cations in the octahedral position, also increase release of potassium from micas.

Release of potassium from mica is not a simple diffusion-controlled process, particularly for the smaller clay size fraction, but involves many complex structural implications. The rate of release of potassium from biotites is often most rapid with smaller particles in sand- and silt-sized materials, although at greater lengths of weathering more potassium may eventually be lost from coarser particles. The rate and total quantity of potassium released from muscovite, however, often decreases with decreasing particle sizes within the silt- and clay-sized materials. It is speculated that removal of potassium from "thicker" particles loosens the remaining potassium-oxygen bonds in the ditrigonal cavity for an entire interlayer, thus promoting greater potassium removal from these particles, particularly for dioctahedral micas. Consequently mica type and particle size are important with regard to potassium in soil fertility as well as soil genesis relationships.

Most of the secondary 2:1 layer silicates present in soils result from the weathering of primary micas or feldspars. These secondary minerals are major sources of exchangeable and nonexchangeable potassium and retain potassium from the leaching action of water. The average potassium content of the lithosphere is about 1.6% with a similar sodium content, although sea water contains only 1.1% potassium compared to 30% sodium. Thus much of the potassium released from primary minerals through weathering is retained by secondary minerals in a form more available to plants.

The potassium in feldspathoids and zeolites is held in true channels in the silicate structure. These minerals are much less dense, and potassium can be readily exchanged from the structure. These structures are also readily weatherable and are present only in coarser fractions of the soil and in young environmental conditions.

Available soil potassium. Forms of soil potassium have been divided into readily, moderately, and difficultly plant-available categories. Potassium in soil solution and exchangeable potassium are readily available to plants. Exchangeable potassium is that held by the negative adsorption sites in soil organic matter and clay minerals, which is easily exchanged with other cations. Fixed potassium is moderately available to plants. It is held between the layers of secondary clay minerals. Potassium in the primary soil minerals is in the difficultly available category. In the average mineral soil, the soil solution potassium makes up about 1 to 3% of the exchangeable potassium, which in turn represents at most a few percent of the total potassium.

Interrelationships exist between the various forms of soil potassium. Fixation of soil solution potassium and exchangeable potassium occurs within clay minerals at higher concentrations of both these readily available potassium forms. This fixation occurs in surface soils when potassium levels are increased by fertilization and in subsoil when potassium levels are increased by downward movement of applied potassium. Release of fixed potassium to exchangeable and soil solution potassium occurs with decreases in levels of these readily available forms. Levels of exchangeable and soil solution potassium are decreased by crop removal and by leaching. A relatively small amount of potassium is also released during weathering of primary minerals to exchangeable and soil solution forms.

These dynamic equilibrium reactions between the solution and exchangeable phases of potassium are generally proposed to be almost instantaneous and occur within seconds. The transformations between exchangeable and nonexchangeable phases are slower, occurring within hours or days, and are strongly affected by prevailing soil conditions. In comparison, the transformations and release of potassium from primary minerals are extremely slow, with release reactions being relatively more rapid than mineral fixation.

The equilibrium between solution and adsorbed potassium is primarily controlled by the degree of potassium saturation and selectivity of adsorption sites. Generally three types of adsorption sites may be distinguished in clay minerals. These include planar sites with a low potassium selectivity, edge sites with a medium potassium selectivity, and interlayer sites with a high potassium selectivity. Adsorption sites of organic matter and kaolinitic minerals are low in potassium selectivity. The 2:1 clay minerals, however, contain adsorption sites that have a higher selectivity for potassium and therefore bind potassium very strongly with mineral differences due to interlayer characteristics.

Only a small fraction of the potassium requirements of plants is obtained by direct contact of the root with the soil. The bulk of the potassium needed by plants has to be transported in soils to the roots. The potassium transport occurs mainly by mass flow of soil solution moving to the plant root, and diffusion of potassium along a concentration gradient that is low at the surface of the adsorbing roots and higher at the surface of the soil particle. Depending on crop, an adequate soil solution concentration of about 0.5 millimole per liter is necessary. Continuous potassium supply to the growing plants is ensured only when the rate of potassium release to the soil solution and transport to the roots keeps pace with the rate of nutrient uptake. Normal uptake for corn and alfalfa is approximately 40 and 190 kg of potassium per hectare, respectively, for a 4-month growing season, whereas for loblolly pine it is about 5 kg per hectare for a 20-year growing season. Theoretically the solution and exchangeable phases of potassium at the beginning of a cropping period and the potassium release from nonexchangeable and primary mineral forms during the cropping period largely control the chemical availability of potassium to plants.

Potassium fertilization. Potassium fertilization has failed to increase corn or soybean yields on certain soils of the Atlantic Coastal Plain region. These soils characteristically have sandy surface horizons and high total potassium concentrations in clayey subsoils. Normally, fertilization recommendations are based on samples collected from the soil surface. Often these samples contain low levels of extractable potassium, but show no yield response to this fertilizer. This lack of response to fertilizer potassium has been related to the availability of subsoil potassium which can amount to 1500 to 4000 kg of potassium per hectare for each 15-cm increment depth in these soils. This subsoil potassium is available for plant uptake if root growth is not restricted by adverse chemical or physical soil properties and the rate of potassium release is sufficient.

Conversely, potassium fertilization has increased crop yields on soils of the northern Great Plains–Intermountain region. These soils are relatively high in extractable potassium and have cool temperature regimes with cool nights and long hot days and a semiarid soil moisture regime. Under these conditions, the plant demands for potassium during periods of rapid growth is very great, and yet the dry soil conditions greatly decrease potassium mobility. Apparently, the relatively large quantities of extractable potassium in these soils do not ensure an adequate rate of supply to the plant root without the addition of fertilizer potassium.

[LUCIAN W. ZELAZNY]

Movement of heavy metals. Research on the movement of heavy metals in soils has recently been receiving more emphasis because of the increased acceptance of the concept of land application of waste materials, the importance of heavy metals in plant and animal life, and improved analytical techniques. Movement of heavy metals in soils has generally been considered to be minimal, although it has been observed in some research studies. Heavy-metal movement is most likely to occur when large heavy-metal applications are made to a sandy, acid, low-organic-matter soil which receives high rainfall or irrigation or where open soil channels or cracks occur and soil particulates move. In such an environment the metal has no opportunity to react with soil attenuation surfaces.

Heavy metals are generally not applied to soils in large amounts except as may occur with a waste utilization program. With their introduction into the soil, heavy metals may: bond to cation or anion exchange sites; form or combine with inorganic precipitates; combine with organic compounds; and enter into the soil-water solution. The extent of heavy-metal movement in a soil thus depends on particular metal reactions in the soil as well as conditions governing soil water movement.

Movement of heavy metals in soils can occur by diffusion in the solution phase as free ions or complexes, by mass flow with percolation solution, or with particulates through open channels caused by rodent tunneling or extensive soil cracking during wetting and drying cycles. Mass flow with percolation water is probably the principal means by which metals move in soils. Generally, diffusion of metals occurs over short distances and, while this is important for plant availability, the movement distances are not large.

Heavy-metal reactions. To move with percolation water in soils, heavy metals must be soluble or associated with very fine mobile particulates. If applied to soils as soluble species, heavy metals will generally react with soil solid or solution constituents to form relatively insoluble compounds such as sulfides, phosphates, hydroxides, or oxides. Thus, the concentration of heavy metals in soil solution is very low. With most heavy metals, precipitation occurs with a pH increase, thus further reducing the ionic concentration in soil solution. As the soil solution moves through the soil profile, the heavy-metal constituents may also react with solid phases such as clay and iron, aluminum, and manganese oxides which may sorb the species, thus further reducing the metal concentration in solution.

In addition to their reactions with the inorganic constituents in the soil profile, the heavy-metal concentrations in soil solution are related to reactivity with organic constituents. Heavy metals have unfilled *d* orbitals and often chelate or complex with soil organic materials. These chelates, if soluble, may enhance movement of the heavy metals; however, the metals may also react with insoluble organic material, thus restricting metal movement. Fulvic acid has been identified as an important organic matter constituent that reacts with metal ions.

In consideration of the numerous processes which serve to remove heavy metals from soil solution and the studies which have found no movement below the zone of metal incorporation, researchers generally conclude that heavy-metal movement is minimal. The soil generally serves as an effective filter to remove the heavy-metal species.

Differences in heavy-metal movement. Chromium, copper, zinc, nickel, and cadmium have been found to leach in soils treated with large applications of sewage sludge. Several of these studies were completed on soils below sewage sludge lagoons. In one study increases were found in the concentrations of all these metals as deep as 3 m under a sludge disposal pond; in another study elevated levels of cadmium, copper, nickel, and zinc were observed as deep as 61 cm below a sludge holding pond.

In contrast to studies where significant movement of heavy metal was reported, other studies found essentially no movement of cadmium, copper, nickel, or zinc in soil columns treated with sewage sludge. In a Michigan study, under conditions where the potential for leaching sludge-borne metals was great [that is, very sandy soil (cation exchange capacity < 5 meq/100 g), very high sludge loadings (320 metric tons/ha over 3 years), and high water application rates], no appreciable leaching of cadmium, chromium, or lead occurred. Also, cadmium, copper, and zinc movement from sewage sludge trenched in a loamy sand soil was not detected for 2 years after sludge application; however, zinc concentrations deeper in the soil profile increased about 3 years after the study was initiated.

Differences in observations of heavy-metal movement can generally be accounted for by consideration of element chemistry, soil texture, soil pH, soil sorptive properties, loading rates, analytical problems, and sample contamination concerns. Regardless of whether or not heavy-metal movement occurred, total accountability for metals applied to a soil has been less than successful.

[V. V. VOLK; R. H. DOWDY]

Kinetics of metal reactions with soils. The use of kinetics to study adsorption reactions in soils is a relatively recent development in soil chemistry, and is being explored because of limitations in the techniques that have been used during the last 20 to 30 years. Suggestions that kinetics should be used to study phosphorus adsorption began to appear in the early 1970s, but the major difficulty of separating soil from solution in less than about 20 min persisted. Thus, rapid reactions could not be followed kinetically. Around 1974 it was suggested that membrane filters, rather than the traditional centrifugation, be used for rapid phase separation. Membrane filters are more effective than centrifugation for phase separation, and a clear soil solution can be obtained more rapidly. Depending on soil clay content, separation can be effected in about 5 to 20 s by this technique. A major breakthrough for use of kinetics in soil–heavy-metal reactions came in 1978 with the development of a method of using a Titrimeter to control reaction pH, and syringes for sampling, thus circumventing many problems of previous methods by conducting studies in a single reaction vessel. Availability of such instrumentation permitted replacing the more traditional methods with kinetic techniques for all detailed adsorption studies.

When metal ions come in contact with soils, they may react with organic matter, clay minerals, or various other minerals and amorphous components of the soil. Reactions may occur via exchange at charge sites, complexation, precipitation, and other mechanisms. Techniques traditionally used to study reactions cannot differentiate between the various mechanisms and sites of reaction in soils. Therefore, researchers must be content with adsorption isotherms, which only provide information on total adsorption capacity and the partition of metal between soil and solution at equilibrium. While this information is useful, it is limited. Kinetic information has always been desirable because of the ability to differentiate different reactions which occur at different rates. Because of the necessity of separating soil from solution by centrifugation, however, kinetic studies have been practical only for slow reactions, such as nitrogen mineralization or potassium release. Most adsorption reactions occur more rapidly than can be followed by centrifugation separation techniques. With development of membrane filtration techniques, reaction rates for all but the most rapid reactions can be followed.

One current study concerns the kinetics of heavy-metal (for example copper, nickel, zinc) adsorption by soils. A typical adsorption reaction is illustrated in Fig. 1. The initial reaction is very rapid, and the reaction rate slows with elapsed time. Actual equilibrium may take an extended time, but the adsorption reaction is usually at least 90 to 95% complete within an hour or two. When the log of metal adsorbed is plotted as a function of time (Fig. 2), the data may frequently be resolved into linear portions, indicating a series of simple (first-order) reactions. Reaction A is completed within 1 min or less; thus, it occurs so rapidly that it cannot be followed even with membrane filtration. This does not mean, however, that no information about the re-

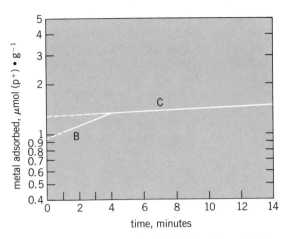

Fig. 1. Typical metal adsorption data that has been plotted log-normal to break curve into constituent linear portions. Reaction A is nearly instantaneous and cannot be plotted as a function of time. The symbol p^+ indicates charge.

action can be obtained. If reaction B is extended to zero time, it does not hit zero adsorption, but approximately 9.5 μmol $(p^+) \cdot g^{-1}$. This is an indication of the amount of metal that has been adsorbed in reaction A. It is further noted that the release of adsorbed cations (Ca^{2+}, Mg^{2+}, K^+) tends to occur rapidly as well, reaching equilibrium within about 1 min, and the amount of cation released corresponds very closely with the amount of metal adsorbed in reaction A. Thus, although reaction A cannot be followed kinetically, information can be derived about the reaction, and it can be confidently attributed to an exchange reaction.

The slower reactions can be readily described in kinetic terms, but a mechanism of the adsorption cannot so easily be assigned. Figure 1 is typical of many results in that desorption of cations is seldom

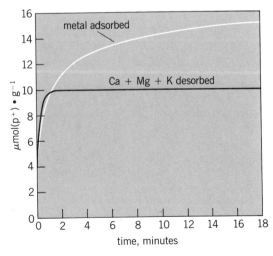

Fig. 2. Typical kinetics curves showing the time dependence of metal adsorption by soil and accompanying cation desorption from soil.

observed after the first minute or so of the reaction. Thus, retention must be by a nonexchange complexation or by physical adsorption. In a few cases, a longer-term increase in cations is observed, which may be attributable to structural replacement, rather than exchange.

Despite the inability to fully characterize reactions, the amounts adsorbed by each reaction and the reaction rates are important, particularly if attenuation of potentially toxic wastes is of interest. Whereas traditional adsorption study techniques can provide only total adsorption information, kinetics can provide time-quantity information. For example, reaction A is nearly instantaneous, so contact time for this reaction is extremely short. Likewise, the rate of reaction C is much slower, so the contact time must be correspondingly longer. Conventional adsorption studies cannot differentiate between these reactions, and if attenuation is of concern, the amount of metal adsorbed by each reaction is of major concern. Two soils having similar total adsorption capacity can differ substantially in the amount of metal adsorbed by each reaction. Therefore, the amount of metal adsorbed by rapid reactions, and not the total adsorption capacity, may be the important factor in keeping metals out of groundwater supplies. On the other hand, slow adsorption reactions are frequently also the least reversible, so metals bound by these reactions may be least available for plant uptake.

For background information *see* ADSORPTION; SOIL CHEMISTRY in the McGraw-Hill Encyclopedia of Science and Technology.

[ROBERT D. HARTER]

Bibliography: R. H. Dowdy and V. V. Volk, Movement of heavy metals in soils, *Chemical Mobility and Reactivity in Soil Systems*, Amer. Soc. Agron. Spec. Publ., 1983; W. E. Emmerich et al., Movement of heavy metals in sludge-treated soils, *J. Environ. Qual.*, 11:174–178, 1982; R. A. Griffin and J. J. Jurinak, Kinetics of the phosphate interaction with calcite, *Soil Sci. Soc. Amer. Proc.*, 38:75–79, 1974; R. D. Harter and G. Smith, Langmuir equation and alternate methods of studying "adsorption" reactions in soils, *Chemistry in the Soil Environment*, Amer. Soc. Agron. Spec. Publ. no. 40, 1981; M. B. Kirkham, Trace elements in corn grown on a long-term sludge disposal site, *Environ. Sci. Technol.*, 9:765–768, 1975; L. J. Lund, A. L. Page, and C. O. Nelson, Movement of heavy metals below sewage disposal ponds, *J. Environ. Qual.*, 5:330–334, 1976; H. M. Selim, R. S. Mansell, and L. W. Zelazny, Modeling reactions and transport of potassium in soils, *Soil Sci.*, 122:77–84, 1976; D. L. Sparks, D. C. Martens, and L. W. Zelazny, Plant uptake and leaching of applied and indigenous potassium in Dothan soils, *Agron. J.*, 72:551–555, 1980; D. L. Sparks, L. W. Zelazny, and D. C. Martens, Kinetics of potassium desorption in soil using miscible displacement, *Soil Sci. Soc. Amer. J.*, 44:1205–1208, 1980; D. L. Sparks, L. W. Zelazny, and D. C. Martens, Kinet-

ics of potassium exchange in a Paleudult from the Coastal Plain of Virginia, *Soil Sci. Soc. Amer. J.*, 44:47–40, 1980; R. J. Zasoski and R. G. Burau, A technique for studying the kinetics of adsorption in suspensions, *Soil. Sci. Soc. Amer. J.*, 42:372–374, 1978.

Soil taxonomy

Recent studies in soil taxonomy are concerned with the development of a rationale for mineral classes and the problems associated with mineral classes in the tropics.

Rationale for mineral classes. Mineralogy enters into United States soil taxonomy mostly in the fifth-highest category (family). It is used as a criterion in

Table 1. Key to mineralogy classes[a]

Determinant size fraction and definition	Class
Applicable to soils of any particle-size class	
Whole soil (< 2 mm[b]); $> 40\%$ carbonates[c] plus gypsum, and the carbonates are $> 65\%$ of the sum	Carbonatic
Whole soil (< 2 mm); $> 40\%$ chemically extractable iron oxide (Fe_2O_3)	Ferritic
Whole soil (< 2 mm); $> 40\%$ hydrated aluminum oxides [$Al(OH)_3$]	Gibbsitic
0.02–2 mm or whole soil (< 2 mm)[d]; $< 90\%$ quartz; $< 40\%$ any other mineral listed below in the table and ratio: $$\frac{\text{extractable } Fe_2O_3 \ (\%) \ + \ \text{gibbsite} \ (\%)}{\text{clay} \ (\%)^e} > 0.2$$	Oxidic
Whole soil (< 2 mm); $> 40\%$ serpentine minerals (antigorite, chrysotile, and so on) and talc	Serpentinitic
Whole soil (< 2 mm[b]); $> 40\%$ carbonates[c] plus gypsum, and the carbonates are $> 35\%$ of the sum	Gypsic
Whole soil (< 2 mm); $> 40\%$ glauconite	Glauconitic
Applicable to soils of fragmental, sandy, sandy-skeletal, loamy, and loamy-skeletal particle-size classes	
0.02–20 mm; $> 40\%$ mica	Micaceous
0.02–2 mm; $> 90\%$ silica minerals (quartz, chalcedony, opal, and so on) and other minerals extremely resistant to weathering	Siliceous
0.02–2 mm; other soils that do not contain $> 40\%$ of any one mineral except quartz and feldspars	Mixed
Applicable to soils of clayey and clayey-skeletal particle-size classes	
< 0.002 mm; $> 50\%$ halloysite (tubular forms)	Halloysitic
< 0.002 mm; $> 50\%$ kaolinite and other 1:1 tabular minerals, and $< 10\%$ smectite	Kaolinitic
< 0.002 mm; $> 50\%$ smectite or more smectite than any other clay mineral	Smectitic[f]
< 0.002 mm; $> 50\%$ illite (commonly $> 4\%$ K_2O)	Illitic
< 0.002 mm; $> 50\%$ vermiculite or more vermiculite than any other clay mineral	Vermiculitic
< 0.002 mm; $> 50\%$ chlorite or more chlorite than any other clay mineral	Chloritic
< 0.002 mm; other soils	Mixed

[a]Adapted from Soil Survey Staff, *Soil Taxonomy*, USDA Handb. no. 436, 1975.

[b]If the < 20-mm fraction contains more carbonates plus gypsum than the < 2-mm fraction, it is used as the determinant size fraction.

[c]$CaCO_3$ equivalent.

[d]The 0.02–20-mm fraction is used for quartz and other minerals; the whole soil (< 2 mm) is used for the ratio of Fe_2O_3 and gibbsite to clay.

[e]Percentage of clay or percentage of 15-bar (1.5-megapascal) water times 2.5, whichever is greater, provided the ratio of 15-bar water to clay is 0.6 or more in at least one-half of the control section.

[f]The class is currently listed as montmorillonitic in the *Soil Taxonomy Handbook* (1975), but smectitic is the recommended usage for this expandable group of minerals.

higher categories of a few taxa because of its overwhelming influence on soil behavior or because of its association with a unique set of soil properties. In such cases, a mineralogy family name would be redundant.

Family classes. In soil taxonomy, family classes are designed primarily for agricultural management and engineering interpretation. In contrast, placement in the higher categories is based on profile morphology and associated measurable properties that are believed to reflect sets of soil-forming processes. Mineral classes are groupings made according to the dominant mineral or minerals present in selected size fractions, and they do not constitute a classification of minerals in soils. In addition to mineralogy, particle size and temperature are used in family placement for most soils. Some other properties, for example, the presence of calcium carbonate, depth, and consistence, are used for family classes in selected taxa of higher categories. Identification of the subgroup (fourth-highest category), along with proper family placement, is essential for interpretation. A soil from the humid northeast and one from the arid southwest may have the same family classification, but the subgroup taxa would indicate great differences between them.

Control section. The same part of the soil profile, termed the control section, is used for both mineralogy and particle-size family placement. The part of the profile designated as the control section varies in depth and thickness according to the presence of specified horizons and features. It is part or all of the subsoil in most soils, but it is the entire profile, from the surface to a root-restricting layer, in shallow soils. A soil must first be placed in the proper particle-size class before the mineralogy class can be determined.

Class criteria. Seventeen mineralogy classes are recognized in the system for mineral (inorganic) soils, as shown in the key (Table 1). Seven classes are applied to any particle-size class. A soil is tested sequentially against the criteria in the key; it is then placed in the first class for which it meets the requirements, regardless of whether it is sandy, loamy, or clayey. If a soil meets the criteria for any of the classes in this group, it is presumed that many of its important properties will be dependent on the mineral or minerals that define the class regardless of its particle-size class. Moreover, the minerals characteristic of these seven classes can occur in any particle size fraction.

If a soil does not have the composition to meet the criteria of any of the first seven classes, and if it has a sandy or loamy texture, it will be placed in one of the next three classes, according to the mineralogy of the size fraction indicated in the table. The rationale in developing the mineral classes for these particle-size groupings is to define taxa by minerals that greatly influence soil chemical and physical behavior and tend to be concentrated in the coarser fractions (0.02–2 mm, or 0.02–20 mm in the case of micas). Soils may exist that do not contain the necessary quantity of mica or silica minerals to meet their respective class requirements and that also contain more than 40% of some mineral other than quartz or feldspars. If such is the case, an additional mineral class could be established if it is deemed to be important from a management standpoint.

The remaining seven mineralogy classes are restricted to clayey soils (those containing more than 35% clay). Only clay-size (less than 0.002 mm effective diameter) particles are used in the class determination. Each of the minerals, which characterizes an individual class, is a phyllosilicate (layer silicate) mineral and as such tends to be more physicochemically active than most other minerals of similar size. All tend to be concentrated in the clay fraction and some, smectite, for example, occur almost exclusively there. It is presumed that once the clay content reaches 35%, and if it is composed mostly of phyllosilicate minerals, it will tend to dominate the behavior of a soil. Admittedly, the percentage limit is somewhat arbitrary because of differences in the physicochemical activity of the individual minerals. Certainly, a lower percentage of smectite will dominate a soil's behavior more than will kaolinite.

Volcanic soils. Seven additional family classes, which substitute for both particle size and mineralogy, are recognized for soils that have been strongly influenced by volcanic materials (Table 2). Mineralogy classes are inappropriate because the soils consist mostly of glass, cinders, or amorphous constituents. The amorphous constituents often have a gel-like consistence. The behavior of such soils is

Table 2. Combined mineralogy and particle-size classes

Class	Description
Cindery	$\geq$ 60%* of the whole soil is volcanic ash, cinders, and pumice; $\geq$ 35%† is cinders $\geq$ 2 mm in diameter
Ashy and ashy-skeletal (feels sandy after prolonged rubbing)	
Ashy	$\geq$ 60%* of the whole soil is volcanic ash, cinders, and pumice; < 35%† is $\geq$ 2 mm in diameter
Ashy-skeletal	Rock fragments‡, other than cinders, constitute $\geq$ 35% of the soil
Medial and medial-skeletal (feels loamy after prolonged rubbing)	
Medial	< 60%* of the whole soil is volcanic ash, cinders, and pumice; < 35%† is $\geq$ 2 mm in diameter; amorphous materials dominate the exchange complex
Medial-skeletal	Rock fragments, other than cinders, constitute $\geq$ 35%† of the soil
Thixotropic and thixotropic-skeletal	
Thixotropic	Particles $\geq$ 2 mm in diameter constitute < 35%†; the fine-earth fraction exhibits thixotropic properties and amorphous materials dominate the exchange complex
Thixotropic-skeletal	Rock fragments, other than cinders, constitute $\geq$ 35%† of the soil; the class otherwise meets the requirements of the thixotropic class

*By weight.   †By volume.   ‡Particles $\geq$ 2 mm in diameter.

mostly determined by the proportion of primary volcanic particles and by secondary amorphous constituents and by their physicochemical properties, such as thixotropy.

Organic soils. Criteria used in the determination of mineral classes for organic soils differ appreciably from those used in inorganic soils. Four classes are recognized: ferrihumic, coprogenous, diatomaceous, and marly. Perhaps some term other than mineralogy classes would be more appropriate, since a variety of materials, which are not minerals in a strict sense, are included. Although it is implicit in the definition of organic soils that the organic material is most influential in the determination of soil behavior, the inorganic components may markedly affect some properties, for example, acidity. Moreover, these components tend to become more important with time because of decomposition of the organic fraction during agricultural use.

Adding classes. Mineral classes are, as indeed is the case of all the taxa in the system, open-ended, that is, classes can be added if the need arises. For example, if clayey soils are found to occur in which fibrous silicate clays (sepiolite or palygorskite) dominate the clay fraction, an appropriate class will be established. Problems have already been identified in applying the combined mineralogy and particle-size classes for volcanic soils, and additional classes will probably be established.

[B. L. ALLEN]

Tropical mineral classes. As new information and experience have been acquired by classifying soils located within the tropical areas of the world, some problems have arisen. Most of the problems relate to low-activity clays, interpretation of gibbsite, and amorphous materials associated with volcanic-ash parent materials. These concerns are more significant in the tropics than they are in the temperate zone only because of their greater areal extent in the tropics. The establishment of a new order, Andisols, is one probable consequence of this fact.

The mineralogical composition of soils is used as criteria for placement at several levels in the taxonomic system. It is specifically, and more or less uniformly, used for all soils at the fifth, or family, level of the system. Where mineralogical criteria are used in categories above the family level, the identification of mineralogy at the family level often becomes redundant. For example, soils placed in the Oxisol order must contain few weatherable minerals and have low cation-exchange capacity. This limits the number of families that can presently be identified. Gibbsite, long believed to be indicative of extreme weathering and thus used to identify oxidic and gibbsitic families, has been found to be a significant component in some relatively unweathered soils.

Oxisols. These can be considered the classical soils of the humid tropics. Although their extent is much more limited than was formerly believed, they occupy about 23% of the tropical land area. By definition at the order level, Oxisols cannot contain more than trace amounts of weatherable minerals. Thus, essentially all Oxisols containing less than 35% clay in their control section are in the siliceous family.

Oxisols also must have an apparent cation-exchange capacity, by the ammonium acetate pH 7 method, of less than 16 milliequivalents (meq) per 100 grams of clay in some part of the oxic horizon within 2 m of the surface. The cation-exchange capacity values associated with various kinds of clay minerals are as follows; kaolinite, 3–15 meq/100 g clay; illite or clay-sized mica, 10–40 meq/100 g clay; smectites, including montmorillonite, 80–150 meq/100 g clay; vermiculite, 100–150 meq/100 g clay; and chlorite, 10–40 meq/100 g clay. Mica and chlorite are easily weathered and not present in appreciable amounts in Oxisols. A clay mixture, to meet the less than 16 meq/100 g cation-exchange capacity limit, would have to have less than 10% of either montmorillonite or vermiculite. By definition a kaolinitic family must be dominated by kaolinite and contain no more than 10% montmorillonite. Thus, in effect, all but a few Oxisols in clayey families are kaolinitic.

Surface area. At the time many of the decisions were being made about how to develop soil taxonomy, the cation-exchange capacity was considered the major role of clay-sized particles in the soil. Soils that were thought to be composed almost entirely of low-activity clays were not separated by particle-size class when clay contents were above 35%. Oxisols and Ultisols with more than 35% clay were placed only in clayey families and not classed at the family level as either fine, (35–60% clay) or very fine (more than 60% clay) as were soils in the other orders.

The more clay, and consequently greater surface area, present in a soil, the greater the amount of phosphate required to overcome the phosphorus-fixation capacity of the soil (Figs. 1 and 2). It now appears desirable to utilize fine and very fine fami-

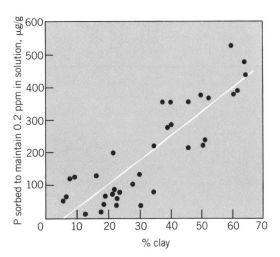

Fig. 1. Relationship between P sorption at 0.2 ppm P in solution and percent clay in kaolinite-dominated systems. (*After R. A. Pope, Use of Soil Survey Information to Estimate Phosphate Sorption by Highly Weathered Soils, Ph.D. thesis, Soil Sci. Dept., N.C. State Univ., 1976*)

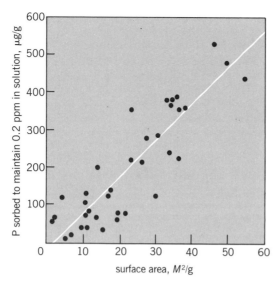

Fig. 2. Relationship between P sorption at 0.2 ppm P in solution and surface area in kaolinite-dominated systems. (*After R. A. Pope, Use of Soil Survey Information to Estimate Phosphate Sorption by Highly Weathered Soils, Ph.D. thesis, Soil Sci. Dept., N.C. State Univ., 1976*)

lies or perhaps surface-area criteria in the classification of Oxisols and Ultisols.

Sand mineralogy. In defining the oxic horizon it was considered desirable to limit the horizon to soil materials that contained only very small amounts of weatherable minerals in the sand and coarse silt fractions. The oxic horizon definition states that only traces of primary aluminosilicates such as feldspars, micas, glass, and ferromagnesian minerals be present. The cambic horizon must contain more than 3% weatherable minerals other than muscovite, or more than 6% muscovite. The cambic horizon limits are usually taken to define traces in the oxic horizon definition. However, some soils that meet all the clay requirements for cation-exchange capacity, and so on, defined in the oxic horizon, contain very little sand or silt. A few instances have been reported where mineral counts in the sand fraction of such soils have found weatherable mineral percentages well in excess of the oxic horizon limit. Thus, they can not be classified as Oxisols. Considering the original intent of the weatherable mineral limit to be one that excluded soils that had the potential to produce clay upon further weathering, it appears more logical to set the weatherable mineral content at a total amount limit, that is, % weatherable minerals $\times$ % sand + silt > some quantity. Such a rationale is presently being considered, but no firm criteria have yet been advanced. One suggestion uses an absolute upper limit of total Ca, Mg, and K content in the soil minus the exchangeable forms. This appears to be a plausible solution.

Oxidic families. Oxidic families are defined as soils that contain less than 90% quartz in the 0.02-to-2 mm size fraction; less than 40% of either serpentine minerals, carbonates plus gypsum, or glauconite; and a ratio of whole soil extractable Fe_2O_3 + gibbsite/% clay ≥ 0.2.

Although the rationale for this family appears lost in committee archives, it seems plausible to assume it was an attempt to set out families of Alfisols and Ultisols that had mineralogies considered tending toward Oxisols. Recent studies, both in the tropics and in temperate areas, have shown gibbsite to form from the direct alteration of feldspar, and subsequently to silicate to form halloysite and kaolinite. Thus, the presence of oxidic families has not been found to be in geomorphic positions bordering Oxisols, but quite the opposite. A substantial number of soils with oxidic mineralogy occur among the Inceptisols of the northeastern United States and only infrequently in older and more developed Paleudults and Paleudalfs. Also, since phosphorus fixation is shown to relate to clay content, a ratio of iron to clay has poor predictability of phosphorus fixation (Fig. 3). Clay-content limits or surface-area criteria may prove to be more satisfactory criteria for defining some soil families.

Volcanic ash. Soils formed from volcanic ash, while not confined to tropical areas, present unique mineralogical challenges. Many of them contain no discrete clay-sized particles but rather amorphous or gellike compounds of silicon and aluminum with varying amounts of cations and anions. Such minerals are highly soluble and chemically reactive.

The most common method of analysis is to subject the soil to dissolution with empirically developed concentrations of strong bases and short boiling times. The amounts or ratio of silicon and aluminum brought into solution has been used to characterize different kinds of volcanic ash soils. It has been observed for several years that when there was a high concentration of aluminum in the soluble amorphous fraction the pH value of the soil in 1 N NaF solution was higher than when the amount of soluble aluminum was low. Efforts are presently being made to utilize NaF pH measurements, which are easy and reli-

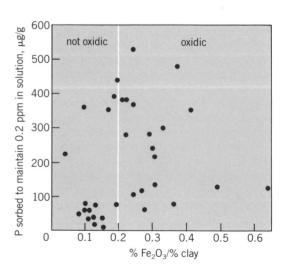

Fig. 3. Relationship between P sorption at 0.2 ppm P in solution and the free iron-to-clay ratio. (*After R. A. Pope, Use of Soil Survey Information to Estimate Phosphate Sorption by Highly Weathered Soils, Ph.D. thesis, Soil Sci. Dept., N.C. State Univ., 1976*)

able, as criteria to classify Andepts, or the proposed new order equivalent, Andisols. The most probable limit for the new order will be pH > 9.4 in 1 N NaF. This value also seems to serve well in predicting which soils developed from volcanic ash are capable of fixing large quantities of fertilizer phosphorus.

Summary. No unique problems of soil mineralogy classification are anticipated in tropical areas. The major concern is that some of the present groups are too broad to satisfactorily separate some extensive soils in tropical areas. Any of the anticipated changes in soil taxonomy will affect like soils in the temperate zone, but the area of soils affected will be less than in the tropics.

For background information *see* SOIL; SOIL CHEMISTRY in the McGraw-Hill Encyclopedia of Science and Technology.

[STANLEY W. BUOL]

Bibliography: B. L. Allen, Mineralogy and soil taxonomy, in J. B. Dixon and S. B. Weed (eds.), *Minerals in Soil Environments*, pp. 771–795, 1977; C. S. Calvert, S. W. Buol, and S. B. Weed, Mineralogical characteristics and transformations of a vertical rock-saprolite-soil sequence in the North Carolina Piedmont: I. Profile morphology, chemical composition and mineralogy; II. Feldspar alteration products—Their transformations through the profile, *Soil Sci. Soc. Amer. J.*, 44:1096–1112, 1980; H. Eswaran and W. C. Bin, A study of a deep weathering profile on granite in Peninsula Malaysia: I. Physicochemical and micromorphological properties, *Soil Sci. Soc. Amer. J.*, 42:144–149, 1978; R. A. Pope, *Use of Soil Survey Information to Estimate Phosphate Sorption by Highly Weathered Soils*, Ph.D. Thesis [Univ. Microfilms Int., 77–29, 631], Soil Sci. Dept., N.C. State Univ., 1976; Soil Survey Staff, *Soil Taxonomy Handbook*, USDA Handb. no. 436, 1975.

Soliton

In the theory of elementary particles, it is widely accepted that the fundamental building blocks of matter are fractionally charged particles termed quarks. Relative to the charge of an electron $-e$, the quarks have charge $\pm e/3$ or $\pm 2e/3$. Elementary particles, such as the proton and neutron, are formed as bound states of three quarks, while mesons are formed as a bound quark-antiquark pair. It has been found that only quark combinations which have net integer charge appear in nature, so that it is impossible to isolate a free quark. At present, no fundamental theory accounts for the fractional quark charges; however, two recent developments in apparently unrelated areas of research have shown how particles or excitations of fractional charge can arise in systems composed of integer charge constituents. In 1976, study of relativistic field theory models led to the discovery that under suitable circumstances excitations termed solitons could carry a fractional fermion number 1/2, analogous to a fractional charge $e/2$. In 1979, it was discovered that a model proposed to account for the properties of the linear conducting

polymer polyacetylene, $(CH)_x$, shows precisely the same behavior discovered in relativistic field theory, except that the fractional charge $\pm e/2$ is masked by a doubling of charge due to the two spin states of the valence electrons. Recently, it has been shown that this model, when applied to a one-dimensional conductor having an average of two electrons per three monomers, leads to solitons of charge $\pm 1/3e$, $\pm 2/3e$, and $\pm 4/3e$. Experiments are under way to establish the existence of such excitations.

Commensurate charge density waves. According to quantum theory, it is impossible to split an electron into two distinct pieces. If elementary particles are indivisible and have integer charge, how is it possible to create excitations of fractional charge from a system of such particles? To understand this phenomenon, consider a one-dimensional chain like $(CH)_x$ having one partly occupied π orbital per site. The system has an average of one electron per monomer. R. F. Peierls has shown that such a one-dimensional system is unstable with respect to a distortion of the chain such that bond lengths alternate along the chain. By translational symmetry of the chain, it is clear that the system has two ground states, termed A and B phases, of equal energy, corresponding to the bonds to the right of even-numbered sites being shorter than those to the right of odd-numbered sites, or vice versa. In chemical terms this corresponds to alternating double and single bonds, although the distortion in $(CH)_x$ is not large enough to form discrete π electron states corresponding to truly localized double bonds.

In Fig. 1, a region on the left is shown in the A phase and on the right in the B phase. The boundary between these two domains is a soliton, a term denoting an excitation which preserves its shape in a nonlinear system. As the soliton moves to the right, it converts A into B phase. A qualitative under-

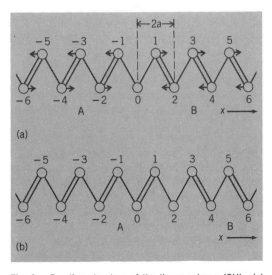

Fig. 1. Bonding structure of the linear polymer $(CH)_x$. (*a*) Configuration in which region on left is in A phase and region on right is in B phase. The center is a soliton acting as a domain wall. (*b*) Bonding configuration after the soliton has moved one unit cell to the right.

standing of the origin of fractional charge follows by considering the charge flow accompanying a displacement of the soliton from site 0 to site 2, that is, a distance $2a$, where a is the average horizontal spacing between sites. The double bond between 1 and 2 flips to become a double bond between 0 and 1. Since the bond which has flipped carried two electrons, one of up spin and one of down spin, the net motion corresponds to a charge $2e$ being transported over a distance a, giving a dipole moment specified by Eq. (1). Since the effective charge Q of a particle is defined as the rate of change of dipole moment with displacement of the particle, the effective charge of a soliton in $(CH)_x$ is given by Eq. (2). If only electrons of up spin are considered, p becomes ea and Eq. (3) is valid. Thus, fractional charge arises from a rearrangement of the charge in the ground state (vacuum) which occurs as the soliton moves through space. In the above simplified view, the width of the soliton is taken to be of order a, while for $(CH)_x$ the width is actually of order $14a$. Nevertheless, the charge-per-spin orientation continues to be given by Eq. (3).

$$2ea = p \qquad (1)$$

$$Q = \frac{p}{2a} = e \qquad (2)$$

$$Q = \frac{ea}{2a} = \frac{e}{2} \qquad (3)$$

Energy states and vacuum charge flow. A complementary view is illustrated in Fig. 2. For either the A or B phase, the allowed electronic energies E_k are shown in Fig. 2a, with the valence band v having all states doubly occupied (one up-spin and one down-spin electron) and the conduction band c unoccupied. For an infinitely long chain the allowed energies E_k form a continuum in each band. Since there is one π electron per site whose charge is balanced by the charge of the $(CH)^+$ ion, the charge and spin density in the perfect A or B phase is 0, when averaged over a unit cell, that is, a length a along the chain.

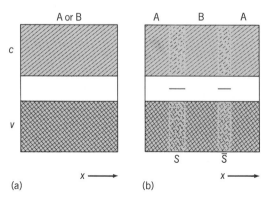

(a)   (b)

Fig. 2. Allowed electronic energy states of $(CH)_x$. (a) A or B phase. Valence band v is fully occupied by two electrons per state, while conduction band c is empty. The spectrum is the same for the A or B phase. (b) Electronic states distorted by the presence of a soliton S and an antisoliton $\bar{S}$.

Consider a long $(CH)_x$ chain in the A phase with the nuclear displacements deformed so that a segment near the center of the chain is placed in the B phase. This process forms a soliton S on the left and an antisoliton $\bar{S}$ on the right, separating the B segment from the rest of the A-phase chain. In the vicinity of S, the nuclear displacements appear as in Fig. 1a, while near $\bar{S}$, the displacements look like the mirror image of this pattern, the mirror plane passing through 0. The electronic states are distorted by the creation of S and $\bar{S}$, as illustrated in Fig. 2b. In the vicinity of S, the amplitude of each electronic wave function, $\psi_k(x)$, is slightly reduced, both in the valence and in conduction bands. Explicit calculations show that the net deficit in the number of π electrons in the valence band in the vicinity of S is 1/2 for a given spin orientation. Therefore, summing over spin orientations, the charge on the soliton is $Q = 2 \times (\frac{1}{2}e) = e$. The same situation holds for the antisoliton, and its charge is also e.

Because of the requirements imposed by quantum mechanics on such systems, the total number of electronic states is conserved when S and $\bar{S}$ are created. For a given spin orientation, one half a state is lost from the valence band near S and one half near $\bar{S}$. Thus, a total of one state is missing from v, and similarly one state is missing from c when S and $\bar{S}$ are present. The two missing states appear as states near the center of the energy gap, as shown in Fig. 2b. Since these are the lowest-energy unoccupied states, if an electron is added to the system, it will occupy the gap center state at S or at $\bar{S}$, changing the charge from $+e$ to 0. The added electron has spin 1/2 so a neutral soliton carries spin 1/2. Finally, if a second electron is added to the gap center state at S or at $\bar{S}$, the charge becomes $-e$ and the net spin is again 0. The soliton charge and spin assignments are given by Eqs. (4). This is the reverse of the relationships for conventional electrons and holes in solids, given by Eq. (5). Experiments on $(CH)_x$ are consistent with the peculiar solitonic charge-spin assignments given by Eqs. (4).

$$Q = 0, \, S = 1/2 \qquad (4)$$
$$Q = \pm e, \, S = 0$$

$$Q = \pm e, \, S = 1/2 \qquad (5)$$
$$Q = 0, \, S = 0$$

The above results also agree with conclusions derived from studies of relativistic field theory models, except for the restriction to one spin orientation in those studies. Also, while the nonrelativistic theory can be treated in a fully self-consistent manner (so-called dynamically broken symmetry), it has not been possible to achieve this to date in the relativistic theory.

Trimerized chains and fractional charge. To circumvent the spin masking of fractional charge, consider a chain having two π electrons per three monomers on average, as shown in Fig. 3a. Peierls's theorem again states that the chain is unstable with respect to length distortions, with the lowest energy

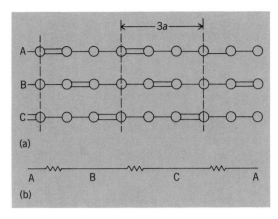

Fig. 3. Trimerized chain having two electrons per three monomers. (a) Three bonding configurations A, B, and C having equal energy. (b) Three solitons S_{AB}, S_{BC}, and S_{CA} separating A, B, C, and A regions of the chain. These solitons have fractional charge.

state having three monomers per repeat distance $3a$ along the chain. It follows from symmetry that the three bonding configurations denoted by A, B, and C have equal energy. Consider a long chain, initially in the A phase. The system has zero charge and spin in each unit cell. Suppose the bonding pattern is slowly distorted so that the left-hand portion of the chain remains A, while a domain wall or soliton is distorted into the B-phase domain on the right, as shown in Fig. 3b. Further, suppose that the system is distorted to the C phase and finally to the A phase again, on the far right. In essence, referring to Fig. 3a, the double bond has moved from the first location in the unit cell in A, to the second location in B, to the third location in C, and to the first location again in A. However, the double bond on returning to the A phase has undergone a net displacement of one unit cell. Since a double bond consists of two electrons, each of charge $-e$, a total charge $-2e$ has been removed from the chain segment containing the three solitons S_{AB}, S_{BC}, and S_{CA}. By conservation of electric charge it follows that the charges of the solitons satisfy Eq. (6).

$$Q_{AB} + Q_{BC} + Q_{CA} = +2e \qquad (6)$$

Finally, since the three solitons are equivalent from translational symmetry, it follows that their charges are equal, and the charge of a soliton is $Q = 2e/3$. As for $(CH)_x$, gap states occur, and if one adds one or two electrons to a gap center state the charge becomes $-e/3$ or $-4e/3$. Similarly, if instead of the distortion in Fig. 3b, the sequence ACBA is created from left to right, then the double bond is moved to the left one unit cell in the sequence so the charge is $Q = -2e/3$ for these antisolitons. Thus, Eqs. (7) are valid. This charge-counting type of proof has

$$Q = \frac{+2e}{3}, \frac{\pm 4e}{3}; S = 0 \qquad (7)$$

$$Q = \frac{\pm e}{3}; S = 1/2$$

been used to derive noninteger charges for solitons in a variety of relativistic field theories, for which the original proof based on charge-conjugation symmetry does not apply.

The question of whether it is possible to account for the fractional charge of quarks by a theory of integer-charged fields which have fractionally charged solitons arises in this regard. While this question is unresolved at present, it appears unlikely that such a goal can be achieved in a straightforward manner. An obstacle to such a program is the difficulty of constructing solitonlike solutions of the proper type having finite energy without introducing extra gauge fields which would probably produce unobserved interactions.

Another issue is whether the fractional charge Q of solitons is a sharp quantum observable such that each experiment would measure the same value Q. Alternatively, the charge might be a fluctuating quantity whose statistical average is fractional, with the result of any given experiment being an integer. It has now been proved that solitonic fractional charge is in fact a sharp, nonfluctuating observable.

There are indications that the methods used to deduce noninteger values of charge may lead to nontraditional values of other quantum numbers such as orbital and spin angular momentum.

For background information *see* MOLECULAR ORBITAL THEORY; QUARKS; SOLITON in the McGraw-Hill Encyclopedia of Science and Technology.

[J. ROBERT SCHRIEFFER]

Bibliography: J. Goldstone and F. Wilczek, Fractional quantum numbers on solitons, *Phys. Rev. Lett.*, 47:986–989, 1981; R. Jackiw and C. Rebbi, Solitons with fermion number 1/2, *Phys. Rev. D*, 13:3398–3409, 1976; M. J. Rice, Charged π phase kinks in lightly doped polyacetylene, *Phys. Lett.*, 71A:152–154, 1979; W. P. Su, J. R. Schrieffer, and A. J. Heeger, Solitons in polyacetylene, *Phys. Rev. Lett.*, 42:1698–1701, 1979, and Soliton excitations in polyacetylene, *Phys. Rev. B*, 22:2099–2111, 1980; W. P. Su and J. R. Schrieffer, Fractionally charged excitations in charge-density-wave systems with commensurability 3, *Phys. Rev. Lett.*, 46:738–741, 1981.

Solution geochemistry

Solution geochemistry is concerned with the chemical interaction of aqueous fluids with mineral assemblages, that is, with rocks. The effect of such interactions is dynamic, in that selective leaching or precipitation of chemical elements by the fluid can change or alter a mineral assemblage, whereas the nature of the mineral assemblage can in turn exert control over the chemical composition of the fluid. Interaction between rocks and fluid is quite common and widespread in nature and occurs in sedimentary, diagenetic, metamorphic, and to some extent, igneous environments. Aside from plastic flow of solids and the flow of liquid magma or lava, transport of material in most geologic systems is accomplished by the movement of dissolved species in aqueous fluid.

Important applications of solution geochemistry are in the formation of hydrothermal ore deposits, metasomatism, precipitation of chemical sediments, alteration of rock textural properties during diagenesis, pollution of ground and surface water by toxic waste, and development and utilization of geothermal energy. It should be evident, therefore, that the science of solution geochemistry encompasses a very broad range of endeavors. This article will be concerned with some recent advances in the study of high-temperature systems, that is, those systems exclusive of sedimentary environments.

Experimental studies. The first experimental studies of the solubilities of common rock-forming minerals in pure water at elevated temperatures (T) and pressures (P) were carried out in the early 1950s. It was demonstrated that most minerals have very low solubilities in pure water, on the order of 0.5% or less total dissolved solids, even at P-T conditions up to 2000 bars (200 megapascals) and 600°C. In the late 1950s it was established experimentally that mineral solubilities were greatly enhanced in chloride-bearing aqueous fluids, and the majority of experimental studies since that time have been carried out in chloride-bearing systems. Natural analogs of such fluids are sea water, oil field brines, and some hydrothermal fluids.

Experimental studies undertaken since the mid-1970s have relied on two different approaches that can be classified roughly as the mineral solubility/speciation approach and the water-rock interaction approach. Both approaches provide valuable information and have certain advantages and disadvantages, depending upon the type of information desired.

Mineral solubility/speciation. These studies are carried out in simple systems that are chemically and mineralogically well characterized, usually involving only a single mineral or simple mineral assemblage. The determination of the concentrations of a small number of elements in the fluid that have achieved chemical equilibrium with the mineral assemblage and the identification of the chemical species by which the elements occur in the fluid are the primary goals of such studies. For example, concentrations of total aqueous magnesium in chloride-bearing solutions that equilibrated with the mineral assemblage talc [$Mg_3Si_4O_{10}(OH)_2$] and quartz (SiO_2) in the P-T range 400–700°C, 1–2 kilobars (100–200 MPa) have been determined experimentally. One of the major goals of the study was to determine the chemical species (often called complexes) in which Mg exists in the fluid. Several species, including Mg^{2+}, $MgCl^+$, and $MgCl_2^0$ are possible. Without knowledge of the identity and amount of each complex present in the fluid at given P-T conditions, the applicability of the results are very limited. If the species can be identified, however, the results can be used to calculate solubilities of other magnesium-silicate, -oxide, -carbonate, and -hydroxide minerals.

It is generally known that a large group of chemical substances, termed electrolytes, dissociate into ions surrounded by a sheath of water molecules when dissolved in water, and are thereby capable of conducting electricity. At room temperature and pressure, a wide variety of alkali, alkaline earth, and transition metal chlorides, nitrates, and hydroxides behave in this manner (for example, NaCl, KCl, KOH, CaCl$_2$). With increasing temperature, however, the changing electrostatic properties of water tend to favor the formation of complexes by mechanisms such as reaction (1) or, in the case of 2:1 electrolytes, a stepwise mechanism, reaction (2). The

$$Na^+(aq) + Cl^-(aq) \rightarrow NaCl^0(aq) \qquad (1)$$

$$Mg^{2+}(aq) + Cl^-(aq) \rightarrow MgCl^+(aq) \qquad (2a)$$
$$MgCl^+(aq) + Cl^-(aq) \rightarrow MgCl_2^0(aq) \qquad (2b)$$

existence of such reactions is clearly documented by measurement of electrical conductances of solutions at elevated pressure and temperature. As a general rule, uncomplexed, ionic species dominate at temperatures below approximately 300°C, whereas at temperatures above approximately 500°C, neutral complexes dominate. The effect of increasing pressure on complexing is opposite that of temperature; that is, increased pressure tends to favor dissociation to ionic species. The magnitude of the pressure effect, however, is considerably less than that of temperature, so that under most geologic conditions the major effect is that of temperature.

Knowledge of the complexing behavior of a given element, combined with knowledge of its overall concentration in a solution as determined by mineral solubility experiments, is sufficient to determine the absolute concentration of each species in the fluid. From Fig. 1, which shows the distribution of Mg species in equilibrium with talc and quartz over a

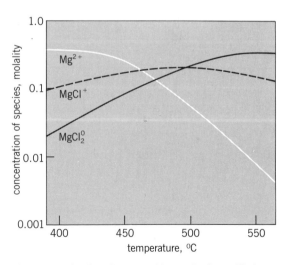

Fig. 1. Distribution of aqueous Mg species in equilibrium with the mineral assemblage talc plus quartz, as a function of changing temperature. Pressure = 2 kilobars (200 kilopascals); total concentration of chloride in the fluid is fixed at 2 moles per kilogram of H$_2$O. (*After J. D. Frantz et al., Mineral-solution equilibria:V. Solubilities of rock-forming minerals in supercritical fluids, Geochim. Cosmochim. Acta, 45:69–77, 1981*)

range of temperatures at fixed pressure, it is evident that the proportions of the species are strongly temperature-dependent. At low temperatures the uncomplexed, ionic species dominate, at intermediate temperatures the monochloride complex dominates, and at high temperatures the dichloride complex dominates. Once the concentrations of the individual species are known, data concerning the thermodynamic properties of the species can be extracted and used to calculate mineral solubilities and species distributions for a large number of other Mg minerals.

Additional mineral-solubility studies in simple systems containing the other major rock-forming elements—Ca, Na, K, Fe, Si, and Al—have been carried out in several laboratories and have provided information for the calculation of diagrams for complex mineral systems (Fig. 1). The chemical nature of chloride-bearing fluids in many rock systems can thereby be reasonably deduced.

Hydrothermal ore deposits form by precipitation of ore minerals from aqueous fluid. The mineral solubility/speciation approach has been directed toward a number of major ore minerals (mainly sulfides and oxides). As a result, understanding of complexing and transport of metals such as Zn, Cu, Cd, Pb, Fe, and Hg in solution has increased significantly in recent years. Complexing of metals in chloride-bearing fluids that contain sulfur is more difficult to understand because of the possibility of sulfur complexes in addition to chloride complexes. In the case of Cu, for example, species such as $Cu(HS)_3^{2-}$, $Cu(HS)_2^-$, and $Cu(HS)_2(H_2S)^-$ might be dominant under certain conditions. Experimental studies have shown, however, that with the exception of Hg, chloride complexes probably dominate over those of sulfur for the metals listed above, under most geologically reasonable conditions of temperature, pH, and total sulfur content of the fluid. The role of organic complexing in metal transport is still poorly understood.

Water-rock interaction. These studies are carried out on systems of much greater chemical and mineralogical complexity than are mineral solubility studies. In a typical experiment, a natural rock would be mixed with natural sea water and subjected to conditions in the range 100–500°C, 0.5–1 kbar (50–100 MPa). Chemical equilibrium cannot be documented in systems of such complexity, but the experiments can delimit the general trends of element precipitation and leaching, as well as the mineralogical changes in the solid. Thus, the ability of these experiments to predict the reactions of other rock types is minimal compared to mineral solubility studies, but because they deal with natural materials, they are more representative of natural processes.

Water-to-rock ratio. Experimental results from a number of research groups have documented that striking changes in both fluid chemistry and mineralogy occur when sea water interacts with hot rocks. A critical variable in determining the final fluid chemistry is the so-called water-to-rock ratio (W/R),

which represents the relative proportions, by mass, of water to rock in the experiments. Sea-water-dominated systems have been defined as those in which W/R ≥ 50, whereas rock-dominated systems are those in which W/R < 50. The two types of systems are characterized by distinctly different fluid compositions. Figure 2 depicts the changes in concentrations of several major elements between normal sea water and sea water that has reacted with basalt at 300°C, 1 kbar (100 MPa). In the rock-dominated system, the fluid changes from a slightly alkaline, oxygenated Na, Mg, SO_4, Cl solution (that is, normal sea water) to a reducing, neutral-to-alkaline Na, Ca, Si, Cl solution. The final fluid in the sea-water-dominated system is acidic and is considerably richer in the minor metals, with the exception of Ca.

The chemical differences between the two types of systems lie in the role of Mg. During interaction

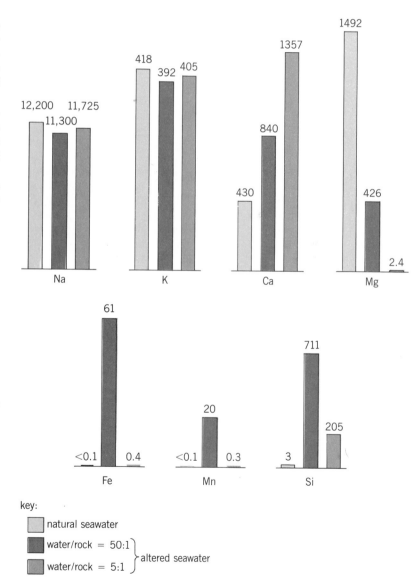

Fig. 2. Comparison of fluid compositions between natural sea water and sea water that has been reacted with basalt at 300°C and 1 kilobar (100 kilopascals) for lengths of time sufficient to produce steady-state fluid compositions. Units are in parts per million. (*After P. A. Rona and R. P. Lowell, Seafloor spreading centers: Hydrothermal systems, Benchmark Papers in Geology, no. 56, Dowden, Hutchinson, and Ross, Inc., 1980*)

with the rock, Mg^{2+} from the original sea water, in the form of $Mg(OH)_2$, enters the crystal structure of certain clay minerals (mainly smectites) formed by alteration of the solid. The OH required by the reaction is derived from the H_2O in the system, thereby liberating H^+ to solution. The H^+ is then consumed by hydrolysis reactions which leach metals from the solid. In rock-dominated systems, Mg in the sea water is rapidly exhausted, and the liberated H^+ is rapidly consumed by hydrolysis reactions. In the sea-water-dominated systems, however, the supply of Mg is much greater, so that the amount of H^+ liberated by the formation of smectite keeps pace with the amount consumed by hydrolysis, thereby maintaining an acidic pH and liberating larger amounts of metals from the solid.

The application of water-rock experiments to natural systems is often complicated by the uncertainty of the water-to-rock ratio for natural processes. The effective ratio depends upon factors which are often unknown, such as rock permeability, rate of fluid flow, reaction rates, and exposed surface area of the affected rock body. However, the general conclusion that circulation of sea water through hot rocks results in profound mineralogical and chemical changes has been well demonstrated.

Natural examples. Experimental sea water–rock interaction studies suggested that elements such as Cu, Fe, and Zn could be liberated from typical igneous rocks and concentrated to ore-forming levels by simply reacting water and rock at elevated temperatures. There was little direct evidence from nature, however, to verify the hypothesis. That such processes do occur naturally was demonstrated in the late 1970s and 1980s when submersible vessels of the FAMOUS and RISE oceanographic projects investigated midocean ridges (sites of submarine basaltic volcanism) in the Pacific Ocean at the Galapagos spreading center and the East Pacific Rise.

On the East Pacific Rise, two types of vents, characterized as warm-water vents and sulfide-mound hot-water vents, were observed discharging hydrothermal fluids directly from the basaltic sea floor. The warm-water vents emit warm, clear water from small cracks and fissures in the sea floor. Exit temperatures as high as 23°C have been recorded for water from these vents, compared to the 2°C temperature of the surrounding sea water. The hydrothermal activity at the hot-water vents is much more dramatic. Fluid is discharged from a number of discrete chimneys, 1–5 m in height, built up on seafloor mounds. Black-smoker chimneys vent the hottest fluids (up to nearly 400°C) at velocities of several meters per second. The "smoke" consists of suspended particulate matter, such as pyrrhotite (FeS), pyrite (FeS_2), sphalerite (ZnS), and chalcopyrite ($CuFeS_2$), precipitated from the fluid. The chimneys and mounds consist of similar material that has settled from the venting fluid. White smokers are generally covered with worm tubes and vent cooler (32–330°C), milky-white fluid at lower velocities. Worm tube particles, pyrite, barite ($BaSO_4$), and amorphous silica (SiO_2) impart the milky-white appearance to the water.

Both warm- and hot-water vents are characterized by a unique and unusual fauna which includes galatheid crabs, large (25-cm-diameter) white clams, white-tubed serpulid polychaetes that often encrust the surfaces of white smokers, and giant tube worms up to 3 m in length. The fauna are of considerable interest to biologists in that the ecosystem is totally independent of photosynthesis and is driven by geothermal energy provided by the hydrothermal fluid.

The "plumbing" system that drives the hydrothermal vents is reasonably well understood. Cold sea water enters fractured basalt on the flanks of midocean ridges and percolates downward to depths of at least several kilometers. The sea water is heated to relatively high temperatures (at least 400°C in the case of fluid from black smokers) and reacts with hot basalt in a manner analogous to the water-rock experiments. Thermal convection drives the altered sea water back toward the surface along the ridge axis, where it is discharged through the sea-floor vents. Mixing with cooler, descending sea water may occur near the surface. Even though the submarine vents have only recently been discovered, geologic and geophysical evidence indicates that this is a large-scale process, occurring at midocean spreading centers throughout the world. The volume of sea water interacting with rock in this manner may be so large that the process may have considerable control on the chemical balance of the oceans.

[ROBERT K. POPP]

Bibliography: H. L. Barnes, Solubilities of ore minerals, in H. L. Barnes (ed.), *Geochemistry of Hydrothermal Ore Deposits*, 1979; J. D. Frantz, R. K. Popp, and N. Z. Boctor, Mineral-solution equilibria:V. Solubilities of rock-forming minerals in supercritical fluids, *Geochim. Cosmochim. Acta*, 45:69–77, 1981; P. A. Rona and R. P. Lowell, *Seafloor Spreading Centers: Hydrothermal Systems*, Benchmark Papers in Geology, no. 56, 1980; F. N. Spiess, East Pacific Rise: Hot springs and geophysical experiments, *Science*, 207(4438):1421–1432, 1980.

Sound

In a medium in which the electrons are not free to flow, the only motion possible is a vibration of the constituent ions or molecules. Such substances are called dielectrics, and their oscillations can be thought of as sound waves bouncing around inside the medium. Typical dielectrics are quartz and liquid and solid helium. Temperature is a measure of the random internal energy of a material. In a dilute gas the temperature scales to the mean kinetic energy of the randomly colliding gas molecules, whereas in a dielectric the temperature is a measure of the intensity of the sound waves randomly propagating inside the medium in equilibrium. The fact that the equilibrium heat capacity follows the Debye form given by the cube of the absolute temperature is experimental evidence that the internal energy of dielectrics is indeed determined by sound waves.

An understanding of the details of sound propagation could lead to an explanation of the energy flow (or transport properties) in such a substance. The basic quantity of interest is the attenuation coefficient of sound or the rate at which energy in an imposed wave is absorbed by the medium. From this coefficient, as a function of frequency, the thermal conductivity and other transport properties can be determined. As transport occurs only in the presence of interactions, nonlinearities become important. It has recently become apparent that by use of the methods of macroscopic nonlinear acoustics the absorption coefficient can be calculated in the quantum as well as the classical limits.

Role of nonlinearities. For waves of sufficiently small amplitude or intensity the equations of motion of the sound field are linear so that the principle of superposition applies. In this limit sound waves run through each other without interacting, and thus the propagation of one sound wave is not affected by the presence of other waves. In general, however, the equations of state and the requirements imposed by galilean relativity cause the equations of motion to become nonlinear at finite amplitude. Nonlinearities cause a propagating wave to interact with other waves to scatter energy out of the original channel (or wavelength), and with itself to distort and form a shock front. Dynamic or transport properties such as the attenuation of sound and thermal conduction are due to the interaction of sound waves with the random background of oscillation, and thus knowledge about the first of these two phenomena is essential to understanding the off-equilibrium behavior of dielectrics.

Scattering of sound by thermal noise. The distance which a thermally excited or driven sound wave propagates before decaying to $1/e$ ($= 1/2.7$) of its original amplitude determines the mean free path of the excitation, a quantity of fundamental interest. In the absence of nonlinearities there is no decay. In the presence of nonlinearities a given wave will interact with other waves to create sum and difference frequencies of the original wave and the various components of the random thermal motion. This happens because nonlinear effects lead to new sound waves whose strength is determined by the product of a given wave, the wave with which it interacts, and the coefficient of nonlinearity. It is well known from trigonometry that a product of two cosines can be written as a superposition of waves at the sum and difference of the original arguments, as in Eq. (1) where the arguments are ωt and $\omega' t$. For an isotropic one-component fluid the coefficient of nonlinearity G is given by Eq. (2), where ρ and c are the equilibrium density and speed of sound respectively. The energy created at the sum and difference frequencies is removed from the original wave and leads to an attenuation γ of sound due to high-frequency noise (in this case the thermal internal energy) given by Eq. (3). Here, $k = 2\pi/\lambda$, where λ is the wavelength of sound, and $U(T)$ is the total

$$\cos \omega t \cos \omega' t = \tfrac{1}{2}[\cos (\omega + \omega')t + \cos (\omega - \omega')t] \tag{1}$$

$$G = 1 + (\rho/c) \, dc/d\rho \tag{2}$$

$$\gamma(k) = (\pi/2\rho c)G^2 U(T)k \tag{3}$$

thermal internal energy. A more involved equation describes attenuation due to an arbitrary spectrum of noise. Equation (3) was derived by L. D. Landau and G. Rumer in 1937 using the approach of quantum perturbation theory. P. Westervelt (1976) emphasized that these results are an immediate consequence of classical nonlinear acoustics.

The above results apply to semidispersive systems where frequency ω is linear in wave number k, so that $\omega = ck$. In the event that ω increases faster than the first power of k, the dispersion is anomalous and the nonlinear acoustics yields the above attenuation formula multiplied with a factor of 2. For the opposite case, or so-called normal dispersion, the attenuation of sound to this order is zero. In liquid ^{4}He the dispersion switches from anomalous to normal as the pressure is increased through about 18 atm (1.8 megapascals) and a large drop in attenuation is observed.

Scattering of sound by quantum noise. In addition to thermally excited vibrations, any oscillator has a zero-point motion which is due to the fundamental quantum of action h as required by the uncertainty principle. This random motion, which persists even at 0° absolute, is also capable of scattering sound. However, in applying nonlinear acoustics to calculate the attenuation of a sound wave due to this mechanism those processes which remove energy from the zero-temperature noise must be deleted, as the energy at this temperature is a minimum. The only nonlinear process which meets this criterion involves sound scattering from zero-point noise of lower frequency through the production of difference frequencies with the incident wave. The attenuation due to this effect, γ_0, represents a spontaneous decay of sound waves into a spectrum of subharmonic frequencies.

Restituting collisions. While sound is being scattered out of a given wave number due to nonlinear effects, interactions within the background are continuously pumping energy back into that channel. For instance, by focusing on the sound energy in the channel with wave number k, two sound waves with wave numbers $k-q$ and q in the background of random motion will interact through sum processes so as to restitute energy into channel k. The equilibrium distribution of sound energy is achieved when the rate at which energy is scattered out of a channel is balanced by the rate at which it is restored to that channel. This determines the equilibrium spectral intensity of sound, as given by expression (4). Here $\beta = 1/k_B T$, where k_B and T are Boltzmann's constant and temperature, and α is an otherwise undertermined constant, which this experiment sets equal to h. The spectral intensity is the sound energy in a unit range of wave number. In this fashion the classical nonlinear acoustics possesses a configuration whose stationary distribution is that of Planck's law plus harmonic zero-point motion.

$$\frac{4\pi\alpha k^3 c}{(2\pi)^3}\left(\frac{1}{2} + \frac{1}{e^{\beta\alpha\omega} - 1}\right) \tag{4}$$

Spontaneous decay of sound. The term involving ½ in Eq. (4) is the zero-point energy. It determines an attenuation of sound at low temperatures ($\beta hkc \gg 1$) given by Eq. (5). Though first derived in 1937 through use of quantum perturbation theory, such a formula was tested for the first time in 1981 in an experiment in which high-frequency sound was generated in doped calcium fluoride (CaF_2) through the use of laser pulses to stimulate nonradiative transitions. The sound waves were detected through the fluorescence they induced in the stress-split energy levels of the Eu^{2+} doping ions. Phonon lifetimes were obtained from the time dependence of the fluorescence radiation and followed the k^{-5} law predicted by Eq. (5).

$$\gamma_0 = \frac{G^2 hk^5}{960\rho\pi^2} \tag{5}$$

Superfluidity and other applications. The macroscopic theory of acoustics possesses solutions of both long and short wavelengths. In addition to the collisions of waves, the nonlinearities lead to a refraction (geometrical acoustics) of the high-frequency waves by the slow (or background) modulations. Finally, the background feels a reaction force due to the bending of the high-frequency waves. A theory which includes all of these effects for nonisotropic distributions of noise has been developed. The stationary distribution of noise for the general case possesses a new additive conserved property of the motion which is related to the momentum of the noise relative to the background. In acoustics this quantity is sometimes referred to as the Stokes drift. Although the basic equations of continuous media obviously conserve the total momentum of the noise plus background, the nonlinear theory developed from it possesses this extra conservation law which can be viewed as a symmetry breaking.

When the thermal noise is sufficiently weak that the Stokes drift can be nonzero, classical hydrodynamics determines the time development of this quantity. The resulting equations are those of the two fluid hydrodynamics of superfluid ^{4}He below the lambda temperature (2.17 K). Superfluids are of particular interest because temperature pulses travel by waves rather than by conduction, and because of their similarities to superconductors.

Nonlinear methods have been extended to include systems which are nonlocal as well as dispersive. They have also been applied to nonlinear Maxwell's equations (vector fields). The renormalizations which must be applied to nonlinear scattering from quantum noise may be important for understanding situations in superconductivity where quantum mode mixing occurs. In any event it is apparent that macroscopic nonlinear acoustics can be used for many systems to understand properties previously thought to be in the realm of microcopic quantum theory.

For background information *see* HEAT RADIATION; HYPERSONICS; LATTICE VIBRATIONS; LIQUID HELIUM; SPECIFIC HEAT OF SOLIDS; UNCERTAINTY PRINCIPLE in the McGraw-Hill Encyclopedia of Science and Technology.

[SETH PUTTERMAN]

Bibliography: R. Baumgartner et al., Spontaneous decay of high frequency acoustic phonons, *Proceedings of the 16th International Conference on Low Temperature Physics, Physica B + C,* 107:109–110, 1981. M. Cabot and S. Putterman, Renormalized classical non-linear hydrodynamics, quantum mode coupling and quantum theory of interacting phonons, *Phys. Lett.,* 83A:91–94, 1981. S. Putterman and P. Roberts, Non-linear hydrodynamics and a one fluid theory of superfluid He^4, *Phys. Lett.,* 89A:444–447, 1982. P. Westervelt, Absorption of sound by sound, *J. Acoust. Soc. Amer.,* 59:760–767, 1960.

Space flight

The contrasting approaches of the American and the Soviet crewed space programs intensified in 1982, as the United States moved into routine, commercial operations with its reusable space shuttle, *Columbia,* while the Soviet Union orbited a new, improved space station. The growing confidence in crewed operations was reflected in changes such as NASA's naming of prime crews for specific missions without naming the customary backup crews. The Soviet Union's long-standing rule of replacing the entire prime crew if one of its members could not fly also changed. Both nations have reached the stage where a cadre of space veterans is available for substitutions.

The international flavor of crewed flight increased with the inclusion of a French cosmonaut on a Soviet launch and the designation for *Spacelab 1* of a physicist from West Germany and one from the Netherlands. Another example of such cooperation was the offering of space launcher services to other nations by the People's Republic of China.

A broadening of the segment of the population which has access to space is evident from such actions and also from the variety of sponsors who provide payloads to both the United States and the Soviet Union for launch. Notable are the opportunity for individuals to fly experiments in a shuttle-carried canister called a "Getaway Special" and the opportunity to use unpurchased flight space for student experiments. The student experiments are chosen in national competition (grades 9–12) under the Shuttle Student Involvement Project, coordinated by the National Science Teachers Association.

A growing number of United States private companies are preparing to offer commercial launch services. They are convinced from market studies that profits will be possible despite the huge pre-launch investments required. NASA projections show that government launch capabilities will be saturated by the mid-1980s, perhaps falling 20% short of anticipated demand. The Soviet Union negotiated its first commercial launch agreement (with India), sparking speculation of a reimbursable service.

Significant space launches in the year ending November 15, 1982, are listed in the table.

Space Transportation System (STS). After earning operational status with its fourth flight, the shuttle flew a paying flight for customers on mission 5. Experience has led to changes in operating procedures and hardware which are resulting in greater payloads and safety. Lighter, stronger materials replaced parts of the thermal protection system. One new material, to be used on orbiters *Discovery* and *Atlantic*, will save about 1000 lb (450 kg) of weight. A new, lightweight external tank, freed of 6000 lb (2700 kg) of material, will fly on mission 6. Further weight savings lie ahead with the use of filament-wound casings for the solid rocket boosters. In addition, there are proposals to more than double the payload volume of the shuttle by adding a cargo compartment to the bottom end of the external tank. There are also possibilities for in-orbit uses of the tank itself, including habitation or as space station modules for assembly in orbit.

Shuttle mission 3. Liftoff on March 22, 1982, took astronauts Jack R. Lousma and C. Gordon Fullerton on a steeper path than that of previous flights, achieving for the first time the nominal trajectory for the shuttle, a continuously ascending path into orbit. The first two flights used an initial climb, an earthward dive, and a final climb as a means of assisting any need for an abort that would require return to launch site.

Liftoff weight was about 4,478,000 lb (2,031,000

kg), nearly 17,000 lb (8000 kg) heavier than mission 1. In orbit, the remote manipulator arm was used for the first time to remove a payload from the shuttle bay, perform dynamic motion tests so loaded, and reseat the cargo for successful reanchoring.

Flooding on the lake bed runway of the primary landing strip at Edwards Air Force Base, CA, forced a change to a touchdown site at the wind-whipped White Sands Missile Range, NM. A new reentry profile had to be used, and astronaut Lousma flew 11 attitude maneuvers as he managed the flight energy that let him glide to touchdown at about 220 knots (113 m/s).

A key objective of mission 3 was successfully achieved: long-term exposure of various parts of the orbiter to sun heat and shadow cold during the 8-day flight. The orbiter and its systems remained at temperature levels below design maximums.

The primary payload was called OSS-1 (Office of Space Sciences) and was mounted on a pallet designed and provided by the European Space Agency (ESA). It checked factors that cannot be pretested on Earth, including reactions of equipment to thermal variations and extremes, the effects of a cloud of particles and gases coming from the materials of the orbiter, and the degree of electrical charging on the shuttle.

Shuttle mission 4. Columbia took off and landed on schedule for the first time. Astronauts Thomas K. Mattingly and Henry W. Hartsfield launched into a 28.5° orbital incline. A mishap on ascent occurred when the solid-fuel boosters experienced stuck parachutes and sank to an ocean depth of 3500 ft (1100 m). NASA used an ocean robot to photograph the debris at depth for analysis. The solid-propellant rockets performed within specifications, but they burned more slowly than expected, causing a depressed trajectory which was compensated for by longer main engine burn time. As a result, the ex-

Significant space launches in the year ending November 15, 1982

Payload name	Date	Payload country or organization	Purpose and comments
Satcom 3R	11/19/81	RCA Corp.	Cable TV feeder; United States Delta launch
Navstar 7	12/18/81	U.S. Air Force	Intended to join six orbiting navigation satellites; Atlas E crashed from 200 ft (60 m)
Ariane IV	12/19/81	ESA	*Marecs A* communications satellite to synchronous orbit
Intelsat V-D	3/4/82	International Consortium	Communications; United States Atlas Centaur launched
Columbia	3/22/82	NASA	Shuttle Mission 3 with OSS 1 experiment
Insat 1A	4/5/82	India	India communications and weather satellite; United States Delta launch
Salyut 7	4/19/82	Soviet Union	Space station; replacing *Salyut 6*
Soyuz T-5	5/13/82	Soviet Union	Two-member record-breaking crew for *Salyut 7*
Soyuz T-6	6/24/82	Soviet Union	Three-member crew visits *Salyut 7*; one French
Columbia	6/27/82	NASA	Shuttle mission 4 with *DOD-82-1*
Landsat D	7/16/82	NASA	Earth resources satellite on Delta 3920
Soyuz T-7	8/19/82	Soviet Union	Three cosmonauts (one female) visit *Salyut 7*
Anik D-1	8/26/82	Canada	Canadian communications satellite; United States Delta launch
ETS 3	9/3/82	Japan	Engineering test satellite; mercury ion attitude thrusters
Ariane V	9/9/82	ESA	Third stage failed; two spacecraft lost
RCA-E	10/27/82	Alascom, Inc.	Alaskan communications satellite; United States Delta launch
DSCS-2/3	10/30/82	U.S. Air Force	First Titan 34D/IUS launch; two communications satellites
Columbia	11/11/82	NASA	Shuttle mission 5; deployed *SBS 3* and *Anik C*

pected burn rate for the next mission was revised downward from 0.368 to 0.365 in./s (9.35 to 9.27 mm/s). The "great depression" caused *Columbia* to fly 8000 ft (2400 m) below its planned trajectory line, costing a theoretical 2000 lb (900 kg) loss in payload capacity.

The shuttle for the first time carried a Department of Defense payload with a classified cluster of instruments, such as infrared and ultraviolet sensors. This was in keeping with the presidential policy that the United States' use of space for peaceful purposes include activities involving national security.

Mission 4 carried many experiments involving the shuttle itself, including its responses to aerodynamic and other loads, plus its generation of electromagnetic, particulate, and other pollutants which could affect sensitive instruments on science flights. A survey of lightning in the Earth's clouds was conducted by using film and photocells. The 50-ft (15-m) arm was exercised extensively. A continuous-flow electrophoresis system experiment became the shuttle's first commercial payload. Designed and funded by the McDonnell Douglas Astronautics Company, it successfully separated protein samples according to their surface electrical charges and demonstrated that space-based equipment could provide in microgravity about 500 times more output than Earth-based equipment. The process is expected to lead to ultrapure medications. Similar medical hopes rest with the monodisperse latex reactor experiment which grew identically sized particles twice the size obtainable on Earth. These may be used as calibration standards for medical measuring devices, and as carriers of drugs and radioactive isotopes for cancer treatment.

Mission 4 carried the first private-sector payload to be flown as a Getaway Special. The 5-ft³ (140-liter) canister was purchased by Gilbert Moore of Utah for $10,000 and donated to Utah college students, who flew nine experiments involving the fields of physics, metallurgy, biology, and thermodynamics.

At one point in the flight, the huge doors to the payload bay could not be closed, a critical requirement for reentry. By rotating the shuttle slowly for several hours in a "barbecue mode," the astronauts equalized temperatures and obtained closure. This and other difficulties added up to only 16 orbiter-related problems, compared with 61 logged on mission 1, 50 on mission 2, and 47 on mission 3.

Columbia ended its 169-hour flight by achieving two key objectives of reentry: expanding the cross-range capability of the orbiter and achieving higher reentry heating for tests. It also made the first shuttle landing on a finite, hard-surface runway and received no height-above-runway calls from chase planes during the last few feet above ground. This routine practice was omitted in the shift to operational procedures in which chase planes will not be used during landings. Rollout distance was 9660 ft (2944 m), achieved with moderate braking.

The astronauts conducted the four hypersonic S

turns common to shuttle reentries and interspersed them with nine flight test maneuvers which were intended to increase test temperatures on the vehicle. The data may lead to lower angles of attack for future landings, providing greater cross-range capability, though the heating is more severe. The current limit is 720 mi (1160 km) and a desired range is 1000 mi (1610 km). This could be vital for launches from California, where there may develop a need to return to Earth within one orbit before overflying Soviet territory.

Shuttle mission 5. For the first time in space history, four astronauts were launched simultaneously. Liftoff weight was about 28,000 lb (12,700 kg) heavier than for mission 1. The ascent trajectory was slightly depressed from the extra weight, from upper winds of 70 knots (36 m/s), and from solid-propellant boosters which burned a little slower than expected. Automatic landing was deleted from the flight requirements before launch because of unreconciled differences of flight characteristics among training facilities.

The flight crew consisted of the commander Vance Brand, pilot Robert Overmyer, and mission specialists Joseph Allen and William Lenoir. The use of a three-man cockpit provided more eyes and hands for coping with launch or landing emergencies. Within 8 h of the November 11 launch the crew had deployed the *SBS 3* communications satellite into the shuttle's 160-mi (257-km) orbit, the first deployment of a commercial satellite from an orbiting space vehicle (see illustration). This 14.6-ft-tall (4.45-m) spacecraft stack, weighing 7211 lb (3271 kg) and spinning at 52 revolutions per minute, was spring-ejected from the payload bay at 2.8 ft/s (0.85m/s). Later, its Thiokol Star-48 solid rocket was fired by a timer and boosted the stack to geosynchronous orbit. The 5-ft-diameter (1.5-m) satellite was 9 ft long at ejection, but in synchronous orbit it extended its circumferential drop skirt of solar cells for a total length of 21 ft (6.4 m). Shortly after ejection, *Columbia* backed off about 18 mi (29 km) and was oriented bottom to the *SBS 3* to prevent its firing debris from degrading the shuttle's windows. Similar procedures were used for the Telesat Canada communications spacecraft, *Anik C,* deployed about a day later. Each of these satellites has 14,000 solar cells and generates 1000 W of direct current power.

Customers paid about $8 million for the *SBS 3* launch and $9 million for the heavier *Anik C.* The launch costs were less than one-third those charged for an expendable booster launch. In addition, the shuttle allows heavier loads of hydrazine fuel to be carried for orbit and attitude maintenance, thus extending anticipated mission lifetime for up to 2 years (from 7 to 9 years) for these two Hughes HS-376 model spacecraft. Each customer carried $500 million in launch liability insurance, costing between $70,000 and $100,000 in premiums. The low premiums are evidence of the insurers' belief that the shuttle is safely operational.

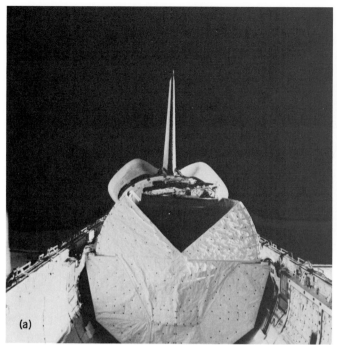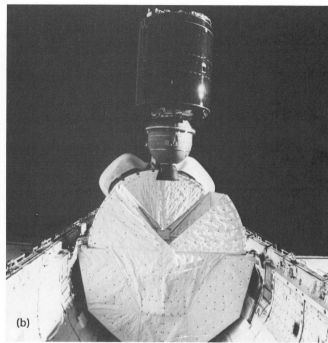

Launch of the *SBS 3* communications satellite during shuttle mission 5. (*a*) Clamshell shroud of *SBS 3* satellite has been opened, and spacecraft rotates on its spin table in the shuttle payload bay. Behind it, still shroud-covered, is the *Anik C* satellite, deployed a day later. (*b*) Spinning spacecraft stack of *SBS 3* is spring-ejected. (*NASA*)

Other items carried in the cargo bay included the development flight instrumentation package and a Getaway Special containing a metals miscibility experiment from the materials processing program of the German Ministry of Research and Technology. On the orbiter's middeck were the monodisperse latex reactor experiment and three Student Involvement experiments. Several orbiter experiments (OEX) were flown to verify the accuracy of wind tunnel simulations and theoretical calculations for further certifying the shuttle and expanding its known capabilities. The remote manipulator arm and the induced environment contamination monitor were removed. The first space walk/work session of the shuttle flights had to be canceled when Allen's suit suffered a malfunctioning fan and Lenoir's suit developed a stuck regulator.

Tests to characterize space-motion sickness symptoms were conducted. One records eye motion at the onset of weightlessness. Doctors believe that 30–50% of persons will experience some symptoms in zero gravity. Soviet crews have also reported early-onset problems, but their long-term missions allow for light work and reduced motion during the first few days. The 5–7-day shuttle missions do not allow such an adaptive period, which would conflict with the required early high activity.

Upon return, *Columbia* made a 2½-min retrograde burn above the Indian Ocean to position itself at about 80 mi (129 km) altitude when nearly 4000 mi (6400 km) from California in order to set up the desired angle of attack on the atmosphere for reentry. Touchdown took place on November 16 and included the first maximum braking test. The craft achieved a stopping distance of 9553 ft (2912 m) after a landing at 140 knots (72 m/s).

Shuttle archeology and geology. What were called "astounding revelations" resulted from radar images obtained on mission 2. Not only did the shuttle imaging radar (SIR) show anticipated surface features, but it also disclosed subsurface features of ancient terrain under the Sahara Desert sands in Egypt and the Sudan over hundreds of miles. The U.S. Geological Survey sees the findings as a major scientific discovery with broad implications for oil and water exploration and for archeology. It was known that radar could theoretically penetrate dry sand, but no one had expected any place on Earth to be dry enough for such results. The areas involved are hyperarid, with rainfall perhaps every 30 years. Images show that under the sand lie large, dry river valleys which extend for perhaps hundreds of miles. Also seen are braided tributary and alluvial fan features, terraces, extensive faults, and other rugged terrain. Human artifacts, perhaps 200,000 years old, have been found atop the sand in many places, reinforcing the prospect that subsurface features which look like lakes could have attracted settlement. Planners are now considering specialized radar sounder missions to study the arid subsurface of Mars. Also, because frozen water allows the passage of radar energy, some are considering penetrating the ice caps of Antarctica and Greenland to disclose the terrain.

Soviet space activity. In the year ending November 15, 1982, there were 100 Soviet launches, grouped generally as follows: communications satellites, 13; military radio store/transmit, 7; electronic ferret, 4; weather, 1; navigation, 9; earth resources, 8 (including one from India); recoverable military photographic, 26; early warning, 5; ocean surveillance, 7; ocean resources, 2; crewed vehicles, 3; new space station, 1; automatic rendezvous supply vessels for the space station, 4; amateur radio, 1; geodesy, 1; minor military, 5; antisatellite, 2 (one target and one interceptor); and the possible test of a manned reusable vehicle prototype, 1.

Possible Soviet shuttle. Speculation that the Soviet Union may be developing a version of the United States' shuttle grew firmer with the launch on June 3 of *Cosmos 1374*. Some analysts believe it achieved a successful wingborne reentry over the Indian Ocean just one earth revolution after launch by an expendable booster. Estimates place the subscale vehicle in the 2000-lb (900-kg) class, leading toward a 40,000 lb (18,000 kg) crewed version. The Soviet effort drew comparisons to the United States Asset program, in which six test gliders made suborbital flights in 1963 down the Eastern Test Range, launched by Douglas Thor boosters.

Soviet Venus landings. The first four-color image of the Venus surface was taken by the *Venera 13* after its landing on March 1, 1982. Only 4 days later, *Venera 14* landed 600 mi (965 km) away in an area which showed different-appearing features. Chemical samples of surface material were placed in a chamber for x-ray fluorescence analysis. Highly alkaline potassium basalts, extremely rare on Earth, are believed to be the prevalent components of rock on Venus.

Soviet space stations. With the Salyut space station design as a core, the Soviets experimented with "building block" modular assembly, in orbit, of large habitats. Trials began in 1981 when the large (30,000-lb or 13,600-kg) *Cosmos 1267* was docked with the slightly larger (42,000 lb or 19,000 kg) *Salyut 6* space station for several months of tests, including the regulation of temperature levels in the large, combined structure.

Later, the *Cosmos 1267* power plant was used to carry both vessels into a destructive reentry. The *Cosmos 1267* has been described by some analysts as armed with antisatellite homing vehicles. Its reentry ended almost 5 years of service by the *Salyut 6*. The station had logged 676 days of crewed operations by five long-duration crews and 11 short-duration crews. Habitation by long-duration teams was for periods of 96, 140, 175, 185, and 75 days (in chronological order).

A replacement space station arrived with the April launch of *Salyut 7*, a spacecraft which retained the overall dimensions and mass of *Salyut 6*, but introduced crew comfort improvements, a strengthened docking unit, and an improved navigation system.

On May 13, cosmonauts Anatoliy Berezovoy and Valentin Lebedev were launched in *Soyuz T-5* to operate the new station. Their resulting 211-day stay broke the previous crewed flight record of 185 days.

They were visited twice during the mission, first by *Soyuz T-6*, launched June 24, carrying cosmonauts Vladimir Djanibekov and Alexander Ivanchenkov, accompanied by French researcher Jean-Loup Chrétien. During launch, television from the cockpit was broadcast live throughout France and the Soviet Union. Prior to flight, the crew had washed with alcohol and slept on beds sterilized with ultraviolet light, a standard Soviet procedure. The five member team in the three-spacecraft complex conducted experiments in medicine, materials processing, biology, and astronomy. A French-built ultrasonic echography device provided cross-sectional views of organs functioning in zero gravity. All three visitors later complained that the number and complexity of experiments caused an overwhelming work load. On July 2 they returned to Earth, using their original spacecraft.

A second set of visitors arrived in *Soyuz T-7*, launched August 19. They were female cosmonaut Svetlana Savitskaya and males Leonid Popov and Aleksandr Serebrov. Some analysts had predicted such a flight would occur before the United States launches astronaut Sally Ride on shuttle mission 7, partly as a reminder to the world that the Soviet Union had women in space before the United States. Vestibular and cardiovascular tests were done on Savitskaya for data on the adaptation of females to zero gravity. Results showed no differences for women versus men.

This visiting team left the newer *Soyuz T-7* spacecraft behind for the return journey of their hosts and rode home on August 27 in the older *Soyuz T-5* transport, because the Soyuz apparently is not considered a qualified reentry vehicle after it has spent more than 100 days in orbit.

Far East activity. India, Japan, and China were active in their budding space programs. A Chinese spacecraft launched September 9 and recovered 5 days later has the characteristics of a reconnaissance mission. The Japanese, on September 3, launched the *Engineering Test Satellite 3*. It features two Japanese-made mercury ion engines for experiment with three-axis attitude control. India acted to modify the *Insat 1B* which will replace its multiservice satellite, *Insat 1A*, which failed September 4 after only 5 months in orbit. India's ground segment includes about 2000 terminals of 10-ft (3-m) diameter which are useful for direct broadcasts to remote villages. An Indian satellite called *Bhaskara 2* was launched on the Soviet SS-5 Skean booster on November 20, 1981. It carried television cameras and microwave radiometers for monitoring sea, wind, and moisture conditions.

European space activity. European plans for operational commercial use of the Ariane launcher were set back when its fifth flight crashed in the Atlantic on September 9, 1982, due to malfunction of a third stage turbopump. Ariane carried ESA's *Marecs B* communications spacecraft for Inmarsat, the global maritime communications system. Also aboard was *Sirio 2*, a European satellite intended as a meteorological distribution link. Both were bound for synchronous orbit. The Ariane marketing strategy calls for transfer of sales, manufacturing, and launch responsibilities to Arianespace, a French company, after seven successful flights. The failure came after a successful fourth flight on December 19, 1981, from French Guiana, which allowed ESA to declare the launcher operational. That flight put *Marecs A* into synchronous orbit. *See* LAUNCH VEHICLE.

International search and rescue. The United States, Canada, France, and the Soviet Union jointly have formed a worldwide search and rescue system which was credited with several life-saving actions within months of its first launch. On July 1, 1982, a standard Soviet navigation spacecraft carried to orbit a piggy-backed search and rescue transponder, which it uses to relay emergency locator signals to ground stations several times daily. The Goddard Space Flight Center in Greenbelt, MD, has demonstrated 1 mi (1.6 km) accuracy in pinpointing test transmitters. *Cosmos 1383* was crucial to the rescue of sailors on a capsized catamaran in 25-ft (8-m) waves in the Atlantic and of crashed fliers in British Columbia. Canadian officials credited the craft not only with saving the plane crew, but with an unusually inexpensive search effort.

Comet exploration. Although the United States is not planning a probe to Halley's Comet, as are ESA, Japan, and the Soviet Union, it nevertheless has initiated comet observation with one of its spacecraft. In September, scientists began maneuvering an old Sun-monitoring satellite called *ISEE 3* from its Earth/Sun libration point orbit about 1 million mi (1.6×10^6 km) from Earth to start it on an intercept course with the comet Giacobini-Zinner. In September 1985, the satellite should come within 2000 mi (3200 km) of the comet's head while passing through its 500,000-mi-long (800,000-km) tail of gas and dust. It will then be some 44 million mi (71×10^6 km) from Earth. *ISEE 3* has little fuel for such a trip, so it will get there by lunar gravity assistance. Controllers will send it into a looping orbit of the Moon, and on its fifth pass *ISEE 3* will be hurled to intercept the comet. The last pass will come within 60 mi (100 km) of the Moon's surface. The targeted comet passes by the Sun every 13 years. Its encounter with *ISEE 3* will occur about 6 months before the other spacecraft encounter Halley's Comet. The information obtained will be made available to the Europeans, Japanese, and Soviets for comparative studies.

Earth resources. The powerful new Delta 3920 carried *Landsat 4*, weighing 4273 lb (1938 kg) into a polar orbit at about 430 mi (690 km) on July 16, 1982. The new satellite was designed to improve the monitoring of renewable resources beyond the abilities of the earlier Landsats. It is also capable of rock discrimination, information not previously available to petroleum geologists. Its main instruments were the multispectral scanner and the thematic mapper. The latter immediately obtained images of higher resolution (30 m) and spectral fidelity than are possible with current or earlier multispectral scanners.

The seven spectral bands of the thematic mapper should allow determination of healthy vegetation; differentiation of plant species; water body delineations; biomass surveys; distinguishing between cloud and snow cover; detection of plant heat stress; acquisition of thermal data on geologic formations; and discrimination of rock type for geologists.

The Soviets continued earth resources activity using the MKF-6 camera on *Salyut 7*, a series of free-flying missions, and experiments with the Meteor-Priroda spacecraft.

Navigation. Using a Proton booster, the Soviets group-launched the first three Navstar-type satellites for their intended GLONASS navigation system.

For background information *see* APPLICATIONS SATELLITES; COMET; COMMUNICATIONS SATELLITE; REMOTE SENSING; SPACE FLIGHT; SPACE PROCESSING; SPACE SHUTTLE; VENUS in the McGraw-Hill Encyclopedia of Science and Technology.

[CHARLES BOYLE]

Bibliography: Aviat. Week Space Technol., issues from November 23, 1981 through November 22, 1982; *NASA Activ.*, issues from November 1981 through November 1982; *Sci. News*, issues from November 21, 1981, through December 4, 1982; *Space World*, issues from December 1981 through December 1982.

Sputtering

Sputtering is the ejection of atoms or clusters of atoms from a surface that is bombarded by projectiles of atomic dimensions; it is sandblasting on an atomic scale. Figure 1 is a schematic diagram of a beam of atoms (or ions) striking the surface of a solid target. It illustrates a number of the phenomena that are initiated by particle bombardment. Most of the projectiles penetrate and stop within the material, a process known as ion implantation; some, however, are reflected, or backscattered. A penetrating projectile may damage or alter the structure of the material, may induce electron or photon emission from the surface, or may sputter away neutral atoms or ions. These closely related impact phenomena are of increasing importance for both basic research and technological development. Perhaps more surprisingly, though, they are natural phenomena whose relevance in understanding extraterrestrial systems is just beginning to be fully appreciated.

This article introduces the physical principles and applications of sputtering. It is limited to a discussion of physical sputtering in which chemical interactions between the projectiles and the target are unimportant. The effects of bombardment by projectiles of energy from about 10^2 to 10^6 eV are considered because this range is best understood and currently of greatest technological importance.

Experimental techniques. As early as the 1850s researchers noticed the slow accumulation of metal deposits on the inside of gas discharge tubes which were being used to study the passage of electrical currents through gases. They found that these mir-

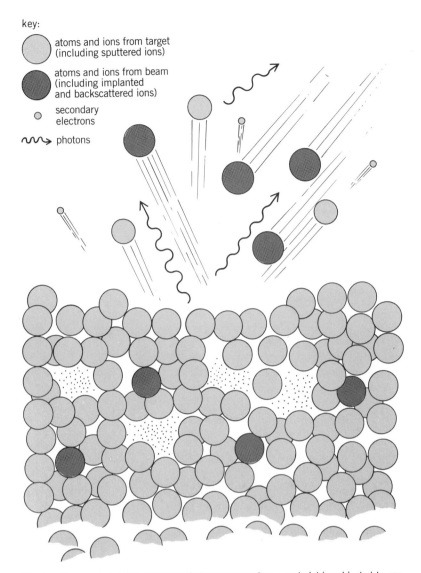

key:

⬤ atoms and ions from target (including sputtered ions)

⬤ atoms and ions from beam (including implanted and backscattered ions)

• secondary electrons

∿→ photons

Fig. 1. Diagram indicating processes that can occur when a material is subjected to particle bombardment. Shown are implanted ions, structural defects, backscattered and sputtered ions, secondary electrons, and photons.

finely focused, narrowly collimated, monoenergetic ion beams that they produce are superior to gas discharges for sputtering studies. Moreover, much higher energies can be achieved. The energy range through which sputtering has been most thoroughly studied is from about 10^3 to 10^6 eV. Sputtering is also greatest within this range for most target-projectile combinations.

Ultrahigh vacuum is essential for a sputtering experiment because the quantity and composition of material that is sputtered from a surface is strongly influenced by absorbed gas. The residual gas in a vacuum system maintained at a pressure of 10^{-9} torr or 10^{-7} pascal (1 atmosphere is 760 torrs or 10^5 pascals) will completely cover an atomically clean surface in about 1 h. Thus, for experiments in which the rate of erosion of the target is low, such pressures are necessary in order that targets remain atomically clean during the period of bombardment. For satisfactory data to be obtained, the rate of removal of material by sputtering should be several times the rate of redeposition of residual gas.

In order for impact-induced particle emission to be termed sputtering, it must occur in the limit that a single projectile strikes a surface. This definition suggests a natural measure of sputtering, called the sputtering yield; it is the average number of sputtered particles emitted per incident projectile. A large majority of all sputtering experiments measure a sputtering yield or one of its various refinements. Early sputtering experiments often measured the yield by observing the change in weight of the target before and after it had been sputtered by an ion beam. The total quantity, or fluence, of projectiles was determined by measurement of the electrical charge deposited on the target. This technique, although still used occasionally, has a number of disadvantages. It certainly requires that a relatively large amount of material be removed; more importantly, though, nothing is learned about the sputtered particles themselves. Their structure, direction, energy, charge, and internal excitation all remain a mystery. To measure these properties it is necessary to look directly at the sputtered particles, either by observing them at the instant of ejection or by collecting them for subsequent analysis.

A number of instruments are used to study the sputtered particles and the targets. The composition of sputtered material may be determined with a mass spectrometer or by collecting material for subsequent analysis by electron spectroscopy or other techniques. Lasers and electrostatic, magnetic, and time-of-flight spectrometers are used to obtain sputtered particle energy spectra, while optical spectrometers give information about the internal states of sputtered atoms. In addition, various forms of electron spectrometers, including electron microscopes, are used to inspect targets for changes in composition and topography.

A typical ultrahigh-vacuum system with instruments for making sputtering measurements is shown in Fig. 2. The projectiles enter from the upper right

rorlike deposits were always of the same material as the cathode of the discharge tube. It is now known than these deposits were the result of bombardment of the cathode by positively ionized atoms of the gas within the tube. These positive ions, which are produced by electron bombardment of the gas, are accelerated toward the negatively charged cathode, strike it, and sputter material which then collects on the tube walls. Although this hypothesis was advanced around 1900, comprehensive investigations of sputtering awaited two more technological developments, the widespread availability of particle accelerators and the development of techniques for producing ultrahigh vacuum.

Low-energy particle accelerators capable of producing stable, well-characterized beams of ions with energies up to about 1 MeV are now common in universities and industrial research laboratories and are also used in semiconductor manufacturing. The

and strike a target in the center of the upper portion of the vacuum chamber. A large number of access ports are provided for the instruments to be mounted. The chamber shown is wrapped with flexible heaters used for baking the system, a process essential for the production of ultrahigh vacuum.

Observations. Much of the basic systematic behavior of the sputtering of amorphous elemental targets is now well established. From a threshold at a few electronvolts, the sputtering yield increases monotonically to a maximum and then decreases to near zero as the energy of the projectile is increased still further. For many target-projectile combinations the maximum occurs at a projectile energy of 10^4 to 10^6 eV. Over a wide range of projectile energies the sputtering yield is found to be proportional to the nuclear stopping power of the target. This is the energy lost by the projectile in elastic (billiard ball–like) collisions per unit length along its path. The sputtering yield is also inversely proportional to the effective strength of the binding of target atoms to the surface, which is usually taken to be the cohesive energy of the substance expressed in electronvolts per atom. For most substances the surface binding energy is between 2 and 7 eV.

Most sputtered particles are single, electrically neutral atoms which originate from the top one or two atomic layers of the target. A typical energy distribution of these atoms has a peak at about one-half of the surface binding energy and decreases approximately as E^{-2} for energies E above a few tens of electronvolts. The angular distribution of sputtered atoms is peaked about the target normal approximately independently of the direction of the incident projectile and is frequently close to a cosine function.

When the average energy per atom deposited in the target becomes large, the sputtering deviates from the description just presented. This is seen most dramatically when a target is bombarded with atomic and molecular beams of the same element with the same energy per atom. The simultaneous impact of two or more projectiles in the same location produces more sputtered particle emission than if each had hit while separated in space or time from the others. Further, the energy distribution of sputtered material contains more particles at lower energies. These data suggest that the underlying physical mechanism must be nonlinear.

If the target is a crystal, atoms are sputtered preferentially in directions defined by densely packed rows of atoms within the crystal. The resulting patterns of sputtered material superficially resemble x-ray diffraction patterns. Also, the sputtering yield depends upon the relative orientation of the crystal and the beam. The yield may fall dramatically if the beam is aligned with crystal directions in which there are relatively open channels. In these directions the projectiles penetrate with fewer large deflection collisions, a phenomenon known as channeling. Since less energy is deposited in the target near the surface, less sputtering occurs.

These are the observations most often made in sputtering experiments. Another worth mentioning is that sputtered surfaces often develop incredibly complex microtopography which is best seen with the scanning electron microscope. The surface structures can be particularly striking when the target is a crystal.

Techniques for the experimental investigation of sputtering are drawn from a wide variety of disciplines, including surface physics, nuclear physics, and laser chemistry. To date, most experimental data on sputtering have been obtained by sputtering metallic elemental targets with projectiles with energy below 1 MeV. The sputtering of heterogeneous materials such as alloys, insulators, and organic or biological materials and sputtering by very energetic projectiles are much less well understood.

Theoretical interpretation. The debate that lasted for many years over whether sputtering is a collision or evaporation phenomenon has been (almost) settled in favor of the collisional picture. High-quality measurements of neutral sputtered atom energy distributions confirm this conclusion. Three qualitatively different events are now recognized as causes

Fig. 2. Ultrahigh-vacuum chamber for studying sputtering. The body of the chamber is about 30 cm in diameter. The ion beam enters from the upper right.

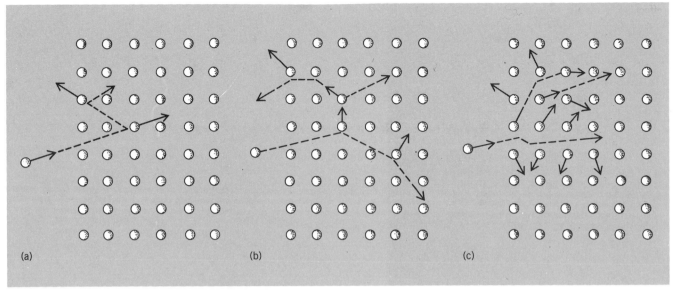

Fig. 3.   Sputtering mechanisms. (*a*) Single knockon. (*b*) Linear collision cascade. (*c*) Spike. (*After P. Sigmund, Sputtering by ion bombardment: Theoretical concepts, in R.* Behrisch, ed., *Sputtering by Particle Bombardment I, pp. 9–71, Springer-Verlag, 1981*)

of sputtering. These are shown schematically in Fig. 3. For projectile energies near the threshold for sputtering, target atoms may be ejected either directly by the projectiles, as shown in Fig. 3*a*, or after a small number of collisions. These events are called single knockons. More commonly, however, sputtering occurs when the projectile generates a linear collision cascade (Fig. 3*b*). In this event, the projectile collides directly with a small number of target atoms, which themselves collide with others. This sharing of energy them proceeds through many generations. The defining requirement for a linear collision cascade is that at any time the density of atoms in motion must be sufficiently small that collisions between moving atoms can be ignored. A theory proposed by Peter Sigmund in 1969 has been quite successful in explaining sputtering from linear collision cascades. Sigmund's theory is based upon a linearized form of the Boltzmann transport equation, which describes the nonequilibrium behavior of large numbers of colliding objects. It is reasonably successful in accounting for most of the effects in amorphous targets described above. One of the results of using a linear theory to describe sputtering is the prediction that when collision cascades produced by different particles overlap, the sputtering yield is the sum of the yields for nonoverlapping cascades. When, however, a projectile deposits a large amount of energy in a small region, the assumptions that underlie the linear theory are not adequate.

A collision cascade which ceases to be linear because so many particles are in motion is called a spike. A spike is awkward to describe mathematically because of its size and complexity. Although the sputtering yield from a spike can be as large as several hundred, a spike is too compact and too en-

ergetic for convincing treatment as a thermal phenomenon. It also contains too few particles for transport theory to be fully satisfactory; yet, it is too large for detailed computer simulation to be generally practical. Seeking the mathematical description of the phenomenon is an active area of research, with several complementary approaches under study. These include models based upon heat flow, nonlinear transport theory, shock-wave propagation, and Monte Carlo simulation.

In summary, sputtering is now known to be basically a collisional phenomenon but to vary in detail as the density of energy deposited by the projectile is increased. Sputtering yields have been measured for various target-projectile combinations from as low as about 10^{-5} to nearly 10^3, with the upper end of the range being for heavy molecular ions bombarding heavy targets. Linear collision cascades are relatively well understood both qualitatively and quantitatively, while the more energetic spikes, which sometimes lead to sputtering which resembles evaporation, are less well described. Some other mechanisms than those discussed can also lead to sputtering, but they are usually restricted to particular targets.

Applications. Originally, sputtering in a gas discharge was viewed as an inconvenient side effect which eroded the cathode. In 1877, however, A. W. Wright demonstrated that cathode sputtering could be used to produce metallic film mirrors of exceptional quality. Sputtering is still sometimes an asset and sometimes a liability. The gas discharge tubes of the 1850s have now been superseded by large particle accelerators. Sputtering, however, continues to limit the lifetime and performance of many internal structures of accelerators and even high-voltage electron microscopes. Unwanted sputtering effects are not, however, just confined to laboratories and esoteric research equipment. Increasingly, integrated circuits, and especially microcomputers, are being manufactured by ion implantation. Electrical structures can be made by implantation that cannot be made by any other process. Unfortunately, the beam also sputters away some of the semiconductor's surface. As a result, the maximum density of implanted atoms that can be produced is limited. Sputtering thus places an intrinsic limit on the amount of change of a surface that can be obtained by ion implantation. It is, however, in the proposed nuclear fusion reactor that sputtering could have its greatest economic impact.

Nuclear fusion reactors produce energy by heating a mixture of deuterium and tritium, both heavier isotopes of hydrogen, to temperatures many times higher than those in the center of the Sun. Such a hot gas, called a plasma, obviously cannot be contained in any conventional vessel. As a result, magnetic fields are being used to provide the principal confinement. Only charged particles are bent by magnetic fields, though, and one of the principal products of the deuterium-tritium nuclear reaction is a neutron whose energy is 14 MeV. When this neutron, or any other neutral particle, escapes magnetic confinement, it strikes the innermost wall of the reactor vessel and can sputter material from it. This not only limits the working life of the wall, but also injects heavy neutral atoms back into the magnetic bottle, contaminating the plasma and leading to excessive losses of energy from the bottle by radiation. Although sputtering of the first wall is not the principal obstacle in the path of commercial fusion reactors, it is important and must be minimized for them to be economically practical.

The effects of sputtering are not always detrimental. Sputtering is still widely used for the production of thin films for scientific and commercial purposes. Sputtered films can be quickly and easily produced with inexpensive equipment, and are often superior to films produced by other techniques. Additional applications of sputtering include ion-beam milling (the use of ion beams to machine surfaces, often on a very small scale), producing ions for particle accelerators in sputter ion sources, and preparing atomically clean surfaces by sputter erosion via low-energy ion guns. By combining an ion gun to slowly erode the surface with a mass spectrometer to analyze sputtered ions, the composition of a material as a function of depth below its surface may be studied. Such an instrument is called a secondary-ion mass spectrometer (SIMS) and is widely used for research in many fields.

In the future, the most important application of sputtering may be to replace chemical etching in the production of large-scale integrated circuits. As the dimensions of circuit elements are reduced below approximately 2 micrometers, etching with liquids becomes difficult. Variable etch rates result in undercutting of some circuit elements and, in general, it is impossible to retain control over the geometric boundaries of removed material. With ion-beam-induced sputtering, in contrast, these geometric boundaries can be delineated with high precision. *See* INTEGRATED CIRCUITS.

Sputtering in nature. Shielded by the Earth's thick atmosphere and extensive magnetic field, people generally are unaware that most of the solar system is awash in an ocean of radiation. The Sun emits a constant stream of energetic particles called the solar wind that reaches to the most distant planets and beyond, and from outside the solar system comes a barrage of cosmic rays, heavy ions often with extremely large energies. This cosmic sandstorm exacts a small but inescapable toll from every object in the solar system. Recent calculations indicate that the bombardment of the atmospheres of both Mars and Venus may have produced observable effects by a process conceptually identical to the sputtering of a solid surface. Both chemical alteration and destruction of tiny interstellar grains are possible as well as the erosion of the ice mantles widely believed to coat these small objects. In the solar system, however, there is no place where conditions are more favorable for sputtering than near Jupiter. Charged particles become trapped in Jupiter's enormous magnetic fields and relentlessly sputter its satellites. Though the effect is small, this sputtering, along with volcanic eruptions and meteorite impacts, is an important element in shaping the landscapes of those distant bodies.

For background information *see* ION IMPLANTATION; ION SOURCES; NUCLEAR FUSION; RADIATION DAMAGE TO MATERIALS; SECONDARY ION MASS SPECTROMETRY (SIMS); SPUTTERING; SURFACE PHYSICS in the McGraw-Hill Encyclopedia of Science and Technology.

[ROBERT A. WELLER]

Bibliography: R. Behrisch (ed.), *Sputtering by Particle Bombardment I*, 1981; P. Sigmund, Theory of sputtering, I. Sputtering yield of amorphous and polycrystalline targets, *Phys. Rev.*, 184:383–416, August 10, 1969.

Steel manufacture

With the increased emphasis on higher quality and greater productivity in steel manufacture, there have been many recent developments in the primary production of steel. These developments have been in the areas of analyzing raw materials, incorporating new technology for converting raw materials to liquid, processing the liquid in steelmaking furnaces, treating the liquid before pouring, and converting the liquid to a solid shape.

Raw materials. The principal raw materials required for steel manufacture in modern basic oxygen furnaces are molten pig iron, steel scrap, lime, and fluorspar. Many varieties of these ingredients are available, and meeting required weights and chemical specifications is important in controlling the variables of the operation.

Before being loaded or charged into the furnaces, the materials must be weighed and tested. By using a computer to solve heat and material equations simultaneously, the amounts of oxygen and other necessary materials can be calculated, automatically weighed or measured, and added to the furnace. The temperature, chemistry, and amount of pig iron and scrap determine the heat balance, and the quantities of carbon, silicon, phosphorus, and sulfur they contain determine the amount of lime and fluorspar needed to meet the final specifications of the liquid steel produced.

Furnace melting. The melting time is determined by the oxygen injection rate relative to the quantity of iron-bearing material in the furnace, the carbon level, and the pouring temperature of the steel. Oxygen is injected through a water-cooled lance maintained at the height which maximizes decarburization of the pig iron and melting of the scrap. At the same time, control of the lance's height minimizes the amount of iron (Fe) which is oxidized and becomes part of the slag by-product.

The chemical metallurgy of the melting process and subsequent formation of the slag are important to ensure good quality and smooth operation. The melting of the lime, the formation of silicon and iron oxides, and the addition of a certain percentage of magnesium oxide neutralize the erosion on the furnace's refractory lining and promote the retention of phosphorus and sulfur in the slag. Either lasers or refractory monitors can measure lining wear during furnace operation. To make the melting process efficient, control of the process by use of a computer program is necessary to regulate the melting of the steel to the required chemical and temperature specifications.

An important in process control is made by using a sensor lance connected to a mass spectrograph to provide analytical data of the melting steel in terms of the elements present in the pig iron. These data, along with the temperature, weight, and composition of the initial charge, are fed into the main computer, which calculates in-line adjustments. This dynamic correction shortens the overall melting time.

Furnace refining. Mixed blowing is the newest refining process in basic oxygen steelmaking. This technique combines oxygen blown from the top lance and an inert gas injected through the bottom of the furnace, which decreases the total Fe (iron units) in the slag and keeps the manganese content in the melt higher than normal. Low carbon steels (lower than 0.02% C) can also be produced by argon bubbling through the bottom after the main oxygen blow. The benefits are improved product yield and lowered consumption of lime and ferroalloys.

Ladle refining. When the liquid steel is ready for further processing, it is poured or tapped from the furnace into a holding vessel or steel ladle. In many cases, the slag in the furnace is prevented from entering the ladle. During the tapping operation, alloying elements are added to the liquid steel to ensure that the steel exhibits the required physical and mechanical properties after solidification and processing. To improve the characteristics of the steel to meet increasing quality demands, additional refining is necessary in the steel ladle. These ladle-metallurgy steps are generally aimed at improving steel cleanliness, chemistry control, and inclusion shape control, and lowering sulfur levels. In ladle metallurgy, vacuum-degassing processes are used which result in cleaner steels because carbon under vacuum becomes a deoxidizer and the oxygen leaves the system as a gas. Also, vacuum degassing results in lower hydrogen and nitrogen contents. Other ladle-metallurgy steps include the addition of calcium or rare-earth compounds which desulfurize the steel and change the shape of the oxide and sulfide inclusions. These practices can impart isotropic properties with as few as 50–70 ppm Ca or rare earths. Steels treated in this manner exhibit improved weldability and high impact strengths at low temperatures.

Pouring ingots. Liquid steel of proper chemical composition and temperature is poured into cast-iron ingot molds in which it solidifies into a shape suitable for subsequent reheating and rolling. The temperature must be controlled in order to maintain a good stream of molten steel and to achieve proper solidification characteristics. The stream of molten steel can be protected from reacting with air and forming oxide inclusions by shrouds of nonoxidizing gases. The surfaces of the cast-iron ingot molds are cleaned and coated to avoid surface defects that can result from rough mold surfaces. After solidification, the solidified ingot is removed from the mold while it is still hot, and is transported to reheating furnaces called soaking pits. When properly heated, the ingots are rolled to a semifinished shape.

Continuous casting. The liquid steel can be cast directly into a semifinished shape by using a continuous casting machine (see illustration). The metal is poured into an intermediate vessel known as a tundish that supplies molten metal to several strands in a controlled manner. The metal stream from tundish to mold is also protected to avoid reaction with air, and the liquid steel begins to solidify in the water-cooled copper molds.

At the caster the skin, or shell, of solidified steel begins to form in the mold, and the solidification front continues to grow as the semifinished shape leaves the mold and is subjected to water sprays and contact with the casting-machine rolls. The casting machine is designed to contain the partially solidified section until it becomes a completely solid structure. The containment parameters of the machinery and the method of cooling while controlling the speed of the process are important factors in determining the quality of the cast steel's internal structure and ultimate application. There have been numerous refinements in casting technology to increase the productivity of the process: (1) automatic control of the liquid from ladle to tundish to mold; (2) automatic control of the spray water; (3) ability to change the width of the mold while casting; (4) ability to rapidly and accurately change the thickness of the machine's support rolls; and (5) ability to cast high-strength steels faster. All these developments improve the

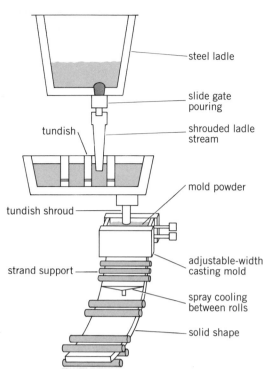

Schematic section of a continuous casting machine.

quality of the steel so that the hot, semifinished product (as it is continuously cast) can be rolled into the final product without further conditioning and excessive reheating.

For background information *see* STEEL; STEEL MANUFACTURE in the McGraw-Hill Encyclopedia of Science and Technology.

[L. HARBOLD]

Stem

Plant biologists are applying cinematographic techniques on a broad scale to investigate the three-dimensional stem anatomy of monocotyledons (such as palms, grasses, and orchids). Recently, the aroid family (Araceae, for example, *Pothos*, *Philodendron*, and *Syngonium*) was examined by using these techniques, and an unprecedented wealth of vascular organization was uncovered.

Three-dimensional reconstructions of dicotyledonous stems have been available for many years. They can be made without cinematography because of the relative simplicity of their stem vascular systems. In the dicotyledons a hollow cylinder of vascular bundles is present in the stem. At each node a small number of vascular bundles, called leaf traces, depart and enter a leaf. The leaf traces serve to connect the leaf with the stem, and thus provide a continuous pathway for food and water throughout the plant. The xylem in a vascular bundle is specialized for transport of water, which moves from the root through the stem to the leaves, where it is transpired. The phloem in a vascular bundle is specialized for the conduction of dissolved food, which moves in either direction,

depending on the location of food sources and sinks.

In monocotyledons the stem vasculature is more complex than in dicotyledons for several reasons. First, vascular bundles are often much more numerous (over 1000 at any level in the small palm *Rhapis*, and about 20,000 in a coconut palm), and second, the vascular bundles are not organized into a ring when seen in a transverse section. Instead, they may be dispersed throughout the entire stem or restricted to a central region, the central cylinder, and they appear to be in a random arrangement. Because of the large number of bundles, the three-dimensional course of an individual vascular bundle is difficult to follow when sections are examined one at a time. Early attempts, without microscopes, were made in the nineteenth century to understand the course of vascular bundles through the stems of palms, which were chosen because of their size. No attempts to study palms were successful until the 1960s, however, when cinematographic methods were developed by M. Zimmermann and P. B. Tomlinson. As a result of their efforts, considerable advances in understanding of vascular organization in the stems of monocotyledons have been made.

Cinematic studies. Films are made routinely over the course of several days by using a movie camera mounted on a compound microscope. Serial transverse stem sections are photographed individually with the movie camera in order to build up a frame-by-frame record of the stem anatomy. A smooth film is produced by precisely lining up one section after another on the stage of the microscope, using a camera lucida drawing of one section as a frame of reference for the next section. The resulting films can be studied repeatedly at variable speed to follow the precise course of bundles from the leaf over long distances

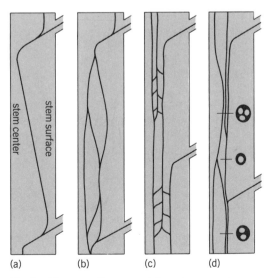

Fig. 1. Diagrammatic representation of the vascular organization in stems of various Araceae: (*a*) type 1; (*b*) type 2; (*c*) type 3; (*d*) type 4. Leaf traces enter the stem at right and fuse with axial bundles of the stem. In type 4, simple and compound bundles are shown as they would appear in transverse section.

(up to 100 cm) to their point of attachment in the stem vascular system. The projection of films on graph paper makes the plotting of the course of bundles through the series of sections possible. Diagrammatic representations are constructed from these plots to show the basic pattern of vascular organization (Fig. 1).

The outcome of the original cinematic study of palms was the recognition of the palm-type three-dimensional course of vascular bundles (type 1, Fig. 1). Film sequences that are going in a downward direction show that each leaf trace seems to enter the stem from a leaf and then to fuse with one peripheral bundle from the stem. The resulting bundle bends sharply, first toward the center of the stem, and then gradually back toward the outside of the stem. Near the periphery the bundle fuses with an incoming leaf trace, and the pattern is repeated. The films revealed for the first time a predictable pattern underlying the seemingly chaotic appearance of palm vascular systems.

The obvious question was whether the vascular bundles of other monocotyledons behaved according to the palm-type principle. In this regard, the most extensive cinematic survey of stem vasculature to date has focused on the Araceae. This family is of particular interest because of its great diversity in both anatomical and morphological features, and it may be expected to produce a corresponding diversity in three-dimensional vascular patterns. The Araceae were known to have three different kinds of vascular bundles: (1) collateral (xylem next to phloem), (2) amphivasal (xylem surrounding phloem), and (3) compound bundles. The course of compound bundles is of particular interest because they consist of several bundles migrating together as a unit, then separating from each other at some point. The Araceae also show a variety of different habits, including tuberous forms, erect herbs, and climbing vines with long internodes.

Types of vascular organization. The films of stem anatomy in the Araceae revealed a wider range of variation in vasculature than was expected. Three general patterns of vascular organization can be recognized in the Araceae with simple vascular bundles. Type 1 is the simplest pattern. It closely resembles the *Rhapis* type in its basic features, and occurs in both *Acorus*, which has amphivasal bundles, and *Anthurium*, which has collateral bundles, as well as in some other genera. Type 2 is more common in the Araceae, occurring for example in *Colocasia*, *Syngonium*, and some *Philodendron* species. In type 2 each entering leaf trace fuses with one or more axial bundles, and the resulting bundle undergoes an unpredictable sequence of splitting and anastomosis so that its identity is quickly lost. This pattern occurs only in species with amphivasal bundles. In tuberous Araceae and species with shortened internodes the type 2 vascular system becomes highly condensed, and more frequent branches are formed.

Type 3 is complex and is unlike any found previously in monocotyledons. It occurs in *Monstera* and closely related genera of root-climbing vines, as well as in *Pothos* and *Pothoidium*. An individual leaf trace does not fuse with a vascular bundle upon entering the central cylinder. Instead, each trace receives numerous branches from neighboring axial bundles. This is not by itself unusual because short branches (bridges) are common in palms. The unusual feature of this pattern is that the axial bundles giving off branches to leaf traces become progressively narrower, and finally fuse with a leaf trace or neighboring axial bundle. The occurrence of type 3 in both *Pothos* and *Monstera* is consistent with other anatomical and morphological evidence, which suggests that a closer relationship of these genera exists than present taxonomic treatments acknowledge.

Type 4 includes those species with true compound vascular bundles, for example, *Rhodospatha* and *Stenospermation*. A compound bundle consists of two or more collateral vascular bundles in close association, so that the entire assemblage migrates as a unit. The bundle sheaths of the component bundles are partly fused together, although a bundle sheath may be absent. Compound bundles are not always easily recognized; sometimes amphivasal bundles with separate strands of xylem are confused with compound bundles. This condition occurs in *Zamioculcas*, *Rhektophyllum*, and *Dieffenbachia*, for example, and is intermediate between compound bundles and amphivasal bundles.

The course of compound vascular bundles in *Rhodospatha* (type 4) does not follow the palm pattern or any other previously recognized pattern. A leaf trace enters the stem and associates with a peripheral axial bundle forming a bipolar bundle. Additional axial bundles contact the leaf trace and form compound bundles with two, three, or more components. Within a compound bundle some components are transitory and form no bridges before bending away and dissociating, while in other instances all the component bundles fuse together, producing a simple bundle again. Thus, in *Rhodospatha* there is no regular pattern of bundle behavior after a leaf trace associates with an axial bundle. The closely related *Stenospermation* shows a similar organization.

The Araceae can now be added to the list of families of monocotyledons with the palm-type pattern of vascular organization. In addition, several patterns occur in the family that have not been reported elsewhere. The Araceae have lived up to their reputation as one of the most diverse groups of monocotyledons with respect to anatomy and morphology.

Bud traces. The Araceae have other novel patterns of vascular organization, namely the way their branches are vascularized by the main axis. Like most monocotyledons, they are faced with the problem of adequately connecting and supporting a branch which grows out from a bud after the main stem is mature. In many Araceae the buds grow out and form a new main shoot after the parent shoot tip is damaged or becomes reproductive. Thus, an extensive vascular interconnection with the main axis is important. In the Araceae (and most monocotyledons) secondary

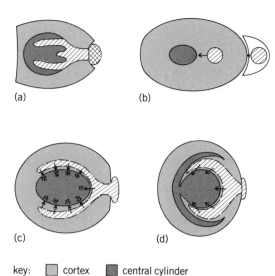

key: ▢ cortex ▪ central cylinder

Fig. 2. Diagrammatic representation of bud-trace organization in various Araceae: (a) *Pothos*; (b) *Acorus*; (c) *Anthurium*; (d) *Culcasia*. Bud traces are hatched and are shown entering the stem from the bud at right. Bud traces traverse the cortex and enter the central cylinder, where they attach to axial bundles. (*After J. C. French, and P. B. Tomlinson, Vascular patterns in stems of Araceae: Subfamily Pothoideae, Amer. J. Bot., 68:213–229, 1981*)

growth from a vascular cambium is absent. This secondary growth could provide an effective way of linking the vascular system of a branch with the rest of the plant and supporting the considerable weight of a branch. The absence of a vascular cambium places a severe growth limitation on monocotyledons. One way certain Araceae have solved this problem is to form extensive vascular connections very early between young developing buds and the main axis, involving numerous bud traces. The course of these bud traces through the stem is remarkably different in various genera, and has already proved to be a useful character in studying the taxonomy of the family. Four different bud-trace patterns are shown in Fig. 2. In *Acorus* the bud traces are arranged like a narrow central cylinder of the main axis; in *Pothos* and in root-climbing vines related to *Monstera*, the bud traces form two prominent arcs inside the main central cylinder; in *Anthurium* the bud traces form two arcs outside the central cylinder; and in *Culcasia* narrow flaps of vascular tissue bend outward to form a pocket into which the bud traces enter.

For background information *see* STEM in the McGraw-Hill Encyclopedia of Science and Technology. [JAMES C. FRENCH]

Bibliography: A. Bell, The vascular pattern of a rhizomatous ginger (*Alpinia speciosa* L. Zingiberaceae) 2. The rhizome. *Ann. Bot.* 46:213–220, 1980. J. C. French and P. B. Tomlinson, Vascular patterns in stems of the Araceae: Monsteroideae, *Amer. J. Bot.*, 68:1115–1129, 1981; J. C. French and P. B. Tomlinson, Vascular patterns in stems of Araceae: Subfamilies of Calloideae and Lasioideae, *Bot. Gaz.*, 142:366–381, 1981; J. C. French and P. B. Tom-linson, Vascular patterns in stems of Araceae: Subfamily Philodendroideae, *Bot. Gaz.*, 142:550–563, 1981; J. C. French and P. B. Tomlinson, Vascular patterns in stems of Araceae: Subfamily Pothoideae, *Amer. J. Bot.*, 68:213–229, 1981.

Stromatolite

Initially known only from the fossil record, stromatolitic blue-green algal communities were first thought to have risen to dominance in the photic benthos of the Precambrian ocean, subsequently declining in abundance and diversity and becoming extinct by the present (Holocene). However, recent discoveries of modern (Holocene) stromatolites in a number of select marine and lacustrine habitats, including Antarctic lakes, have stimulated great interest because these modern stromatolites can serve as models for understanding paleoenvironmental conditions and early processes in diagenesis and fossilization.

Antarctic lake stromatolites. In the last few years, stromatolitic blue-green algal mats have been discovered in five lakes located within a 1250-km^2 (450-mi^2) area of the ice-free southern Victoria Land oasis of Antarctica. Lakes Bonney, Chad, Hoare, Fryxell, and Vanda each contain two or more macromorphological types in the benthos beneath their perennial 4- to 6-m thick (13 to 20 ft) ice covers. Evidence suggests that other lakes in this region contain sim-

Fig. 1. A 0.7-m core of a prostrate anaerobic stromatolite from Lake Fryxell, Antarctica, showing internal laminae or sediment layers. Stromatolite structure also resembles prostrate aerobic stromatolites in these Antarctic lakes.

ilar, if not additional, stromatolite types. These are the only known cold fresh-water modern (Holocene) stromatolites on Earth found at such high latitudes, and may represent analogs of certain Precambrian stromatolites.

Macromorphology and species composition. The four presently known macromorphological types of Antarctic stromatolites are prostrate aerobic mats, prostrate anaerobic mats (Fig. 1), columnar lift-off mats (Fig. 2), and pinnacle mats (Fig. 3). All four types and variations thereof, while differing somewhat in their algal species composition and abundance, are more striking by the universal presence of the dominant photosynthetic blue-green filamentous alga *Phormidium frigidum*, which apparently represents a key microorganism in the stromatolite formation. Present also are several photosynthetic pennate diatoms abundant in the uppermost layers (except prostrate anaerobic), occasional coccoid green algae, and

Fig. 3. Pinnacle stromatolites, Lake Vanda, Antarctica, resembling the common fossil form *Conophyton*.

Fig. 2. Columnar stromatolites from the shallow, highly oxygenated waters of: (*a*) Lake Hoare, Antarctica; some are lifting off and floating and others are remaining in place while precipitating calcite. (*b*) Lake Fryxell, Antarctica; all are remaining in place and precipitating calcite.

numerous nonphotosynthetic bacteria and yeasts. The rarity or absence of grazing or burrowing animals is striking, as microfauna are rarely observed in any of these lakes.

In all cases, the stromatolitic mats show slow seasonal growth; photosynthetically mediated precipitation of calcite (and perhaps other minerals); trapped seasonal and shorter-term influxes of fairly fine sediment from glacial meltstreams; and bound, coarser sand and gravel derived from the benthos or occasional, nonannual passage of gravels through or between cracks in the perennial lake ice. All stromatolites begin as prostrate, spreading *P. frigidum* filaments with associated algae and microorganisms. In the deeper portions of the lakes where photosynthetic oxygen production is light-limited (aerobic) or lacking altogether (anaerobic), prostrate mats form which have fine parallel internal laminae or sediment layers (Fig. 1). Stromatolite thicknesses of up to 2.5 m (8.2 ft) have been obtained with no indication that the maximum depth was achieved.

In the shallower, less light-limited benthos of all five lakes, greater photosynthetic oxygen production accompanies the trapping of bubbles of oxygen 1 cm (0.4 in) or less in diameter within the stromatolitic mat matrix, causing upward growth and lift-off. In four lakes (except Vanda) the macromorphology achieved is columnar up to 20 cm (8 in.) high and 10 cm (4 in.) wide at the base (Fig. 2). Some of these lift-off columnar mats tear free of the gravelly substrate and float upward under the lake ice; ultimately some of these floating stromatolitic mats freeze into the ice, eventually reaching the upper lake surface where they are blown away. However, at least a proportion of the lift-off mats remain in place and precipitate sufficient calcite and trap sediments to ensure their permanent anchorage to the benthos (Fig. 2). Pinnacle stromatolitic mats are distributed widely in the more brightly lit depths of Lake Vanda (Fig. 3). These are conical with pointed tips, up to 13 cm (5 in.) high and 5 cm (2 in.) wide at the base. Their macromorphology is clearly reminiscent of the common Precambrian stromatolite *Conophyton*, a modern form of which also has been described from hot springs and marine habitats.

Ecology of Antarctic stromatolites. While other microorganisms make up the stromatolites of Antarctic oasis lakes and warrant study, the physiological ecology of the abundant and ubiquitous filamentous blue-green alga *P. frigidum* is best understood. Features possibly significant to this alga's formation of modern stromatolites which are counterparts to Precambrian ones include: the ability to photosynthesize and grow at very low light intensities; active growth at low temperatures (-2 to $20°C$ or 28 to $68°F$); apparent preference for large amounts of calcium (calciphilic), which probably is related to calcite production and lithification; poor tolerance to high salinity (hence it is confined to fresher waters); and the ability of the thin-sheathed, stiff filaments to undergo limited movement, such that movement toward light (phototaxis) allows perpetual building of the stromatolite structure. In other words, once it is formed within these Antarctic lakes, the stromatolite persists. The conditions governing the slow growth of stomatolites also allow slow rates of decomposition and no disturbance by turbulence or animal grazing or burrowing. Modern stromatolites are of more than academic interest, for it is recognized generally that oil shales are associated with stromatolite deposits. Indeed, the hydrocarbons in oil shales probably derive from the stromatolite organic carbon.

For background information *see* CYANOPHYCEAE; OIL SHALE; STROMATOLITE in the McGraw-Hill Encyclopedia of Science and Technology.

[BRUCE C. PARKER; ROBERT A. WHARTON]

Bibliography: S. M. Awramik, L. Margulis, and E. S. Barghoorn, Evolutionary processes in the formation of stromatolites, in M. R. Walter (ed.), *Stromatolites*, pp. 149–162, 1976; F. G. Love et al., Modern *Conophyton*-like algal mats discovered in Lake Vanda, Antarctica, *Geomicrobiol. J.*, vol. 3, no. 1, 1983; B. C. Parker et al., Living stromatolites in Antarctic oases lakes, *BioSci.*, 31(9): 656–661, 1981; R. A. Wharton, Jr., et al., Biogenic calcite structures forming in Lake Fryxell, Antarctica, *Nature (London)*, 295:403–405, 1982.

Sun

The Sun supplies virtually all the energy consumed on Earth and that energy sustains human existence. Until very recently the Sun has been considered a steady and stable source of energy. There is now compelling evidence obtained from historical records and recent measurements from various observation platforms that the Sun's energy output is not constant but varies on time scales from minutes to hundred of years and possibly longer. These variations are related to many things, including global oscillations of the Sun, the rotational period of the Sun, the existence of sunspots, the 11-year solar cycle, and, of course, the eruptions called solar flares.

An active area of theoretical research concerns the impact of external forces such as volcanic eruptions, industrial-age carbon dioxide production, and variations in solar luminosity on weather and climate. The meteorological or climatic effects of a changing Sun can be quite severe. The current stage of this research predicts changes in the Earth's climate ranging from a little ice age to melting of the polar ice caps and flooding of coastal areas. Although these effects would probably be on a minimum time scale of hundreds of years, it would be important to be aware of such trends at the earliest possible time. Smaller changes in the weather are also possible on a much shorter time scale, such as changes in mean annual temperatures large enough to have a sociological impact.

There is also academic interest in a changing Sun. A change in the quantity or spectrum of solar radiation provides clues and signatures of physical processes taking place on or under the Sun's surface. In addition, the things that are learned about the Sun can be applied in various forms to understanding the workings of other stars.

Effects of solar radiation on weather. The radiation from the Sun covers the full electromagnetic spectrum from radio waves to gamma rays and also includes the emissions of plasma in the form of the solar wind and energetic particles from solar flares. The solar wind in turn modulates the number and intensity of galactic cosmic rays striking the Earth. The impact of this radiation on the terrestrial environment varies markedly from one part of the spectrum to another. The two ends of the electromagnetic spectrum, radio waves and gamma rays, carry little energy and have a minimal effect on the Earth's atmosphere. These parts of the spectrum do, however, serve as indicators of solar activity and solar flares which, as discussed below, are related to measurable changes in the total solar energy flux. The higher-energy parts of the spectrum, that is, x-rays and extreme ultraviolet radiation, affect the higher levels of Earth's atmosphere. These forms of radiation, like gamma rays and radio emission, are also highly variable following solar activity. The x-ray and extreme ultraviolet radiations, in addition to the effects of a variable solar wind, produce ionization in the Earth's ionosphere and distortion of the magnetosphere, which affect telecommunications and the coupling of the lower-altitude neutral atmosphere to the heavily ionized magnetosphere. The effects of the neutral atmosphere-magnetosphere coupling are most pronounced at the poles, where the atmosphere has its best access to interplanetary space. The energy deposited in the Earth's atmosphere by x-rays, extreme ultraviolet radiations, and plasma emissions is much less than 1% of the total. This energy in itself cannot produce major changes in the Earth's climate, as the amount is small and is deposited far above the level of the atmosphere that contains the weather. However, the response of the atmosphere to solar variations in these portions of the spectrum and the response of the magnetosphere to variations in the solar wind and energetic particles are rapid enough to produce daily changes in the weather. A cascading and amplifying process would be required for such small power fluctuations to produce significant weather effects at much lower altitudes, and the existence of this process is highly controversial.

The bands of the solar spectrum containing the most power and having the most recognizable effect on the environment are the ultraviolet, optical, and near infrared. The ultraviolet has its strongest effects on the stratosphere and mesosphere, where it serves as a prime agent in the photochemical reactions involving nitrogen, oxygen, ozone, and oxides of nitrogen. The ultraviolet contains about 1% of the total solar power and produces measurable heating in the stratosphere.

By far the most solar power (99%) is carried in the optical and near-infrared bands of the spectrum. This radiation penetrates to the troposphere below 10 km in altitude, or essentially ground level, providing heat to the Earth's surface. This effect is modulated by the presence of clouds which reflect the incoming light; by the concentration of carbon dioxide, which traps Earth's escaping radiation by the greenhouse effect; by the evaporation and condensation cycle of water; and by other meteorological processes. If there were a general heating or cooling of the Sun, its effects upon the Earth's climate would be most pronounced in this band of the spectrum.

Measurement of solar flux. Ironically, it is the optical and near-infrared portions of the spectrum which have been the most difficult to measure to the required precision and accuracy. Ground observations of this radiation are strongly affected by the overlying atmosphere. A difficult experimental problem arising in short rocket flight observations is the accurate calibration of different instruments, or of the same instrument at different times. The possible terrestrial effects of a changing Sun require in some theories variations in the total solar power of as little as 0.1%, and absolute instrument calibrations to this accuracy are difficult.

Perhaps the best experiment to measure the integrated solar flux is the active cavity radiometer irradiance monitor (ACRIM) on the Solar Maximum Mission satellite. The instrument was designed and built by R. C. Willson of the Jet Propulsion Laboratory in Pasedena, California, and the spacecraft is operated by NASA Goddard Space Flight Center in Greenbelt, Maryland. The instrument consists of three identical and redundant cavities which absorb

solar radiation. The amount of solar heat absorbed in a cavity is adjusted with a calibrated heat source to measure the flux with respect to a standard level. The system has demonstrated exceptional stability since satellite launch in February 1980, with short-term precision of about 10 parts per million, long-term precision of 0.02%, and absolute accuracy of roughly 0.1%. The cavities are sensitive to radiation from infrared to low-energy x-rays, thus amply covering the dominant portions of the solar spectrum.

Variations in solar power output. Since satellite launch the data collected have allowed investigation of solar power output fluctuations ranging from a few minutes' duration upward to a long-term drift over a few years' span. The results of this investigation provide proof that the Sun does vary in its illuminated power over the entire frequency range, but the variations are far from uniform. The power fluctuations are synchronized with different features or physical processes, on or below the solar surface, which occur at different rates or frequencies.

Longer-term variations and the solar cycle. The first 2 years of data are shown in illustration *a*. These data represent the solar energy flux above or below the average value of 1367.9 W/m^2. This quantity shows significant fluctuations on the order of 0.1% from the average, and the fluctuations are superposed on a definite downward trend of about 0.04% per year. At this rate the Sun will be shining at one-half its current level in 1700 years. This rate of decline cannot continue, as there is little evidence for total fluctuations of more than 0.1% over recorded history. It could, however, represent a real trend over the time scale of an 11-year solar cycle.

A cause for the large fluctuations in solar power may be conjectured from the similarity of the flux measurements with the projected sunspot area (illustration *b*). The Solar Maximum Mission satellite was launched during the maximum of the sunspot cycle, and it was expected that a large number of dark sunspots would reduce the toal amount of light radiated from the Sun and result in low solar power when the number of sunspots was the greatest. This, in fact, accounts for the majority of the fluctuations seen in the ACRIM data. The total area of the dark sunspots (relative to the rest of the Sun) facing the Earth times a number which accounts for the reduced luminosity of the relatively cool sunspots predicts many of the fluctuations.

The possibility of this correlation was predicted in earlier literature, but it prompts the following question: If the production of energy in the core of the Sun is constant and the sunspots block some of it from escaping into space, does this energy get bottled up within the Sun or does it emerge by going out around the sunspot immediately? The answer seems to be that the energy flow blocked by a sunspot or group of sunspots emerges from around the sunspot a few days later. This new, emerging radiation can be seen in the solar phenomenon called plages, which are seen as diffuse bright areas surrounding sunspot groups. The plage area on the Sun

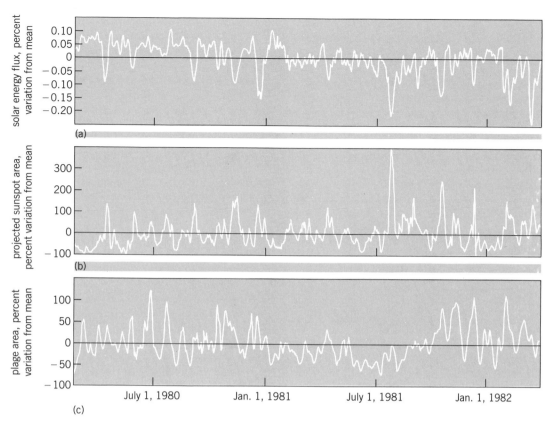

Variation of solar parameters in the years 1980–1981. (*a*) Solar energy flux as measured by ACRIM. (*b*) Projected sunspot area. (*c*) Solar plage area.

during the mission is shown in illustration *c*. Every dip in the ACRIM data has been identified with the emergence of a new sunspot group or the return of an old sunspot group as the Sun rotates in its nominal 27-day period. However, these dips are often smaller than would be expected from the size of the sunspot group. This discrepancy is virtually eliminated when the measured area of the bright plages present on the Sun is taken into account. In fact, this effect predicts the peaks which are often seen before and after a dip.

Going one step further, the main solar feature which shows a downward trend in the years 1980–1981 is the plage area. Thus, it can be conjectured that the total solar luminosity is decreasing, because there is progressively less plage area on the Sun as the solar cycle progresses toward solar activity minimum. This is the opposite of what was predicted earlier, based only upon sunspot blocking being small at solar minimum. Thus, in the picture that has been developed in this discussion, the solar luminosity temporarily decreases as a sunspot group emerges until the surrounding plage area grows, counteracting the decrease; and, as solar minimum approaches in 1987, the total plage area approaches a minimum, minimizing the energy flux during the 11-year cycle of solar activity.

These new observations then tell something about the structure of the outer parts of the Sun, in partic-

ular, how deep sunspots go into the surface. Below the solar surface the transfer of energy outward takes place by convection or by a boiling motion, where hot cells of solar material bubble up from below. The delay between the emergence of a sunspot group and the emergence of the surrounding plage shows how deep the energy blockage goes if the speed with which the energy can bubble up around the sides of a sunspot is known.

Rapid variations and solar oscillations. A deeper probe of the solar structure involves fluctuations of the solar luminosity on time scales of minutes. Here the effect of the Sun's oscillations can be observed. The Sun is capable of oscillating in a variety of radial motions, similar to the motions of a guitar string or a drum membrane. The only difference is that the Sun is a three-dimensional body, instead of a one-dimensional body like the guitar string or a two-dimensional body like the drum, and the restoring forces are gravity and pressure rather than tension. Different modes of a 5-minute oscillation have been observed by telescopic means for small sections of the Sun. The ACRIM data provide the first observation of this effect on a global scale. The only oscillations expected to affect the solar luminosity, however, are those called low-order p-mode oscillations. With these oscillations, a major fraction of the Sun, as it oscillates, is moving toward or away from the observer. Higher-order oscillations have some of the solar surface moving in one direction relative to the observer while the remainder of the visible Sun moves in the other direction, giving on the average little or no effect at all. The power in these oscillations represents only between 1 and 2 ppm of the total solar flux; however, these oscillations serve as a valuable tool in probing the solar interior. They penetrate deeply into the Sun, and the relative strengths and frequency separations of the different modes of oscillation reveal properties of the Sun's core. This new form of research has been named solar seismology and will provide a new window to the solar interior, just as traditional seismology has provided a window to the Earth's interior. The new information contained in these and other similar measurements should help scientists deduce the details of the stratification and energy production and transport within the Sun.

Plans are under way to continue the Solar Maximum Mission observations for several more years. This will provide an extensive set of data for detailed studies of the phenomena described above. Such measurements, coupled with vigorous theoretical work and observations on other platforms, will provide a new level of understanding of the internal and external workings of the Sun. This detailed history of solar luminosity can then be compared with the existing detailed meteorological data on Earth to help unravel the subtle, intricate, and important relationship of the Earth's climate with the Sun.

For background information *see* SUN in the McGraw-Hill Encyclopedia of Science and Technology.

[JAMES M. RYAN]

Bibliography: J. A. Eddy (ed.), Historical and arboreal evidence for a changing Sun, *The New Solar Physics*, pp. 11–13, 1978; H. A. Hill, Seismic sounding of the Sun, in J. A. Eddy (ed.), *The New Solar Physics*, pp. 135–214, 1978; R. Revelle, Carbon dioxide and world climate, *Sci. Amer.*, 247(2):35–43, August 1982; R. C. Willson, Solar irradiance variations and solar activity, *J. Geophys. Res.*, 87(A6):4319–4326, 1982.

Superconducting computer

Superconductivity and the Josephson effect are being explored as potentially new ideas to apply to the design of ever-improving high-performance computers, supercomputers.

Computer performance. A stored-program digital computer generally consists of a central processing unit (CPU), memory, and input/output (I/O) system. All problems that have to be solved on the computer are reduced to a series of elementary instructions, stored in memory, that govern the various operations to be performed on the data, such as add, multiply, and compare. The instructions are interpreted and the execution is controlled by the CPU, which contains the arithmetic-logical unit (ALU) which performs the operations on the data, in addition to registers which are used to store intermediate results. Ultimately the operating performance of the computer is

governed by how the machine is organized to deal with the various problems and how fast the circuit technology used in the computer operates. Presently, high-performance synchronous digital computers have operating cycle times in the range of 12 to 60 nanoseconds, and it normally takes from 1 to 10 cycles to execute an elementary instruction. High-performance computers are usually measured in terms of how many million instructions per second (MIPS) they can execute. This figure typically runs from a few to over 100 MIPS, depending on the machine, its technology, its organization (architecture), and the problem set (specific or general purpose).

Josephson technology and superconductivity are being explored as ways of reducing the cycle time further; initially the goal is to reduce the cycle time to below 4 ns, and ultimately it is thought that a subnanosecond cycle time will be possible. The reasons that a change from the well-established silicon technology may be needed to enter the subnanosecond region have to do with the nature of the limitations on the computer's cycle time.

Limitations on cycle time. There are principally two components to the cycle time of a computer: the circuit delay and the package delay. The circuit delay, reflecting the time it takes to perform the logical operations, is a strong function of the chosen technology (CMOS, NMOS, bipolar, or Josephson), the power dissipated per circuit, and the level of integration (for example, 100 circuits per chip, 1000 circuits per chip, and so forth). Usually for a given technology, higher circuit speed can be obtained by increasing dissipation or reducing circuit dimensions, or both. The reduction in device size, paced by available lithographic tools, is also very desirable from the point of view of reliability and cost. However, there are two principal limitations to available circuit speed, both of which are related to power requirements.

The first is associated with the wasted power: the power dissipated in the form of heat. This has to be removed by a cooling medium, and the level of heat removal from circuit chips is limited on the one hand by the cost and complexity of the heat removal structures, and on the other hand by the allowable rise in temperature that the circuits can sustain while still operating correctly. Heat removal structures usually also limit the volumetric efficiency of packing circuit chips.

The second limitation to the power comes from the circuit engineers' ability (or inability) to provide regulated power to the circuit chips. A large fraction of the chip input/output connections is devoted to power input—to carry the current, but even more importantly to reduce the power voltage swings caused by varying current drain on the chip as the circuits change their binary-encoded state under program control. This inductive variation of chip voltage (sometimes called the Δi noise problem) becomes an increasingly difficult problem to solve as the circuit chip performance is improved. Of the two principal

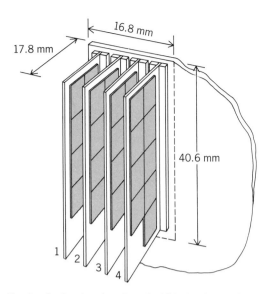

Fig. 1. Card-on-board package for high-density packing of chips. Typical dimensions expected for a small computer are indicated. (*After F. F. Tsui, JSP: A research signal processor in Josephson technology, IBM J. Res. Develop., 24:243–252, 1980*)

power limitations to circuit performance this one appears the most demanding and severe.

The package delay component of the cycle time depends on the way the integrated circuit chips are arranged. A high-performance supercomputer usually consists of a large number of high-speed logic and memory chips that are required to be able to communicate with one another within a single cycle time. Signals propagate at typically one-third to three-fourths of the velocity of light in high-performance machines, and yet the total path length can easily become the principal limit to how small a cycle time can be realized. The chips must be packed closer and closer together—not only in a two-dimensional arrangement but most effectively in a three-dimensional arrangement. In other words, the ultimate goal of high-performance computer packaging schemes is to achieve the highest possible density of circuits in a given volume, and this not only means high circuit density on each chip, but also high-density packing of chips. It is necessary to approach as closely as possible the ideal of a three-dimensional arrangement, and a good approximation is the card-on-board package (Fig. 1), where chips are attached to both sides of the card. The cards in turn are plugged closely together into the board.

Advantages of Josephson technology. The Josephson effect, predicted in 1962 and observed very soon thereafter, was quickly identified as being the basis of a very promising computer technology. Initially, the very fast switching speed and low power dissipation of this device were the main attractions, but it soon developed that many other aspects of the physics and engineering were very favorably inclined

toward superconducting computer technology. Principal among these was the essentially lossless-transmission-line characteristics of all superconducting circuit interconnection lines. Intercircuit and interchip signals could be transmitted on matched lossless lines, allowing communication in a single pass with no reflections and with a speed of approximately one-third the speed of light. The characteristic wave impedances of such lines suitable from both fabrication and circuit criteria are sufficiently low (for example, 10 ohms) that interline crosstalk and disturbs are very small and, for all practical purposes, negligible. The interconnection lines have very small cross sections; they are nearly lossless, and so there is no concern with respect to series resistance losses which significantly lower the projected and actual limits on the conductor cross-sectional area that semiconductor technology can employ. Furthermore, the very serious limit of electromigration is also absent in superconducting lines.

It soon became apparent that circuits with such low power dissipation would not, for a very long time in the development of this technology, be concerned about fundamental cooling limits; the power removal problem was no longer a concern. This has enabled particularly attractive and novel techniques to be applied to the power supply and regulation concerns faced by the circuit designer; active power supply and regulation circuits can be incorporated into the

Fig. 2. Scanning electron micrograph of high-speed pluggable connections: array of platinum micropins, line of right-angle solder joints from mid-left to top right, and larger guidepin in foreground. (*From M. B. Ketchen et al., A Josephson technology system level experiment, IEEE Electr. Device Lett., EDL-2:262–265, 1981*)

on-chip design, thereby removing the Δi problem.

These advantages naturally accrue from the basic physics of the superconducting state and the Josephson effect. The energy gap that characterizes these devices has shrunk from the hundreds of meV of the semiconductor electronic devices to just a few meV with the superconducting Josephson devices.

The next significant advantage that Josephson technology realized was in the techniques and technologies that are available to the circuit package engineer. With little or no cooling constraint, simple attachment of high-performance chips to the package is feasible; there are no heat removal structures required, and the package can be fabricated in a high-density three-dimensional card-on-board arrangement such as that shown in Fig. 1, where the total power dissipation is estimated at only 400 mW. Josephson technology also makes possible the high-density superconducting interconnection lines and the ability, for the first time, to make a monolithic package in the sense that the package components are all made from the same substrate material: silicon. This has significant importance in that it allows the engineer

to work with little concern for differential thermal expansion in the package design. The use of silicon not for its electrical properties, but rather its crystal perfection and machining properties, as well as mechanical strength and thermal properties, opens up a new door to micromachine fabrication techniques that can be applied to this novel package.

As package dimensions are shrunk and chip densities increase, it becomes increasingly difficult to provide an adequate number of very small, high-speed, pluggable connectors for each card. The techniques presently being explored consist of the use of very small solder joints to take signals around corners, and of platinum micropins that plug into extremely small sockets each containing a captured mercury drop, 200 μm in diameter. The pins are only 80 μm in diameter and 200 μm long, and they are arranged in a square array of a thousand or more pins with 300 μm between centers (Fig. 2).

Finally, there is a significant anticipation that one of the advantages in Josephson technology will be the low-temperature operation. Most of the failure-inducing mechanisms that plague room-temperature computers will be frozen out and will not affect an operating computer at 4.2 K; thus, computer operation at this temperature holds the promise of high reliability.

Disadvantages of Josephson technology. This significant optimism generated by the attractions of Josephson technology and superconductivity is tempered by the fact that a lot of detailed engineering and materials work remains to be done before a superconducting computer becomes a reality. Josephson devices are not a simple extension of transistor circuits. They involve a much more significant break with the past than the transition from the triode vacuum tube to the transistor did. The change involves much more than reducing power supply voltage and reducing impedances. Current and magnetic flux become of prime importance, rather than potentials and charge. Duality has been helpful only at providing physicists and engineers with hints of what must be done.

The biggest challenge is perhaps the sheer magnitude of the tasks involved in introducing a fundamentally new technology: chip technology, circuit design, package technology, design techniques, testing, and so forth. Good control over the prime circuit parameters has to be shown on a scale that allows superconducting computers to be contemplated. In this respect the control of the device threshold current presents a significant challenge. The superconducting pair tunneling current depends exponentially on the barrier thickness to such an extent that average oxide thickness control to an accuracy of better than 0.1 nm is required to control the tunneling current to better than 10 or 20%. Also, the methods needed to ensure sufficient device isolation and power gain for reliable logic-circuit and memory design are still evolving.

There are obvious difficulties associated with the

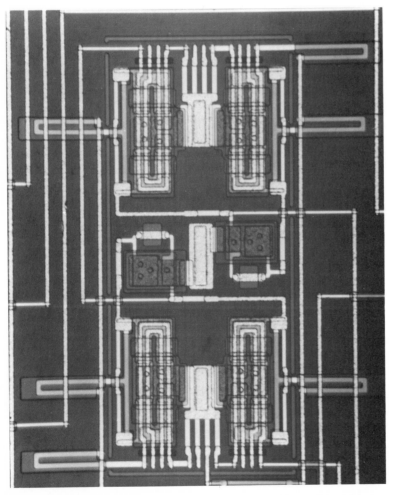

Fig. 3. Josephson chip circuit consisting of six Josephson devices configured into two logical AND gates. (*IBM*)

use of liquid helium. Cooling the computer from 300 to 4.2 K will undoubtedly produce stresses. These have to be understood and taken care of; they affect the choice of materials both for the devices and for the package. The choice of more familiar but relatively ductile lead-alloy Josephson devices has to be weighed against the more robust refractory-based devices, such as niobium. More is known with respect to lead-alloy processes and device design, and this technology also provides a faster device since the device capacitance represented by lead oxide (PbO) is approximately one-third that of niobium oxide (Nb_2O_5). It is only very recently that sufficient progress has been made with the refractory-based devices that they have begun to challenge the position held for the last decade by lead alloy, particularly with respect to low capacitance and low device leakage in the nonzero voltage state.

Perhaps the most obvious disadvantage of a superconducting computer, that of the liquid helium environment, is in fact not a particularly significant one. Admittedly, small liquid helium refrigerators are inefficient at converting kilowatts of electrical energy to only a few watts of cooling power at 4.2 K; however, this is not the bottleneck in the conventional cooling problem. That bottleneck, which is in the transistor technology of getting the heat away from the circuit without affecting correct operation and impacting packaging efficiency, has been broken with the direct cooling of Josephson chips in liquid helium.

Recent progress and prospects. Logic and memory test vehicles are receiving very active attention at a number of locations. Recent years have seen a significant growth in these fields by the Japanese. Figure 3 shows an example of recent progress in Josephson chip technology: six Josephson devices configured into two logical AND gates. This circuit was designed and fabricated by using a 2.5-μm design rule. Figure 4 is a portion of a Josephson logic chip with a regular array of logic gates, such as those shown in Fig. 3, connected together via wires running in channels configured in a rectangular array. Across the center of the figure runs a utility line or the form of the on-chip power regulation and distribution system.

In 1981 some results associated with a cross-section model of a prototype machine were announced. This experiment, which attempts to explore all of the fundamental aspects of the packaging and circuit technologies, was designed to reveal potential problems associated with the fabrication of this card-on-board design. A representative path through the package was exercized with a minimum cycle time of 3.7 ns.

There is still much to be done before a superconducting supercomputer will be involved in calculations on weather forecasting, hydrodynamics, and so forth, but the direction of development appears to be set. Fundamental considerations involving information processing and thermal noise argue that ultimately 4.2 K operation will be used for computers.

Fig. 4.   Portion of Josephson logic chip showing regular array of logic gates. (*IBM*)

For background information *see* DIGITAL COMPUTER; INTEGRATED CIRCUITS; JOSEPHSON EFFECT; SUPERCONDUCTING DEVICES in the McGraw-Hill Encyclopedia of Science and Technology.

[DENNIS J. HERRELL]

Bibliography: W. Anacker, Computing at 4 degrees Kelvin, *IEEE Spectrum*, 16(5):26–37, May 1979; *IBM J. Res. Develop.*, vol. 24, no. 2, March 1980 (issue devoted entirely to Josephson computer technology); M. B. Ketchen et al., A Josephson technology system level experiment, *IEEE Electr. Device Lett.*, EDL-2(10):262–265, October 1981; J. Matisoo, The superconducting computer, *Sci. Amer.*, 242(5):50–65, May 1980.

Switching systems (communications)

The most widely used technique of voice and data transmission for general application in telecommunications employs circuit switching, in which a specific path is established for each desired call or connection. Recent wide application of integrated circuits in switching have brought forth new circuit switching system architectures and capabilities. Two new generations of switching systems have evolved.

The first comprised circuit switching systems that could interface efficiently with multiplex transmission systems carrying a plurality of voice signals by the more robust digital, rather than analog, techniques. Analog voice signals may be repetitively sampled and coded into 64,000 bit/s digital signals. Most systems for switching digitized voice signals employ a form of circuit switching that provides paths on a time-division basis; that is, part of the selected path or portions of paths through the switch are shared in time by many simultaneous calls.

Central-control digital TDM systems. Circuit switching systems consist of two basic functions, the switching center network (SCN) and the control. The first-generation digital switching systems were time-division systems that included a single central stored-program control (SPC; Fig. 1). Digital time-division multiplex (TDM) transmission facilities (lines) need only be in synchronism internally. Since a plurality of digital TDM line facilities terminates on a time division switch, these lines, and indeed the entire network of nodes and lines, must be in synchronism. There are several basic synchronization methods for telecommunication networks. The one most frequently employed bases synchronization upon the timing of signals arriving over the multiplexed lines from several offices. Since only the rate and not the phase of synchronism is achieved, the digital signals arriving at the switch must be buffered (stored) in order to pass through the time-division SCN in phase.

The channels of TDM transmission and switching systems are known as time slots. Buffering may also be required to place signals emerging from the SCN onto digital transmission lines into selected time slots. Time slots are assigned to calls internal to a TDM switching system at the input to TDM transmission systems. A form of buffering known as time slot interchange (TSI), employing integrated circuit mem-

ory, is used to momentarily store successive digitized speech samples so that they may be delayed for release in different time slots, possibly on different TDM transmission lines.

Another type of switching technique required in TDM systems is to switch between TDM lines. This technique uses high-speed gates in one or more stages known as time multiplex switching (TMS). Coordinate (matrix) configurations of these gates are similar to those found in non-TDM switches and are therefore called space division stages. The TMS space stages switch at time slot rates, whereas in non-TDM space stages the connections are established through crossbar or reed switching devices for the duration of each connection.

In TDM switching systems the type of SCN is delineated by the order of TSI or T (time) stages and TMS or S (space) stages through which the path is established. Typically, SCN configurations such as T-S-S-S-S-T or S-T-S may be found. The particular choice is determined by engineering considerations, such as growth increments and system size range.

Initially, most TDM switching systems were used to switch calls between trunks to and from intermediate offices or between trunks and internal links. In the first-generation local offices, the conversion of signals from analog to digital (A/D), and vice versa, took place in trunk or link interface circuitry. As the cost of interface circuitry has decreased, newer systems have been designed to terminate ordinary telephone lines carrying analog voice signals directly on A/D conversion circuits. To use TDM switching, voice signals are digitized, assigned time slots, and then multiplexed. The line interface circuit, colloquially known as BORSCHT, provides the system with these capabilities, as well as those normally expected within local space-division switching systems. BORSCHT is an acronym for Battery (to feed the analog telephone transmitter), Overvoltage protection, Ringing, Supervision, digital Coding, Hybrid to separate directions of transmission (a requirement for digital TDM systems), and Test access (to reach both the in-office and outside plants).

Most early local switching systems (Fig. 1a) employed space-division concentrators using metallic crosspoints, such as crossbar or sealed reed switches between the lines and the interface circuits. These were located in the links so that their cost could be spread over more lines. The first-generation switches of this type were pioneered by the French in 1970. Semiconductor crosspoint concentrators are in use, and in the United States include a variety that has electrical characteristics much like metallic devices, in that they can carry the high energy required by analog telephone lines for talking and ringing.

In anticipation of the rapidly decreasing cost of integrated circuits and the prospects for including much of the interface in them, some first-generation local systems introduced into North America were designed with a BORSCHT circuit on each line (Fig. 1b). Local time-division digital systems are also

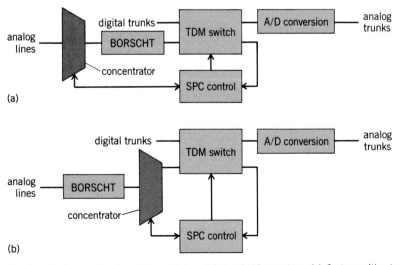

Fig. 1. First-generation local time-division digital switching system. (a) System without interface (BORSCHT) circuitry on each line. (b) System with interface (BORSCHT) circuitry on each line.

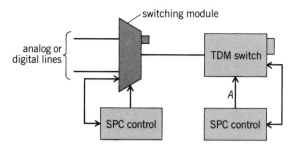

Fig. 2. Second-generation local time-division digital switching system. If the TDM switch is self-directing, the path labeled *A* is omitted.

suitable for application where there is considerable growth of lines and where digital multiplex transmission facilities are the economic choice. Since much local trunking employs digital TDM, the use of TDM digital switching for intermediate offices (tandem, toll, and transit) has also been popular. A system of this type with a capacity of 107,000 trunks is in use in the United States.

Distributed-control TDM systems. A new generation of local digital switching systems is now appearing. Not only is the interface circuit highly integrated, but line and trunk terminations are grouped into modules that permit the application of distributed control techniques. Some or all of the routine call-processing functions are performed by programs stored in microprocessors associated with the modules. In addition, the first switching stage, usually a time stage, is included in the module. The module therefore includes all of the basic switching functions of the network, interface, and control (Fig. 2). Systems with varying degrees of distributed control and network are known as distributed switching systems. The close association of these functions in a unit will lead to further opportunities for the use of very-large-scale integrated (VLSI) circuits.

Distributed switching has also renewed interest in separating the modules from the TMS and central control system core, sometimes called the host. The modules may be located away from the system core, closer to the customers they serve, thereby saving line facilities. Typically these remote units serve from a hundred to several hundred lines. For the larger units, the microprocessor controls used in the switch module are sufficiently autonomous to enable them to function on intralocal calls should the trunk linking the module with the host be disrupted.

Changes in the system core are also beginning to appear as a result of the application of distributed switching and large-scale circuit integration (LSI). These changes affect both the network and the control. The network employs both TMS and TSI techniques, interchanging channel information between multiplexed links to and from the modules and between time slots. It is now possible to design LSI chips with sufficient intelligence that the modules themselves may control the establishment of paths through the switching network. This technique is known as self-selection, since the switching actions in the chip are controlled by signals arriving over the links.

Alternative digital switching techniques. The time-division digital switching technique will be used in newly installed systems as it becomes economically competitive with space division. Distributed switching is applicable to space division as well as time division. These applications may also use powerful low-cost microprocessors for call processing in switching modules remote from the host. A remote switch employing a nonmetallic switching network has been deployed in the United States.

Another technique for the switching of digital signals that is about to be introduced is the use of existing metallic space-division switching systems. Most modern switching systems will pass 64,000 bit/s digitized voice signals, highlighting the fact that a digital switching system may employ time or space division. But there is interfacing synergy between time-division multiplex transmission and time-division switching systems. Furthermore, a time-division system provides an opportunity to transmit digital signals from nonvoice sources, such as computers and terminals. This capability gives rise to the concept for the future of networks integrating digital multiplex transmission, switching, and services.

Existing and planned systems. Table 1 lists the time-division digital central-office switching systems developed to date. Most of these consist of a central

Table 1. Central stored-program-control time-division multiplex switching systems

Country	Code	Manufacturer	First service	Type	Size (1K = 1000)
United States	No. 4 ESS	Western Electric	1976	Toll	107K trunks
United States	ITS 4/5*	TRW-VIDAR	1976	Local/toll	7K lines
Canada	DMS 10*	Northern Telecom	1977	Local	7K lines
Canada	DCO	Stromberg-Carlson	1977	Local	32K lines
United States	1210*	ITT-North	1978	Local/toll	26K lines
Sweden	AXE†	L. M. Ericsson	1978	Local/toll	64K lines
United States	No. 3 EAX	GTE-Auto Elec.	1978	Toll	60K trunks
Japan	NEAX-61	Nippon Electric	1979	Local	100K lines
Canada/United States	DMS 100/200*	Northern Telecom	1979	Local/toll	100K lines

*Local systems have a BORSCHT circuit on each line.
†1982 design has a BORSCHT circuit on each line.

Table 2. Distributed stored-program-control time-division multiplex switching systems

Country	Code	Manufacturer	First service	Type
West Germany	EWS-D	Siemens	1980	Local/toll
United Kingdom	System X	British Telecom.	1981	Local/toll
Italy	UT 10/3	Italtel	1981	Local
Europe	1240	ITT-BTM-SE1	1982	Local/toll
United States	No. 5 ESS	Western Electric	1982	Local/toll
United States	No. 5 EAX	GTE-Auto Elec.	1982	Local/toll
France	MT 20/*25	Thomson-Tele.	1982	Local/toll
United States/France	E10S/TSS	TSS-Alcatel	1983	Local
Netherlands	PRX-D	Philips	1983	Local/toll

*Local systems have a BORSCHT circuit on each line.

time-division digital network and central stored program processor. Table 2 lists the distributed time-division switching systems expected within the next few years.

For background information *see* SWITCHING SYSTEMS (COMMUNICATIONS) in the McGraw-Hill Encyclopedia of Science and Technology.

[AMOS E. JOEL, JR.]

Bibliography: A. E. Joel, Jr., *Electronic Switching*: *Digital Central Offices of the World*, 1982; *Proceedings of the International Switching Symposium*, Montreal, September 1981 (IEEE Publ. CH 1736–8).

Tectonophysics

Seismic investigations have shown that the Earth's interior is composed of three main concentric layers: the crust, mantle, and core. The distinction between the crust and the underlying mantle was discovered early in this century by A. Mohorovičić, a Yugoslavian geophysicist, whose analysis of earthquake-generated seismic waves indicated that a few tens of kilometers below the Earth's surface, seismic velocities increase rapidly to a value of a little over 8 km/s. This seismic discontinuity, known as the Mohorovičić discontinuity, or Moho, typically lies at a depth of 30–40 km (19–25 mi) beneath the continents, although beneath major continental mountain systems, such as the Alps, Andes, and Himalayas, the crust may reach a maximum thickness approaching 80 km (50 mi). The Moho shallows significantly beneath the ocean basins, where crustal thicknesses are typically only 5–7 km (3–4 mi). Recent studies, however, have shown that near large transform-fault zones, in areas such as the North Atlantic, the oceanic crust is unusually thin. In places this anomalously thin crust may be less than 2 or 3 km (1–2 mi) thick, making it the thinnest crust yet found on Earth. The processes responsible for the formation of this anomalously thin crust are not well understood, but it may be caused by the thermal effect of juxtaposing at ridge-transform intersections older, colder lithosphere against the spreading center forming oceanic crust.

Oceanic crustal structure. Seismic investigations in the 1950s and early 1960s suggested a relatively simple three-layered model for the oceanic crust. More recent studies have modified this picture of the oceanic crust somewhat by suggesting that seismic velocity increases more continuously with depth. From these seismic investigations, studies of rocks recovered from the sea floor by dredging or drilling, and investigations of sequences of rocks on land thought to represent uplifted sections of oceanic crust and upper mantle, the following picture of the oceanic crust has emerged (Fig. 1). The shallowest part of the oceanic crust (layer 2) is believed to consist of a disrupted sequence of pillow lavas, massive basalt flows, and rubble zones, about 500 m (1500 ft) thick, grading downward into a 1–1.5-km-thick (0.6–1-mi) sequence of sheeted dikes. These rocks overlie layer 3, an assemblage of 3–5 km (2–3 mi) of gabbro, below which lie the ultramafic rocks of the upper mantle. The uppermost 2 km (1.2 mi) of the oceanic crust is characterized seismically by steep velocity gradients of 1.5–2.0 s^{-1}, while velocities increase

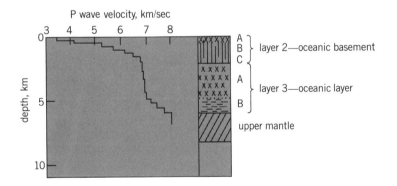

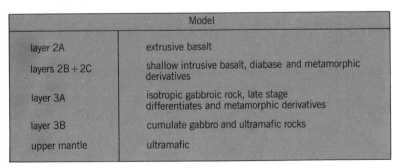

Model	
layer 2A	extrusive basalt
layers 2B + 2C	shallow intrusive basalt, diabase and metamorphic derivatives
layer 3A	isotropic gabbroic rock, late stage differentiates and metamorphic derivatives
layer 3B	cumulate gabbro and ultramafic rocks
upper mantle	ultramafic

Fig. 1. Generalized model of the velocity structure of the oceanic crust and an inferred geological model. (*After P. J. Fox, R. S. Detrick, and G. M. Purdy, Evidence for crustal thinning near fracture zones: Implications for ophiolites, in A. Panayiota, ed., Ophiolites: Proceedings of the International Ophiolite Symposium, pp. 161–168, 1980*)

much more slowly (0.1–0.2 s^{-1}) in the underlying gabbros. The crust/mantle boundary appears to be regionally variable in the ocean basins. In some areas there is a sharp velocity contrast between the crust and mantle, while in other places the transition is much smoother, occurring over a 1–2 km (0.6–1.2 m) depth interval with a basal high-velocity layer overlying the Moho. At still other places it has been argued that a low-velocity zone overlies the Moho.

While this picture of the oceanic crust is generally correct, a closer examination of available data reveals significant variability in both the velocity structure and total thickness of the crust. The nature and origin of these variations in crustal structure are still not well understood. Some are apparently related to changes in the properties of the upper part of the crust that vary systematically with age. Other variations in crustal structure may be due to the nature of the volcanic processes that form the oceanic crust. However, one particularly interesting result of recent studies is the indication that anomalously thin oceanic crust may be present along large offset, slow-slipping transform-fault zones in areas such as the North Atlantic.

Thin crust in fracture zones. Marine geological and geophysical surveys over the past two decades have clearly shown that the axis of the mid-ocean ridge system is frequently offset by large fault systems known as transform faults. The seismically inactive extensions of these faults out onto the ridge flanks are known as fracture zones. In well-surveyed areas, such as the central North Atlantic, transform faults occur on the order of one every 50–100 km (30–60 mi), making them among the most common tectonic features in the ocean basins.

Geological evidence. Morphologically, transform faults on slowly (1–2 cm/year or 0.4–0.8 in./year) spreading ridges display many similar characteristics (Fig. 2a). A distinctive closed-contour depression is usually present at ridge axis–transform intersections. The axis of the fault zone is typically defined by a deep, linear trough bounded by escarpments rising several kilometers above the transform-valley floor. Linear ridges frequently border the transform valley for distances of up to several hundred kilometers. This lineated terrain, composed of the axial trough and flanking ridges, ranges from several kilometers to more than 30 km (20 mi) in width, with the scale of the relief apparently related to the length of the ridge-axis offset.

A diverse suite of basaltic and ultramafic rocks have been recovered from transform-valley walls, suggesting that rocks representative of the lower part of the oceanic crust are exposed. Many early investigators assumed that the great relief found in fracture-zone escarpments was the product of a single large fault with a throw of several thousands of meters providing a structurally controlled "tectonic window" into the lower crust and upper mantle. However, recent high-resolution bathymetric surveys of several major North Atlantic transforms and submersible investigations of several other fracture zones have shown clearly that the walls flanking transform valleys consist

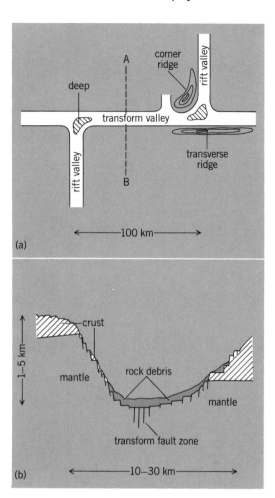

Fig. 2. Some features of transform faults. (a) Generalized morphology of a transform fault offsetting a portion of a slowly spreading ridge. (b) Schematic geological cross section (A–B in a) across a slowly slipping transform based on submersible and high-resolution bathymetric studies of North Atlantic fracture zones. (*After P. J. Fox and D. G. Gallo, 1982*)

of a large number of small throw faults (less than 500 m or 1500 ft) linked together by terraces and talus ramps (Fig. 2b). The outcropping of deep-crustal and upper-mantle rocks on these escarpments would be difficult to explain if normal crustal thicknesses (5–7 km or 3–4.5 mi) existed here. If, however, the oceanic crust near transform faults and fracture zones is anomalously thin, then this type of rock distribution is easily accounted for, especially if the basaltic layer (layer 2 in Fig. 1) is no more than a few hundred meters thick. Recent detailed submersible investigations of parts of the Kane and Oceanographer fracture zones, two large central North Atlantic transforms, have provided additional support for the presence of a very thin crustal section in these transforms.

Geophysical evidence. The best direct geophysical evidence for the existence of thin crust in a large Atlantic fracture zone is from a seismic refraction line shot along the trough of the Kane fracture zone (Fig. 3). Total crustal thicknesses of only 2–3 km (1.2–2 mi) were found along a 50-km-long (30-mi) segment of this fracture zone. These crustal thick-

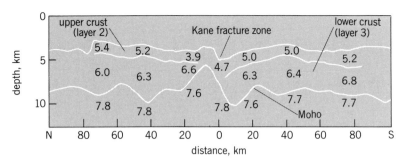

Fig. 3. Schematic crustal section across the Kane fracture zone in the central North Atlantic showing the thin, low-velocity crust associated with the fracture zone. Layers 2 and 3 correspond to those in Fig. 1. The boundary between the two layers is indicated. The numbers along the curves represent compressional wave velocities determined from seismic refraction studies. (*After R. S. Detrick and G. M. Purdy, 1980*)

nesses are only about half the thickness of the normal oceanic crust found on either side of the fracture zone. This anomalously thin crust is also characterized by lower compressional wave velocities than are typical of most oceanic crust. A similar refraction experiment on the Oceanographer fracture zone, southwest of the Azores, also suggests that the crust beneath the transform valley thins in places by about 50%, predominantly within the main crustal layer (layer 3).

It is still not known how extensive this anomalous fracture-zone crust is. Analysis of refraction data from the Vema transform valley in the central, equatorial Atlantic and a recent interpretation of gravity data across the Kane fracture zone indicate that more normal oceanic crustal thicknesses are sometimes present. However, there is good evidence that along at least parts of these fracture zones crustal thicknesses are probably 2–3 km (1.2–2 mi) or less.

One common characteristic of this fracture-zone crust, despite the differences in apparent crustal thicknesses, is the presence of relatively low crustal seismic velocities compared to most oceanic crust. These low seismic velocities have been interpreted as a result of the intense fracturing and brecciation of crustal rocks from faulting within the active transform. A similar low-velocity, low-density region is present along large strike-slip faults on land, such as the San Andreas Fault in California. Other factors which may contribute to the low seismic velocities observed in fracture zones are significant thicknesses of basaltic rubble which may accumulate in transform valleys and serpentinization (hydration) of the ultramafic components of the lower crust and upper mantle.

Origin of thin crust. The origin of this anomalously thin fracture-zone crust is still poorly understood. It has been proposed that at ridge-transform intersections the thermal effect of juxtaposing old, relatively cold lithosphere against a spreading center alters the volcanic processes forming the oceanic crust. The thickness of oceanic crust is dependent on the volume of basaltic melt that is segregated at depth from the mantle and delivered to shallow-level magma cham-

bers. Near the thick, cold wall of the transform boundary, heat conduction across the fault will reduce mantle temperatures, resulting in either less partial melting of mantle material or impairment of the ability of melt that has formed to reach the surface, or both. As a result, the volcanic budget decreases as the transform boundary is approached, and the crust thins correspondingly.

If this model is correct, the thinnest crust should be found at large-offset transforms on slow-spreading ridges where the age and thermal contrast across the fault are largest. The effect may be less significant at smaller-offset transforms or along transforms on faster-spreading ridges such as in the eastern Pacific. There are, as yet, not enough geological and geophysical data from transform faults to adequately test this hypothesis; however, studies now in progress are expected to provide the necessary information. The results to date strongly suggest that in at least some cases the presence of transform faults can significantly alter the volcanic and tectonic processes responsible for the formation of oceanic crust.

For background information *see* FAULT AND FAULT STRUCTURES; MOHO (MOHOROVIČIĆ DISCONTINUITY); PLATE TECTONICS; TECTONOPHYSICS; TRANSFORM FAULT in the McGraw-Hill Encyclopedia of Science and Technology.   [ROBERT S. DETRICK, JR.]

Bibliography: R. S. Detrick and G. M. Purdy, The crustal structure of the Kane fracture zone, *J. Geophys. Res.*, 85:3759–3778, 1980; P. J. Fox, The geology of the Oceanographer transform, *EOS*, 61:1105, 1980; P. J. Fox, R. S. Detrick, and G. M. Purdy, *Ophiolites: Proceedings of the International Ophiolite Symposium*, pp. 161–168, 1980; P. Spudich and J. Orcutt, A new look at the seismic velocity structure of the oceanic crust, *Rev. Geophys. Space Phys.*, 18:625–645, 1980.

Telephone service

For many years the entrance link of the telecommunications network, called the loop, was provided exclusively by twisted copper pairs. Now, however, several new digital transmission and processing technologies are also being used to provide the entrance.

The most widespread of these technologies is called digital loop carrier, which uses digital multiplexers to provide 24 circuits on two pairs of wires. The use of the digital carrier reduces costs and improves voice quality for conventional telephone circuits. It also brings channels capable of carrying information at rates of 64 kilobits/second (kb/s) and 1.5 megabits/second (Mb/s) within reach of users at a much lower cost than is possible with twisted pair loops. These rates are high enough for high-speed computer terminals, for bulk data transfer, and for limited-motion video.

Most digital loop carrier systems are presently connected to the telephone central office over existing copper pairs. However, some systems are beginning to use optical fibers, increasing the information rate available to users to 45 Mb/s. A 45 Mb/s channel can transport a video picture of standard broadcast quality.

Supplementing these technologies, other new digital transport systems increase the data rates possible on existing copper loops, further reducing the cost of bringing digital transmission to users.

The increased digital capability of the loop, coupled with rapid evolution toward digital capabilities of the remainder of the telecommunications network, makes end-to-end digital transmission a rapidly growing reality. International standards for interfaces and services are being set, and by the end of this century it is expected that computer and video services will be widespread.

Loop network. As shown in Fig. 1, the present loop network is devided into feeder and distribution regions. The feeder region uses large cables containing pairs of 26, 24, 22, or 19 gage (American Wire Gage) wire, whose sizes range up to nearly 4000 pairs. Feeder pairs are connected at various points to smaller distribution cables, which connect directly to users. New feeder cables are added every 5 to 10 years, as demand for circuits approaches the capacity of a section. Distribution cables, in contrast, contain sufficient pairs for a much longer period, on the order of 30 years, in order to avoid frequent construction activity near residences and businesses.

The techniques described above have served for many years to control costs while providing high-quality voice-frequency circuits. They have several drawbacks, however, including high labor cost for planning and construction, and large variability in transmission characteristics. The use of digital transmission techniques can go a long way toward eliminating these drawbacks.

Digital and optical transmission. Figure 2 shows the principles of digital and optical transmission.

In a digital, or PCM (pulse-code modulation), transmission system, each analog input signal is sampled at the so called Nyquist rate of twice the highest frequency present, which for telephone speech is 8 kHz. Each sample of instantaneous signal level is then encoded in an 8-bit binary word. (If the input signal is digital data, it is already in digital form and need not be sampled and coded.) Binary words for 24 users are time-multiplexed into a single 1.544-megabit stream, which is transmitted to the distant terminal. Although the bandwidth required for the 1.544-megabit signal is many times the combined analog bandwidth of the 24 voice circuits, requiring signal regenerators to be added every mile (1.6 km), digital transmission has significant advantages over other alternatives:

(1) Since the signal is regenerated rather than being amplified at each repeater, it can transverse hundreds of repeaters without degradation. With analog amplifiers, such as those used in coaxial television systems, impairments accumulate with each added amplifier. (2) Digital terminals and repeaters make extensive use of integrated logic circuits, which over

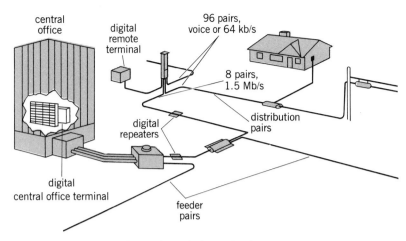

Fig. 1. Application of digital carrier to the loop network.

the past decade have been reduced in cost by factors of 10 to 100. (3) The user port of a digital transmission system operates at 64 kb/s, giving it a much higher information capacity than a voice-frequency circuit, which is limited to lower rates (1.2 to 4.8 kb/s) without circuit tailoring and requires more costly terminal equipment. Thus, digital transmission systems are much better suited to computer and data transmission.

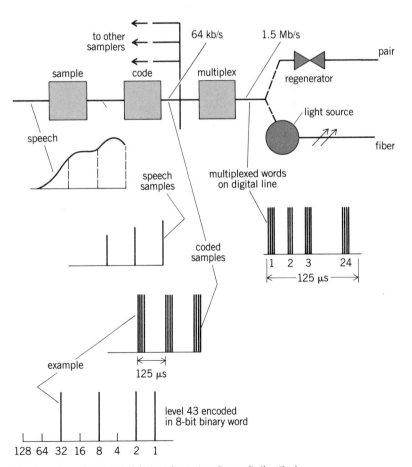

Fig. 2. Block diagram of digital carrier system (transmit direction).

In an optical transmission system, photons replace electrons as the information carriers, and glass replaces metallic conductors as the transport medium. The chief advantage of optical transmission systems over metallic systems is their extremely wide bandwidth; a fiber pair used for loop transmission can carry 45 megabits of information—sufficient capacity for 672 digitally encoded voice signals—for up to 10 mi (16 km) without intermediate repeaters. To obtain the same capacity with twisted pairs, 560 digital repeaters are required on 56 pairs.

Application to loop network. In loop applications, digital loop carrier terminals are placed both at a remote terminal near the user and in the central office (Fig. 1). Feeder circuits provided by a digital carrier have a lower labor content and higher material costs than circuits using multipair cables. The material costs continue to be reduced by large-scale integration and other innovations, however, so that carrier costs have started to approach cable costs. In individual cases, where distances or construction costs are large, carrier-derived circuits are considerably less expensive than cable alternatives. Also, because of its modularity and flexibility, the use of a digital carrier reduces both inventory and planning costs.

From a transmission point of view the path from the terminal near the user to the central office has low loss and distortion. Since the path from the terminal to the user is short, usually 1 mi (1.6 km) or less, the overall variation in transmission performance is small. For certain types of circuits, such as those with stringent transmission requirements, and for digital circuits, digital loop carriers provide greater capability at less cost than twisted pairs.

Although the majority of circuit demand is for voice channels, the growth rate of data circuits is several times that of voice circuits. The location of the digital terminal near the user brings not only the 64 kb/s port within a short distance, but also the underlying 1.544 Mb/s line and the 45 Mb/s optical line rate. The relationship of these bit rates to various service capabilities is given in the table.

The use of digital carriers for loops is based primarily on minimizing the cost of basic telephone service. The increase in the ability of the network to carry data and video signals is therefore achieved at low cost as a natural consequence of network growth.

Other digital transport systems. From remote terminals and central offices to nearby users, twisted pairs will remain the prevalent transmission medium. Pairs have a much wider bandwidth than that needed for analog voice, but suffer from transmission irregularities. They can, however, be made to carry 64 kb/s and higher bit rates by using recently developed transmission equipment. This equipment, which uses various transmission techniques such as time-compression multiplexing, is placed at the user location and at the central office or remote terminal. For example, a recent system can transport 64 kb/s in two directions simultaneously on a single, unmodified pair for distances up to 3 mi (4.8 km). Prior to the introduction of this system, two modified pairs were required for the same capability. Similarly, techniques are available for allowing a pair to share independent telephone and low-bit-rate digital circuits, or to carry several independent data signals.

End-to-end network implications. The telecommunications network as a whole is becoming increasingly digital. Most new intercentral office links (trunks) in metropolitan areas are now provided over digital transmission systems. Existing switching equipment is being modified to switch 64 kb/s digital signals as well as voice signals, and new digital switches are beginning to be deployed. Long-distance links remain largely voice, but digital optical fiber systems are beginning to be deployed. The result is an increasing number of networks in which the path between the central offices nearest end users is capable of carrying digital signals.

With the extension of digital capability directly to the telephone user by using the techniques just described, end-to-end digital transmission is becoming a reality. Simultaneously, various telecommunications administrations have begun to define digital network standards and services. The name given the resulting network is the Integrated Services Digital Network (ISDN). Utilizing ISDN user interface and services standards, a variety of bit rates suitable for interactive data base, home security, bulk data, and video services can be readily provided. By the end of this century, computer and video services will be widespread and become as commonplace as telephones are today.

For background information *see* OPTICAL COMMUNICATIONS; PULSE MODULATION; TELEPHONE SERVICE; in the McGraw-Hill Encyclopedia of Science and Technology.

[NORWOOD G. LONG]

Bibliography: Bell Telephone Laboratories, *Engineering and Operations in the Bell System*, 1977; A. J. Ciesielka and N. G. Long, New Technology for Loops: A Plan for the 80's, *IEEE Trans. Commun.*, Com-28(7)923–930, July 1980; S. E. Miller and A. G. Chynoweth, *Optical Fiber Telecommunications*, 1979; H. Taub and D. L. Schilling, *Principles of Communications Systems*, 1971.

Bit rate capabilities

Bit rate	Voice channels	Data	Video
64 kb/s	1	12–48 voice-frequency modems, 1 facsimile page per second	Exploratory
1.5 Mb/s	24	Bulk data	1 limited-motion video channel
45 Mb/s	672	Computer nets	1 broadcast video channel

Tissue culture

A great deal is now known about the requirements, in the form of nutrients and special supplements like growth regulators (both natural and synthetic), for rapid growth of explanted organs and pieces of higher plants both on semisolid (agar) media and in liquid. Such cultures traditionally have been called tissue cultures, although it is recognized by all who work with them that they rarely constitute or derive from true tissues, such as cambium. In recent years the methods have improved to the point that they are playing a useful role in horticulture and agriculture, holding still greater promise for their use in combination with genetic engineering techniques.

Development of procedures. Plant tissue cultures were originally defined as aseptic preparations of somatic plant cells and tissues which grow and function without serious physiologic derangement but which, at the same time, do not differentiate into distinct organs. In the 1930s Pierre Nobécourt, R. J. Gautheret, and P. R. White probably had the first true cultures which fulfilled this definition. The main criterion of successful tissue cultures was an unlimited capacity for undifferentiated growth as a callus mass (Fig. 1a).

As a further development of the aseptic culture of isolated pieces of plant materials as callus, it became possible in the latter 1950s to obtain, maintain, and grow isolated free cells (Fig. 1b). In the

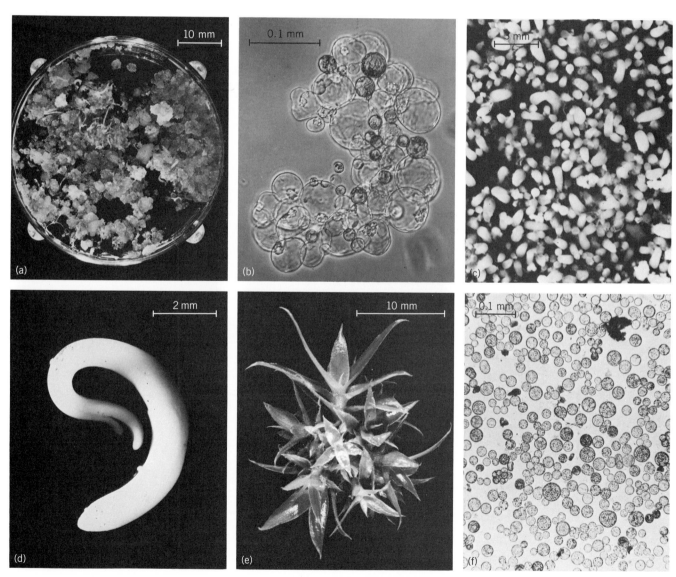

Fig. 1. Higher plant cultures. (a) A petri dish containing a predominantly unorganized mass of callus tissue. (b) Cells growing in suspension culture. (c) Somatic embryos of carrot developed from freely cultured somatic cells. (d) Excised zygotic embryo of *Datura stramonium* grown in culture from a much earlier stage of development. (e) Proliferated culture of a bromeliad showing precocious axillary branching; each of the branches can give rise to a whole plant. (f) Population of protoplasts from the day lily, *Hemerocallis*; these can regenerate walls, divide, and go on to grow into plants much like the somatic embryos of carrot in c. (*Courtesy of A. D. Krikorian*)

further study of these cells in liquid suspension culture in the late 1950s and early 1960s it was discovered that in some cases such free cells could redevelop into plants and, in this sense, behave like embryos (Fig. 1c). The procedures permitted the normal sexual process to be bypassed, eventually producing a population of plants with characteristics of the original plant from which the so-called primary explant derived. It was appreciated long ago that these procedures held promise for the production in large numbers of identical copies of individual plants with desirable qualities or characteristics. The principle behind such clonal micropropagation, or cloning, is called totipotency; this is the innate potential of every living nucleated cell of the plant body to give rise to a complete, new plant.

Somatic embryogenesis. Over the years, it has become possible, as a laboratory procedure, to successfully grow plants from the free cells of a number of flowering species. However, to date the numbers of species so grown have been relatively few and, although theoretically informative, only recently have cloning procedures started to provide the means with which to compete with more conventional methods of multiplicative propagation. Culture procedures have

potential for practical applications, but they also provide the opportunity to investigate the stimuli that release the otherwise suppressed totipotency of mature quiescent cells as they exist in the intact plant body, and the factors that control the direction and pace of their subsequent development. A continuing objective is to learn to control the development of somatic cells growing in isolation from the plant body, so that they will recapitulate the behavior of zygotes perfectly. If this were feasible, it would not only greatly add to the knowledge of development but would provide the long-sought means for clonal, asexual propagation of plant species. Figure 2 provides a schematic representation of propagation of plants from aseptically cultured cells. Emphasis is given to the stages and intervening events in the induction of growth and morphogenesis in somatic cells and tissue. The scheme represents the situation in a so-called model system. The complexity of the interactions and stages are emphasized to show that it is no easy matter to stimulate free body cells to grow like embryos.

Not enough is presently known about the control of somatic embryogenesis so that the techniques of cell culture and the induction of somatic embryoge-

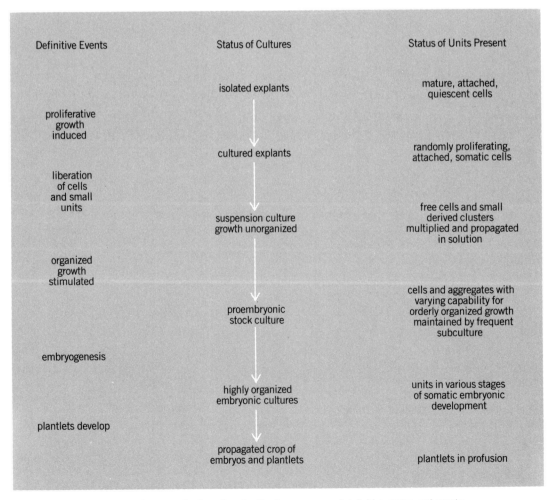

Definitive Events	Status of Cultures	Status of Units Present
	isolated explants	mature, attached, quiescent cells
proliferative growth induced		
	cultured explants	randomly proliferating, attached, somatic cells
liberation of cells and small units		
	suspension culture growth unorganized	free cells and small derived clusters multiplied and propagated in solution
organized growth stimulated		
	proembryonic stock culture	cells and aggregates with varying capability for orderly organized growth maintained by frequent subculture
embryogenesis		
	highly organized embryonic cultures	units in various stages of somatic embryonic development
plantlets develop		
	propagated crop of embryos and plantlets	plantlets in profusion

Fig. 2. Propagation of plants from aseptically cultured cells: the sequence of definitive stages and events.

nesis may be applied with confidence to all cases. However, it would be useful if cloning from cells could be extended to plants in which propagation by seeds involves many years, or results in great variability, or to plants in which conventional breeding procedures are not feasible because of failures in the normal reproductive process via flowers and seeds. Similarly, it could be advantageous to apply cloning procedures to the multiplication of desirable but sterile hybrids and to plants which are difficult to propagate by conventional means. Much valuable time could be saved by the rapid multiplication of very large numbers of new hybrids or individual mature plants known to have especially desirable characteristics (elite specimens). Cloning from cells would also be useful in plant pathology. For instance, clonal micropropagation from cells could permit production of virus-free or specific pathogen-free plant stocks or allow direct isolation or even production by mutation of strains resistant or tolerant to pathogens.

Although the free-cell culture of higher plants and their manipulation to produce somatic embryos is still at the stage of laboratory experimentation, other aseptic culture techniques are more or less routinely used in commercial and agricultural settings for exactly the same reason. Since these methods have also come to be categorized under the broad heading of tissue culture, it is important to distinguish them.

Embryo culture. It was shown as far back as the early 1920s that stimulation, otherwise unobtainable or erratic, of the growth of certain embryos could be achieved in aseptic culture. In some cases, embryos with poorly developed food reserves do not germinate because they are very dependent on externally available nutrient sources. For instance, the seeds of orchids contain a very small embryo comprising only a simple mass of cells. The embryo is totally dependent upon exogenous organic food such as sugar to germinate. In nature this sugar is provided by a symbiotic mycorrhizal (root fungus) relationship. Another example where embryos fail to germinate involves the formation of inhibitors in the seed. In this case embryos can often germinate only after an appropriate period of dormancy. In some plants (for example, the iris) one can eliminate both the dormancy requirement and the effect of germination inhibitors present in the seed of some hybrids by excising embryos and rearing them in sterile culture, in the absence of constraints, to a size sufficient for transplanting to soil. Aseptic culture has become a widely used and routine procedure for rescuing embryos that would normally not grow into plantlets (Fig. 1d).

Meristem, shoot, or stem tip culture. The discovery in the 1960s of the ability of explanted meristems (apical growing points) and shoot tips of the orchid *Cymbidium* grown in aseptic culture to produce, when appropriately cut, protuberances resembling normal protocorms which can grow to plantlets provided the most dramatic impetus for further development of procedures for multiplying and maintaining plants in aseptic culture. Shoot-tip cultures from many other plants have since been exploited in the procurement, maintenance, and multiplication of stocks. In some cases, a plant is generated from one cultured shoot tip; in others, multiple shoots can be stimulated to form a plant. As long as development of shoots emerging from the proliferated area at the base of a shoot-tip explant can be maintained at a rate consistent with their removal by excision, an open-ended system is possible. A balance is maintained which favors the continued formation of undifferentiated growth that will organize in culture by adjusting the medium. As these shoots (with or without roots) are removed from the proliferating mass and are transferred to an environment or different medium conducive to further root development, new proliferations grow to replace them.

A variation on this theme involves the stimulation of some plants by exogenous growth regulators of the cytokinin class to form precocious axillary shoots in profusion. Since axillary shoots can, in turn, produce additional axillary branches, theoretically in perpetuity, as each newly formed shoot is subcultured, the method is a good one for rapid clonal multiplication (Fig. 1e). This technique has been applied to a great variety of species, ranging from herbaceous foliage plants to bulbous monocotyledons and woody species.

Other organ cultures. A laboratory research tool used for some 60 years is to grow isolated plant parts such as root tips under aseptic condition in order to study their nutrition, growth, and development. When immature organs, such as leaves and fruits, are isolated for culture, their performance in isolation can be evaluated. When it is fully understood what developing organs receive in terms of stimuli and nutrients and when it is possible to re-create the environment in which they originate in place, researchers should be able to grow them separately, if not from the initiating cells then from the primordia. At present the achievements of classical organ culture apply more to roots than to other organs.

Anthers as sources of haploid cells. In the mid 1960s it was discovered that cultured anthers of several plants were capable of yielding haploid plantlets. Cultures can be initiated from anthers containing immature pollen grains, which are actually microspores prior to the development of the pollen grain or mature male gametophyte. In tobacco, for instance, the vegetative nucleus divides to give rise to the proembryo while still within the original wall of the pollen grain. This method has great potential for raising many haploid plants and has been heralded as a major breakthrough with great consequences for genetics, plant breeding, and agriculture. This is especially so since homozygous diploids can be raised by the use of colchicine or by taking advantage of the fact that many cultured cells and tissues spontaneously undergo endopolyploidization or doubling of the chromosome complement.

Protoplast cultures. Higher plant cells are different from animal cells in that they have a cellulose

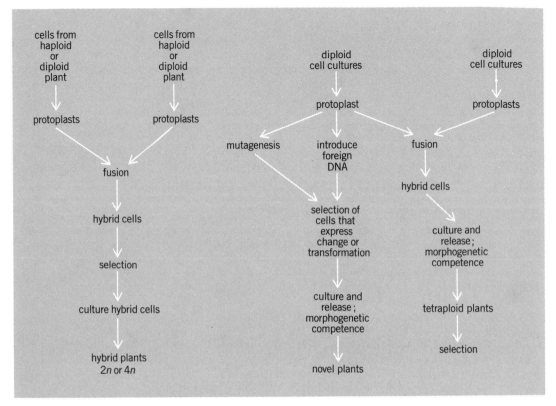

Fig. 3. Use of protoplasts for novel plant production. The left- and right-hand sides of the scheme involve different levels of parasexual hybridization. The middle column outlines the use of protoplasts in mutagenesis procedures and in genetic engineering.

wall. Without their walls, the cells could not exist in an ambient medium which is osmotically weaker than their contents. Enzymatic isolations, en masse, of still-viable wall-less cells (protoplasts) have been known since 1960. Cellulases and pectinases, enzymes generally derived from certain wood-degrading fungi, are capable of dissolving the intercellular components and cell walls. These enzymes are now widely used to produce large numbers of protoplast preparations from different plants and organs and are even used to produce aseptically derived tissues and cells in culture (Fig. 1f). Having reconstituted their surrounding walls, they are able to divide, proliferate, and in some instances eventually give rise to plants. In still other cases, by obtaining naked protoplasts, and in this way overcoming the barrier to cell fusion inherent in the presence of a cell wall, fusion of protoplasts can lead to fused nuclei and production of reconstituted cells from which new plants can grow. If the protoplasts were from haploid cells (as from anther culture), new diploid cells could be produced in a sort of artificial fertilization or syngamy. This is called parasexual or somatic hybridization.

In select cases the exploitation of these methods in the production of novel plants, even between evolutionarily disparate organisms, by so-called fusion breeding techniques can be envisioned. Figure 3 shows how this might be achieved and also outlines how cells might be modified by using mutagens and genetic engineering techniques in combination with protoplast methodology and cell culture. Nevertheless, the full exploitation of these ideas and methods, despite numerous claims for their routine usefulness, still awaits a more complete understanding and control of protoplasts and cells in culture.

For background information *see* Breeding (plant); Culture technique; Plant growth; Seed germination; Somatic cell genetics; Tissue culture in the McGraw-Hill Encyclopedia of Science and Technology.

[A. D. Krikorian]

Bibliography: B. V. Conger (ed.), *Cloning Agricultural Plants via in vitro Techniques*, 1981; J. H. Dodds and L. W. Roberts, *Experiments in Plant Tissue Culture*, 1982; A. D. Krikorian, Cloning higher plants from aseptically cultured tissues and cells, *Biol. Rev.*, 57:157–218, 1982; J. Reinert and Y. P. S. Bajaj (eds.), *Plant Cell, Tissue, and Organ Culture*, 1977.

Tooth disorders

Tooth decay (dental caries) is one of the most prevalent infectious diseases caused by bacterial accumulations (dental plaques) on teeth. Recent findings suggest that some of the different plaque organisms (for example, *Streptococcus mutans*) play a unique etiological role. Effective methods are now available for inhibiting dental caries.

Development of caries. Dental caries is the localized destruction of the tooth surface beneath adherent dental plaques. It may affect the outer enamel surface of the crowns of teeth as well as the cemental surface covering their roots. Whereas enamel caries begins to occur at an early age, caries of cementum generally develops at older age after exposure of the roots due to periodontal, or gum, disease. The present discussion will be restricted to enamel caries.

The mechanism of destruction is generally believed to entail an acid dissolution of the mineral structure of enamel or cementum $[Ca_{10}(PO_4)_6(OH)_2]$, resulting in formation of soluble calcium and phosphate ions. Acid formation and accumulation in plaque is due to fermentation of common dietary carbohydrates by plaque bacteria; among the end products formed are: lactic, formic, pyruvic, acetic, and propionic acids. Remineralization of slightly demineralized areas of the tooth surface by uptake of calcium and phosphate ions can also occur after "acid attacks," when the plaque milieu becomes neutral due to exhaustion of dietary carbohydrate. Retentive tooth surface areas, such as fissures in molars or areas on the contacting surfaces of neighboring teeth, are particularly prone to plaque formation and therefore caries development. Continued demineralization may lead to an actual cavity in enamel or cementum which may progress into the underlying dentin and eventually reach the inner pulp of the tooth.

Major etiologic factors. The development of caries requires a susceptible tooth, plaque, and dietary carbohydrate. Common dietary carbohydrates are sucrose, glucose, fructose, and various starches. Sucrose has been considered the major culprit because it is the most widely used readily fermentable carbohydrate and is present in many food items eaten as between-meal snacks. It is interesting to note that hereditary fructose-intolerant people who learn to avoid sucrose early in life, but who frequently eat other carbohydrates, generally have a negligible caries experience.

There are other factors that influence caries development; foremost among these is saliva. The removal of salivary glands in rodents can greatly increase their caries activity. Humans with xerostomia (dry mouth) due to irradiation-induced malfunctioning of the salivary glands will also develop rampant caries unless their teeth are adequately treated with topical fluoride solution. The effect of saliva may be mediated in part by factors which influence the acidity of plaque, including rate of flow, buffer capacity, or concentration of urea, ammonia, and so on, which can yield basic products and counterbalance bacterial acid. Calcium and phosphate ions are also important; at higher ion concentrations in saliva, plaque demineralization will occur less readily and remineralization after an acid attack may commence sooner.

Plaque flora. Plaques consist of a variety of bacteria. They are characteristically very heterogeneous and generally differ considerably from person to person, from tooth to tooth in the same mouth, and even between different areas on the same tooth surface. These differences are due to variables such as the type of tooth surface, diet, caries experience, and pH of the plaque. Plaques covering the crowns of teeth are generally predominated by streptococci (gram-positive, chain-forming cocci), *Actinomyces* species (gram-positive rods), and *Veillonella* (gram-negative cocci). Among the streptococci, the proportions of *Streptococcus mutans* may vary widely from undetectable to 50% or more of the total cultivable flora; the proportions of *S. sanguis* and *S. mitis* average about 5 and 10%, respectively, whereas *S. salivarius*, enterococci, or anaerobic streptococci constitute generally less than 1%. *Actinomyces* species, such as *A. viscosus*, *A. naeslundii*, *A. odontolyticus*, and *A. israelii*, may average a total of 40% or more of the total cultivable flora. Other organisms, such as staphylococci, species of *Candida*, *Neisseria*, and *Lactobacillus*, and a variety of other gram-positive or gram-negative rods, constitute generally less than 1% of the flora.

Bacterial specificity. The ability of plaque bacteria to produce acid is of critical importance in the formation of caries. Acid formation can be affected by changing conditions. For example, acidification of plaque will cause a slowdown of bacterial metabolism, and lead to a lower rate of bacterial growth and acid production. Numerous studies have indicated that plaque bacteria vary greatly in their ability to produce acids or alkaline substances, or in their tolerance to an acidic milieu. Another important aspect of plaque cariogenicity is related to the plaque matrix. This material, which is located between the bacteria, is responsible for cohesion of the bacterial mass. Major plaque matrix components are various salivary glycoproteins and bacterial extracellular polysaccharides, such as the glucans (polymers of glucose such as dextran), which are produced specifically from sucrose (glucose linked to fructose) by *S. mutans*.

Dental plaque can best be visualized as a conglomerate of tightly packed microcolonies of different bacteria surrounded by matrix material of varying chemical composition. Colonies of *S. mutans* may be surrounded by dextran, whereas colonies of other bacteria may be surrounded by other types of polysaccharides, salivary glycoproteins, or other materials. Experimental evidence suggests that the diffusion of carbohydrates, other bacterial substrates, or bacterial metabolic products in and out of plaque is limited to varying degrees; that is, plaque acts as a diffusion-limiting system. It may be expected, therefore, that plaques with a different bacterial composition differ in their cariogenicity because they possess different abilities to produce acids or other bacterial products and to retain products formed within different microcolonies.

Ever since the late 1800s, when L. Pasteur, R. Koch, and others demonstrated that single microbial agents were the cause of different infectious diseases, there has been a search for a single causative agent of dental caries. Lactobacilli, studied since the early 1900s, were the first organisms to be implicated. Interest in these organisms as the sole etiologic agents began to wane after World War II, however, when it was demonstrated that their numbers in plaque were often extremely low and that caries lesions could develop in their absence. More recently, another species, *S. mutans*, has been strongly implicated and is now considered by many to be of major etiologic importance. However, the concept that caries is due to one type or only a small spectrum of plaque organisms, rather than to the combined metabolic activity of all plaque organisms, is still controversial.

The role of different plaque bacteria in caries development has been studied in laboratory cultures as well as in animals and humans. However, conclusive evidence about the contribution of individual organisms can be obtained only in studies of caries development in which a certain type or group of organisms has been completely eliminated. This kind of proof is lacking because such studies have been impossible to perform.

The contribution of each type of plaque organism to caries initiation is related to its plaque concentration and to its metabolic activity. Limited data from studies of laboratory cultures suggest that lactobacilli and streptococci have a higher acid production rate or acid tolerance than many other plaque organisms, including the numerous *Actinomyces* species. Some organisms, such as *Veillonella*, cannot ferment carbohydrate to acid but instead can decompose acids, such as lactic acid, and thus may counterbalance plaque acidity. Many plaque organisms can produce basic substances with the same effect.

Studies with animals. Studies with animals suggest considerable differences in cariogenicity among plaque organisms. For example, caries in rodents can be inhibited by antibiotics active against grampositive bacteria but not by antibiotics which inhibit only gram-negative organisms. Studies with rats that are fed a cariogenic diet and have been artificially infected with single plaque organisms show that *S. mutans* strains almost invariably induce caries involving tooth fissures as well as other, less retentive tooth surfaces. In contrast, other types of streptococci, lactobacilli, or actinomyces are often not cariogenic. Moreover, if such organisms are cariogenic, they induce a much less severe type of caries which is nearly always restricted to the fissures. Studies with monkeys (*Macaca fascicularis*) also suggest a specific role of *S. mutans* in caries development. These monkeys, which possess a microbial plaque flora composed of many of the same organisms found in human plaque, remain virtually free of caries when fed a cariogenic high-sucrose diet, unless they are artificially infected with *S. mutans*. In some cases, test organisms form little plaque and are present in grossly inadequate quantities for caries development in animals, even if acid production by individual cells is extremely high. On the other hand, bacterial metabolism may also play a critical role in cariogenicity. In one study, *S. mutans* as well as *A. viscosus* were found to form copious amounts of plaque on the teeth of inoculated rodents, but only the *S. mutans*–infected animals developed caries.

Studies with humans. Studies with humans have shown a positive correlation between the levels of lactobacilli or *S. mutans* in plaque or saliva and caries activity. Such a correlation is absent in the case of *S. sanguis*, *S. mitis*, *Veillonella*, or different *Actinomyces* species and has not been determined for many other, less predominant plaque organisms. An important role for lactobacilli in the initiation of caries seems unlikely. Lactobacilli are often undetectable or present only in very low numbers in plaques covering sound surfaces or even surfaces with soon-to-develop or beginning caries. Instead, these organisms are often found in high concentrations in caries lesions, which appear to be their major and indispensable oral habitat because they can be eliminated from most mouths merely by cleaning out and filling of cavities. The localization of lactobacilli in cavities may be explained partly by the known acidity of cavities, which may provide these acid-tolerant organisms with a selective growth advantage over other, less acid-tolerant bacteria. In sum, lactobacilli appear to be more important in the progression of cavities than in their actual initiation.

By far the strongest evidence for a single group of organisms as a major causative agent in the initiation and progression of caries has been obtained in the case of *S. mutans*. Many caries-free persons have low or negligible plaque levels of *S. mutans*, whereas caries-active persons generally have much higher levels. Similarly, the presence of *S. mutans* in plaques on sound surfaces that remain caries-free is generally undetectable or much lower than on sound surfaces that become carious. Proportions of *S. mutans* in plaques covering caries lesions are generally much higher than in plaques on sound surfaces, and the organisms are nearly always found in cavities on all types of tooth surfaces.

Although *S. mutans* should be considered an important etiologic agent because of its highly acidogenic and acid-tolerant properties, its unique ability to cause caries in animals and its positive correlation with human caries development are not entirely understood. Analysis of research findings, however, suggests the following: (1) An increased sucrose intake is known to lead to higher *S. mutans* concentrations in plaque because extracellular polysaccharide formation from sucrose will enhance the organism's accumulation by strongly binding growing cells together, thereby preventing them from being removed by oral cleansing; (2) a frequent high acidity of plaque will favor selectively the growth of acid-tolerant *S. mutans* cells; and (3) plaques of caries-free (caries-inactive) persons become much less acid after exposure to carbohydrates than plaques of caries-active persons. Collectively, these findings suggest that caries development is caused by an increase in the metabolic activity of plaque which is due to a shift in its bacterial composition toward organisms such as *S. mutans* as the result of a more frequent carbohydrate consumption. In fact, children with so-called nursing bottle syndrome hold a bottle in their mouths for long periods during the day or night, exposing their teeth very intensively to the carbohydrate in the bottle fluid, and have rampant caries and plaques which often consist of *S. mutans* alone.

Caries inhibition. Control of caries has been attempted in different ways, including protection of the tooth from bacterial acid attack, antiplaque (bacterial) methods, or dietary changes. Protection of fissures against acid attack by so-called sealants, which consist of a thin layer of hardened acrylic material painted like a nail varnish in liquid form onto the tooth surface, can be very effective and their use is on the increase. Fluorides in water or pills, or applied locally to the teeth by various methods, are thought to render tooth surfaces less acid-soluble, thus producing a caries-inhibiting effect. Antiplaque methods include oral hygiene (tooth brushing, use of dental floss, and so forth), which is very effective only when done meticulously. Disinfectants and antibiotics, although proved effective, are not as yet used on a large scale for the specific purpose of combating plaque bacteria. The use of enzymes to break down extracellular glucans by *S. mutans* or to inhibit their formation, thus dispersing *S. mutans* colonies in plaque, has not been very successful. The development of a suitable caries vaccine directed against *S. mutans* or other potentially cariogenic plaque bacteria is still to be achieved. Removal of all sucrose from the diet is obviously impossible. Attempts to minimize the effect of sucrose have been aimed so far principally at beverages or chewing gum in which sucrose has been replaced by artificial nonfermentable sweeteners. Recently, however, there have been reports of a significant decrease in the prevalence of caries in various parts of the world. This may be at least partly due to the use of fluorides in various forms.

For background information *see* DENTISTRY; TOOTH DISORDERS in the McGraw-Hill Encyclopedia of Science and Technology.   [JOHANNES VAN HOUTE]

Bibliography: G. W. Burnett and G. S. Schuster, *Oral Microbiology and Infectious Disease*, student ed., pp. 141–212, 1978; J. van Houte, Bacterial specificity in the etiology of dental caries, *Int. Dental J.*, 30(4):305–326, 1980: D. H. Leverett, Fluorides and the changing prevalence of dental caries, *Science*, 217:26–30, 1982.

Toxin

Some of the most poisonous nonproteinaceous substances known are found in marine organisms. One that has received considerable attention is tetrodotoxin, the pufferfish poison. In Japan this toxin is a public health concern since the pufferfish, in particular *Fugu rubripes rubripes*, is a highly prized delicacy, but the viscera, which contain tetrodotoxin, must be removed before the fish can be eaten. The structure of tetrodotoxin was discovered by American and Japanese scientists in the mid-1960s and is the first powerful marine toxin to be identified. In the last decade the structures of many other nonprotein toxins have been discovered.

Saxitoxin, gonyautoxins, and brevitoxins. Toxic red tides are large blooms of certain dinoflagellates.

When red tides appear, the shellfish frequently become unsuitable for human consumption since they ingest the toxic dinoflagellates and concentrate the toxin in their digestive tracts. Clams and mussels along the western coast of temperate North America become toxic during the summer months from ingestion of *Gonyaulax catenella*. The Alaska butter clam, *Saxidomas giganteus*, however, is toxic throughout the year, even when dinoflagellates are absent from the water. The reason for this is unclear. Saxitoxin (I) is the major toxin in *S. giganteus*. Other shellfish, as well as *G. catenella* and *G. tamarensis*, the dinoflagellate responsible for paralytic shellfish poisoning along the coasts of the temperate North Atlantic Ocean, elaborate a complex array of saxitoxin-related compounds, frequently referred to as gonyautoxins. Saxitoxin and the gonyautoxins, which act on the neuromuscular systems, are sodium channel blockers. Tetrodotoxin has essentially the same pharmacology.

The dinoflagellate *Ptychodiscus brevis* (*Gymnodinium breve*) is responsible for massive fish kills and mollusk poisoning along the Florida coast and in the Gulf of Mexico. Three ichthyotoxins, brevitoxins A, B, and C, were isolated from unialgal cultures of *P. brevis* obtained during an outbreak in Florida in 1953. The structure of brevitoxin B (II) has recently been discovered, and its absolute stereochemistry has been determined. The extraordinary fused, polycyclic ether structure of brevitoxin B has no precedent.

Aplysiatoxin and lyngbyatoxin A. Sea hares are gastropod mollusks that frequently secrete poisonous substances when molested. The toxins are located in the digestive gland, where they are accumulated through diet. Two toxins, aplysiatoxin (III, R = Br) and debromoaplysiatoxin (III, R = H), both of which are lethal to mice, have been isolated from the sea hare *Stylocheilus longicauda*. The origin of these unusual bislactones has been traced to *Lyngbya majuscula*, a blue-green alga that is one of the favorite foods of this mollusk.

Lyngbya majuscula is also the causative agent of a severe contact dermatitis that frequently affects. swimmers and bathers in Hawaii during the summer months. Aplysiatoxin and debromoaplysiatoxin, each highly inflammatory, have been isolated from varieties shown to be responsible for outbreaks of the dermatitis on the windward side of Oahu. *Lyngbya majuscula* from the leeward side of Oahu, however, contains a different inflammatory agent, lyngbyatoxin A (IV), which is an indole alkaloid. Varieties containing lyngbyatoxin A as the major irritant have so far not been implicated in seaweed contact dermatitis. It is interesting that the digestive tract of *S. longicauda* is not affected by these powerful vesicants.

Aplysiatoxin and lyngbyatoxin A are potent skin tumor promoters. Since skin cancer proceeds in two major stages, initiation and promotion, the skin cell must first of all be exposed to a carcinogen (initiator) which reacts irreversibly with the DNA of the cell. This damaged or initiated cell is then transformed

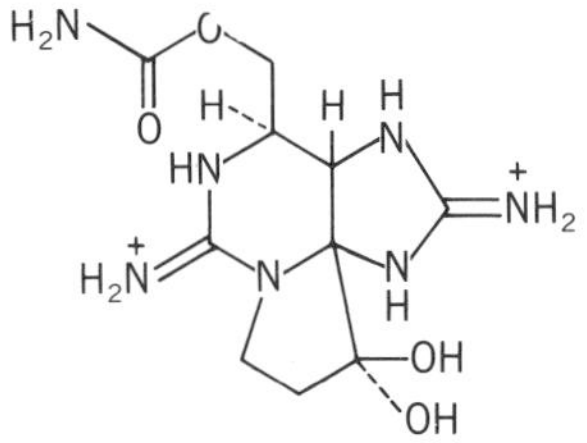

(I)

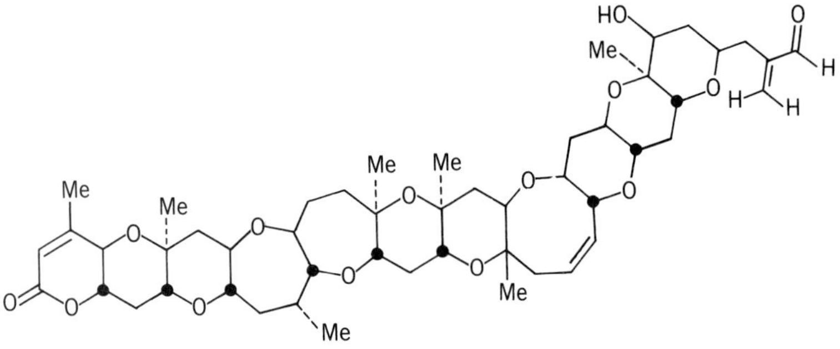

(II)

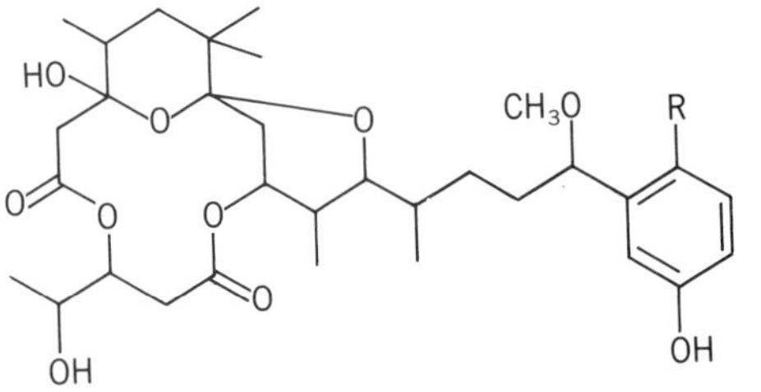

(III)

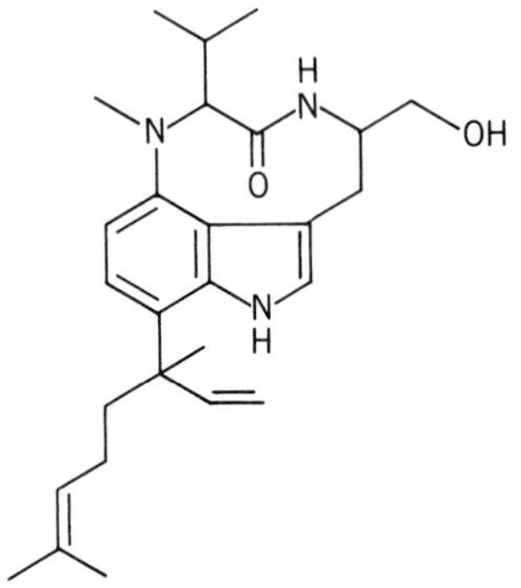

(IV)

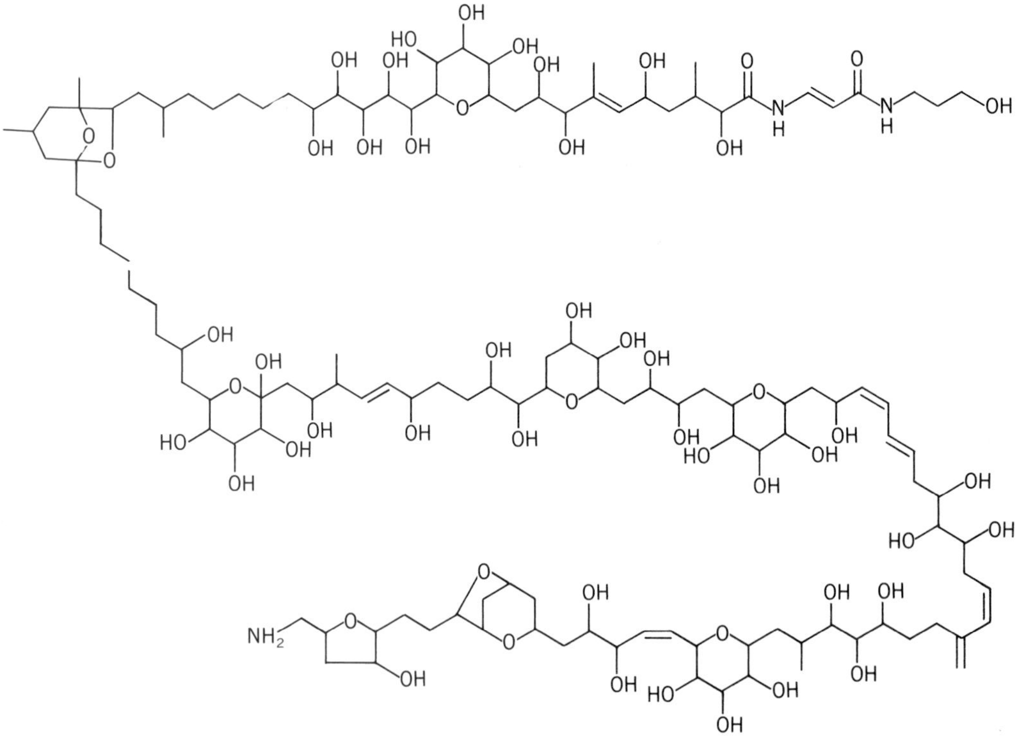

(V)

into a tumor cell in a complex second stage by the noncarcinogenic aplysiatoxin or lyngbyatoxin A (promoter). Unlike the initiator, the promoter reacts in a reversible manner, at least in the early phase of the second stage, with the membrane of the initiated cell. There must be repeated contact of the cell with the promoter, however, before tumors will develop. Tumors therefore are produced on mouse skin when a single application of 7,12-dimethyl-benz[a]anthracene as the initiator is followed by repeated application of aplysiatoxin or lyngbyatoxin A. The bodily effects are identical to those produced by

12-O-tetradecanoylphorbol-13-acetate (TPA), a well-known tumor promoter found in croton oil. Curiously, debromoaplysiatoxin is a weaker promoter of skin tumors.

Aplysiatoxin and debromoaplysiatoxin share many effects with TPA in cell-culture systems; for example, all three compounds inhibit the binding of epidermal growth factor to its membrane receptor in a rat embryo cell line with equal potencies. Debromoaplysiatoxin, however, requires a tenfold greater concentration to achieve the same effect. Some effects, however, are identical for all four compounds; for example, aplysiatoxin, debromoaplysiatoxin, lyngbyatoxin A, and TPA increase ornithine decarboxylase (ODC) activity in mouse skin and stimulate the release of arachidonic acid from mouse embryo cells to the same extent. The latter two processes appear to be important in the development of tumors from initiated cells, since vitamin A analogs (a blocker of ODC activity) and indomethacin (which prevents inflammation by inhibiting the synthesis prostaglandin E_2 from the released arachidonic acid) lower tumor incidence in mouse skin when coapplied with the promoter.

Palytoxin. Except for certain toxic proteins and polypeptides such as botulinus toxin, palytoxin (V) is the most poisonous substance known. This water-soluble toxin was first isolated from *Palythoa toxica*, a coelenterate which grows in a single, small tidepool near Hana on the island of Maui. The same toxin is found in smaller amounts in other species of *Palythoa*. Palytoxin is also found occasionally in the filefish *Alutera scripta* and results from feeding on *P. tuberculosa*. *Palythoa toxica* is highly toxic throughout the year, but toxicity levels of individual animals varies markedly. Other species of *Palythoa* appear to be highly toxic only during the summer months when the female polyps produce eggs. Recent studies suggest that the toxin is produced by a symbiotic bacterium, probably a *Vibrio* species. Palytoxin, ($C_{129}H_{223}N_3O_{54}$), is unusual because it totally lacks the familiar amino acid, sugar, and fatty acid units found in other biomolecules of this size (molecular weight is 2681 daltons). The biogenesis of palytoxin is not known.

The lethality of palytoxin in several animals has been determined. The primary cause of death is congestive heart failure. It produces profound vasoconstriction and induces intense cardiac and smooth muscle contractions that are accompanied by an increased uptake of calcium into the myocardial tissue. The mechanism is unknown, but it is suspected that the toxin interferes with the operation of the slow calcium channel. The two ends of the molecule, namely the β-amidoacrylamide and primary amine functionalities, are necessary for toxicity.

Ciguatoxin. This is a toxin that could present a serious public health hazard and is responsible for ciguatera, a human illness that results from the ingestion of a number of tropical and subtropical coral reef fishes. Ciguatera outbreaks, which are currently widespread in the Pacific, are sporadic and unpredictable, and the factors that control their sudden appearances are not well understood. In recent years ciguatoxin has been traced to a benthic dinoflagellate, *Gambierdiscus toxicus*, which grows epiphytically on certain macroalgae and is initially consumed by herbivorous fishes. The toxin, which is lipophilic, accumulates in the flesh and viscera of the herbivorous and carnivorous fishes as it is passed through the food chain. Pure ciguatoxin has been isolated, and it is anticipated that its molecular structure will be determined in the near future.

For background information *see* DINOFLAGELLIDA; POISON in the McGraw-Hill Encyclopedia of Science and Technology.

[RICHARD E. MOORE]

Bibliography: Y. Y. Lin et al., Isolation and structure of brevetoxin B from the red tide dinoflagellate Ptychodiscus brevis (Gymnodinium breve), *J. Amer. Chem. Soc.*, 103:6773–6775, 1981; R. E. Moore, Toxins, anticancer agents, and tumor promoters from marine prokaryotes, *Pure Appl. Chem.*, 54:1919–1935, 1982; C. F. Wichmann et al., Structures of two novel toxins from Protogonyaulax, *J. Amer. Chem. Soc.*, 103:6977–6979, 1981; N. W. Withers, Ciguatera fish poisoning, *Annu. Rev. Med.*, 33:97–111, 1982.

Two-dimensional systems

Perhaps the most fundamental phase change which occurs in condensed matter is the transformation of a disordered liquid into a periodic crystalline solid. Research in this field has recently been extended to the crystallization of truly two-dimensional systems, and the new physics which has emerged underscores the importance of spatial dimensionality in determining even the most qualitative properties of matter. Understanding the physics of two-dimensional systems has impact on various research fields, such as membrane biology, catalytic chemistry, and surface physics, as well as on applied technology. For example, the rapid progress in microminiaturization of electronic components will ultimately face the fundamental limits and possibilities of the two-dimensional world.

Recent theoretical predictions have indicated that the crystallization, or equivalently the melting behavior, of a two-dimensional solid may be entirely different from the abrupt, unpredictable phase transformation which generally occurs in bulk three-dimensional systems. These theories suggest an entirely new mechanism for the destruction of order at melting, predicting that melting occurs in two distinct steps as a function of temperature. At the stage between the two steps the two-dimensional system is predicted to be in a qualitatively new state. Experimental work has given new insights into, and provided substantial reinforcement of, some of the general theoretical concepts, but the agreement between experimental and theoretical results is not complete.

Theoretical concepts. A three-dimensional crystalline solid is by definition both crystalline in that it has a well-defined periodic lattice with long-range order, and solid in that it resists shear. A two-di-

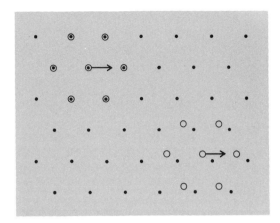

Fig. 1. Schematic representation of a two-dimensional solid. The atoms (open circles) on the left are coincident with the black dots. The atoms in the right-hand region are not in registration with those on the left, as seen by the position of the dots. The two regions do have the same orientation of the crystal axes. (*After W. F. Brinkman, D. S. Fisher, and D. E. Moncton, Melting of two-dimensional solids, Science, 217:693–700, 1982*)

mensional solid is similar to the more familiar three-dimensional counterpart, but if differs in one important way. Thermal fluctuations in a solid cause displacements of the atoms from their perfect lattice positions. In three dimensions the fluctuations in position of atoms separated by the average distance r remain finite as r gets very large. However, in two dimensions the fluctuations continue to grow as r grows. For macroscopic samples the fluctuations prevent the system from having rigorous, perfect periodic order. Nevertheless the order is considerably greater than the short-range order of a liquid, and so the term quasi-long-range order is used. A schematic representation of this situation is shown in Fig. 1, where the atoms in a two-dimensional lattice are shown to fluctuate from their ideal lattice position. Although a two-dimensional system has only quasi-long-range periodic order, it exhibits true long-range order in the orientation of its crystal axes, as shown in Fig. 1.

The question arises as to the nature of the melting phase transition which must occur as the system energy is increased by heating. It was suggested in 1973 that the long-wavelength fluctuations might control the behavior at the melting transition. These fluctuations are controlled, in turn, by the behavior of defects in the system. For a two-dimensional crystal the important defects are lattice dislocations; a dislocation for a hexagonal lattice is illustrated in Fig. 2a. This dislocation involves an additional half row of atoms; it has only a small effect on the local orientation of the lattice. If the temperature is above a critical temperature, free dislocations form spontaneously. Once this occurs, the system loses its quasi-long-range periodicity and it can no longer resist shear, because stresses can be relaxed simply by moving free dislocations. The two-dimensional crystal thus melts at temperature T_M, and its shear modulus jumps discontinuously to zero.

In 1978 it was recognized that the defect theory of melting implied an additional result. As mentioned above, the planar crystal exhibits long-range order of its crystal axes in addition to its quasi-long-range periodicity. Dislocations violently disrupt the quasi-long-range periodicity but have only a negligible effect on the direction of the crystal axes, as Fig. 2a shows. A second transition is predicted to disrupt the orientational order of the crystal axes. Like the previous transition, it involves defects, but this time they are defects in the lattice orientation called disclinations. An example is shown in Fig. 2b, where the crystal axes rotate by 60° around the central site, which has fivefold symmetry. There are also disclinations with sevenfold symmetry, which produce rotations of $-60°$. The state between the two transitions is called the intermediate phase, or hexatic phase, because it has hexagonal lattice orientational order, even in the absence of periodic order.

Experimental studies. The first attempts to study the melting of two-dimensional crystals exploited the remarkable ability of some liquid crystals to form freely suspended films (like soap-bubble films). These can be made as thin as two molecular layers (5 nm) with an area of 1 cm². This technique has been exploited in x-ray scattering experiments. The film is drawn across an open hole in a glass cover slide. In the crystalline phase, the elongated molecules order with their long axes perpendicular to the plane of the film and form a simple close-packed hexagonal lattice on the plane. From x-ray diffraction measurements on $\overline{14}S5$ (4-*n*-pentylbenzenethio-4'-*n*-tetradecyloxybenzoate) taken with synchrotron radiation, there is certainly no doubt about the two-dimensional solid nature of these materials. The data are very sensitive to the fluctuations which are unique to two-dimensional systems, and these fluctuations are clearly seen.

Studies of thick films (greater than 100 layers) have produced indirect evidence in support of the existence of the hexatic phase between the solid and fluid phases in a material called 650BC (*n*-hexyl-4'-*n*-pentyloxybiphenyl-4-carboxylate). In this work, a

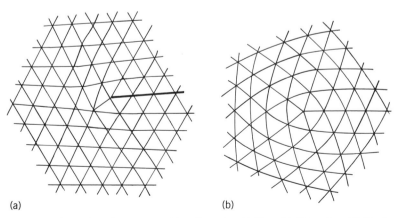

(a)   (b)

Fig. 2. Lattice defects in a hexagonal two-dimensional lattice. (*a*) Dislocation. (*b*) Disclination. (*After W. F. Brinkman, D. S. Fisher, and D. E. Moncton, Melting of two-dimensional solids, Science, 217:693–700, 1982*)

new liquid crystal phase has been discovered which can be thought of as a stack of two-dimensional hexatic layers.

Another two-dimensional system, consisting of a monolayer of xenon atoms adsorbed on the (001) surface of pyrolytic graphite, has enabled substantial progress on the nature of the two-dimensional melting problem. It has been shown theoretically that the weak graphite potential (3.2 meV) should not qualitatively alter the behavior of the melting phase transition from that on a smooth substrate, even though the system differs slightly from the ideal two-dimensional system. Data obtained by x-ray diffraction demonstrate remarkable quantitative consistency with detailed predictions of the dislocation theory of the melting at T_M. Above T_M, a free two-dimensional system would presumably be in the hexatic phase, with an isotropic liquid phase at higher temperatures. However, a physisorbed system is expected to exhibit only one phase above T_M, which will be orientationally ordered at all temperatures because of the presence of the substrate. Further study of the role of the substrate is necessary, but it is already clear that the behavior of two-dimensional solids can be quite different from that of ordinary matter.

For background information *see* CRYSTAL DEFECTS; LIQUID CRYSTALS; PHASE TRANSITIONS in the McGraw-Hill Encyclopedia of Science and Technology.

[DAVID E. MONCTON]

Bibliography: W. F. Brinkman, D. S. Fisher, and D. E. Moncton, Melting of two-dimensional solids, *Science*, 217:693–700, 1982; R. Pindak and D. E. Moncton, Two-dimensional systems, *Phys. Today*, 35(5):56–62, May 1982; S. K. Sinha (ed.), *Ordering in Two Dimensions*, 1980.

Venus

Among the inner planets (Mercury, Venus, Earth, Mars) and their satellites, Venus appears to be the most Earth-like in many ways. It is the closest planet

Basic characteristics of terrestrial planetary bodies

Planetary body	Mass, in Earth masses	Mean radius, km*	Density, g/cm³
Mercury	0.055	2439	5.43
Venus	0.815	6051	5.24
Earth	1.000	6371	5.51
Mars	0.107	3390	3.93
Moon	0.012	1738	3.34

*1 km = 0.6214 mi.

to Earth and is similar to Earth in size and density (see table). The Earth differs from the Moon, Mars, and Mercury, which are all approximately one-half the diameter or less of this larger body. The surfaces of these small planetary bodies are very ancient, dating back to the first half of solar system history, and are covered with impact craters and related deposits. They are characterized by a single, global rigid lithospheric plate (the outer mechanical layer of a planet usually composed of the crust and upper mantle), which has remained stable since earliest solar system history. In contrast, the surface of the Earth is very dynamic. The lithosphere is divided into a series of laterally moving plates. In a process known as plate tectonics, these plates form at ocean ridges, spread laterally at rates measured in centimeters per year, collide, and are subducted back into the interior of the planet. Thus two-thirds of the surface of the Earth (the ocean basins) is less than 200 million years old. The continental regions, by virtue of their lower density, escape complete subduction and preserve a part of the record of earlier Earth history.

The extremely dynamic plate tectonic system of the Earth is fundamentally different from the single, rigid, ancient lithospheric plate of the Moon, Mars, and Mercury. The planet Venus offers an opportunity to test hypotheses about the differences among the inner planets, such as: why these bodies are so different; planetary size as an important control on

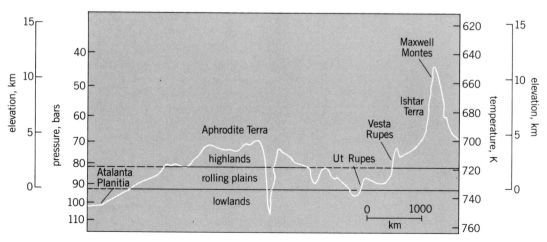

Fig. 1. Composite cross section of Venus topography as revealed by the Pioneer Venus altimeter data. Features not mentioned in the text are Vesta Rupes (scarp), Ut Rupes (scarp), and Atalanta Planitia (plain). Information on the range of temperatures and pressures is also shown. 1 bar = 10⁵ pascals.

heat loss and thus on the thickness and stability of the lithosphere; the importance of other factors, such as position in the solar system and initial conditions; and whether plate tectonics is unique in the solar system.

Although Venus has many similarities to the Earth, one of the major differences is its hot, dense, carbon dioxide atmosphere, which results in surface temperatures of approximately 730 K and pressures of 9.5 megapascals (95 bars). Unfortunately, this dense cloud layer precludes visual observations of surface features from Earth-based or orbiting spacecraft. Thus, investigations of Venus have had to rely on observations at radar wavelengths, where topographic and imaging data could be obtained, and on the few normal images obtained by spacecraft that have descended through the atmosphere and landed.

Pioneer observations. The most comprehensive data set comes from the Pioneer mission, in which a spacecraft was placed into orbit around Venus and multiple probes were sent into the atmosphere. The spacecraft carried a radar altimeter that provided topographic data over more than 95% of the surface at horizontal resolution of about 100 km (60 mi) and vertical resolution of greater than 200 m (650 ft). This basic data set provided a fundamental, but low-resolution, view of Venus topography.

The terrain of Venus can be subdivided into lowlands (about 27% of the surface), rolling plains (65%), and highlands (about 8%) [Fig. 1]. The highlands are concentrated in three main areas, Ishtar Terra, Aphrodite Terra, and Beta Regio. Ishtar Terra stands several kilometers above the mean planetary radius and is characterized by relatively steep slopes. The western part of Ishtar (Lakshmi Planum) is a vast plateau 2500 km (1500 mi) in diameter, which is surrounded and cut by linear mountain ranges rising up to 3 km (2 mi) above the plain. A massive mountain range (Maxwell Montes) occurs in Eastern Ishtar, rising 11 km (7 mi) above mean planetary radius—higher than Mount Everest is above sea level on Earth. The topography of Ishtar Terra is unlike any seen on the smaller terrestrial planets, but other highland areas show more predictable characteristics. Beta Regio contains a central linear depression and several equidimensional mountains which have been interpreted as a rift zone and volcanoes, respectively. Aphrodite extends along the equator for more than 10,000 km (6000 mi) and contains numerous riftlike depressions in its central interior. The lowland areas of Venus are relatively smooth and are concentrated in roughly circular and linear areas. Midlands or rolling uplands are the most extensive terrain type and contain a diversity of topographic features, including possible rift zones, volcanoes, and impact craters.

Thus, present knowledge of Venus strongly hints at a geological diversity comparable to some aspects of both the Earth and the smaller terrestrial planets, including impact cratering, volcanic activity, tectonic activity, and perhaps continents. Evidence for plate tectonics on Venus even at low resolution was a major topic of interest in the Pioneer project. To investigate this topic, several scientists have asked whether the presence of plate tectonics could be proved for Earth with Pioneer-type spacecraft data of similar resolution. The answer is ambiguous. Many of the major features indicative of plate tectonics (trenches at subduction zones, folds, graben along oceanic spreading centers) are not detectable at this resolution. Other broad features, however, such as the oceanic rises typical of spreading centers, are visible. But the question then arises as to whether these could be interpreted without prior knowledge of their origin. An additional question concerns how the Venus environment will change the nature of the

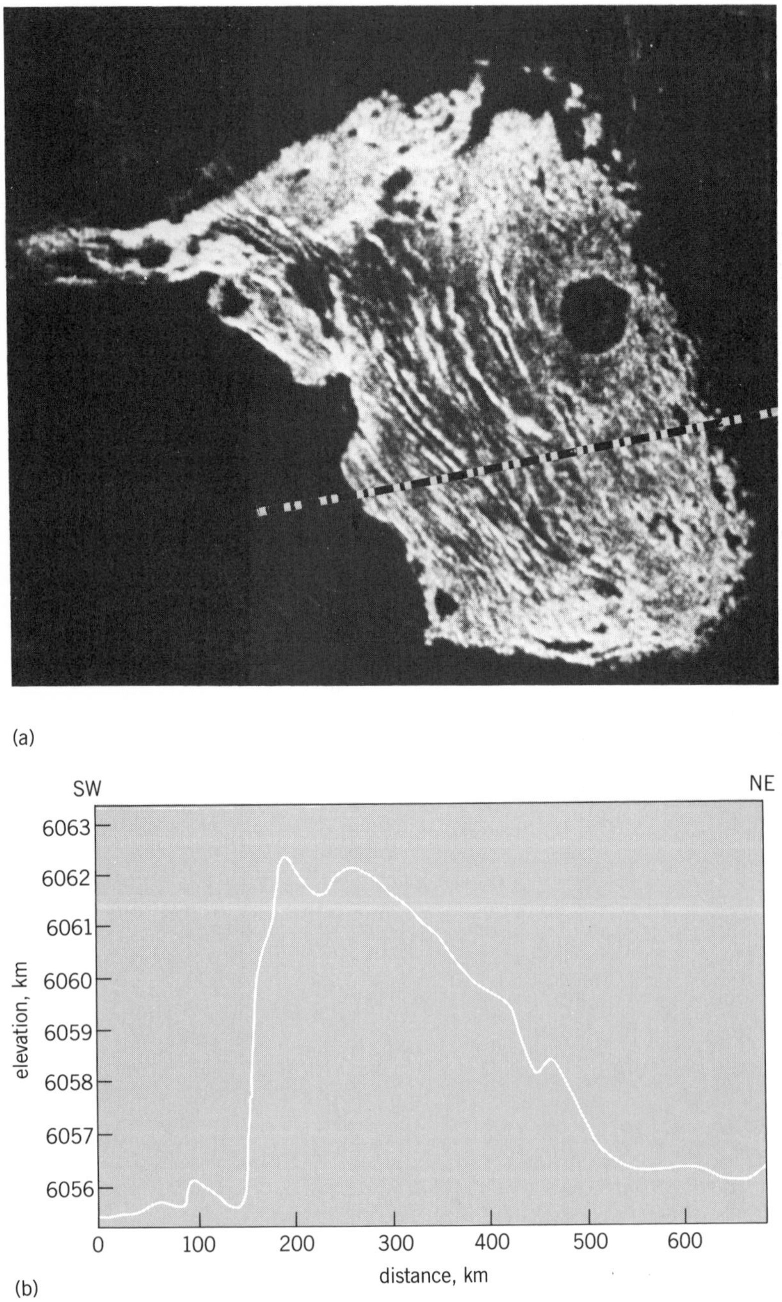

(a)

(b)

Fig. 2. Maxwell Montes in the Ishtar Terra region of Venus. (a) Radar image showing the banded texture of the mountain range. (b) Topographic profile of the location shown by the broken line in the radar image.

Fig. 3.   Images of the surface of Venus from the Soviet Venera missions. (a) *Venera 13.* (b) *Venera 14.* Distance between points of the teeth at base of spacecraft is 5 cm (2 in.).

plate tectonics process and the topographic characteristics of this process. Thus, at the Pioneer Venus radar resolution, only the physiography can be characterized, and characterization of geological processes responsible for this physiography must await higher-resolution images. The question of the possible presence of plate tectonics on Venus does not yet have an answer.

Earth-based observations. Higher-resolution images have recently been obtained for a portion of the surface of Venus by using Earth-based radar telescopes located at Arecibo (Puerto Rico) and Goldstone (California). Recent data from Arecibo for Ishtar Terra show details of the mountain ranges at approximately 6 km (4 mi) horizontal resolution. Within these mountain ranges are seen parallel bands, approximately 15–20 km (9–12 mi) wide, which show many similarities to compressional fold belts associated with plate tectonic activity on Earth. In the Maxwell Montes Region, the bands are parallel to the long axis of the mountains and are best developed at high elevations (Fig. 2). The resolution is still too coarse and the coverage too incomplete to draw definitive conclusions about whether these bands and mountain ranges represent plate tectonic activity on Venus.

Venera 13 and 14 observations. In March 1982, two Soviet spacecraft, *Venera 13* and *14*, descended through the Venus atmosphere and landed on the eastern flanks of Beta-Phoebe Regio. These two spacecraft provided images of the area around the landing point (Fig. 3), as did their predecessors, *Venera 9* and *10*. The landing site at *Venera 13* shows evidence of abundant soil and rock fragments, while the *Venera 14* site is characterized by platy layered units with little associated soil. Each spacecraft carried an x-ray fluorescence spectrometer to analyze the composition of Venus surface materials. A drill obtained samples of surface materials which were brought inside the spacecraft for analysis. The composition at the *Venera 14* site was comparable to a typical basalt obtained from the Earth's ocean basins. At the *Venera 13* site the sample was also basaltic in nature but contained an anomalously high percentage of potassium, a composition found only locally on Earth. At the present time Soviet and American scientists are studying these data to try to unravel the nature of processes responsible for the formation and erosion of surface rocks on Venus.

Future studies. At present there are two types of data available for the Venus surface. Venera spacecraft provide a significant view of Venus on a very local scale, while Pioneer Venus altimetry and Earth-based radar data provide a view at relatively coarse resolution, characterizing the physiography but not the nature of geological processes. What is required are images of the surface at resolutions of 1 km (0.6 mi) or better in order to bridge the gap between these two data sets and answer fundamental questions about the nature and evolution of Venus and its relationship to the Earth. Such a mission is under active consideration by the National Aeronautics and Space Administration.

For background information *see* PLATE TECTONICS; SOLAR SYSTEM; VENUS in the McGraw-Hill Encyclopedia of Science and Technology.

[JAMES W. HEAD, III]

Bibliography: D. B. Campbell and B. A. Burns Earth-based radar imagery of Venus, *J. Geophys. Res.*, 85:8271–8281, 1980; J. W. Head and S. C. Solomon, Tectonic evolution of the terrestrial planets, *Science*, 213:62–76, 1981; H. Masursky et al., Pioneer Venus radar results: Geology from images and altimetry, *J. Geophys. Res.*, 85:8232–8260, 1980; G. H. Pettengill et al., Pioneer Venus radar results: Altimetry and surface properties, *J. Geophys. Res.*, 85:8261–8270, 1980; Yu. A. Surkov et al., X-ray fluorescence spectrometry on the surface of Venus, *Anal. Chem.*, 54:957A–966A, 1982.

Vesicle membranes

About 20 years ago, A. D. Bangham discovered that artificial cells could be produced in aqueous media by natural phospholipids such as egg lecithin. These cells, called liposomes, consist of spherical bilayer lipid membranes held together by hydrophobic forces. They are immediately and irreversibly destroyed in the presence of organic solvents, such as ethanol. Liposomes are viewed as the simplest models of biological cells and have been used by biochemists for studies of immobilized proteins, ions, and polar molecules. They have also been of interest to pharmacologists as possible vehicles for organ-selective transport of pharmaceuticals in the bloodstream. In addition, biophysical investigations have elucidated the thermodynamic behavior of liposomes in phenomena such as ion transport, phase transitions, aggregation, and fusion. However, for years there were no reports involving synthesis of liposomes independently from natural amphiphiles. Consequently, the mechanism of membrane formation and the stereochemistry of bilayer lipid membranes were not well understood.

The first totally synthetic bilayer membrane vesicle was formed from didodecyldimethylammonium bromide (DODAB) in 1977. This work confirmed that such membranes could be produced from a water-soluble amphiphile with a polar head group and a large hydrophobic part of equal diameter. It was demonstrated that the nature of the head group, which could be any of a number of groups (for example, carboxylate, sulfonate, carbohydrate, or tetraalkylammonium) was unimportant. The hydrophobic part usually consisted of two oligomethylene chains with 14–18 carbon atoms.

Figure 1 shows two typical amphiphiles. Several of these amphiphiles have been shown to form planar bilayer lipid membranes (BLMs) in aqueous suspensions. Application of ultrasound disrupts the liquid crystals and causes a rearrangement of the planar bilayer or membrane with equal numbers of molecules in both layers, to a curved bilayer, and finally to formation of a sphere with fewer molecules on the

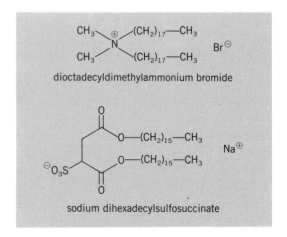

dioctadecyldimethylammonium bromide

sodium dihexadecylsulfosuccinate

Fig. 1. Two typical amphiphiles.

inside than on the outside. After 2 hours of ultrasound, small unilamellar vesicles are obtained (Fig. 2), which have rather uniform diameters of 35 ± 10 nanometers. The thickness of the curved lipid membrane bilayers ranges from 4 to 6 nm, the critical vesicle concentration from $< 10^{-10}$ to 10^{-4} molar. The ratio of the number of molecules in the outer layer to the number in the inner layer, obtained by [1]H-nuclear magnetic resonance (NMR) experiments, is usually about 1.8.

Unsymmetric bilayer membranes. A bilayer vesicle obtained by ultrasound tends to be symmetric. Two different amphiphiles, for example, a positively charged ammonium group and a negatively charged sulfonate, will be found in both the inner and outer layers. However, vesicle membranes with reactive head groups, for example, phenylenediamine or benzenediazonium derivatives, can be made fully unsymmetric by application of water-soluble reagents that do not penetrate the membrane. Thus, the outer phenylenediamine can be oxidized and coupled with a dye-forming reagent. Such reactions are known in color photography development. The inner molecules remain unchanged, and an outer benzenediazonium group can undergo selective photolysis by sunlight as well by using water-soluble, membrane-insoluble sensitizers. Such processes are illustrated in Fig. 3.

Some of these unsymmetric membranes provide an electron acceptor at the outer surface and an electron donor at the inner surface. Attempts to use such membranes for photochemical charge separation by irradiating porphyrin chromophores dissolved in the hydrophobic membrane have so far been unsuccessful. This is because efficient electron trapping requires an optimization of redox potentials and distances, and this has proved impossible with present techniques.

Monolayer membranes. Vesicle membranes have also been obtained from water-insoluble amphiphiles with a long hydrophobic part and hydrophilic head groups on both ends. Such molecules yield extremely thin membranes consisting of only one molecular layer (monolayer lipid membrane; Fig. 4). If two identical head groups are present, a spacer, or wedge, is needed to fill the gaps at the outer surface of the vesicle, as shown in Fig. 5. Such treatment is unnecessary when macrocyclic amphiphiles with one large head group (for example, thiosuccinic acid) are used with one smaller head group (for example, sulfonate), which produces uniform vesicles by ultrasound methods. The large head group is outside, the small group inside. The most versatile monolayer lipid membrane–forming system reported so far is derived from maleic acid anhydride, α,ω-bisfunctional alcohols, and sulfur nucleophiles, which can be added stepwise to the activated double bond of maleic esters.

Monolayer lipid membrane vesicles provide several advantages over bilayer vesicles. Monolayers can be synthesized with an unsymmetrical arrangement of both surfaces, and they are stable for longer periods (typically months) than bilayer vesicles (typ-

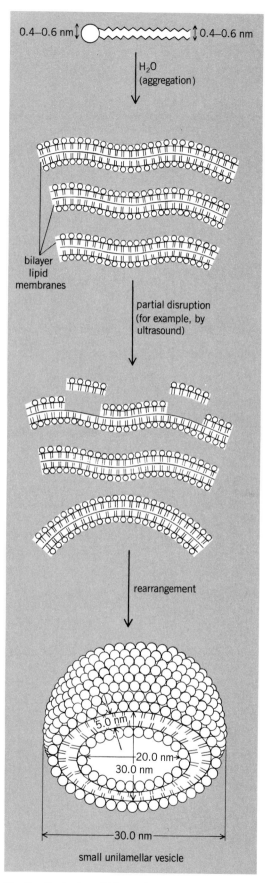

Fig. 2. Mechanism of bilayer lipid membrane vesicle formation from water-insoluble bilayers. (*After J.-H. Fuhrhop, Synthetische Vesikel mit Mono- oder Doppelschichtmembranen, Nachr. Chem. Tech. Lab., 28:792, 1980*)

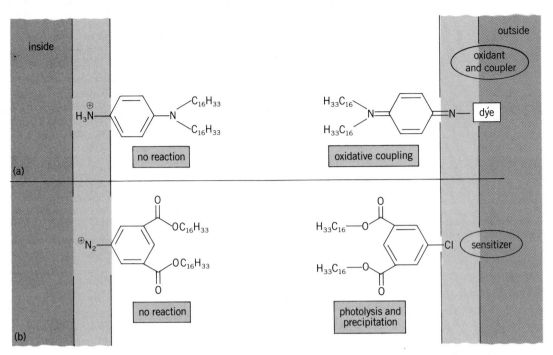

Fig. 3. Surface reactions of unsymmetric bilayer membranes. (a) Redox-active membrane. (b) Photolabile membrane.

ically a day) because the double-headed amphiphile cannot leave the monolayer, thus hindering migration of molecules to neighboring vesicles and preventing fusion of vesicles. Furthermore, the thinness of monolayers allows channels to be constructed into the membrane with molecules of much smaller size than the polypeptides or polyalcohols needed for bilayers.

Polymeric membranes. Polymeric vesicle membranes have been prepared from lipids with butadiene, diacetylene, or vinyl units in the hydrophobic part. The amphiphile is first treated with ultrasound and then polymerized by ultraviolet irradiation. Polymerization preserves the structure of the monomer vesicle and stabilizes it against organic solvents, for example, 50% ethanol. Polymeric vesicle membranes show low leakage rates for entrapped glucose, exhibit improved osmotic shock resistance, and retain their spherical structure after precipitation.

Chiral membranes. Amphiphiles with bulky head groups containing a center of asymmetry, for exam-

ple, glutamic acid or a carbohydrate, may produce vesicle surfaces with a chiral superstructure. This leads to circular dichroism more than a thousand times stronger than that exhibited by isolated molecules in solution. Such chiral membranes recognize the difference between a molecule and its mirror-image counterpart. It has been found that in a racemic mixture of chiral nitrophenylesters one enantiomer is hydrolyzed much faster if the catalyst is a vesicle with L-histidine head groups. The observed enantioselectivity ratio of 4:1 is higher than for any other chiral imidazole catalyst or corresponding aggregate.

Adsorption, solution, entrapment. Since most synthetic vesicles are highly charged, they tend to adsorb water-soluble organic compounds of opposite charge tightly to their surfaces. If the adsorbed molecule is a porphyrin or similar dye, it will be protonated by acids and metallated by zinc salts which are added to the aqueous vesicle solution. However, the water-soluble dye may also be evenly distributed between the inner and outer surfaces of the vesicle membrane by ultrasound. In this case, the outside pigments will react faster with acids than the entrapped dye molecules. This effect can be improved by dissolving cholesterol within the membrane, which lowers the proton permeability.

Water-insoluble porphyrins which are dissolved within the hydrophobic membrane are not protonated down to pH 1.5. In micellar solution the pK_a of the same porphyrins is about 4.5, and quantitative diprotonation is observed at pH 3. Although protons migrate through vesicle membranes, their concentration within the membrane is exceedingly low. This fact also explains the observed stability of

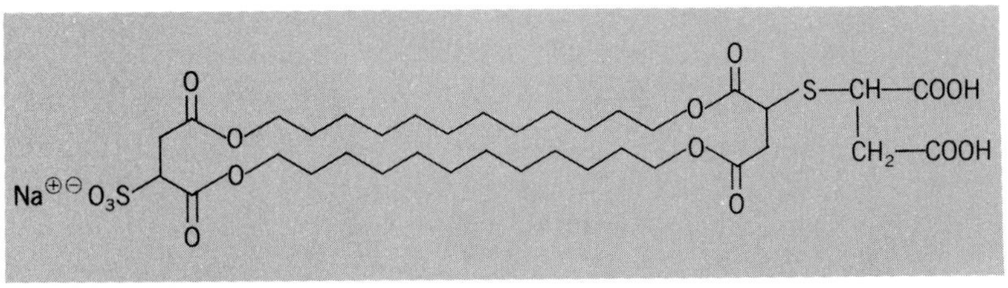

Fig. 4. Typical monolayer lipid membrane. On the left the small head group is inside of vesicle membrane, and on the right the large head group is outside of vesicle membrane.

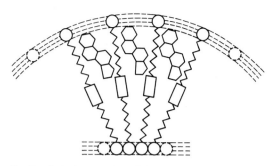

Fig. 5.   Part of a monolayer lipid membrane with a steroid wedge. (*After J.-H. Fuhrhop, Nachr. Chem. Tech. Lab.,* 28:792, 1980)

ester bonds within the membrane against strong acids in the aqueous phases.

Future directions and applications. The current emphasis in vesicle research is on synthesis and analysis of well-defined vesicles in terms of size (both surfaces) and permeability, and also on the localization of entrapped materials. A major problem is the concentration of material in the inner encapsulated water volume, which is usually at least 10^2 times smaller than the outer aqueous volume. Active transport through nonbiological membranes has not been reported.

Well-characterized and stable vesicles could have numerous applications, three of which are particularly interesting. Vesicles with chiral surfaces could transport dissolved pharmaceuticals in the bloodstream and release them selectively to cells whose surfaces interact strongly with the vesicle surface. Photoexcited dyes within the membrane could reduce electron acceptors localized on one surface and oxidize donors on the other surface, and light-induced charge separation similar to that in biological photosynthesis could be achieved. Finally, photolysis of benzenediazonium head groups to electrically neutral benzene-derivative head groups, for example, chlorobenzene, leads to precipitation of the vesicle. Since a vesicle contains about 10^4–10^5 molecules, and since only a few of the head groups must undergo photolysis to achieve precipitation, a quantum yield of 10^3 might be achieved in diazo-type photographic processes.

For background information *see* INTERFACE OF PHASES; LIPID; MICELLE; MONOMOLECULAR FILM in the McGraw-Hill Encyclopedia of Science and Technology.

[J.-H. FUHRHOP]

Bibliography: J. H. Fendler, *Membrane Mimetic Chemistry,* 1982; J.-H. Fuhrhop et al., Monomolecular membranes from synthetic macrotetrolides, *Angew. Chem. Int. Ed. Engl.,* 21:440, 1982; G. Gregoriadis and A. C. Alison, *Liposomes in Biological Systems,* 1980; L. Gros, H. Ringsdorf, and H. Schnupp, Polymere antitumormittel auf molekularer und zellulärer basis?, *Angew. Chem.,* 93:311, 1981 (*Angew. Chem. Int. Ed. Engl.,* 20:305, 1981).

Viral chemotherapy

Vaccination has proved efficacious in the prevention of many human viral diseases, although there are still a number of such diseases for which no vaccine is available. The treatment for herpes keratoconjunctivitis and the prophylaxis of influenza A virus have been available for almost 20 years; however, it has been only within the last decade that effective therapy for other viral infections has become available.

The therapeutic accomplishments of these drugs have paved for the way for compounds with greater activity and specificity, as determined by their higher therapeutic index (maximum tolerated dose divided by minimum effective dose) in cell culture and in animals. The identification of viral enzymes with unique properties that are not shared by corresponding enzymes in noninfected cells was a significant discovery. Thus it seemed possible that antiviral agents could be developed that would selectively interact with viral-induced enzymes and interfere with viral replication at an early or late metabolic stage. Several of these compounds, which approach the ideal antiviral agent, have now been synthesized and are being tested for clinical usefulness in various types of viral diseases.

This article will review some of these antiviral agents (Fig. 1), and it is limited to those that are currently in clinical use in the United States (see table) and in Europe. Isatin-3-thiosemicarbazone (Methisazone), a drug used for the treatment of smallpox and vaccinia infections, will not be discussed since smallpox has been eradicated.

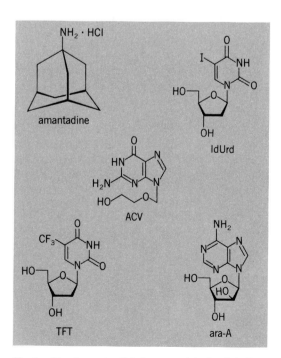

Fig. 1.   Structures of antivirals approved by the U.S. Food and Drug Administration.

Usage and mechanism of action of FDA-approved antiviral drugs

Antiviral drug	Use	Mechanism
Amantadine (1-adamantanamine, Symmetrel, Mantadix)	Influenza A	Inhibition of virus penetration or uncoating
5-Iodo-2'-deoxyuridine (IdUrd, IDU, idoxuridine, Stoxil)	Herpes keratitis	Activated by cellular and virus thymidine kinase; inhibition of virus DNA polymerase and incorporation into virus DNA
Adenine arabinoside (ara-A, Vidarabine, vira-A)	Herpes keratitis Herpes encephalitis (adults) Neonatal herpes*	Inhibition of virus DNA polymerase and incorporation into viral DNA
5-Trifluorothymidine (TFT, Viroptic, Trifluridine)	Herpes keratitis	Activated by cellular and virus thymidine kinase; incorporation into virus DNA; inhibition of thymidylate synthetase
Acyclovir (ACV, Zovirax, BW 248U)	Primary genital herpes Mucocutaneous herpes in the immuno-compromised	Activated by virus thymidine kinase; inhibition of viral DNA·polymerase and incorporation into DNA as a chain terminator

*Pending FDA approval.

Amantadine (1-adamantanamine). This drug prevents infection by influenza A and C, parainfluenza 1, Sendai, rubella virus, and pseudorabies virus in cell culture. Amantadine does not inhibit the virus, and has no effect on the adsorption of virus to host cells, but it does inhibit virus penetration or uncoating, and thus the viral genome is not expressed.

Amantadine was licensed by the Food and Drug Administration in 1966 for prophylaxis of Asian influenza. In 1976, the FDA-approved use of amantadine was changed to include therapy as well as prophylaxis of all influenza A strains. While the first and most effective line of protection against influenza A still remains vaccination, amantadine is recommended for use in an epidemic of influenza A. In addition, patients with cancer, or those who receive immunosuppressive therapy and might not respond fully to vaccine, could be better protected by amantadine in addition to vaccine. The side effects of amantadine on the central nervous system are primarily insomnia, jitteriness, and difficulty in concentration. Symptoms generally clear within 48 hours after therapy is stopped. Amantadine has been reported to be embryotoxic and teratogenic in rats but not in rabbits. Resistance of virus to amantadine has developed in culture and in humans, and indiscriminate use could result in the outgrowth of resistant strains of influenza A. Rimantadine, a congener of amantadine, is reported to have a lower frequency of side effects on the central nervous system.

5-Iodo-2'-deoxyuridine (IdUrd). This compound has potent activity against most human DNA viruses in cell culture, including herpes simplex virus type 1 (HSV-1) and type 2 (HSV-2), varicella-zoster virus (VZV), cytomegalovirus (CMV), pseudorabies virus, vaccinia virus, polyoma virus, adenovirus, and simian virus 40. IdUrd selectively inhibits viral replication by being more rapidly anabolized to the iodo analog of the thymidylate in the infected cells than in the noninfected cells, and is subsequently incorporated into viral DNA as counterfeit thymidine triphosphate. Along the way, the drug or its metabolites effect competitive inhibition and also cause feedback inhibition of regulatory enzymes. A direct correlation has been found between the extent of incorporation of IdUrd into HSV-DNA and inhibition of viral replication.

IdUrd is the first effective topical antiviral drug approved by the FDA for the treatment of ocular herpes simplex virus infection. IdUrd prepared in dimethyl sulfoxide (DMSO) is available in Europe and is used in the treatment of cold sores, genital herpes, herpetic whitlow, and varicella-zoster skin lesions. However, the use of this regimen is controversial. At high dosages, IdUrd or DMSO is teratogenic in several animal species. Topical applications of IdUrd may result in signs and symptoms of drug toxicity, such as punctate keratitis, follicular and papillary conjunctivitis, punctal occlusion, and allergic reaction. Resistance to IdUrd by herpesvirus develops rapidly in cell culture and is attributed to a decrease or lack of activity of the viral enzyme (thymidine kinase) necessary to phosphorylate this compound. Resistance to IdUrd in clinical applications has also been encountered. The systemic use of IdUrd for viral infection is also limited because it causes bone marrow suppression and is not transported to the brain.

5-Ethyl-2'-deoxyuridine (Aedurid) and 5-iodo-2'-deoxycytidine (Cebe-Viran) are two analogs of IdUrd which are available in Europe for the treatment of herpes keratitis. Clinical efficacy has not been confirmed in the United States.

Adenine arabinoside (ara-A). This compound shows extensive antiviral activity against several clinically important DNA viruses, such as HSV, VZV and CMV, as well as against animal RNA viruses that induce an RNA-dependent DNA polymerase (reverse transcriptase). The antiviral activity of ara-A is dependent on the levels of the enzyme adenosine deaminase in the cell line, since these probably dictate ara-A's activity. Ara-A is rapidly deaminated by this enzyme in humans to arabinosylhypoxanthine (ara-H), which is about 5 to 50 times less active, but 25 times less toxic, than ara-A in cell culture. There is still some uncertainty about the primary site of inhibition of ara-A (or ara-H). Its triphosphate, ara-ATP, is a competitive inhibitor of HSV-induced DNA polymerase with respect to ATP. The affinity of ara-ATP for viral DNA polymerase is greater than for the host cell DNA polymerase; however, this specificity depends on the source of the viral enzyme. Recent studies also indicate that ara-ATP is incorporated in internucleotide linkage of viral as well as cellular DNA.

Ara-A is approved by the FDA for topical therapy of HSV dendritic and geographic corneal ulcers and for parenteral treatment of herpetic encephalitis in adults. Approval for treatment of neonatal herpes with ara-A is pending. By timely therapy of herpes encephalitis with ara-A, the mortality has been decreased from 70 to 28%. It is not effective in its present formulation for the treatment of deep ocular disease, orolabial or genital herpes, and CMV infections. The drug may be effective in the therapy of VZV in patients receiving immunosuppressive drugs for the treatment of cancer or for procedures such as organ transplantation, and for the therapy of chronic hepatitis B virus infection.

Several approaches have been pursued to circumvent the problem of deamination and low solubility of ara-A in order to decrease the fluid volume input required for parenteral administration. Studies of compounds that are resistant to deamination, such as 2-fluoro-ara-A, cyclo-ara-A, and 2'-azido-ara-A, are in progress. Also under consideration are the coadministration of adenosine deaminase inhibitors and the 5'-phosphate of ara-A (ara-AMP), which is over 50 times more soluble than ara-A. Whether ara-AMP and ara-A are equally effective remains to be resolved.

Nausea and other gastrointestinal symptoms are the most common adverse effects of parenteral ara-A. Topically applied ara-A may result in signs and symptoms of drug toxicity similar to those observed with IdUrd. Ara-A has teratogenic effects in rats and rabbits and is carcinogenic in mice and rats. Ara-A–resistant HSV variants have been reported in cell culture and in humans. This resistance is probably mediated by mutations in the DNA polymerase region of the virus genome.

Trifluorothymidine (TFT). The spectrum of antivirus activity in cell culture of this compound is similar to that of IdUrd. After phosphorylation by either cellular or viral thymidine kinase, TFT is further phosphorylated to the triphosphate and is then incorporated into DNA. The uptake of TFT into DNA is responsible for its antiviral activity. Fragmentation of DNA or chemical hydrolysis of TFT-DNA to 5-carboxy-2'-deoxyuridine-DNA could contribute to the antiviral activity. The monophosphate of TFT is also a potent inhibitor of thymidylate synthetase, an enzyme necessary for pyrimidine synthesis. This inhibition is probably responsible for the cytotoxicity of the drug, making it too toxic for systemic use.

TFT is approved by the FDA for topical therapy of epithelial keratitis caused by HSV. It is probably the most effective drug for this infection since it readily penetrates the eye, and because resistant mutants are rare, although clinical resistance has been reported. TFT is also a potent inhibitor of cytomegalovirus in cell culture, but it is unlikely that this drug will be used clinically because of its systemic toxicity.

TFT ophthalmic solution, unlike IdUrd or ara-A, rarely causes hypersensitivity, although burning or stinging and palpebral edema occur frequently and contact dermatitis can also occur. TFT is mutagenic in culture and teratogenic in certain, but not all, animals. This drug is relatively stable under acidic conditions, but is rapidly hydrolyzed under physiologic and alkaline conditions to 5-carboxy-2'-deoxyuridine, a metabolite that is devoid of antiviral, but not cytotoxic, activity.

Acyclovir (ACV). This drug exhibits potent activity against HSV, VZV, and Epstein-Barr virus (EBV) in culture. Its activity against human CMV is marginal, and it has no activity against pseudorabies, adeno, or RNA viruses. Acyclovir is a member of an important new class of specific antiviral agents which depend on their preferential activation by the viral-induced thymidine kinase. The selective antiviral activity of acyclovir is, in part, a function of a higher affinity for viral than for cellular thymidine kinase. In addition, the triphosphate form of the drug (ACV-TP) may either inhibit viral DNA polymerase or act as a substrate for incorporation into viral DNA as a chain terminator, or both (Fig. 2). This class

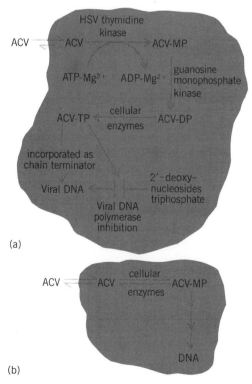

Fig. 2. Action mechanism of acyclovir. (a) In HSV-infected cell and (b) in noninfected cell.

of compounds includes 5-iodo-5'-amino-2',5'-dideoxyuridine (AldUrd), 5-ethyl-2'-deoxyuridine, arabinosylthymine (ara-T), E-5-(2-bromovinyl)-2'-deoxyuridine (BVDU), and 2'-deoxy-2'-fluoroarabinosyl-5-iodocytosine (FIAC) and its thymine analog (FMAU). These experimental compounds are not yet approved by the FDA.

ACV ointment (5%) has been approved by the FDA for the treatment of primary genital herpes (that is, for patients in whom genital herpes represents the first experience with either HSV-1 or HSV-2). A formulation of 5% ACV in polyethylene glycol was not effective in patients with recurrent genital or orolabial herpes. A remarkably large number of controlled clinical trials have already been completed with this drug, and it is likely that the use of ACV will not be limited to the treatment of primary genital herpes. ACV ointment (3%) has been shown to be as effective as ara-A for the treatment of dendritic corneal epithelial ulcers in humans. Favorable results have been noted in controlled trials of ACV in patients with acute herpes zoster and mucocutaneous HSV infections, as well as ACV prophylaxis of HSV infections in bone-marrow recipients. Other studies in patients infected with CMV or EBV are ongoing. ACV is also being evaluated in parallel with ara-A for the treatment of herpes encephalitis and neonatal herpes. The FDA recently approved intravenous ACV for treating initial and recurrent mucosal and cutaneous HSV infections in immunocompromised adults and children and for treating patients with severe initial genital herpes who are not immunocompromised.

Side effects of ACV ointment are similar to those described for IdUrd and ara-A. Intravenous ACV can produce a reversible increase in urea and creatine. Patients receiving bolus systemic ACV can develop nephrotoxicity caused by deposition of ACV crystals. Slow infusion or reduced dosage of the drug has now been recommended, especially for patients with altered renal function.

Herpes simplex viruses resistant to ACV emerge rapidly in culture; already three cases of ACV-resistant mutants in humans have been reported. The use of combinations of antiviral drugs may overcome the problem of viral resistance and has the potential added advantage of reducing toxic effects.

Ribavirin (Virazole). In culture, this drug has a broad antiviral spectrum, affecting DNA and RNA viruses, including herpes, adeno, vaccinia, influenza, parainfluenza, measles, and rhino viruses. This compound structurally resembles guanosine and is converted to its active form by phosphorylation. In the cell, ribavirin is metabolized to the 5'-monophosphate; this product inhibits inosinate dehydrogenase, which ordinarily converts inosinate to xanthylate. In other words, ribavirin interferes with the formation of guanosine monophosphate, upon which both DNA and RNA synthesis depend. The 5'-triphosphate derivative is a potent inhibitor of influenza virus RNA polymerase, and this may be responsible for the antiviral activity against this virus.

Although this drug is marketed in many countries for the treatment of viral respiratory infections and other viral childhood diseases (for example, measles), the FDA has not yet licensed this drug for the treatment of any viral infection. Controlled clinical trials of oral ribavirin for influenza A and B virus, hepatitis B virus, and genital herpes infections have resulted in conflicting reports: the drug either had no beneficial effect or resulted in slight clinical improvement. In a recent controlled study, an aerosol preparation had some effectiveness in reducing the duration of virus shedding and symptoms in patients infected with influenza A virus. Limited open trials with oral preparations against herpes zoster, orolabial herpes, measles, and Lassa fever have yielded favorable, but still unconfirmed, results.

The major toxicological effect of ribavirin in humans at high dose levels is the development of anemia; this effect is reversible on termination of therapy. Ribavirin is teratogenic in rats and hamsters but not in baboons, and is known to cause fetal resorption in rats and rabbits.

The following antivirals are still being tested for either herpetic or rhinoviral (common cold) infections: phosphonoformate (PFA), 5-n-propyl-2'-deoxyuridine (PdUrd), E-5-(1-propenyl)-2'-deoxyuridine; 5-methoxymethyl-2'-deoxyuridine (MMUdR), (S)-9-(2,3-dihydroxypropyl)adenine [(S)-DHPA]; 2-deoxy-D-glucose, L-lysine, and arildone for HSV infections; and enviroxime, zinviroxime, and 4',6-dichloroflavan for rhinovirus infections.

Conclusions. All of these agents attempt to reduce viral load during acute disease in the hope that the immune mechanisms of the host may be able to complete a cure. None of these compounds is effective in preventing reactivation of latent viruses or in reducing the inflammatory response caused by the viruses. Antiviral chemotherapy aimed at controlling recurrences by keeping the virus in, that is, by interfering with the mechanisms of reactivation of the virus, will probably be the only truly effective way of controlling recurrences. Until this is done, the presently available effective antiviral treatments that curtail the duration of the disease, particularly initial infections, will have to be optimized. The discovery of selective antiviral drugs of proven efficacy will undoubtedly hasten efforts at understanding viral metabolism and the identification of new viral-induced enzymes. This will, in turn, lead to the development of more effective drugs for the treatment of various viral infections.

For background information see ANIMAL VIRUS; DRUG RESISTANCE; INTERFERON; VIRUS; VIRUS CHEMOPROPHYLAXIS in the McGraw-Hill Encyclopedia of Science and Technology.

[RAYMOND F. SCHINAZI; WILLIAM H. PRUSOFF]

Bibliography: G. J. Galasso, An assesment of antiviral drugs for the management of infectious diseases in humans, Antiviral Res., 1:73–96, 1981; A. J. Nahmias, W. R. Dowdle, and R. F. Schinazi (eds.), The Human Herpesviruses: An Interdisciplin-

ary Perspective, 1981; R. F. Schinazi and W. H. Prusoff, Antiviral drugs: Modes of action and strategies for therapy, *Hosp. Prac.*, 16:113–124, 1981.

Volcano

After lying dormant for 123 years, Mount St. Helens spewed gas and dust into the atmosphere during much of 1980. Its more explosive phases devastated neighboring forests, lakes, and towns, and sent plumes into the stratosphere that gradually spread over much of the Northern Hemisphere. Extensive measurements performed on the ground, from aircraft and balloons, and from spacecraft have permitted the first comprehensive characterization of material emitted by explosive volcanoes, and a determination of the environmental and climatic consequences of this emitted material.

Temperatures beneath the Earth's surface steadily increase with depth, and at about 80 km (50 mi) beneath the continents, some rocks will melt. However, in special locations, molten rock (magma) exists much closer to the surface. Volcanoes are localized places where such magma occasionally reaches the surface. In some cases, magma simply flows out of a volcano and spills across the neighboring surface. In other cases, gas pressure builds up and eventually is released explosively, with gases and particles being thrown high into the air. The gases responsible for explosive volcanism are derived either from the magma itself, bubbling out of it as pressure drops with decreasing depth, or from groundwater that seeps close to the magma and is heated to steam.

Mount St. Helens' eruption. Gases and ash particles from the eruption of Mount St. Helens on May 18, 1980, were injected as high as 22 km (14 mi) into the air, thus penetrating the lowest 10 km (6 mi) of the stratosphere (Fig. 1). The larger, lower-altitude ash rained down on eastern Washington, northern Idaho, and western Montana, producing ash deposits as thick as 70 mm (3 in.). Smaller-sized ash particles and gases were carried greater distances, depending on the wind speed and direction at their altitude (Fig. 2). For example, material injected near the base of the stratosphere at an altitude of 12 km (8 mi) was caught by the strongest winds and traveled in an easterly direction at a speed of 30 m/s (70 mph), ultimately traversing the globe in 15 days. Material located close to the top of the volcanic plume was carried initially in a southeasterly direction, but then reversed direction to move toward the northwest. Within a period of about 2 months, the volcanic material in the stratosphere had spread fairly uniformly in an east-west direction. In a more gradual fashion it moved north and south to cover much of the Northern Hemisphere. Material injected into the lower atmosphere, or troposphere, beneath the stratosphere, was removed in a period of days because of vigorous vertical motion and rainfall. Several subsequent major explosions, such as the ones on May 25 and June 13, added more material to the stratosphere, but not as much or to as high an altitude as the event on May 18.

Observations. Extensive observations made during 1980 from the ground, air, and space not only have provided in-depth information about Mount St. Helens, but have also permitted generalizations to be made about the nature and consequences of other explosive volcanoes. Geologists monitored seismic activity associated with the movement of magma within the volcano and studied deposits of volcanic debris; atmospheric scientists sampled the gases and particles of the volcanic plumes in both the troposphere and stratosphere from aircraft and balloon platforms, including a NASA-operated U-2 aircraft that flew in the lower stratosphere to investigate the climatic consequences of the volcanic debris; and space scientists monitored the gradual spreading of the volcanic cloud in the stratosphere with satellite experiments, such as NASA's Satellite Aerosol and

Fig. 1. Plumes of gas, ash (darker clouds), and water (lighter clouds) injected into the atmosphere by Mount St. Helens on May 18, 1980. (*NASA*)

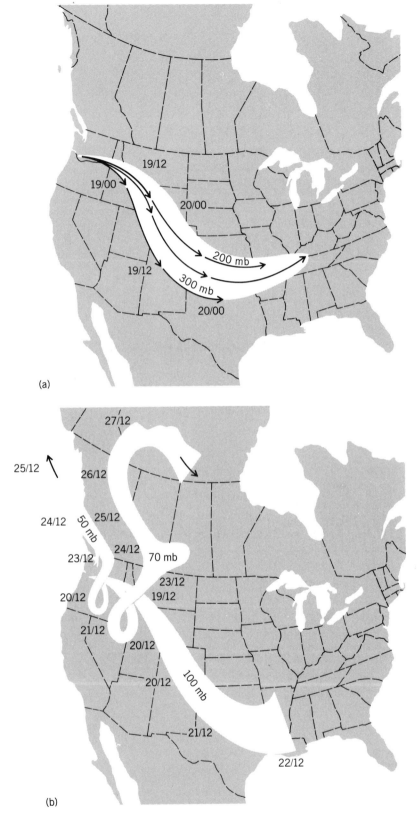

(a)

(b)

Fig. 2. Paths followed by material injected into the atmosphere at different altitudes or pressure levels by Mount St. Helens on May 18, 1980. The trajectories in *a* refer to the upper troposphere and lower stratosphere, while those in *b* refer to somewhat higher altitudes in the stratosphere. The numbers flanking the arrows denote the day and hour at which the material reached given places, while the numbers within arrows show the pressure level in units of millibars (1 mb = 100 pascals). The pressure at sea level is 1000 mb and at the boundary between the troposphere and stratosphere 200 mb.

Gas Experiment (SAGE) experiment.

Mount St. Helens spewed ash, water vapor, and magmatic gases into the atmosphere. The ash particles consisted of a mixture of volcanic glass, created when tiny liquid droplets of magma froze, and crystalline material, made chiefly of the mineral plagioclase, that resulted from the shattering of rocks overlying the magma chamber. Some of the particles from this debris cloud were collected on fine wires by a U-2 aircraft. Electron microscope photographs of them are shown in Fig. 3. In the stratosphere, ash grains were typically about 1 μm (1/25,000 in.) in size, with the grain size varying from about 0.1 to 10 μm. The average grain dimensions increased somewhat with decreasing altitude in the stratosphere and troposphere: big grains fell more quickly than smaller ones.

The chief gases injected into the atmosphere were water vapor and sulfur-containing compounds. Measurements obtained in the stratosphere a few days after the May 18 explosion showed enhancements within the volcanic cloud over corresponding values in the undisturbed stratosphere of up to a factor of 10 in water vapor, 1000 in sulfur dioxide (SO_2), and 3 in carbon disulfide (CS_2) and carbonyl sulfide (COS). In addition, both methyl chloride (CH_3Cl) and hydrogen chloride (HCl) were found to be enhanced by as much as a factor of 10 in the stratospheric debris. Within the lower atmosphere, sulfur gases were also prominent members of the volcanic plumes, with hydrogen sulfide (H_2S) and sulfur dioxide (SO_2) alternating as the most abundant sulfur-containing gas species.

Chemical transformations. A concentrated aqueous solution of sulfuric acid almost always constitutes the most abundant type of particle (or aerosol) in the stratosphere. These particles are produced by a series of oxygen- and water-adding reactions that convert sulfur gases into sulfuric acid aerosols. For example, hydroxyl radicals (OH) that result from the breakup of water vapor molecules by ultraviolet sunlight react with SO_2 to produce sulfur trioxide (SO_3). In turn, SO_3 may combine with water vapor to produce sulfuric acid vapor (H_2SO_4) which subsequently condenses.

The sulfur gases injected by Mount St. Helens into the stratosphere—particularly SO_2—resulted in the creation of a significant amount of new sulfuric acid particles. This material appeared very rapidly. The abundance of sulfuric acid was enhanced by factors of 10 and 200 over its background amount in the 1-day-old and 4-day-old clouds, respectively, produced by the May 18 explosion. The initial conversion of SO_2 gas into sulfuric acid particles may have taken place on the surfaces of the ash grains, which thus acted as catalysts to greatly accelerate the chemical transformation. The oxidants for these reactions may have been derived from background air mixed into the volcanic cloud, while the cloud's own water vapor initially supplied much of this ingredient.

Sulfur dioxide was only incompletely exhausted during the early rapid conversion phase. Later transformations to sulfuric acid proceeded at a much slower pace over a period of months because of the poisoning of the catalytic surfaces of the ash grains, which then fell out of the stratosphere. The sulfuric acid produced during the two phases formed new, initially tiny particles, and also coatings around ash particles. The net result of this conversion was an enhancement by a factor of 3 of sulfuric acid aerosols in the Northern Hemisphere stratosphere during the fall and winter of 1980. These particles also gradually fell out of the stratosphere, but except for the first few days after the May 18 explosion, sulfuric acid and not ash was the dominant aerosol in the stratospheric volcanic plume.

Climate effects. Aerosols affect the Earth's climate by altering the amount of sunlight absorbed by the Earth and the amount of heat it radiates to space (Fig. 4). Such alterations can modify temperatures at the ground and in the atmosphere, and may thereby also alter atmospheric circulation patterns, ultimately producing changes in regional weather patterns. For these effects to be significant, aerosols must be sufficiently abundant to intercept a nontrivial fraction of sunlight and heat radiation.

Prior to the measurements made of the stratospheric debris from Mount St. Helens, it was difficult to determine whether volcanic aerosols cause a net cooling or warming of the Earth's surface and its lower atmosphere, since this depended critically on several poorly known properties: their ability to absorb sunlight as opposed to their ability to scatter sunlight, and their mean size. Certain (highly absorbing) aerosols absorb a larger fraction of sunlight, and therefore the Earth tends to warm; other (poorly absorbing) aerosols produce the opposite effect. Even in the latter case, however, the troposphere and ground could still experience a net warming if the aerosols were effective in blocking the heat these regions radiate toward space. This effectiveness, in turn, depends on the size of the aerosols because both very small and very large particles are good blockers of heat relative to their ability to intercept sunlight. In all cases, stratospheric aerosols produce a rise in temperature in the region where they reside since their thermal and solar interactions work together. Measurements conducted from the U-2 showed that the volcanic aerosols from Mount St. Helens were very poor absorbers of sunlight and that they had a mean size of a few tenths of a micrometer during much of the year that they remained in the stratosphere. Therefore, these volcanic aerosols, and probably those produced by other volcanic explosions, cause a net cooling of the ground and troposphere, but a warming of the lower stratosphere.

By the time the volcanic material from Mount St. Helens had spread throughout much of the Northern Hemisphere, the aerosol concentration had become sufficiently diluted so that only about 2% of the incident sunlight was intercepted by the volcanic aerosols. This is too small a percentage to have significantly affected the Earth's climate during 1980. However, there have been times during which much more volcanic debris was present in the stratosphere.

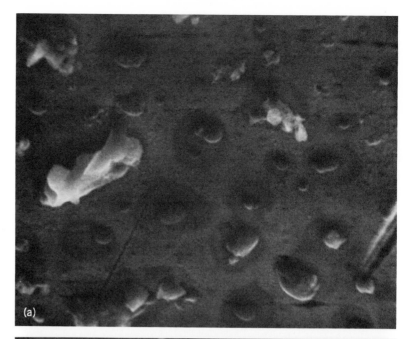

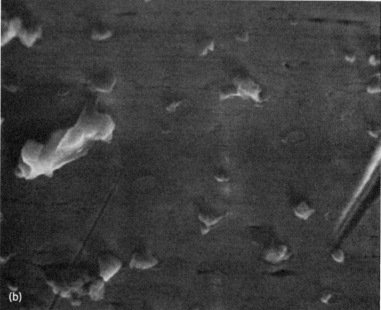

Fig. 3. Electron micrographs of particles that were part of the debris cloud injected by Mount St. Helens into the stratosphere. The changes in the appearance of some particles (a) before and (b) after heating is due to the evaporation of sulfuric acid that coated some ash particles and that formed separate particles. The particles in b are very fine ash particles. The biggest particle has a long dimension of about 7 μm (1/3000 in.). (*Courtesy of Ken Snetsinger*)

At these times, the Earth's climate may have been altered in a significant fashion. The period from 1880 to 1915, when a number of major volcanic explosions, including Krakatoa, occurred, was cooler than the several subsequent decades, in accordance with the inferences later drawn about Mount St. Helens' aerosols. It is also interesting to note that the Chichón volcano of Mexico introduced a massive amount of material in April 1982—possibly enough to cause short-term changes in the climate.

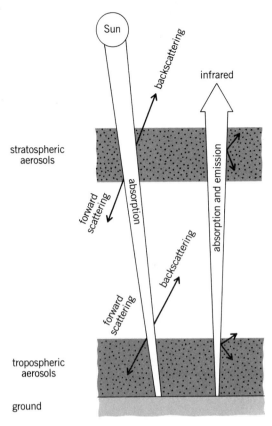

Fig. 4. Major interactions between aerosols and radiation in the stratosphere and troposphere.

For background information *see* ATMOSPHERIC CHEMISTRY; CLIMATIC CHANGE; VOLCANO in the McGraw-Hill Encyclopedia of Science and Technology.   [JAMES B. POLLACK]

Bibliography: R. Findley, Eruption of Mount St. Helens, *Nat. Geog.*, 159:3–65, January 1981; J. B. Pollack et al., Radiative properties of the background stratospheric aerosols and implications for perturbed conditions, *Geophys. Res. Lett.*, 8:26–28, January 1981; Reports (on Mount St. Helens), *Science*, 211:815–838, February 1981.

Water conservation

Low dissolved-oxygen concentration in releases from reservoirs has been recognized for some time as a significant contributor to water quality problems in many areas. Although many reservoirs were constructed years ago, technology is only now being developed and applied to correct some of these difficulties.

Reservoir stratification. During summer months, many reservoirs separate, or stratify, into distinct water layers (Fig. 1). Although reservoirs can stratify in northern climates under extremely cold conditions, the stratification condition that causes the most significant dissolved-oxygen problems occurs in warmer months. In this case the top layer (epilimnion) is exposed to sunlight and warm inflows, making it warmer and less dense than the unexposed bottom layer (hypolimnion). This density differential between the epilimnion and hypolimnion starts to develop during spring, with very little subsequent mixing of the two layers taking place during the summer and fall.

Respiration of living organisms, decomposition of naturally occurring organic materials and organic pollutants, and chemical reactions—all processes which deplete dissolved oxygen in reservoirs—are at work in both the epilimnion and hypolimnion. Because of its exposure to air and sunlight, however, the epilimnion receives dissolved oxygen through diffusion and photosynthesis, enabling it to replenish its supply. The hypolimnion, isolated from these dissolved-oxygen-replenishing processes, continues to have depressed levels of dissolved oxygen.

It is important to recognize the distinction between reservoirs and lakes. Lakes occur in natural depressions and do not have outlets that can draw from beneath the surface as reservoirs do. Although lakes can also stratify, outflows come from the surface. Therefore, the stratification process does not affect the dissolved oxygen of the outflow.

Water drawn from lower levels of a stratified reservoir, such as during hydroturbine operation, often has a low concentration of dissolved oxygen during the summer and fall. Releases from the reservoir during this period may reduce the dissolved oxygen downstream to such an extent that there is no longer enough of it to adequately assimilate wastes discharged into the river or to support fish and other aquatic life.

Not all reservoirs stratify, and, of those that do, not all are deficient in hypolimnetic dissolved oxygen. The problem, however, is particularly severe

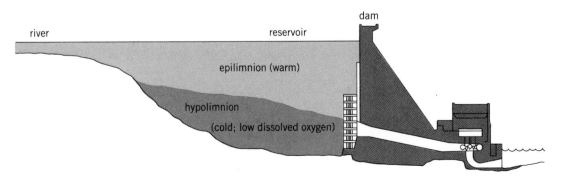

Fig. 1. Reservoir stratification which is typical during summer months.

in the southeastern United States, where high water turbidity and high summer temperatures concentrate heat input near the surface of the reservoir. This characteristic, combined with the high levels of biological activity associated with lengthy stratification periods, often results in severely depressed dissolved-oxygen concentrations. The solution to this problem is to increase the amount of dissolved oxygen in reservoir releases through methods such as reaeration of these releases.

Aeration technology. Reaeration of reservoir releases can be accomplished in a number of ways: multilevel intakes can be installed to take water from the epilimnion to mix with water from the hypolimnion before release downstream; water can be released through spillways, where it is aerated by the turbulence caused by the spillway structure; or special aeration valves can be installed on reservoir outlets. All of these devices are more suited for in-

corporation in new reservoirs or for existing reservoirs which do not generate hydropower. If they could be incorporated into an existing hydropower facility, they would be expensive to construct or, in some cases, cause a significant reduction in power-generating capacity. Other methods to aerate water within reservoirs before it is released or as it passes through hydroturbines are in various stages of development.

Diffusion. The most efficient means of adding oxygen to reservoirs is to bubble high-purity oxygen through small-pore diffusers placed near the bottom of the reservoir. The small size, and hence large surface area, of these bubbles—coupled with the reservoir's generally substantial depth, which allows high pressures and a long contact period—contributes to excellent oxygen transfer efficiency that sometimes exceeds 90%. Oxygen is used rather than air (which is only 20% oxygen) because gas-handling and diffusion equipment to accommodate air

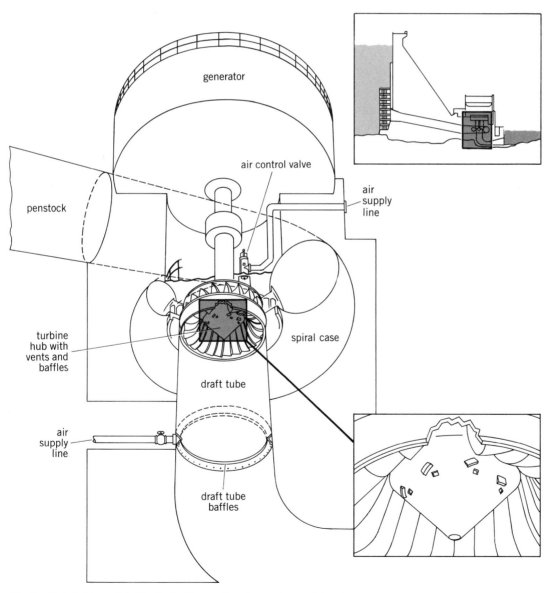

Fig. 2. Hydroturbine with baffles for turbine aeration. Insets show position of the turbine in relation to the reservoir (upper right) and an enlarged view of the turbine hub (lower right).

would have to be five times as large as that used for high-purity oxygen. Also, when air is added to the water, the large nitrogen component would contribute to gas supersaturation of the water, possibly resulting in gas bubble disease in fish.

Diffused air has been used in small reservoirs to improve dissolved-oxygen conditions through aeration but primarily by mixing the hypolimnion and the epilimnion, and thus breaking up stratification. Large pumps have also been employed to mix the two layers.

Baffle technology. Another method of aeration involves introduction of air below the hydroturbine. Because of the water-flow characteristics of some hydroturbines, a low-pressure area exists immediately downstream of the turbine in the draft tube. Sometimes, the low pressure is sufficient to draw or aspirate small amounts of air into the draft tube, if the tube is vented to the atmosphere. Through the use of baffles, which induce a localized area of low pressure, large quantities of air can be drawn into the draft tube. High turbulence and high pressures farther down the draft tube provide excellent oxygen transfer efficiency. This technique has been successfully tested at the U.S. Army Corps of Engineers' Clarks Hill Dam (Georgia and South Carolina), Tennessee Valley Authority's Norris Dam (Tennessee), and Alabama Power's Bankhead Dam and Logan Martin Dam (Alabama).

Figure 2 shows two baffle arrangements demonstrated on a hydroturbine at Norris Dam. In this hydroturbine, a type known as a Francis, water enters through the penstock. The water is then distributed around the periphery of the hydroturbine wheel by the spiral case, and as the water travels through the hydroturbine it passes over the runner cone (hub). For the Francis unit, it is sometimes possible to mount baffles on the runner cone over existing vent holes, which are connected to air-supply pipes, as shown in Fig. 2. Often these pipes prove to be too small, and additional air-supply lines must be added.

Another location for baffles is on the draft tube wall with an air-supply pipe feeding into the draft tube. This configuration is suitable for both Francis and propeller turbines. For this arrangement, baffles can be continuous around the draft tube wall, as in the ring design shown in Fig. 2, but it is not necessary that they be so. Where air-supply pipes could be placed behind the draft tube, as many as eight baffles have been used.

One of the disadvantages of baffle technology is that it reduces the power generation capability of the turbines. Although this reduction is small, generally 1–3%, the cost represents the most significant portion of the expense of the aeration installation.

Compressors. Other methods, which have been tested at Idaho Power's American Falls Dam (Idaho), Tennessee Valley Authority's Norris, Cherokee, Douglas, and South Holston dams (all in Tennessee), and several European locations, make use of low-pressure compressors to blow air through the hydroturbine mechanism or directly into the draft tube. An advantage of these methods over baffles is that they limit power reduction to the period of aeration rather than all the time. Compressor installations are also being developed for certain hydroturbines which are not well suited for baffles.

Dissolved oxygen levels. Desirable minimum limits of dissolved oxygen in streams generally are 6 mg/liter to support salmonoids and 4–5 mg/liter for other fish. Even though some reservoirs have virtually no dissolved oxygen in their releases during certain parts of the year, a reservoir oxygen-injection system can be designed to supply enough oxygen to meet stream-water quality standards even under extremely high water-flow conditions. The best turbine aeration systems have increased dissolved-oxygen levels a maximum of only about 4 mg/liter in reservoirs where little dissolved oxygen was present initially and water temperature was below 20°C. The amount of dissolved oxygen transferred to the water in a turbine aeration system decreases as the initial dissolved-oxygen concentration and the temperature increase. Work is continuing to increase the amount of oxygen transferred under a full range of operating conditions.

For background information *see* WATER CONSERVATION in the McGraw-Hill Encyclopedia of Science and Technology.

[CHARLES E. BOHAC]

Bibliography: C. E. Bohac et al., Methods of reservoir release improvement, *Proceedings of the 37th Purdue Industrial Waste Conference,* Purdue University, 1982; C. E. Bohac et al., *Techniques for Reaeration of Hydropower Releases,* U.S. Army Engineer Waterways Experiment Station, Vicksburg, MS, September 1982; Tennessee Valley Authority, *Improving Reservoir Releases,* TVA/ONR/WR-82/6, Knoxville, TN, 1981.

Wind ship

Pure sailing ships were used for all ocean cargo transport until the early 1900s. With the advent of steam propulsion, powered cargo vessels replaced pure sailing vessels in virtually all ocean cargo transportation systems. Fossil fuels were inexpensive, and powered vessels were more reliable than their sailing counterparts. Since powered vessels were not at the mercy of wind and weather conditions, they could maintain strict schedules and sail at higher average voyage speeds.

Recently, as oil prices have steadily increased, naval architects and marine engineers have looked toward the use of wind power once again as a means of reducing ship operating costs. Most research has focused on sail-equipped motor vessels rather than on pure sailing ships. A sail-equipped ship is a hybrid vessel which combines the reliability and constant speed of fossil fuel propulsion with the economy of wind energy afforded by the use of sails.

Wind propulsion for merchant vessels is being explored by many nations. The Japanese are among the leaders in the development of this technology. Their first such ship, the *Shin Aitoku Maru,* is a cargo vessel which takes advantage of the technological and engineering developments of this century to utilize wind propulsion for merchant marine fuel conservation.

Performance. The 699-gross-ton (1979-m³ capacity), steel-hulled ship measures 66 m (217 ft) long, has a beam of 10.6 m (34.8 ft) and a draft of 5.2 m (17.1 ft). During the ship's first year of commercial service, operating in coastal waters between Japan, China, and Korea, the sail rig was used in conjunction with the main diesel engine about 60% of the time. During the year, the total fuel savings attributed to use of the sails was about 10%. An additional benefit of the sails is that they contribute to a reduction in rolling motions, thereby aiding the vessel in seakeeping.

Research and development. The hybrid *Shin Aitoku* was built by the Aitoku Company at the Inamura Shipbuilding Company after nearly 2 years of research by Nippon Kokan Kaisha (NKK), under consignment from the Japanese Marine Machinery Development Association. Early research and development activities included wind tunnel tests and analyses and the construction of an experimental sail-equipped ship to test sail rigs at sea. The ship was rigged with various sail designs, and was used to pinpoint optimum sail construction and shape—the best method for the automation of sail handling—and to test a ship's stability and maneuverability with large sails.

After wind tunnel tests, three types of sails were chosen for sea trials: soft sails, because of their low construction costs; fore-and-aft sails, because of superior performance sailing into the wind; and rigid sails, because of their best performance overall. After sea trials, rigid sails with a laminar flow, a maximum camber, and a large aspect ratio were chosen for construction and installation aboard the *Shin Aitoku*. Two sets of these sails were built at a cost of $250,000. The rectangular sails measure 8 m (26.2 ft) across and 12.15 m (39.9 ft) high, for a total sail area of about 194 m² (20.88 ft²). The sails are constructed of canvas stretched across steel frames (Fig. 1).

Sail rig operation. The two sails are controlled automatically from the ship's bridge by the use of a microprocessor tied into a hydraulic power system. No additional crew members are needed to operate the sail system (Fig. 1).

Wind speed and relative direction are constantly monitored and processed by the computer. Normally, sail movements are accomplished automatically by the computer as wind direction and course changes dictate. Unlike sails used on traditional sailing craft, the rigid sails on board the *Shin Aitoku* are not taken down from their masts. Instead, the sails are actually folded vertically in half when not in use (Figs. 2 and 3).

If the wind is blowing from an unfavorable direction, for example, directly ahead, the computer will automatically fold the sails and feather them into the wind, or turn them so that they will produce the

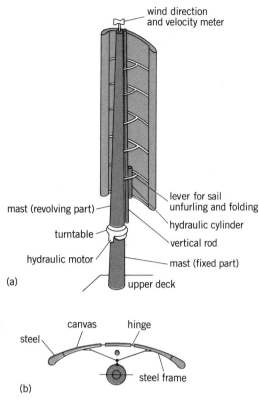

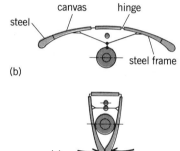

Fig. 1. Diagram of sail components and hydraulic equipment used to open, close, and rotate the sails. (*a*) Plan view. (*b*) Cross section of sail unfurled. (*c*) Cross section of sail folded.

least amount of wind resistance. When the winds become favorable, the sails are stretched open again.

Designed to be used for propulsion in winds up to 20 m/s (45 mi/h), the sails can revolve through 360° and can be stretched or folded in about 2 min.

As an auxiliary propulsion system, the sails create thrust which reduces the amount of power needed from the main diesel propulsion system. When the winds are at a speed and from a direction which can be efficiently utilized by the sails, the output of the main diesel engine is automatically reduced by the amount of thrust obtained through the sails. This allows the ship to travel at a constant speed while the main engine's fuel consumption is reduced.

The main diesel engine speed is regulated by an automatic load controller. To increase or decrease the main engine output, the engine rotation rate and the blade angle on the ship's controllable pitch propeller are adjusted for maximum fuel economy.

Other types of sails. Research into the use of wind energy for marine propulsion systems is not limited to the type of sails used on the *Shin Aitoku*. Other research has centered on the use of wing sails, which

Fig. 2. Ship cruising with sails fully stretched. Hydraulic sail-furling components and mast and turntable connection are visible in foreground.

Fig. 3. Sails fully closed and aligned with ship's longitudinal axis.

create thrust using the same principle that allows an airfoil to create lift; and wind turbines, which translate wind energy into mechanical motion in the same manner as a land-based windmill.

For background information *see* MERCHANT SHIP; MICROPROCESSOR; WIND POWER in the McGraw-Hill Encyclopedia of Science and Technology.

[DAVID W. ROBB]

Bibliography: D. Robb, Japanese wind ship proves successful, *Sea Technol. Mag.*, pp. 37–40, April 1982; *Sail Equipped Motor Ships: Interim Summary*, NKK (Nippon Kokan Kaisha) no. 271–014, March 1980; M. Sudo, *Operating Performance of Sail Equipped Small Tanker*, NKK Overseas Rep. no. 33, 1981; World's first commercial motor tanker equipped with sails, *Jap. Ind. Technol. Bull.*, pp. 2–6, September 1980.

Yersiniosis

Yersiniosis, in a narrow definition, refers to infections caused by two members of the genus *Yersinia*, namely *Y. pseudotuberculosis* and *Y. enterocolitica*, but exclusive of the plague caused by the notorious *Y. pestis*. While many clinical and microbiological aspects relative to the disease-producing potential of the long-recognized *Y. pestis* and *Y. pseudotuberculosis* have been resolved, much research centers today on *Y. enterocolitica*.

Described in the United States barely 40 years ago, *Y. enterocolitica* entered into scientific prominence in 1976 when it was linked to a major outbreak of gastrointestinal illness in upper New York State (Holland Patent), involving 222 school-age subjects, 16 of whom were subjected to appendectomies. The study of this epidemic provided the foundation upon which many current methodologies embodying molecular biology were coupled with standard microbiological techniques. This led to greater understanding of virulence mechanisms underlying this microbial species' ability to invade the gastrointestinal mucosa to produce gastroenteritis, mesenteric lymphadenitis, terminal ileitis, and rarely, blood infections. Paramount among the endeavors were: taxonomic differentiation of the heterogenous species, comprising *Y. enterocolitica*; refinement of methods for the isolation and identification of this species from human clinical sources; the identification of plasmids associated with virulence markers; and the definition of epidemiologic trends among isolates, especially in the United States.

Taxonomically, based on deoxyribonucleic acid (DNA) hybridization studies and biochemical characteristics, *Y. enterocolitica* was differentiated from a group of isolates resembling *Y. enterocolitica*. These latter isolates were accorded distinct species status within the genus *Yersinia* (Table 1). The taxonomic differentiation clearly delineated the pathogenic *Y. enterocolitica* from nonvirulent species derived mainly from environmental sources, such as water, animals, or food.

Microscopic and culture characteristics. Members of the genus *Yersinia* are gram-negative, asporogenous, facultatively anaerobic, coccobacillary to ba-

Table 1. Biochemical characterization of _Yersinia_ species*

Test	_enterocolitica_	_fredriksenii_	_intermedia_	_kristensenii_	_pseudotuberculosis_	_pestis_
Urease production	+	+	+	+	+	0, rare +
Ornithine decarboxylase	+	+	+	+	0	0
Lysine decarboxylase	0	0	0	0	0	0
Arginine dihydrolase	0	0	0	0	0	0
Motility: 25°C	+	+	+	+	+	0
37°C	0	0	0	0	0	0
Voges-Proskauer: 25°C	+	+	+	0	0	0
37°C	0	0	0	0	0	0
β-Galactosidase (_O_-nitrophenol-β-D-galactoside)	+ +(W)37°C	+	+	+	+	+
Simmons citrate	0	V	+(25°C)	0	0	0
Glucose	+	+	+	+	+	+
Mannitol	+	+	+	+	+	+
Maltose	+	+	+	+	+	V
Cellobiose	+	+	+	+	0	0
Sorbitol	+	+	+	+	0	0
Sucrose	+	+	+	0	0	0
Trehalose	+	+	+	+	+	+
Melibiose	0	0	+	0	+	V
Rhamnose	0	+	+(O)	0	+	0
Raffinose	0	0	+	0	V	0, rare +
α-Methyl glucoside	0	0	+(O)	0	0	0
Salicin	V†	+	+	+	V	V
Esculin	V†	+	+	+	+	+
Xylose	V†	+	+	+		0
Lecithinase	V†	+	+	+(W),0	Unknown	Unknown
Indole	V†	+	+	+(W)	0	0

*+ = positive. +(W) = weak positive. 0 = negative. (O) = occasional strain negative. V = variable.
  †Biotype 1, strains (+ + + + +); example serogroup 0:8.
Biotype 2, strains (0 0 + 0 +); example serogroup 0:5,27; 0:9.
Biotype 3, strains (0 0 + 0 0); example serogroup 0:1,2,3.
Biotype 4, strains (0 0 0 0 0); example serogroup 0:3.
Biotype 5, strains (0 0 0 0 0);
additional negative for trehalose, sorbitol, sucrose, β-galactosidase, ornithine decarboxylase, and Voges-Proskauer.

cillary microorganisms. They grow well, albeit slowly and variably, on a variety of bacteriological culture media. Isolation, especially from fecal specimens, of the more commonly occurring strains (serogroups) of _Y. enterocolitica_ that cause human infection (for example, serogroups 0:3; 0:5,27; 0:8; and 0:9) may be achieved through cultivation on a variety of enteric media and on newly developed selective media designed for the purpose (Table 2). Awareness of the similarity of _Y. enterocolitica_ colonies on some enteric agars to those of other enteric organisms comprising the fecal flora, such as _Escherichia coli_, will prevent

Table 2. Colonial characteristics of _Yersinia enterocolitica_ and _Y. pseudotuberculosis_ on enteric isolation media*

Medium	_enterocolitica_	_pseudotuberculosis_
Hektoen-enteric	Pinpoint, salmon colored	Pinpoint, smooth, colorless (green) to yellow (48 h)
Eosin methylene blue	Pinpoint, lavender, green metallic sheen (48 h)	Pinpoint, colorless (lavender)
MacConkey	Pinpoint, pink hue	Pinpoint, pink (48 h)
Salmonella-Shigella	Pinpoint, colorless	Pinpoint, pink hue (48 h)
Xylose-lysine-deoxycholate	Pinpoint, yellow	Pinpoint, colorless to yellow (48 h)
Pectin agar	Depressed colonies (48 h)	Not tested
Cellobiose-arginine-lysine (CAL)	Burgundy, red; clear with pinpoint red center	Not tested
Cefsulodin-irgasan-novobiocin (CIN)	Deep red center, sharp borders	Not tested

*24 h incubation at 37°C, unless otherwise indicated.

overlooking this species when present.

Patients with acute gastroenteritis due to *Y. enterocolitica* shed large numbers of this microorganism in stools. Cultures are thus usually positive during acute syndromes by direct plating of the specimen to primary isolation media, such as MacConkey agar. During convalescence, however, fecal shedding of *Y. enterocolitica* may be sparse. In these instances, the inoculation of stools into phosphate-buffered saline at pH 7.2, with the subculturing of these broths 2 to 3 days after incubation at 4°C (cold enrichment), increases the isolation of *Y. enterocolitica* during this stage of the disease. Presently, however, there is disagreement on the practical value of cold enrichment for the recovery of *Y. enterocolitica*. Several investigators contend that strains of *Y. enterocolitica* recovered after prolonged incubation (up to 28 days) at 4°C are usually of biotype 1 and of a serogroup seldom associated with human infections. It is stressed that serogroup 0:8 *Y. enterocolitica*, although of biotype 1 (Wauters schema), is highly pathogenic and considered distinct from other biotype-1 isolates which apparently lack many of the phenotypic markers associated with virulence (Table 3). Although cold enrichment does increase the recovery rate of *Y. enterocolitica*, the question of the clinical significance of isolates, especially nonserogroup 0:8 biotype-1 strains, obtained during postconvalescence by such enhancement is still being resolved.

Once isolation of a suspect yersinia has been accomplished, the biochemical identification ensues more logically since the so-called *Y. enterocolitica*–like strains have been accorded species designation. It is emphasized, however, that since test results are a function of incubation temperature, determinations should be evaluated at both 37 and 25°C. Reactions varying with temperature are indicated in Table 1. Intimately associated with the biochemical behavior of *Y. enterocolitica* are correlates of pathogenicity and ecology. Human pathogenic serogroups are found among biochemically well-characterized subsets (biotypes) of *Y. enterocolitica*. Tests necessary for biotyping and representative serogroups in these biotypes are indicated in Table 1.

Virulence factors. Most research centers on the virulence-associated attributes of *Y. enterocolitica* that cause the various clinical syndromes linked to yersiniosis. The presence of a series of plasmid species associated with a number of temperature-dependent phenotypic markers of virulence (Table 3) were discovered. Among such markers expressed after growth at 37°C, but not at 25°C, and associated with the presence of a 40–48-megadalton (Mdal) plasmid species are: (1) autoagglutination in tissue-culture medium; (2) calcium requirement for growth as indicated by lack of growth on magnesium oxalate (MOX) agar at 37°C; (3) synthesis of V and W antigen complex necessary for antiphagocytic activity and intracellular survival; and (4) tissue invasiveness as assessed by the intraperitoneal and oral inoculation of gerbils, mice, and rabbits, or by eliciting conjunctivitis upon instillation of a dense (10^{10} cells per milliliter) suspension into the guinea pig eye (serogroup 0:8). Recently a new plasmid of 82 Mdal has been recognized in serogroup 0:8, 0:13, and 0:20 strains, the presence of which correlated with mouse lethality. This plasmid species may have been overlooked in virulent strains containing the 42-Mdal plasmid.

Internalization of *Y. enterocolitica* in HeLa and HEP-2 tissue-culture cells does not require the presence of a plasmid species; however, detachment of these cells (a monolayer) from the surface of tissue-culture tubes is apparently a function of a plasmid-mediated factor. Human pathogenic and avirulent strains also produce a heat-stable (ST) enterotoxin

Table 3. Virulence markers associated with presence of plasmid species*

	Plasmid species, Mdal				
Phenotypic marker	<35, >70 Biotype 1†	42 (0:8)	44 0:9	47 0:3	82‡ 0:8
VW antigen	0	+	+	+	
Autoagglutination at 37°C	0	+	+	+	
Calcium dependency for growth at 37°C	0	+	+	+	
HeLa, HEP-2 cell§ internalization	0	0	0	0	
Monolayer-detachment guinea pig conjunctivitis	0	+	0	0	
Serum resistance¶ after growth at 37°C	0	0	0	0	
Mouse lethality	0	+	NT	D	+
Gerbil lethality	0	+	NT	0	
Enterotoxin production	0	0	0	0	

* + = positive, expressed in presence of plasmid species. NT = not tested. D = diarrhea. 0 = absent or not correlated with presence of a plasmid, for example, heat-stable (ST) enterotoxin production.

†Serogroups 0:5; 0:6; 0:7,13; 0:10.

‡Also present along with 42 Mdal in serogroups 0:20 and 0:13. Strains containing only 82 Mdal plasmid not tested for other plasmid (42 Mdal) associated markers such as autoagglutination and so on.

§Internalization is not plasmid-mediated.

¶May also be expressed by plasmid-containing strains.

that is not plasmid-encoded and whose role in gastrointestinal disease is questionable. Serum resistance is another virulence marker found among human pathogenic groups independent of other phenotypic markers of virulence. Many of these *Y. enterocolitica* plasmid-associated phenotypic markers have also been found in *Y. pseudotuberculosis*.

Epidemiologically, there still appears to be a geographic loculation regarding the occurrence of different *Y. enterocolitica* serogroups. While yersiniosis has a global distribution and most infections are caused by serogroups 0:3; 0:5,27; 0:8; 0:9—and more rarely by 0:1,2,3; 0:5; and 0:21—these strains are not encountered widely. For example, serogroup 0:3, phage type 9b, found commonly in Canada, is just emerging in the United States; serogroup 0:8, the American strain, has rarely been encountered outside continental North America; and serogroup 0:9, common in Europe, has not been isolated from patients in the United States. Factors accounting for these ecological nuances are still unexplained. Additionally, serogroups 0:3 (biotype 4) and 0:1,2,3 (biotype 3) are apparently host-bound in animals, and are recovered almost exclusively from the pig and chinchilla, respectively.

Transmission, diagnosis, and treatment. Transmission of *Y. enterocolitica*, and *Y. pseudotuberculosis* also, from their natural habitat (animate or inanimate) to humans is still unresolved. With the exception of the Holland Patent *Y. enterocolitica* outbreak, which was definitely linked to contaminated chocolate milk, the mode of acquisition of these *Yersinia* species is speculative and thought to occur from healthy and sick human or animal carriers, especially pigs and dogs, and through contaminated water or food.

The diagnosis of gastrointestinal yersiniosis is achieved by recovery of *Y. enterocolitica* or *Y. pseudotuberculosis*, especially during the acute and early convalescent stages of infection. In the absence of cultural evidence of yersiniosis, diagnosis may be attempted through assessment of an antibody response to one of the more frequently occurring serogroups in a given geographic locale. This method is especially valuable during outbreaks. Outside this setting, however, numerous factors, such as age of patient, underlying disease, administration of antibiotics and immunosuppressive agents, and nature of infecting strain affect the immunologic response and hence the interpretation of antibody titers. Technical factors, such as prozone phenomena (in which visible evidence, such as agglutination or precipitation, of an antigen-antibody reaction does not occur because of antigen or antibody excess, or of the presence of incomplete antibodies), also contribute to the difficulty of establishing a serologic diagnosis. Hence, serology as a diagnostic method is a secondary alternative to cultural diagnosis.

Gastrointestinal yersiniosis is usually a self-limited disease not requiring antimicrobial therapy. In more protracted cases or in patients with mesenteric lymphadenitis, terminal ileitis, or extraintestinal involvement of visceral organs and blood, antibiotic therapy may be indicated. As most *Y. enterocolitica* (but not *Y. pseudotuberculosis*) produce a broad-spectrum beta-lactamase active against natural and semisynthetic penicillins and cephalosporins, the aminoglycosides (gentamicin, tobramicin, and so forth), trimethoprim-sulfamethoxazole, and chloramphenicol appear to be the antibiotics of choice when treatment is indicated.

For background information *see* EPIDEMIOLOGY; MEDICAL BACTERIOLOGY; MICROBIOLOGICAL METHODS; YERSINIA in the McGraw-Hill Encyclopedia of Science and Technology.

[EDWARD J. BOTTONE]

Bibliography: R. E. Black et al., Epidemic *Yersinia enterocolitica* infection due to contaminated chocolate milk, *New Engl. J. Med.*, 298:76–79, 1978; E. J. Bottone, *Yersinia enterocolitica*: A panoramic view of a charismatic microorganism, *CRC Crit. Rev. Microbiol.*, 5:211–241, 1977; D. J. Brenner et al., Deoxyribonucleic acid relatedness in *Yersinia enterocolitica* and *Yersinia enterocolitica*–like organisms, *Current Microbiol.*, 4:195–200, 1980; D. A. Schiemann and J. A. Devenish, Relationship of HeLa cell infectivity to biochemical, serological, and virulence characteristics of *Yersinia enterocolitica*, *Infect. Immun.*, 35:497–506, 1982.

McGRAW-HILL YEARBOOK OF SCIENCE AND TECHNOLOGY

List of Contributors

List of Contributors

A

Agerwala, Dr. Tilak. *Manager, Architecture and System Design, IBM T. J. Watson Research Center, Yorktown Heights, New York.* DATA FLOW SYSTEMS.

Aharony, Prof. Amnon. *Department of Physics and Astronomy, Tel Aviv University.* FRACTALS—coauthored.

Albaugh, Dr. G. P. *Department of Botany, University of Maryland.* SEED GERMINATION—in part.

Allen, Prof. B. L. *College of Agricultural Sciences, Department of Plant and Soil Science, Texas Tech University.* SOIL TAXONOMY—in part.

Altamuro, Vincent M. *President, Management Research Consultants, Yonkers, New York.* PARETO'S LAW.

Ames, Prof. David R. *Head, Department of Animal Sciences, Colorado State University.* AGRICULTURAL SCIENCE (ANIMAL).

Ashby, Prof. Eugene C. *Department of Chemistry, Georgia Institute of Technology.* ORGANIC REACTION MECHANISM.

B

Baker, Dr. Victor R. *Department of Geosciences, University of Arizona.* MARS.

Barbacid, Dr. Mariano. *Laboratory of Cellular and Molecular Biology, National Cancer Institute, National Institutes of Health, Bethesda.* ONCOGENE, HUMAN.

Bardack, Dr. David. *Department of Biological Sciences, University of Illinois.* OSTEICHTHYES.

Barger, Prof. Vernon D. *Department of Physics, University of Wisconsin.* ELEMENTARY PARTICLE.

Barron, Dr. Eric J. *National Center for Atmospheric Research, Boulder, Colorado.* PALEOCLIMATOLOGY.

Betts, Dr. Russell. *Chemistry Division, Argonne National Laboratory.* NUCLEAR MOLECULE.

Bloch, Dr. Salman. *ARCO Oil and Gas, Dallas.* SHALE.

Bohac, Charles E. *Water Quality Branch, Tennessee Valley Authority, Chattanooga.* WATER CONSERVATION.

Bonfiglio, Dr. Thomas A. *Director, Surgical Pathology, University of Rochester, Medical Center.* CLINICAL PATHOLOGY.

Bottone, Dr. Edward J. *Director, Department of Microbiology, Mount Sinai Hospital, and Professor of Clinical Microbiology, Mount Sinai School of Medicine.* YERSINIOSIS.

Boxer, Dr. Steven G. *Department of Chemistry, Stanford University.* CHLOROPHYLL.

Boyle, Charles P. *NASA Goddard Space Flight Center, Greenbelt, Maryland.* SPACE FLIGHT.

Buol, Dr. Stanley W. *Department of Soil Science, North Carolina State University.* SOIL TAXONOMY—in part.

Burke, Dr. Deborah. *Department of Psychology, Pomona College.* MEMORY—coauthored.

C

Cassen, Dr. P. M. *Theoretical and Planetary Studies Branch, NASA-Ames Research Center, Moffett Field, California.* JUPITER—coauthored.

Chaffin, Dr. Roger J. *Solid State Device Physics Division, Sandia National Laboratories, Albuquerque.* ELECTRONICS.

Chamberlain, Dr. John A., Jr. *Department of Geology, Brooklyn College, and Research Associate, Osborn Laboratories of Marine Sciences, New York Aquarium, New York Zoological Society.* EVOLUTION.

Chandra, Dr. G. Ram. *Chief, Seed Research Laboratory, U. S. Department of Agriculture, Beltsville, Maryland.* SEED GERMINATION—in part.

Chincarini, Prof. Guido. *Department of Physics and Astronomy, University of Oklahoma.* COSMOLOGY.

Conger, Prof. Bob V. *Institute of Agriculture, Department of Plant and Soil Science, University of Tennessee.* CLONING AGRICULTURAL PLANTS—feature.

Connell, Dr. Joseph H. *Department of Biological Science, University of California, Santa Barbara.* ECOSYSTEM.

Converti, V. *Manager, Computer Services, Arizona Public Service Company, Phoenix.* ELECTRICAL UTILITY INDUSTRY—in part.

Cooke, Dr. Todd. *Department of Botany, University of Maryland.* FERN—in part.

Couch, Dr. Richard B. *Ship Hydrodynamics Laboratory, University of Michigan.* SHIP POWERING.

Cowles, Dr. Joe R. *Department of Biology, University of Houston.* LIGNIN.

Cozart, Jerome X. *Graphic Arts Technical Foundation, Inc., Pittsburgh.* PRINTING.

Crouch, Dr. Martha L. *Department of Biology, Indiana University.* MOLECULAR BIOLOGY.

Currey, Prof. John D. *Department of Biology, University of York, England.* MOLLUSCA.

D

Dalgarno, Dr. Alexander. *Harvard College Observatory Center for Astrophysics, Cambridge, Massachusetts.* SCATTERING EXPERIMENTS (ATOMS AND MOLECULES).

Davenport, Dr. Demorest. *Emeritus Professor of Zoology and the Humanities, University of California, Santa Barbara.* SEA ANEMONE.

Dayanandan, Dr. P. *Division of Biological Sciences, University of Michigan.* PLANT PHYSIOLOGY—coauthored.

Delevoryas, Dr. Theodore. *Department of Botany, University of Texas.* PALEOBOTANY—in part.

Detrick, Dr. Robert S., Jr. *Graduate School of Oceanography, University of Rhode Island, Narragansett Bay Campus.* TECTONOPHYSICS.

Dewar, Donald L. *President, Quality Circle Institute, Red Bluff, California.* QUALITY CIRCLES—feature.

Dodington, Dr. Sven H. *Mountain Lakes, New Jersey.* AIR NAVIGATION—in part.

Donoghue, Prof. John F. *Department of Physics and Astronomy, University of Massachusetts.* GLUONS.

Doty, Prof. Keith L. *Department of Electrical Engineering, University of Florida.* INTELLIGENT MACHINES—feature.

Dougherty, Prof. Ralph C. *Department of Chemistry, Florida State University.* MOLECULAR CHIRALITY.

Dowdy, Dr. R. H. *Department of Soil Science, University of Minnesota.* SOIL CHEMISTRY—in part.

Drennon, B. J. *Environmental Science Group, Los Alamos National Laboratory.* SOIL—in part.

Dunn, Prof. Floyd. *Department of Electrical Engineering, Bioacoustics Research Laboratory, University of Illinois at Urbana-Champaign.* BIOPHYSICS.

E

Eder, Dr. Jerome M. *Medical Adviser, Los Angeles Herpes Resource Center, Palo Alto.* HERPES.

Egger, Dr. Hans. *Department of Physics, University of Illinois at Chicago.* LASER—coauthored.

Elderfield, Dr. Henry. *Department of Earth Sciences, University of Cambridge.* SEA WATER—coauthored.

Ernst, Dr. Wallace Gary. *Department of Earth and Space Sciences, University of California, Los Angeles.* EVOLUTION OF THE EARTH'S CRUST—feature.

Eswaran, Dr. Hari. *National Coordinator, International Soils Program, USDA, Washington, D.C.* SOIL—in part.

Etherington, Cliff. *Research Manager, Process Engineering Technology, Springfields Nuclear Power Development Laboratories, United Kingdom Atomic Energy Authority, Lancashire.* EXTRUSION.

Everest, Charles E. *Everest Interscience, Tustin, California.* PYROMETER.

F

Farkas, Dr. Daniel F. *Department of Food Science and Human Nutrition, University of Delaware.* FOOD MANUFACTURING.

Feigenbaum, Prof. Mitchell J. *Department of Physics, Cornell University.* PERIOD DOUBLING.

Feldman, Dr. Jerry F. *Thimann Laboratories, University of California, Santa Cruz.* FUNGI.

Fisher, Dr. Dennis F. *U. S. Army Human Engineering Laboratory, Aberdeen Proving Ground, Maryland.* HUMAN-FACTORS ENGINEERING.

Fletcher, Dr. Ronald D. *Professor and Associate Chairman, Department of Microbiology, School of Dental Medicine, University of Pittsburgh.* PERIODONTAL DISEASE.

Franklin, Dr. C. L. *Division of Biological Sciences, University of Michigan.* PLANT PHYSIOLOGY—coauthored.

Franzini, Prof. Paolo. *Department of Physics, Columbia University.* MESON.

Frederickson, Dr. Robert C. A. *Lilly Research Laboratories, Indianapolis.* CELLULAR RECEPTORS—in part.

French, Dr. James C. *Department of Biological Sciences, Mississippi State University.* STEM.

Fuhrhop, Prof. Jurgen H. *Institut für Organischen Chemie der Freien Universität Berlin.* VESICLE MEMBRANES.

G

Gage, Dr. J. D. *Scottish Marine Biological Association, Dunstaffnage Marine Research Laboratory, Oban.* ECHINODERMATA.

Gallo, Dr. Robert C. *Chief, Laboratory of Tumor Cell Biology, Department of Health and Human Services, National Cancer Institute, National Institutes of Health, Bethesda.* LEUKEMIA—coauthored.

Geary, Dr. Robert B. *Director of Engineering, General Dynamics, Quincy Shipbuilding Division, Quincy, Massachusetts.* MARINE ENGINEERING.

Gefen, Dr. Yuval. *Department of Physics and Astronomy, Tel Aviv University.* FRACTALS—coauthored.

Giannetta, Prof. Russell. *Joseph Henry Laboratories of Physics, Princeton University.* LIQUID HELIUM.

Glover, Dr. David M. *Eukaryotic Molecular Genetics Research Group, Department of Biochemistry, Imperial College of Science and Technology, London.* MOLECULAR GENETICS.

Goodstein, David H. *Director, Inter/Consult, Cambridge, Massachusetts.* ELECTRONIC PUBLISHING—feature.

H

Harbison, Dr. G. R. *Woods Hole Oceanographic Institution, Woods Hole, Massachusetts.* MARINE ECOSYSTEM—coauthored.

Harbold, L. M. *Superintendent, Steelmaking Department, Bethlehem Steel Corporation, Burns Harbor Plant, Chesterton, Indiana.* STEEL MANUFACTURE.

Harper, Dr. Judson M. *Head, Department of Agricultural and Chemical Engineering, Colorado State University.* FOOD ENGINEERING—in part.

Harter, Dr. Robert D. *Institute of Natural and Environmental Resources, University of New Hampshire.* SOIL CHEMISTRY—in part.

Haxton, Dr. W. C. *Los Alamos National Laboratory.* DOUBLE BETA DECAY.

Hayes, William C. *Editor in Chief, "Electrical World," McGraw-Hill Publications Company, New York.* ELECTRICAL UTILITY INDUSTRY—in part.

Head, Dr. James W., III. *Department of Geological Sciences, Brown University.* VENUS.

Heath, Dr. Clark W., Jr. *Director, Chronic Diseases Division, Department of Health and Human Services, Centers for Disease Control, Atlanta.* POLLUTION.

Heckman, Dr. Harry H. *Lawrence Berkeley Laboratory, University of California, Berkeley.* ANOMALONS.

Herberman, Dr. Ronald B. *Chief, Biological Development Branch, National Cancer Institute, Frederick Cancer Research Facility, Frederick, Maryland.* IMMUNOLOGY.

Herrell, Dr. Dennis J. *IBM, Thomas J. Watson Research Center, Yorktown Heights, New York.* SUPERCONDUCTING COMPUTER.

Heskel, Dr. Dennis L. *Department of Anthropology, University of Utah.* ARCHEOLOGICAL METALLURGY—feature.

Hopkins, Dr. John J. *President, Navigation Development Services, Inc., Northridge, California.* AIR NAVIGATION—in part.

Houk, Dr. Kendall N. *Department of Chemistry, University of Pittsburgh.* FRONTIER MOLECULAR ORBITAL THEORY.

Hu, Dr. Evelyn L. *Bell Laboratories, Murray Hill, New Jersey.* INTEGRATED CIRCUITS—coauthored.

Hunter, Prof. Stanley D. *Department of Physics, Louisiana State University.* ACOUSTICS.

Hutchins, Carleen Maley. *Catgut Acoustical Society, Inc., Montclair, New Jersey.* MUSICAL INSTRUMENTS.

Hutter, Dr. Kolumban. *Leader, Physical Limnology Group, Laboratory of Hydraulics, Hydrology and Glaciology, Federal Institute of Technology, Zurich.* GLACIOLOGY.

Hwang, Dr. Sun-Tak. *Department of Chemical Engineering, University of Iowa.* MEMBRANE SEPARATION TECHNIQUES.

J

Jablonski, Dr. David. *Department of Ecology and Evolutionary Biology, University of Arizona.* LARVA.

Jackel, Dr. Lawrence D. *Member, Technical Staff, Electronics Research Laboratory, Bell Laboratories, Holmdel, New Jersey.* INTEGRATED CIRCUITS—coauthored.

Jaeger, Dr. Ralph W. *Assistant Director for Technical Marketing, Arianespace, Evry, France.* LAUNCH VEHICLE.

Janssen, Dr. John. *Woods Hole Oceanographic Institution, Woods Hole, Massachusetts.* MARINE ECOSYSTEM—coauthored.

Joel, Amos E., Jr. *Switching Consultant, Bell Laboratories, Holmdel, New Jersey.* SWITCHING SYSTEMS (COMMUNICATIONS).

Jones, Dr. Meredith L. *Curator of Worms, Smithsonian Institution, Washington, D.C.* POGONOPHORA.

Jorgensen, Dr. Timothy J. *Radiobiology Laboratory, Johns Hopkins Oncology Center, Baltimore.* PHOTOREACTIVATING ENZYME.

K

Karel, Dr. Marcus. *Department of Nutrition and Food Science, Massachusetts Institute of Technology.* FOOD ENGINEERING—in part.

Kaufman, Dr. Peter B. *Division of Biological Sciences, University of Michigan.* PLANT PHYSIOLOGY—coauthored.

Kayser, Prof. Fritz. *Institute of Medical Microbiology, University of Zurich.* ANTIMICROBIAL RESISTANCE.

Kellermann, Dr. Kenneth I. *National Radio Astronomy Observatory, Green Bank, West Virginia.* QUASARS.

Khasawneh, Dr. F. E. *Research Soil Chemist, Agricultural Research Branch, Division of Agricultural Development, National Fertilizer Development Center, Tennessee Valley Authority, Muscle Shoals, Alabama.* FERTILIZER.

Klostermeyer, Dr. E. C. *Emeritus Professor of Entomology, Washington State University.* AGRICULTURAL SCIENCE (PLANT)—in part.

Kossowsky, Dr. Ram. *Manager, Physical Metallurgy, Research and Development Center, Westinghouse Electric Corporation, Pittsburgh.* ION IMPLANTATION.

Krikorian, Dr. Abraham D. *Department of Biochemistry, State University of New York, Stony Brook.* TISSUE CULTURE.

Kuhn, Dr. H. *Max-Planck Institut für Biophysikalische Chemie, Göttingen.* LIFE, ORIGIN OF.

Kukacka, Dr. Larry. *Leader, Process Materials Group, Department of Energy and Environment, Brookhaven National Laboratory, Associated Universities, Inc., Upton, New York.* CONCRETE.

L

Laughlin, Dr. Robert B. *Lawrence Livermore National Laboratory, University of California, Livermore.* HALL EFFECT.

Light, Prof. Leah. *Department of Psychology, Pitzer College, Claremont, California.* MEMORY—coauthored.

Lipps, Prof. Jere H. *Department of Geology, University of California, Davis.* MICROPALEONTOLOGY.

Lockyer, Dr. Nigel. *Stanford Linear Accelerator Center, Stanford, California.* PARITY (QUANTUM MECHANICS).

Loehlin, Prof. John C. *Chairman, Department of Psychology, University of Texas at Austin.* PSYCHOLOGY.

Long, Norwood G. *Director, Loop Systems Engineering Center, Bell Laboratories, Whippany, New Jersey.* TELEPHONE SERVICE.

Luther, Dr. Gabriel G. *Quantum Metrology Group, Center for Absolute Physical Quantities, National Bureau of Standards, U. S. Department of Commerce, Washington, D.C.* GRAVITATION.

M

McCammon, Prof. J. Andrew. *Department of Chemistry, University of Houston.* PROTEIN—coauthored.

McCave, Dr. I. N. *Department of Environmental Sciences, University of East Anglia, Norwich.* MARINE SEDIMENTS.

McElroy, Dr. Michael W. *Project Manager, Air Quality Control, Electric Power Research Institute, Palo Alto, California.* COAL.

McGehee, Prof. O. Carruth. *Department of Mathematics, Louisiana State University.* FOURIER SERIES AND INTEGRALS.

Mao, Dr. Boryeu. *Department of Chemistry, University of Houston.* PROTEIN—coauthored.

Marlatt, Dr. Robert B. *University of Florida Agricultural Research and Education Center, Homestead, Florida.* PLANT PATHOLOGY—in part.

Martinelli, Dr. Mario, Jr. *Project Leader, USDA Rocky Mountain Forest and Range Experiment Station, Fort Collins, Colorado.* AVALANCHE.

Maynard, Dr. Julian D. *College of Science, Department of Physics, Pennsylvania State University.* QUANTUM SOLIDS AND LIQUIDS.

Meinel, Dr. Aden. *Optical Sciences Center, University of Arizona.* OPTICAL TELESCOPE—coauthored.

Meinel, Dr. Marjorie. *Optical Sciences Center, University of Arizona.* OPTICAL TELESCOPE—coauthored.

Meisel, Dr. Dan. *Chemistry Division, Argonne National Laboratory.* PHOTOELECTROLYSIS.

Moncton, Dr. David E. *Department of Physics, Brookhaven National Laboratory, Associated Universities, Inc., Upton, New York.* TWO-DIMENSIONAL SYSTEMS.

Moore, Dr. Randy. *Biology Department, Baylor University.* GRAFTING IN PLANTS.

Moore, Prof. Richard E. *Department of Chemistry, University of Hawaii at Manoa.* TOXIN.

Morton, Arthur. *Senior Staff Engineer, Production Engineering Services, Conoco Inc., Houston.* OIL AND GAS, OFFSHORE.

Mueller, Dr. Richard J. *Department of Biology, Utah State University.* FERN—in part.

Munoz, Dr. James L. *Department of Geological Sciences, University of Colorado.* GEOCHEMICAL PROSPECTING.

Murphy, Dr. Robert C. *Department of Pharmacology, University of Colorado Health Sciences Center.* LEUKOTRIENES.

N

Nagaoka, Dr. Sakae. *Evaluation Division, Electronic Navigation Research Institute, Tokyo.* RADAR—in part.

Neter, Dr. Erwin. *Department of Pediatrics, State University of New York at Buffalo.* BACTERIAL ENDOTOXIN.

Niklas, Dr. Karl J. *Division of Biological Sciences, Section of Plant Biology, Cornell University.* PALEOBOTANY—in part; REPRODUCTION (PLANT).

Norris, Dr. James R., Jr. *Senior Chemist/Group Leader, Chemistry Division, Argonne National Laboratory.* PHOTOSYNTHESIS.

Nye, Dr. Peter H. *Department of Agricultural Science, Soil Science Laboratory, University of Oxford.* SOIL—in part.

O

Owen, Dr. Denis. *Department of Biology, Oxford Polytechnic, Headington, Oxford, England.* DESERTIFICATION.

Owen, Dr. Tobias C. *Earth and Space Science Department, State University of New York, Stony Brook.* ORIGIN OF PLANETARY ATMOSPHERES—feature.

P

Palmer, Dr. Martin. *Department of Earth Sciences, University of Leeds.* SEA WATER—coauthored.

Parker, Prof. Bruce C. *Department of Biology, Vir-*

ginia Polytechnic Institute and State University. STRO-
MATOLITE.

Parks, Robert E. *Optical Sciences Center, University of Arizona.* DIAMOND-TURNED OPTICS.

Patrick, Prof. William H. *Center for Wetland Resources, Louisiana State University.* SOIL—in part.

Peale, Prof. Stanton J. *Department of Physics, University of California, Santa Barbara.* JUPITER—coauthored.

Peercy, Dr. P. S. *Supervisor, Ion Implantation Physics, Sandia National Laboratories, Albuquerque.* SEMICONDUCTOR MEMORIES.

Pistole, Dr. Thomas G. *Department of Microbiology, University of New Hampshire.* LECTINS.

Pollack, Dr. James B. *Space Science Division, NASA Ames Research Center, Moffett Field, California.* VOLCANO.

Pope, Dr. Daniel H. *Director, Fresh Water Institute, and Associate Professor of Biology, Rensselaer Polytechnic Institute.* METALLURGICAL ENGINEERING.

Popp, Dr. Robert K. *Department of Geology, College of Geosciences, Texas A&M University.* SOLUTION GEOCHEMISTRY.

Power, Dr. J. F. *USDA, SEA, Soil and Water Conservation Research Unit, University of Nebraska.* AGRICULTURAL SOIL AND CROP PRACTICES.

Prusoff, Dr. William H. *Department of Pharmacology, Yale University School of Medicine.* VIRAL CHEMOTHERAPY—coauthored.

Putterman, Prof. Seth. *Department of Physics, University of California, Los Angeles.* SOUND.

R

Racusen, Richard H. *Department of Botany, University of Maryland.* FERN—in part.

Raju, Dr. M. R. *Fellow, Toxicology Group, Life Sciences Division, Los Alamos National Laboratory.* RADIATION BIOLOGY.

Ratcliffe, Stanley. *Malvern, Worcestershire, England.* AIR-TRAFFIC CONTROL.

Reynolds, Dr. R. T. *Theoretical and Planetary Studies Branch, NASA-Ames Research Center, Moffett Field, California.* JUPITER—coauthored.

Rhodes, Prof. Charles K. *Research Professor, Department of Physics, University of Illinois at Chicago.* LASER—coauthored.

Rickords, Tom J. *Senior Member Technical Staff, G & E Division, Litton, Inc., Woodland Hills, California.* LAND NAVIGATION SYSTEM.

Ritter, Dr. Robert C. *Department of Veterinary and Comparative Anatomy, Pharmacology and Physiology, Washington State University, and Department of Veterinary Medicine, University of Idaho.* CELLULAR RECEPTORS—in part.

Robb, David W. *Associate Editor, "Sea Technology," Compass Publications, Inc., Arlington, Virginia.* WIND SHIP.

Rochow, Prof. W. F. *Department of Plant Pathology, Cornell University.* PLANT PATHOLOGY—in part.

Roos, Dr. Eric E. *Plant Physiologist, USDA National Seed Storage Laboratory, Colorado State University.* SEED GERMINATION—in part.

Rubin, Dr. Emanuel. *Chairman, Department of Pathology and Laboratory Medicine, Hahnemann Medical College and Hospital, Philadelphia.* CELLULAR RECEPTORS—in part.

Rudman, Dr. W. B. *Department of Malacology, Australian Museum, Sydney South.* ECOLOGICAL INTERACTIONS.

Ryan, Dr. James M. *Space Science Center, University of New Hampshire.* SUN.

S

Sakazaki, Dr. Riichi. *Chief, Enterobacteriology Laboratories, National Institute of Health, Tokyo.* BACTERIAL TAXONOMY.

Sarngadharan, Dr. M. G. *Laboratory of Tumor Cell Biology, National Cancer Institute, National Institutes of Health, Bethesda.* LEUKEMIA—coauthored.

Scheer, Dr. Hugo *Botanisches Institut der Universität München.* BILIPROTEIN.

Schiffer, Dr. John. *Argonne National Laboratory.* NUCLEAR PHYSICS.

Schinazi, Dr. Raymond F. *Department of Pediatrics, Division of Infectious Diseases, Allergy and Immunology, Emory University School of Medicine.* VIRAL CHEMOTHERAPY—coauthored.

Schrieffer, Prof. J. Robert. *Department of Physics, University of California, Santa Barbara.* SOLITON.

Schull, Dr. William J. *Director, Center for Demographic and Population Genetics, University of Texas Health Science Center, Houston.* HUMAN GENETICS.

Shaw, Dr. H. J. *Edward L. Ginzton Laboratory, Stanford University.* GYROSCOPE.

Soderblom, Dr. Laurence A. *U.S. Geological Survey, Flagstaff, Arizona.* SATELLITE (ASTRONOMY).

Stanbridge, Dr. Eric J. *Department of Medical Microbiology, College of Medicine, University of California, Irvine.* ONCOLOGY.

Summers, Dr. William C. *Department of Therapeutic Radiology, Yale University School of Medicine.* ANIMAL VIRUS.

Sureau, Dr. Jean-Claude. *Technical Director, Radant Systems, Inc., Stow, Massachusetts.* RADAR—in part.

Szczesniak, Dr. Alina S. *Principal Scientist, Central Research Department, General Foods Corporation, White Plains, New York.* FOOD ENGINEERING—in part.

T

Tappan, Prof. Helen. *Department of Earth and Space Sciences, University of California, Los Angeles.* EXTINCTION (BIOLOGY).

Teixeira, Dr. Arthur A. *Institute of Food and Agricultural Sciences, Agricultural Engineering Department, University of Florida.* COGENERATION.

Tilton, Dr. George R. *Department of Geological Sciences, University of California, Santa Barbara.* EARTH, AGE OF.

Tomasetti, Richard L. *Senior Vice President, Lev Zetlin Associates, Inc., New York.* BUILDINGS.

Turner, Dr. B. E. *National Radio Astronomy Observatory, Charlottesville, Virginia.* INTERSTELLAR MATTER.

V

Vandermeer, Dr. John. *Division of Biological Sciences, University of Michigan.* MATHEMATICAL ECOLOGY.

van Houte, Dr. Johannes. *Forsyth Dental Center, Boston.* TOOTH DISORDERS.

Volk, Prof. V. V. *Department of Soil Science, Oregon State University.* SOIL CHEMISTRY—in part.

W

Weller, Dr. Robert A. *A.W. Wright Nuclear Structure Laboratory, Physics Department, Yale University.* SPUTTERING.

Wernicke, Dr. Brian. *Department of Geology, Syracuse University.* EXTENSIONAL TECTONICS.

White, Prof. William B. *Department of Geochemistry, Pennsylvania State University.* MINERALOGY.

Wiech, Raymond E., Jr. *Witec California, Inc., San Diego.* POWDER METALLURGY.

Witham, Clyde L. *SRI International, Menlo Park, California.* AEROSOL.

Woese, Prof. Carl R. *Department of Genetics and Development, University of Illinois, Urbana.* ARCHAEBACTERIA.

Wyn-Jones, Dr. Evan. *Department of Chemistry and Applied Chemistry, University of Salford, England.* ACOUSTIC RELAXATION.

Y

Youngberg, Prof. Harold. *Crop Science Department, Oregon State University.* AGRICULTURAL SCIENCE (PLANT)—in part.

Z

Zelazny, Dr. Lucian W. *Department of Agronomy, Virginia Polytechnic Institute and State University.* SOIL CHEMISTRY—in part.

McGRAW-HILL YEARBOOK OF SCIENCE AND TECHNOLOGY

Index

Index

Asterisks indicate page references to article titles.

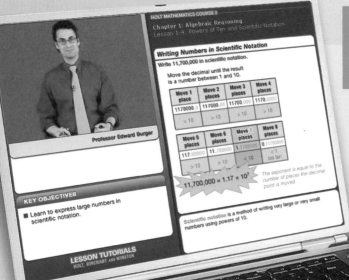